WASTE MANAGEMENT PRACTICES

PRACTICES

Municipal, Hazardous, and Industrial

WASTE MANAGEMENT PRACTICES

Municipal, Hazardous, and Industrial

John Pichtel

Taylor & Francis
Taylor & Francis Group

Boca Raton London New York Singapore

A CRC title, part of the Taylor & Francis imprint, a member of the
Taylor & Francis Group, the academic division of T&F Informa plc.

Published in 2005 by
CRC Press
Taylor & Francis Group
6000 Broken Sound Parkway NW
Boca Raton, FL 33487-2742

International Standard Book Number-10: 0-8493-3525-6 (Hardcover)
International Standard Book Number-13: 978-0-8493-3525-9 (Hardcover)
Library of Congress Card Number 2004058570

Library of Congress Cataloging-in-Publication Data

Pichtel, John, 1957-
Waste management practices : municipal, hazardous, and indistrial / John Pichtel.
    p. cm.
  Includes bibliographical references and index.
  ISBN 0-8493-3525-6 (alk. paper)
   1. Refuse and refuse disposal-Management. 2. Hazardous wastes-Management. 3.
Factory and trade waste-Management. I. Title.

TD791.P46 2005
628.4-dc22   2004058570

Taylor & Francis Group
is the Academic Division of T&F Informa plc.

Visit the Taylor & Francis Web site at
http://www.taylorandfrancis.com

and the CRC Press Web site at
http://www.crcpress.com

Preface

*When written in Chinese, the word 'crisis' is composed of two characters —
one represents danger and one represents opportunity.*

John F. Kennedy

Even if you're on the right track, you'll get run over if you just sit there.

Will Rogers

In his *Laws of Ecology*, Dr. Barry Commoner postulated that "In nature there is no waste; everything is connected to everything else; everything must go someplace; and there is no such thing as a free lunch." These laws have been faithfully followed for eons by all biota on the planet; except for humans. This has become particularly evident over the past few centuries. Worldwide, human population growth continues to increase exponentially. The quantities of nonrenewable natural resources extracted and used, and the consequent degrees of air, water, and soil pollution also follow an upward trend. The "garbage crisis," as it became known in the late 1980s, will not go away; the number of sanitary landfills in the United States continues to decline rapidly, and the amount of waste generated per capita has only recently begun to stabilize. Demands for convenient and disposable consumer products have reached unprecedented levels. Humans are producing numerous substances that nature simply does not possess the capability to decompose. Payment for our "lunch" is indeed due.

In the United States, regulators, scientists, policy makers, and the general public have belatedly recognized that the context in which we have managed our wastes, whether household, industrial, commercial, or hazardous, has been inadequate if not outright flawed. In the 1970s, disasters including Love Canal (NY), Times Beach (MO), and Valley of the Drums (KY) underscored the lack of a comprehensive strategy for hazardous waste management. In the 1980s, the Islip (NY) "Garbage Barge" made headlines along with the washing of medical waste on to New Jersey, New York and California beaches. The *Khian Sea*, transporting incinerator ash from Philadelphia, experienced a lengthy and frustrating odyssey in hopes of finding a home for its toxic cargo. The Fresh Kills landfill, located in Staten Island, New York, is now the world's largest landfill, constructed without a liner on porous sandy soils. It has become apparent that our earlier mindset on management and disposal of wastes was neither adequately serving public health nor protecting the environment.

In response to the above and similar events, federal and state legislation has been enacted addressing the proper storage, collection, transportation, processing, treatment, recovery, and disposal of wastes from many sources. The Resource Conservation and Recovery Act (RCRA) established a comprehensive framework for the overall management of existing and future hazardous waste generation, transportation, treatment, storage, and disposal activities. The Act also called for the more effective management of both hazardous and nonhazardous wastes, by way of reduction, reuse, and recycling. Amendments to RCRA now cover the management of used oil, industrial waste, and other residues that do not fit conveniently into either category.

There is a need for well-trained scientists, regulatory personnel, and policy makers to appreciate and integrate the technical and regulatory complexities of waste management. The public must make well-informed decisions concerning the allocation of resources toward future management efforts. They must accept the consequences of their lifestyle choices on local and large-scale environments. Complacency has brought us to where we are today; we now need comprehensive knowledge combined with committed action to establish a new framework in managing wastes.

There are few references which collectively address the management of the above listed wastes although the engineer, scientist, or regulatory person may ultimately be responsible for the proper disposition of one or more types. This book is intended to serve as a comprehensive manual for the identification and management of a wide range of wastes ranging from those that are merely a nuisance to extremely hazardous. This is an introductory manual for waste management as mandated by the RCRA and related statutes, with an emphasis on basic environmental science and related technical fields.

The first part of this book (Part I) provides an overview of the historical and regulatory development of waste management. The second part (Part II) delineates the management of municipal solid wastes, i.e., those we encounter on a daily basis. Both conventional (e.g., sanitary landfill, aerobic composting) and innovative (bioreactor landfill, high-solids anaerobic digestion) technologies are discussed. The third part (Part III) addresses hazardous wastes and their management, from the perspectives of identification, transportation, and requirements for generators and for treatment, storage, and disposal facilities. Disposition via incineration, chemical treatment, and land disposal is also presented. The final part (Part IV) is devoted to special categories of waste that cannot find a regulatory "home" under either RCRA Subtitle D (Solid Wastes) or Subtitle C (Hazardous Wastes). These include used motor oil, medical waste, and electronics waste, among others.

In addition to end-of-chapter problems provided in all chapters to this book, Chapters 4 (Characterization of Solid Waste), 8 (Composting MSW), 9 (Incineration of MSW) and 10 (The Sanitary Landfill) contain exercises using data from field situations. Data are supplied in Microsoft Excel format. The student is to analyze the data, answer the questions and provide conclusions. The exercises and data can be accessed on the Web by going to: www.crcpress.com/e_products/downloads/download.asp? cat_no=3525

Regardless of how passionately some Americans may adhere to the "reduce, reuse, and recycle" mantra, wastes of varying toxicity and mobility in the biosphere will continue to be produced in incomprehensible quantities. It is therefore critical that Americans become aware of the hazards and potential benefits of wastes in order to manage them in the most appropriate fashion.

Author's Biography

John Pichtel is a Professor of Natural Resources and Environmental Management at Ball State University in Muncie, Indiana, where he has been on the faculty since 1987. He received the Ph.D. degree in Environmental Science from Ohio State University, the M.S. degree in Soil Chemistry/Agronomy from Ohio State University, and the B.S. degree in Natural Resources Management from Rutgers University. His primary research and professional activities have been in management of hazardous and municipal wastes, remediation of contaminated sites, reclamation of mined lands, and environmental chemistry. He teaches courses in management of solid and hazardous wastes, environmental site assessment, site remediation, and emergency response to hazmat incidents.

Dr. Pichtel is a Certified Hazardous Materials Manager and a Certified Professional Soils Specialist. Dr. Pichtel holds memberships in the Institute of Hazardous Materials Managers, Sigma Xi Scientific Society, the American Society of Agronomy, and the Indiana Academy of Science. He was selected as a Fulbright Scholar in 1999.

In addition to *Waste Management Practices*, Dr. Pichtel has written one book on cleanup of contaminated sites (*Site Remediation Technologies*), and has been the author or co-author of approximately 30 research articles. He has served as a consultant in hazardous waste management projects and has conducted environmental assessments and remediation research in the United States, the United Kingdom, Finland, and Poland.

Acknowledgments

The author wishes to acknowledge Matt Lamoreaux, Suzanne Lassandro, and Mike Masiello of CRC Press who have been instrumental in preparing this work for publication. Their assistance and professionalism have helped to make this book a rewarding experience. Thanks also to the technical editors at Macmillan India Limited for their fine work in editing the manuscript.

I am indebted to Mr. Brian Miller, Ms. Kendra Becher and Mr. Grant Daily for preparing the figures in this text.

Thanks to those loved ones who provided their unfailing support and encouragement throughout the course of this project: Rose and Ed Gonzalez; Theresa, Leah and Yozef Pichtel; and my respected colleagues Paul Weller and Joseph Timko.

Finally, special thanks to my students: your desire to understand the underlying mechanisms, trends, and other concepts of environmental science provided me with the incentive to pursue this work.

Table of Contents

Chapter 3

Part II
Municipal Solid Wastes

Chapter 4

Chapter 7
Municipal Solid Waste Processing; Materials Recovery Facilities..............................169

Chapter 21
Construction and Demolition Debris...609

Part I

Historical and Regulatory Development

The term *solid waste* is a rather generic one, used to describe those materials that are of little or no value to humans; in the same vein, disposal may be preferred rather than use. Solid wastes have been categorized by citizens and governments alike as municipal solid waste, domestic waste, and household waste. As we shall see, however, the regulatory definition of solid waste is a highly inclusive one, incorporating hazardous wastes, nonhazardous industrial wastes, sewage sludges from wastewater treatment plants, along with garbage, rubbish, and trash. However, not all of the above wastes are necessarily managed in the same manner or disposed in the same facility. The definition only serves as a starting point for more detailed management decisions.

Until recently, waste was given a low priority in the conference rooms of municipal, state, and federal offices responsible for public health and safety. Waste management has now become a pressing concern for industrial societies because they produce large volumes of waste as a result of economic growth and lifestyle choices. There have been concomitant concerns regarding the inherent hazards of many such materials, as well as the cost of their overall management and disposal.

Over the past decade and a half, significant legislation has been enacted for the purpose of protecting humans and the local environment from the effects of improper waste management and disposal. Additionally, a wide range of economic incentives (e.g., grants and tax breaks) have been made available to municipalities, corporations, and universities to support waste reduction, recycling, and other applications of an integrated waste management program. Some have proven highly successful.

Part 1 provides the reader with a framework within which to establish a context for the management of many types of wastes. Following the Introduction is a history of waste management and then a discussion of regulatory development in waste management.

1 Introduction

Conspicuous consumption of valuable goods is a means of reputability to the gentleman of leisure.

Thorstein Veblen
The Theory of the Leisure Class, 1899

As recently as three to four decades ago in the United States, the chemical, physical, and biological properties of the municipal solid waste stream were of little or no concern to the local hauling firm, the city council, or the citizens who generated the waste. Similarly, little thought was given to the total quantities of waste produced. Waste volumes may have appeared fairly consistent from year to year, since few measurements were made. Wastes were transported to the local landfill or perhaps the town dump alongside the river for convenient final disposal. The primary concerns regarding waste management were, at that time, aesthetic and economic, i.e., removing nuisance materials from the curb or the dumpster quickly and conveniently, and at the lowest possible cost.

By the late 1980s, however, several events were pivotal in alerting Americans to the fact that the present waste management system was not working. When we threw something away, there was really no "away":

1. *The Islip Garbage Barge.* The *Mobro 4000* left Islip, Long Island, with another load of about 3100 tons of garbage for transfer to an incinerator in Morehead City, North Carolina. Upon learning that the barge may be carrying medical waste, concerns were raised by the receiving facility about infectious materials on board, and the *Mobro* was refused entry. From March through July 1987, the barge was turned away by six states and countries in Central America and the Caribbean (Figure 1.1) The Mexican Navy intercepted the barge in the Yucatan Channel, forbidding it to enter Mexican waters. The ongoing trials of the hapless barge were regular features on many evening news programs. A municipal solid waste incinerator was eventually constructed in the Islip area to receive the aged wastes for final disposal.
2. *Beach washups.* In 1988, medical wastes began to wash up on the beaches of New York and New Jersey. In 1990 the same phenomenon occurred on the West Coast. Popular beaches along the East Coast and in California closed because of potentially dangerous public health conditions. Outraged officials sought the causes of this pollution, arranged for clean-ups, and attempted to assure the public that the chances of this medical debris causing illness were highly remote; however, public fears of possible contact with hepatitis B and HIV viruses led to a concomitant collapse in local tourist industries.
3. *The Khian Sea.* This cargo ship left Philadelphia in September, 1986, carrying 15,000 tons of ash from the city's municipal garbage incinerator for transfer to a landfill (Figure 1.2). It was soon suspected that the ash contained highly toxic chlorinated dibenzodioxins; as a result, the ship was turned away from ports for 2 years, during which it wandered the high seas searching for a haven for its toxic cargo (Holland Sentinel, 2002). A total of 4000 tons of the toxic ash was dumped on a beach in Haiti near the port of Gonaives. An agreement was arranged 10 years later for the eventual return of the ash to the United States.
4. *The plight of the sanitary landfill.* The mainstay for convenient waste disposal in the United States was becoming increasingly difficult and costly to operate and keep

FIGURE 1.1 The ill-fated "garbage barge" from Islip, Long Island. (Greenpeace/Dennis Capolongo. No archiving; not for resale.)

    functioning Increasingly stringent regulations for landfill construction, operation, and final closure were forcing underperforming landfills to shut down. Those that remained in operation were compelled to charge higher tipping (i.e., disposal) fees, often in the form of increased municipal taxes.

5. *Love Canal*. This event galvanized American society into an awareness of the acute problems that can result from mismanaged wastes (particularly hazardous wastes). In the 1940s and 1950s, the Hooker Chemical Company of Niagara Falls, New York, disposed over 100,000 of hazardous petrochemical wastes, many in liquid form, in several sites around the city. Wastes were placed in the abandoned Love Canal and also in a huge unlined pit on Hooker's property. By the mid-1970s chemicals had migrated from the disposal sites. Land subsided in areas where containers deteriorated, noxious fumes were generated, and toxic liquids seeped into basements, surface soil, and water. The incidence of cancer, respiratory ailments, and certain birth defects was well above the national average. A public health emergency was declared for the Love Canal area, and many homes directly adjacent to the old canal were purchased with government funds and the residents were evacuated. Numerous suits were brought against Hooker Chemical, both by the U.S. government and by local citizens. At the time, however, there was simply no law that assigned liability to responsible parties in the event of severe land contamination.

    With greatly enhanced environmental awareness by U.S. citizenry, and with public health, environmental as well as economic concerns a paramount focus of many municipalities, a proactive and *integrated* waste management strategy has evolved. The new mindset embraces waste reduction, reuse, resource recovery, biological processing, and incineration in addition to conventional land disposal. Given these new priorities, therefore, the importance of documenting the composition and quantities of municipal solid wastes (MSWs) produced, and ensuring its proper management (including storage, collection, segregation, transport, processing, treatment, disposal, recordkeeping, and so on) within a community, city, or nation cannot be overstated.

(a)

(b)

FIGURE 1.2 Legacy of the *Khian Sea*: (a) ash pile dumped on Haitian beach; (b) *Khian Sea* sailor eating ash on the beach, attempting to disprove any hazard. (Greenpeace/Dennis Capolongo. No archiving; not for resale.)

1.1   DEFINITION OF A SOLID WASTE

We can loosely define solid waste as a solid material possessing a negative economic value, which suggests that it is cheaper to discard than to use. Volume 40 of The U.S. Code of Federation Regulations (40 CFR 240.101) defines a solid waste as:

> garbage, refuse, sludges, and other discarded solid materials resulting from industrial and commercial operations and from community activities. It does not include solids or dissolved material in domestic sewage or other significant pollutants in water resources, such as silt, dissolved or suspended solids in industrial wastewater effluents, dissolved materials in irrigation return flows or other common water pollutants.

1.2 CATEGORIES OF WASTES

American consumers, manufacturers, utilities, and industries generate a wide spectrum of wastes possessing drastically different chemical and physical properties. In order to implement cost-effective management strategies that are beneficial to public health and the environment, it is practical to classify wastes. For example, wastes can be designated by generator type, i.e., the source or industry that generates the waste stream. Some major classes of waste include:

- Municipal
- Hazardous
- Industrial
- Medical
- Universal
- Construction and demolition
- Radioactive
- Mining
- Agricultural

In the United States, most of the waste groupings listed above are indeed managed separately, as most are regulated under separate sets of federal and state regulations.

1.2.1 MUNICIPAL SOLID WASTE

MSW, also known as domestic waste or sometimes household waste, is generated within a community from several sources, and not simply by the individual consumer or a household. MSW originates from residential, commercial, institutional, industrial, and municipal sources. Examples of the types of MSW generated from each major source are listed in Table 1.1.

Municipal wastes are highly heterogeneous and include durable goods (e.g., appliances), nondurable goods (newspapers, office paper), packaging and containers, food wastes, yard wastes, and miscellaneous inorganic wastes (Figure 1.3). For ease of visualization, MSW is often divided into two categories: garbage and rubbish. Garbage is composed of plant and animal waste generated as a result of preparing and consuming food. This material is putrescible, meaning that it can decompose quickly enough through microbial reactions to produce bad odors and harmful gases. Rubbish is the component of MSW excluding food waste, and is nonputrescible. Some, but not all, of rubbish is combustible. Table 1.2 lists materials that constitute MSW.

TABLE 1.1
Municipal Solid Waste Generation as a Function of Source

Residential (single- and multi-family homes)	Food scraps, food packaging, cans, bottles, newspapers, clothing, yard waste, old appliances
Commercial (office buildings, retail companies, restaurants)	Office paper, corrugated boxes, food wastes, disposable tableware, paper napkins, yard waste, wood pallets
Institutional (schools, hospitals, prisons)	Office paper, corrugated boxes, cafeteria waste, restroom wastes, classroom wastes, yard waste
Industrial (packaging and administrative; *not* process wastes)	Office paper, corrugated boxes, wood pallets, cafeteria wastes
Municipal	Litter, street sweepings, abandoned automobiles, some construction and demolition debris

Adapted from Franklin Associates, EPA530-R-98-010, 1999.

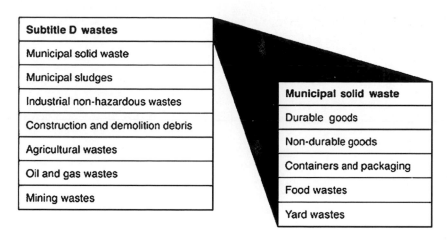

FIGURE 1.3 MSW as a component of Subtitle D wastes. (Reproduced with the kind permission of Franklin Associates, Prairie Village, KS,1998.)

TABLE 1.2
Physical Composition of Municipal Solid Waste

Chemical Class	General Composition	
Organic	Paper products	Office paper, computer printout, newsprint, wrappings
		Corrugated cardboard
	Plastics	Polyethylene terephthalate (1)[a]
		High-density polyethylene (2)
		Polyvinyl chloride (3)
		Low-density polyethylene (4)
		Polypropylene (5)
		Polystyrene (6)
		Multi-layer plastics (7)
		Other plastics including aseptic packaging
	Food	Food (putrescible)
	Yard waste	Grass clippings, garden trimmings, leaves, wood, branches
	Textiles/rubber	Cloth, fabric
		Carpet
		Rubber
		Leather
Inorganic	Glass	Clear ("flint")
		Amber, green, brown
	Metals	Ferrous
		Aluminum
		Other non-ferrous (copper, zinc, chromium)
	Dirt	Dirt
		Stones
		Ash
	Bulky wastes	Furniture, refrigerators, stoves, etc. ("white goods")

[a] Plastics coding system, Society of the Plastics Industry, Inc.

The heterogeneity of the waste stream is further demonstrated at the disposal site. As we shall see in Chapter 3, Subtitle D of the Resource Conservation and Recovery Act (RCRA) regulates the management of wastes other than hazardous wastes. As shown in Table 1.3, RCRA Subtitle

D landfills accept many kinds of wastes, and not just MSW. Household items such as old newspapers, food waste, plastic packaging and refrigerators may be present along with construction and demolition debris. It has been a common practice to landfill nonhazardous industrial wastes, such as those from the oil and gas industry, as well as residues from automobile salvage operations, along with MSW. About 16% of municipal landfills are used for the disposal of sewage sludges ("biosolids") from wastewater treatment plants (Tammemagi, 1999). Such co-disposal practices are fairly common, because wastewater treatment plants are often owned and operated by the same municipal body, thus encouraging cooperation between the two facilities. Finally, MSW landfills contain a significant proportion of potentially hazardous materials from homes. Households generate used motor oil, pesticide and paint containers, batteries, household solvents, and many other hazardous wastes, albeit in small volumes.

1.2.2 HAZARDOUS WASTE

Hazardous wastes are produced by most, if not all, of the sources listed in Table 1.1. However, when the monthly quantity generated exceeds a certain limit, both the wastes and the generator are subject to compliance with federal and state regulations. RCRA defines hazardous waste as (40 CFR 240.101):

> Any waste or combination of wastes which pose a substantial present or potential hazard to human health or living organisms because such wastes are non-degradable or persistent in nature or because they can be biologically magnified, or because they can be lethal, or because they may otherwise cause or tend to cause detrimental cumulative effects.

In other words, an RCRA hazardous waste is a solid waste that, owing to its quantity, concentration, or physical, chemical or infectious characteristics, may (1) cause or contribute to an increase in mortality, serious illness, or incapacitation or (2) pose a substantial hazard to human health or the environment when improperly treated, stored, transported or disposed of, or otherwise managed.

Solid wastes are classified as hazardous under the RCRA regulations if they exhibit one or more of the following characteristics:

- Ignitability
- Corrosivity
- Reactivity
- Toxicity

Examples of hazardous wastes include residues from solvent manufacture, electroplating, metal treating, wood preserving, and petroleum refining. Regulations require that these wastes be managed much more stringently as compared with ordinary MSW. For example, an extensive "paper trail" is required, indicating the status of the waste from the point of generation, through interim storage, treatment (if any), transportation, and ultimate disposal. Requirements are stringent for facilities that generate waste, as well as for transporters, and treatment, storage and disposal facilities. This "cradle to grave" approach to handling hazardous wastes has been central to promoting sound management.

1.2.3 INDUSTRIAL WASTE

Billions of tons of industrial solid waste are generated and managed on-site at industrial facilities each year; the amount generated is approximately four times greater than the amount of MSW produced (Tammemagi, 1999). Generated by a broad spectrum of U.S. facilities, industrial wastes are by-products from manufacturing and other processes. Many (but not all) of these wastes are of low toxicity and are usually produced in fairly large quantities by an individual generator. Examples of an industrial waste stream are coal combustion solids, including bottom ash, fly ash, and flue gas desulfurization sludge. Other common sources of industrial wastes are the pulp and paper industry, the iron and steel industry, and the chemical industry.

Industrial waste is usually not classified directly by federal or state laws as either municipal waste or hazardous waste. If an industrial waste stream, based on a knowledge of the processes involved and laboratory testing, is designated as hazardous waste, the waste must be managed as such and shipped to a licensed treatment, storage, and disposal facility. Wastes designated as non-hazardous are placed in landfills or land application units (typically installed on company property), or incinerated. A large proportion of industrial waste is composed of wastewater, which is stored or treated in surface impoundments. Treated wastewaters are eventually discharged into surface waters under Clean Water Act permits issued by the U.S. EPA or state governments via the National Pollutant Discharge Elimination System (NPDES).

State and some local governments have regulatory responsibility for ensuring appropriate management of industrial waste. Regulatory programs will therefore vary widely.

1.2.4 MEDICAL WASTE

Medical waste is generated during the administration of healthcare by medical facilities and home healthcare programs, or as a result of research by medical institutions. The U.S. institutions generating most of the medical waste include hospitals, physicians, dentists, veterinarians, long-term healthcare facilities, clinics, laboratories, blood banks, and funeral homes. The majority of regulated medical waste, however, is generated by hospitals. Although not all waste generated by the above sources is considered infectious, many facilities choose to handle most or all of their medical waste streams as potentially infectious.

Specific classes of regulated medical wastes include: cultures and stocks of infectious agents (e.g., cultures from medical, pathological, research, and industrial laboratories); pathological wastes (tissues, organs, body parts, body fluids); waste human blood and blood products; sharps (both used and unused hypodermic needles, syringes, scalpel blades, etc.) for animal or human patient care or in medical, research, or industrial laboratories; animal waste (contaminated carcasses, body parts, and the bedding of animals exposed to infectious agents); and isolation wastes (discarded materials contaminated with fluids from humans who are isolated to protect others from highly communicable diseases) (40 CFR Part 259).

Congress passed the Medical Waste Tracking Act in November, 1988, which directed the U.S. EPA to develop protocols for the comprehensive management of infectious waste. RCRA was amended to include medical waste management. The Act established a cradle-to-grave medical waste tracking program. The medical waste tracking program had limited participation, and the program expired in June 1991 without being reauthorized by Congress; however, the course of U.S. medical waste management changed significantly as a result of this legislation.

1.2.5 UNIVERSAL WASTE

Universal wastes include: (1) batteries such as nickel–cadmium and small lead–acid batteries found in electronic equipment, mobile telephones, and portable computers; (2) agricultural pesticides that have been recalled or banned from use, or are obsolete; (3) thermostats that contain liquid mercury; and (4) lamps that contain mercury or lead.

Universal wastes are generated by small and large businesses regulated under RCRA; these businesses had been required to classify the above materials as hazardous wastes. The Universal Waste Rule, first published in the May 1995 Federal Register, was implemented to ease the regulatory burden on businesses that generate these wastes. Specifically, the Rule simplifies requirements related to notification, labeling, marking, prohibitions, accumulation time limits, employee training, response to releases, offsite shipments, tracking, exports, and transportation. Universal wastes are also generated by households, which are not regulated under RCRA and are permitted to dispose of these wastes in the trash.

Many industries strongly support the Universal Waste Rule because it facilitates company efforts to establish collection programs and participate in manufacturer take-back programs required by a number of states. Also appealing to industry are substantial cost savings when the above wastes do not have to be managed as hazardous. The implementation of universal waste programs varies from state to state; for example, some states have included their own universal wastes in addition to those listed by federal regulations.

1.2.6 Construction and Demolition Debris

Construction and demolition (C&D) debris is waste material produced during construction, renovation, or demolition of structures. Structures include residential and nonresidential buildings as well as roads and bridges. Components of C&D debris include concrete, asphalt, wood, metals, gypsum wallboard, and roofing. Land-clearing debris such as tree stumps, rocks, and soil are also included in C&D debris.

1.2.7 Radioactive Waste

Radioactive wastes are a specialized category of industrial wastes. The main generators are electricity-producing nuclear plants, nuclear waste reprocessing facilities, and nuclear weapons facilities. Radioactive wastes are also produced by research and medical (e.g., pharmacological) procedures. Radioactive wastes are, by definition, unstable; they contain atoms with nuclei that undergo radioactive decay. Energy is naturally released from the nucleus to convert it into some stable form. Energy can be emitted as particles or electromagnetic waves. Particles include alpha particles, which are composed of two protons and two neutrons (the equivalent of a helium atom stripped off its planetary electrons), and beta particles, essentially identical to electrons. Gamma radiation is a form of electromagnetic energy similar to light or x-rays.

A major concern with radioactive materials (including wastes) is their capability of causing effects from a distance; in other words, particles, and particularly gamma radiation, can travel for a measurable distance. Gamma waves can penetrate matter including living tissue. The alpha, beta, and gamma forms of radioactive energy are designated "ionizing radiation" because they can ionize other matter, i.e., create a charge on a previously uncharged atom or molecule. This effect is potentially hazardous to health, as ionized nucleic acids (DNA and RNA) can lead to genetic mutations and cancer.

High-level radioactive wastes are generated in nuclear plants by the fission of uranium nuclei in a controlled reaction. The Nuclear Regulatory Commission (NRC) defines high-level radioactive waste as (10 CFR Part 72):

1. the highly radioactive material resulting from the reprocessing of spent nuclear fuel, including liquid waste produced directly in reprocessing and any solid material derived from such liquid waste that contains fission products in sufficient concentrations; and
2. other highly radioactive material that the Commission, consistent with existing law, determines by rule requires permanent isolation.

Spent uranium fuel is an example of a highly radioactive waste and contains many other radionuclides. Generators of this waste include commercial nuclear plants that produce electricity, nuclear waste reprocessing facilities, and nuclear weapons facilities. These wastes are highly regulated and rigorously managed; there are strict licensing requirements for the storage of spent nuclear fuel and high-level radioactive waste (10 CFR Part 72).

Due to the inherent hazards, the disposal of high-level wastes is fraught with controversy. For most nuclear-technology countries, the primary disposal choice involves some form of sophisticated burial in deep, stable geologic formations. In the United States, the Yucca Mountain site, located about 90 miles north of Las Vegas, Nevada, is under consideration as the primary choice for a repository. The site appears to have the approval of engineers due to the presence of volcanic tuff deposits, substantial depth to groundwater, and an arid environment. The Yucca Mountain site had been approved by both President Bush and the U.S. Senate to serve as the nation's high-level nuclear waste repository (*New York Times*, 2002a, 2002b). Legal and political battles continue over this decision, however.

Low-level radioactive wastes comprise many diverse materials generated from industrial, research, educational, and other processes. Sources include private and government laboratories, industry, hospitals, and educational and research institutions. The Nuclear Regulatory Commission (NRC) defines low-level radioactive waste as radioactive material that (10 CFR Part 62):

1. is not high-level radioactive waste, spent nuclear fuel, or by-product material as defined in the Atomic Energy Act of 1954 (42 U.S.C. 2014(e)(2)); and
2. the NRC, consistent with existing law and in accordance with 10 CFR Part 61, classifies as low-level radioactive waste.

Low-level radioactive wastes consist of trash and other materials that have come into contact with radioactive materials, and may have become measurably radioactive themselves. Such wastes include cleanup items like mops and rags, lab gloves, protective clothing, filters, syringes, tubing, and machinery. Hundreds of different radionuclides can occur in low-level waste (Tammemagi, 1999). Approximately two million cubic feet of low-level radioactive wastes are disposed at commercial disposal sites annually (Liu and Liptak, 2000).

Several techniques are available for the disposal of low-level radioactive wastes. In the United.States, some wastes are buried in trenches situated in thick clay formations. Some are permitted for disposal in a Subtitle D sanitary landfill. In France, low-level wastes are stored in heavily reinforced concrete vaults (Tammemagi, 1999).

1.2.8 MINING WASTE

Mine waste includes the soil or overburden rock generated during the physical removal of a desired resource (coal, precious metals, etc.) from the subsurface. Mine waste also includes the tailings or spoils that are produced during the processing of minerals, such as by smelting operations. In addition, heap wastes are produced when precious metals such as gold, silver, or copper are recovered from piles of low-grade waste rock or tailings by spraying with acid or cyanide solutions.

In mining operations, overburden wastes and tailings are returned to the surrounding environs (Figure 1.4). Due to the enactment of federal and state mining reclamation laws—for example the

FIGURE 1.4 Mining wastes are typically returned to the site of operation.

Surface Mining Control and Reclamation Act (SMCRA) of 1977—mine operators are required to return the affected site to its previous contours and land use, and must post a sufficient bond until all operations are satisfactorily completed.

Quantitative estimates of mine wastes produced in the United States are limited; estimates range from 1 to 2 billion tons annually. Approximately one half occurs as overburden spoils, and the remaining one half as heap leach waste (Rhyner et al., 1995).

1.2.9 AGRICULTURAL WASTE

The largest proportion of agricultural wastes occurs as animal manures and crop residues; however, other wastes, such as pesticide containers and packaging also contribute to this category.

In the United States, agricultural wastes are produced in much greater quantities than are municipal solid wastes. Much of this waste goes unnoticed by most Americans, however, as the sources are more diffuse and wastes are generated in areas of low population density. In small-scale agricultural operations, animal and plant wastes can be recycled directly on to the soil surface. Used on-site, this process can be viewed as the application of an inexpensive soil amendment (Figure 1.5). However, when large numbers of animals are concentrated in a relatively small area, for example in livestock feedlots and poultry operations, the accumulation and management of the wastes become a more acute concern. Manures may need to be moved off-site for disposal; cost and feasibility issues become significant, as manures are composed mostly of water and are therefore only a dilute source of plant nutrients. Problems related to odor, pathogen content, salt concentration, and ammonia production are also present. In such cases, more sophisticated management techniques may be required to reduce the volume and potential toxicity of the wastes (e.g., anaerobic digestion or composting), thereby rendering the material more cost-effective for transport as well as hygienically safe.

FIGURE 1.5 Animal manures applied to agricultural fields serve as a low-cost soil conditioner and source of nutrients.

The management of municipal, hazardous, medical, universal, C&D, and other special wastes will be discussed in detail in subsequent chapters. The management of radioactive, mining, and agricultural wastes is not covered in this book.

1.3 GENERATION OF MSW

At the close of the Second World War, economic activity increased markedly for many Americans. Following the fulfillment of basic material needs, expenditures for personal consumption increased substantially. Americans have enjoyed a growing amount of discretionary spending dollars. As a result of this growth in personal consumption expenditure (PCE) dollars, otherwise referred to as consumer spending, we have increasingly become a nation of consumers. Waste generation is inevitably correlated with this increased consumption.

Advertising has been central to stoking the current level of overconsumption in American society. In addition, new marketing and production practices, such as disposable products and the planned obsolescence of various goods, have been introduced (Tammemagi, 1999). To exacerbate the situation, packaging has become important in marketing practices for consumer goods. Packaging now comprises more than one-third of the U.S. waste stream. The overall result of these trends has been an explosive growth in the variety and volume of consumer goods, and concurrently in the volumes and heterogeneity of solid wastes (Figure 1.6). The need for adequate management of wastes, therefore, continues to grow in urgency.

Table 1.3 and Figure 1.7 show trends in MSW generation, materials recovery, and disposal in the United States from 1960 to 1999. The generation of MSW has increased steadily, from 80 million metric tons (88 million tons) in 1960 to 208 million metric tons (229 million tons) in 1999 (U.S. EPA, 2001). Per capita waste generation increased from 1.2 kg (2.7 lb) per person per day in 1960 to 2.1 kg (4.6 lb) per person per day in 1999. Only recently have annual per capita waste generation rates begun to stabilize. Such trends have occurred partly because the public is more informed of environmental concerns and responsibility (i.e., awareness of reduce, reuse, and recycling), and partly because disposal costs have increased markedly.

1.4 SOLID WASTE MANAGEMENT

Solid waste management is concerned with the generation, on-site storage, collection, transfer, transportation, processing and recovery, and ultimate disposal of solid wastes.

FIGURE 1.6 Reproduced with the kind permission of King Features Syndicate.

TABLE 1.3
Materials Generated (thousands of tons) in MSW, 1960 to 1999

Material	1960	1970	1980	1990	1999
Wastes from Specific Products					
Paper and paperboard	29,990	44,310	55,160	72,730	87,470
Glass	6,720	12,740	15,130	13,100	12,560
Metals					
Ferrous	10,300	12,360	12,620	12,640	13,320
Aluminum	340	800	1,730	2,810	3,130
Other nonferrous	180	670	1,160	1,100	1,390
Total Metals	10,820	13,830	15,510	16,550	17,840
Plastics	390	2,900	6,830	17,130	24,170
Rubber and leather	1,840	12,970	4,200	5,790	6,220
Textiles	1,760	2,040	2,530	5,810	9,060
Wood	3,030	3,720	7,010	12,210	12,250
Other	70	770	2,520	3,190	4,010
Total	54,620	83,280	108,890	146,510	173,580
Other Wastes					
Food wastes	12,200	12,800	13,000	20,800	25,161
Yard wastes	20,000	23,200	27,500	35,000	27,730
Miscellaneous inorganic wastes	1,300	1,780	2,250	2,900	3,380
Total other Wastes	33,500	37,780	42,750	58,700	56,270
Total MSW generated	88,120	121,060	151,640	205,210	229,850

Source: U.S. EPA, 2001.

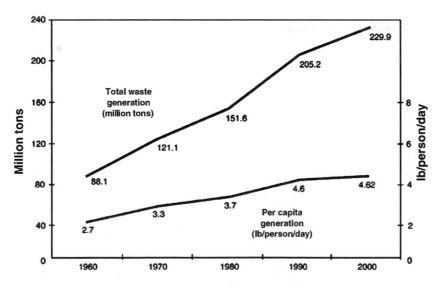

FIGURE 1.7 Trends in total solid waste generation and per capita waste generation in the United States. (U.S. EPA, 2001.)

The U.S. EPA Agenda for Action of 1989 (U.S. EPA, 1989) promoted an innovative and comprehensive program for *integrated waste management*, i.e., the utilization of technologies and management programs to achieve waste management objectives. The U.S. EPA integrated waste management hierarchy includes the following components, in order of preference:

- Reducing the quantity and toxicity of waste
- Reusing materials
- Recycling materials
- Composting
- Incineration with energy recovery
- Incineration without energy recovery
- Sanitary landfilling

Strategies that emphasize the top of the hierarchy are encouraged whenever possible; however, all components are important within an integrated waste management system. The integrated waste management program is customized to meet a particular community's capabilities and needs based on criteria such as population size, presence of industry and business, existing infrastructure, and financial resources. The integrated approach has made great strides within the past decade in educating the American consumer about individual responsibility in waste management, in fostering industry cooperation in waste reduction, and, ultimately, in reducing some of the massive volumes of wastes targeted for landfill disposal.

In 1999, 64 million tons of MSW (a recovery of 27.8%) was recycled (including composting). A total of 34 million tons was combusted (14.8%), and 131.9 million tons (57.4%) was landfilled. Relatively small amounts of this total were littered or illegally dumped. Figure 1.8 shows MSW recovered for recycling (including composting) and disposed of by combustion and landfilling in 1999.

Most states have aggressively encouraged recycling and have established goals for rates of various components of the waste stream (e.g., paper wastes, metals, yard waste) to be recycled. Many businesses and industries have responded by establishing goals for reducing wastes from manufacturing processes. Through such participation, businesses have discovered that reducing the amount of hazardous and non-hazardous materials from product manufacture actually results in substantial cost savings. Many states have responded to the integrated waste management initiative by providing financial incentives for source reduction and recycling.

These committed approaches using state-mandated recycling targets and financial incentives have experienced their share of difficulties, however. At the initiation of such programs, the supply

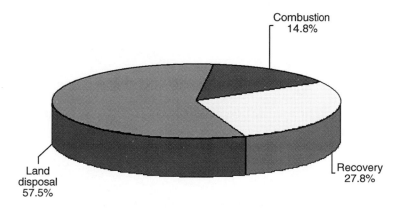

FIGURE 1.8 Current waste management priorities in the United States. (U.S. EPA, 2001.)

of diverted materials grows, but often without a corresponding demand. Consequently, collected materials may have to be stockpiled or sometimes dumped until new markets are created through incentives, legislation, or the market system. Such pitfalls are addressed in later chapters.

1.4.1 SOURCE REDUCTION

Source reduction or waste prevention includes the design, manufacture, purchase, or use of materials such as products and packaging, in a manner that reduces their amount or toxicity before they enter the waste management system. In other words, by not producing the waste, there is no longer a concern over storage, collection, disposal costs, and liability. Examples of source reduction activities include (U.S. EPA, 2001):

- Designing products or packaging to reduce the quantity or the toxicity of the materials used, or to make the materials easy to reuse.
- Reusing existing products or packaging, for example, refillable bottles, reusable pallets, and reconditioned barrels and drums.
- Lengthening the lives of products such as tires to postpone disposal.
- Using packaging that reduces the amount of damage or spoilage to the product.
- Managing nonproduct organic wastes (e.g., food scraps and yard waste) through on-site composting or other alternatives to disposal (e.g., leaving grass clippings on the lawn).

The U.S. EPA has only recently been estimating source reduction based on national production and disposal data. In 1999, the U.S. public and businesses prevented more than 50 million tons of MSW from entering the waste stream. Containers and packaging represent approximately 24% of the materials source reduced in 1999, in addition to nondurable goods (e.g., newspapers) at 18%, durable goods (e.g., appliances, furniture, tires) at 11%, and other materials (e.g., yard and food wastes). Almost half (47%) of the total waste prevented since 1992 includes organic waste materials such as yard and food wastes. This is the result of many locally enacted bans on the disposal of yard waste from landfills, as well as successful programs promoting backyard composting. Table 1.4 shows the progress made in waste reduction since 1992. The rate of source reduction for selected materials in the waste stream is presented in Table 1.5.

The disposal of some materials have increased over the past decade. In particular, clothing shows significantly increased disposal rates, as do plastic containers. Some of the rise in plastics use is attributed to the trend of manufacturers substituting glass packaging with plastic. Another waste category experiencing explosive growth is that of electronic wastes, such as personal computers. The management of electronic wastes is dealt with in Chapter 22.

TABLE 1.4
The Progress of Source Reduction in the United States, 1992–1999

Year	Amount Source Reduced, tons
1992	630,000
1994	7,974,000
1995	21,418,000
1996	23,286,000
1997	32,019,000
1998	40,319,000
1999	50,042,000

Source: U.S. EPA, 2001.

TABLE 1.5
Source Reduction by Major Categories, 1999

Waste Stream	Amount Source Reduced (tons)
Durable goods (e.g., appliances, furniture)	5,289,000
Nondurable good (e.g., newspapers, clothing)	8,956,000
Containers and packaging (e.g., bottles, boxes)	12,004,000
Other MSWs (e.g., yard wastes and food wastes)	23,793,000
Total source reduction from 1990 baseline	50,042,000

Source: U.S. EPA, 2001.

Over the past several decades, there had been few incentives for industry to manufacture more durable products, reduce the amount of material used in the product, design products that could be easily repaired, use minimal packaging, use potentially recyclable packaging materials, or purchase post-consumer wastes as raw materials for manufacturing processes. Many of these approaches are now supported enthusiastically (Rhyner et al., 1995). Government incentives and mandates, legislated recycling targets, public support, and concern for the "bottom line" (via curbing waste removal and disposal costs) have all contributed to the growing interest and participation in integrated waste management.

1.4.2 RECYCLING

Recycling (including community composting programs) recovered 28% (62 million tons) of the total of 229 million tons of MSW generated in 1999 (U.S. EPA, 2001). The percentage recycled was up from 16% in 1990 and 10% in 1980. There were over 9300 curbside recycling programs in the United States in 1998. About 3800 yard waste composting programs were reported in 1998. Waste recycling will be discussed in detail in Chapter 6.

1.4.3 INCINERATION

Incineration is defined as the controlled burning of solid, liquid, or gaseous wastes. "Controlled" conditions may include an oxygen-enriched combustion chamber under elevated temperatures, the use of auxiliary fuel, and vigorous agitation of the incoming waste. About 15% of all MSW generated is disposed via incineration.

The main purpose of incineration is volume reduction, with the ultimate result of extending the lifetime of a land disposal facility. A second purpose has been labeled "waste to energy," i.e., the recovery of heat energy from combustion for water or space heating or electricity generation. A third benefit of incineration is detoxification — the destruction of microbial and other pathogenic organisms — within the waste. Incineration of MSW, hazardous waste, and medical waste will be discussed in Chapters 9, 15, and 20, respectively.

1.4.4 LAND DISPOSAL

Presently, about 55% of all MSW generated is disposed in landfills. Figure 1.9 shows that the number of municipal solid waste landfills has decreased substantially from about 8000 in 1988 to 2216 in 1999. Average landfill size increased during this period. At the national level, capacity does not appear to be an issue, although problems of insufficient capacity have been experienced in certain regions of the United States, for example in the northeastern states (U.S. EPA, 2001).

With recovery rates increasing and combustion remaining relatively constant, the percentage of MSW discarded to landfills has decreased from 1980 to the present, and has remained relatively

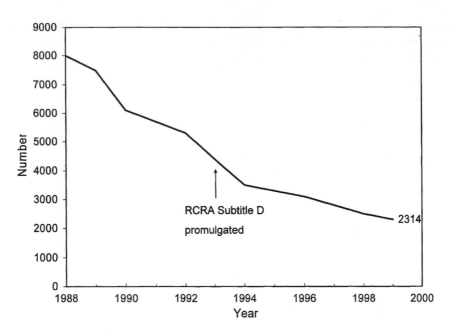

FIGURE 1.9 Decline of the sanitary landfill over the past decade (U.S. EPA, 2001.)

steady. Sanitary landfilling will be presented in Chapter 10, and secure landfilling of hazardous waste in Chapter 17.

1.4.5 GOALS AT THE FEDERAL LEVEL

The U.S. EPA set a goal for the nation to recycle at least 35% of MSW by the year 2005, while reducing the generation of solid waste to 1.9 kg (4.3 lb) per person per day. Because economic growth results in the generation of more products and materials being generated, there will be an increased need to further develop recycling and composting infrastructure, buy more recycled products, and create opportunities for source reduction activities such as reuse of materials and products and modify manufacturing processes (e.g., making containers with less materials) in order to meet these goals.

REFERENCES

Code of Federal Regulations, Vol. 10, Part 61, Licensing Requirements for Land Disposal of Radioactive Waste, U.S. Government Printing Office, Washington, DC, 2004.

Code of Federal Regulations, Vol. 10, Part 62, Criteria and Procedures for Emergency Access to Non-federal and Regional Low-level Waste Disposal Facilities., U.S. Government Printing Office, Washington, DC, 2004.

Code of Federal Regulations, Vol. 10, Part 72, Licensing Requirements for the Independent Storage of Spent Nuclear Fuel and High-level Radioactive Waste, U.S. Government Printing Office, Washington, DC, 2004.

Code of Federal Regulations, Vol. 40, Part 240, Guidelines for the Thermal Processing of Solid Wastes, U.S. Government Printing Office, Washington, DC, 2004.

Code of Federal Regulations, Vol. 40, Part 259, Medical Wastes, U.S. Government Printing Office, Washington, DC.

Franklin Associates, Characterization of Building-related Construction and Demolition Debris in the United States, EPA530-R-98-010, Prepared for the U.S. Environmental Protection Agency, Municipal and Industrial Solid Waste Division, Office of Solid Waste, Prairie Village, KS, June 1998.

Franklin Associates, Characterization of Municipal Solid Waste in the United States: 1998 Update, EPA 530-R-01-014, Prepared for the U.S. Environmental Protection Agency, Office of Solid Waste and Emergency Response, Washington, DC, 1999.

Holland Sentinel, Wandering ash arrives home in Pennsylvania, June 28, 2002, see: http://hollandsentinel.com/stories/062802/new_062802029.shtml

Liu, D.H.F. and Liptak, B.G., *Hazardous Waste and Solid Waste*, Lewis Publishers, Boca Raton, FL, 2000.

New York Times, Bush Signs Bill For Nevada Nuclear Dump, July 24, 2002a, p. 13.

New York Times, Senate Approves Nuclear Waste Site in Nevada Mountain, July 10, 2002b, p. 14.

Rhyner, C.R., L.J. Schwartz, R.B. Wenger, and M.G. Kohrell, *Waste Management and Resource Recovery*, Lewis Publishers, Boca Raton, FL, 1995.

Tammemagi, H., *The Waste Crisis., Landfills, Incinerators, and the Search for a Sustainable Future*, Oxford University Press, Oxford, UK, 1999.

U.S. Environmental Protection Agency, The Solid Waste Dilemma: An Agenda for Action, Final Report of the Municipal Waste Task Force, EPA 530 SW-89-019, Office of Solid Waste and Emergency Response, Washington, DC, 1989.

U.S. Environmental Protection Agency, Municipal Solid Waste in the United States: 1999 Facts and Figures, EPA 530-R-01-014, Prepared for the U.S. Environmental Protection Agency, Office of Solid Waste and Emergency Response, Washington, DC, 2001.

SUGGESTED READINGS AND WEB SITES

Brown, H., Ash use on the rise in the United States, *World Waste*, 1997. p. 16.

Conn, W.D., Reducing municipal solid waste generation: lessons from the seventies, J. Resour. Manage. Technol., 16, 24–27, 1995.

Essential Action, Philadelphia Ash Dumping Chronology. see http://www.essentialaction.org/return/chron.html.

Goldberg, D., The magic of volume reduction, *Waste Age*, 21, 98–104, 1980.

Glenn, J., The state of garbage in America, Biocycle, 32, 65–72, 1998.

Hannon, B.M., Bottles, cans, and energy, *Environment,* 14, 11–21, 1972.

Jones, K.H, Risk assessment: comparing compost and incineration alternatives, *MSW Manage.*, 29-32, 36-39, 1991.

Lober, D., Municipal solid waste policy and public participation in household source reduction. Waste Manage. Res., 14, 129–145, 1996.

Matsuto, T. and Ham, R.K., Residential solid waste generation and recycling in the U.S.A. and Japan, *Waste Manage.* Res., 8, 229–242, 1990.

O'Leary, P.R., Walsh, P.W., and Ham, R.K., Managing solid waste, *Sci. Am.*, 256, 36–42, 1988.

Rathje, W. and Murphy C., Rubbish! *The Archaeology of Garbage,* Harper Collins Publishers, New York, NY, 1992.

Reed, J., *Love Canal*, Chelsea House, Philadelphia, PA, 2002.

Sherman, S., Local government approaches to source reduction, *Resour. Recycling*, 9, 112, 119, 1991.

Shop Earth Smart: A Consumer Guide to Buying Products That Help Conserve Money and Reduce Waste,. The City and County of San Luis Obispo, Cal, 1999.

Streissguth, T., Nuclear and Toxic Waste. Greenhaven Press, San Diego, CA, 2001

U.S. Congress Office of Technology Assessment, Facing America's Trash: What Next for Municipal Solid Waste? OTA-O-424, U.S. Government Printing Office, Washington, DC, 1989.

QUESTIONS

1. The Love Canal (Niagara Falls, New York) disaster was considered by some public health officials and environmental regulators and scientists to be a "blessing in disguise". Explain.
2. Municipal solid waste is generated within a community from several sources, not just the household. List and discuss these sources.
3. "Sanitary landfills and incinerators are no longer adequate to serve America's waste management concerns." Do you agree or disagree with this statement? Discuss.

4. According to the U.S. EPA definition, "solid waste" include: (a) discarded solids; (b) discarded semisolids (sludges); (c) contained gases; (d) materials from commercial, industrial, and domestic sources; (e) all of the above.

5. Under the RCRA regulations, how can a solid waste become designated as a hazardous waste?

6. Industrial waste may or may not include hazardous waste. Explain and provide examples.

7. Livestock feedlot wastes are an environmental concern due to: (a) the low water content of the wastes; (b) potential for eutrophication of water by nitrogen; (c) high pathogen content; (d) low fertilizer value; (e) all of the above

8. What is the primary purpose of the Medical Waste Tracking Act? What event(s) catalyzed this legislation?

9. Why was the Universal Waste Rule enacted? What are its benefits and whom does it serve?

10. Why is ionizing radiation a human health hazard? Explain its mode of action on the body.

11. How is the majority of mining waste disposed? Agricultural waste? Low-level radioactive waste?

12. How are high-level nuclear wastes to be managed and disposed in the near future? Discuss the pitfalls, in your opinion, associated with the single repository approach of nuclear waste disposal.

13. MSW generation is influenced by both population size and consumer lifestyle. How do these two factors differ for the United States and a less developed country, for example India or Mexico?

14. Describe the waste management hierarchy under RCRA. What is the ultimate goal of such a hierarchy?

15. What factors have occurred over the past 50 years to increase significantly the quantities of MSW generated in the United States?

16. Explain why the manufacturing of consumer products using more durable, long-lived materials had been slow to catch on by both industry and the American consumer. If you have traveled to Europe or Japan, did you notice a difference in the quality of consumer products or in the amount of packaging? Explain.

17. In the United States and elsewhere, formal and comprehensive solid waste management programs have been slow to develop. Explain the reasons for this.

18. Based on parameters such as population and economic trends and personal lifestyles, identify the issues you feel will significantly affect waste management in the coming decade.

19. What is your state's position on integrated waste management? If there is a formal program, discuss its administration and practical application including the waste management hierarchy.

2 A Brief History of Waste Management

Man is everywhere a disturbing agent. Wherever he plants his foot, the harmonies of nature are turned to discords.

George Perkins Marsh, 1874

2.1 INTRODUCTION

When early man roamed the Earth, solid waste was probably composed of the remains from hunting, gathering, and food preparation. Human fecal matter comprised the other category of the prehistoric waste stream. When wastes accumulated, nomadic people would simply move to another location. Natural processes of scavenging and microbial decomposition easily absorbed and incorporated such wastes. As a result, and also due to the very low human populations extant at the time, the characteristic problems associated with wastes such as disease, air pollution, and groundwater contamination were probably insignificant.

When man began inhabiting caves, wastes were piled near entrances, and when the heap became too large, inhabitants would simply move on to another dwelling. In about 9000 B.C.E., people began to abandon nomadic life and created permanent communities. Humans advanced from hunters and gatherers to farmers and craftsmen, and became civilized and urbanized. Waste quantities increased and began to accumulate for longer periods. As a result, wastes became more harmful to health and to natural environments. Stationary human societies have since had to confront the logistical problem of how to manage their residues.

The types of materials predominantly in use by early societies, such as tools, weapons, and handiworks, have helped to identify various eras: for example, the Stone Age, the Bronze Age, Iron Age, etc. When archaeologists excavate and examine the villages of ancient peoples, they search the waste piles, cooking hearths, tombs, and structures of the former inhabitants. By sorting through the refuse of ancient habitations, archaeologists have gained insights into the lifestyle, diet, and social order of the inhabitants of early societies. For example, Stone Age humans left behind nondegradable items such as tools, weapons, and utensils. During times of economic decline, the Mayans of Central America buried defective utensils, ornaments, and other household items no longer useful in homes in their royal tombs. Some wastes appear to have been recycled as well — debris such as broken pots and ceramics have been found within the high platforms and walls of some temples (Alexander, 1993).

2.2 EARLIEST CIVILIZATIONS

When civilizations arose in Mesopotamia, Egypt, and elsewhere, the effects of solid wastes became significant; as a result, certain rules and practices emerged to encourage some rudimentary programs of waste management. As early as 8000 to 9000 B.C.E., dumps were established away from settlements, probably located so that wild animals, insects, and odors would not migrate to populated areas (Bilitewski et al., 1997). The Minoans (3000 to 1000 B.C.E.), placed their wastes, covered periodically with layers of soil, into large pits, thus operating the first proto-sanitary land-fills (Priestley, 1968; Wilson, 1977). By 2100 B.C.E., cities on the island of Crete had trunk sewers connecting homes to carry away many wastes (Melosi, 1981; Vesilind et al., 2002). In the Egyptian city of Heracleopolis (founded about 2100 B.C.E.), the wastes in the "non-elite" section were ignored, while in the elite and religious sections efforts were made to collect and dispose of all wastes, which usually ended up in the Nile River (Melosi, 1981). At Kouloure in the ancient Crete capital of Knossos (ca. 1500 B.C.E.), an effective composting effort was established in pits (Kelly, 1973). By 800 B.C.E., old Jerusalem established sewers and had installed a primitive water supply. In the Indus valley, the city of Mohenjo-daro had houses equipped with waste chutes and trash bins, and may have had waste collection systems (Melosi, 1981). Harappa, in the Punjab region, now a part of modern-day India, installed toilets and drains in the bathrooms. Many Asian cities collected waste in clay containers which were hauled away (Vesilind et al., 2002).

The first recorded regulations for the management of solid wastes were established during the Minoan civilization (Tammemagi, 1999). Around 2000 B.C.E., Israel provided guidelines as to how to manage wastes; instructions for the management of human waste are provided in the Bible (Deuteronomy 23:12–13). By 200 B.C.E., many cities in China employed "sanitary police," who were responsible for the enforcement of waste disposal laws (Vesilind et al., 2002).

2.3 GREECE

During the fifth century B.C.E., Greek municipalities began to establish town dumps that were main-tained in a relatively orderly condition. Garbage normally consisted of food waste, fecal matter, pot-sherds, and abandoned babies (e.g., malformed or illegitimate) (Kelly, 1973). In Athens (ca. 320 B.C.E.), each household was responsible for collecting and transporting its wastes (Tammemagi, 1999). Residents were required by law to sweep the streets daily, and it was mandated that wastes be transported to sites beyond the city walls (Bilitewski et al., 1997).

During the early Bronze Age, it was common for the Trojans to allow many of their wastes (e.g., bones, rubbish) to accumulate on floors, which were eventually covered by a layer of soil and packed into a new surface. It has been speculated that floor levels may have been raised by as much as 20 in., possibly requiring inhabitants to raise the roofs and doors of their dwellings periodically (Blegen, 1958; Alexander, 1993). Putrescible and bulky garbage was thrown into the streets, where scavengers such as pigs or geese were allowed to forage among the piles. In some locations, slaves and other "underclass" inhabitants were given the right to pick through the wastes that they carried away (Alexander, 1993). For the most part, however, city dwellers lived amid waste and squalor. Direct action for waste management was implemented only when the volume of wastes affected local defense. For example, in Athens in 500 B.C., a law was passed that required all wastes to be deposited at least 2 km outside of town limits because piles next to the city walls provided an oppor-tunity for invaders to scale over them (Bilitewski et al., 1997).

Greek and Persian scholars were among the first to suggest an association between personal hygiene, contaminated water, spoiled food, and disease outbreaks and epidemics. Hippocrates (ca. 400 B.C.E.) and the Persian Ibn Sina (980–1037 C.E.) suggested a relationship between waste and infectious disease (Bilitewski et al., 1997).

2.4 ROME

In ancient Rome, wastes were dumped into the Tiber River, tossed into the streets, or dumped into open pits on the outskirts of the city. Rome was the first civilization to create an organized waste collection workforce in 14 C.E. (Vesilind et al., 2002). To handle the piles of wastes left on the streets, teams of sanitation workers shoveled the materials into horse-drawn wagons. The collection team transported the refuse to a pit, located either outside the city gates or at some distance from the community. The city's inhabitants, however, often preferred the convenience of a more local, neighborhood dump. Administrators replied by posting signs reading, "Take your refuse further out or you will be fined." The signs included arrows showing the way out of the city (Kelly, 1973).

During the rule of the Caesars (27 B.C.E. to C.E. 410), thousands of carcasses from gladiatorial combats (both human and animal) were disposed in open pits at the city's outskirts. The only known law in existence at that time concerning waste disposal regarded the management and disposal of fecal matter. The sanitation subcommittee of the Roman Senate decreed that fecal matter was not to be disposed of in carts or open pits (Kelly, 1973).

The Romans had gods for every occasion, and they unwittingly had a goddess dedicated to the consequence of their indiscriminate waste disposal, the Goddess of Fever. In spite of their sacrifices at the altars, Rome was a victim to plagues in 23 B.C.E., C.E. 65, 79, and 162. The Romans did not yet fully grasp the connection between waste and infectious diseases. During the first century C.E., Roman emperors began to realize that municipal solid wastes posed a significant public health concern. Emperor Domitian (81–96 C.E.) ordered pest control, because his advisors noticed that lack of cleanliness in the city was associated with an increase in the population of rats, lice, bedbugs, and other vermin (Bilitewski et al., 1997). Emperor Vespasian (69–79 C.E.) ordered the installation of public toilets, which were designed to have running water beneath (Kelly, 1973). By 300 C.E., there were 144 public toilets in Rome (Bilitewski, 1997).

Some researchers claim that over time, waste accumulation may have contributed to the burying of cities, which subsequently were rebuilt. The old Roman section of the city of Bath, England, is 12 to 20 ft beneath the existing city (Wilson, 1977).

The population of Rome eventually grew to over one and a quarter million. At this point, municipal wastes could no longer be handled adequately. Some historians have suggested that the intense odor of these wastes may have driven the aristocracy from the city into the mountains or along the seaside. It is speculated that such a decentralization of power may have precipitated the decline of the empire (Alexander, 1993). Additionally, the growing mounds of wastes outside the city walls are thought to have compromised the defense of the city (Vesilind et al., 2002).

2.5 EUROPE

With the end of the Roman Empire came the loss of any semblance of order and discipline that had been instituted by the imposition of laws and the presence of an organized, active military. Equally significant were the loss of technical knowledge and the science of basic hygiene. As a result, from the Dark Ages through the Renaissance, there was no organized method of waste disposal, with street dumping among the most common practices (Kelly, 1973). Routine procedure was to simply dump wastes, including fecal matter, directly out of a window (Figure 2.1). These materials would decompose naturally and eventually become incorporated into the unpaved street. In some locations, a centralized receptacle was established directly in front of homes for the general dumping of sewage and other wastes.

As the population in Europe surged and also became urbanized, the impact of wastes became more acute. In London, each household established its own waste heap outdoors. According to one report on London's sanitation (Greater London Council, 1969):

> As the population density rose and pressure on land within the urban area increased a street system evolved. The pattern of refuse disposal changed accordingly. Everything from domestic refuse to cinders from foundries, offal from shambles [slaughterhouses] and manure from stables went to the streets where it was placed in the central kennel or gutter.

It was unsafe to burn wastes within the city due to the proximity of countless wooden structures (Wilson, 1977). As a result, wastes remained in place. In 1297, an order was issued that required all tenants to maintain a clear pavement in front of their dwelling. The order was largely ignored. However, much waste was burned in household open fires. During the mid-1300s, scavenging kites

FIGURE 2.1 Medieval woodcut showing fecal matter being dumped from a window.

and ravens were protected by law because they fed upon the waste heaps. According to one report of the period, "The pigs which roamed about grew fat on the offal in the streets. Dogs were innumerable." (Rawlinson, 1958)

The city of Paris had a unique experience associated with their waste problems. In 1131, a law was passed prohibiting swine from running loose in the streets after young King Philip, son of Louis the Fat, was killed in a riding accident caused by a loose pig. The monks of a local abbey protested the law and were granted a dispensation. The controversy on allowing animals to run free in the streets, however, continued for years (Melosi, 1981).

During the medieval period, sufficient fodder to feed winter animals was typically unavailable near large cities like London; hence, many farm animals were slaughtered during the fall when grazing was no longer possible. Smoking and salting meats did not preserve meats for the entire winter, which created a strong demand for spices. Spices were used to mask the foul tastes and odor of partially spoiled meat, fish, and other foods. Even with these efforts, spoiled food was a large component of medieval European wastes (Alexander, 1993).

In 1354, an order was issued that "filth" deposited in front of houses was to be removed weekly (Wilson, 1977). London wards were assigned a beadle or bailiff, who hired assistants called "rakers." Once a week, rakers would collect rubbish and dung from the middle of the streets and from the fronts of houses, following which it was to be carted away, outside the city (Alexander, 1993; Harris and Bickerstaffe, 1990). As quoted by Rawlinson (1958):

> The refuse was raked together and loaded onto tumbrels [farmer's wagons], drawn by two horses. London maintained 12 of these specially designed carts. ... A number of laystalls were [sic] established in the city suburbs and on the banks of the river. Special days were appointed when refuse was to be put outside doors for the rakers to scoop up and trundle away to the laystalls.

Another common practice during the medieval period was to discard wastes into surface water. The plague of 1347 may have been precipitated by waste disposal into watercourses such as the Thames River. Also known as "The Black Death," the plague claimed the lives of 25 million of 80 million European citizens over the period 1347 to 1352. The epidemic was rapidly spread by fleas whose hosts (Norway rats) flourished in the abominable sanitary conditions of the period (Alexander, 1993). Edward III notified the Mayor and Sheriffs of London to discontinue the practice of dumping into waterways, after experiencing an unpleasant trip down the fouled Thames. In 1383, an ordinance was passed against river disposal by people living on the Walbrook watercourse because the preponderance of garbage plugged the river. The English Parliament prohibited dumping filth and garbage into rivers, ditches, and watercourses in 1388. It was also ordered that refuse be carried away to selected sites so that it would not become the source of nuisance (Wilson, 1977). The practice of dumping in water, however, continued illegally into the 19th century with the consequent contamination of roads, rivers, and groundwater by human and animal waste (Bilitewski, 1997).

In 1407, inhabitants of London were instructed to keep their refuse indoors until rakers could carry it away (Wilson, 1977). "Refuse collected was sold to farmers and market gardeners; that from the riverside laystalls was taken downstream in boats to be dumped on the Essex marshes." (Rawlinson, 1958) The paving of streets was also required, so that inhabitants would not have to wade through fecal matter and other wastes (Bilitewski et al., 1997). In 1408, Henry IV ordered that refuse be removed or else forfeits would have to be paid. Garbage cans were introduced during this time. The streets were cleaned, animal carcasses were collected, and the possessions of people who had died from the plague were burned (Bilitewski et al., 1997).

Despite acts, ordinances, and threats, however, the mounds of solid waste persisted as a nuisance and health hazard in Europe. London city officials began paying informers to report offenders who threw their garbage into the streets and who were later fined. As an example, one Londoner in 1421 (Rawlinson, 1958):

> was arraigned for making a great nuisance and discomfort to his neighbours by throwing out horrible filth onto the highway, the stench of which was so odious, that none of his neighbours could remain in their shops.

Paris and some medieval German cities required that wagons, which had brought goods and supplies into the city, must depart with a load of wastes to be deposited in the countryside (Wilson, 1977; Gerlat, 1999).

The waste issue reached a crisis stage in Europe in about 1500. Populations continued to surge into the cities. Garbage from households, animal manure, and industrial debris continued to be dumped into the central gutters in the street. As was the case with Athens 2000 years earlier, municipal wastes were piled so high outside the gates of Paris as to potentially interfere with the defense of the city (Tammemagi, 1999).

Even with the increased efforts of English rakers, whose work included cleaning large public spaces and market squares, lawmakers were still offended by the filth in the streets (Wilson, 1977). Paris was somewhat ahead of London in its institution of municipal street cleaning, paid for by public funds, in 1506 (Wilson, 1977; Hosch, 1967).

During the mid-1600s the population of London reached about 400,000. The journals of Daniel Defoe and Samuel Pepys described the stench of the garbage and documented the plague that was afflicting Europe during this time. About 100,000 inhabitants of London died during the plague of 1665. Even the fashion of that period was affected by the squalor. Doublet and hose for gentlemen and pin-up skirts for ladies were designed to keep their clothing out of the filth of the city thoroughfares; scented handkerchiefs and snuff were also used to help mask the stench (Alexander, 1993). The Great Fire of London in 1666 had some cleansing effect on the city environs, and complaints about refuse in the streets eased to some extent (Wilson, 1977).

During the 1700s, it was ruled that London's inhabitants could not bury dung within the city limits, and could not take out their garbage after 9:00 p.m. By this time of night, lawmakers reasoned, any honest person was home in bed, and those roaming the streets were presumably up to no good, i.e., sneaking somewhere to dump garbage (Kelly, 1973). This and other proposals such as "the removal of ordure (filth, dung, manure) and rubbish lying in the streets" and a suggestion to place the entire London area under a uniform public management so that all filth would be taken by boat on the Thames to "proper distances in the country" were made to no avail (Wilson, 1977).

Profitable uses could be found, however, for virtually every type of waste generated during this period, up to the beginning of the Industrial Revolution. The British were rather enterprising when dealing with certain wastes. Rush-covered floors of some houses during the Tudor period contained debris up to 3 ft thick. This debris was rich in nitrates, and in the early 17th century was "mined" for saltpeter, which was used in the manufacture of gunpowder (Wilson, 1977). Around 1815, the dust from a century-old refuse heap at the bottom of Grays Inn Lane was extracted and sold to Russia to make brick for the rebuilding of Moscow after Napoleon's invasion. The refuse yards of Edinburgh, Scotland, remained the same size for one hundred years because much of the waste that was brought in was sorted and eventually sold (Wilson, 1977). The general composition of wastes of the period tended to be high in ash, dust, and cinder. The composition of London's wastes for over a century is shown in Table 2.1.

The Industrial Revolution had its beginnings in the 18th century, when the availability of raw materials and increased trade and population stimulated new inventions and an intense reliance on machine labor. Increased production led to greatly increased waste generation.

Charles Dickens and other writers have chronicled the living conditions of the working poor in European cities during the 19th century. Industrial production was high priority for governments and businesses, with public health and environmental quality being of lesser importance. Water supply and wastewater disposal were, by modern standards, totally inadequate. For example, Manchester, England, had on average one toilet per 200 people. About one sixth of the city's inhabitants lived in cellars, frequently with walls oozing human waste from nearby cesspools. People often lived around small courtyards where human waste was piled, and which also served

TABLE 2.1
Composition of London's Solid Wastes, 1888–2000

Component	1888	1892	1926	1967	2000
Fine dust and cinder	81.7	83.2	54.8	19.3	—
Vegetable, putrescible and bone	13.2	8.3	14.7	19.2	38.8
Paper	—	4.3	15.0	34.0	19.5
Metals	0.4	1.0	3.6	10.6	3.6
Rags	0.4	0.4	1.8	2.4	—
Glass	1.3	1.4	3.0	10.9	8.4
Plastic	—	—	—	1.3	8.1
Miscellaneous	3.0	1.4	7.0	2.3	21.7[a]

Adapted from Wilson, D.G., *Handbook of Solid Waste Management,* Van Nostrand Reinhold Company, New York, 1977. With permission.

[a]"Fines/miscellaneous" plus textiles.

as children's playgrounds (Vesilind et al., 2002). In 1741, Lord Tyrconnel described the streets of London as "abounding with such heaps of filth as a savage would look on in amazement." In 1832, citizens complained that the streets near Westminster Abbey were "the receptacle of all sorts of rubbish which lay rotting and corrupting, contaminating the air and affording a repast to a herd of swine." (Tammemagi, 1999) (Figure 2.2).

In 1842, Sir Edwin Chadwick drafted the Report from the Poor Law Commissioners on an Inquiry into the Sanitary Conditions of the Labouring Population of Great Britain. The report described the sanitary conditions as follows:

> Many dwellings of the poor are arranged round narrow courts having no other opening to the main street than a narrow covered passage. In these courts there are several occupants, each of whom accumulated a heap. In some cases, each of these heaps is piled up separately in the court, with a general receptacle in the middle for drainage. In others a pit is dug in the middle of the court for the general use of all the occupants. In some the whole courts up to the very doors of the houses were covered with filth.
>
> ... defective town cleansing fosters habits of the most abject degradation and tends to the demoralization of large numbers of human beings, who subsist by means of what they find amidst the noxious filth accumulated in neglected streets and bye-places.

The report included a recommendation that "public authorities undertake the removal of all refuse from habitations, streets and roads, and the improvement of the supplies of water."

In the middle to late 19th century, the research of physicians and scientists such as Frenchman Louis Pasteur, German Robert Koch, and German-Hungarian Ignaz Semmelweis revealed the

FIGURE 2.2 London slum, 19th century.

connection between bacteria and viruses and the incidence of specific diseases. Public health offi-cials eventually linked sanitation practices, including improper waste disposal to the incidence of dis-ease and other health complaints. Thus was born the "Great Sanitary Awakening". The understanding of the pathology of infectious disease may very well have been the incentive behind modern sanita-tion practices such as wastewater treatment and sanitary landfilling (Bilitewski et al., 1997).

By the mid-1830s, London became more strict about its enforcement policies regarding waste disposal. The Metropolitan Police Act of 1839 was enacted, which penalized those who "cut tim-ber or stone; threw or lay coal, stone slates, lime, bricks, timber, iron or other materials; or threw or laid any dirt, litter or ashes, or any carrion, fish, offal or rubbish" into any thoroughfare (Wilson, 1977). London's Public Health Act of 1875 mandated the removal of refuse by the Sanitary Authority on appointed days. All tenants were required to place their wastes into a mobile recepta-cle (this was, incidentally, the first legal recognition of the trash container). The Public Health Act of 1891 directed the Sanitary Authority to "employ or contract with a sufficient number of scav-engers to ensure the sweeping and the cleansing of the several streets within their district and the collection and removal of street and house refuse." Until 1965, the disposal of Greater London's refuse was handled by about 90 local authorities (Wilson, 1977).

2.6 UNITED STATES

As early as 1657 the residents of New Amsterdam (later New York City) prohibited the throwing of garbage into streets; furthermore, keeping streets clean was the responsibility of the individual homeowner (Gerlat, 1999). Garbage was piled high near elegant homes and, reminiscent of Ancient Rome, hogs, geese, dogs, and vultures rummaged for food within the heaps. In 1834, Charleston, West Virginia, enacted a law protecting garbage-eating vultures from hunters (Vesilind et al., 2002).

In early American cities, the collection of MSW was rare. Benjamin Franklin is considered to be one of the first to organize a crude form of sanitation in any of the Colonial cities. In 1792, Franklin hired servants to remove waste from the streets of Philadelphia (Alexander, 1993), which had already expanded to a population of 60,000. According to a plan developed by Franklin, slaves carried loads of wastes on their heads and waded into the Delaware River for waste disposal down-stream from the city (Kelly, 1973). In 1795, the Corporation of Georgetown adopted the first ordi-nance on record in America concerning waste management. The regulations forbade long-term storage of wastes on private property or dumping in the street (APWA, 1976; Wilson, 1977). The ordinance did not, however, provide details on collection or removal of waste. In 1800, Georgetown and Washington, DC, contracted with "carriers" to clean streets and alleys periodically.

Urban solid waste problems caused by rapid industrialization and overcrowding were acute in the northeastern United States, and were probably among the worst worldwide at that time. A flood of immigrants from Europe and Asia exacerbated the on-going population migrations from the countryside to the city (Alexander, 1993). New York slums were, at that time, the most densely pop-ulated acreage in the world, worse than even those in Bombay (Melosi, 1981). Sidewalks were piled high with garbage, and roadways were crowded with carts, horses, and people (Figure 2.3). According to Zinn (1995), the cities:

> were death traps of typhus, tuberculosis, hunger, and fire. In New York, 100,000 people lived in the cel-lars of the slums … the garbage, lying 2 feet deep in the streets, was alive with rats.

As in the colonial period, pigs were allowed to run free because they scavenged some of the garbage (Alexander, 1993).

Other parts of the country were not without problems of poor sanitation and inadequate waste management. Zinn (1995) noted conditions of urban populations in the south after the Civil War:

> And the slums of the southern cities were among the worst, poor whites living like the blacks, on unpaved dirt streets "choked up with garbage, filth and mud", according to a report of one state board of health.

FIGURE 2.3 The waste problem in New York City. Garbage dumped on sidewalks impeded both pedestrian and vehicular traffic.

FIGURE 2.4 Sanitation worker and the horse problem, New York City, 1900

The horse was a major contributor to the urban U.S. waste load (Figure 2.4). There were more than three million horses in U.S. cities at the turn of the century, 120,000 in New York City alone (Melosi, 1981). Each generated about 20 lb of manure per day. Also at that time, over 80,000 horses, mules, and cows were maintained in the city of Chicago. It is estimated that these animals produced about 600,000 tons of manure annually (Melosi, 1973; Wilson, 1977). In 1900, 15,000 horses in Rochester, New York, "produced enough manure to cover an acre of ground to a height of 175 feet"

(Bettmann, 1974; Alexander, 1993). Another difficult issue of this period, owing to their numbers and size, was the disposal of carcasses of dead horses and cattle.

Waste collection at the public's expense began in 1856, and, in 1895, the District of Columbia passed a bill for the construction of incinerators. The incinerators, however, were employed only during winter months. During summer, wastes were placed onto flat-bottomed boats called scows, and transported to a site south of Alexandria, Virginia, for final disposal (Figure 2.5).

During the mid-1800s, health conditions in major American cities were declared deplorable. A New York City citizen, George Strong, noted in his diary in 1852 (Kelly, 1973):

> Such a ride uptown! Such scalding dashes of sunshine coming in on both sides of the choky, hot railroad car. … Then the feast of fat things that come reeking filth that Center Street provided in its reeking, fermenting, putrefying, pestilential gutter! I thought I should have died of the stink, rage, and headache before I got to Twenty-first Street.

That New Yorker, however, actually loved the city despite its overall "civic filthiness."

U.S. public health officials, observing the progress in Europe, made requests for improved disposal of garbage and "night soil" (i.e., human excrement). Despite the complaints of sanitary engineers, journalists, and others, the state of refuse collection and disposal in the United States in the 1880s and early 1890s remained poor. Methods were inconsistent, technology was primitive, and the public, as a whole, did not seem to be concerned (Wilson, 1977). In Chicago, St. Louis, Boston, and Baltimore, much of the wastes were simply carted to open dumps. In New York, street teams collected the garbage where it was carted away by open horse-drawn wagons (with the horses fouling the streets during collection) to barges fated for dumping 25 miles offshore. This practice was still an improvement over older methods. Prior to 1872, the city used simple dumping platforms built over the East River to unload the city's wastes. Given the relatively closed position of the

FIGURE 2.5 Unloading garbage from scows off the Atlantic Coast. According to George Waring: "About twenty Italians unload the cargo of a deck-scow in about two and one half hours. In 1896 over 760,000 cubic yards of refuse were disposed of in this manner, on 1531 scows."

Lower Manhattan Bay, currents and tides dispersed little of the waste into ocean waters; much of it washed up on the beaches of Long Island and New Jersey (Figure 2.6) (Bettmann, 1974; Alexander, 1993). Another waste disposal practice involved a public facility called a "dispose," which was used to convert animal carcasses, meat by-products, and other waste food products into raw materials for industrial products ranging from soap to explosives. These facilities disappeared with the decline in the supply of raw materials and an increase in local ordinances regulating the foul-smelling runoff they generated (Melosi, 1981).

In the late 1800s, enterprising individuals scoured the streets and trash piles searching for material of value, essentially carrying out a simple form of recycling (Figure 2.7). Scavengers, also known as "rag pickers," removed much unwanted material in cities. For example, in the city of Chicago, rag pickers collected over 2000 yd^3 daily (Gerlat, 1999). Partly because of such efforts and partly because of the simpler lifestyles of the period, in 1916, the municipal collection crews collected only about 0.23 kg (0.5 lb) of refuse per capita per day, compared with about ten times that which is collected today (Vesilind et al., 2002).

FIGURE 2.6 Coney Island beach pollution from disposal off the New York City coast.

FIGURE 2.7 Rag pickers removing materials of value from waste. (Reproduced with the kind permission from the New York City Municipal Archives.)

Due to the sanitary problems generated by the intensive industrialization and urbanization of the United States in the latter half of the 19th century, modern solid waste management programs emerged in the 1890s (Blumberg and Gottlieb, 1989). Reformers called for city control over the collection of urban wastes. Prior to that point, waste was considered primarily the individual's responsibility with minor government participation. This lack of governmental input served to exacerbate the expanding accumulation of garbage in the streets. Europe, had, by this time, already developed relatively sophisticated disposal systems and technologies. As late as 1880, only 43% of all U.S. cities provided some minimum form of collection (McBean et al., 1995). Just 24% of the cities surveyed maintained municipally operated garbage collection systems, and an additional 19% contracted out for the service (Blumberg and Gottlieb, 1989). A 1902 MIT survey of 161 U.S. cities showed that 79% provided regular collection of wastes. By 1915, 89% of major American cities had some kind of waste collection system, and by 1930 virtually all large cities had waste collection services (Tammemagi, 1999).

New York City took the lead in handling municipal waste management and promoting overall civic cleanliness. Colonel George E. Waring, Jr., a Civil War veteran and the individual responsible for the establishment of a municipal sewer system in Memphis, Tennessee, served as the city's Commissioner of the Department of Street-Cleaning from 1895 to 1898. One of Waring's first steps in managing the city's wastes was to establish a systematic classification scheme. He encouraged, at homes and businesses, segregation of organic refuse, ash, and general rubbish fractions into separate bins. He then contracted for as much recovery of salable materials from the wastes as possible, making a profit from this phase of the operations (Figure 2.8). Reduction processes were developed, for example, for the extraction from the wastes of by-products such as ammonia, glue, grease, and dry residues for fertilizer (Figure 2.9). The city obtained substantial revenue by salvaging these materials (Melosi, 1973).

FIGURE 2.8 Early waste separation activities.

To make street cleaning more efficient and thorough, Waring raised the competence and status of the workers. The street sweepers were made to wear white uniforms, to associate them with the medical profession, which also became the department's trademark. The workers were eventually dubbed the "White Wings" (Figure 2.10).

Waring's reform efforts made a positive impression on city dwellers. Collection became more efficient, and the cost to clean the streets dropped to about half of 1895 figures. Public health also improved. According to the Board of Health, the city's death rate and sick rate declined substantially. The average annual death rate in New York was 19.63 per 1000 during the first half of 1897, down from 26.78 per 1000 in 1882 to 1894. Similarly diarrheal diseases decreased significantly (Melosi, 1973).

In dealing with the final disposal of refuse, Waring employed both innovations and established techniques. Most of the dry waste was still dumped at sea, until experiments with controlled incineration had been completed. For ocean disposal, Waring recommended the use of the new catamaran-type vessel, the Delehanty Dumping Scow, which was self-emptying and self-propelled (Melosi, 1973).

Waring encouraged experimentation to find more efficient and economical methods of waste reduction and utilization. His goal, far-sighted by today's standards, was to reduce the amount of the city's wastes for ultimate disposal, and place the entire program under city management. A waste reduction plant was eventually built on Barren Island (Figure 2.11). A land-reclamation program later began on Riker's Island, using ashes and other incinerated materials as fill material (Figure 2.12).

Waring's enthusiasm for reform had an impact well beyond his brief tenure as commissioner.

His organization of the Department of Street-Cleaning as a quasi-military outfit drew much ridicule from the press. Criticism soon turned to lavish praise when New Yorkers, for the first time in many years, could walk along uncluttered sidewalks and drive through streets free of garbage and manure. High praise of Waring's handiwork became widespread (Melosi, 1973; Wilson, 1977).

(a)

(b)

FIGURE 2.9 Waste reduction plant, Barren Island, ca. 1897.

Around the turn of the century, the average waste generated per Manhattan citizen was 160 lb of garbage (food wastes and debris), 1230 lb of ash, and 97 lb of rubbish. Total annual waste generation was about 675 kg (1487 lb), slightly higher than today's national average (Melosi, 1981; Alexander, 1993). The ash fraction, from the burning of coal or wood in home furnaces, comprised the major component. Coal-burning home furnaces remained in common use in much of the United States until the end of World War II.

FIGURE 2.10 One of New York City's White Wings, 1905.

FIGURE 2.11 Unloading garbage from scows at Barren Island.

FIGURE 2.12 Convicts unloading scows of ashes at Riker's Island. The ashes were used for fill at the site of the future prison.

2.7 RECENT WASTE MANAGEMENT INITIATIVES

The management of municipal refuse has, fortunately, changed substantially over the years; the composition of the U.S. waste has also changed. Some events that significantly altered the characteristics of solid wastes over the past century are shown in Table 2.2.

By the turn of the 20th century, a variety of waste disposal practices were adopted by municipalities, ranging from land disposal, water disposal (including ocean dumping), incineration, reduction, or some combination of methods (Table 2.3). With an increase in public awareness, ocean dumping received the greatest criticism. Dumping wastes in surface waters was seen as merely shifting one community's waste to another, with no regard for public health. The pollution of East Coast and West Coast Beaches forced the passage of federal legislation in 1934 making the dumping of municipal refuse into the sea illegal. Industries and commercial establishments were exempted from the regulations, however, and continued dumping into offshore waters (Vesilind et al., 2002).

From the 1880s to the 1930s, land dumping remained the most common method of waste disposal, regardless of opposition by public health officials and many sanitary engineers (Figure 2.13). Already by the 1890s, concerns were being raised about the health risks posed by large open dumps. Sanitary engineers at the time preferred either of two methods — incineration or reduction (Blumberg and Gottlieb, 1989).

TABLE 2.2
Significant Milestones in MSW Generation and Management in the United States

1868	Celluloid, the first commercial synthetic plastic, is invented
1903	Corrugated paperboard containers are in commercial use
1907	First paper towels developed
1908	Paper cups replace tin cups in vending machines, in public buildings, and on trains
1913	Corrugated cardboard becomes popular as packaging
1924	Kleenex facial tissues first marketed
1930s	Kimberly-Clark markets the Kotex brand disposal sanitary pad for feminine hygiene protection
1930	Invention of nylon
1935	First beer can is manufactured
1939	Arrival of 25-cent paperback books, "cheap enough to throw away"
1944	Dow Chemical invents Styrofoam™
1949	Johnson and Johnson introduces disposable diapers to United States (invented in Sweden)
1950s	In-house garbage disposals become popular. In some cities, 25-30% of wastes are ground up
1953	Swanson introduces the TV dinner
1960	Pop-top beer cans are invented
1960	Plastic gains popularity as a packaging material
1963	Aluminum beverage cans are developed
1972	Oregon passes the first refundable deposit bottle law
1972	The Intel MCS-4-based SIM4 is the first microcomputer (but not the first personal computer)
1976	There are over 50 million microwave ovens in U.S. households
1977	The Apple][is the first highly successful mass-produced personal computer
1977	PETE soda bottles begin to replace glass
1981	The IBM PC is introduced
1985	Mass-marketing begins of the Swatch® watch, a disposal wristwatch
1986	Fuji introduces the disposable camera
1986	The Fresh Kills Landfill on Staten Island, New York, becomes the world's largest landfill
1988	An estimated 20 million personal computers have become obsolete
2003	The disposable DVD is introduced
2004	The disposable cellular telephone is introduced

TABLE 2.3
Trends of Waste Disposal Practices in the United States, Turn of the 20th Century Compared with Current Data. Values given in Percent

Method	1899	1902	1913	1999[a]
Dumped on land	70	46.5	61	—
Dumped in water	3	2.5	3	—
Incineration	16	29.5	7	15
Sanitary landfill	—	—	7	57
Combination of methods	—	1.5	11	—
No systematic method	11	0.5	—	—
No data	—	19.5	11	—

Adapted from Truini, 2003; Gerlat, 1999.

[a]28% recovered for recycling and composting.

(a)

(b)

FIGURE 2.13 An unsanitary open dump. Photos provided by Wright Environmental Management, Inc. Original source unknown.

2.8 SOLID WASTE INCINERATION AND OTHER THERMAL PROCESSES

England and Germany were at the vanguard in developing effective solid waste incineration systems for the purposes of both volume reduction and energy production. The first municipal waste inciner-ation system began operations in Nottingham, England, in 1874 (Murphy, 1993). In 1892, a cholera epidemic swept Hamburg, Germany. Communities surrounding the city refused to accept the city's cholera-tainted waste, thus forcing the city to build and operate one of Germany's first waste incinerators, designed with the cooperation of English engineers. The incinerator suffered from a range of initial operating problems. One problem related to the significantly different composition of household waste in Hamburg compared with that of England (Erhard, 1991; Bilitewski et al., 1997).

During the same period in the United States, construction of mass-burn facilities was not considered economically justifiable. Allegheny, PA, installed the first municipal incinerator in 1885, followed by Pittsburgh and Des Moines in 1887 and Yonkers, New York, and Elwood, Indiana, in 1893 (Figure 2.14) (Kelly, 1973). In designing waste incinerators, including mobile and stationary systems, engineers applied methods that were under development in Europe. It was not until after 1910, however, that incineration came into widespread use in the United States. The so-called "garbage crematories" appeared throughout the United States (Figure 2.15). Chicago experimented with both a stationary facility and a traveling incinerator. The latter rolled through the city's alleys disposing of refuse as it passed. An intense competition developed between the designers of the mobile and the stationary furnaces. Out of this "picturesque rivalry grew a startlingly clean condition of alleys in the city" (Figure 2.16) (Lane, 1894; Melosi, 1973).

The early application of incinerators in the United States, however, included a long list of problems and failures. Faulty design and construction in addition to inadequate preliminary studies contributed to widespread system malfunctions. Often, U.S. incinerators burned only wet wastes without the organic materials necessary to maintain combustion. Partly as a result of such initial errors, 102 of the 180 incinerators built in the United States between 1885 and 1908 were abandoned by 1904 (Wilson, 1986; Blumberg and Gottlieb, 1989). Shortly afterwards, however, a new generation of incinerators was being promoted by engineers, and in the decade after 1910 incineration returned to widespread use. By this time some sanitation experts thought that incinerators would replace open dumps in smaller communities. A 1924 report indicated that out of 96 cities surveyed, 29% burned or incinerated their wastes. This compared with 17% that dumped, filled with, or buried wastes; 38% that used wastes as fertilizer or animal feed; 2% that used reduction; and the remainder that used no systematic method at all. At its peak in the 1930s to 1940s, between 600 and 700 U.S. cities constructed incineration plants. Avoiding some of the earlier design problems, incineration from a stationary source became a significant method of disposal of municipal wastes (U.S. EPA, 1973; Blumberg and Gottlieb, 1989).

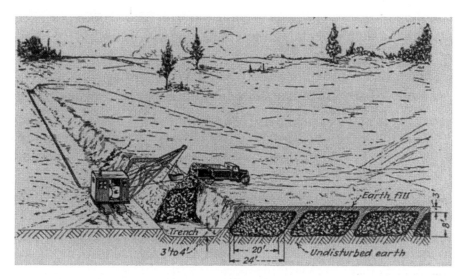

FIGURE 2.14 Diagram of a California sanitary landfill, 1939. (Reproduced with the kind permission from Engineering News-Record, Copyright Oct. 26, 1939, The McGraw-Hill Companies, Inc., All rights reserved.)

FIGURE 2.15 Early MSW incinerator. (Reproduced with the kind permission from the New York City Municipal Archives.)

FIGURE 2.16 Traveling municipal waste burner.

Another innovative thermal disposal method at the turn of the century involved the technology of "reduction." This essentially entailed "cooking" the garbage to extract a wide range of marketable by-products, including grease and "tankage," i.e., dried animal solids which could be sold as fertilizers (Blumberg and Gottlieb, 1989).

Beyond simple mass-burn incineration as a waste reduction system, the British and Germans developed technologies to recover energy from incineration (Figure 2.17). The first plant to generate electricity from incineration was developed in Great Britain in the mid-1890s. By 1912, 76 plants in Great Britain produced energy as did 17 more in the rest of Europe. A pilot project was built in New York City in 1905. Interest in using incinerators to convert waste into energy, however, was low in an era of cheap energy alternatives. During this period only two cities in North America — Westmount, Quebec, and Milwaukee — derived any revenue from steam produced by incinerators (Marshall, 1929; Melosi, 2000). The "waste-to-energy" technology failed to become established in the United States for another 60 years (Blumberg and Gottlieb, 1989).

Up through the 1960s, in many large U.S. cities, household wastes were incinerated in apartment units in order to reduce waste volumes. These incinerators, unfortunately, burned unsorted wastes, operated at relatively low temperatures, and lacked air pollution control. As a result, metals, soot, and

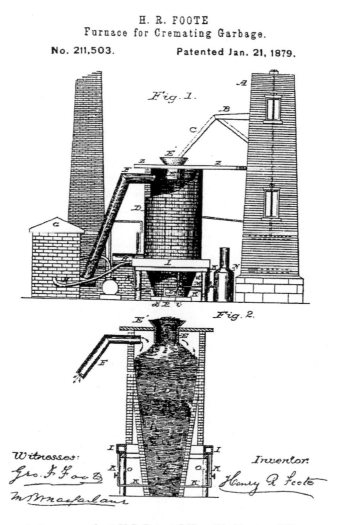

FIGURE 2.17 Early waste-to-energy plant. U.S. Patent Office, Washington, DC.

other products of incomplete combustion were released in abundance via the flue (stack). A recent study of sediments in Central Park Lake, New York City, correlates the accumulation of lead, tin, and zinc with the use of incinerators (Chillrud et al., 1999). By the late 1960s to the early 1970s, public concerns with regard to management of both domestic and toxic chemical wastes increased. The Air Quality Act of 1967 introduced new emissions standards that required the retrofitting of air pollution control devices, such as scrubbers and precipitators (see Chapter 9) to older incinerators. Since incinerators were already more expensive and technology-intensive than landfills, the Act essentially priced incineration out of the market. Within five years of the Act, 100 large-scale incinerators had been shut down (Tammemagi, 1999).

The energy crises of the 1970s created a resurgent interest in the possibility of obtaining inexpensive energy from the thermal decomposition of MSW. The so-called "waste-to-energy" plants and "refuse-derived fuel" systems were designed and developed. Given the continued closing of sanitary landfills across the United States, incineration with the possible production of energy began to appear as an attractive alternative. In the early to mid-1980s, approximately 100 new plants were committed and another 200 planned in the United States (Tammemagi, 1999).

2.9 LAND DISPOSAL AND THE SANITARY LANDFILL

Until the 1900s, "land disposal" of solid wastes involved nothing more than direct dumping on to the land surface followed by abandonment. On the outskirts of many cities, wetlands, often considered "nuisance areas," were filled using layers of household refuse and ash. Early in the century, however, disposal methods prescient of sanitary landfills began to evolve. Simple burying was employed in the United States in 1904 (Public Administration Service, 1970; McBean et al., 1995). The first excavated site that was periodically covered with soil, a precursor of today's modern sanitary landfill, opened in 1935 in California (Figure 2.14). In addition to MSW, the landfill accepted industrial wastes. As a result, the site has secured a ranking on the U.S. EPA Superfund list due to its content of hazardous materials (Gerlat, 1999; Vesilind et al., 2002).

Up to the 1950s, however, open-pit dumping of wastes remained a standard procedure (California State Water Pollution Control Board, 1954; McBean et al., 1995). Due to the incompatibility of wastes (e.g., disposal of hot ashes with paper products), fires were a frequent hazard. In many municipalities, controlled burning was allowed for the purpose of volume reduction. There were considerable problems with odor, smoke, insects, noise, and seagulls. Up to this point, most disposal sites were obviously not designed or constructed with much engineering input. Planning to address environmental protection remained inadequate. Siting of the landfill was based on convenience and efficiency, rather than practical technical concerns such as proximity to surface water and groundwater, and soil and geologic considerations. When such landfills were completed, they were often covered with a thin cap of soil, and the growth of surface vegetation was encouraged. Land subsidence was common and many sites leaked for decades after closure. Subsurface liners were rarely used and the layers of waste were usually only a few meters thick. As landfills expanded, growth occurred laterally and covered large tracts of land. Many were situated near expanding urban areas and their water supplies. As a result, public opposition to these landfills became increasingly contentious.

Alternative waste disposal techniques were attempted in the United States; however, landfills remained the most common method due to the appealing costs of land and labor, and the simple, inexpensive technology involved. To address the growing criticism, the concept of the "sanitary landfill" was introduced in the 1950s. Also known as a "cut and cover" or controlled tipping system, the sanitary landfill was touted as an engineered system for disposing solid wastes on land by spreading them into thin layers, compacting to the smallest practical volume, and covering with a layer of soil at the end of each working day (Stone, 1977). The concept of controlled tipping, in which solid wastes are sealed in "cells" formed from soil or other cover material at regular intervals, was devised in order to keep wastes relatively free from odor, less attractive to vectors, and less of an overall hazard to public health. This simple step was significant as it greatly alleviated problems such as uncontrolled fires, windblown refuse, and rodent infestations. Landfill practices gradually improved over time although proper planning, engineering, operations, and staff training were slow to evolve. By 1959, the sanitary landfill was the primary method of solid waste disposal for U.S. communities (ASCE, 1959; McBean et al., 1995). The American Society of Civil Engineers in 1959 published the first engineering guide to sanitary landfilling which detailed the compaction of refuse and the placement of a daily cover to reduce hazards of fire, odor, and rodents.

Although design and operation were a substantial improvement over earlier land disposal efforts, the sanitary landfill still suffered from numerous deficiencies. It did not sufficiently address groundwater contamination, surface runoff, odor and gas emissions, and related public health concerns. At this time, there was only limited knowledge regarding the infiltration of surface water through a covered landfill, and the consequent reactions of this water upon contact with wastes. There also came the belated recognition of the potential impact of escaping contaminants (leachates) on groundwater quality. Eventually, it was decided that, to ensure minimal protection of the local environment, the installation of an engineered cover system to reduce long-term leachate generation and an impermeable or slowly permeable liner across the base of a landfill to prevent the escape of any leachates to the environment, were essential (Tammemagi, 1999).

Many communities, however, continued open burning and open dumping into the 1960s. According to the U.S. Federal Bureau of Solid Waste Management, 94% of all land disposal operations in the mid-1960s was inadequate in terms of air and water pollution, insect and rodent problems, and physical appearance (Tammemagi, 1999).

Public awareness of the potential hazards posed by MSW landfills increased. By the 1970s and 1980s, there was growing concern about the effects of landfills on contamination of groundwater. Groundwater, stored in underground strata and tapped by wells, provides drinking water for approximately 90% of the U.S. population (U.S. EPA, 2001b). Once groundwater becomes contaminated, it is very difficult, slow, and expensive to remediate. The concentration and toxicity of some landfill leachates were considered capable of increasing the risk of cancer (Brown and Donnelly, 1988; Tammemagi, 1999). A number of toxic materials from households (e.g., batteries, pesticide containers and paints) were found in leachates along with a wide range of industrial wastes, previously not restricted from municipal landfills.

As a result of the enactment of stringent federal regulations such as the Resource Conservation and Recovery Act and its amendments (see Chapter 3), numerous modifications were made in sanitary landfill design. Caps constructed of clay or impermeable synthetic materials, such as high-density polyethylene, were placed over landfills undergoing closure to decrease the infiltration of precipitation, thus limiting the formation of leachate. Bottom liners, constructed of similar materials, were introduced to capture any leachate that formed within the landfill. In addition, subsurface and surface collection systems were installed to capture and remove leachate and landfill gas. Monitoring of groundwater quality and gaseous emissions became a required component of proactive landfill operation.

In spite of these technical advances, there was continuing concern about groundwater contamination. Studies during the late 1970s indicated that leaking leachate was a problem facing all landfills. The U.S. EPA estimated that in 1990 more than 75% of U.S. landfills were polluting groundwater with leachate (Lee and Jones, 1991). There was also concern that even state-of-the-art municipal landfills with double liners and other modern leachate containment systems would eventually fail. In other words, the increased use and sophistication of engineering techniques could only postpone the onset of groundwater contamination.

By the 1980s, the importance of selecting a site that minimized the environmental impact of a landfill was recognized. New siting criteria emphasized the importance of sites that were well above the groundwater table, that did not occur in groundwater recharge zones, that were not in natural flood areas, that occurred where soil water moved only very slowly, and that had natural imperme-able clay formations to prevent contaminant migration. Over the past two decades, landfill siting has become a sophisticated process that incorporates technical as well as political and social con-cerns. Continuing improvement in the siting process has significantly restricted locations where landfills can be placed (Tammemagi, 1999).

The number of MSW landfills has dropped substantially from about 20,000 in 1979 to 2216 in 1999 (U.S. EPA, 2001a) (see Chapter 1). One of the major causes for the decline is the NIMBY (not-in-my-backyard) syndrome, a result of the "dump" stigma that persists in the minds of many Americans. The NIMBY phenomenon evolved as a result of increased awareness and the affluence of citizens combined with increased public education and attention by the media. There have also been tremendous improvements in the ability to detect contaminants with improved technology and instrumentation. Another cause for the decline of landfills was the enactment of stringent new guidelines, both at the federal and state levels, for landfill construction, operation, and closure. Such guidelines have essentially put many older landfills out of business.

2.10   RECYCLING/REUSE

Recycling in the late 1800s was by individuals who scoured the streets and trash piles looking for material of value. The first organized municipal recycling program was attempted in 1874 in Baltimore, but did not succeed (Gerlat, 1999; Vesilind et al., 2002).

From the late 1800s through World War I, raw garbage was fed to pigs on farms as a means of increasing food production. By 1917, 35% of all cities monitored in one survey utilized this method. The figure increased to 44% in 1925, and then leveled off at 39% by 1930 (Hering and Greeley, 1921; Blumberg and Gottlieb, 1989). Scientists discovered that this practice contributed to the infection of animals by *Trichinella spiralis* and *Vesicular exanthema*, which could be passed on to humans who ate undercooked pork. When a series of swine epidemics occurred in the 1950s and several operations were shut down, public health regulations were issued to prevent the feeding of raw garbage to pigs. The cost of cooking the garbage prior to feeding to pigs was expensive and so the practice gradually disappeared (Alexander, 1993).

In 1898 the first materials recovery facility (MRF)(see Chapter 7) was built in New York City. The facility processed the waste of over 116,000 residents, and recovered up to 37% (by wt) of the wastes. Soon MRFs were constructed in Berlin, Hamburg, and Munich. Assisted by trommel screens and conveyor belts, the Munich MRF processed over 275 metric tons (300 tons) of waste per day (Bilitewski et al., 1997).

Europe led the United States in recycling. By 1939, as war approached, German householders were expected to separate rags, paper, bottles, bones, rabbit skins, iron, and other metals from their discards (APWA, 1941; Alexander, 1993). Prewar Japan's needs for imported scrap iron is well documented.

Regardless of advances in sanitation technology, waste composition and volumes will change, public and governmental attitudes will vary, and there will always be new challenges in the field of waste management. A health official noted the sense of frustration experienced by sanitation offi-cials when he wrote (Alexander 1993):

> Appropriate places for [refuse] are becoming scarcer year by year, and the question as to some method of disposal... must soon confront us. Already the inhabitants in proximity to the public dumps are begin-ning to complain. . . . I can not urge too strongly upon the Commissioners the necessity for action in this direction. The waste that is taken from yards and dwelling places must be provided for, and that provision should not longer be delayed.

This warning was ignored. The letter was sent to the attention of the Commissioners of Washington, DC, in 1889.

REFERENCES

Alexander, J. H., in *Defense of Garbage*, Praeger, Westport, CT, 1993.

American Public Works Association, Refuse Collection Practice, Chicago, IL, 1941.

American Public Works Association, in *History of Public Works in the United States*, 1776–1976. Armstrong, E.L., Robinson, M.C. and Hoy, S.M., (Eds.), Chicago, IL, 1976.

ASCE Committee on Sanitary Engineering Research, Refuse volume reduction in a sanitary landfill, *ASCE J.Sanitary Eng. Div..*, 85, 37–50, 1959.

Bettmann, O., *The Good Old Days — They Were Terrible!* Random House, New York, NY, 1974.

Bilitewski, B.B., Hardtle, G., and Marek. K., *Waste Management*, Springer, Berlin, 1997.

Blegen, C.W., *Troy,* Vol. 1, Princeton University Press, Princeton, NJ, 1958.

Blumberg, L., and Gottlieb, J., *War on Waste*, Island Press, Washington, DC, 1989.

Brown, K.W. and Donnelly, K.C., An estimation of the risk associated with the organic constituents of hazardous and municipal waste landfill leachates, *Hazardous Waste Hazardous Mater.,* 5, 1–30, 1988.

California State Water Pollution Control Board, Investigation of Leaching of a Sanitary Landfill, Sacramento, CA, 1954.

Chillrud, S.N., Bopp, R.F., H.J. Simpson, J.M. Ross, E.L. Shuster, D.A. Chaky, D.C. Walsh, C.C. Choy, L. Tolley, and A. Yarme. Twentieth century atmospheric metal fluxes into Central Park Lake, New York City, *Environ. Sci. Technol.*, 33, 657–661, 1999.

Gerlat, A., Garbage: the long view, *Waste News*, May 3, 1999, pp. 22–25, 1999.

Greater London Council, *Refuse Disposal in Greater London,* London, UK, 1969.

Harris, C. and Bickerstaffe, J., *Finding out about Managing Waste. A Resource Book for National Curriculum: Science, Geography and Technology,* Hobsons, London, 1990.

Hering, R. and Greeley. S.A., *Collection and Disposal of Municipal Refuse,* McGraw Hill, New York, NY, 1921.

Hosch, K., From donkey-drawn wheelbarrows to sanitary landfills, *Stradhygiene,* 18, 228–231,1967.

Kelly, K., Garbage: *The History and Future of Garbage in America*, Saturday Review Press, New York, NY, 1973.

Lane, R., Chicago garbage burning, *Harper's Weekly*, 38, 408, 1894.

Lee, G.F. and Jones, R.A., Landfills and ground-water quality, *Groundwater*, 29, 482–486, 1991.

Marshall, C. E., Incinerator knocks out garbage dump in Long Island town, *American City*, 40:129, 1929.

McBean, E.A., Rovers, F.A., and Farquhar, G.J., *Solid Waste Landfill Engineering and Design,* Prentice-Hall, Englewood Cliffs, NJ, 1995.

Melosi, M.V., Out of Sight, Out of Mind: The Environment and Disposal of Municipal Refuse, 1860–1920, *The Historian,* 35, 621–640, Kingston, RI (August), 1973.

Melosi, M. V., *Garbage in the Cities,* Texas A&M Press, College Station, TX, 1981.

Melosi, M.V., *The Sanitary City: Urban Infrastructure in America from Colonial Times to the Present,* The Johns Hopkins University Press. Baltimore, MD, 2000.

Miller, B., *Fat of the Land. Garbage of New York the Last Two Hundred Years,* Four Walls Eight Windows, New York, NY, 2000.

Murphy, P., The Garbage Primer: A Handbook for Citizens, League of Women Voters, New York, NY, 1993.

Priestley. J.J., *Civiliation, Water and Wastes. Chemistry and Industry,* 1968, pp. 353–363,

Public Administration Service, Municipal Refuse Disposal, Interstate Publishers and Printers, Illinois, 1970.

Rawlinson, J., *A History of Technology,* Vol. IV, Singer, C. et al., Eds., Clarendon Press, Oxford, UK, 1958.

Stone, R., Sanitary landfill, *in Handbook of Solid Waste Management,* Wilson, D.G., Ed., Van Nostrand Reinhold, New York, NY, 1977.

Tammemagi, H., *The Waste Crisis. Landfills, Incinerators, and the Search for a Sustainable Future,* Oxford University Press, New York, NY, 1999.

U.S. Environmental Protection Agency, Municipal Solid Waste in the United States: 1999 Facts and figures, EPA 530-R-01-014, Office of Solid Waste and Emergency Response, Washington, D.C, 2001a.

U.S. Environmental Protection Agency, Safe Drinking Water Act, Underground Injection Control (UIC) Program, EPA 816-H-01-003, Office of Water, Washington, DC, August 2001b.

U.S. Environmental Protection Agency, An Environmental Assessment of Gas and Leachate Problems at Land Disposal Sites, 530/SW–110–OF, Office of Solid Waste, Washington, DC, 1973.

Vesilind, P.A., Worrell, W. and Reinhart, D., *Solid Waste Engineering,* Brooks/Cole, Pacific Grove, CA, 2002.

Truini, J., 3,2,1: Self-destructing DVDs may stir waste issues, *Waste News,* Jul. 21, 2003, p. 1.

Waste Watch, National Analysis of Household Waste, www.wastewatch.org.uk.

Wilson, D. G., Ed., *Handbook of Solid Waste Management,* Van Nostrand Reinhold Company, New York, NY, 1977.

Wilson, D.G. Ed., History of solid waste management, in *The Solid Waste Handbook: A Practical Guide,* Wiley, New York, NY, 1986.

Winslow, C.E.A. and Hansen, P., Some statistics of garbage disposal for the larger American cities in 1902. Public Health: Papers and Reports, 29, 141–153. American Public Health Association, October 1903.

Zinn, H., *A People's History of the United States: 1492 – Present,* Harper Collins, New York, NY, 1995.

QUESTIONS

1. Explain how the chemical and physical properties of solid wastes have evolved over the millennia. Keep in mind the specific sources that contributed to the waste stream of the period.

2. What were the most common waste disposal methods of earlier societies? Approximately when did these modes of 'management' change? What was (were) the root cause(s) for the change?

3. Who were the first known individuals to associate improper waste disposal and adverse health effects?

4. What was the first civilization to create an organized waste collection workforce? The first 'landfills'?

5. After the fall of the Roman Empire, how did waste management change? What specific-factors were responsible for this change?

7. Explain how a society's waste problems could actually compromise the security of a community.

8. Explain how the Industrial Revolution changed both the quantity and composition of solid wastes. How were human populations affected by the change in waste composition?

9. U.S. cities of the late 19th century experienced deplorable health conditions as a result of improper waste management. What were some initiatives that drastically altered urban waste management?

10. Discuss the efforts of George Waring on urban waste management. Were his efforts of long-lasting benefit or only superficial?

11. What is the benefit of 'controlled tipping' of wastes, compared with prior land disposal practices?

12. Discuss how landfills from 1920 to the mid-1970s were managed with regard to (a) siting, (b) day-to-day operations, (c) leachate collection and removal, (d) methane recovery, and (e) closure.

13. "Routine open burning of municipal wastes, in the open landfill cell, serves effectively as an expedient and cost-effective means of extending landfill lifetime." Is this statement accurate? Give reasons.

14. In the United States, the number of MSW landfills has dropped substantially in the past 30 years, despite the fact that there seems to be ample space available to construct new landfills or expand existing ones. What are the primary reasons for this decline?

15. Discuss the evolution of MSW incineration from 19th century Europe to the present. What factors were responsible for its decline in popularity in the early 20th century? Why have incinerators increased in popularity today?

16. Search local historical records and old news articles to draft a chronology of waste management in your community. What were the specific events that catalyzed some of the more important changes? Can you locate the long covered-over disposal sites in your town or city? What is the current land use at these sites?

3 Regulatory Development

'This country is planted thick with laws, from coast to coast—man's laws not God's—and if you cut them down—do you really think you can stand upright in the winds that would blow then?'

Robert Bolt
A Man For All Seasons

3.1 INTRODUCTION

As discussed in Chapter 2, the modern system of solid waste collection and disposal by a local authority had its beginning with the British Public Health Act of 1875. Since then the complexity and reach of waste management laws have increased along with the complexity and volumes of wastes generated.

3.2 SIGNIFICANT U.S. LEGISLATION

National legislation addressing the management of wastes dates back to the Rivers and Harbors Act of 1899. The Act prohibits the unauthorized obstruction or alteration of any navigable waters of the United States. Examples of activities requiring an Army Corps of Engineers permit include constructing a structure in or over any waters of the United States, excavation or deposit of material in such waters, and various types of work performed in such waters, including filling (33 CFR Part 322).

3.2.1 THE SOLID WASTE DISPOSAL ACT

Modern U.S. solid waste management legislation dates from 1965 when the Solid Waste Disposal Act, Title II of Public Law 89–272, was enacted as Title II of the Clean Air Act of 1965 by the U.S. Congress. The intent of the act was to:

- Promote the demonstration, construction, and application of solid waste management and resource recovery systems that preserve and enhance the quality of air, water, and land resources.
- Provide technical and financial assistance to state and local governments and interstate agencies in conducting surveys of waste disposal practices and problems, and in the planning and development of resource recovery and solid waste disposal programs.
- Promote a national research and development program for improved management techniques; more effective organizational arrangements; new and improved methods of collection, separation, recovery, and recycling of solid wastes; and the environmentally safe disposal of nonrecoverable residues.
- Provide for the promulgation of guidelines for solid waste collection, transport, separation, recovery, and disposal systems.
- Provide for training grants in occupations involving the design, operation, and maintenance of solid waste disposal systems.

The U.S. Public Health Service and the U.S. Bureau of Mines were responsible for enforcement of this act. The former agency had the responsibility for regulating Municipal Solid Waste (MSW) generation and the latter was charged with the supervision of solid wastes generated from mining and from fossil fuel combustion (e.g., coal ash at an electric generating utility). The primary thrust of this legislation was on the development of more efficient disposal methods rather than on the protection of public health and the environment.

3.2.2 THE RESOURCE RECOVERY ACT

The Resource Recovery Act of 1970 (Public Law 95–512) was considered a shift in federal legislation from waste disposal efficiency to efforts to recover energy and materials from solid waste. The Act authorized grants for demonstrating new resource recovery technologies.

By 1970 the U.S. Environmental Protection Agency was established by presidential order under Reorganizational Plan No. 3 of 1970. All solid waste management activities were transferred from the U.S. Public Health Service to the U.S. Environmental Protection Agency (EPA). The Resource Recovery Act subsequently required annual reports from the U.S. EPA on methods of promoting recycling and reducing the overall generation of solid waste.

3.2.3 THE RESOURCE CONSERVATION AND RECOVERY ACT

The acts promulgated up to this point did little to establish firm regulations; rather, guidelines were formulated. The federal government became engaged in a more active regulatory role, manifested in the Resource Conservation and Recovery Act (RCRA) of 1976, passed by Congress as Public Law 94–580. For the first time, comprehensive federal regulations were established to regulate many categories of waste. As of this writing, RCRA consists of ten subtitles, and are listed in Table 3.1.

3.2.4 SOLID WASTE MANAGEMENT UNDER RCRA

The RCRA solid waste management program, Subtitle D, encourages environmentally sound solid waste management practices that maximize the reuse of recoverable material and promote resource recovery. The term "solid waste" as used in Subtitle D is broad and includes waste materials beyond ordinary MSW, which is typically collected and disposed in municipal solid waste landfills; for example, hazardous waste generated by conditionally exempt small quantity generators (CESQGs) (see Chapter 11) are included, as are hazardous wastes that are excluded from the Subtitle C regulations (e.g., household hazardous waste). The solid waste management program also addresses MSW generated by businesses.

TABLE 3.1
Outline of RCRA Subtitles

Subtitle	Provisions
A	General Provisions
B	Office of Solid Waste; Authorities of the Administrator and Interagency Coordinating Committee
C	Hazardous Waste Management
D	State or Regional Solid Waste Plans
E	Duties of the Secretary of Commerce in Resource and Recovery
F	Federal Responsibilities
G	Miscellaneous Provisions
H	Research, Development, Demonstration, and Information
I	Regulation of Underground Storage Tanks
J	Standards for the Tracking and Management of Medical Waste

Within the context of RCRA, the U.S. EPA promotes an integrated, hierarchical approach to managing MSW that includes source reduction, recycling, incineration, and landfilling. Waste reduction and recycling are the preferred elements of the system while landfilling is the lowest priority.

Subtitle D includes technical criteria for MSW landfills to ensure that their routine operation will be protective of public health and the environment. A major provision in RCRA involving MSW management is the prohibition of open dumps. This prohibition is implemented by the states, using EPA criteria, to determine which facilities qualify as sanitary landfills and therefore remain in operation. The EPA criteria were originally promulgated in 1979; open dumps were to close or be upgraded by September 1984. An open dump is defined as a disposal facility that does not comply with one or more of the 40 CFR Part 257 or Part 258 Subtitle D criteria. Using the Part 257, Subpart A criteria as a benchmark, each state evaluated its solid waste disposal facilities to determine which facilities were open dumps that needed to be closed or upgraded. For each open dump, the state completed an Open Dump Inventory Report form that was sent to the Bureau of the Census.

In the 1984 amendments to RCRA (see below), the U.S. EPA was required to revise the sanitary landfill criteria for facilities that received small quantity generator hazardous waste (see Chapter 11) or hazardous household waste. Under this authority, the Agency promulgated revised regulations applicable to MSW landfills in October 1991, to take effect in October 1993 for most provisions. The new criteria required the installation of liners, leachate collection and removal systems, groundwater monitoring, and corrective action at MSW sanitary landfills.

Other provisions authorized in RCRA for MSW management include: financial and technical assistance for states and local governments (most of which was ended in 1981 due to budget cutbacks); research, development, and demonstration authority; and a procurement program, whose goal was to stimulate markets for recycled products by requiring federal agencies to purchase recycled materials. Consistent with its strong emphasis on recycling, RCRA contains provisions for the EPA to encourage recycling and promote the development of markets for items with recovered materials content. To help achieve this goal, the EPA publishes federal procurement guidelines that set minimum recovered materials content standards for designated items. RCRA requires federal procuring agencies to purchase items manufactured with the highest percentage of recovered materials practicable. These requirements are specified in Comprehensive Procurement Guidelines (CPG) and Recovered Materials Advisory Notices (RMAN)(U.S. EPA, 2002). While EPA is the lead agency under RCRA, the Department of Commerce is given several responsibilities for promoting the greater commercial use of resource recovery technologies.

The U.S. EPA has established a number of programs to encourage sound waste management, for example Wastewise, the Jobs Through Recycling program, unit pricing, and full cost accounting for MSW (U.S. EPA, 2002).

3.2.5 REGULATION OF HAZARDOUS WASTE UNDER RCRA

Subtitle C of RCRA embraces the hazardous waste management program. A waste is declared "hazardous" if it appears on a list of about 100 industrial process waste streams and more than 500 discarded commercial products and chemicals. Beyond these lists, a waste may still be deemed hazardous if it is ignitable, corrosive, reactive, or toxic, measured via specific test protocols. These requirements are discussed in detail in the Code of Federal Regulations and in Chapter 11. The 1976 law expanded the definition of solid waste, which includes hazardous waste, to include:

> sludge . . . and other discarded material, including solid, liquid, semi-solid, or contained gaseous material.

This expanded definition is significant with respect to hazardous wastes because approximately 95% occurs as liquids or sludges. Some hazardous wastes are specifically excluded, however, from this definition, for example industrial point source discharges (regulated under the Clean Water Act) and nuclear wastes (regulated under the Atomic Energy Act).

RCRA grants the U.S. EPA (and, ultimately, relevant state agencies) broad enforcement authority to require all hazardous waste management facilities to comply with the regulations. The hazardous waste management program is intended to ensure that hazardous waste is managed safely from the moment it is generated to the moment it is ultimately disposed of (the "cradle to grave" program).

For generators of hazardous waste, the Subtitle C program includes procedures for proper identification and measuring ("counting") of hazardous waste. Under RCRA, hazardous waste generators must comply with regulations concerning recordkeeping and reporting; the labeling of wastes; the use of appropriate containers; providing information on waste chemical composition to transporters, treatment, storage and disposal facilities; and the use of a hazardous waste manifest system. Initially, facilities generating less than 1000 kg of waste per month were exempt from the regulations. The 1984 amendments to RCRA lowered this exemption to 100 kg per month. Generator requirements are presented in Chapter 12.

Under Subtitle C, transporters of hazardous waste must comply with numerous specific operating standards. The RCRA regulations were integrated with existing Department of Transportation regulations, which address the transport of hazardous materials. Requirements include the use of the hazardous waste manifest system, hazard communication, appropriate waste packaging and waste segregation, and handling incidents during transport. Details of transporter requirements appear in Chapter 13.

The RCRA program includes standards for facilities that treat, store, or dispose of hazardous waste. These standards include requirements for general facility management and specific hazardous waste management units (e.g., landfills, incinerators). RCRA requires treatment, storage and disposal (TSD) facility owners and operators to obtain a hazardous waste permit from the EPA or relevant state agency. All TSD facilities are required to meet financial requirements in the event of accidents and to close their facilities in compliance with EPA regulations.

The 1984 amendments imposed a number of new requirements on TSD facilities with the goal of minimizing land disposal. Bulk or noncontainerized hazardous liquid wastes were prohibited from disposal in any landfill, and severe restrictions were placed on the disposal of both hazardous and nonhazardous liquids in hazardous waste landfills. Landfill disposal of several specific highly hazardous wastes was phased out from 1986 to 1990. The U.S. EPA was directed to review the characteristics of all wastes defined as hazardous and to determine the suitability of their disposal to land. These safeguards became known as the land disposal restrictions (LDR). Minimum technological standards were set for new landfills and surface impoundments, for example requiring the installation of double liners, a leachate collection system, and groundwater monitoring. In the 1984 amendments the federal government set deadlines for the closure of TSD facilities not meeting minimum standards.

With the understanding that the routine management of hazardous waste may result in spills or other releases to the environment, RCRA Subtitle C contains provisions governing corrective action for the clean-up of contaminated air, groundwater, and soil. Requirements for TSD facilities under RCRA appear in Chapter 14.

The Subtitle C program also contains provisions that allow the U.S. EPA to authorize and financially assist state governments to implement and enforce the hazardous waste program. In order to receive final authorization from the federal government, the state program must be equivalent to, no less stringent than, and consistent with the federal program. As the EPA develops new regulations, the state's program must be reviewed to determine whether the state has the authority to enforce the relevant requirements. In cases where states do not have authorization, they often participate in running the program under the so-called Cooperative Arrangements. The Cooperative Arrangements provide financial assistance and allow states to participate in specific aspects of the program (e.g., assisting in permit evaluation, conducting inspections), while working to achieve full authorization by the EPA (McCarthy and Tiemann, no date).

3.2.6 UNDERGROUND STORAGE TANK MANAGEMENT

At the same time when regulators and the public were expressing concerns regarding environmental contamination from improper waste management practices, concerns were heightened about the underground storage of fuels and hazardous substances. To address the issue of leaking underground storage tanks (USTs), Congress established a leak prevention, detection, and cleanup program through the 1984 RCRA amendments and the 1986 Superfund Amendments and Reauthorization Act (SARA).

The RCRA Subtitle I UST program regulates underground tanks that store either petroleum or hazardous substances. The UST regulations govern tank design, construction, installation, operation, release detection, release response, corrective action, closure, and financial responsibility. Similar to RCRA Subtitle C, Subtitle I contains provisions that allow the EPA to authorize state government implementation and enforcement of the UST regulatory program.

The provisions of Subtitle I created a Leaking Underground Storage Tank (LUST) Trust Fund to finance the clean-up of leaks from petroleum USTs in cases where the UST owner or operator does not remediate a site or when a release from the tank requires emergency response. The Trust Fund provides money for the EPA to administer the program and for states to direct the clean-up operations, take enforcement actions, and conduct clean-ups when necessary. The money to support the fund is obtained via a 0.1 cent-per-gallon federal tax on motor vehicle fuels and other petroleum products.

UST management is not covered in this book.

3.2.7 AMENDMENTS TO RCRA

RCRA has been amended nine times. Some of the amendments have been relatively minor, for example those involving clarification to portions of the law. The most significant amendments came into effect in 1980, 1984, and 1992.

3.2.7.1 The 1980 Amendments

The RCRA Amendments of 1980 provided the U.S. EPA with greater enforcement capabilities to handle illegal dumpers of hazardous waste. Funds were authorized to conduct an inventory of hazardous waste sites, and RCRA authorizations for appropriations were extended through 1982. Also under the 1980 amendments, the EPA's authority to regulate certain high-volume, low-hazard wastes (initially designated "special wastes") was restricted.

3.2.7.2 The Comprehensive Environmental Response, Compensation and Liability Act of 1980

Public law 96–510, 42 U.S.C. Article 9601, The Comprehensive Environmental Response, Compensation and Liability Act (CERCLA) of 1980 was promulgated in order to address the management of uncontrolled and abandoned hazardous waste sites in a timely fashion. CERCLA, commonly known as Superfund, imposed a tax on chemical and petroleum industries and provided broad federal authority to respond directly to releases or threatened releases of hazardous substances that may endanger public health or the environment. Since CERCLA's creation, several billion dollars in taxes have been collected and held in a trust fund for cleaning up the nation's most hazardous sites. Of great practical importance, CERCLA established liability criteria for persons or facilities responsible for the disposal and release of hazardous waste at affected sites.

The law under CERCLA authorizes two kinds of response actions at uncontrolled hazardous waste sites:

- Short-term removals, where actions are taken to address releases or threatened releases that require a prompt response.
- Long-term remedial response actions that permanently and significantly reduce the dangers associated with releases of hazardous substances that are serious, but not immediately life-threatening. These actions can be conducted only at sites listed on EPA's National Priorities List (NPL).

CERCLA also allowed for the revision of the National Contingency Plan (NCP), which provides the guidelines and procedures needed to respond to releases and threatened releases of hazardous substances. CERCLA was amended by the Superfund Amendments and Reauthorization Act (SARA) on October 17, 1986 (U.S. EPA, 2003a).

3.2.7.3 The Hazardous and Solid Waste Amendments of 1984

The most significant set of amendments to the RCRA were the Hazardous and Solid Waste Amendments of 1984 (HSWA), a complex law with many detailed technical requirements. Some of its major provisions were:

- Restrictions on the land disposal of hazardous waste.
- The inclusion of small-quantity hazardous waste generators (those producing between 100 and 1000 kg of waste per month) in the hazardous waste regulatory program.
- A new regulatory program for underground storage tanks (described above).
- The U.S. EPA was mandated to issue regulations governing those facilities that produce, distribute, and use fuels produced from hazardous waste, including used oil.
- Hazardous waste facilities owned or operated by federal, state, or local government agencies must be inspected annually, and privately owned facilities must be inspected at least every 2 years.
- Each federal agency was required to submit to the EPA an inventory of hazardous waste facilities it ever owned or operated.

The 1984 law also required that the EPA establish a timetable for issuing or denying permits for treatment, storage, and disposal facilities; required permits to be set for fixed terms not exceeding 10 years; required permit applications to contain information regarding the potential for public exposure to hazardous substances from facility operations; and authorized the EPA to issue experimental permits for facilities demonstrating new technologies. The EPA's enforcement powers were increased, the list of prohibited actions constituting crimes was expanded, penalties were increased, and citizen suit provisions were expanded.

Other provisions of the 1984 amendments prohibited the export of hazardous waste unless the government of the receiving country formally consented to accept it; created an ombudsman's office in the EPA to handle RCRA-associated complaints and requests for information; and reauthorized RCRA through fiscal year 1988 at a level of about $250 million per year. HSWA called for a National Ground Water Commission to assess and report to Congress on groundwater issues and contamination from hazardous wastes. However, the commission was never funded or established (McCarthy and Tiemann, no date).

3.2.7.4 The Medical Waste Tracking Act

RCRA has focused on waste concerns beyond MSW and hazardous waste. It established a medical waste tracking program to ensure that such waste is properly handled from the moment it is generated to the moment it is disposed. Congress passed House Bill 3515, the Medical Waste Tracking Act in November 1988, which directed the EPA to develop protocols for dealing with infectious waste disposal. The EPA was required to publish an interim final rule for a 2-year demonstration of the medical waste management and tracking program. RCRA was amended by adding a Subtitle J.

The medical waste tracking program ended in June 1991 and no federal EPA tracking requirements are currently in effect; however, some states have instituted their own medical waste requirements.

3.2.7.5 The Federal Facility Compliance Act

The Federal Facility Compliance Act of 1992 handled the legal dispute as to whether federal facilities are subject to enforcement actions under RCRA. The Act waived governmental immunity from prosecution with regard to the improper management of hazardous wastes. As a result, the EPA, the Department of Justice, and the states can enforce the provisions of RCRA against federal facilities, and federal departments and agencies can be subjected to injunctions, administrative orders, and penalties for noncompliance. Additionally, federal employees may be subject to both fines and imprisonment under any federal or state solid or hazardous waste law. The Act also contains provisions applicable to mixtures of radioactive and hazardous waste at Department of Energy facilities and to munitions, military ships, and military sewage treatment facilities handling hazardous wastes.

3.2.7.6 The 1996 Amendments

The Land Disposal Program Flexibility Act (P.L. 104–119), passed by the 104th Congress, exempts hazardous waste from RCRA regulation if the waste is treated to a point where it no longer exhibits the characteristics that made it hazardous, and is disposed in a facility regulated under the Clean Water Act or in a Class I deep injection well regulated under the Safe Drinking Water Act (see Chapter 17). A second provision exempted small landfills sited in arid or remote areas from groundwater monitoring requirements, provided there was no prior evidence of groundwater contamination.

The chronology of the Solid Waste Disposal Act, Resource Conservation and Recovery Act, and major amendments are given in Table 3.2.

TABLE 3.2

Solid Waste Disposal Act, Resource Conservation and Recovery Act and Major Amendments

Year	Act	Public Law Number
1965	Solid Waste Disposal Act	P.L. 89–272, Title II
1970	Resource Recovery Act of 1970	P.L. 91–512
1976	Resource Conservation and Recovery Act of 1976	P.L. 94–580
1980	Used Oil Recycling Act of 1980	P.L. 96–463
1980	Solid Waste Disposal Act Amendments of 1980	P.L. 96–482
1984	Hazardous and Solid Waste Amendments of 1984	P.L. 98–616
1988	Medical Waste Tracking Act of 1988	P.L. 100–582
1992	Federal Facility Compliance Act of 1992	P.L. 102–386
1996	Land Disposal Program Flexibility Act of 1996	P.L. 104–119

3.3 OTHER RECENT LAWS AFFECTING MSW MANAGEMENT

Several other solid and hazardous waste-related measures have been enacted by Congress over the past decade. Although these are technically not amendments to RCRA, they are implemented at the federal level with authority for enforcement subsequently provided to states.

3.3.1 THE PUBLIC UTILITY REGULATION AND POLICY ACT OF 1978

The Public Utility Regulatory and Policy Act of 1978 (PURPA) was enacted in response to the energy crises of the 1970s. PURPA was intended to increase the diversity of fuel use and increase the production and efficiency of electricity generation while providing better prices to customers. The new legislation was designed to boost domestic supplies of energy, which includes directing public and private utilities to purchase power from waste-to-energy facilities.

Under PURPA a new class of electricity generators called qualifying facilities (QFs) was created. QFs were composed of cogenerators using natural gas and small power producers that used renewable resources such as wind, solar, municipal waste, or biomass. PURPA required utilities to connect QFs to transmission grids and to purchase their power at a price that did not exceed the avoided cost of installing and operating new capacity. At the same time the Power Plant and Industrial Fuel Use Act, enacted concurrently with PURPA, prohibited the use of oil and natural gas in new power plants (Levitan and Nezam-Mafi, 1998).

3.3.2 SANITARY FOOD TRANSPORTATION ACT

Some waste haulers travel long distances, sometimes to other states, in order to transport MSW. In order to economize shipments some waste hauling companies would carry produce or other farm products on their return trip, which understandably raised concerns regarding food safety. The Sanitary Food Transportation Act of 1990 (P.L. 101–500) required the regulation of trucks and rail cars that haul both food and solid waste. The Act directed the Departments of Agriculture, Health and Human Services, and Transportation to promulgate regulations specifying the following (McCarthy and Tiemann, no date):

- Recordkeeping and identification requirements
- Decontamination procedures for refrigerated trucks and rail cars
- Materials for construction of tank trucks, cargo tanks, and ancillary equipment

3.3.3 CLEAN AIR ACT

The Clean Air Act (CAA) Amendments of 1990 (Section 305 of P.L. 101–549) contain a provision mandating more stringent federal standards for solid waste incinerators, known as the Maximum Achievable Control Technology (MACT) emission standards (see Chapter 15). The CAA amendments require the EPA to issue new source performance standards to control air emissions from municipal, hospital, and other commercial and industrial incinerators, including hazardous waste burning cement kilns and lightweight aggregate kilns. The MACT standards set emission limitations for polychlorinated dibenzodioxins (PCDDs), polychlorinated dibenzofurans (PCDFs), metals, particulate matter, total chlorine, hydrocarbons, carbon monoxide, and destruction and removal efficiencies (DRE) for organic emissions.

New facilities must comply with the EPA requirements within 6 months of the time they are issued and existing units must comply within 5 years of issuance.

3.3.4 POLLUTION PREVENTION ACT

The Pollution Prevention Act of 1990 (sections 6601–6610 of P.L. 101–608) was enacted as part of the Omnibus Budget Reconciliation Act of 1991. The Act declared pollution prevention to be a national policy for waste management and directed the EPA to conduct activities aimed at preventing the generation of pollutants, rather than managing them after they are created. The Pollution Prevention Act focused industry, government, and public attention on reducing the amount of pollution through cost-effective changes in production, operation, and raw materials use.

Matching grants were authorized for states to establish technical assistance programs for businesses, and the EPA was directed to establish a Source Reduction Clearinghouse to disseminate information. The Act also imposed new reporting requirements on industry. Firms that were required to file an annual toxic chemical release form under the Emergency Planning and Community Right-to-Know Act of 1986, must also file a report detailing their source reduction and recycling efforts over the previous year.

3.3.5 INDIAN LANDS OPEN DUMP CLEANUP ACT

Public Law 103-399, The Indian Lands Open Dump Cleanup Act of 1994, acknowledged concerns that solid waste open dump sites located on American Indian or Alaskan native lands threatened the health and safety of local residents. The purpose of the Act was to identify the location of open dumps on Indian lands, assess the health and environmental hazards posed by those sites, and provide financial and technical assistance to Indian tribal governments to close such dumps in compliance with Federal regulations or standards promulgated by tribal governments or native entities (U.S. EPA, 1998).

The Act required the Director of the Indian Health Service (IHS) to develop an inventory of all open dump sites on Indian lands. In addition, the IHS was to submit annual reports to Congress indicating a priority for addressing waste management deficiencies and progress made in addressing those needs. The Act also called for the IHS to identify the level of funding necessary to bring those open dump sites into compliance with all regulations and to develop comprehensive waste management plans for every tribal entity.

According to the IHS, prior to the law's enactment, only two of more than 600 waste dumps on Indian lands met current U.S. EPA regulations.

3.3.6 MERCURY-CONTAINING AND RECHARGEABLE BATTERY MANAGEMENT ACT

The Mercury-Containing and Rechargeable Battery Management Act of 1996 (Battery Act) (P.L. 104–142) was enacted in 1996 to phase out the use of mercury in batteries, and to provide for the efficient and cost-effective disposal of used nickel-cadmium batteries, used small sealed lead-acid batteries, and certain other regulated batteries. The act applies to battery and product manufacturers, battery waste handlers, and certain battery and product importers and retailers. The law also places uniform national labeling requirements on regulated batteries and rechargeable consumer products and encourages battery recycling programs.

Additionally, the collection, storage, and transportation of used rechargeable batteries, used consumer products containing batteries that are not easily removable, and certain other batteries are subject to regulation under the Universal Waste Rule (60 F.R. 25492) (see Chapter 18). The rule applies to battery and product manufacturers, battery waste handlers, and certain battery and product importers and retailers. Other types of batteries not covered by the Battery Act, such as the larger lead-acid batteries found in automobiles, trucks, and other equipment, are regulated as universal wastes under the RCRA hazardous waste regulations, 40 CFR Subpart 273 (U.S. EPA, 2003b).

3.4 RELATIONSHIP OF RCRA WITH OTHER ENVIRONMENTAL STATUTES

RCRA is only one of the several federal regulatory programs in place to protect environmental quality. The RCRA regulations work in concert with other environmental statutes such as the Clean Air Act (CAA); the Clean Water Act (CWA); the Emergency Planning and Community Right-to-Know

Act (EPCRA); the Federal Insecticide, Fungicide, and Rodenticide Act (FIFRA); the Marine Protection, Research, and Sanctuaries Act (MPRSA); the Occupational Safety and Health Act (OSHA); the Safe Drinking Water Act (SDWA); and the Toxic Substances Control Act (TSCA).

3.5 LAWS, REGULATIONS AND OTHER ACTIONS AT THE FEDERAL LEVEL

3.5.1 THE LAWMAKING PROCESS

The primary function of Congress is the making of laws, and the legislative process comprises a number of formal steps. The work of Congress is initiated by the introduction of a proposal in one of four principal forms: the bill, the joint resolution, the concurrent resolution, or the simple resolution. The bill is introduced to the appropriate committee for consideration. Following public hearings and markup sessions, the bill is forwarded to the House floor for consideration. After a measure passes in the House, it moves on to the Senate for consideration. A bill must pass both bodies in the same form before it can be presented to the President for signature into law. After both the House and Senate have passed a measure in identical form the bill is considered "enrolled." It is sent to the

Friday,
January 31, 2003

Part VI

Environmental Protection Agency

40 CFR Part 62
Federal Plan Requirements for Small Municipal Waste Combustion Units Constructed on or Before August 30, 1999; Final Rule

FIGURE 3.1   Federal Register.

President, who may sign the measure into law, veto it and return it to Congress, let it become law without signature, or, at the end of a session, pocket-veto it.

3.5.2 REGULATIONS

The Resource Conservation and Recovery Act, passed by Congress, includes a mandate that directs the U.S. EPA to develop regulations. Regulations, or rulemakings, are issued by an agency such as the EPA, DOT, or OSHA, which translate the general mandate of a law into a set of requirements for the agency and the regulated community.

Environmental regulations are formulated by the U.S. EPA with the support of public participation. When a regulation is proposed, it is published in the *Federal Register*, a government document, to notify the public of the EPA's intent to create new regulations or modify existing ones (Figure 3.1). EPA provides the public, which includes the potentially regulated community, with an opportunity to submit comments. Following a comment period, the EPA may revise the proposed rule based on both an internal review process and public comments. The final regulation is published, or promulgated, in the *Federal Register*. Included with the regulation is a discussion of the agency's rationale for the regulatory program. The final regulations are compiled annually and incorporated in the Code of Federal Regulations (CFR) (Figure 3.2). This process is called codification and each CFR title corresponds to a different regulatory authority. For example, the

code of federal regulations

Protection of Environment

40

PARTS 260 to 265

Revised as of July 1, 1998

CONTAINING
A CODIFICATION OF DOCUMENTS
OF GENERAL APPLICABILITY
AND FUTURE EFFECT

AS OF JULY 1, 1998

With Ancillaries

Published by
the Office of the Federal Register
National Archives and Records
Administration

as a Special Edition of
the Federal Register

FIGURE 3.2 Code of Federal Regulations.

EPA regulations appear in Title 40 of the CFR and the RCRA regulations are found in Title 40 of the CFR, Parts 240–282. These regulations are often cited as 40 CFR, with the part (e.g., 40 CFR Part 262), or the part and section (e.g., 40 CFR §262.40) listed after the CFR title.

The above relationship between an act and the regulations is fairly typical; one exception, however, is the relationship between HSWA and its regulations. Congress, through HSWA, provided the EPA with a mandate to promulgate regulations but also placed explicit instructions in the statute to develop specific regulations. Many of these requirements are so specific that EPA incorporated them directly into the regulations. HSWA is also significant because Congress established ambitious schedules for the implementation of the Act's provisions. Another unique aspect of HSWA is that it established "hammer provisions" or statutory requirements that would go into effect automatically with the force of regulations, if EPA failed to issue regulations by certain dates (U.S. EPA, 2002).

The interpretation of statutory language does not end with the codification of regulations. EPA further clarifies the requirements of an act and its regulations through guidance documents and policy.

3.5.3 GUIDANCE AND POLICY

Policy statements specify operating procedures that should generally be followed by a facility or agency. They are mechanisms used by EPA program offices to outline the manner in which the RCRA program is implemented. For example, EPA's Office of Solid Waste may issue a policy out-

THE OFFICE OF PESTICIDE PROGRAM'S GUIDANCE DOCUMENT

ON

METHODOLOGY
for DETERMINING the DATA NEEDED and the
TYPES of ASSESSMENTS
NECESSARY to MAKE FFDCA SECTION 408
SAFETY DETERMINATIONS for
LOWER TOXICITY PESTICIDE CHEMICALS

OFFICE OF PESTICIDE PROGRAMS
U.S. ENVIRONMENTAL PROTECTION AGENCY

June 7, 2002

FIGURE 3.3   Example of a Guidance Document.

lining what actions should be taken to achieve RCRA-corrective action cleanup goals. Policy statements are usually addressed to the staff working on implementation, but they may also be addressed to the regulated community.

Guidance documents are issued by the EPA to provide direction for implementing and complying with regulations (Figure 3.3). These are not strict requirements but are "how to" documents. For example, the regulations in 40 CFR Part 270 detail what is required in a permit application for a hazardous waste management facility, while the guidance for this part suggests how to evaluate a permit application to ensure that all information has been included. Guidance documents also elaborate on the agency's interpretation of the requirements of the act (U.S. EPA, 2002).

3.5.4 PUBLIC INVOLVEMENT IN RCRA

RCRA encourages public participation and involvement provisions to facilitate permitting, corrective action, and state authorization processes. The EPA, consistent with the requirements of the Administrative Procedures Act (APA), involves the public every time the agency issues a rulemaking that establishes or changes regulatory provisions. Because the RCRA program is a complex regulatory framework, EPA has established several public outreach and assistance mechanisms to foster public involvement. These include access to information through training grants, the Freedom of Information Act (FOIA), EPA's Office of Ombudsman, the EPA Docket Center, the EPA Dockets (EDOCKET) Web site, and the RCRA, Superfund & EPCRA Call Center.

REFERENCES

Levitan, R.L. and Nezam-Mafi, N., A Business Perspective on the Competitive Transition of the Electric Utility Industry, Levitan and Associates, Inc. *American Bankruptcy Institute Fifth Annual Northeast Bankruptcy Conference* July 1998. See: http://www.levitan.com/pdf/ABIspeech98.PDF

McCarthy, J.E. and Tiemann, M., Solid Waste Disposal Act/Resource Conservation and Recovery Act. Summaries of Environmental Laws Administered by the EPA, Congressional Research Service Report RL 30022, National Council for Science and the Environment, Washington, DC. See: http://www.ncseonline.org/nle/crsreports/briefingbooks/laws/h.cfm

U.S. Environmental Protection Agency, The Nation's Hazardous Waste Management Program at a Crossroads, Report No. EPA/530-SW-90-069, Washington, DC July 1990.

U.S. Environmental Protection Agency, Report on the Status of Open Dumps on Indian Lands, 1998 Report, Waste Management in Indian Country, 1998. See: http://www.epa.gov/tribalmsw/pdftxt/98report.pdf

U.S. Environmental Protection Agency, Office of Solid Waste, *RCRA Orientation Manual*, U.S. Government Printing Office, Washington, DC, 2002.

U.S. Environmental Protection Agency, CERCLA Overview, 2003a. See: http://www.epa.gov/superfund/action/law/cercla.htm

U.S. Environmental Protection Agency, Mercury-Containing and Rechargeable Battery Management Act (Battery Act) Enforcement, 2003b. See: http://www.epa.gov/Compliance/civil/programs/ba/index.html

SUGGESTED READINGS

Butler, J.C., III, Schneider, M.W., Hall, G.R., and Burton, M.E., Allocating superfund costs: cleaning up the controversy, *Environ. Law Reporter*, 23, 10133–10144, 1993.

Carlson, R.L., RCRA overview, in *Handbook on Hazardous Materials Management*, Cox, D.B., Ed., Institute of Hazardous Materials Management, Rockville, MD, 1995.

Case, D.R., Resource Conservation and Recovery Act, in *Environmental Law Handbook*, Sullivan, T.F.P., Ed., 14th ed., Government Institutes, Rockville, MD, 1997.

Landfair, S.W., Toxic substances Control Act, in *Environmental Law Handbook*, Sullivan, T.F.P., Ed., 14th ed., Government Institutes, Rockville, MD, 1997.

Lopez-Cepero, B.D., Emergency Planning and Community Right-To-Know Act of 1986, in *Handbook on Hazardous Materials Management*, Cox, D.B., Ed., Institute of Hazardous Materials Management, Rockville, MD, 1995.

Nardi, K.J., Underground storage tanks, in *Environmental Law Handbook*, Sullivan, T.F.P., Ed., 14th ed., Government Institutes, Rockville, MD, 1997.

Paschal, E.F., Jr., Clean Water Act, in *Handbook on Hazardous Materials Management*, Cox, D.B., Ed., Institute of Hazardous Materials Management, Rockville, MD, 1995.

Phillips, J.W. and Lokey, J.D., Toxic air pollution control through the Clean Air Act, in *Handbook on Hazardous Materials Management*, Cox, D.B., Ed., Institute of Hazardous Materials Management, Rockville, MD, 1995.

Shimberg, S.J., The Hazardous and Solid Waste Amendments of 1984: What Congress Did ... and Why, *The Environmental Forum*, March 1985, pp. 8–19.

Steinzor, R.I. and Lintner, M.F., Should taxpayers pay the cost of Superfund? *Environ. Law Reporter*, 22, 10089–10090, 1992.

Wagner, T.P., *The Complete Guide to the Hazardous Waste Regulations*, 2nd ed., Van Nostrand Reinhold, New York, NY, 1991.

Williams, S.E., Safe Drinking Water Act, in *Environmental Law Handbook*, 14th ed., Sullivan, T.F.P., Ed., Government Institutes, Rockville, MD, 1997.

QUESTIONS

1. Identify the primary state and local regulations and agencies involved in MSW management in your city or county. Who is primarily responsible for MSW recycling? Waste reduction? Management of a landfill or incinerator?

2. Prepare a chronology of the development of MSW legislation in your state.

3. Discuss how the general public can become involved in the promulgation of RCRA regulations.

4. Explain the difference between a law, a regulation, a policy, and a guidance document. How do they differ in terms of enforcement capability?

5. Discuss the evolution of the waste regulatory process in the United States. Based on industry trends and public concerns, how do you think the laws and regulations may evolve in the next 10 years?

6. Define a solid waste and a hazardous waste in general terms. What materials do they include? Then refer to Volume 40 of the Code of Federal Regulations. Review the definitions of a solid waste (Part 261.2) and a hazardous waste (Part 261.3). What materials are included? What are some of the exemptions to the definitions of each?

7. What is the general relationship between RCRA and CERCLA? How do the two acts differ in terms of when the waste is generated and when it is disposed?

8. Using the World Wide Web, compare the regulations of three different states with regard to management of MSW. How do they differ in terms of sanitary landfill siting, landfill operation, siting of a composting facility, and recycling of electronics waste?

Part II

Municipal Solid Wastes

Waste management embraces the application of techniques and systems that ensure the proper storage, collection, transportation, and disposal of a waste stream. State governments, industry, and citizens continue to seek avenues to reduce the volume of waste, to reuse it, or to manage and dispose of it properly. Integrated management of solid waste is one strategy that has been firmly embraced by the U.S. EPA and other nations, for example, the European Union, for the appropriate management of solid waste. In this management hierarchy, reuse and recycling are given high priority and landfilling is considered to be the least favored option. Up to the present time, however, landfilling continues to serve as the primary destination for the majority of U.S. solid waste.

The chemical, physical, and biological characterization of municipal solid waste is presented in the following chapter. Most of Part II, however, addresses the processing and ultimate disposition of the waste, including recycling, composting, incineration, and sanitary landfilling. Both conventional and innovative technologies in these applications are presented. Discussion is primarily limited to wastes generated from residential and commercial sources.

4 Characterization of Solid Waste

4.1 INTRODUCTION

In order for a community to formulate an integrated solid waste management program, accurate and reliable data on waste composition and quantities are essential. Such data will encourage well - organized and smoothly functioning recycling programs; foster the optimal design and operation of materials recovery facilities and municipal incinerators; and, ultimately, reduce the amount of waste generated and keep the overall waste management costs low.

Knowledge of the chemical composition of MSW will guide engineers and scientists of its utility as a fuel and will also help in predicting the makeup of gaseous emissions after incineration as well as of possible hazardous substances occurring in the ash. Waste composition will provide information on the utility of the material for composting or for biological conversion into biogas fuel. In addition, given that the majority of U.S. municipal solid waste (MSW) is disposed of in landfills, knowledge of chemical composition will help in predicting leachate composition and necessary treatment options. The physical properties of MSW will indicate ease of transport, processing requirements, combustion characteristics, and a rough prediction of landfill lifetime.

4.2 SAMPLING PROTOCOLS FOR MSW

MSW consists of a wide range of materials that vary depending on the community and its consumers' income and lifestyles, its degree of industrialization, institutionalism, and commercialism. Given these variables, several protocols can be followed to estimate the MSW composition for an area.

In order to compile accurate data, several issues must be addressed (Rhyner et al., 1995):

- How to obtain representative samples of the MSW?
- What is the desired sample size?
- How many samples are needed to achieve a desired level of accuracy?

4.2.1 DIRECT SAMPLING

Direct sampling is useful on a small scale for obtaining information about MSW composition. The direct sampling method involves physically sampling and sorting MSW at the source of generation. Although MSW can be extremely heterogeneous, direct sampling is one of the more accurate characterization methods. In order to make accurate judgments as to composition, sorting and analysis should be conducted in several randomly selected locations within the community. Waste sampling from single- and multi-family homes, commercial establishments (restaurants and businesses), and institutions (schools, hospitals) is encouraged, as these inputs create local variations.

Another direct sampling approach is to sample the waste after it has arrived at a centralized collection point or a tipping (i.e., unloading) area. This may include a transfer station or disposal facility. ASTM Method D5231-92 (ASTM, 1998) calls for a sample size of 91 to 136 kg (200 to 300 lb) to be manually sorted at the disposal facility. Whether at the source or a disposal facility, the degree of sorting is a function of the number of product categories desired. For example, if a composting program is to be instituted, a sorting scheme might include organic and inorganic materials only. Alternatively, food and yard wastes, the highest quality compost feedstock, can be separated from all other MSW. If a comprehensive materials recovery program is being considered, however, more detailed data about waste categories will be needed — for example, wastes may have to be separated into aluminum, ferrous metals, glass, and paper. In some cases, paper products are further subdivided into old newspaper (ONP), old corrugated cardboard (OCC), laser-quality office paper, and colored paper.

One disadvantage of direct sampling programs based on a limited number of samples is that data may be misleading if unexpected circumstances occurred during the sampling period. These circumstances could include the delivery of infrequent and exotic wastes, a severe wet or dry season, or errors in sampling methods (U.S. EPA, 1999). Such errors will be compounded when a small number of samples are collected to represent the community waste stream. Sampling studies do not provide accurate information about trends unless they are performed in a consistent manner over a long period of time (U.S. EPA, 1999). Another disadvantage of direct sampling is that it would be prohibitively expensive for making estimates on a national scale.

4.2.2 MATERIAL FLOWS

Another approach to determining waste composition is to assess material flows. This method is useful for estimating waste stream composition and trends on a regional basis. The U.S. EPA uses materials flow estimation for the compilation of waste data for the United States (U.S. EPA, 2001). The methodology is based on production data (by weight) for materials and products in the waste stream. For a particular municipality, inputs and outputs are recorded and compared. For example, if a community purchases 500,000 aluminum beverage cans in 1 week, it can be expected that about 500,000 aluminum cans will end up in the waste stream some time soon afterward. This model is, of course, an oversimplification; and one must also consider that the community is an open system having numerous imports and exports (U.S. EPA, 1999).

4.2.3 SURVEYS

Waste quantity and composition can be estimated by distributing questionnaires to producers of the waste. This system typically applies to generators of commercial and industrial wastes, and does not work effectively for domestic sources. A questionnaire is distributed to companies in an area, with detailed questions concerning the quantities of waste generated and its composition. Waste types may be listed in relation to product or material categories; for example, a county building may be asked to quantify the laser-quality office paper, mixed, colored papers, ONP, and ONP boxes. Other questions may pertain to seasonal variations in waste generation and any recycling programs already in operation (Williams, 1998). In many cases, however, companies do not maintain accurate records of the amount of waste generated. Data on composition may also be difficult to obtain due to concerns over the release of company and proprietary information.

Yu and MacLaren (1995) compared the accuracy of direct waste analysis with the survey for determining waste stream composition. Table 4.1 demonstrates that there is substantial variability in material estimates between the two methods.

4.2.4 MULTIPLIERS FOR PROJECTING WASTE QUANTITIES

Waste generation multipliers are used for estimating waste quantities from sources in a particular region. These multipliers express the relationship between the amount of waste produced and an identifiable parameter, for example a household or a specific industry. The value of the multiplier is based upon surveys, published data, and direct sampling for an area. For example, for a county in the midwest United States, a household waste multiplier may be derived based on the size of the population. Agricultural multipliers may be formulated based on the number and type of livestock and the total land area available for grazing. Industrial waste multipliers may be based on the number of employees at a facility. The population of the area in question is multiplied by the appropriate value

TABLE 4.1

Waste Composition as Estimated by Direct Analysis and Surveys (wt%; $n = 78$)

Waste Type	Direct Sampling	Survey
Paper	24.7	33.2
Paperboard	22.3	9.0
Ferrous metal	5.9	3.3
Non–ferrous metal	0.9	0.7
Plastics	13.3	6.9
Glass	2.8	8.4
Rubber	0.4	0.5
Leather	0.0	0.0
Textiles	4.5	0.7
Wood	7.5	10.3
Vegetation	1.4	0.4
Fines	0.3	2.2
Special wastes	0.6	0.7
Construction materials	4.6	2.2
Food	10.7	20.9

Source: From Yu, C. and McLaren, V., *Waste Manage. Res.,* 13, 343–361, 1995. Reproduced with kind permission of Elsevier Academic Press.

TABLE 4.2

Typical U.S. Waste Generation Rates as a Function of Generator Type

Waste Generation Sector	Average	Units
Single family residential	1.22	kg/person/day
Apartments	1.14	kg/person/day
Offices	1.09	kg/employee/day
Eating and drinking establishment	6.77	kg/employee/day
Wholesale and retail trade[a]	0.009	kg/$ sales
Food Stores	0.015	kg/$ sales
Educational facilities	0.23	kg/student/day

[a] Except food stores

Source: From Savage., G.M. *Warmer Bull. J. World Resour. Found.,* 49, 18–22, 1996. Reproduced with kind permission of the Warmer Bulletin, www.residua.com.

to obtain an estimate of waste production. Table 4.2 presents waste generation multipliers based on generator type.

In efforts to develop more accurate waste generation multipliers, some surveys have taken into account numerous factors, including the size of the local population in a region, the type and age of residence occupied, season of the year, and the types of businesses in an area. Also useful are economic data such as industrial output and number of employees (Rhyner and Green, 1988; Savage, 1996; Williams, 1998).

Household waste generation multipliers have varied widely. Estimates of household waste production have varied between 1.08 and 1.22 kg/person/day (2.37 and 2.68 lb/person/day) (Rhyner

TABLE 4.3
Household Waste Multipliers Based on the Community
Population Size

Population	Waste Generation Multiplier (kg/person/day)
<2,500	0.91
2,500–10,000	1.22
10,000–30,000	1.45
>30,000	1.63

Reproduced with kind permission of Wisconsin Department of Natural Resources.

and Green, 1988). More accurate estimates can be generated for household waste using multipliers based on the population size of the community. Smaller communities produce a lower waste generation per person per day compared with larger communities (Table 4.3)(Yu and MacLaren, 1995).

The multipliers used for predicting future waste production quantities have significant implications for planning. If waste quantities are expected to increase or if composition is expected to change (e.g., due to the arrival of new businesses or industries), changes may be needed to accommodate the new waste stream, for example, the establishment of a MRF or expansion of a landfill.

4.3 VARIABILITY AFFECTING WASTE SAMPLING

4.3.1 SEASONAL

To ensure accurate waste generation estimates, wastes should be sampled regularly over a defined period (e.g., one calendar year) to account for seasonal variations. The season of the year strongly affects the amounts of yard waste generated, for example. During spring, summer and fall months, the volume of grass clippings from low-density residential neighborhoods sharply increases. The quantities generated are also highly dependent on the yard area per living unit (Pfeffer, 1992). In the fall, leaves will add to the waste load, and the number and types of trees in the community affect the total amounts. Many states have banned grass clippings, leaves, twigs, and branches of certain sizes from landfills. Given that the burning of these materials is often prohibited, an alternative means must be provided for their disposal. The generation of other wastes are also affected by season. There will be a greater percentage of construction and demolition waste and waste tires during warmer months (Table 4.4).

In areas that are heavily industrialized or support diverse commercial activity, patterns of the community's waste generation will be significantly affected by season. For example, industries heated with coal or utilities burning coal for heat or electric generation will produce significantly more ash in the winter months. Also, food-processing plants will produce more waste during harvest months (Rhyner, 1992; Rhyner et al., 1995).

In addition to seasonal variations, the quantity and overall composition of MSW will vary over the course of a week. More yard waste is produced on weekends; in contrast, more commercial and industrial wastes are produced on weekdays.

Municipal waste processing and disposal systems must take into account the shifting quantities and composition of waste over the course of a day, a week, or a year. For example, a city recycling program with a composting system should expect substantial quantities of potential feedstock in the spring, and therefore allow for sufficient space for initial storage, for the establishment of the compost piles, and for the stockpiling of the final, cured product. Municipal incinerator will be affected by large inputs of wet grass or leaves in the summer. These materials will

TABLE 4.4
Seasonal Variations for Various MSW Components (Expressed as %)

Waste Component	Autumn	Winter	Spring	Summer	Average
Organics	86.0	86.5	87.7	89.8	87.5
Paper	44.7	45.7	47.5	40.3	44.5
Plastic	6.1	6.5	7.0	6.0	6.4
Yard waste	15.0	15.1	7.4	15.0	13.1
Wood	1.4	0.8	1.0	0.8	1.0
Food	15.2	14.3	18.0	21.5	17.3
Textiles	1.5	1.3	1.3	2.0	1.5
Other	2.3	2.9	5.4	4.2	3.7
Inorganics	13.4	12.8	12.2	9.8	12.1
Metals	3.0	3.9	4.0	3.1	3.5
Glass	7.0	7.1	7.1	5.6	6.7
Soil	0.9	0.4	0.0	0.0	0.3
Other	2.5	1.4	1.1	1.2	1.5
Special wastes	0.5	0.6	0.1	0.3	0.4
Total	100	100	100	100	100

Adapted from Savage, G.M., *Warmer Bull. J. World Resour. Found.*, 49, 18–22, 1996. Reproduced with kind permission of the Warmer Bulletin, www.residua.com.

lower the heat content of the waste, and an auxiliary fuel may be required to maintain combustion temperatures.

4.3.2 REGIONAL

Different parts of the country produce markedly different types and amounts of wastes. Communities along the Gulf coast, by virtue of the warm, moist climate during most of the year, will produce substantially more yard and garden waste than would communities in central Arizona. Moisture will also induce specific effects; the MSW of Louisiana or western Florida would be expected to possess a higher moisture content compared to that of Tucson or Santa Fe, all other factors being equal. Much of this higher moisture content may occur in yard waste, but the relatively higher humidity along the Gulf will also permeate stored wastes. Finally, certain marketing/community/grassroots activities in a county or state will affect waste composition. A striking example is the enactment of recyclable bottle bills in various states. A financial incentive for the reuse of soda and beer bottles will sharply limit their appearance in the local waste stream.

4.3.3 HOUSEHOLD

A study conducted in 1976 found that the number of persons per household have a significant influence on waste production (Rhyner, 1976). In Figure 4.1, daily household waste production is plotted against the number of persons per household. For two persons, total waste generation is about 2.5 kg (5.5 lb) per day, which is equivalent to 1.3 kg (2.8 lb) per person per day; for ten people in a household, the amount of waste produced is 4 kg (8.8 lb) per day, which converts to 0.4 kg (0.9 lb) per person per day. These results are not surprising; economies of scale occur with a larger number of residents. Large families will purchase food and beverages in larger containers and will share newspapers and other consumables. Except for very low-income households, the data from this study were found to be independent of family income level.

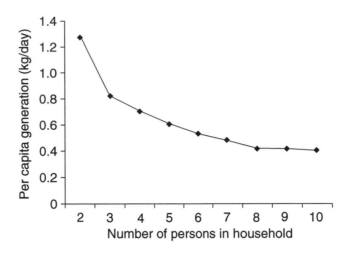

FIGURE 4.1 Variability in quantities of MSW according to size of household. (From Rhyner, C.R., *Waste Age,* 1976, pp. 29–50. With permission.)

4.3.4 NATIONAL ECONOMY

There is some correlation between waste generation rates and the overall economy of a country. Figure 4.2 presents waste generation and gross domestic product data for several developed countries. Up to a certain point, per capita waste generation does not change significantly with increase in per capita gross domestic product. However, beyond about $20,000 per capita GDP, waste generation varies sharply. The highest incomes, however, correlate with the highest waste generation rates.

4.4 COMMON COMPONENTS IN MUNICIPAL SOLID WASTE

The predominant components of the U.S. municipal waste stream are discussed below. Some details on their generation are summarized in Table 4.5 and Figure 4.3. Information on manufacturing and recycling processes for these materials is given in Chapter 6.

4.4.1 PAPER PRODUCTS

Paper and paper products comprise the largest component of the U.S. municipal waste stream. The products that comprise paper and paperboard wastes are shown in Table 4.6. Total generation of paper products in MSW has grown from 30 million tons in 1960 (34% of the waste stream) to 87.5 million tons (38.1%) in 1999 (U.S. EPA, 2001).

4.4.2 GLASS

Glass occurs in MSW primarily in the form of containers (Table 4.7), as beer and soft drink bottles, wine and liquor bottles, and bottles and jars of food and other consumer products. Glass is also common in durable goods such as appliances and consumer electronics.

Glass accounted for 6.7 million tons of MSW in 1960 or 7.6% of total generation. The generation of glass grew over the next two decades; however, aluminum and plastic containers gained a strong foothold in the food container market and replaced some glass containers. Thus, the tonnage of glass in MSW declined in the 1980s. Glass composed about 10% of MSW generation in 1980, declining to 5.5% in 1999 (U.S. EPA, 2001).

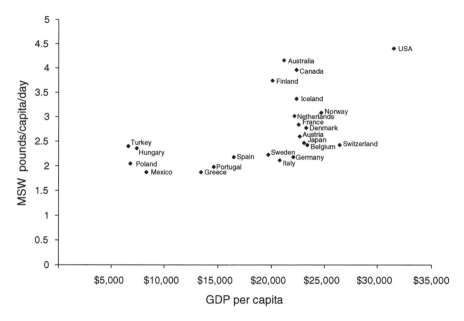

FIGURE 4.2 Relationship between GDP and per capita MSW generation for various countries.

TABLE 4.5
Materials Generated (thousands of tons) in MSW, 1960 to 1999

Material	1960	1970	1980	1990	1999
Waste from Specific Products					
Paper and paperboard	29,900	44,310	55,160	72,730	87,470
Glass	6,720	12,740	15,130	13,100	12,560
Metals					
Ferrous	10,300	12,360	12,620	12,640	13,320
Aluminum	340	800	1,730	2,810	3,130
Other nonferrous	180	670	1,160	1,100	1,390
Total Metals	10,820	13,830	15,510	16,550	17,840
Plastics	390	2,900	6,830	17,130	24,170
Rubber and Leather	1,840	12,970	4,200	5,790	6,220
Textiles	1,760	2,040	2,530	5,810	9,060
Wood	3,030	3,720	7,010	12,210	12,250
Other	70	770	2,520	3,190	4,010
Total Materials in Products	54,620	83,280	108,890	146,510	173,580
Other wastes					
Food Waste	12,200	12,800	13,000	20,800	25,161
Yard Trimmings	20,000	23,200	27,500	35,000	27,730
Miscellaneous Inorganic					
Wastes	1,300	1,780	2,250	2,900	3,380
Total Other Wastes	33,500	37,780	42,750	58,700	56,270
Total MSW Generated	88,120	121,060	151,640	205,210	229,850

Source: U.S. EPA, 2001; Franklin Associates, 1999. Reproduced with kind permission of Franklin
Associates.

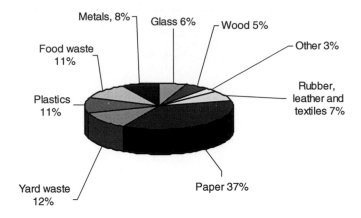

FIGURE 4.3 Composition of MSW in the United States, 1999 (U.S. EPA, 2001).

TABLE 4.6
Paper Products in Municipal Solid Waste, 1999

Product Category	Generation (thousands of tons)
Nondurable Goods	
Newspapers	
Newsprint	11,330
Groundwood inserts	2,630
Total newspapers	13,960
Books	1,120
Magazines	2,310
Office papers	7,670
Telephone directories	680
Third class mail	5,560
Other commerical printing	5,940
Tissue paper and towels	3,360
Paper plates and cups	930
Other nonpackaging paper	4,790
Total paper and paperboard nondurable goods	46,320
Containers and Packaging	
Corrugated boxes	31,230
Milk cartons	490
Folding cartons	5,780
Other paperboard packaging	290
Bags and sacks	1,690
Other paper packaging	1,670
Total paper and paperboard containers and packaging	41,150
Total paper and paperboard	87,470

[a] Includes tissue in disposable diapers, paper in games and novelties and cards.

Source: U.S. EPA, 2001.

TABLE 4.7
Glass Products in MSW, 1999

Product Category	Generation (thousands of tons)
Durable goods[a]	1150
Containers and packaging	
Beer and soft drink bottles	5450
Wine and liquor bottles	1830
Food and other bottles and jars	3770
Total glass containers	11,050
Total glass	12,560

[a] Glass as a component of appliances, furniture and consumer electronics.

Source: U.S. EPA, 2001.

4.4.3 ALUMINUM

The largest sources of aluminum in MSW are used beverage containers (UBCs) and other packaging (Table 4.8). In 1999 about 2 million tons of aluminum were generated in containers and packaging, while 1 million tons were found in durable and nondurable goods. The total, 3.1 million tons, comprises 1.4% of total MSW generation in 1999. This compares with the generation of 340,000 tons (0.4%) in 1960.

4.4.4 FERROUS METALS

Ferrous metals (iron and steel) are the largest category of metals in MSW on a weight basis (Table 4.8). The majority of ferrous metals in MSW are found in appliances, furniture, and other durable goods. Containers and packaging are the other primary source of ferrous metals in MSW (U.S. EPA, 2001).

Approximately 10.3 million tons of ferrous metals were generated in 1960. Weights increased during the 1960s and 1970s, but later decreased as lighter materials like aluminum and plastics replaced steel in some applications. The percentage of ferrous metals generation in MSW declined from 11.7% in 1960 to 5.3% in 1999.

4.4.5 OTHER NONFERROUS METALS

Nonferrous metals such as copper, zinc, and lead are found in durable products such as appliances and consumer electronics. The generation of nonferrous metals in MSW totaled 1.4 million tons in 1999. Lead in automotive batteries is the most prevalent of the nonferrous metals in MSW. The generation of nonferrous metals has increased slowly, up from 180,000 tons in 1960. As a percentage of total generation, nonferrous metals remain below 1% (U.S. EPA, 2001).

4.4.6 PLASTICS

Plastics are used in durable and nondurable goods and in containers and packaging, with the latter being the largest category of plastics in MSW (Table 4.9). In durable goods, plastics are found in appliances, furniture, carpets, and other products.

There are hundreds of different resin formulations used in appliances, carpets, and other durable goods. Plastics are found in such nondurable products as disposable diapers, trash bags, cups, eating utensils, sporting goods, and household items such as shower curtains.

TABLE 4.8
Metal Products in MSW, 1999

Product Category	Generation (thousands of tons)
Durable Goods	
Ferrous metals	10,390
Aluminum	960
Lead	970
Other nonferrous metals	420
Total metals in durable goods	12,740
Nondurable Goods	
Aluminum	200
Containers and Packaging	
Steel	
Food and other cans	2,690
Other steel packaging	240
Total steel packaging	2,930
Aluminum	
Beer and soft drink cans	1,540
Food and other cans	50
Foil	380
Total aluminum packaging	1,970
Total metals in containers and packaging	4,900
Total Metals	17,840
Ferrous	13,320
Aluminum	3,130
Other nonferrous	1,390

Source: U.S. EPA, 2001.

Plastic food service items are generally made of clear or foamed polystyrene while trash bags are made of high- or low-density polyethylene. A wide variety of other resins are used in other nondurable goods. Plastic resins are used in container and packaging products such as polyethylene terephthalate (PET) soft drink bottles, high-density polyethylene (HDPE) bottles for milk and water, and a wide variety of other resin types are used in containers, bags, sacks, wraps, and lids (U.S. EPA, 2001).

Plastics are a rapidly growing segment of MSW. In 1960, plastics comprised an estimated 390,000 tons (< 1%) of MSW generation. The quantity has increased to 24.2 million tons (10.5%) in 1999.

4.4.7   RUBBER AND LEATHER

Automobile and truck tires are the predominant sources of rubber in MSW (Table 4.10). Other sources include clothing and footwear and other durable and nondurable products. The generation of rubber and leather in MSW has increased from 1.8 million tons in 1960 to 6.2 million tons in 1999. One reason for the relatively slow rate of growth is that tires have been made smaller and longer-wearing than previously. As a percentage of total MSW generation, rubber and leather have remained steady at 3.0%.

4.4.8   TEXTILES

Textiles in MSW occur in discarded clothing, although other sources include furniture, carpets, tires, footwear, and other nondurable goods such as towels. A total of 9.1 million tons of textiles were generated in 1999, comprising 3.9% of total MSW generation.

TABLE 4.9
Plastics in MSW, 1999

Product Category	Generation (thousands of tons)
Durable goods	
Total plastics in durable goods	7,180
Nondurable goods	
Plastic plates and cups	910
Trash bags	950
All other nondurables[a]	3,970
Total plastics in nondurable goods	5,830
Plastic containers and packaging	
Soft drink bottles	900
Other containers	2,650
Bags, sacks, wraps	4,240
Other packaging[b]	2,680
Total plastics in containers and packaging, by resin	
PET	1,850
HDPE	4,180
PVC	480
LDPE/LLDPE	3,320
PP	1,100
PS	250
Other resins	70
Total plastics in containers and packaging	11,160
Total plastics in MSW by resin	
PET	2,430
HDPE	5,340
PVC	1,450
LDPE/LLDPE	6,020
PP	3,060
PS	2,280
Other resins	3,590
Total plastics in MSW	24,170

HDPE = high-density polyethylene; PET = polyethylene terephthalate; PS = polystyrene; LDPE = low-density polyethylene; PP = polypropylene; PVC = polyvinylchloride; LLDPE = linear low-density polyethylene.
[a]All other nondurables include plastics in disposable diapers, clothing and footwear.
[b]Other plastic packaging includes coatings, closures, caps, trays and shapes.
Source: U.S. EPA, 2001.

4.4.9 FOOD WASTES

Food wastes include uneaten food and food preparation wastes from residences, commercial establishments (restaurants, fast food establishments), institutional sources such as school cafeterias, and industrial sources. The generation of food wastes was estimated at 25.2 million tons in 1999 (U.S. EPA, 2001).

TABLE 4.10
Rubber and Leather in MSW

Product Category	Generation (thousands of tons)
Durable goods	
Rubber in tires[a]	2,890
Other durables[b]	2,430
Total rubber and leather	
Durable goods	5,410
Nondurable goods	
Clothing and footwear	540
Other nondurables	250
Total rubber and leather	
Nondurable goods	790
Containers and packaging	20
Total rubber and leather	6,220

[a] Automobile and truck tires. Does not include other materials in tires.
[b] Includes carpets and rugs and other miscellaneous durables.
Source: U.S. EPA, 2001. With permission.

4.4.10 YARD WASTE

Yard waste includes grass clippings, leaves, and tree trimmings from residential, institutional, and commercial sources. The average composition by weight is estimated to be about 50% grass, 25% leaves, and 25% tree trimmings (U.S. EPA, 1999). Quantities and relative proportions will vary widely according to geographic region and climate. Yard waste is the second largest component of MSW, at 12.1% of total generation.

In the past, the generation of yard waste increased steadily as the U.S. population and amount of residential housing grew, although per capita generation remained relatively constant. In recent years, however, the amounts of yard waste have declined substantially in many areas, as a result of local and state legislation (usually in the form of bans) on the disposal of such wastes in landfills. With such so-called "flow control" in place, homeowners are adjusting by establishing backyard composting and using mulching lawnmowers that allow grass clippings to remain on the lawn surface. In 1992, 11 states had legislation banning or discouraging yard waste disposal in landfills. By 1999, 23 states and the District of Columbia, representing more than 50% of the nation's population, had legislation affecting the disposal of yard waste (U.S. EPA, 2001).

4.4.11 HOUSEHOLD HAZARDOUS WASTES

The portion of MSW referred to as "household hazardous waste" (HHW) refers to those hazardous materials occurring in MSW regardless of their source. Most hazardous material in household waste occurs either as heavy metals, organic compounds, or asbestos. These items are considered hazardous because they may contain materials that are ignitable, corrosive, reactive, or toxic. Metallic wastes such as lead, cadmium, and mercury are hazardous due to direct toxicity. Many of the organic wastes are deemed hazardous solely because they are flammable, although some pose inhalation hazards (e.g., paint strippers and other solvents) and others may damage or penetrate the skin (solvents and pesticides). Most of the asbestos occurring in solid waste occurs as old vinyl-asbestos floor tiles and asbestos shingles. Since the primary asbestos hazard is in the form of respirable particles, asbestos tiles and shingles are typically not a significant hazard. Table 4.11 lists

TABLE 4.11

Common Household Hazardous Wastes

Batteries (Ni–Cd, Pb, Hg)
Drain openers
Oven cleaners
Metal cleaners and polishers
Used motor oil
Automotive fuel additives
Grease and rust solvents
Carburetor and fuel injection cleaners
Air-conditioning refrigerants
Starter fluids
Paints
Paint thinners
Paint strippers and removers
Adhesives
Herbicides
Insecticides
Fungicides and wood preservatives
Asbestos-containing materials

TABLE 4.12

Hazardous Elements and Compounds Occurring in Common Household Products

Ingredient	Types of Products Found in
Acrylic acid	Adhesives
Aniline	Cosmetics (perfume), wood stain
Arsenic (III) oxide	Paint (nonlatex anti-algae)
Benzene	Household cleaner (spot remover, oven cleaner) stain, varnish, adhesives cosmetics (nail polish remover)
Cadmium	Ni–Cd batteries, paints, photographic chemicals
Chlordane	Pets (flea powders)
Chlorinated phenols	Latex paint
Chlorobenzene	Household cleaners (degreaser)
Hexachloroethane	Insect repellents
Lead	Stain/varnish, auto batteries, paint
Mercury	Batteries, paint (nonlatex anti-algae), fluorescent lamps
Methylene chloride	Household cleaners, paint strippers, adhesives
Nitrobenzene	Polish (shoe)
Silver	Batteries, photographic chemicals
Warfarin	Rodent control
Xylene	Transmission fluid, engine treatment (degreaser), paint (latex, nonlatex, lacquer thinners), adhesives, microfilm, fabric, cosmetics (nail polish)

Source: U.S. Congress Office of Technology Assessment, 1989.

many common HHWs and Table 4.12 presents selected hazardous compounds occurring in common household products.

The U.S. Congress Office of Technology Assessment (USCOTA, 1989) determined that the amount of HHW generated ranges between 0.2 and 0.4% of the residential waste stream. The average

U.S. household generates more than 20 lb of HHW per year. As much as 100 lb can accumulate in homes (U.S. EPA, 1993). Overall, Americans generate 1.6 million tons of HHW per year.

Estimates of the abundance of HHW vary, the possible reasons for the discrepancies being (Liu and Liptak, 2000):

- Some estimates include less toxic materials such as latex paint.
- Most estimates include the weight of the containers, and many estimates include the containers even if they are empty.
- Some estimates include materials that were originally in liquid or paste form but have dried, such as dried paint and adhesives. Toxic substances can still leach from these dried materials, but drying reduces the potential leaching rate.

As discussed in Chapter 11, the U.S. EPA has established stringent requirements for the management of hazardous waste generated by industry. Congress chose not to regulate HHW, however, due to the impracticality of regulating every household. Some of the concern surrounding such HHWs, therefore, is that unwitting consumers may dispose of these wastes, many of which are toxic and nonbiodegradable, directly into the sewer or household trash. As a result, such wastes will eventually find their way into the biosphere, either in the form of sewage solids or landfill leachate. During the 1980s, many communities started special collection days (Figure 4.4) or permanent collection sites for handling HHW. There are more than 3000 permanent HHW programs and collection events throughout the United States (Figure 4.5; U.S. EPA, 1993).

4.4.11.1 Toxic Metals

Lead is widespread in the municipal waste stream; it occurs in both the combustible and noncombustible portions of MSW. Discards of lead in MSW are substantially greater than discards of cadmium, mercury, and other toxic metals. Of the lead products entering the waste stream, lead-acid batteries (primarily for automobiles) rank first (U.S. EPA, 2000b). Trends in quantities of lead discarded in products in MSW are shown in Table 4.13. Lead discards in batteries are growing steadily as are discards in consumer electronics. Discards of leaded solder in cans and lead in pigments, however, virtually disappeared between 1970 and 1986.

Similar to lead, cadmium is widespread in products discarded into MSW although it occurs in much smaller quantities overall. Since 1980, nickel-cadmium household batteries have been the primary contributors of cadmium in MSW. Trends in quantities of cadmium discarded in products in MSW are shown in Table 4.14. Discards of cadmium in household batteries were low in 1970 but then increased dramatically. Cadmium discards in plastics are relatively stable. Discards of cadmium in consumer electronics have decreased over time, while amounts in the other categories listed are relatively low.

There are a number of sources of mercury in MSW, with total discards of mercury in 2000 estimated to be 173 tons, a substantial decline from the 1989 estimate of 709 tons (Table 4.15). Some of the common items using or containing mercury include household batteries, electric lighting, paint residues, fever thermometers, thermostats, pigments, dental uses, special paper coatings, mercury light switches, and film pack batteries (U.S. EPA, 1992).

4.4.11.2 Organic Compounds

The organic components of household hazardous waste include volatile organic compounds (VOCs) and persistent nonvolatile organics (POCs). The VOCs occur in household products such as cleaners and solvents, lawn and garden products (including pesticides), fuel products, and oil-based paints. VOCs are sometimes very toxic and may be carcinogenic, mutagenic or teratogenic. For example, benzene, a common component of automotive gasoline, has been declared a human carcinogen, mutagen, and possible teratogen. It damages the central and peripheral nervous system, is linked with blood cell disorders, and irritates the eyes and skin. Methylene chloride, an

(a)

(b)

(c)

FIGURE 4.4 Tox-away day for local citizens to properly dispose of HHW.

active ingredient in some paint strippers, is a possible carcinogen and is linked with central nervous system, respiratory, and cardiovascular disorders.

A number of common automotive products are considered hazardous and many states require their management as HHW. For example, antifreeze is toxic by virtue of its content of ethylene glycol; transmission fluid is toxic due to its content of hydrocarbons and mineral oils; brake fluid is toxic due to glycol as well as heavy metals; used motor oil is toxic due to the presence of polycyclic aromatic hydrocarbons and other hydrocarbons and heavy metals. Gasoline is toxic due to its content of benzene, toluene, ethylbenzene, and xylene as well as from being extremely flammable.

FIGURE 4.5 HHW drop-off and treatment center.

TABLE 4.13
Lead (tons) in Products in MSW, 1970–2000

Products	1970	1986	2000	Percentage
Lead-acid Batteries	83,825	138,043	181,546	Variable
Consumer electronics	12,233	58,536	85,032	Increasing
Glass and ceramics	3,465	7,956	8,910	Increasing; stable after 1986
Plastics	1,613	3,577	3,228	Fairly stable
Soldered cans	24,117	2,052	787	Decreasing
Pigments	27,020	1,131	682	Decreasing
All others	12,567	2,537	1,701	
Total	164,840	213,652	281,887	

Source: U.S. EPA, 2000b.

Persistent nonvolatile organics include pesticides, herbicides, and fungicides in common lawn and garden products. Others include hydraulic fluids and lubricants. Some POCs are probable or known carcinogens and some damage the liver, kidneys, central nervous system, lungs and reproductive system.

4.5 CHEMICAL PROPERTIES OF MSW

Accurate information on the chemical composition of the components of MSW is important for a number of reasons. First, the composition of landfill leachate (see Chapter 10) is directly affected by MSW composition. Excluding from or otherwise managing materials within the

TABLE 4.14
Cadmium (tons) in Products in MSW, 1970–2000

Products	1970	1986	2000	Percentage
Household batteries	53	930	2,035	Increasing
Plastics	342	502	380	Variable; decreasing after 1986
Consumer electronics	571	161	67	Decreasing
Appliances	107	88	57	Decreasing
Pigments	79	70	93	Variable
Glass and ceramics	32	29	37	Variable
All others	12	8	11	Variable
Total	1196	1788	2684	

Source: U.S. EPA, 2000b. With permission.

TABLE 4.15
Discards[a] of Mercury (tons) in Products in MSW, 1970 to 2000

Product	1970	1980	1989	2000
Household batteries	310.8	429.5	621.2	98.5
Electric lighting	19.1	24.3	26.7	40.9
Paint residues	30.2	26.7	18.2	0.5
Fever thermometers	12.2	25.7	16.3	16.8
Thermostats	5.3	7.0	11.2	10.3
Pigments	32.3	23.0	10.0	1.5
Dental uses	9.3	7.1	4.0	2.3
Special paper coating	0.1	1.2	1.0	0.0
Mercury light switches	0.4	0.4	0.4	1.9
Film pack batteries	2.1	2.6	0.0	0.0
Total Discards	421.8	547.5	709.0	172.7

[a] Discards before recovery.
Source: U.S. EPA, 1992. With permission.

waste stream (e.g., solvents, nickel–cadmium household batteries) will potentially improve leachate properties and prevent groundwater contamination. Secondly, composition must be known for evaluating alternative MSW processing and recovery options. For example, if the organic fraction of MSW is to be composted or is to be used as feedstock for the production of other biological conversion products, information on the major elements (e.g., ultimate analysis) that comprise the waste is important. Information will be required on trace element composition in the waste as well; for example, even modest concentrations of cadmium, arsenic, or lead may be detrimental to efficient composting or biogas production. Finally, the feasibility of MSW combustion is directly affected by chemical composition. Wastes can be considered a combination of semi-moist combustible and noncombustible materials. If solid wastes are to be used as a fuel, some important properties to determine include ultimate analysis, proximate analysis, energy content, and particle size distribution.

4.5.1 ULTIMATE ANALYSIS OF SOLID WASTE COMPONENTS

The ultimate analysis of a material is defined as its total elemental analysis, i.e., the percentage of each individual element present. The results of the ultimate analysis are typically used to characterize the chemical composition of the organic fraction of MSW. Such a determination is essential for assessing the suitability of the waste as a fuel and predicting emissions from combustion. The data are also used to define the proper mix of MSW materials to achieve suitable nutrient ratios (e.g., C/N) for biological conversion processes such as composting.

The ultimate analysis involves the determination of the percent values of carbon, hydrogen, oxygen, nitrogen, sulfur, and ash in a sample. Due to concerns over emissions of chlorinated compounds during combustion, the determination of halogens is often included in an ultimate analysis (Tchobanoglous et al., 1993; Liu and Liptak, 2000). The percent values of carbon, hydrogen, nitrogen, sulfur, and chlorine are measured directly by established procedures. The oxygen value is calculated by subtracting the other components, including ash and moisture, from 100%.

Data on the ultimate analysis of individual combustible materials are presented in Table 4.16. The majority of MSW is composed of carbon, hydrogen, and oxygen. Five materials tend to predominate in the organic portion of MSW: cellulose, lignins, fats, proteins and hydrocarbon polymers. Cellulose accounts for the majority of the dry weight of MSW and is the predominant compound in paper, wood, food waste, and yard waste (Masterson et al., 1981, Liu and Liptak 2000;). The relatively low sulfur and nitrogen contents are significant, as both are precursors to acid rain. Sulfur is not a component of any solid waste category, except perhaps building materials (gypsum panels) or yard waste. Nitrogen occurs mainly in food waste, grass clippings, and textiles (e.g., wool and nylon)(Liu and Liptak, 2000). Chlorine occurs in the organic form as polyvinyl chloride (PVC) and vinyl, and paper products bleached with chlorine. Chlorine may also occur in the inorganic form as sodium chloride and other simple salts.

The ash fraction is the residual remaining after combustion and is primarily inorganic although some organics may remain as well. Ash can impart significant environmental and public health effects if improperly managed. Ash may exit an incinerator and enter the atmosphere via the flue,

TABLE 4.16
Ultimate Analysis of the Combustible Components in Household MSW

Component	% by Wt (dry basis)					
	Carbon	Hydrogen	Oxygen	Nitrogen	Sulfur	Ash
Organic						
Paper	43.5	6.0	44.0	0.3	0.2	6.0
Plastics	60.0	7.2	22.8	—	—	10.0
Food Wastes	48.0	6.4	37.6	2.6	0.4	5.0
Yard Wastes	47.8	6.0	38.0	3.4	0.3	4.5
Textiles	55.0	6.6	31.2	4.6	0.15	2.5
Rubber	78.0	10.0	—	2.0	—	10.0
Wood	49.5	6.0	42.7	0.2	0.1	1.5
Inorganic						
Glass	0.5	0.1	0.4	<0.1	—	98.9
Metals	4.5	0.6	4.3	<0.1	—	90.5
Dirt, ash	26.3	3.0	2.0	0.5	0.2	68.0
MSW	15–30	2–5	12–24	0.2–1.0	0.02–0.1	—

Adapted from Kaiser, E.R., *Proceedings of the National Incinerater Conference*, ASME, New York, 1969; U.S. Department of Health, Education, and Welfare, 1969. Data reproduced with kind permission of the American Society of Mechanical Engineers.

TABLE 4.17
Composition of a Sample of MSW Ash

Material	% by Wt
Metals	16.1
Combustibles	4.0
Ferrous metal	18.3
Nonferrous metal	2.7
Glass	26.2
Ceramics	8.3
Mineral, ash, other	24.1

Source: Chesner, W.H. et al., From Hasselriis, F. *Handbook of Solid Waste Management,* in Keith, R., Ed., McGraw-Hill, New York, 1994. Reproduced with kind permission of the McGraw-Hill Companies.

or it may be retained within the solid residue. The composition of ash is largely influenced by the composition of the charge, i.e., the MSW entering the incinerator. Ash from unprocessed, unsorted MSW typically contains a much higher content of potentially toxic metals such as cadmium, lead, and mercury. Some of these metals may be readily leached if placed in a landfill and therefore require segregation from other wastes along with specialized treatment. A number of nontoxic metals also occur, such as iron, copper, magnesium, calcium, and sodium. Table 4.17 presents representative data from an MSW ash fraction.

EXAMPLE 4.1

Estimate the chemical composition of the organic fraction of a sample of MSW. Some data on waste properties is shown below.

	Wet Weight (kg)	Dry Weight (kg)
Paper	19	16
Plastics	3.7	3.5
Food wastes	5.1	1.9
Yard wastes	8.4	2.6
Textiles	1	0.8
Rubber	0.22	0.22
Wood	1.3	0.9

SOLUTION

Determine the percentage distribution of C, H, O, N and S occurring in the waste sample. Use the percent values of these elements from Table 4.16 (ultimate analysis).

	Dry Weight (kg)	Percent by Weight (Dry Basis)					
		C	H	O	N	S	Ash
Paper	16.0	43.5	6.0	44.0	0.3	0.2	6.0
Plastics	3.5	60.0	7.2	22.8	—	—	10.0
Food wastes	1.9	48.0	6.4	37.6	2.6	0.4	5.0
Yard wastes	2.6	47.8	6.0	38.0	3.4	0.3	4.5
Textiles	0.8	55.0	6.6	31.2	4.6	0.15	2.5
Rubber	0.22	78.0	10.0	—	2.0	—	10.0
Wood	0.9	49.5	6.0	42.7	0.2	0.1	1.5

Determine the percentage distribution of the elements in the sample.

	Wet Weight (kg)	Dry Weight (kg)	Composition (kg)					
			C	H	O	N	S	Ash
Paper	19	16	6.96	0.96	7.04	0.048	0.032	0.96
Plastics	3.7	3.5	2.1	0.252	0.798	0	0	0.35
Food wastes	5.1	1.9	0.912	0.122	0.7144	0.0494	0.0076	0.095
Yard wastes	8.4	2.6	1.243	0.156	0.988	0.0884	0.0078	0.117
Textiles	1	0.8	0.44	0.053	0.2496	0.0368	0.0012	0.02
Rubber	0.22	0.22	0.172	0.022	0	0.0044	0	0.022
Wood	1.3	0.9	0.446	0.054	0.3843	0.0018	0.0009	0.0135
Total			12.27	1.618	10.174	0.2288	0.0495	1.5775

	Weight (kg)
C	12.27
H	1.62
O	10.17
N	0.22
S	0.05
Ash	1.58

Determine the molar composition of the elements. Ignore the data for the ash.

Element	Atomic Weight (g/mol)	Mole
C	12.01	1.022
H	1.01	1.604
O	16.0	0.636
N	14.01	0.016
S	32.07	0.002

Calculate an approximate chemical formula. Determine mole ratios (sulfur = 1).

	Mole Ratio
C	655.32
H	1028.84
O	407.71
N	10.07
S	1.00

The chemical formula for the waste mixture given above is $C_{655.3}H_{1028.8}O_{407.7}N_{10.1}S$.

4.5.2 PROXIMATE ANALYSIS OF MSW

Proximate analysis is more specific compared with ultimate analysis, and is used to estimate the capability of MSW as a fuel. Proximate analysis includes the following tests (Drobney et al., 1971; Singer, 1981):

- Moisture content, determined by loss of moisture after heating at 105°C for 1 h.
- Volatile combustible matter, the additional loss of weight after ignition at 950°C for 7 min in a covered crucible (oxygen is excluded).

- Fixed carbon, the combustible residue left after volatile matter is removed; ignition at 600 to 900°C.
- Ash, the weight of residue after combustion in an open crucible.

Moisture content and ash represent the noncombustible component of the MSW. Moisture is undesirable in MSW as it adds weight to the fuel without adding to the heating value. In addition, moisture content will adversely affect heat release from the fuel. Ash similarly adds weight without providing heat energy. Furthermore, ash retains heat when removed from the furnace; as a result, potentially useful heat is lost to the environment.

The volatile matter and the fixed carbon content are the preferred indicators of the combustion capability of MSW. The volatile matter is the portion of MSW converted into gas as the temperature increases. Gasification occurs before the onset of combustion. In many incineration systems these carbonaceous gases are drawn away from the heating mass and carried to a secondary combustion chamber where combustion of the fuel gas occurs (see Chapter 9). Heat release is rapid and combustion is complete within a short time (Pfeffer, 1992).

Fixed carbon is the solid carbon residue that has settled on the furnace grates. Combustion occurs in the solid state, i.e., on the surface of this "char" material. The rate of combustion is affected by the temperature and surface area of the char. A waste fuel with a high percentage of fixed carbon will require a longer retention time in the combustion chamber to achieve complete combustion as compared with a fuel low in fixed carbon (Pfeffer, 1992).

The value for fixed carbon from laboratory results is calculated as follows (Liu and Liptak, 2000):

$$\% \text{ fixed carbon} = 100\% - \% \text{ moisture} - \% \text{ ash} - \% \text{ volatile matter} \quad (4.1)$$

A limitation of proximate analysis is that it does not provide an indication of possible pollutants emitted during combustion. These data are determined by conducting ultimate analysis. Proximate analysis data for the combustible components of MSW and bulk samples of MSW are presented in Table 4.18.

TABLE 4.18
Typical Proximate Analysis of MSW and MSW Components

Waste Type	Proximate analysis (% by wt)			
	Moisture	Volatiles	Fixed Carbon	Noncombustable (ash)
Food mixed	70.0	21	3.6	5.0
Paper mixed	10.2	76	8.4	5.4
Newspapers	6.0	81	11.5	1.4
Cardboard	5.2	77	12.3	5.0
Plastics mixed	0.2	96	2	2
Polyethylene	0.2	98	<0.1	1.2
Polystyrene	0.2	99	0.7	0.5
PVC	0.2	87	10.8	2.1
Textiles	10	66	17.5	6.5
Yard wastes	60	30	9.5	0.5
Wood mixed	20	68	11.3	0.6
Glass	2			96–99
Metals	2.5			94–99
Domestic MSW	10–40	30–60	3–15	10–30

Source: Kiely, G., *Environmental Engineering*, McGraw-Hill, New York, 1997. Reproduced with kind permission of the McGraw-Hill Companies.

4.5.3 ENERGY CONTENT OF MSW

The energy content of the organic components of MSW can be determined by (1) combusting samples in a full-scale boiler and measuring steam output (Figure 4.6), (2) using a laboratory bomb calorimeter (Figure 4.7), or (3) calculation from elemental composition (i.e., ultimate analysis). Most data on the energy content of MSW are based on the results of bomb calorimeter tests. This test measures the heat release at a constant temperature of 25°C (77°F) from the combustion of a dry sample. The value of 25°C is used as a standard reference temperature for heat balance calculations.

The energy stored within the chemical bonds of a material is known as the *heat of combustion*. This heat is released when the material is burned. The heat generated by the combustion of a material in a calorimeter may be determined by measuring the temperature rise that occurs upon its combustion:

$$U = C_V \Delta T / M \qquad (4.2)$$

where U is the heat value (cal/g) of the unknown material, ΔT the rise in temperature (°C) from thermogram, M the mass (g) of the unknown material, and C_V the heat capacity (cal/°C) of the calorimeter (measured using a standardized material).

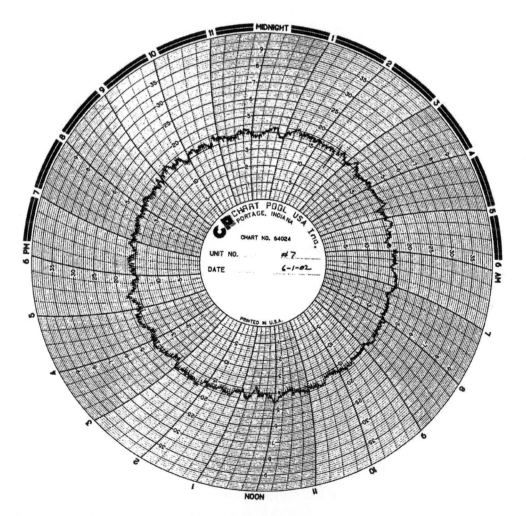

FIGURE 4.6 Chart showing steam production at a heating plant.

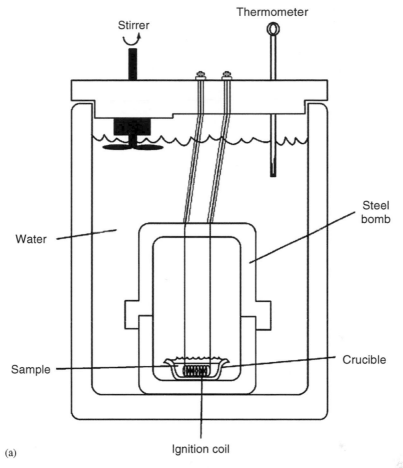

(a)

(b)

FIGURE 4.7 Bomb calorimeter: (a) schematic showing major components; and (b) laboratory unit.

EXAMPLE 4.2

A 10 g sample of mixed MSW is combusted in a calorimeter having a heat capacity of 8850 cal/°C. The temperature increase on combustion is 3.35°C. Calculate the heat value of the sample.

SOLUTION

$$U = C_V \, \Delta T \, / \, M = (8850 \times 3.35) \, / \, 10.00 = 2965 \text{ cal/g} = 5278 \text{ Btu/lb}$$

It should be obvious by now that the heat content of a MSW sample is essentially a function of composition; specifically, the percentage of materials having high Btu values such as paper, plastics, food, and yard wastes will provide the highest heat release. Moisture and inorganics (e.g., ash) will diminish the heat of combustion in a sample.

Heat values for the individual waste materials can be approximated by using Equation 4.3, known as the modified Dulong formula:

$$\text{MJ/kg} = 337C + 1419(H_2 - 0.125O_2) + 93S + 23N \qquad (4.3)$$

where C, H_2, O_2 S, and N are given in percent by weight.

Using a more direct approach, Khan et al. (1991) estimated the energy content from MSW with the equation

$$E = 0.051[F + 3.6(CP)] \cdot 0.352(PLR) \qquad (4.4)$$

where E is the energy content in MJ/kg, F the percent by weight food in the waste, CP the percent cardboard and paper and PLR the percent plastic and rubber.

EXAMPLE 4.3

Determine the energy content of the MSW sample presented in Example 4.1.

SOLUTION

The chemical formula for the waste mixture given in Example 4.1 was $C_{655.3} H_{1028.8} O_{407.7} N_{10.1} S$ Using the Dulong formula

$$\begin{aligned}
\text{MJ/kg} &= 337C + 1419(H_2 - 0.125O_2) + 93S + 23N \\
&= 337 \, (50.4) + 1419 \, (6.6 - 0.125 \times 41.8) + 93 \, (0.21) + 23 \, (0.90) \\
&= 18{,}975 \text{ MJ/kg}
\end{aligned}$$

EXAMPLE 4.4

Estimate the energy content using the Khan equation, for MSW having the following properties:

Component	Percent by Weight
Paper products	37
Plastics	7
Glass	9
Metals	6
Food waste	24
Textiles	2
Misc.	15
Total	100

(handwritten) ✳ Khan Equation
$$E = 23(F + 3.6(CP)) + 160\ PLR = \frac{BTU}{lb}$$

SOLUTION

$$E = \underset{0.053}{\cancel{0.051}}\ [F + 3.6\ (CP)] + \underset{0.372}{\cancel{0.352}}\ (PLR)$$
$$= 0.051\ [24 + 3.6\ (37)] + 0.352\ (7)$$
$$= 10.48\ \text{MJ/kg}$$

Two heat of combustion parameters are of significance: *high heating value* and *low heating value*. The higher heat of combustion includes the latent heat of vaporization of water molecules generated during the combustion process. The reaction for the combustion of cellulose and the consequent formation of water is

$$(C_6H_{10}O_5)_n + 6nO_2 \rightarrow 6nCO_2 + 5nH_2O \tag{4.5}$$

This water results solely from the combustion process, i.e., hydrogen is oxidized to form a water molecule. Therefore, even a seemingly dry sample of MSW will generate moisture, and this free water must be evaporated. The energy required may be substantial and may result in an inefficient combustion process. Subtracting the latent heat of vaporization of water provides a lower heat of combustion; this value represents the net heat available during the incineration of MSW.

The high and low heating values can be estimated from composition data of the material. The higher heat value (HHV) is calculated using the equation

$$\text{HHV (MJ/kg)} = 0.339\ (C) + 1.44\ (H) - 0.139\ (O) + 0.105\ (S) \tag{4.6}$$

The lower heat value (LHV) is calculated as

$$\text{LHV} = \text{HHV (in MJ/kg)} - 0.0244\ (W + 9H) \tag{4.7}$$

where W represents the mass % of water and H the wt % of H in the waste.

The as-received heat value of a waste is approximately proportional to the carbon content of the waste. The heat values of plastics and, to a lesser extent, of paper, are among the highest because of their high carbon content and relatively low ash and moisture contents. In contrast, yard waste and food waste, although mostly organic, possess lower heat values because of their high moisture contents (Liu and Liptak, 2000).

The following four factors must be considered when evaluating MSW as a potential fuel (Pfeffer, 1992):

- Only dry organic matter yields energy.
- Ash reduces the proportion of organic fuel per pound of MSW.
- Ash retains heat when removed from the furnace, therefore wasting heat.
- Water reduces the amount of organic fuel per pound of MSW and requires a significant amount of energy for removal (evaporation).

The heat contents for various fractions of MSW appear in Table 4.19.

4.5.4 FUSION POINT OF ASH

The fusion point of ash provides information on its physical behavior under high temperatures, i.e., softening and melting. The fusion point of MSW ash is the temperature at which the ash from waste combustion forms "clinker" by fusion and agglomeration. The fusion point should, ideally, correlate with the potential for boiler fouling by the ash.

TABLE 4.19
Typical Values for Inert Residue and Energy Content of Residential MSW

Component	Inert Residue[a] (%) Range	Energy Content (kJ/kg) Range	Energy Content (Btu/lb) Range
Organic			
Food wastes	2–8	3,350–6,700	1,500–3,000
Paper	4–8	11,200–18,000	5,000–8,000
Cardboard	3–6	13,400–16,800	6,000–7,500
Plastics	6–20	26,800–35,750	12,000–16,000
Textiles	2–4	14,500–17,900	6,500–8,000
Rubber	8–20	20,125–26,800	9,000–12,000
Leather	8–20	14,500–19,000	6,500–8,500
Yard wastes	2–6	2,225–17,900	1,000–8,000
Wood	0.6–2	16,770–19,000	7,500–8,500
Misc. organics	—	—	—
Inorganic			
Glass	96–99+	110–225	50–100
Tin cans	96–99+	225–1100	100–500
Aluminum	90–99+	—	—
Other metal	94–99+	225–1120	100–500
Dirt, ashes.	60–80+	2230–11,175	1,000–5,000
Municipal solid wastes		8950–13,400	4,000–6,000

[a]After complete combustion.

Adapted from Tchobanoglous, G. et al., *Integrated Solid Waste Management: Engineering Principles and Management Issues*, McGraw-Hill, New York, 1993. Data reproduced with kind permission of the McGraw-Hill Companies, Inc.; Kaiser, E.R., *Proceedings of the National Incinerator Conference, ASME,* New York, 1969. Data reproduced with kind permission of the American Society of Mechanical Engineers.

Fusion temperatures are often measured under both reducing and oxidizing conditions. Typical fusion temperatures for clinker formation from MSW range from 1100 to 1200°C (2000 to 2200°F) (Tchobanoglous et al., 1993).

4.5.5 CONTENT OF NUTRIENTS AND OTHER SUBSTRATES

In applications where the organic fraction of MSW is used as feedstock for compost or biological conversion into methane and ethanol, information on the essential nutrients in the waste materials is important. Both composting and biogas production are carried out by diverse consortia of heterotrophic microorganisms. Therefore, the microbial nutrient balance of the MSW should be assessed to allow for maximal conversion for final uses. The composition of essential nutrients and elements in the organic fraction of MSW is shown in Table 4.20. Nitrogen content, both as nitrate and ammonium, is highest in food and yard wastes by virtue of their higher protein contents (see below). Sulfur, potassium, calcium, and magnesium are also markedly higher in food and yard waste.

The organic fraction of most MSW (i.e., food waste, yard waste, paper products, textiles) can be classified according to their relative degree of biodegradability as follows:

- Sugars
- Starches and organic acids
- Proteins and amino acids

- Hemicellulose
- Cellulose and lignocellulose
- Lignin
- Fats, oils, and waxes

4.5.6 CARBOHYDRATES

The main sources of carbohydrates are putrescible garbage and yard wastes. Carbohydrates are designated by the general formula $(CH_2O)_x$ and include a range of sugars and their polymers such as starch and cellulose (Figure 4.8). Some polymers vary markedly in their resistance to hydrolysis. The starch polymers readily hydrolyze to glucose (Figure 4.9), which is a water-soluble and highly biodegradable simple sugar. Such polymers may therefore attract pests such as flies and rats. Sugars account for 4 to 6% and starches 8 to 12% of the dry weight of MSW (Pfeffer, 1992).

TABLE 4.20
Elemental Analysis of Organic Materials Used as Feedstock for Biological Conversion Processes

Component	Newspaper	Office Paper	Yard Waste	Food Waste
K (%)	0.35	0.29	2.27	4.18
Ca (%)	0.01	0.10	0.42	0.43
Mg (%)	0.02	0.04	0.21	0.16
NH_4–N (mg/kg)	4	61	149	205
NO_3–N (mg/kg)	4	218	490	4278
SO_4–S (mg/kg)	159	324	882	855
P (mg/kg)	44	295	3500	4900
B (mg/kg)	14	28	88	17
Zn (mg/kg)	22	177	20	21
Mn (mg/kg)	49	15	56	20
Fe (mg/kg)	57	396	451	48
Cu (mg/kg)	12	14	7.7	6.9
Ni (mg/kg)	—	—	9.0	4.5

Source: Tchobanoglous, G. et al., *Integrated Solid Waste Management: Engineering Principles and Management Issues,* McGraw-Hill, New York, 1993. Data reproduced with kind permission of the McGraw-Hill Companies, Inc.

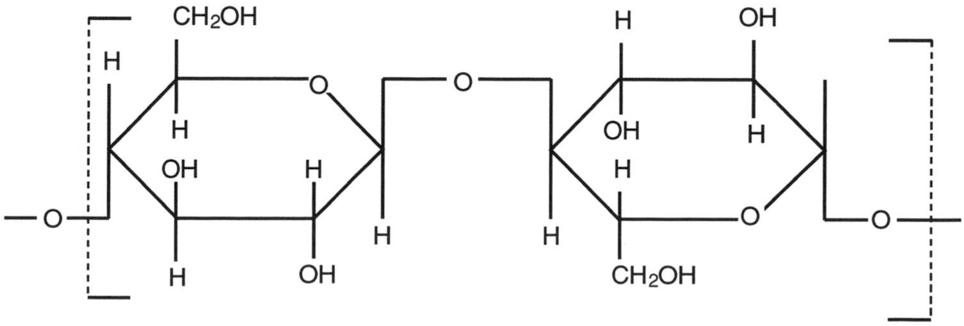

FIGURE 4.8 General structure of a carbohydrate molecule.

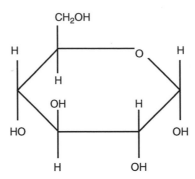

FIGURE 4.9 Structure of a glucose molecule.

4.5.7 CRUDE FIBERS

This category includes natural fibers such as cotton, wool, and leather, which are generally resistant to degradation. The major polymers are cellulose and lignin. Cellulose is a polymer of glucose and due to the nature of its chemical bonds, it is only slowly biodegradable (Figure 4.10). Lignin is composed of a number of monomers, with benzene being the most common (Figure 4.11). The benzene ring is resistant to biodegradation. Both cellulose and lignin contribute to the formation of soil humus, also a very resistant organic compound. Natural fibers found in paper products, food waste, and yard waste are the major source of these polymers. Cellulose may account for 25 to 30% of the dry weight of MSW, while lignin may comprise 8 to 10% (Pfeffer, 1992).

4.5.8 PROTEINS

All proteins possess a backbone of an amine group ($-NH_2$) and an organic acid (R–COOH) (Figure 4.12). Food wastes and yard wastes are sources of proteins, which comprise about 5 to 10% of the dry solids in MSW. Proteins are important in the biodegradation of MSW, as they are an important N source for heterotrophic microorganisms. Efficient microbial degradation of carbonaceous wastes requires a sufficient supply of N. Partial decomposition of proteins can result in the production of amines, which produce intense odors. Common names for some of these amines include "putrescine" and "cadaverine" (Pfeffer, 1992).

4.5.9 LIPIDS

Also known as fats, oil, and grease, these may comprise approximately 8 to 10% of MSW on a dry weight basis. A generalized structure of a lipid molecule in shown in Figure 4.13. The main sources of lipids are putrescible garbage, fat, and cooking oils. Lipids typically possess a high energy value, in the range of 35,775 to 38,000 MJ/kg (16,000 to 17,000 Btu/lb) (Pfeffer, 1992). Solid wastes, high in lipid content, are well suited for energy recovery processes. Lipids become fluid at slightly above ambient temperatures. This can add to the liquid content of MSW and will change physical properties due to wetting of paper products. Lipids have a low solubility in water that renders them slowly biodegradable.

4.5.10 BIODEGRADABILITY OF MSW FRACTIONS

The above compounds serve as a substrate for a wide range of micro- and macroorganisms important in composting and other biological processing. All of these organic components can be biologically converted into gases and relatively stable organic and inorganic solids.

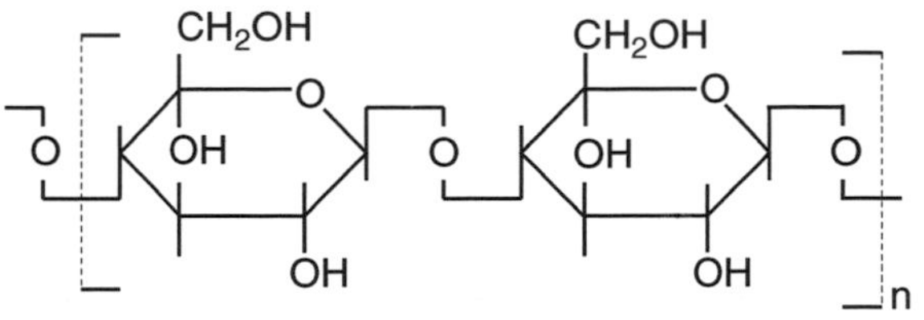

FIGURE 4.10 Structure of cellulose.

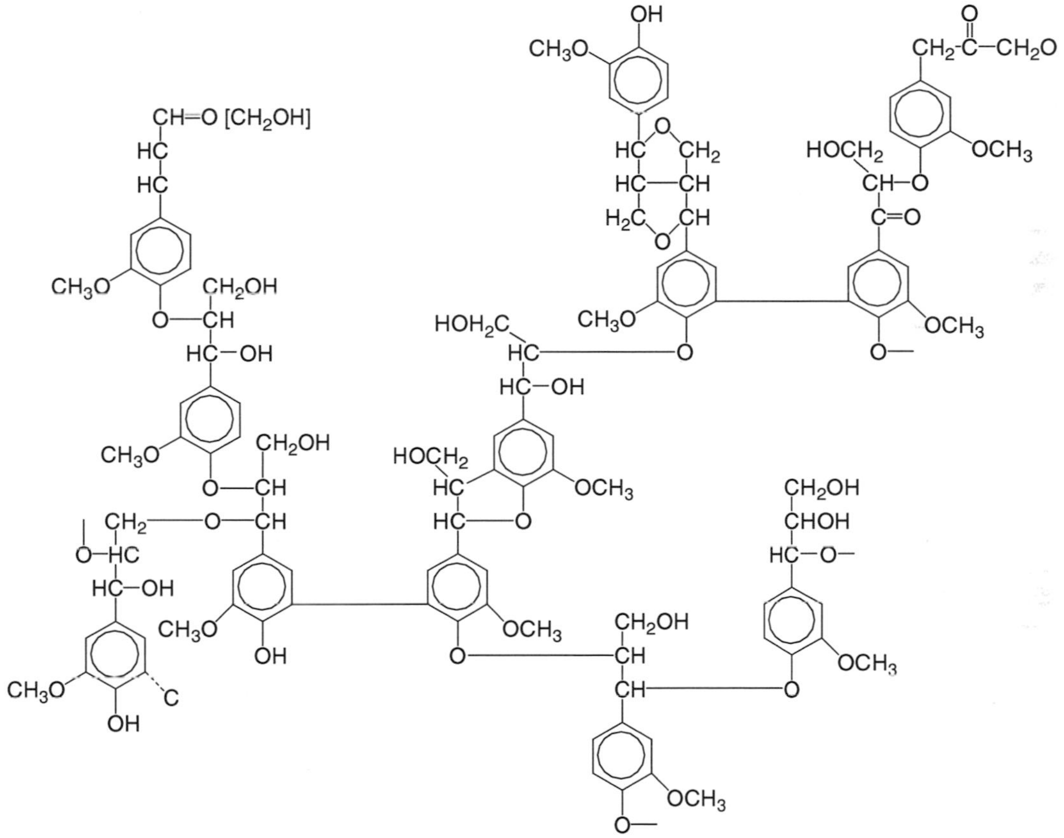

FIGURE 4.11 Structure of lignin. (Pearl, I.A., *Chem. Eng. News,* July 6, 1964, pp. 81–92. Reprinted with kind permission from the American Chemical Society.)

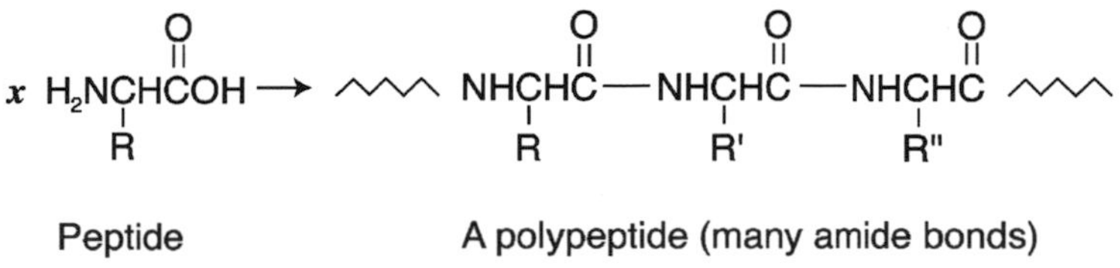

Peptide   A polypeptide (many amide bonds)

FIGURE 4.12 Generic structure of a protein molecule.

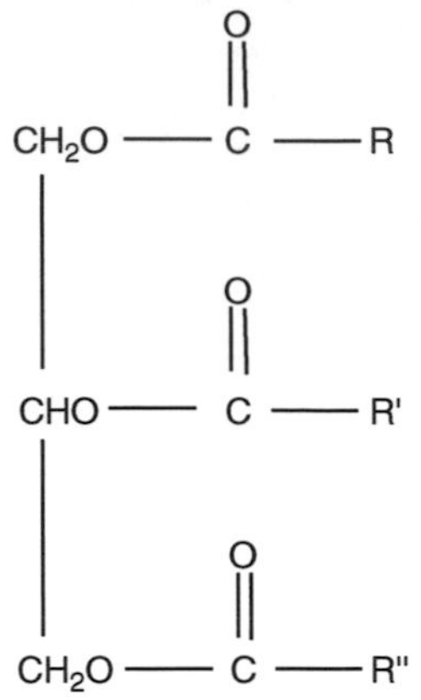

FIGURE 4.13 Structure of a lipid molecule.

The biodegradability of the organic fraction of MSW can be determined via simple laboratory tests for volatile solids and lignin content. The biodegradability factor can be calculated by the equation (Tchobanoglous et al., 1993)

$$BF = 0.83 - 0.028LC \qquad (4.8)$$

where BF represents the biodegradable fraction expressed on a volatile solids basis and LC the lignin content of the volatile solids expressed as a percent of dry weight.

The biodegradability of several organic compounds in MSW is shown in Table 4.21. Wastes with high lignin contents such as newspaper and cardboard tend to be of low biodegradability. Materials with a low lignin content, for example food wastes and grass clippings, tend to have a high biodegradability.

4.6 PHYSICAL PROPERTIES OF MSW

4.6.1 DENSITY

Density is a useful parameter in waste characterization as it provides information for predicting storage volume, including as-discarded at a residence or commercial facility, after compaction in a collection truck, and after compaction within a landfill cell.

The density of raw, uncompacted solid waste will vary as a function of composition, moisture content, physical shape, and degree of compaction. With the increase in the proportion of glass, ceramics, ashes and metals, the density also increases. Moisture will replace the air occurring in voids, thus increasing density until it becomes saturated. Excessive water contents may actually displace solids, which will eventually lower the overall density.

Raw wastes range in density from about 115 to 180 kg/m³ (200 to 300 lb/yd³). This low density is partly a function of the shape of the material in the waste stream. Corrugated boxes, bottles, and cans contain large void spaces which greatly decrease the density. If these materials were crushed, waste density would sharply increase. Some compaction occurs during storage in piles. Shredding, baling, and other size-reduction techniques also decrease irregularity and increase density (Liu and Liptak, 2000). MSW compacted in a landfill ranges in density from 300 to 900 kg/m³ (Sincero and Sincero, 1996; Kiely, 1997).

Volume reduction has a significant impact on the cost of collection and hauling MSW. Collection trucks are space-limited; therefore, greater compaction capabilities will result in a greater density of MSW and more cost-effective hauling. High-pressure compaction using stationary balers can greatly

TABLE 4.21

Biodegradability of Selected Organic Components in MSW

Component	Lignin Content (as % of VS)	BF (as % of VS)
Food waste	0.4	0.82
Newsprint	21.9	0.22
Office paper	0.4	0.82
Cardboard	12.9	0.47
Yard waste	4.1	0.72

BF (Biodegradable fraction) = 0.83 – (0.028) × LC, where LC = % of VS (volatile solids).

Source: Tchobanoglous, G. et al., *Integrated Solid Waste Management: Engineering Principles and Management Issues,* McGraw-Hill, New York, 1993. Data reproduced with kind permission of the McGraw-Hill Companies, Inc.

increase MSW density for long-distance transport, for example in rail cars. An upper limit of baled density is approximately 900 kg/m³ (1500 lb/yd³) (Pfeffer, 1992). Values for waste density are shown in Table 4.22.

EXAMPLE 4.5

During a sampling event at a tipping floor of a MRF, MSW is found to contain the following components:

Component	Density (kg/m³)	Amount in Sampled Waste (% by Wt)
Food waste	290	22
Mixed plastics	60	12
Glass	200	8
Ferrous and aluminum	200	12
Textiles	60	5
Dust, dirt	500	28

What is the average density of this solid waste mixture?

SOLUTION

$$\text{Average density} = (0.22)(290) + (0.12)(60) + (0.08)(200) + (0.12)(200)$$
$$+ (0.05)(60) + (0.28)(500)$$
$$= 254 \text{ kg/m}^3$$

Efficient use of landfill volume is an essential parameter in landfill management. During routine waste management operations (e.g., landfilling, tipping at a transfer station), trucks are weighed when entering and exiting the facility. With knowledge of the compacted density, the volume of land required at a landfill can then be calculated.

The density of MSW can be calculated on an as-compacted or as-discarded basis. The compaction ratio r is defined as the ratio of the as-compacted density ρ_c to the as-discarded density ρ_d and is given by

$$r = \rho_c / \rho_d \qquad (4.9)$$

TABLE 4.22
Density and Moisture Content of MSW

Waste Source	Component of Waste	Density (kg/m³)	Moisture Content (% by Wt)
Domestic	Food	290	70
	Paper products	70	5
	Plastic	60	2
	Glass	200	2
	Metals	200	2
	Clothing and textiles	60	10
	Ashes, dust	500	8
Municipal	Uncompacted	60–120	20
	Baled waste	470–900	—
	Compacted in collection truck	300–400	20
	Compacted in landfill	300–890	25

Adapted from Vesilind, P.A. et al., *Environmental Engineering,* 2nd ed., Butterworths, Boston, MA, 1998. Reproduced with kind permission of Elsevier Publishing.

A final disposal compaction ratio is calculated for landfills, and a compactor machine ratio is used for equipment such as a baler, which is used to increase MSW density prior to disposal. A common compaction ratio for a compacter machine may range from 2 to 4 (Sincero and Sincero, 1996).

If materials having different densities are expressed in terms of their weight fraction, the equation for calculating the overall bulk density is

$$\frac{(M_a+M_b)}{M_a/P_a+M_b/P_b} = \rho_{(a+b)} \tag{4.10}$$

where M_a is the mass of A, M_b the mass of B, ρ_a the bulk density of A and ρ_b the bulk density of B. When there are more than two materials to be considered, the above equation is extended.

The degree of volume reduction that occurs as a result of waste compaction, whether in a baler or landfill, is an important design variable. Waste volume reduction is calculated by the equation (Vesilind et al., 2002)

$$V_c / V_o = F \tag{4.11}$$

where F is the fraction remaining of initial volume as a result of compaction, V_o the initial volume and V_c the compacted volume.

EXAMPLE 4.6

For the following waste mixture,

Component	% by Wt	Uncompacted Bulk Density (kg/m³)
Corrugated cardboard	25	30
Paper products	15	61
Aluminum	9	38
Food waste	29	368
Yard waste	22	7.1

(a) What is the bulk density for the waste mixture prior to compaction? Assume that the compaction in the landfill cell is 500 kg/m³.
(b) Estimate the volume reduction (expressed as %), during compaction in the landfill.
(c) If the food and yard waste is diverted for composting, what is the uncompacted bulk density of the remaining waste?

SOLUTION

(a) Bulk density prior to compaction:

$$\frac{(25 + 15 + 9 + 29 + 22)}{35/30 + 15/61 + 9/38 + 29/368 + 22/7.1} = 22.2 \text{ kg/m}^3$$

(b) Percent volume reduction resulting from compaction:

$$22.2/500 = 0.04 \text{ or } 4\%$$

In other words, the landfill volume required is 4% of that required without compaction.

(c) When food waste and yard waste is removed, uncompacted bulk density is

$$\frac{(9 + 29 + 22)}{9/38 + 29/368 + 22/7.1} = 29.2 \text{ kg/m}^3$$

4.6.2 MOISTURE CONTENT

The moisture content of solid wastes is useful for estimating heat content, landfill sizing, and transport requirements. Moisture content is expressed either as a percentage of the wet weight or as a percentage of the dry weight of the material. The wet-weight method is more commonly used and is expressed as follows:

$$M = (w - d)/w \times 100 \tag{4.12}$$

where M is the moisture content (%), w the initial weight of sample as delivered (lb [kg]) and d the weight of sample after drying at 105°C (lb [kg]).

Typical data on the moisture content for solid waste components are given in Table 4.22. For most MSW in the United States the moisture content will vary from 15 to 40% depending on composition, season of the year, and weather conditions (Tchobanoglous et al., 1993; Kiely, 1997).

EXAMPLE 4.7

Using the data for a MSW sample provided below, determine the average moisture content of the sample. Base your calculations on a 100 kg sample size.

Component	Moisture Content (%)	Wt%	Discarded Weight (kg)
Paper waste	7	25	25
Yard waste	55	18	18
Food waste	65	20	20
Plastic	2	5	5
Wood	20	8	8
Glass	3	7	7
Metals	3	9	9
Textiles	12	8	8
Total		100	

SOLUTION

The dry weight of each MSW component is calculated using the equation

Dry weight = [(moist weight)(100 − % moisture)]/100

Component	Moisture Content (%)	Wt%	Moist Weight (kg)	Dry Weight (kg)
Paper waste	7	25	25	(1.0 − 0.07)(25) = 23.25
Yard waste	55	18	18	(1.0 − 0.55)(18) = 8.10
Food waste	65	20	20	(1.0 − 0.65)(20) = 7.00
Plastic	2	5	5	(1.0 − 0.02)(5) = 4.9
Wood	20	8	8	(1.0 − 0.2)(8) = 6.4
Glass	3	7	7	(1.0 − 0.03)(7) = 6.79
Metals	3	9	9	(1.0 − 0.03)(9) = 8.73
Textiles	12	8	8	(1.0 − 0.12)(8) = 7.04
Total		100		72.21

Totaling the values in the final column, the average percent moisture content of the MSW is equal to [(100 − 72)/100](100%) = 28%.

4.6.3 PARTICLE SIZE DISTRIBUTION

The size distribution of solid waste components is important for improving the rate of chemical reactions; in other words, smaller particle sizes provide greater surface area and thus more rapid reaction with microorganisms in a compost pile, or more rapid combustion in an incinerator. Size distribution is also an important consideration in the recovery of materials, for example, with the use of processing equipment such as a trommel screen or a magnetic separator (see Chapter 7).

MSW tends to stratify vertically when mixed, with smaller and denser components migrating to the bottom of a pile and lighter, bulkier objects migrating to the top. Such stratification has implications for efficient combustion on a traveling grate in a boiler or for materials separation in a MRF.

Size distribution is measured by passing samples of MSW over a series of screens, beginning with a coarse screen and continuing down to a fine screen. As discussed earlier, MSW is extremely heterogeneous; therefore, neither MSW nor any of its components are considered to possess a characteristic particle size (Liu and Liptak, 2000).

The size (i.e., "diameter") of a waste component may be calculated by any of the following equations:

$$D = 1 \tag{4.13}$$

$$D = (l + w + h)/3 \tag{4.14}$$

$$D = (l + w)/2 \tag{4.15}$$

$$D = (lw)^{1/2} \tag{4.16}$$

$$D = (lwh)^{1/3} \tag{4.17}$$

where D is the diameter, l the length, w the width and h the height.

Particle size distributions of various MSW components are given in Table 4.23.

EXAMPLE 4.8

A mixture of nonspherical waste particles are uniformly sized as follows: $l = 4$ units; $w = 1.2$ units, and $h = 1.5$ units.

TABLE 4.23
Typical Particle Size Distribution of MSW

Component	Size Range (mm)	Typical (mm)
Food	0–200	100
Paper and cardboard	100–500	350
Plastics	0–400	200
Glass	0–200	100
Metals	0–200	100
Clothing and textiles	0–300	150
Ashes, dust	0–100	25

Source: Kiely, G., *Environmental Engineering,* McGraw-Hill, New York, 1997. Reproduced with kind permission of the McGraw-Hill Companies.

Using the five equations provided above, calculate the particle diameter D. What is the range of variation in the calculated values?

$$D = l = 4$$
$$D = (l + w + h) / 3 = 2.23$$
$$D = (l + w) / 2 = 2.6$$
$$D = (lw)^{1/2} = 2.19$$
$$D = (lwh)^{1/3} = 1.93$$

Particle diameters range from 1.93 to 4 units, i.e., by a factor of 2.1.

Obviously, MSW will contain particles having a wide range of individual sizes. Under such circumstances, the particle size is often expressed as mean particle diameter. A number of calculations are possible. Some examples include:

arithmetic mean $\quad D = \dfrac{D_1 + D_2 + D_3 + D_4 + \cdots + D_n}{n}$

geometric mean $\quad D = (D_1 \times D_2 \times D_3 \times D_4 \times \cdots \times D_n)^{1/n}$

weighted mean $\quad D = \dfrac{W_1 D_1 + W_2 D_2 + W_3 D_3 + W_4 D_4 + \cdots + W_n D_n}{W_1 + W_2 + W_3 + W_4 + \cdots + W_n}$

number mean $\quad D = \dfrac{M_1 D_1 + M_2 D_2 + M_3 D_3 + M_4 D_4 + \cdots + M_n D_n}{M_1 + M_2 + M_3 + M_4 + \cdots + M_n}$

where W is the weight of material in each sieve size, M the total number of particles in each sieve size and n the number of sieve sizes (diameters).

EXAMPLE 4.9

Given the data for the following waste sizes,

	Sieve Size (mm)				
Particle diameter (mm)	100	75	50	25	5
Weight of fraction (kg)	2	6	12	4	4
Number of particles	225	310	500	2000	5750

calculate the arithmetic mean, the geometric mean and the weighted mean.

arithmetic mean $\quad D = \dfrac{100 + 75 + 50 + 25 + 5}{5} = 51$ mm

geometric mean $\quad D = (100 \times 75 \times 50 \times 25 \times 5)^{1/5} = 34.2$ mm

weighted mean $\quad D = \dfrac{(2 \times 100) + (6 \times 75) + (12 \times 50) + (4 \times 25) + (4 \times 5)}{2 + 6 + 12 + 4 + 4}$

$\qquad\qquad\quad = 48.9$ mm

number mean $\quad D = \dfrac{(225 \times 100) + (310 \times 75) + (500 \times 50) + (2000 \times 25) + (5750 \times 5)}{225 + 310 + 500 + 2000 + 5750}$

$\qquad\qquad\quad = 17.0$ mm

Note: The term *diameter* is defined to reflect a spherical particle shape; therefore, the above equations must serve only as an approximation.

Other calculations of particle size distribution incorporate particle surface area and volume as well.

4.6.4 FIELD CAPACITY

Field capacity may be defined as the total amount of moisture retained by mixed solids against the force of gravity. Water in excess of field capacity will be released by the force of gravity as leachate. The field capacity of a waste stream is of critical importance for two reasons; first, aerobic microbial activity is optimized at or slightly below the field capacity. Therefore, biological processing such as composting is optimized near this point. Secondly, field capacity is important in predicting leachate formation in landfills, compost piles, or storage piles.

Field capacity varies with the degree of pressure applied to the waste and the state of decomposition of the waste. The field capacity of uncompacted commingled wastes from residential and commercial sources may range from 50 to 60% (Tchobanoglous et al., 1993; Kiely, 1997).

One equation for field capacity of MSW is

$$FC = 0.6 - 0.55(W/[4500 + W]) \tag{4.18}$$

where FC is the field capacity (percent of dry weight of waste) and W the overburden weight calculated at midheight of the waste in lift (kg).

4.6.5 HYDRAULIC CONDUCTIVITY OF COMPACTED WASTE

The hydraulic conductivity, designated K, of compacted wastes is a physical property that strongly influences the movement of liquids (especially leachate) and gases in a landfill. Dense materials such as sludges tend to resist rainfall infiltration and promote runoff from a landfill cell. In contrast, paper and yard waste, by virtue of having large particles and therefore large void space, exhibit little resistance to rainfall infiltration.

Loose samples of MSW have a hydraulic conductivity value of 15×10^{-5} m/s, while dense baled waste may have a K of 7×10^{-6} m/s. The hydraulic conductivity for shredded waste ranges from 10^{-4} to 10^{-6} m/s (Kiely, 1997). Since MSW is very heterogeneous, these values serve only as an approximation.

REFERENCES

American Society for Testing and Materials, Test Method D5231-92 (1998) Standard Test Method for Determination of the Composition of Unprocessed Municipal Solid Waste, West Conshohocken, PA, 1998.

Chesner, W.H., Collins, R.J., and Fung, T., Assessment of the potential stability of Southwest Brooklyn incinerator residue in asphaltic concrete mixes. From Hasselriis, F., Ash disposal, in *Handbook of Solid Waste Management,* Keith, R., Ed., McGraw-Hill, New York, NY, 1994.

Drobney, N.L., Hull, H.E., and Testiiu, R.F., Recovery and Utilization of Municipal Solid Waste, U.S. Environmental Protection Agency, SW-10c, Washington, DC, 1971.

Franklin Associates, Characterization of municipal solid waste in the United States: 1998 update. EPA 530-R-01-014. Prepared for U.S. Environmental Protection Agency, Office of Solid Waste and Emergency Response, Washington, DC, 1999.

Kaiser, E.R., Chemical analyses of refuse compounds, *Proceedings of the National Incinerator Conference,* ASME, New York, NY, 1969.

Khan, M.Z.A. and Abu-Ghararah, Z.H., New approaches for estimating energy content in MSW, *J. Environ. Eng.,* 117, 376–380, 1991.

Kiely, G., *Environmental Engineering,* The McGraw-Hill Companies, New York, NY, 1997.

Liu, D.H.F. and Liptak, B.G., *Hazardous Waste and Solid Waste,* Lewis Publishing, Boca Raton, FL, 2000.

Masterson, W.L., Slowinski, E.J., and Stanitski, C.L., *Chemical Principles,* 5th ed., Sauders College Publishing, Philadelphia, PA, 1981.

Pearl, I.A., Lignin chemistry, *Chem. Eng. News,* July 6, 1964, pp. 81–92.

Pfeffer, J.T., *Solid Waste Management Engineering,* Prentice-Hall, Englewood Cliffs, NJ, 1992.

Rhyner, C.R., Domestic solid waste and household characteristics, *Waste Age,* pp. 29–50, 1976.

Rhyner, C.R., The monthly variations in solid waste generation, *Waste Manage. Res.,* 10, 67, 1992.

Rhyner, C.R. and Green, B.D., The predictive accuracy of published solid waste generation factors, *Waste Manage. Res.,* 6, 329–338, 1988.

Rhyner, C.R., Schwartz, L.J., Wenger, R.B., and Kohrell, M.G., *Waste Management and Resource Recovery,* Lewis Publishers, Boca Raton, FL, 1995.

Savage, G.M., Assessing waste quantities and properties: a vital requirement for successful solid waste management planning, *Warmer Bull, J. World Resour. Foun.,* High Street, Tonbridge, Kent, UK, 49, 18–22, 1996.

Sincero, A.P. and Sincero, G.A., *Environmental Engineering: A Design Approach,* Prentice-Hall, Upper Saddle River, NJ, 1996.

Singer, J.G., Ed., *Combustion: Fossil Power Systems,* Combustion Engineering, Inc., Windsor, CT, 1981.

Tchobanoglous, G., Theisen, H., and Vigil, S., *Integrated Solid Waste Management: Engineering Principles and Management Issues,* McGraw-Hill, New York, NY, 1993.

U.S. Department of Health, Education, and Welfare, *Incinerator Guidelines,* Washington, DC, 1969.

U.S. Environmental Protection Agency, Characterization of products containing mercury in municipal solid waste in the United States, 1970 to 2000, PA530-R-92-013, Office of Solid Waste, Washington, DC, 1992.

U.S. Environmental Protection Agency, *Household hazardous waste: steps to safe management*, EPA530-F-92-03, Office of Solid Waste and Emergency Response (OS-305), 1993.

U.S. Environmental Protection Agency, Characterization of municipal solid waste in the United States: 1998 update, EPA 530, Office of Solid Waste and Emergency Response, Washington, DC, 1999.

U.S. Environmental Protection Agency, Municipal solid waste generation, recycling and disposal in the United States: Facts and figures for 1998, EPA 530-F-00-024, Solid Waste and Emergency Response, Washington, DC, 2000a.

U.S. Environmental Protection Agency, Characterization of products containing lead and cadmium in municipal solid waste in the United States, 1970 to 2000, EPA/5S30-SW-015A. Office of Solid Waste, Washington, DC, 2000b.

U.S. Environmental Protection Agency, Municipal solid waste in the United States: 1999 Facts and figures, EPA 530-R-01-014, Office of Solid Waste and Emergency Response, Washington, DC, 2001.

U.S. Congress Office of Technology Assessment, Facing America's trash: what next for municipal solid waste? OTA-O-424, U.S. Government Printing Office, Washington, DC, 1989.

Vesilind, P.A., Peirce, J.J., and Weiner, R.F., *Environmental Engineering,* 2nd ed., Butterworths Publishing, Boston, MA, 1988.

Vesilind, P.A., Worell, W.A., and Reinhart, D.A., *Solid Waste Engineering,* Brooks/Cole, Pacific Grove, CA, 2002.

Williams, P.T., *Waste Treatment and Disposal,* Wiley, New York, NY, 1998.

Wisconsin Department of Natural Resources, The state of Wisconsin solid waste management plan, Madison, WI, 1981.

Yu, C. and MacLaren, V., A comparison of two waste stream quantification and characterization methodologies, *Waste Manage. Res.,* 13, 343-361, 1995.

SUGGESTED READINGS AND WEB SITES

Ali Khan, M. and Buney, F., Forecasting solid waste compositions, *Resource Conserv. Recycling,* 3, 1–17, 1989.

Alter, H., The origins of municipal solid waste: the relations between residues from packaging materials and food, *Waste Manage. Res.,* 7, 103–114, 1989.

Black, R.J., Muhich, A.J., Klee, A.J., Hickman, H.L., and Vaughn, R.D., The National Solid Wastes Survey: An Interim Report, National Technical Information Service, NTIS Report PB 260 102, 1968.

Boyd, G. and Hawkins, M., Methods of Predicting Solid Waste Characteristics, U.S. Environmental Protection Agency, Office of Solid Waste, SW-23c, Washington, DC, 1971.

Cailas, M.D., Kerzee, R., Swager, R., and Anderson, R., Development and Application of a Comprehensive Approach for Estimating Solid Waste Generation in Illinois, The Center for Solid Waste Management and Research, University of Illinois, Urbana-Champaign, IM, 1993.

Davidson, G. R., Residential Solid Waste Generation in Low Income Areas, U.S. Environmental Protection Agency, Office of Solid Waste, Washington, DC, 1972.

Dayal, G., Yadav, A., Singh, R. P., and Upadhyay, R., Impact of climatic conditions and socioeconomic status on solid waste characteristics: a case study, *the Sci. Total Environ.,* 136, 143–153, 1993.

DSM Environmental Services, Inc., Vermont Waste Composition Study. Prepared for Vermont Department of Environmental Conservation Solid Waste Program, Ascutney, VT, 2002.

Franklin, W. and Franklin, M., Solid waste stream characteristics, in *Integrated Solid Waste Management: Options for Legislative Action,* Kreith, F., Ed., Genium Publishing Corporation Schenectady, NY, 1990.

Frossman, D.J., Hudson, F., and Marks, D. H., Waste generation models for solid waste collection, *J. Environ. Eng. Div. ASCE,* 100, 1219–1230, 1974.

Henricks, S.L., 1994. Socio-economic Determinants of Solid Waste Generation and Composition in Florida, MS thesis, Duke University School of the Environment. Durham, NC, 1994.

Hockett, D., Lober, D. J., and Pilgrim, K., Determinants of per capita municipal solid waste generation in the southeastern United States, *J. Environ. Manag.,* 45, 205–217, 1995.

McCauley-Bell, P. and Reinhart, D., Development of a Waste Composition Study Planning and Analysis Tool (MSW-XPert), Report #97-3, Florida Center for Solid and Hazardous Waste Management, Gainesville, FL, 1997.

Mantell, C. L., Ed. Solid Wastes: Origin, Collection, Processing, and Disposal. Wiley-Interscience, New York, NY, 1975.

Miller, C., Garbage by the numbers, National Solid Wastes Management Association Research Bulletin 02-02, 2002. See: http://www.nswma.org/General%20Issues/Garbage%20By%20The%20Numbers%202002.pdf

Oregon Department of Environmental Quality, Waste Composition 2000 Report and 2002 Study Update, 2000. See: http://www.deq.state.or.us/wmc/solwaste/wcrep/wcrep2002.htm

Toronto, City of, No date, Solid Waste Quantity and Composition. How much? What is it? See: www.city.toronto.on.ca/wes/techservices/involved/ swm/net/pdf/presentation_March05_03.pdf

QUESTIONS

1. Since the late 1980s many municipalities have invested large sums in order to obtain accurate and reliable data on local waste composition and quantities. What is the significance to a community to obtain such data, i.e., how are the data used?

2. Explain the different methods for sampling MSW. What are the advantages and disadvantages of each? Consider accuracy, feasibility, and cost.
3. How do waste generation multipliers work? In your opinion, are they accurate predictors of waste generation?
4. Fluctuations in waste composition are affected by both geographic region and season of the year. Explain.
5. What specific attributes of MSW are the preferred indicators of the combustion capability of MSW?
6. Explain the difference between fixed carbon and volatile matter.
7. List three methods for determining the energy content of the organic components in MSW. Which method most accurately reflects 'real-world' energy production?
8. What factors must be considered when evaluating MSW as a potential fuel?
9. Why are lignin and cellulose only slowly biodegradable? Refer to their chemical structures.
10. For a city seeking to better control waste management costs, a thorough assessment of waste generated from various sources is necessary. From the city's perspective, would waste measurements by weight or by volume be the most accurate and efficient? Explain.
11. How does total moisture content affect overall management of MSW? Are there environmental or other implications to a high-moisture content waste stream?
12. Packaging makes up what percentage of the U.S. waste stream? Has this value increased, decreased or stabilized over the past decade?
13. The heating value of raw MSW is (a) less than, (b) approximately equal to, (c) higher than that for Midwest bituminous coal (approx 12,000–14,000 Btu/lb)?
14. Define: heat value, putrescible.
15. The majority of U.S. domestic solid wastes occur as:
    (a) plastics, especially PVC and polyethylene; (b) metals; (c) animal manures and yard wastes; (d) paper and paper products; (e) none of the above.
16. What are the primary sources of aluminum in the U.S. waste stream? Glass? Paper?
17. What is (are) the primary source(s) of lead in MSW? Of mercury? Of cadmium?
18. In your community, are MSW quantities being routinely measured? If yes, compile data on the total population of the area being served; next, calculate the total weight of MSW being collected, and convert to number of kilograms (or pounds) of wastes generated per person per day.
19. For your community, are data being collected regarding waste composition? Compute the percentage distribution of each waste component in the local waste stream. How does this distribution compare with the data of Figure 4.3? What factors may account for some of these differences? If possible, perform calculations for daily generation of specific waste types, e.g., kg. or pounds of food waste per person per day.
20. In your community, what are possible future trends (10 years, 20 years) in the production of food wastes, yard wastes, paper, plastic, and hazardous wastes? Consider population trends, personal lifestyles, the movement of businesses in or out of the area, and urban sprawl.
21. MSW composition is a critical factor in formulating waste management programs. What changes in waste composition do you predict for the next 10 years in U.S. waste composition? Justify your reasoning.
22. Three decades ago, it was predicted that the use of personal computers would result in a 'paperless society.' While this prediction obviously did not become reality, has computer use affected waste paper production? In what ways? Be specific.

23. Estimate the chemical composition of the organic fraction of the MSW sample described below.

Component	Wet weight (kg)	Dry weight (kg)
Food waste	63.2	24.5
Yard waste	95.5	29.7
Paper products	174.0	152.2
Plastics	29.7	28.2

24. For the waste sample in Question 23, calculate the heat content (MJ/kg) using the modified Dulong formula.

25. Estimate the energy content from MSW, using the Khan equation, having the following properties:

Component	% by wt
Paper products	25
Corrugated cardboard	15
Plastics	6
Glass	8
Metals	12
Food waste	15
Ash, dirt, misc.	19
Total	100

26. For the following waste mixture,

Component	% by Wt	Uncompacted Bulk Density (kg/m³)
Paper products	39.5	61
Ferrous	10	44
Food waste	26	375
Yard waste	25	10.4

what is the bulk density for the waste mixture prior to compaction? Assume that the compacted density in the landfill cell is 575 kg/m³. Estimate the volume reduction (expressed as %) during compaction in the landfill.

27. Given data for the following waste sizes,

	Sieve Size (mm)			
Particle diameter (mm)	100	75	50	25
Weight of fraction (kg)	9	12	28	8
Number of particles	450	1200	2500	5250

calculate the arithmetic mean, the geometric mean and the weighted mean of particle size distribution.

28. The higher heating value for cellulose, $C_6H_{10}O_5$, is 32,500 kJ/kg. Determine the lower heating value.

EXCEL EXERCISE

WASTE CHARACTERIZATION

FILE NAME: CHARACTERIZ.XLS

Background

For this exercise, you will work with waste sampling data for the town of Goat Cheese, WI from January 2001 through April 2002. The data are stored on a Microsoft Excel spreadsheet.

Waste is collected and the trucks are brought to the transfer station for unloading. Once a month, a truck is randomly selected and dumps its contents on to a secluded portion of the tipping room floor. Three university students were hired to sort the wastes into designated fractions and then weigh the fractions.

At each sampling date, a subsample of this waste was fed into a micronizing mill (a small shredder) and the shredded mixture was weighed and combusted in a tabletop furnace. The ash was collected and weighed. A separate subsample was shredded and placed in a bomb calorimeter. The heat content, measured in BTU/lb, was determined.

The data for this exercise can be located at www.crcpress.com/e_products/downloads/download.asp?cat_no=3525

Tasks

1. Calculate the percentages of each waste component for each month. What is the predominant fraction in the waste?
2. Determine the average values for each component over 2001.
3. Plot the data for total paper, plastics, food waste and yard waste, over the year 2001. What trends do you observe?
4. Plot the data for BTU and ash content at each sampling date. Are any seasonal trends observed?
5. Finally, perform a simple regression analysis of the heat content vs. ash content data. Is there a significant correlation between the data sets? What is the correlation coefficient?

5 Municipal Solid Waste Collection

To take ... out of the compact cities as you pass through,
To carry buildings and streets with you afterward wherever you go ...

Walt Whitman
Song of the Open Road

5.1 INTRODUCTION

Collection is one of the first steps of a solid waste management system; therefore, its proper planning and implementation can serve as a foundation for a sound waste management system. Disposal costs continue to grow rapidly across the United States and the costs of waste collection actually exceed those of ultimate disposal; collection costs range between 40 and 60% of a community's solid waste management system costs (U.S. EPA, 1999). Therefore, an efficient collection program can ultimately hold down waste management costs.

5.2 DEVELOPING A WASTE COLLECTION SYSTEM

Collection services are provided to residents in most urban and suburban areas in the United States as well as in some rural areas either by municipal governments or private haulers. Collection programs in different communities vary greatly depending on waste types collected, community characteristics, economics, and the desires of their residents. In recent years, collection services have expanded in many communities to include the collection of recyclable materials, yard wastes, and even household hazardous wastes. Different collection equipment and hauling companies are utilized in a single community to serve different customers (e.g., single-family, multi-family, commercial) or to collect different materials (MSW, recyclables, bulky waste) from the same customers.

Because collection and waste transfer systems may be complex, many factors and options must be considered in their planning and design. When a community is considering the implementation of a new collection program, some of the most immediate variables to identify and address are waste types, the service area, the level of desired service, public vs. private hauling, how to fund the program, creating and meeting waste reduction goals, and handling labor contracts. Some of the more salient issues for a community planning its MSW collection program are discussed below.

5.2.1 CHARACTERIZATION OF WASTES

Intuitively, data concerning waste sources, waste composition, and total volumes are critical for the proper planning of a collection program. Accurate and current data on the characteristics of municipal wastes will not only encourage well-organized and smoothly functioning collection, but will also enhance recycling programs and possibly reduce the amount of waste generated, thus holding down overall waste management costs. Waste characterization was discussed in Chapter 4.

5.2.2 SERVICE AREA AND LEVEL OF SERVICE

City street and block maps should be evaluated by program planners to determine street configurations, including the number of houses, location of one-way and dead-end streets, and traffic patterns. The ultimate goal is to formulate an efficient system where dead time (e.g., U-turns, detours, delays at railroad crossings) for vehicles and collection crews is minimized.

The level of services to the community includes specifying the materials to be collected and any requirements for separate collection (e.g., picking up recyclables in a separate vehicle). The frequency of pickup and the set-out requirements for residents must also be determined.

5.2.3 PUBLIC VS. PRIVATE COLLECTION

The collection system for a municipality may be operated by a municipal department, one or more competing private firms, or a combination of public and private haulers. In municipal collection, a city or county agency such as the local waste management office hires its own employees and equipment to collect solid waste. With private collection, the municipal agency contracts with a private collection firm. Larger communities may issue multiple collection contracts, each for a different geographic area, type of customer (single-family vs. multi-family units), or material collected (e.g., MSW vs. recyclables). Private collection relies on competition to control prices and quality of service. Some U.S. communities allow haulers to bid competitively to provide a specified level of service to residents within an area. Residents then contract directly with the designated hauler for their area.

5.2.4 FUNDING THE COLLECTION SYSTEM

The municipality must formulate a funding plan to generate the money necessary to pay for collection services. The three alternatives for funding solid waste services are property tax revenues, flat fees, and variable-rate fees. Property taxes are the conventional means of funding solid waste collection, especially in communities where municipal employees are the waste collectors and haulers. The property tax method is preferred for its administrative simplicity; no separate system is necessary to bill and collect payments, since funds are derived via the collection of personal and corporate property taxes. Funding waste collection from property taxes, however, provides no incentive for waste reduction by residents (U.S. EPA, 2003).

In recent years, many municipalities have shifted away from covering costs through property taxes and are instead instituting user fees, primarily a result of imposed caps on property tax increases. With the property tax method of payment, customers usually never see a bill and generally have no idea how much it costs to remove their wastes. Flat fees are a common method for funding collection in communities served by private haulers and in municipalities where a separate authority is used for solid waste services. As with the property tax method, the flat-fee method provides no incentive for reducing wastes by residents.

The variable-rate fee system (also known as "pay as you throw") requires waste generators to pay in proportion to the amount of wastes they set out for collection. Variable-rate systems typically require that residents purchase special bags or stickers (Figure 5.1); a range of service levels are made available to generators. The purchase price of bags or stickers is set sufficiently high to cover program costs. The use of such bags and stickers helps citizens to become more aware of how much waste they are producing; thus, there is an incentive to reduce waste volumes. In addition, by using smaller or fewer bags or fewer stickers, residents can generate savings from source reduction efforts. Another option is to charge different rates for various sizes of cans or other containers. Some communities will collect recyclables at reduced cost to residents as a financial incentive for recycling instead of disposal. In a study of eight communities (Miranda and Aldy, 1996), significant increases in recycling tonnages were reported when a pay-as-you-throw pricing system was established. San Jose, California, and Lansing, Michigan, experienced more than a doubling of recycling

FIGURE 5.1 Bags and stickers may be used in a "pay as you throw" waste collection program.

levels over a 2- and 3-year period, respectively. Communities in Illinois experienced recycling rate increases between 41 and 64% over 5 years. Pasadena and Santa Monica, California, experienced recycling rate increases of approximately 70 and 30%, respectively.

Many communities have chosen to combine elements of the above funding methods to form a "hybrid system" which is best suited to their community.

5.2.5 Labor Contracts

Any conditions in existing contracts with labor unions that would affect the types of collection equipment or operations must be evaluated. The degree of any such constraints must be assessed early (U.S. EPA, 2003).

5.3 LOGISTICS OF THE COLLECTION PROGRAM

5.3.1 Storage Container Requirements

Specific solid waste storage containers are often required for a particular collection program. Containers should be appropriate for the collection vehicles used. For example, a community may decide to use self-loading compactor trucks in certain neighborhoods. Residents will therefore have to place wastes in containers that fit the container-lifting devices of the trucks. Containers should also be easy to handle, durable, resistant to corrosion, weather, and animals. In areas where waste is collected manually, standard-sized metal, plastic containers or plastic bags are usually specified for waste storage. Many municipalities limit the size of cans to 30 to 35 gal or to a maximum total weight. If plastic bags are used, a minimum thickness may be required. Some programs require the use of bags because they do not have to be emptied and returned to the curb and are therefore quicker to collect than cans. Many cities prohibit the use of other containers because they may be difficult to handle and increase the risk of worker injury. Some municipalities also limit the total number of containers collected at a single residence. Additional fees may be charged for additional containers.

5.3.2 Set-Out Requirements

To establish uniform collection, communities usually formulate guidelines and enact ordinances that specify how residents are to prepare solid waste and recyclables for collection. Set-out requirements address the types of containers to be used, the segregation of recyclables or other wastes for separate collection, how frequently materials are collected, and where residents are to place materials for collection.

5.3.3 WASTE SEPARATION

Many communities have arranged to collect some portions of solid waste separately; for example, recyclable materials or yard wastes may be collected on a different day from ordinary MSW. Residents will therefore be required to separate wastes before collection. Residents may be expected to separate recyclable materials such as paper, cardboard, glass, aluminum, and plastic; similarly, yard waste, bulky items, and household hazardous wastes may have to be segregated for separate collection. Bulky items such as white goods (e.g., large appliances) and furniture are usually placed at the same point of collection as other solid wastes. Some U.S. communities have tested the so-called wet and dry collection systems, in which "wet" organic wastes suitable for composting are collected separately from "dry" wastes, which may be sorted for the recovery of recyclables (U.S. EPA, 2003).

5.3.4 FREQUENCY OF COLLECTION

The greater the frequency of collection in a community, the more costly will be the collection system. Factors to consider when establishing collection frequency include total cost, desires of the residents, storage limitations, and climate. Collection once or twice per week is common for most U.S. municipalities, with collection once per week being prevalent. Crews collecting once per week can collect more tons of waste per hour, but make fewer stops per hour than the twice-a-week collection vehicles. Some communities in hot, humid climates use twice-per-week service due to health and odor concerns. In one study, once-per-week systems were found to collect 25% more waste per collection hour while serving 33% fewer homes during that period. Personnel and equipment requirements were 50% higher for once-per-week collection (U.S. EPA, 1974a). In Montgomery County, Maryland, one part of the county received weekly MSW pickup, while other areas received twice-per-week pickup. Twice-per-week collection was almost 70% more costly than once-per-week collection (U.S. EPA, 1999).

EXAMPLE 5.1

In the town of Livengood, Ohio, it is determined that the per capita waste generation rate is 1.4 kg (3.1 lb) per person per day. Collection is conducted once per week by the municipality. If the density of MSW in a typical trash container is 150 kg/m³, how many 120 L (30 gal) containers would be needed for a family of four?

1.4 kg/person/day × 7 days/week = 9.8 kg MSW

9.8 kg lb/person × 4 persons = 39.2 kg/family

39.2 kg / 150 kg/m³ = 0.26 m³

0.26 m³ × 1000 L/m³ = 260 L

Thus, three 120 L containers are required.

EXAMPLE 5.2

From the above example, collection trucks have a capacity of 11.5 m³ (15 yd³), which can compact the waste to a density of 420 kg/m³. How many customers can a truck handle in a single run, before departing for the transfer station?

11.5 m³ × 420 kg/m³ = 4830 kg capacity

4830 kg / 39.2 kg/household = 123 households

5.3.5 Waste Pickup Locations

In urban and suburban communities, waste is typically collected using curbside or alley pickup. Backyard service, more common in the past, is still used by some communities. Curbside or alley service is more economical but requires greater resident participation than backyard service. The productivity of backyard systems is about one half than that of curbside or alley systems (Hickman, 1986). Therefore, with smaller municipal budgets and increased service costs, more municipalities have switched to curbside or alley collection. Some municipalities also offer collection services to larger apartment buildings and commercial establishments. In other communities, service to these customers is provided by private collection companies. In general, wastes from such buildings are stored in dumpsters or roll-off containers.

Regarding collection services in rural areas, residents are usually required to place containers near their mailboxes or other designated pickup points along major routes. Other municipalities require a drop-off arrangement, where wastes are brought to a facility known as a transfer station (see below). A drop-off service is obviously much less expensive than a collection service but is less convenient for residents. Table 5.1 lists various waste collection methods and their advantages and disadvantages.

TABLE 5.1
Advantages and Disadvantages of Various Pickup Points for Collecting MSW

Curb-side/Alley Collection
Residents place containers to be emptied at curb or in alley on the collection day. Collection crew empties containers into collection vehicle. Residents return the containers to their storage location until next scheduled collection time.

Advantages
- Crew can move quickly
- Crew does not enter private property, so fewer accidents and trespassing complaints arise
- This method is less costly than backyard collection because it generally requires less time and fewer crew members
- Adaptable to automated and semiautomated collection equipment

Disadvantages
- On collection days waste containers are visible from the street
- Collection days must be scheduled
- Residents are responsible for placing containers at the proper collection point

Backyard Setout – Setback Collection
Containers are carried from backyard to curb by a special crew and emptied by the collection crew. The special crew then transports the containers back to their original storage location

Advantages
- Collection days need not be scheduled
- Waste containers are not usually visible from the street
- Use of additional crew members reduces loading time as compared to backyard collection method

Disadvantages
- Because crews enter private property, more injuries and trespassing complaints are likely
- The method is more time-consuming
- Residents are not involved and more crew members than for curb-side and alley collection are required
- This is more costly than curb-side and alley collection because additional crews are required

Backyard Collection
In this method, collection crews enter property to collect refuse. Containers may be transported to the truck, emptied, and returned to their original storage location, or emptied into a tub or cart and transported to the vehicle so that only one trip is required

(continued)

TABLE 5.1 (*continued*)

Advantages
- Collection days need not be scheduled
- Waste containers are not usually visible from the street
- Residents are not involved with container setout or movement
- This method requires fewer crew members than setout/ setback method

Disadvantages
- Because crew enters private property, more injuries, and tresspassing complaints are likely
- This approach is more time-consuming than curb-side and alley or setback method
- Spills may occur where waste is transferred

Drop-Off at Specified Collection Point
Residents transport waste to a specified point. This point may be a transfer station or the disposal site

Advantages
- Drop-off is the least expensive of methods
- Offers reasonable strategy for low population densities
- This method involves low staffing requirements

Disadvantages
- Residents are inconvenienced
- There is increased risk of injury to residents
- If drop-off site is unstaffed, illegal dumping may occur

Source: American Public Works Association, *Solid Waste Collection Practice*, 4th ed., Chicago, IL, 1975. (Reproduced with kind permission of the American Public Works Association.)

5.3.6 COLLECTION EQUIPMENT

A wide range of collection vehicles are available to a municipality. Collection equipment is continually being redesigned to meet changing needs and to incorporate advances in technology. Trends in the collection vehicle industry include the increased use of computer-aided equipment, mechanical lifting devices, and electronic controls. Some trucks are equipped with onboard computers for monitoring truck performance and collection operations.

Collection vehicles used for MSW transport in various countries include (Kiely, 1997):

- Traditional compacter-type trucks taking loose and bagged waste
- Modern single-compartment trucks taking wheeled bins from single-unit dwellings
- Single-compartment trucks taking wheeled bins from multi-unit apartment buildings and commercial establishments
- Multi-compartment trucks that remove source-separated waste
- Trucks taking container loads, either closed- or open-topped
- Vacuum trucks, used in areas with limited accessibility, with tube lengths up to 100 m
- Traditional open-top trucks, commonly used in low-income countries

Truck chassis and bodies are usually purchased separately and can be combined in a variety of arrangements. When selecting truck chassis and bodies, municipalities must consider regulations regarding truck size and weight. Truck selection must address maximizing the amount of wastes that can be collected while remaining within legal weights for the vehicle.

Compactor trucks are by far the most prevalent waste collection vehicles in use (Figure 5.2). Compactor vehicles are equipped with hydraulically powered rams that compact wastes and later push the wastes out of the truck at the disposal or transfer facility. Compactor trucks are classified as front-, side-, or rear-loading with capacities varying from 7.5 to 35 m^3 (10 to 45 yd^3).

FIGURE 5.2 Compactor-type truck common in urban and suburban neighborhoods.

Prior to the development of compactor trucks, open and closed noncompacting trucks were used to collect solid waste. These vehicles are inefficient for MSW collection because they carry a relatively small amount of waste and workers must lift waste containers high to place wastes into the truck. Noncompacting trucks are still used for collecting bulky items like furniture and white goods or other materials that are collected separately, such as recyclables and yard waste. Noncompacting trucks may also be suitable for servicing small communities and rural areas.

5.3.7 AUTOMATED WASTE COLLECTION

Waste collection is a labor-intensive business, often requiring as many as three workers per vehicle to lift and dump containers. With the advent of automated lifting systems, however, collection requires fewer workers, thereby reducing labor costs and workers' compensation claims.

Semiautomated and fully automated systems are two innovative approaches to MSW collection. Both systems rely on special trucks with mechanical or hydraulic lifting systems and require customers to use special wheeled carts. With semiautomated vehicles, crews wheel the carts to the collection vehicle and line them up with hydraulic lifting devices mounted on the truck body, activate the lifting mechanism, then return empty containers to the collection point. In fully automated vehicles, drivers control hydraulic arms or grippers from the vehicle cab. Unless there are problems such as the overflow of materials, improperly prepared materials, or obstructed setouts, the driver can service a route without leaving the collection vehicle (Figure 5.3). The benefits of automated waste collection include the following (U.S. EPA, 1999):

- *Reduced injury risk*: Increased automation typically reduces work-related lifting injuries as well as puncture wounds and lacerations.
- *Reduced vehicle needs*: Fully automated collection increases (by up to 300%) the number of households served per hour. This increased productivity typically results in a smaller fleet of vehicles.
- *Decreased labor needs*: Automated collection reduces crew size per truck. For semiautomated collection, one- or two-person crews are typical. With fully automated systems, the driver typically works alone.
- *Reduced environmental impacts*: Automated collection means fewer trucks, lower fuel usage, fewer air emissions, and fewer traffic and safety impacts on community streets.
- *Reduced tipping fees*: Carts with lids help keep water, ice, and snow from setouts, which also helps control the weight of setouts and decreases tipping fees.

(a)

(b)

FIGURE 5.3 Automated collection vehicles for (a) residential wastes, (b) commercial wastes.

- *Improved neighborhood aesthetics*: Uniform containers eliminate unsightly setouts. Containers with lids are less likely to be tipped over or torn apart by animals, reducing litter potential.
- *Reduced public health risks*: Containers with lids help mitigate odor and health concerns.

Waste setout requirements, waste quantities, and the characteristics of the collection routes are important considerations in the selection of collection vehicles. For example, suburban areas with wide streets and little on-street parking may be ideally suited to side-loading automatic collection systems. Conversely, urban areas with narrow alleys and tight corners may require rear loaders and shorter wheelbases. For large apartment buildings and complexes and for commercial and industrial applications, hauled-container systems are often used.

Table 5.2 provides criteria that should be used to determine the most appropriate collection equipment. Municipalities can use these criteria to determine the requirements their equipment must meet.

TABLE 5.2
Factors to Consider in Selecting Waste Collection Equipment

Loading Location

Compactor trucks are loaded on the side, back, or front. Front-loading compactors are often used with self-loading mechanisms and dumpsters. Rear loaders are often used for both self and manual loading. Side loaders are more likely to be used for manual loading and are often considered more efficient than back-loaders when the driver does some or all of the loading

Truck Body or Container Capacity

Compactor capacities range from 10 to 45 yd^3. Containers associated with hauled systems generally have a capacity range of 6 to 50 yd^3. To select the optimum capacity for a particular community, the best trade-off between labor and equipment costs should be determined. Larger capacity bodies may have higher capital, operating, and maintenance costs

Heavier trucks may increase wear and tear, and corresponding maintenance costs for residential streets and alleys

Design Considerations

- The loading speed of the crew and the collection method used
- Road width and weight limits (consider weight of both waste and vehicle)
- Capacity should be related to the quantity of wastes collected on each route. Ideally, capacity should be an integral number of full loads
- Travel time to transfer station or disposal site, and the probable life of that facility
- Relative costs of labor and capital

Chassis Selection

Chassis are similar for all collection bodies and materials collected

Design Considerations

- Size of truck body. Important for chassis to be large enough to hold truck body filled with solid waste
- Road width and weight limitations (also need to consider waste and truck body weight)
- Air emissions control regulations
- Desired design features to address harsh treatment (e.g., driving slowly, frequent starting and stopping, heavy traffic and heavy loads) include the following: high-torque engine, balanced weight distribution, good brakes, good visibility, heavy-duty transmission, and power brakes and steering

Loading and Unloading Mechanisms

Loading mechanisms should be considered for commercial and industrial applications, and for residences when municipalities wish to minimize labor costs over capital costs. A variety of unloading mechanisms are available

Design Considerations—Loading

- Labor costs of collection crew
- Time required for loading
- Interference from overhead obstructions such as telephone and power lines
- Weight of waste containers

Design Considerations—Unloading

- Height of truck in unloading position. Especially important when trucks will be unloaded in a building
- Reliability and maintenance requirements of hydraulic unloading system device

Truck Turning Radius

Radius should be as short as possible, especially when part of route includes cul-de-sacs or alleys. Short wheelbase chassis are available when tight turning areas will be encountered

Watertightness

Truck body must be watertight so that liquids from waste do not escape

Safety and Comfort

Vehicles should be designed to minimize the danger to solid waste collection crews

Design Considerations

- Carefully designed safety devices associated with compactor should include quick-stop buttons. In addition, they should be easy to operate and convenient
- Truck should have platforms and good handholds so that crew members can ride safely on the vehicle
- Cabs should have room for crew members and their belongings
- Racks for tools and other equipment should be supplied
- Safety equipment requirements should be met
- Trucks should include audible back-up warning device
- Larger trucks with impeded backview should have a video camera and a cab-mounted monitor screen

(continued)

TABLE 5.2 (*continued*)

Loading Height

The lower the loading height, the more easily solid waste can be loaded into the truck. If the truck loading height is too high, the time required for loading and the potential of injuries to the crew members will increase because of strain and fatigue

Design Considerations
- Weight of full solid waste containers
- If higher loading height is being considered, consider an automatic loading mechanism

Speed

Vehicles should perform well at a wide range of speeds

Design Considerations
- Distance to disposal site
- Population and traffic density of an area
- Road conditions and speed limits of routes that will be used

Adaptability to Other Uses

Municipalities may wish to use solid waste collection equipment for other purposes such as snow removal

Source: Pferdehirt, W., 1994. Reproduced with kind permission of W.P Pferdehirt, University of Wisconsin-Madison.

Additionally, certain cost data should be compared for each truck being considered, including initial capital cost, annual maintenance and operation costs, and expected service life (U.S. EPA, 2003).

If the number of households that a truck can service in a single day has been determined, the number of collection vehicles needed for a community can be estimated by the equation (Vesilind et al., 2002)

$$N = SF/XW \tag{5.1}$$

where N is the number of collection vehicles needed, S the total number of households serviced, F the number of collections per week, X the number of customers a truck can service per day and W the number of workdays per week.

EXAMPLE 5.3

From the data for the town of Livengood (Example 5.1), determine the number of collection vehicles needed if 8250 households must be serviced once per week. The trucks collect wastes 4 days per week, with 1 day for routine truck maintenance and other projects.

We will assume that an average truck can service 1.4 households per minute. The actual time spent collecting is 5 h. The total number of households served per day is

$1.4/1 = x/(5 \text{ h} \times 60 \text{ min}) = 420$ customers per day

$N = SF/XW$

$N = 8250 \times 1/420 \times 4 = 4.9$ trucks

5.3.8 DEVELOPING COLLECTION ROUTES

Thorough collection routes and schedules must be developed for the proposed collection program. Efficient routing and rerouting of collection vehicles hold down costs by reducing the labor expended for collection. Routing procedures usually comprise two separate components: microrouting and macrorouting (U.S. EPA, 2003). Macrorouting consists of dividing the total collection area into routes of a size sufficient for a 1 day collection for a single crew. The size of a route is a function of the amount of waste collected per stop, distance between stops, loading time, and traffic conditions. Barriers such as railroad embankments, rivers, and roads with heavy competing traffic can be used to divide route areas.

For large areas, macrorouting is best accomplished by first dividing the entire community into districts. Each district is subsequently divided into routes for individual crews. Using the results of the macrorouting analysis, microrouting will designate the specific path that each crew and collection vehicle will take on a given day. Results of microrouting analyses can then be used to readjust macrorouting decisions. Microrouting analyses and planning can accomplish the following (U.S. EPA, 2003):

- Increase the likelihood that all streets will be serviced equally and consistently.
- Help supervisors locate crews quickly because they know the specific routes that will be taken.
- Provide potentially optimal routes that can be tested against the driver's experience to provide the best actual routes.

The method selected for microrouting must be simple enough to be used for route adjustments; for example, seasonal variations in waste generation will require adjustments. Seasonal fluctuations in waste generation can be accommodated by providing fewer, larger routes during low-generation periods (typically winter) and increasing the number of routes during high-generation periods (spring and fall).

5.3.9 ROUTE DEVELOPMENT

The Office of Solid Waste Management Programs of the U.S. Environmental Protection Agency has developed a simple, noncomputerized "heuristic" (i.e., manual) approach to collection vehicle routing based on certain logical principles. The EPA developed the method to promote efficient routing layout and to minimize the number of turns and dead space encountered (U.S. EPA, 1974b). This method relies on developing, recognizing, and using certain patterns that repeat in every municipality. Using this approach, route planners can place tracing paper over a large-scale block map. The map should show collection service garage locations, disposal or transfer sites, one-way streets, natural barriers, and areas of heavy traffic flow. Routes should then be traced onto the tracing paper using the rules presented below:

1. Routes should not be fragmented or overlapped. Each route should be compact, consisting of street segments clustered in the same geographical area.
2. Total collection plus haul times should be reasonably constant for each route in the community (equivalent workloads).
3. The collection route should be started as close to the garage or motor pool as possible, taking into account heavily traveled and one-way streets (see rules 4 and 5).
4. Waste from heavily traveled streets should not be collected during rush hours.
5. In the case of one-way streets, it is best to start the route near the upper end of the street, working it through the looping process (see Figure 5.4).
6. Services on dead-end streets can be considered as services on the street segment that they intersect, since the waste can only be collected by passing down the street segment. To keep left turns to a minimum, wastes should be collected from the dead-end streets when they are to the right of the truck. They must be collected by walking down, backing down, or making a U-turn.
7. When practical, service stops on steep hills should take place on both sides of the street while the vehicle is moving downhill, for safety, ease, speed of collection, reduction of the wear on vehicle, and conservation of fuel and oil.
8. Higher elevations should be at the start of the route.
9. For collection from one side of the street at a time, it is generally best to route with many clockwise turns around blocks.
10. For collection from both sides of the street at the same time, it is generally best to route with long, straight paths across the grid before looping clockwise.
11. For certain block configurations within the route, specific patterns should be applied.

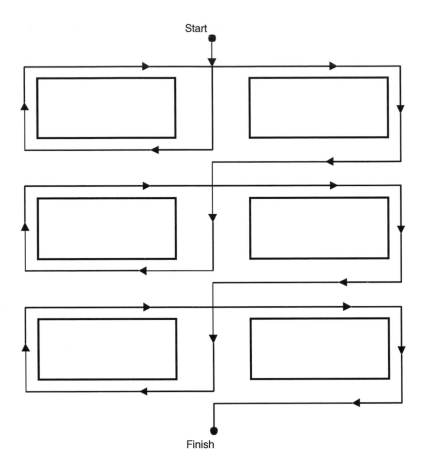

FIGURE 5.4 Routing patterns for one-way street collection (U.S. EPA, 1974b).

Figures 5.5 and 5.6 are examples of heuristic tools that can be applied depending on the block patterns within a collection area.

5.3.10 COMPUTER-ASSISTED ROUTING

Computer programs are becoming popular in establishing route design, especially when routes need to be adjusted periodically. Programs can be used to develop detailed microroutes or to adjust existing routes. To program detailed microroutes, planners require information similar to that needed for heuristic routing, for example block configurations, waste generation rates, distances between residences, distances between routes and disposal or transfer sites, and loading times (U.S. EPA, 2003). Municipalities that have a geographic information system (GIS) database can utilize data for their area to facilitate computerized route balancing.

5.3.11 WASTE TRANSFER

Waste transportation costs will be substantial if the distance between a collection zone and the final destination (e.g., landfill, incinerator) is significant. In the interest of economics, many municipalities choose to transfer waste from neighborhood collection trucks or stationary containers to larger vehicles before transporting it to the disposal site. A transfer station may be established between the waste collection sources and the final destination to serve in this capacity.

The primary objective in using a transfer station is to reduce the traffic of smaller vehicles to the disposal site, ultimately resulting in reduced transport costs including labor (crews spend less

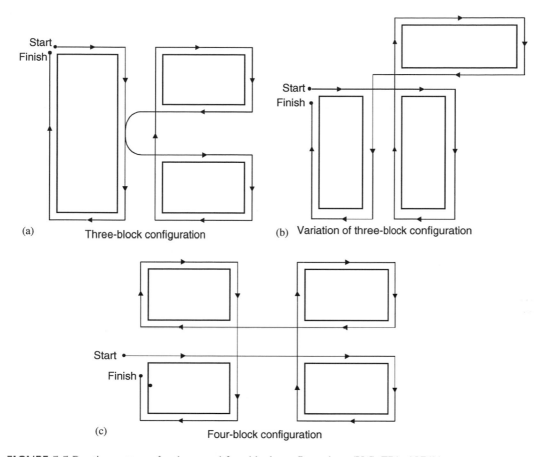

Start
Finish
(a) Three-block configuration

Start
Finish
(b) Variation of three-block configuration

Start
Finish
(c) Four-block configuration

FIGURE 5.5 Routing patterns for three- and four-block configurations (U.S. EPA, 1974b).

time traveling to the disposal site) and fuel. In addition to lower collection costs, transfer stations offer benefits including reduced maintenance costs for collection vehicles, increased flexibility in the selection of disposal facilities, the opportunity to recover recyclable materials at the transfer site, and the opportunity to process wastes (shred or bale) prior to disposal. In determining whether a transfer station is appropriate, municipal decision makers should compare the costs and savings associated with the construction and operation of the facility with costs for the direct shipping of the wastes from local neighborhoods to the landfill.

Transfer stations are often difficult to site and permit, particularly in urban areas. The farther the ultimate disposal site is from the collection area, the greater the savings attained from the use of a transfer station. The disposal site is typically at least 10 to 15 miles from the generation area before a transfer station is economically justified (see Figure 5.7). Transfer stations are sometimes used for shorter hauls to complete other duties such as sort wastes or allow the shipment of wastes to more distant landfills (U.S. EPA, 2003).

5.3.12 TYPES OF TRANSFER STATIONS

The type of station that would be most appropriate for a community depends on several design variables, for example (U.S. EPA, 2003):

- Capacity for waste storage
- Types of wastes received

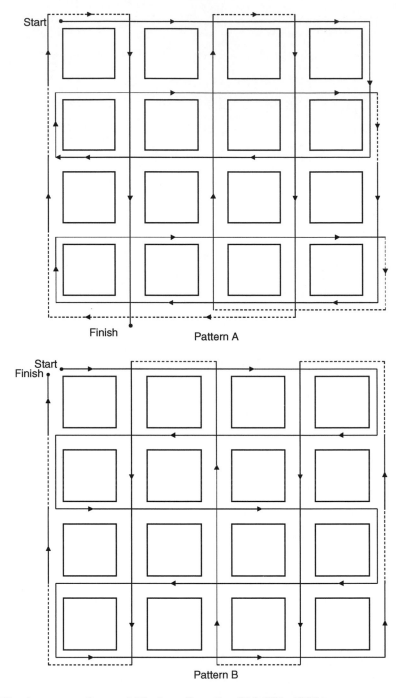

FIGURE 5.6 Routing patterns for a multiblock configuration (U.S. EPA, 1974b).

- Processes necessary to recover material from wastes
- Types of collection vehicles using the facility
- Types of transfer vehicles to be accommodated
- Site access

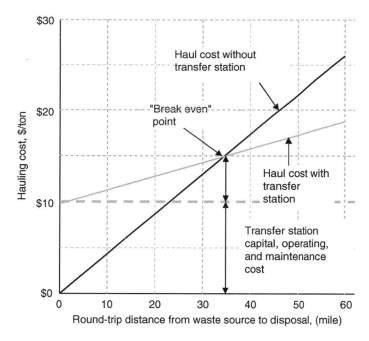

FIGURE 5.7 Comparison of waste hauling costs with and without a transfer station (U.S. EPA, 2002). Assumptions used to create this sample comparison were as follows:

Cost to construct, own and operate transfer station ($/ton)	$10
Average payload of collection truck hauling directly to landfill (tons)	7
Average payload of transfer truck hauling from transfer station to landfill (tons)	21
Average trucking cost (direct or transfer hauling) ($/mile)	$3

5.3.12.1 Small to Medium Transfer Stations (capacity of less than 100 to 500 tons/day)

Small to medium transfer stations are usually "direct-discharge" facilities that provide little area for interim waste storage. Such stations are equipped with operating areas for waste collection trucks and are often provided with drop-off areas for use by the public. Direct-discharge stations are often constructed with two operating floors. A compactor or open-top container is located on the lower level. Users enter the upper level and dump wastes into hoppers attached to these containers.

Some smaller transfer stations used in rural areas may use simple drop-off collection, in which a series of open-top containers are filled by users. The containers are then emptied into a larger vehicle at the station or hauled directly to the disposal site. The number and size of containers at the facility depends on the size and population density of the area served and the frequency of collection.

5.3.12.2 Large Transfer Stations

Large transfer stations are designed for heavy commercial use by private and municipal collection vehicles. When collection vehicles arrive at the site, they are checked in for billing, weighed, and directed to the appropriate dumping area. Check-in and weighing procedures are often automated for regular users. Collection vehicles travel to the dumping area and empty wastes into a trailer, pit, or onto a platform. Transfer vehicles are weighed after loading to just under maximum legal weights; this maximizes payloads and minimizes weight violations.

Several different designs for larger transfer operations are common depending on the transfer distance and vehicle type. Most designs fall into one of three categories: (1) direct-discharge noncompaction stations; (2) platform/pit noncompaction stations; or (3) compaction stations (U.S. EPA, 2003).

Direct-discharge Noncompaction Stations

These stations are generally designed with two operating floors. In the transfer operation, wastes are dumped directly from collection vehicles on the top floor through a hopper and into open-top trailers on the lower floor. The trailers are often positioned on scales so that dumping is halted when the maximum payload is reached. These stations are efficient because waste is handled only once. However, some provision for waste storage must be provided at peak drop-off times or during system interruptions.

Platform or Pit noncompaction Stations

In platform or pit stations, collection vehicles dump their wastes onto an area where wastes are temporarily stored and sorted for recyclables or unacceptable materials. The waste is then pushed into open-top trailers by front-end loaders (Figure 5.8). Platform stations are also constructed with two levels. Temporary storage is provided that can accommodate peak inflow of wastes. Construction costs may be higher with this type of station because of the increased floor space; however, the ability to temporarily store wastes results in a need for fewer trucks and trailers. Also, facility operators can haul wastes at night or during other slow traffic periods (U.S. EPA, 2003).

Compaction Stations

Compaction transfer stations use mechanical equipment to compact wastes before they are transferred. A hydraulically powered compactor is commonly used to compress wastes. Wastes are fed into the compactor through a chute either directly from collection trucks or after storage in a pit. The hydraulic ram pushes waste into the transfer trailer, which is mechanically linked to the compactor. Compaction stations are used when (1) wastes must be baled for shipment (e.g., rail haul) or for delivery to a balefill; (2) open-top trailers cannot be used because of size restrictions such as viaduct clearances; and (3) the site layout does not accommodate a multilevel building conducive to loading open-top trailers (U.S. EPA, 2003).

Transfer Station Design Considerations

The main objective in designing a transfer station should be to facilitate efficient operations. The operating program should be as simple as possible; waste handling should be minimized and the

FIGURE 5.8 Pit-type noncompaction transfer station.

facility should be sufficiently flexible to be modified as needed. Equipment and building durability are essential to minimize maintenance costs.

Site Location and Design

Establishment of a transfer station can be hindered by the NIMBY syndrome. Local residents are more likely to accept a new transfer station if the site is carefully selected and the buildings are designed appropriately for the site. Other factors to be addressed when considering a potential site are discussed below.

Proximity to waste collection area. Proximity to the collection area helps to maximize savings from reduced hauling time and distance. In some cases municipalities may consider the construction of more than one transfer station. For example, two transfer stations may be economically preferable if travel time from one end of the city to the other are excessive.

Accessibility of haul routes to disposal facilities. Transfer trucks should be able to easily enter major truck routes, which reduces haul time and potential impacts on nearby homes and businesses. Planners may have to determine whether improvements to local roads are necessary. The availability of rail lines and waterways may allow the use of rail cars or barges for transfer to disposal facilities.

Traffic. Transfer stations will generate additional amounts of traffic in its immediate area, which can contribute to increased road congestion, air emissions, noise, and wear on roads. For this reason, waste transfer stations are often located in industrial areas that have ready access to major roadways. Travel routes and resulting traffic impacts typically receive significant attention during transfer station siting and design (U.S. EPA, 2001).

Site zoning and design requirements. Municipalities must be certain that the proposed use meets site zoning requirements. In addition, the local site plan ordinance should be reviewed to identify any restrictions that could affect design, such as building height and setback (U.S. EPA, 2003).

Availability of utilities. A typical municipal transfer station will probably require full electricity and gas, water (for domestic use and fire fighting), telephones, and sanitary and storm sewers. Transfer station designers should determine the cost of connecting to these utilities and the regular service charges associated with them.

Visual impacts and aesthetics. The transfer station should be oriented so that transfer operations and vehicle traffic are not visible to local residents. Visibility can be restricted if the site is sufficiently large. Landscaping, installation of berms, and other site improvements will greatly improve the aesthetic quality of the entire facility.

Informing the community. When initiating a siting process, education must be extended beyond the siting committee and include a community-wide outreach initiative. Components of this type of public outreach typically include (U.S. EPA, 2002):

- Public meetings
- Interviews with local newspapers, media editorial boards, and broadcast media
- News conferences and press releases
- Paid advertising
- Internet sites
- Informational literature and direct mail
- City council or county commission presentations
- Presentations to civic, environmental, religious, professional, and neighborhood groups
- Community education programs and workshops
- Files located in public libraries or community centers

Building Design

Whenever putrescible wastes are being handled, larger transfer stations should be enclosed. Typically, transfer station buildings are constructed of concrete, masonry, or metal. Wood is not commonly used because it is difficult to clean, is less durable, and is more susceptible to fire damage.

Major considerations in building design include durability of construction, adequate size for tipping and processing, minimization of overhead obstructions to trucks, and flexibility of layout.

Transfer Station Sizing
The transfer station should have a sufficiently large capacity to manage the wastes that are expected to be received at the facility throughout its entire operating life. Factors to be considered in determining the appropriate size of a transfer facility include (U.S. EPA, 2003):

- Capacity of collection vehicles using the facility
- Number of days of storage on tipping floor
- Time required to unload collection vehicles
- Number of vehicles that will use the station and their expected days and hours of arrival (design to accommodate peak requirements)
- Waste sorting or processing to be accomplished at the facility
- Transfer trailer capacity
- Hours of station operation
- Availability of transfer trailers awaiting for loading
- Time required to attach and disconnect trailers from tractors or compactors
- Time required to load trailers.

Transfer stations are usually designed to have 1/2 to 2 days of storage capacity. The collection vehicle unloading area is usually the waste storage area and a waste sorting area. When planning the unloading area, adequate space should be provided for vehicle and equipment maneuvering. To minimize the space required, the facility should be designed so that collection vehicles back into the unloading position. Adequate space should also be available for offices, employee facilities, and other facility-related activities. Formulas for determining transfer station capacity are presented below (U.S. EPA, 2003).

Stations with Surge Pits
Based on the rate at which wastes can be unloaded from collection vehicles:

$$C = P_C (L/W)(60H_W/T_C)F \tag{5.2}$$

Based on the rate at which transfer trailers are loaded:

$$C = (P_t N \times 60H_t)/(T_t + B) \tag{5.3}$$

Direct Dump Stations

$$C = N_n P_t F \times 60H_W/[(P_t/P_C)(W/L_n)T_C] + B \tag{5.4}$$

Hopper Compaction Stations

$$C = (N_n P_t F \times 60H_W)/(P_t/P_C \times T_c) + B \tag{5.5}$$

Push-Pit Compaction Stations

$$C = (N_p P_t F \times 60H_W)/[(P_t/P_C)(W/L_p)T_C] + B_c + B \tag{5.6}$$

Where C = station capacity (tons/day)
P_C = collection vehicle payloads (tons)
L = total length of dumping space (ft)

W = width of each dumping space (ft)
H_w = hours per day that waste is delivered
T_C = time to unload each collection vehicle (min)
F = peaking factor (ratio of number of collection vehicles received during an average 30-min period to the number received during a peak 30-min period)
P_t = transfer trailer payload (tons)
N = number of transfer trailers loading simultaneously
H_t = hours per day used to load trailers (empty trailers must be available)
B = time to remove and replace each loaded trailer (min)
T_t = time to load each transfer trailer (min)
N_n = number of hoppers
L_n = length of each hopper
L_p = length of each push pit (ft)
N_p = number of push pits
B_C = total cycle time for clearing each push pit and compacting waste into trailer

5.3.13 TRANSFER VEHICLES

Most transfer systems use tractor trailers for hauling wastes; however, other types of vehicles may also be used.

5.3.13.1 Trucks and Semitrailers

Trucks and semitrailers are often used to carry wastes from transfer stations to disposal sites. These vehicles are flexible and effective because they can be adapted to serve the needs of individual communities. Truck and trailer systems should be designed to meet the following requirements:

- Wastes must be covered during transport.
- The vehicles should be designed to operate safely in the traffic conditions encountered on the hauling routes.
- Truck capacity should be designed so that road weight limits are not exceeded.
- Unloading methods should be simple and reliable.
- Truck design should prevent the leakage of liquids during hauling.
- The materials used to make the trailers and the design of sidewalls, floors, and suspension systems should be able to withstand the MSW loads.
- The number of required tractors and trailers depends on peak inflow, storage at the facility, trailer capacity, and the number of hauling hours.

Two types of trailers are used to haul wastes from the transfer station to the final disposal facility: compaction and noncompaction trailers. Noncompaction trailers are used with pit or direct-dump stations, and compaction trailers are used with compaction stations. Noncompaction trailers can usually haul higher payloads than compaction trailers because the former do not require an ejection blade for unloading. Based on a maximum gross weight of 80,000 lb, legal payloads for compaction trailers are typically 16 to 20 tons, while legal payloads for open-top live-bottom trailers are 20 to 22 tons (U.S. EPA, 2003).

Transfer vehicles should be able to negotiate the rough and muddy conditions of landfill access roads.

5.3.13.2 Rail Cars

Railroads carry only about 5% of transferred wastes in the United States (Lueck, 1990). As the distance between sanitary landfills and urban areas increases, however, railroads tend to become more appealing for transporting wastes to distant sites. Rail transfer stations are usually more expensive than similarly sized truck transfer stations because of costs for constructing rail lines, installing special equipment to remove and replace roofs of rail cars to load or bale wastes, and installing special equipment to unload rail cars at the disposal facility. A 60 ft boxcar can transport approximately 90 tons of waste compared with transfer trailers, which usually transport only 20 to 25 tons of waste (U.S. EPA, 2003).

REFERENCES

American Public Works Association, Institute for Solid Wastes, *Solid Waste Collection Practice*, 4th ed., Chicago, IL, 1975.

Hickman, H.L., Collection of residential solid waste, In The Solid Waste Handbook: A Practical Guide, W.D. Robinson Ed., John Wiley and Sons, Inc., New York, 1986.

Kiely, G., *Environmental Engineering*. The McGraw-Hill Companies, New York, NY, 1997.

Lueck, G.W., Elementary lessons in garbage appreciation, *Waste Age*, September 1990.

Miranda, M.L. and Aldy, J.E., Unit pricing of residential municipal solid waste: lessons from nine case study communities, Report prepared for Office of Policy, Planning and Evaluation, U.S. Environmental Protection Agency, Washington, DC, 1996.

Pferdehirt, W., University of Wisconsin–Madison Solid and Hazardous Waste Education Center, Madison, WI, 1994.

U.S. Environmental Protection Agency, *Residential Collection Systems, Volume 1: Report Summary*, SW-97c.1, 1974a.

U.S. Environmental Protection Agency, *Heuristic Routing for Solid Waste Collection Vehicles*, DSW/SW-1123, 1974b.

U.S. Environmental Protection Agency, *Collection Efficiency: Strategies for Success*, EPA530-K-99-007, Solid Waste and Emergency Response (5306W), Washington, DC, December 1999.

U.S. Environmental Protection Agency, *Waste Transfer Stations: Involved Citizens Make the Difference*, EPA530-K-01-003, Solid Waste and Emergency Response (5306W), Washington, DC, January 2001.

U.S. Environmental Protection Agency, *Waste Transfer Stations: A Manual for Decision-Making*, EPA530-R-02-002, Solid Waste and Emergency Response (5306W), Washington, DC, June 2002.

U.S. Environmental Protection Agency, *Decision-Makers Guide to Solid Waste Management*, 2nd ed., EPA530-R-95-023, Solid Waste and Emergency Response (5306W), Washington, DC, June 2003.

Vesilind, P.A., Worrell, W.A., and Reinhart, D.A., *Solid Waste Engineering*, Brooks/Cole. Pacific Grove, CA, 2002.

SUGGESTED READINGS

Bader, C., Where are the collection trucks going? *J. Municipal Solid Waste Professionals*, Sept./Oct. 2001. www.forester.net/mw_0109.where.html

British Columbia Ministry of Environment, Lands, and Parks, Guidelines for Establishing Transfer Stations for Municipal Solid Waste, British Columbia Ministry of Environment, Lands, and Parks, Victoria, BC, 1996.

Bush, S. 2003. Is automation altering refuse collection? *J. Municipal Solid Waste Professionals*, Sept./Oct. 2003. www.forester.net/mw_0309_automation.html

Bush, S. and Luken, K., Automated collection: getting the biggest bang for your buck, *J. Municipal Solid Waste Professionals*, Sept./Oct. 2002. www.forester.net/mw_0209_automated.html

California Integrated Waste Management Board, Completion of Solid Waste Information System Inspection Reports for Disposal Sites and Transfer Stations, Integrated Waste Management Board, Sacramento, CA, 1994.

California Integrated Waste Management Board, Inspection Guidance for Transfer Stations, Materials Recovery Facilities, and Waste-to Energy Facilities, Integrated Waste Management Board, Sacramento, CA, 1995.

California Integrated Waste Management Board, Transfer, Processing, and Material Recovery Facilities, Integrated Waste Management Board, Sacramento, CA, 1998.

Canterbury, J., *Pay As You Throw: Lessons Learned About Unit Pricing of Municipal Solid Waste*, DIANE Publishing Company, Chicago, IL, 1996.

Central and Eastern Europe Business Information Center, Environment Municipal Waste Management, Market Research, May 2002. www.mac.doc.gov/ceebic/countryr/Poland/market/municipalWASTE.html

Chicago, Illinois Department of Streets and Sanitation, City of Chicago Solid Waste Management Plan: Needs Assessment Report, Department of Streets and Sanitation, Chicago, IL, 1990.

Colorado Hazardous Materials and Waste Management Division, Colorado Transfer Stations, Hazardous Materials and Waste Management Division, Denver, CO, 1996.

Cutler, T., The components of comprehensive integrated waste management software systems, *J. Municipal Solid Waste Professionals*, March/April 2003. www.forester.net/mw_0303_componants.html

DIANE Publishing Company, *Bibliography of Municipal Solid Waste Management Alternative*, DIANE Publishing Company, Chicago, IL, 1995.

Illinois Environmental Protection Agency, Active Permitted Waste Transfer Stations. Springfield, IL, 1998.

Jonathan, R. and Manoor, A., *Vehicles for People or People For Vehicles? Issues in Solid Waste Collection in Low-Income Countries*, WEDC Publishing, Cholchester, UK, 2002.

Kansas Bureau of Waste Management, Kansas Directory of Municipal Solid Waste Landfills, Transfer Stations, Construction/Demolition Landfills and Composting Operations, The Department, Topeka, KS, 2000.

Kinhaman, T., *The Economics of Residential Solid Waste Management*, Ashgate Publishing Ltd, Hampshire, UK, 2003. Ashgate.com.

Lin, W.-H., *Solid Waste Transfer Stations in Hong Kong: a Critical Review*, University of Hong Kong, 2000.

Ludwig, C., Hellweg, S., and Stucki, S., *Municipal Solid Waste Strategies and Technology for Sustainable Solutions*, Springer, New York, NY, 2003.

Maine Waste Management Agency, *Transfer Stations and Bulky Waste Management: A Planning Manual*, Maine Waste Management Agency, Augusta, ME, 1993.

Massam, B.H., The Location of Waste Transfer Stations in Ashdod, Israel using a Multi-Criteria Decision Support System, Urban Studies Program, Division of Social Sciences, York University, Toronto, Ontario, 1990.

Merrill, L., Broken springs, busted drivelines and driver duties, *J. Municipal Solid Waste Professionals*, July/Aug. 2001. www.forester.net/mw_0107_refuse.html

Missouri Water Pollution Control Program, Storm Water Permit for Solid Waste Transfer Stations, Water Pollution Control Program, Jefferson City, MO, 2001.

National Center for Resource Recovery, *Municipal Solid Waste Collection: A State of the Art Study*, Lexington Books, Lexington, MA, 1990.

Pasternar, S. and Yanke, S., Municipal solid waste management in Texas: a decade of change? *J. Municipal Solid Waste Professionals*, May/June. 2003. www.forester.net/mw_0305_municipal.html

Schnorf, R.A., Transfer Stations: A Guide for Massachusetts Municipalities, Bureau of Solid Waste Disposal, Boston, MA, 1990.

Sloggett, G., An Economic Analysis of Solid Waste Stations for Coal County, Rural Development and Cooperative Extension Services, Oklahoma State University, OK, 1992.

Sloggett, G., Fitzgibbon, J., and Doesken, G. A., Solid Waste Transfer Stations for Rural Oklahoma, Cooperative Extension Service and Division of Agricultural sciences and Natural Resources, Oklahoma State University, OK, 1993-1994.

Tasmania Department of Environment and Land Management: Environmental Management Division, Guidelines for the Establishment and Management of Waste Transfer Stations, Department of Land Management, Tasmania, 1996.

Tilton, J., Scales and software, *J. Municipal Solid Waste Professionals*, Nov/Dec. 2001. www.forester.net/mw_0209_scales.html

U.S. Environmental Protection Agency, *Decision Makers Guide to Solid Waste Management*, No date. www.epa.gov/epaoser/non-hw/muncpl/dmg2/chapter4.pdf

U.S. Environmental Protection Agency, A Regulatory Strategy for Siting and Operating Waste Transfer Stations Response to a Recurring Environmental Circumstance: the Siting of Waste Transfer Stations in Low Income Communities and Communities of Color, EPA 500-R-00-001, Washington, DC, 2000.

U.S. Environmental Protection Agency, *Waste Transfer Stations: Involved Citizens make the Difference*, EPA 530-K-01-003, Office of Solid Waste and Emergency Response, Washington, DC, 2001.

U.S. Environmental Protection Agency, *Waste Transfer Stations: A Manual for Decision Making*, EPA 530-R-02-002. Office of Solid Waste and Emergency Response, Washington, DC, 2002.

U.S. Environmental Protection Agency, Frequently Asked Questions About Recycling and Waste Management, Municipal Solid Waste, Oct. 2003. www.epa.gov/epaoswer/non-hw/muncpl/faq.htm

QUESTIONS

1. MSW compaction in a collection truck allows for increased volumes of waste to be transported; however, if this compacted material is brought to a materials recovery facility, there are potential disadvantages. Discuss these problems.

2. In your community, observe the different containers used for the storage of MSW. What types of collection trucks are used for residential (single-family) neighborhoods? Apartment complexes? Commercial zones (restaurants, industrial parks, etc.)?

3. In your community, what are the major systems and equipment used for the collection of domestic and commercial solid wastes? Are there programs for the collection of source-separated wastes? For household hazardous wastes? Other?

4. In your community, is waste collection carried out by the municipality or by a private hauler? Who collects the commercial wastes?

5. What recommendations would you make regarding current waste collection in your community in order to reduce costs and improve efficiency?

6. How are yard wastes managed in your community? Are these materials collected separately from other wastes? If yard wastes are currently collected along with MSW, what changes to the collection system would be required in order to collect these wastes separately?

7. In your community, is there a program for collecting recyclable materials? If a program is in place, what materials are collected and how? Where are the recyclables shipped after collection? How much of each material (in tons) is collected per month?

8. In your community, how are household hazardous wastes transported to treatment or disposal facilities? In your opinion, is this management scheme effective or are there problems to address?

9. If your community's wastes are collected by a municipal system, how much of the municipal budget is earmarked to cover these costs? Visit the local waste management office and determine how these costs have changed over the past 10 years or more.

10. You have been hired as a consultant for a small town (population = 12,000) that wishes to begin a municipal program for solid waste collection. The community wants to collect wastes once per week. What type of collection vehicle would you recommend? What capacity truck would be appropriate (available truck capacities are 14, 16 and 20 yd^3)? Is once-per-week collection suitable for this community?

11. Obtain a map of a portion of your community and use the U.S. EPA heuristic routing guidelines to formulate a suitable route for collection vehicles. Contact the waste hauler in your community and determine the method used to route collection vehicles.

12. Locate three sites in your community that could serve as locations for transfer stations. Justify your choices.

13. Discuss the benefits of transfer stations to a community in terms of economics, time savings, and environmental quality.

14. At your City Hall, obtain the accident records for city employees. Determine the relative accident rate for solid waste collection employees.

6 Recycling Solid Wastes

We are not to throw away those things which can benefit our neighbor. Goods are called good because they can be used for good: they are instruments for good, in the hands of those who use them properly.

Clement of Alexandria (ca. 150–220)

6.1 INTRODUCTION

In an era when energy conservation, material cost and availability, and solid waste management are major concerns to municipal administrations, scientists and the general public, it is imperative that all parties appreciate the importance of recycling and of the value of products manufactured from scrap. As we shall see, the benefits from waste recycling are not solely environmental, but economic and aesthetic as well.

As discussed in Chapter 1, integrated waste management embraces a hierarchy of waste management options to achieve maximum economic and environmental returns. Recycling was listed near the top of the hierarchy and will be addressed in this chapter.

As indicated in Chapter 2, recycling is not a new phenomenon. Animal manures, plant debris and "night soil" have been applied to agricultural lands for millennia, and rag pickers were important recyclers in America as recently as the early 20th century. Modern recycling can trace its roots back to the 1960s, following the growing awareness of citizens to a myriad of environmental and public health concerns. At that time, however, recycling programs often only involved the simple *segregation* of materials from the waste stream. Unfortunately, markets were not established for the purchase and reuse of separated materials. Manufacturers were reluctant to invest and participate in new processing technologies and many were not yet equipped to handle these so-called "secondary materials." As a result, many separated materials found their way to the landfill. Many recycling programs failed not only due to the lack of processing, then, but more importantly, due to the lack of established markets for separated materials.

A new environmental awareness arose by the late 1980s, catalyzed by news of washups of medical wastes, decline of landfill space, a possible global greenhouse effect, and atmospheric ozone depletion. At this time it became apparent that sanitary landfills were rapidly closing and new ones faced substantial regulatory and grassroot hurdles to permitting and siting. The cost of disposing wastes correspondingly increased. As a result, interest in recycling by the public and, significantly, also by industries and government increased markedly.

In recent years, numerous community recycling efforts have originated from efforts to reduce the waste load to the local landfill, thus saving tax dollars. There are many examples of municipalities establishing recycling drop-off centers or materials recovery facilities (MRFs) as a result of public pressure. On a larger scale, however, a wide range of legislation at both national and state levels has been promulgated that encourages the recycling of MSW. Many are aimed at waste generators, whether the individual homeowner or business; some take the form of guidelines or requirements for extending the lifetime of the local landfill. On the heels of federal mandates since 1990, most states have set specific guidelines for reducing the tons of waste entering landfills. These quotas were to be met via a combination of source reduction, recycling, and composting. Other legislation, in contrast, addressed the purchase of recycled materials. Some government offices, for example, are now required to purchase paper manufactured containing a specified percentage of recycled fibers. By the early 1990s, most states had established regulations addressing waste recycling. As a result of such incentives and pressures, many industrialized nations have established innovative and proactive recycling strategies.

There are two primary approaches to the segregation of MSW for eventual recycling: source separation and the MRF. Source separation includes the segregation of specific waste components by the individual homeowner and commercial establishment. (i.e., *at the source*). The individual products (e.g., aluminum cans, paper, glass, and plastics) are collected and transported to a facility for further processing such as densifying and shredding. These slightly processed, clean materials are then sold to and removed by brokers or manufacturers. In contrast, the MRF is a centralized and mechanized facility which accepts either raw ("commingled") MSW or source-separated materials. The mixed items are placed on conveyor belts where specific recyclables are removed at various stations, either by hand or by a specialized mechanical device. Both source separation and MRF methods differ drastically in terms of efficiency of separation, capital costs, labor costs, energy use, and other factors. The MRF will be discussed in the next chapter.

6.2 RECYCLING TERMINOLOGY

Several terms related to recycling are often misused; in order to avoid confusion, it is important at the outset to clarify some of the relevant language.

Source separation — Removal of potentially recyclable materials from the waste stream. Conducted by the individual consumer and commercial establishment (Figure 6.1).

Reuse — Using an item for its original purpose. A common example is refilling a returnable soft drink bottle.

Recycling — Use of a material in a form similar to its original use. Newspapers are recycled into cardboard or new newspaper. Plastic is shredded and manufactured into fabric. Aluminum window frames are converted into new beverage containers.

Waste-to-energy — The conversion of MSW (preferably the organic fraction only) into energy by combustion in a controlled incinerator. Energy is recovered as heat and can be utilized directly; however, some facilities convert the heat energy into electrical energy.

FIGURE 6.1 Drop-off centers are one means of segregation of MSW components.

Resource recovery — Extraction of energy or materials from wastes. This term incorporates all of the above. Thus, a waste-to-energy facility will incinerate organic wastes to generate heat energy. Glass and rubber are separated from wastes, processed, and used as road building materials.

With the above terminology in mind, we can address the overall mechanism of recycling as shown by the universal recycling symbol.

One arrow in the figure indicates source separation, i.e., the removal of materials from the waste stream (e.g., setting aside aluminum cans in the homeowner's kitchen) and placement for pickup in specialized bins at the curb or at a designated drop-off facility. The second arrow symbolizes processing of the material. The aluminum cans are collected by a municipal waste hauler or private firm and brought to a broker or distributor where they are compressed into large bales. Given the proper market conditions, the bales are sold and shipped off-site to an aluminum smelter. At the smelter, the bales are melted, the material is drawn into sheets, and new cans are eventually manufactured. The cans are shipped to a soft drink manufacturer and filled, and are then shipped to a retail store for sale to the consumer. Finally, the consumer purchases the soft drink stored in the *recycled* cans (third arrow).

Given the above cycle, therefore, a material is not truly considered recycled until it has proceeded through all three steps, and is ultimately purchased by the consumer.

Over the past several decades, the source separation component of this cycle has become disproportionately large compared with the other two. Problems with the second step have been encountered, as some industries have complained of excessive costs for retooling the equipment and facilities needed to process recycled stock. In other cases, financial incentives, whether in the form of subsidies or market prices, may still be in effect for the use of virgin materials. A bottleneck has also occurred at the third arrow; in other words, there has been insufficient demand by purchasers, particularly the individual consumer, for the purchase of items manufactured from recycled products.

6.3 RECYCLING PROGRESS AND STATISTICS

According to the U.S. EPA (2001), Americans generated approximately 229 million tons of municipal solid waste in 1999. A total of 28% (62 million tons) of MSW was recovered by recycling (including composting) in 1999. There were nearly 9000 curbside recycling programs in the United States in 1997, as well as more than 12,000 drop-off centers for recyclables (Figure 6.1). About 380 MRFs are in operation to process the collected materials. There were also approximately 3500 yard waste composting programs.

EXAMPLE 6.1

Using the data for the MSW composition of the city of Pristine, Illinois, in the table below, calculate the maximum contribution of source separation to the city's solid waste disposal program.

Component	% by wt
Paper	31
Cardboard	5
Ferrous metals	10
Nonferrous metals	1
Glass	9

SOLUTION

The above components make up to total 56% of the MSW stream. Of this 56%, however, not all are potentially recyclable.

Component	Expected Recovery (%)	MSW (% by wt)
Newspaper	25	7.8
Cardboard	100	5.0
Ferrous	75	7.5
Nonferrous	50	0.5
Glass	75	6.75
Total		27.60

MSW comprises about 50% of total solid waste. As discussed in Chapter 1, the remainder is industrial, construction and demolition debris, and so on. Therefore, source separation can handle about 27.6% of 50% or

$$27.6\% \times 50\% = 14\% \text{ of the total solid waste}$$

Unfortunately, no U.S. community has come close to attaining 100% participation in a source separation program. If we optimistically assume 50% participation, then $14\% \times 50\% = 7\%$ is the maximum contribution of source separation to recycling (Adapted from Schwarz and Brunner, 1983).

6.4 RECOVERY AND MARKETS FOR COMPONENTS OF THE WASTE STREAM

To understand the opportunities and challenges, and to formulate a holistic approach to waste recycling, it is useful to have a basic understanding of the individual materials involved, various manufacturing processes, and the nature of secondary material markets. Questions that recycling professionals must address include the following:

- What is the demand for reclaimed materials?
- What are the specific requirements for a raw material used by an industry?
- How can a recovered material compete, technologically and economically, with virgin feedstock?
- Are new processes or equipment necessary in order to process and work with wastes?
- What are the incentives or disincentives that affect the use of recycled materials by manufacturers?
- Can new uses be identified that will increase the demand for recycled materials? (Adapted from Rhyner et al., 1995)

6.5 MARKET ISSUES

Price volatility in recycling markets is inherent in the system. Prices for recycled materials follow the overall demand for manufactured goods. Supply and demand for materials are based on broader issues such as markets for certain raw materials, trade agreements and tariffs between countries, and so on. Handling the fluctuations in revenue will help to secure the long-term success of a recycling program. Some communities manage fluctuations by creating local manufacturing demand for recycled materials. Others negotiate contracts that include price floors.

Recycling programs that collect a wide variety of materials such as mixed paper, newspaper, cardboard, glass, metals, plastic bottles, and lead automotive batteries, may be at an advantage over programs collecting only one or two items. As markets decline for one material, it can be stockpiled until the market improves; meanwhile, other, more profitable materials can continue to be made available for sale.

6.6 PURITY OF MATERIALS

Processors and end users of recovered materials typically require that the materials be homogeneous and free of contamination. A small amount of an unwanted material may negatively affect the quality of a recycled product, and may, in some cases, be a hazard to workers. Some industries comply with strict standards as to composition and will not tolerate even very low levels of contamination. Other industries routinely process materials to remove foreign material.

There is significantly less foreign material in source-separated wastes compared with raw mixed wastes processed via an MRF. However, many citizens (and politicians) prefer the convenience of shipping commingled wastes to a central processing facility for sorting into various fractions.

Buyers may also require that the materials be compacted or established under a specific condition (e.g., bottles are not to be broken, aluminum beverage cans are to be crushed, and HDPE containers must be baled).

Materials that are commonly recycled or are potentially recyclable are discussed in the following sections.

6.7 PAPER

As mentioned in Chapter 4, paper products comprise by far the majority of the municipal solid waste stream — about 38% of U.S. MSW, more than double than any other component. Paper waste has its share of environmental and economic costs, as it occupies substantial volume in collection trucks and landfills. Production of paper waste implies that more trees need to be cut in order to satisfy continued needs for new paper. The large quantities generated along with the associated costs for disposal provide economic incentives for paper recycling.

6.7.1 PAPER MANUFACTURE

The Chinese developed the first known papermaking process as early as 100 C.E. A suspension of bamboo fibers was used as the paper base. In the United States, the first paper mill was constructed in 1690 near Philadelphia. Until the mid-1800s, paper was made exclusively from recycled fiber derived from cotton, linen rags, and waste paper. With an increased demand for paper and paperboard, techniques for utilizing wood fiber in papermaking were developed (American Paper Institute, 1990; Tchobanoglous et al., 1993).

Both coniferous and deciduous wood pulp is used in modern papermaking. Hardwoods and softwoods have very different fiber morphologies, which will therefore result in different paper properties. The fibers of softwoods are longer and stronger than those of hardwoods (Roberts, 1996); however, softwood fibers tend to form flocs of entangled fibers during the sheet-forming process, resulting in problems with appearance. To control this effect, softwood fibers are blended with those of hardwood to provide adequate strength and appearance.

Paper is derived from fibers originating within the cells of land plants; therefore, paper does not have a fixed chemical composition. Specifically, terrestrial plant cells are mostly composed of carbohydrate polymers (polysaccharides) incorporated to some degree with lignin, a complex aromatic polymer. The amount of lignin commonly increases with the age of the plant. The carbohydrate component of the cell contains primarily the structural polysaccharide cellulose. There are other, lower-molecular weight nonstructural polysaccharides known as hemicelluloses, which play an important part in pulp and paper properties. There are also relatively small amount of water-soluble compounds such as alcohols, resin acids and fatty acids, and trace inorganic materials (Rhyner et al., 1995; Roberts, 1996).

Essential to papermaking, whether utilizing wood fibers or recycled fibers as the base, is that the fibers be conformable, i.e., capable of being matted into a uniform sheet. They must also be capable of forming sufficiently strong bonds at the point of contact. Proper conformability and bonding begin with the pulping process, when the bonds in the wood fibers are broken.

There are three methods of pulping virgin fiber: (1) mechanical pulping, where fibers are freed by the application of mechanical energy; (2) chemical pulping, where chemicals are added to dissolve the lignin and retain the cellulose; and (3) semichemical pulping, which is a combination of (1) and (2).

Beyond pulping, paper manufacture operations are as follows (Smook, 1982; Rhyner et al., 1995):

- *Sheet formation.* Pulp slurry (1% pulp, 99% water) is guided into a headbox.
- *Forming.* The pulp is dewatered by about 20%. The fibers are formed into a sheet.
- *Pressing.* Fibers are pressed together. Another 20% of the water is removed from the pulp.
- *Drying.* The sheet is dried to about 90 to 95% solids. Fibers bond together at this point.
- *Converting and finishing.* The sheet is pressed between rolls to reduce thickness and increase uniformity, and is wound on to reels (Figure 6.2).

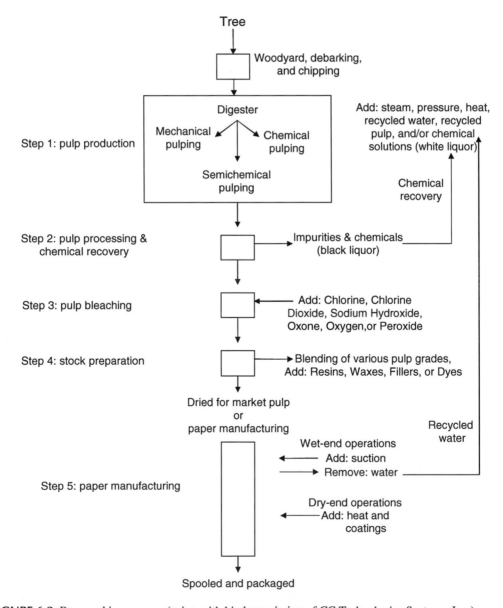

FIGURE 6.2 Papermaking process (using with kind permission of CC Technologies Systems, Inc.).

6.7.2 PAPER RECYCLING

Paper recycling has a long history. Collection drives by the Boy Scouts and other organizations were in place before World War II. Such programs increased greatly during the war.

Recovery of paper and paperboard for recycling is highest overall compared with all other materials in MSW. A total of 65.1% of all corrugated boxes was recovered for recycling in 1999 (Table 6.1) according to the U.S. EPA (2001). Newspapers were recovered at 60.06% and high-grade office papers at 52.7% with lesser amounts of other papers recovered. The American Forest and Paper Association (2002) reports the rate of paper product recycling as follows:

- Corrugated cardboard 70.1%
- Newspapers 68.9%
- Office paper 43.2%
- Printing and writing paper 37.8%

TABLE 6.1
Generation and Recovery of Paper Waste in the Solid Waste Stream, 1999

Product Category	Generation (thousands of tons)	Recovery (thousands of tons)	Recovery (percent of generation)
Nondurable Goods			
Newspapers			
Newsprint	11,330	6,800	60.0
Groundwood inserts	2,630	1,430	54.4
Total Newspapers	13,960	8,230	59.0
Books	1,120	200	17.9
Magazines	2,310	530	22.9
Office papers	7,670	4,040	52.7
Telephone directories	680	110	16.2
Third class mail	5,560	1,230	22.1
Other commerical printing	5,940	1,360	22.9
Tissue paper and towels	3,360	Neg.	Neg.
Paper plates and cups	930	Neg.	Neg.
Other nonpackaging paper[a]	4,790	Neg.	Neg.
Total Paper and Paperboard			
Nondurable goods	46,320	15,700	
Containers and Packaging			
Corrugated boxes	31,230	20,340	65.1
Milk cartons	490	Neg.	Neg.
Folding cartons	5,780	400	6.9
Other paperboard packaging	290	Neg.	Neg.
Bags and sacks	1,690	230	13.6
Other paper packaging	1,670	Neg.	Neg.
Total Paper and Paperboard			
Containers and packaging	41,150	20,970	51.0
Total paper and paperboard	87,470	36,670	41.9

Neg.=negligible

[a] Includes tissue in disposable diapers, paper in games and novelties, and cards.

Source: U.S. EPA, 2001.

Production of paper continues to increase in the United States and, therefore, more paper waste continues to be produced. Over the past decade paper recycling trends have continued to increase. According to the American Forest and Paper Association, between 1987 and 1999, the paper and paperboard recovery rate increased from 28.8 to 45%. The United States used more than 72 million tons of paper products, but only 25.5% is made from recycled paper. This compares with 35% in Western Europe, almost 50% in Japan, and 70% in the Netherlands (Liu and Liptak, 2000).

Waste paper can be classified as bulk or high grade. The highest grade includes manila folders, hard manila cards, and similar computer-related paper products. High-grade waste paper is used as a pulp substitute. Bulk grades consist of newspapers, corrugated cardboard, and mixed paper waste (unsorted office or commercial paper waste). Bulk grades are used to make paperboards, construction paper, and other products. The Institute of Scrap Recycling Industries has established standards and practices that apply to paper stock for repulping in the United States and Canada (ISRI, 2002). Common paper grades are listed in Table 6.2.

Paper manufacture from recovered paper and paperboard utilizes a different pulping process compared with pulping virgin fibers. A significant mechanical component needed for pulping recycled fibers is the continuous pulper. In this unit, the input material is ground into a smooth pulp, and extraneous materials (glues, plastic, metal and clips) are removed. Recovered pulps are then de-inked by a process of chemical disintegration or chemical treatment. In some plants a washing step is incorporated to further clean the pulp. The pulp is washed free of ink and other contaminants on a fine mesh screen. An optional flotation process may be used in which chemicals are added to the pulper to create air bubbles that separate and float ink droplets away from the pulp. The pulp may also be bleached (Rhyner et al., 1995). Apart from washing and flotation, the resultant pulp is screened and thickened. Once processed, the pulp, whether from virgin or recycled fibers, enters the paper manufacturing process.

Each time paper is recycled, some of the longer fibers are shortened, generally losing their flexibility and bonding ability. This is due to a process called "hornification," which is collectively a number of partly irreversible physical changes. Virgin pulp is added to maintain the paper strength required for efficient runnability at fast speeds, both on the paper mill and during conversion (e.g., printing press), as well as for the end-use. Besides strength, brightness also deteriorates each time the paper is recycled. In summary, waste paper tends to downgrade in quality as it is recycled. However, recycled fibers possess some advantages in that the twice-dried stock may drain faster than its virgin equivalent, require less refining, be co-refined with hardwood pulp or combined hardwood and softwood pulps without significant damage, and impart improved opacity (Ferguson, 2001).

As with many other materials, the waste paper market is volatile and strongly influenced by region. Economic conditions continue to affect progress in paper recycling. In some locations, mixed paper waste is of little value. One recycling marketer from the Midwest recently lamented, "Right now we can get $5 per ton for recovered paper. For that amount you can't start the collection truck."

One limitation on the amount of waste paper that can be recycled in a year is the capacity of paper mills. Construction or modification of such mills is capital-intensive; therefore, investors must be assured that there will be an adequate supply of waste paper to the mills and at a competitive price. Much waste paper is shipped to markets along the Pacific rim (e.g., South Korea), where timber resources are scarce. China is a major influence in the mixed waste paper market. Imports from the United States have increased from 209,000 tons in 1993 to 2.2 million tons in 2000. The mixed waste paper is sorted and processed in the recipient country. The demand is expected to grow further (Paper Technology, 2001).

Over the past decade, legislative programs have been developed in several countries that require a certain percentage of recycled fiber content in newspaper, office paper, and other products. Such initiatives increase the demand and the quantity of paper available for recycling. However, there may be resistance on the part of waste paper recyclers to make the large capital investments necessary to increase plant capacity (Pfeffer, 1992).

TABLE 6.2
Selected Paper Grades for Repulping and Recycling

No.	Name	Composition
1	Soft mixed paper	Mixture of various qualities of paper not limited as to type of baling or fiber content
2	Mixed paper	Baled clean, sorted mixture of various quality of paper containing less than 10% of groundwood content
3	—	Grade not currently in use
4	Boxboard cuttings	Baled new cuttings of paperboard used in the manufacture of folding cartons, set-up boxes, and similar boxboard products
5	Mill wrappers	Baled paper used as outside wrap for rolls, bundles, or skids of finished paper
6	News	Baled newspaper as typically generated from news drives and curbside collections
7	News, de-ink quality	Baled sorted, fresh newspapers, not sunburned, containing not more than the normal percentage of rotogravure and colored sections. May contain magazines
8	Special news de-ink quality	Baled, sorted, fresh newspapers, not sunburned, free from magazines, white blank, pressroom overissues, and paper other than news, containing not more than the normal percentage of rotogravure and colored sections. This grade must be tare-free
9	Overissue news	Unused, overrun newspapers printed on newsprint, baled or securely tied in bundles, containing not more than the normal percentage of rotogravure and colored sections
10	Magazines	Baled coated magazines, catalogues, and similar printed materials May contain a small percentage of uncoated news-type paper
11	Corrugated containers	Baled corrugated containers having liners of either test liner, jute, or kraft
12	Double-sorted corrugated	Baled, double-sorted corrugated containers, generated from supermarkets, industrial, or commercial facilities, having liners of test liner, jute, or kraft. Material has been specially sorted to be free of boxboard, off-shore corrugated, plastic, and wax.
13	New double-lined kraft corrugated cuttings	Baled new corrugated cuttings having liners of either test liner, jute, or kraft. Treated medium or liners, insoluble adhesives, butt rolls, slabbed or hogged medium, are not acceptable in this grade
14	—	Grade not currently in use
15	Used brown kraft	Baled used brown kraft bags free of objectionable liners and original contents
16	Mixed kraft cuttings	Baled new brown kraft cuttings, sheets and bag scrap, free of stitched paper
17	Carrier stock	Baled printed or unprinted, unbleached new beverage carrier sheets and cuttings. May contain wet strength additives
18	New colored kraft	Baled new colored kraft cuttings, sheets and bag scrap, free of stitched papers
19	Grocery bag scrap	Baled, new brown kraft bag cuttings, sheets and misprint bags.
20	Kraft multi-wall bag scrap	New brown kraft multi-wall bag cuttings, sheets, and misprint bags, free of stitched papers

Source: ISRI, Reproduced with kind permission of the Institute for Scrap Recycling Industries, Inc., Washington, DC 20005-3104, 2002. The ISRI Scrap Specifications Circular is subject to change. To find the most recent edition, go to www.isri.org.

6.7.3 OTHER PAPER MARKETS

Building manufacturers have used waste paper as a raw material for manufacturing other items. Home insulation has been manufactured from old newspapers. The cellulosic material is sprayed with an antiflammability agent and can be blown directly into wall cavities. Other applications of waste paper include insulation board, fiberboard, roofing, and siding. The molded pulp industry uses waste paper for the manufacture of plant pots, egg cartons, meat trays, and packaging materials. In agriculture, waste paper has been used as animal bedding (NAA, 2002). Old newspaper bedding has replaced straw bedding in some businesses due to its availability and relatively lower price, particularly during years of weather extremes. Old newspaper bedding has been shown to not adversely affect animals using the bedding or the soil where the bedding is applied as a component of manure (Rhyner et al., 1995).

Refuse-derived fuel (RDF) is a mostly carbonaceous material derived from MSW. The non-combustible components of waste such as stones, glass, and metals are removed usually by mechanical means, the organics are shredded, and the resultant "fluff" is either used directly as a fuel or is compressed into pellets. RDF is typically burned as a co-fuel along with coal, which accrues a number of environmental and economic benefits. A detailed discussion of RDF appears in Chapter 9.

6.7.4 BENEFITS OF PAPER RECYCLING

Seventeen trees are required to produce 1 ton of paper. All Sunday newspapers in the United States require the equivalent of one-half million trees every week. When paper is manufactured from waste paper, however, trees are conserved as are considerable amounts of energy. For 1 ton of paper recycled there is a savings of approximately 4100 kWh of energy along with 7000 gal of water and 3 yd^3 of landfill space (Liu and Liptak, 2000). Paper production from recycling also requires less chemicals, including bleaches, and will therefore produce less toxic wastes.

6.8 GLASS

6.8.1 GLASS MANUFACTURE

The ancient Egyptians were the first society known to manufacture glass for containers, by first forming a sand or clay mold and then wrapping strands of molten glass around the mold. A more widespread use of glass as a packaging material was made possible by the development of glass blowing in about 50 B.C.E. Across the Atlantic Ocean, glass manufacturing was the first known industry in Pre-Colonial America, developed more than 10 years before the arrival of the Pilgrims in 1620. Techniques for mass production of glass containers were developed in the 19th century and the first fully automatic bottle machine, producing one million bottles per week, was developed in 1903. Today, approximately 10 million tons of glass containers are manufactured in the United States annually (ISRI, 2001).

In order to manufacture new glass products, relatively inexpensive raw materials are required, including silica (SiO_2), soda ash (Na_2CO_3), and limestone ($CaCO_3$). Sometimes muriate of potash (KCl) is used in place of soda ash. Silica is the basic foundation of the product, however. It is common for recycled glass ("cullet") to be added to the mixture. The mixture is heated to approximately 1480–1570°C (2700–2850°F) and liquified.

The basic structural unit of silica has a tetrahedral shape with silicon in the center linked symmetrically to four oxygen atoms at its corners, giving the chemical formula SiO_4. Upon cooling molten silica quickly, a randomly organized network of these tetrahedra are formed, linked at their corners to give vitreous silica, an amorphous material (Pilkington, 2003):

$$Na_2CO_3 + SiO_2 \xrightarrow{1500°C} Na_2SiO_3 + CO_{2(g)} \tag{6.1}$$

$$Na_2SiO_3 + xSiO_2 \xrightarrow[\text{digestion}]{Na_2SO_4} (Na_2O)(SiO_2)_{(x+1)} \tag{6.2}$$

Soda ash makes the silica melt at a lower temperature. The sodium–oxygen atoms enter the silicon–oxygen network. Limestone imparts strength to the glass. Calcium enters the network structure rendering it more complex so that during the cooling process, it is more difficult for the atoms to arrange themselves for crystallization to occur.

The molten glass is pressed into molds that form bottles and jars. Air is injected into the containers to form the openings within. Cooling in an annealing oven strengthens the new bottles and jars. The cooled product is checked for flaws such as bubbles before shipment. The final step is filling with food, nonfood (medicines, perfume, cosmetics, and cleaning supplies), and other consumer and industrial items.

6.8.2 GLASS RECYCLING

Glass has experienced a more rapid growth in recycling than that of any other commodity except aluminum beverage containers. In recent years the use of recycled glass has increased more than 80% with glass manufacturers melting more than 2.5 million tons in their furnaces in a single year. This growth is the result of both increased collection through curbside recycling programs and accelerated demand from glass manufacturers (CMI, 2002).

About 12.5 million tons of glass is disposed of in the United States annually, which comprises about 8% of the total waste stream (U.S. EPA, 2001). Clear glass ("flint"), green or amber bottle and container glass, comprise 90% and the remaining 10% is plate and similar glass. A total of 2.9 million tons of glass containers were recovered for recycling in 1999. Based on 1999 glass generation rates, an estimated 26.6% of glass containers were recovered for recycling, with a 23.4% recovery rate for all glass in MSW (Table 6.3). The Glass Packaging Institute reported a recovery rate of 35.2% for glass containers in 1997; however, this rate includes refilling of bottles. In contrast with estimates for the United States, Japan recycles about 50% of its total glass production.

Most recovered glass is used for the production of new containers. A smaller fraction is applied to other uses such as fiberglass, bricks, and "glassphalt," a mixture of glass and asphalt that serves as a paving mixture for highway construction. The main purchaser of waste glass ("cullet") is, however, the glass container industry.

6.8.3 THE GLASS RECYCLING PROCESS

Glass recycling begins with the collection of used bottles and other containers. The glass is separated into clear, green, and amber (brown) colors by the consumer, the operators of the collection vehicle, or by the processing facility. At the MRF or other receiving facility, containers are crushed into small pieces (approximately 3/8 to ¾ in. across), which are shipped to manufacturing plants.

TABLE 6.3
Glass Waste Production and Recycling, 1999

Product Category	Generation (thousands of tons)	Recovery (thousands of tons)	Recovery (percent of generation)
Durable goods[a] containers and packaging	1,150	Neg.	Neg.
Beer and soft drink bottles	5,450	1,560	28.6
Wine and liquor bottles	1,830	940440	24.0
Food and other bottles and jars	3,770	940	24.9
Total glass containers	11,050	2,940	26.6
Total Glass	12,560	2,940	23.4

Neg.=Less than 5000 tons or 0.05 %.

[a] Glass as a component of appliances, furniture, consumer electronics, etc.

Source: U.S. EPA, EPA 530-R-01-014, 2001.

This material is known as processed, or furnace-ready cullet. In some situations, the containers are not broken but sent directly to the manufacturer.

Manufacturers purchase whole and broken glass cullet and combine it with soda ash, limestone, and silica to create new glass products for consumer, industrial, and other applications. Recyclers in the United States utilize 10 to 80% cullet in glass manufacture. This compares with 80 to 90% in Switzerland and Germany (Rhyner, 1995; Liu and Liptak, 2000).

Modern glass container manufacturing requires clean and uniform feedstock. There are four requirements for cullet used in recycling:

- Must be separated by color
- Must be contaminant-free
- Must meet market specifications
- Must be container glass

6.8.4 COLOR

On separating glass by color, manufacturers can ensure the desired quality and color consistency of new glass products. The preferred situation is for the individual consumer or business to separate glass by color at the source. This is also the optimum location for removal of contaminants such as food, labels, and dirt. Many community recycling programs will allow for the collection of mixed glass. Although a convenient practice for the consumer, such mixing may actually hinder the marketability of the product.

If separation by color does not occur at the source of generation, colors and contaminants should be sorted out early during processing. At a MRF or transfer station, cullet is almost always color-sorted by hand. In some communities, intermediate processors known as glass benefaction facilities receive glass from recycling programs and use sophisticated optical sorting machines to separate the glass into three color types (CMI, 2002). Optical sorting equipment is capital-intensive; therefore, hand-picking is typically the only feasible sorting mechanism.

If the cullet is not completely color-sorted and becomes thoroughly mixed, undesired colors are difficult to remove and can ruin an entire load. Mixed cullet has significantly lower demand and value. Glass manufacturers set limits on the amount of mixed cullet that is acceptable for manufacturing new containers. Many companies simply prefer not to buy mixed cullet. Other markets for mixed glass are available (see below), but are limited.

6.8.5 CONTAMINATION

Contaminated cullet, probably the single greatest problem for glass manufacturers, is not suitable for the production of new glass containers. Cullet can be contaminated at any point during the recycling process: at home, during collection, processing, or shipping. Contaminated cullet decreases quality and increases costs. Contaminants are a risk to the glass manufacturer and disrupt production, cause injury to workers, damage manufacturing equipment, and produce a poor-quality product.

Virtually all glass food and beverage containers, including food jars, soft drink bottles, juice containers, beer bottles, wine and liquor bottles, are recyclable. However, household glass products, such as light bulbs, drinking glasses, and window panes are not acceptable for producing glass containers. Such glass products vary significantly in chemical composition; furthermore, many possess different melting temperatures. As a result, mixing these products with container cullet may cause defects such as bubbles, cracks, or other weak points and imperfections in new containers.

Common materials that contaminate cullet include:

- Ceramic cups, plates, and pottery
- Crystal and opaque drinking glasses

- Mirrors
- Windshields and window glass
- Heat-resistant cookware (e.g., Pyrex)
- Light bulbs
- Clay garden pots
- Laboratory glass

Other contaminants include:

- Ceramic and wire caps for beer bottles
- Metal rings from wine bottles
- Metal caps, lids, and neck rings
- Food and dirt

Glass benefaction facilities (mentioned above) receive glass from community recycling programs and direct it through a sequence of steps to remove contaminants (stones, ceramics, and metal caps). Metals are removed magnetically. Eddy current separators are used to remove nonmagnetic metal contamination from caps and lids. An air classifier removes lightweight items such as loose paper or plastics. These unit operations for waste separation are discussed in detail in Chapter 7. Some contaminants are removed manually from mixed cullet; however, this is a slow and potentially dangerous activity. The final product is a ground glass feedstock which is uniform in color and free of contaminants, and readily acceptable by container manufacturers (CMI, 2002).

Preprocessors such as glass benefaction facilities provide a valuable market for recycling programs that do not produce the volume or have the ability to produce glass for direct delivery to a mill. A clean feedstock at the outset of processing, however, is strongly preferred by the industry.

6.8.6 GLASS MARKETS

To be a competitive commodity, recycled glass must maintain a price that competes with raw materials. Crushed glass has a fairly strong and consistent market value compared with most postconsumer recycling materials. Based on the discussion so far in this chapter, it is obvious that the price paid for cullet will strongly depend on the color and cleanliness of the recovered product. Marketed as flint (clear), amber (brown), emerald (green), or mixed color glass, the values range from $0 to $65 per metric ton delivered to the glass plant. Clean flint cullet is the most desirable form of glass scrap. Mixed color glass contaminated with food or ceramic fragments is the least desirable grade of cullet and will bring the lowest price (GPI, 2002).

6.8.7 CONTAINER GLASS

Container glass is 100% recyclable, and glass containers can be recycled into new ones repeatedly. There is no change in chemical or physical properties, and therefore no decline in quality with the repeated recycling of cullet.

According to the Glass Packaging Institute (no date), glass collectors, haulers, suppliers, and processors can reduce the risks and increase revenue by following some basic glass recycling guidelines:

- Contact potential buyers for their specifications and acceptance policies, ability to remove contaminants, transport preference (i.e., truck or rail car), and "furnace ready" requirements.
- Ask about buyers' capacity to remove metals.
- Conduct inspections regularly, especially before adding newly collected glass to stored recyclables and during loading for shipment.

FIGURE 6.3 Cullet stored outdoors and unprotected will inevitably become contaminated.

- If stored outdoors place the cullet on a concrete pad, not on the ground or asphalt, to avoid contamination from dirt or gravel during loading (Figure 6.3). Cover the cullet during inclement weather.
- When storing multiple loads of colored cullet, keep the cullet separated so that no intermingling of colors can occur.
- Prior to loading cullet shipments, wash the truck bed. Inspect the truck bed and the tarp used to cover the previous load for any residue.

6.8.8 OTHER USES FOR RECYCLED GLASS

In addition to serving as feedstock for manufacturing new glass containers, recycled glass is used to make other products. Fiberglass, a common alternative market for cullet, is predominantly used in the form of glass wool for thermal and acoustical insulation. Recycled glass used in the manufacture of fiberglass now constitutes the second highest volume of postconsumer glass. Industry standards for product quality and consistency are very high. Another promising alternative market is glassphalt, a road-paving material, which is a mixture of crushed low-grade, mixed-color cullet and asphalt. Some glassphalt mixtures may contain ground glass, sand, gravel, and limestone (Liu and Liptak, 2000). Given the comparatively low costs for road building aggregate, however, the demand for glassphalt is modest. Cullet is also used to produce highway reflectors and signs, sandblasting materials, decorative glass, and drainage aggregate. Cullet has been used in the manufacture of some wastewater plumbing. As glass is a relatively inert material, it can readily withstand the corrosive agents within wastewater. Some of the more innovative uses of recycled glass are listed in Table 6.4.

6.8.9 BENEFITS OF GLASS RECYCLING

Whereas some industrial sectors have been reluctant to utilize scrap material in routine production because of concerns over retooling and excessive costs, glass recycling, particularly that of containers, is an integral component of the glass production industry. Recovered glass waste has a lower melting point (1370°C [2500°F]) than the standard silica, soda ash, and limestone mixture; in container manufacture, each 10% increase in cullet reduces the melting energy by about 2.5%. Less

TABLE 6.4
Other Secondary Uses for Cullet

Abrasives	Finely ground container and noncontainer glass used in sand blasting. Such abrasives contain no silica, which is the causative agent of silicosis.
Aggregate substitute	Container and noncontainer glass utilized as drainage medium, backfill or for landscaping purposes
Bead manufacturing	Container and noncontainer glass is melted into rounded glass pellets or beads and used in reflective paint for highways
Decorative applications	Ceramic tiles, picture frames, costume jewelry, and some household items may include recycled container and noncontainer glass
Frictionators	Recycled glass is used to make frictionators needed for firing ammunition and lighting matches
Fluxex or other additives	Glass powders used as lubricants, core additives and fluxes in metal foundry work and fabrication, as well as flux/finders in the ceramics industry

Source: Glass Packaging Institute, n.d. Reproduced with kind permission of the Glass Packaging Institute

energy to manufacture glass products compared with raw materials will hold down manufacturing costs. There are also fewer gaseous emissions when working with cullet.

New glass containers manufactured from cullet possess the same quality and structural integrity as do containers made with raw materials only. Using recycled glass saves wear on furnaces, resulting in extended furnace life and savings on maintenance. Recycling glass also reduces the amount of solid waste brought to landfills. Lower volumes of solid waste lessen the demand for landfill space and reduce disposal costs. Finally, recycled glass is usually closer to the bottle plants than the sources of the raw materials (CMI, 2002).

6.9 ALUMINUM

6.9.1 Aluminum Manufacturing

The starting material for primary aluminum manufacture is bauxite ore, a mined mineral. Dissolving powered bauxite in sodium hydroxide produces alumina, which serves as the raw material for primary aluminum production. The aluminum industry utilizes the Bayer process to produce alumina from bauxite. The three major stages in the Bayer process are extraction, decomposition, and calcination. During extraction, hydrated alumina is selectively removed from other insoluble oxides by transferring it into a solution of sodium hydroxide (caustic soda) (World-Aluminum, 2000):

$$Al_2O_3 \cdot xH_2O + 2NaOH \rightarrow 2NaAlO_2 + (x + 1)H_2O \qquad (6.3)$$

This product is transferred to a heated pressure digester. Conditions within the digester (concentration, temperature, and pressure) vary according to the properties of the bauxite ore being used. Modern plants typically operate between 200 and 240°C and involve pressures of approximately 30 atm.

After the extraction stage the liquor, containing the dissolved Al_2O_3, is separated from the insoluble bauxite residue, purified and filtered before it is delivered to the decomposer. The mud is thickened and washed so that the caustic soda can be removed and recycled. During the decomposition phase, crystalline alumina trihydrate is extracted from the digestion liquor by hydrolysis (World-Aluminum, 2000):

$$2NaAlO_2 + 4H_2O \rightarrow Al_2O_3 \cdot 3H_2O + 2NaOH \qquad (6.4)$$

The alumina trihydrate crystals are then classified into size fractions and fed into a rotary or fluidized-bed calcination kiln. In the kiln, alumina trihydrate crystals are calcined to remove their water of crystallization and prepare the alumina for smelting.

The basis for aluminum smelting plants is the Hall-Héroult process. Alumina is dissolved in an electrolytic bath of molten cryolite (sodium aluminum fluoride) within a large carbon- or graphite-lined steel container. The bath also contains a small amount of aluminum fluoride and calcium fluoride. An electric current is passed through the electrolyte at low voltage, but very high current, typically 150,000 A. The electric current flows between a carbon anode made up of petroleum coke and pitch, and a cathode formed by the thick carbon or graphite lining of the pot. Molten aluminum is deposited at the bottom of the pot and is siphoned off periodically, taken to a holding furnace, and often alloyed with selected elements to produce required qualities for specific end-uses such as beverage cans, sheet, transportation uses, and building and construction products.

Few U.S. companies refine bauxite into alumina in the United States. Most import alumina from Australia, Jamaica, Suriname, Guyana, and Guinea. The United States is the largest aluminum producer in the world, with containers and packaging accounting for the largest share of shipments.

6.9.2 Aluminum Recycling

Aluminum waste consists of industrial scrap, which is a by-product of aluminum manufacturing processes ("new scrap"), and old scrap consisting of postconsumer items such as used aluminum beverage cans, window frames, building siding, and foil. Nearly 80% of the aluminum in MSW consists of used beverage containers (UBCs).

Nationwide, aluminum cans constitute less than 1% of MSW; in communities having established recycling programs or container deposit laws, the percentage in the local waste stream is negligible. Table 6.5 presents data on the quantity of aluminum recycled. In 1975, about 25% of aluminum cans were recycled. This percentage remained relatively constant until about 1980. The increased rate in the latter half of the 1980s is attributable to additional collection programs and container deposit legislation. During this period, a number of states passed laws requiring deposits of $0.05 to $0.10 per container, thus providing an additional incentive for recycling. Aluminum beverage containers were recovered at a rate of 59.5% of generation (0.9 million tons) in 1997, and 48.5% of all aluminum in containers and packaging was recovered for recycling in 1997. The rate of aluminum recycling has been in decline for a decade, after peaking at 65% in 1992.

There are numerous successful community recycling programs for mixed aluminum scrap and aluminum cans. These programs are generally self-sufficient and, in some municipal programs, provide an income to subsidize other recycling activities (Pfeffer, 1992). Used aluminum cans are collected in curbside pickup programs, at buy-back locations, at recycling centers, and by scrap metal dealers. A number of states have established mandatory deposits for beverage containers and have installed redemption centers at supermarkets.

Cans brought to collection centers are processed in a number of ways. Small, low-volume processors normally flatten cans and sell them to a nearby wholesaler. Larger operations will bale, densify, or shred cans for shipment to aluminum consumers (Figure 6.4). Aluminum manufacturers have established specific criteria as to how aluminum cans should be prepared. The baled or shredded aluminum is shipped by truck, railcar, or sea container to regional mills or reclamation plants.

At the reclamation plant the bales are unloaded and the cans are tested for quality and moisture content. After inspection the bales of cans are shredded to reduce volume. The shredded cans are conveyed to a delacquering oven to remove coatings and moisture. The hot shredded aluminum is then passed over a small screen to remove dirt and contaminants and fed directly into a reverberatory furnace. Heated to 1400°F (650°C), the cans melt and blend in with the molten metal already in the furnace. Alloying elements and primary aluminum are added as needed. A mixture of salt and potassium fluoride is added as a flux to separate any oxides ("dross") that are skimmed off (CMI, 2002).

Molten aluminum is analyzed for the appropriate chemical properties and then tapped (removed) from the furnace and poured into large molds that cast sheet ingots. These large rectangular ingots

TABLE 6.5
Metal disposal and recycling, 1999

Product Category	Generation (thousands of tons)	Recovery (thousands of tons)	Recovery (percent of generation)
Durable Goods			
Ferrous metals[a]	10,390	2,800	26.9
Aluminum[b]	960	Neg.	Neg.
Lead[c]	970	930	95.9
Other nonferrous metals[d]	420	Neg.	Neg.
Total Metals in Durable Goods	12,740	3,730	29.3
NonDurable Goods			
Aluminum	200	Neg.	Neg.
Containers and Packaging Steel			
Food and other cans	2,690	1,150	56.1
Other steel packaging	240	170	70.8
Total Steel Packaging	2,930	1,680	57.3
Aluminum			
Beer and soft drink cans	1,540	840	54.5
Food and other cans	50	Neg.	Neg.
Foil	380	30	7.9
Total Aluminum Packaging	1,970	870	44.2
Total Metals in Containers and Packaging	4,900	2,550	52.0
Total Metals	17,840	6,280	35.2
Ferrous	13,320	4,480	33.6
Aluminum	3,130	870	27.8
Other Nonferrous	1,390	930	66.9

Neg. = Less than 5000 tons or 0.05 %.

[a] Ferrous metals in appliances, furniture, tires, and miscellaneous durables.

[b] Aluminum in appliances, furniture, and miscellaneous durables.

[c] Lead in lead-acid batteries.

[d] Other nonferrous metals in appliances and miscellaneous durables.

Source: U.S. EPA, EPA 530-R-01-014, 2001.

(9100 to 18,200 kg or 20,000 to 40,000 lb each) are allowed to cool and harden. The surface of the sheet ingot is milled to a smooth surface in a process called "scalping." The scalped ingot is then passed between two giant steel rollers in a large mill. The sheet is passed through a few more times until it is about 1.25 cm (0.5 in.) thick and about 300 m (1000 ft) long. This long sheet is then annealed to soften it and passed to a series of rollers in a finishing mill where it acquires the necessary hardness and thickness. The edges are trimmed in a slitter and the sheet is rolled up for shipment to a can manufacturer (CMI, 2002).

The finished sheet may be 3 km (2 miles) long and made from over 1.2 million recycled cans. The sheets are shipped to container manufacturing plants and cut into discs that are ultimately formed into cans. The cans are printed with the company label or logo and are shipped, often with the tops separate, to the filling plant (Tchobanoglous et al., 1993).

FIGURE 6.4 Baled aluminum UBCs ready for shipping to a container manufacturer.

6.9.3 Specifications for Recovered Aluminum Cans

Collection centers and other buyers accept cans that are free of gross contamination such as dirt and food wastes. The buyers then compact and bale the material according to mill specifications regarding dimension and weight. Noncontainer aluminum products purchased by scrap dealers must simply be dry and free of contamination; the dealers collect and bale the material for shipment to users. The Institute of Scrap Recycling Industries has developed a set of standard specifications for a number of recycled commodities. Typical specifications for preparing aluminum beverage can scrap for sale to an aluminum recycling company are presented in Table 6.6 (ISRI, 2002).

There are stringent quality requirements at U.S. mills for aluminum scrap. Aluminum UBCs must be relatively clean and free from dirt, oil, grease, and other surface contaminants. Iron, aluminum foil, and other types of aluminum scrap are unacceptable if mixed in the bales. Any contamination with lead, copper, brass, and other nonferrous metal may result in immediate rejection. All flammables, paper, and plastic should be removed prior to baling and cans must be relatively dry. All incoming material is tested for moisture by the receiving mills. They generally have a threshold of 4% allowable moisture. The mills will accept loads with a higher moisture content but will deduct for any moisture over 2% as a penalty for wet loads (CMI, 2002).

6.9.4 Benefits of Aluminum Recycling

Aluminum manufacturers such as Reynolds and Alcoa have actively promoted recycling since the mid-1960s. The aluminum industry recognized the advantages of a domestic aluminum supply and therefore established the necessary infrastructure for transportation and processing. While other industries have resisted recycling programs and mandatory container deposit legislation, the aluminum industry has developed collection and processing centers, a transportation network, and reclamation plants. Recycling makes economic sense to manufacturers for several reasons:

- Recycling provides a stable, domestic source of aluminum. In contrast, most of the bauxite required to produce new aluminum must be imported, and 4 lb of bauxite is required to produce each pound of new metal.

TABLE 6.6
Specifications for Aluminum Beverage Can Scrap

Baled UBC	Used aluminum beverage cans, magnetically separated and free from all other types of material. Average bale dimensions 30 to 36 in. × 36 to 48 in. × 60 to 72 in. (75 to 90 cm × 90 to 120 cm × 150 to 180 cm). Density 14 to 30 lb/ft². Bales should be kept dry. Most mills will allow 4% maximum water content.
Densified UBC	Cans are compressed to a small block, approximately 12 to 16 in. × 16 to 24 in. by a variable thickness of 6 to 10 in. Density 30 and 50 lb/ft². All bricks should be of the same dimensions and fairly uniform in weight. The individual bricks have slots for banding and are stacked in a uniform fashion and strapped into a bundle of 2 to 3000 lb with 1/2 to 3/4 in. × 0.020 in. steel strapping. This package is only made by specific machines ("Densican") designed to produce these uniform can bricks. Aluminum must be free of steel, aluminum foil, paper, wood, oil and all types of non-UBC metals. Most mills will allow up to 4% water content
Bricked UBC	Cans compressed to a density of 45 to 70 lb/ft² in a high-compression press with two equal dimensions between 12 and 24 in. and one variable dimension up to 48 in. For shipment to U.S. consumers, briquettes should be stacked at least 4 ft high and strapped into bundles without the use of pallets or any support sheets or wrapping other than steel straps. It is important that the bricks be free of iron, dirt, and any other scrap since they are sometimes charged directly into a furnace.

Source: ISRI, no date. Reproduced with kind permission of the Institute for Scrap Recycling Industries, Inc., Washington, DC 20005-3104. The ISRI Scrap Specifications Circular is subject to change. To find the most recent edition, go to www.isri.org.

- Aluminum recycling is profitable and well established because it requires only 5% of the electric power to remelt aluminum as it does to extract it from bauxite ore (Rhyner et al., 1995; Liu and Liptak, 2000).
- Recycled cans are of uniform and known composition and impurities are readily removed.

6.9.5 ONE FINAL NOTE ON ALUMINUM

A recently released report indicates that in 2001 more aluminum cans were littered, landfilled, or incinerated than were recycled. According to Container Recycling Institute (2002), the 50.7 billion aluminum cans wasted last year squandered the energy value equivalent to 16 million barrels of crude oil, or enough energy to supply 2.7 million American homes with electricity for a year.

6.10 FERROUS METALS

Ferrous metals are those containing iron and are used in the manufacture of industrial and consumer goods. Industrial ferrous waste may include aged tanks and silos, obsolete machine tools, retired railway locomotives, dismantled bridges, entire ships, demolished steel-framed buildings, and worn-out motor vehicles. Consumer ferrous waste includes appliances ("white goods"), automobiles (each year about 10 million discarded), food, and nonfood containers.

6.10.1 STEEL MANUFACTURE

Five major groups of activities are involved in steel manufacture: coking, sintering, iron-making, steelmaking, and final rolling and finishing (Russell and Vaughan, 1976). Coking involves heating coal pyrolytically (i.e., in the absence of air) to produce a fuel high (approximately 90%) in carbon. Sintering agglomerates fine ore particles into a porous mass for charging into the blast furnace.

In the blast furnace, molten iron is produced. Workable iron ores tend to be rich in iron (III) oxide (Fe_2O_3). Recovering the iron by removing oxygen from the ore is an early processing step. Industrial iron production involves reducing iron (III) oxide in the blast furnace. Most of the iron (III) oxide is reduced using carbon monoxide as follows:

$$Fe_2O_3 + 3CO \quad \rightarrow \quad 2Fe(I) + 3CO_2 \tag{6.5}$$

Not all the iron (III) oxide is reduced by carbon monoxide, however. A fraction of the iron is reduced directly using carbon as the oxidizing agent:

$$Fe_2O_3 + 3C \quad \rightarrow \quad 2Fe(I) + 3CO \tag{6.6}$$

In the actual steelmaking process, iron is converted into steel by forcing oxygen through the molten metal from the furnace. This oxidizes the impurities in the molten metal.

In steelmaking there are three basic types of furnaces: the open hearth, the basic oxygen furnace, and the electric arc furnace. The open-hearth furnace has declined in popularity in recent years as it is relatively slow in preparing a batch of steel compared with the other furnace types. The primary feedstock in the basic oxygen furnace is molten pig iron, produced in a blast furnace from iron ore (hematite and magnetite), limestone, and coke. Molten pig iron can be combined with steel scrap. The electric arc furnace operates almost exclusively on steel scrap and is discussed below.

A schematic of the steelmaking process is shown in Figure 6.5.

6.10.2 FERROUS RECYCLING

Among all the materials recycled worldwide, iron and steel represent the greatest tonnages. Iron has been manufactured for thousands of years and scrap was recycled even in the earliest times of production. Today, the scrap recycling industry processes an average of 50 million tons of scrap iron and steel annually (ISRI, 1993b).

Recovery of ferrous metals from appliances ("white goods") was estimated at 2.3 million tons of the total ferrous in appliances in 1997. Overall recovery of ferrous metals from durable goods (large and small appliances, furniture, and tires) was estimated to be 29.3% (3.7 million tons) in 1999 (Table 6.5). Steel food cans and other cans were recovered at a rate of 56.1% (1.5 million tons). Approximately 170,000 tons of other steel packaging, mostly steel barrels and drums, was recovered for recycling in 1999 (U.S. EPA, 2001).

Processors buy ferrous scrap from a number of sources including municipalities, demolition operations, automobile dismantlers, shipyards, and industrial plants (Figure 6.6). Steel cans, also known as "tin cans" due to the presence of a corrosion-resistant tin coating, are recovered along with other consumer items at the curbside or an MRF. Cans are often commingled with nonferrous containers and must be separated magnetically. Afterwards they are compacted and shipped to a de-tinning facility. In the de-tinning plant, the cans are shredded and the material is again passed through a magnetic separator to remove aluminum, often from bi-metal cans and other nonferrous metals. The clean steel is then de-tinned either by heating in a kiln to volatilize the tin, or by reaction with sodium hydroxide and an oxidizing agent. Tin is recovered by electrolysis and formed into ingots (Tchobanoglous et al., 1993). This process allows for the production of both high-quality tin and steel. The chemically de-tinned steel is used for the production of new steel. Tin cans de-tinned by heating are not suitable in steelmaking, however, as the heat causes some of the tin to diffuse into the steel and occur as an impurity. In some applications impurities from tin will not interfere with the production of new steel, and the de-tinning process may be skipped altogether (Rhyner et al., 1995).

White goods are large, bulky appliances such as washing machines, refrigerators, freezers, and stoves (Figure 6.7). The annual discard rate is over 12 million tons (Table 6.5) (U.S. EPA, 2001). Appliances contain large amounts of ferrous along with copper and aluminum. Before baling or

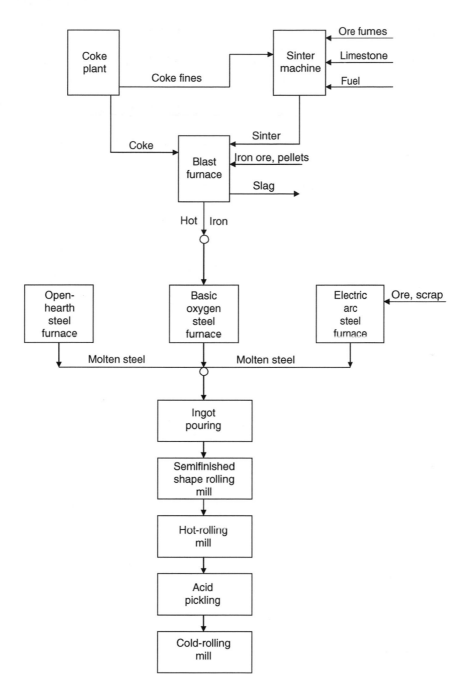

FIGURE 6.5 Schematic of the steelmaking process.

shredding, such appliances must be checked for the removal of potentially useful or hazardous materials. For example, equipment manufactured prior to 1979 may contain polychlorinated biphenyls (PCBs) within electrical capacitors; similarly, CFCs may be present in compressor units in refrigerators or freezers.

Old automobiles are a major source of ferrous scrap. About 75% of an average automobile can be recycled (Rhyner et al., 1995). Historically, most of the recyclable materials were ferrous; however, plastics are becoming more common in automobile manufacture. Prior to processing for

FIGURE 6.6 Metal processing facilities may accept ferrous waste from municipalities, demolition operations, industrial plants, and individual consumers.

FIGURE 6.7 White goods set aside for recycling.

ferrous, all hazardous materials (battery, refrigerants, used oil, and antifreeze) must be removed. Auto salvage operations remove the fuel tank, tires, windshields, radiators, and other items with potential resale value. The remains are placed into an industrial shredder which converts the automobile into small chunks (Figure 6.8) or compressed in a high-capacity compactor. These packages can be fed directly into an electric arc furnace (see below).

Larger ferrous scrap from industry is processed for reuse via cutting and baling. In dealing with industrial ferrous wastes, common machinery include a crane, either mounted or mobile, which

(a)

(b)

FIGURE 6.8 Automobile shredding operation. (Reproduced with kind permission of Sims Group Limited, Figure 6.8b [bottom] only.)

houses a large electromagnet (Figure 6.9); a baling press, used to densify objects such as automobiles; a hydraulic guillotine shear to slice steel I-beams and pipe; and a shredder (Figure 6.10). Several of these unit operations are discussed in more detail in the next chapter.

Industrial consumers purchase ferrous scrap directly or through a materials broker. These mills and foundries remelt the scrap and manufacture new products. Basic oxygen furnaces and electric arc furnaces utilize most of the iron and steel scrap. In the basic oxygen furnace, molten pig iron is combined with 20 to 30% steel scrap (Rhyner et al., 1995). The electric arc furnace operates almost exclusively on steel scrap. Melting is accomplished by supplying energy to the furnace interior. This energy can be electrical or chemical. Electrical energy is supplied via graphite electrodes and is usually the largest contributor in melting operations. The scrap falls

FIGURE 6.9 Electromagnet for moving ferrous wastes onto rail cars.

FIGURE 6.10 Shredder housing at a metal recovery facility.

into the furnace via a crane. The electrodes swing into place over the furnace. The roof is lowered and then the electrodes are lowered to strike an arc on the scrap. The arc consists of a plasma of hot, ionic gases reaching temperatures in excess of 3515°C (6000°F). Contact with the arc initiates the melting of the scrap. Once the charge is melted, the furnace sidewalls are exposed to intense radiation from the arc, melting the entire mass. Once the desired steel composition and temperature are achieved in the furnace, the tap hole is opened, the furnace is tilted, and the steel pours into a ladle for transfer to the next batch operation (usually a ladle furnace or ladle station). During the tapping process, alloy additions are made based on the bath analysis and the desired steel grade (Jones, 2002).

6.10.3 BENEFITS OF FERROUS RECYCLING

Using ferrous waste in place of iron ore to manufacture steel has many advantages beyond cost considerations. Recycled ferrous is nearly 100% metal and is often readily available or readily transported in bulk. In contrast, iron ore has to be mined and milled free from tailings and chemical impurities, and smelted in a blast furnace before it can be converted into steel. The use of ferrous scrap in comparison with ore imparts substantial energy savings, significantly reduces the amount of water needed for processing, and causes less air pollution.

Iron and steel can be processed and remelted repeatedly for the manufacture of industrial and consumer items with absolutely no diminution of quality. Steel made from scrap is chemically and metallurgically equivalent to steel manufactured from virgin ore (ISRI, 1993b).

6.11 PLASTICS

Plastics possess many properties that make them desirable, if not indispensable, for the modern consumer. These synthetic polymers are shatter-resistant, waterproof, lightweight, durable, and strong. As a result, plastics have replaced glass and a number of other materials in packaging, construction, and other uses. The United States is the largest consumer and producer of plastics in the world which is consistent with its large consumer-driven economy, low-cost chemical feedstocks, and a well-developed and organized petrochemical infrastructure.

Prior to 1970, plastics were generally not listed as a component of MSW. Data for 1970 show that plastics comprised about 2 to 3% of the waste stream. Today, plastics occupy 30% of landfill space although their weight percentage ranges between 5 and 9% (Table 6.7)(U.S. EPA, 2001; Liu and Liptak, 2000).

TABLE 6.7
Recovery of Plastics from the Waste Stream, 1999

Product	Generation (thousands of tons)	Recovery (thousands of tons)	Recovery (percent of total)
Total plastics in durable goods	7,180	270	3.8
Total plastics in nondurable goods	5,830	0	0
Total plastics in containers and packaging[a]	11,160	1,080	9.7
Total plastics in MSW, by resin			
PET	2,430	470	19.3
HDPE	5,340	490	9.2
PVC	1,450	Neg.	Neg.
LDPE/LLDPE	6,020	130	2.2
PP	3,060	120	3.9
PS	2,280	10	0.4
Other resins	3,590	130	3.6
Total Plastics in MSW[b]	24,170	1,350	5.6

HDPE=high-density polyethylene; PET=polyethylene terephthalate; PS=polystyrene; LDPE=low-density polyethylene; PP=polypropylene; PVC=polyvinyl chloride; LLDPE=linear low-density polyethylene; Neg.=Less than 5000 tons or 0.05 %.

[a] Plastic plates and cups, trash bags, plastics in disposable diapers, clothing, footwear, etc.

[b] Includes coatings, closures, caps, trays, shapes, etc.

Adapted from U.S. EPA, EPA 530-R-01-014, 2001.

6.11.1 PLASTICS MANUFACTURE

The raw materials for virtually all plastics are natural gas, petroleum, and liquified petroleum gases. Simple hydrocarbon monomers serve as the building blocks for conventional plastics. These monomers are linked together to form long chains of repeating molecules called polymers. In the simplest case, gaseous ethylene monomers ($-CH_2-$) are concatenated to produce a solid polymer measuring tens of thousands of carbons in length (Figure 6.11). Hundreds of these high-molecular-weight polymers are in use for plastics manufacture. Each polymer possesses unique properties such that it will meet the requirements of industry and the consumer. About 80% of plastic used in consumer products is either polyethylene terepthalate (PET), also known as #1 plastic, or high-density polyethylene (HDPE) or # 2 plastic. As we shall see, these are the most commonly recycled polymers as well. The plastics numbering system appears in Figure 6.12.

There are two main categories of synthetic polymers, thermoplastics and thermosets. A thermoplastic is one that consists of individual (nonlinked) chains of the polymer. They can be melted and reformed into the same polymer repeatedly. In contrast, thermosets consist of polymer chains linked to each other by cross-bonding (Figure 6.13). Once a product made of thermoset plastic is melted, it cannot be reformed. Thermoplastics make up about 90% of all plastic products.

The main manufacturing processes used to transform newly formed polymers into a useful form are extrusion, blow molding, and injection molding. Most of these processes begin with plastic resins as pellets. These are subsequently subjected to heat and pressure and melted before processing.

6.11.2 EXTRUSION

Extrusion molding is employed to convert plastics into continuous sheeting, film, tubes, rods, and filaments, and to coat wire and cable. In extrusion, dry plastic beads are loaded into a hopper and then fed into a long heating chamber through which it is moved by the action of a continuously revolving screw. At the end of the heating chamber, the molten plastic is forced out through a small

A radical initiator

$$In\bullet + CH_2 = CH_2 \longrightarrow [In-CH_2CH_2 \bullet] \xrightarrow{CH_2 = CH_2} [In-CH_2CH_2CH_2CH_2\bullet]$$

Repeat

Polyethylene chain

FIGURE 6.11 Ethylene monomers joining to form a polyethylene polymer. (From McMurry, J., *Organic Chemistry*, 3rd ed., 1992. Reproduced with kind permission of Brooks/Cole, a division of Thomson Learning: www.thomsonrights.com.)

| 1 | 2 | 3 | 4 | 5 | 6 | 7 |
| PETE | HDPE | PVC | LDPE | PP | PS | Other |

FIGURE 6.12 Plastics numbering system.

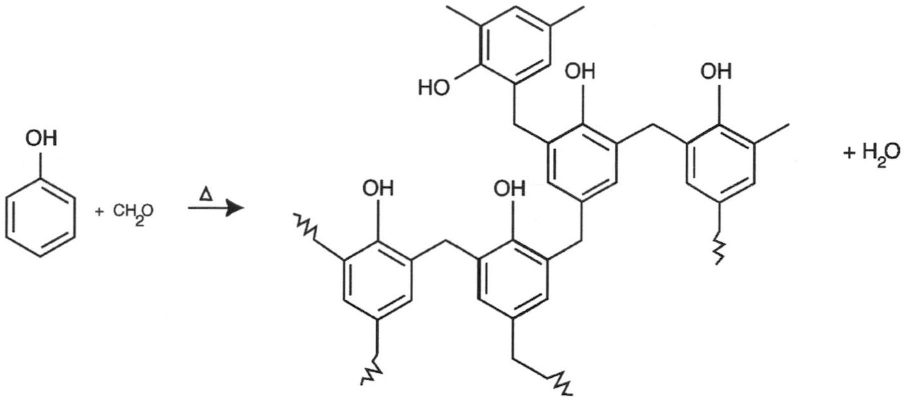

FIGURE 6.13 Thermoset polymer showing cross-links. (From McMurry, J., *Organic Chemistry*, 3rd ed., 1992. Reproduced with kind permission of Brooks/Cole, a division of Thomson Learning: www.thomsonrights.com.)

FIGURE 6.14 Plastics extruder.

opening or die with the desired shape for the finished product (Figure 6.14). As the working piece is removed from the die it is fed onto a conveyor belt where it is cooled, typically by blowers or by immersion in water. In the production of wide film or sheeting, the plastic is extruded in the form of a tube. This tube may be split as it comes from the die and then stretched and thinned to the dimensions desired in the finished film (SPI, 1999). Extruded products include plastic pipe and plastic lumber (Rhyner et al., 1995).

6.11.3 BLOW MOLDING

Blow molding is a method of forming hollow molten tubes out of thermoplastic materials, then with the use of compressed air, blowing up the tube to conform to the interior of a chilled blow mold (SPI, 1999).

6.11.4 Injection Molding

In injection molding (Figure 6.15), plastic is placed into a hopper which feeds a heated injection unit. A reciprocating screw pushes the plastic through this long heating chamber, where the material is softened to a fluid state. At the end of this chamber a nozzle abuts firmly against an opening into a cool, closed mold. The fluid plastic is forced at high pressure through this nozzle into the mold. A system of clamps holds the mold halves shut. As soon as the plastic cools to a solid state, the mold opens and the finished plastic is ejected from the press (SPI, 1999). Food tubs used for yogurt and cottage cheese are made by injection molding.

6.11.5 Compression Molding

Compression molding is simply the squeezing of a material into a desired shape by the application of heat and pressure to the material in a mold. Compression molding is used for forming thermoset materials but not for thermoplastics. Plastic molding powder, mixed with polymer feedstock and fillers such as cellulose to strengthen or impart other qualities to the finished product, is placed directly into the open mold cavity. The mold is then closed, pressing down on the plastic and causing it to flow throughout the mold. While the heated mold is closed, the thermosetting material undergoes a chemical change that permanently hardens it into the shape of the mold (SPI, 1999).

Other methods of plastics formation include thermoforming, transfer molding, and reaction injection molding (SPI, 1999).

6.11.6 Plastics Recycling

The American Plastics Council estimated that about one half of all U.S. communities, nearly 19,400, were collecting plastics for recycling (Testin and Vergano, 1992). PET and HDPE were the primary recycled polymers. About 7400 communities collect plastics at the curb and another 12,000 collect plastics at drop-off centers. In addition, thousands of grocery stores in the United States accept plastic bags for recycling into new trash can liners and other products.

A wide range of consumer products can be manufactured from recovered plastics. Some are listed in Table 6.8. The U.S. Food and Drug Administration (FDA) regulates the use of recycled

FIGURE 6.15 Injection molding apparatus.

TABLE 6.8
Other Products Manufactured from Recycled Plastics

Impact barriers
Docks, decks
Fences
Scuff boards, floor boards
Boundary markers, right of way markers
Signposts
Benches and picnic tables
Fiberfill for sleeping bags
Plastic lumber
Flowerpots
Containers for nonfood products
Mats
Strapping
Scouring pads
Toys
Compost bins
Recycling containers

resins in plastic food containers. The FDA has allowed for only limited use of recycled polymers in food container manufacture due to concerns about possible food contamination. Recycled containers have been used in soda bottles, tubs for butter, and detergent bottles. Recycled resins in nonfood containers continue to increase in popularity (APC, 2002).

Plastic lumber has become popular over the past decade. Plastic lumber is made by extrusion and contains either single or mixed resins. Such lumber possesses physical characteristics similar to those of standard wood lumber. The advantages of plastic lumber are that it is resistant to the elements, water, and insect damage. Less maintenance (painting, etc.) is required. However, plastic lumber is relatively costly compared with its natural counterpart. There are also some concerns that plastic timbers may bend slightly with time.

6.11.7 PROCESSING FOR RECYCLING

Postconsumer plastics can replace or supplement virgin plastic resins. Plastics recycling is difficult because each type of plastic must undergo a different process before becoming a new product. There is a major concern with contamination as well. Plastics are typically segregated by resin type and ideally by production method (e.g., extrusion vs. injection). Different resins possess differing physical characteristics including melting points, tensile strength, shatter resistance, and so on. Buyers often require that plastics are color-separated with no contamination (Rhyner et al., 1995).

There are seven major types of plastics, created by a voluntary labeling system to encourage recycling. Types are indicated by a recycling logo (the three chasing arrows) with a number from 1 to 7 situated in the center (Figure 6.12). As noted above, polyethylene PET and HDPE are the predominantly recycled polymers. Postconsumer items made from PET and HDPE resins have developed stable markets in the United States and Asia.

Postconsumer plastics are recovered from collection centers loose in wire mesh cages but more typically baled to reduce volume. After breaking bales, containers are deposited along a conveyor belt for final sorting. Undesired plastics and extraneous wastes are removed manually. Plastics are also sorted by color.

Plastics can be recycled in several ways. In HDPE recycling, containers are chipped to small flakes by a granulator designed to cut chips without causing excessive heat that might fuse the particles. The flakes measure about 1 cm (3/8 in.) across. The flakes are washed with hot water and detergents to remove labels, adhesives and dirt, and floated to remove any heavy contaminants. The HDPE is placed into a spin dryer to remove free water. Flakes are dried with hot air, reducing moisture content to about 0.5% (Tchobanoglous et al., 1993). The dried flakes may be sold in that form. More sophisticated plants reheat the flakes, add pigment, and pass them through a pelletizer, which produces small plastic beads that are used in injection molding presses to create new products (Figure 6.16)(CMI, 2002).

Resin may also be fluidized using an extruder. Flakes are fed into the extruder and compressed as they are forced forward toward the die. The combined heat from flow friction and supplemental heating causes the resin to melt. Volatile contaminants are vented from the mixture. The melted resin mixture may pass through a fine screen to remove any remaining solid impurities (Tchobanoglous et al., 1993).

PET is a form of polyester that is extremely tough and versatile. Soft drink and water bottles are made from this resin as are many plastic jars and "clamshell" packages (e.g., salad containers). Recycling PET is similar to that for HDPE. Bottles may be color-sorted and are then ground and washed. Unlike polyethylene, however, PET sinks in the wash water while the plastic caps and labels float off. The clean chips are dried and pelletized. PET bottles may contain aluminum caps, and granulated aluminum may contaminate PET chips. Electrostatic precipitation is used to remove the aluminum (Tchobanoglous et al., 1993). Recycled PET has many uses and there are well-established markets for this resin. The largest usage of recovered PET is in textiles. Carpet companies often use 100% recycled resin to manufacture polyester carpets in a variety of colors and textures. PET is also spun into fine filaments to make fiber filling for pillows and jackets. A substantial quantity of recycled PET returns to the bottle market (CMI, 2002).

Regardless of the reassuring numbers of collection programs, however, the overall recovery of plastics for recycling is quite small, totaling 1.3 million tons or 5.6% of plastics generation in 1999 (Table 6.7). However, recovery of some plastic containers has generally increased. PET soft drink

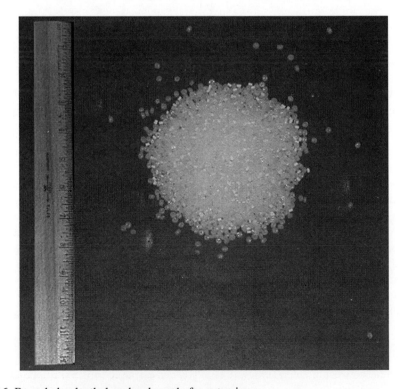

FIGURE 6.16 Recycled polyethylene beads ready for extrusion.

bottles were recovered at a rate of 22.8% in 1999. Milk and water bottles (HDPE) were recovered at an estimated 23.8% in 1999. Significant recovery of plastics from lead-acid battery casings and other containers was also reported (APC, 2002). According to the U.S. EPA (1999), however, the plastics industry recycled 5.2% of postconsumer polymers in 1997 and that rate was expected to increase marginally at best over the next few years. The same report ranks plastics recycling at the bottom of all recycled materials listed. The plastics industry has recently launched an intensive public relations campaign and research program to enhance the image of plastic as an easily recycled material.

6.12 YARD WASTE

The U.S. EPA (1999) defines yard waste as grass, leaves, and tree and brush trimmings from residential, institutional, and commercial sources. There are limited data on the composition of yard wastes; however, it is estimated that the average composition is about 50% grass, 25% leaves, and 25% brush on a weight basis. These numbers will vary as a function of climate, region of the country, and season of the year.

Due to the huge volumes of yard wastes produced, along with concerns over diminishing landfill space, many states have enacted legislation to divert these wastes away from landfills. By 1998, 22 states and the District of Columbia, representing more than 50% of the U.S. population, had enacted legislation banning or discouraging yard waste disposal in landfills. Such legislation has led to an increase in the use of mulching lawnmowers and backyard composting. Many municipalities have also established compost programs near waste transfer stations or landfills.

About 3500 composting facilities for yard waste exist in the United States (U.S. EPA, 2001). These vary in size and sophistication as well as in terms of quality and quantity of finished product.

Based on sampling studies at landfills and transfer stations, a total of 11.5 million tons of yard wastes were recovered for composting in 1997, the second largest fraction of total recovery of the waste stream. The percentage of yard waste composted (41%) has more than doubled since 1992. This is due to increased numbers of yard waste composting facilities, more material being handled at facilities, and bans of yard waste from landfills by states (U.S. EPA, 2001). Within the past few years, however, composting has increased by lesser amounts, suggesting that much of the impact of the states' bans of yard waste from landfills has taken place.

Details of the composting process are discussed in Chapter 8.

6.13 FOOD WASTE

Food wastes include uneaten food and food preparation waste from residences, commercial establishments (restaurants, etc.), institutional sources (school cafeterias, hospital cafeterias), and industrial sources (factory lunchrooms). Food waste generated during the preparation and packaging of food products is not included in the U.S. EPA estimates for food waste. Food waste generation from residential and commercial sources was estimated using data from sampling studies from parts of the country combined with demographic data, grocery store sales, and restaurant sales. The estimated food waste production was 24.6 million tons in 1999 (U.S. EPA, 2001).

As mentioned in Chapter 2, a substantial portion of food wastes generated during the 1940s and 1950s was fed to hogs. Today, however, "recycling" of food waste primarily refers to its incorporation into the composting process for later use as a soil conditioner or landscaping material. The U.S. EPA (1999a) estimates that approximately 285,000 tons of food wastes are recycled (i.e., composted) annually.

6.14 TIRES AND RUBBER

It is estimated that between 2 and 3 billion tires have been disposed in the United States alone, and another 270 million tires (weighing 3.4 million tons) are added to the waste stream every year

TABLE 6.9
Rubber and Leather in MSW and Recycled, 1999

Product	Generation (thousands of tons)	Recovery (thousands of tons)	Recovery (percent of generation)
Durable goods			
Rubber in tires[a]	2,980	7902	6.5
Other durables[b]	2,430	Neg.	Neg.
Total rubber and leather in durable goods	5,410	790	14.6
Nondurable goods			
Clothing and footwear	540	Neg.	Neg.
Other nondurables	250	Neg.	Neg.
Containers and packing	790	Neg.	Neg.
Total rubber and leather	6,220	790	12.7

Neg.=Less than 5000 tons or 0.05 %. Details may not add to totals due to rounding.

[a] Automobile and truck tires. Does not include other materials in tires.

[b] Includes carpets and rugs and other miscellaneous durables.

Adapted from U.S. EPA, EPA 530-R-01-014, 2001.

(Table 6.9) (U.S. EPA, 2002). This number does not include over 30 million tires that are retreaded every year. Until recently, waste tires were simply stockpiled (Figure 6.17), landfilled, or burned. Open, uncontrolled tire fires have resulted in the production of many noxious and hazardous air pollutants and such fires are difficult to extinguish, in some cases lasting months or years. Tires also serve as a breeding ground for insects such as mosquitoes and other pests.

6.14.1 DESIGN AND MANUFACTURE

Tires are constructed from one of two distinct designs, i.e., nonbelted and steel-belted. The latter type dominates the tire market by virtue of its greatly improved lifespan as well as improved gas mileage. A longer lifespan results in less tires ending up in landfills. Unfortunately, however, steel-belted tires are more difficult to recycle, and comprise about 90% of all used tires in the waste stream. Tables 6.10 and 6.11 list the typical types of materials used in tire manufacture.

6.14.2 DISPOSAL AND RECYCLING

Many states have targeted the tire dumping problem by restricting land disposal of tires, setting up recycling programs, and assisting in the development of markets for collected scrap tires. Bans on disposing scrap tires in landfills are in effect in 33 states, and over 30 states collect disposal fees on tires to fund disposal and management and, in some cases, to support research and market development for tire recycling (U.S. EPA, 2002). The fate of scrap tires is outlined in Figure 6.18. The majority is land-disposed (landfilled and stockpiled).

Disposal of waste tires by sanitary landfilling causes problems for operators, such as tires rising to the surface due to their low density. Such rising may eventually damage the integrity of a landfill liner. Some landfill operators cut tires to prevent such "floating" behavior. Others have shredded tires for use as a daily landfill cell cover material. Landfill disposal of tires is generally considered wasteful for the following reasons (Rhyner et al., 1995):

- Tires are relatively inert and may not necessarily need land disposal to limit any hazards.
- Tires have potential value as a recovered material.
- Tires have a potentially high economic value as a fuel.

FIGURE 6.17 Illegal tire dumps pose hazards from fires and insect breeding, and are unsightly.

TABLE 6.10
Typical Chemical Composition of a Tire

Synthetic rubber
Natural rubber
Sulfur and sulfur compounds
Silica
Phenolic resin
Oil: aromatic, naphthenic, paraffinic
Fabric: polyester, nylon, etc.
Petroleum waxes
Pigments: zinc oxide, titanium dioxide, etc.
Carbon black
Fatty acids
Inert materials
Steel wire

Source: Rubber Manufacturers Association, no date. Reproduced with kind permission of the Rubber Manufacturers Association.

Since 1996, the use of scrap tire monofills (i.e., a landfill dedicated to one type of material) has become more prominent in some locations as a means to manage scrap tires. In some cases, monofills are used where there are no other markets available and were MSW landfills are not accepting tires. In other cases, monofills are portrayed as a management system that allows long-term storage of scrap tires without the problems associated with above ground storage. In theory, monofilled processed scrap tires can be "harvested" when markets for scrap tire material improve. Using monofills for scrap tires is preferable to above ground storage in piles, especially if a pile is not well managed (Scrap Tire Management Council, 1999).

Markets for waste tires recovered an estimated 66% (177.5 million out of 270 million scrap tires) in 1999 (Table 6.12). The conversion of scrap tires into fuel increases every year, and is currently the largest single use of scrap tires (Scrap Tire Management Council, 1999). The use of tires as a fuel material is discussed in Chapter 9. Overall 12.7% of rubber and leather in MSW was recovered in 1999 (U.S. EPA, 2001).

TABLE 6.11
Composition (by wt) of Passenger and Truck Tires

Material	Passenger Tire (%)	Truck Tire (%)
Natural rubber	14	27
Synthetic rubber	27	14
Carbon black	28	28
Steel	14–15	14–15
Fabric, fillers, accelerators, antiozonants, etc.	16–17	16–17
Average weight (lb)		
New	25	120
Scrap	20	100

Source: Rubber Manufacturers Association, no date. Reproduced with kind permission of the Rubber Manufacturers Association.

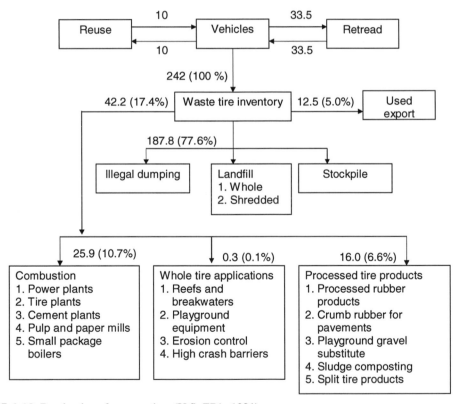

FIGURE 6.18 Destinations for scrap tires (U.S. EPA, 1991).

6.14.3 RECYCLING

Over the past decades many innovative uses have been found for recycled tires. For example, ground rubber from scrap tires is recycled into rubber products such as rubber-modified asphalt, playground cover, and flooring material. Tire material has also been employed as an alternative to pea stone in septic systems. Some facilities use scrap tires as a combustion fuel.

TABLE 6.12
Estimated Total Scrap Tire Market for 1998

Market	Quantity (millions of tires)	
Fuel		114
Cement kilns	38	
Pulp or paper mills	20	
Dedicated tires to energy	16	
Electric utilities	25	
Industrial boilers	15	
Civil engineering		20
Products		23
Ground rubber	15	
Cut, punched or stamped	8	
Miscellaneous and agriculture		5.5
Export		15
Subtotal		117.5
Total generation of waste tires		270

Source: Scrap Tire Management Council, 1999. Reproduced with kind permission of the Scrap Tire Management Council.

6.14.4 CRUMB RUBBER

For several recycling processes, tires are shredded to a fine particle size (about 5×5 cm or 2×2 in.) for eventual processing. The steel is removed magnetically and the particles are often shredded a second time to produce crumb rubber. The remaining rubber is treated to restore its ability to bond with other materials. Recovered rubber can then be combined with virgin rubber or other materials to produce a quality product.

More than 227 million kg (500 million lb) of crumb rubber is used in North America in one year and the rubber manufacturing industry used more than half this amount. Uses for crumb rubber include fillers in rubber compounds and asphalt modifiers, for example athletic tracks. Rubber adds flexibility to the surface, better traction, and increases the lifetime of the material (Rhyner et al., 1995; Kohrell, 1993). However, there are many other applications for recovered crumb rubber (Table 6.13). Manufacturers can achieve substantial savings in material costs. Improved mixing and curing properties are additional benefits that result from the use of crumb rubber.

6.14.5 RETREADED TIRES

Approximately 26.2 million retreaded tires were sold in the United States and Canada in 2000, with the following breakdown (Tire Retread Information Bureau):

- 6.3 million light truck tires
- 18.2 million medium-and heavy-duty truck tires
- 750,000 other tires (aircraft, off-road vehicles, industrial or lift trucks, motorcycles, farm equipment, and specialty uses)
- 1.5 million passenger car tires

Table 6.14 lists various industries using retreaded tires.

U.S. and Canadian retread tire industries used approximately 260 million kg (575 million lb) of tread rubber in 2000. There are approximately 1200 retreading plants in North America, a large

TABLE 6.13
Applications for Recycled Crumb Rubber

Construction and Equipment
Adhesives and sealants
Bin liners
Carpet underlay
Conveyer skirt boarding
Custom molded goods
Dams, silos, ponds, roof liners, and covers
Floor mats
Floor tiles
Foundation waterproofing
Gaskets
Hospital, industrial, and bathroom flooring
Insulation
Livestock stable mats
Nonskid surfaces
Paint
Patio bricks
Raised flooring
Roof shingles
Vibration dampers
Waterproofing compounds for roofs and walls

Geotechnical and Asphalt Applications
Drainage pipes
Fill materials for highway embankments
Porous irrigation pipes
Railroad crossings
Road building and repair
Roadway crack and joiner sealants
Rubberized asphalt for roads and driveways
Soil conditioner and ground cover
Subbase for horse-racing tracks
Subsoil drainage
Traffic cone bases
Traffic and people barricades

Automotive Industry
Belts
Brake disk pads
Brake linings
Bumpers
Car body underseal and rustproofing materials
Floor liners for trucks and vans
Floor mats for cars and trucks
Seals
Shock absorbers
Splash guards and mud guards
Tires and tire inner liners

Continued

TABLE 6.13 *Continued*

Athletic Surfaces
  Running tracks
  Golf tee-off areas
  Kindergarten playgrounds and recreation areas
  Nonslip boat dock surfaces
  School sports areas
  Swimming pool borders
  Walkways and garden paths
  Tennis and basketball courts

Adapted from: Rubberecycle, 2001. Reproduced with kind permission of Rubberecycle.

TABLE 6.14
Industries using Retreaded Tires

Eighty percent of the tires used by the commercial aviation industry are retreaded tires
Nearly 100% of off-road, heavy-duty vehicles
School buses and municipal vehicles
Trucking fleets and overnight delivery vehicles
Taxi fleets, race cars, and industrial vehicles
Fire trucks and other emergency vehicles
Farm tractors and other agricultural equipment
Millions of passenger cars
Federal and military vehicles, including those operated by the U. S. Postal Service, use
  retreaded passenger, truck, and aircraft tires

Source: Tire Retread Information Bureau, 2002. Reproduced with kind permission.

---

percentage of which are owned and operated by independent small businesses. The remaining plants are owned and operated by new tire manufacturers and a major tread rubber supplier (TRIB, 2002).

Approximately 70% of the cost of a new tire is in the tire body. Retreaded tires can be driven at the same legal speeds as comparable new tires with no loss in safety or performance. Retreaded truck tires are manufactured according to rigorous industry-recommended practices. Commercial aircraft retreads are approved by the Federal Aviation Administration. Retreaded passenger car tires are manufactured according to federal safety standards developed by the U.S. Department of Transportation.

Steel-belted radials are routinely retreaded and are available with all types of tread patterns. Retreading greatly reduces solid waste disposal problems and conserves hundreds of millions of barrels of petroleum every year. Truck tires can often be retreaded several times (TRIB, 2002).

Despite the fact that alternative uses for scrap tires exist, it is estimated that roughly 500 million scrap tires were lying in stockpiles as of 1998 (U.S. EPA, 1999b). To further alleviate the scrap tire problem, more actions must be taken up the product chain. Beyond recycling, work is in progress by some tire manufacturers to increase the recycled content of new tires they manufacture to reduce the use of virgin materials and, at the same time, provide a significant end market for scrap tires. Manufacturers also strive to design tires with increased durability, thus prolonging the useful life of tires. Lastly, reuse of scrap tires via retreading gives tires a new useful life (EPA, 2002).

6.15 GOALS FOR THE NATION

The U.S. EPA goal for the United States was to recycle at least 35% of MSW by the year 2005 (compared with 1990 baseline numbers), while reducing the generation of solid waste to 1.95 kg (4.3 lb) per person per day. As shown so far, the United States recycles 28.2% and per capita generation is 2.1 kg (4.62 lb) per person per day. Because economic growth results in the generation of more products and materials, there will be an increased need to further develop the U.S. recycling and composting infrastructure, buy more recycled products, and invest in source reduction activities—such as the reuse of materials and products, and "lightweighting" of products and packaging —in order to meet these goals.

REFERENCES

American Forest and Paper Association, Recovered Paper Statistical Highlights, 2002. www.afandpa.org/recycling/rec-introduction.html.

American Paper Institute, Paper Recycling and its Role in Solid Waste Management, New York, NY, 1990.

American Plastics Council, Recycling Facts from The American Plastics Council, 2002. http://www.plastics-resource.com/recycling/recycling_backgrounder/bk_1998.html.

California Integrated Waste Management Board (CIWMB), Effects of Waste Tires, Waste Tire Facilities, and Waste Tire Projects on the Environment, Sacramento, CA, 1996.

CC Technologies Systems, Inc. No date. Corrosion Cost. See: http://www.corrosioncost.com/pdf/pulppaper.pdf

Connecticut Metal Industries, 2002. Refer to http://www.ctmetal.com/glass.htm.

Container Recycling Institute, Trashed Cans: The Global Environmental Impacts of Aluminum Can Wasting in America, Arlington, VA, 2002.

Ferguson, L., Can Deinked Pulp Dare to Compete with Virgin Pulp? 6th Research Forum on Recycling, Magog, QC, Oct. 2001, p. 181.

Glass Packaging Institute, No date, Glass Handling and Recycling. See: http://www.gpi.org/Handling.html.

Institute of Scrap Recycling Industries, Recycling Paper, Washington, DC, 1993a.

Institute of Scrap Recycling Industries, Recycling Scrap Iron and Steel, Washington, DC, 1993b.

Institute of Scrap Recycling Industries, Glass Recycling: It Just Keeps Going On and On, 2001. See: http://www.isri.org/industryinfo/glass.htm.

Institute of Scrap Recycling Industries, Scrap Specifications Circular 2002. Guidelines for Nonferrous Scrap, Ferrous Scrap, Glass Cullet, Paper Stock, Plastic Scrap, SPECS 2002-1, Washington, DC, 2002.

Jones, J.D., Electric arc Furnace Steelmaking, American Iron and Steel Institute, Steelworks, 2002. http://steel.org/learning/howmade/eaf.htm.

Kohrell. M.G., Business Opportunities in Wisconsin's Postconsumer Waste Stream, University of Wisconsin-Green Bay, WI, 1993.

Liu, D.H.F. and Liptak, B.G., Hazardous Waste and Solid Waste, Lewis Publishing, Boca Raton, FL, 2000.

McMurry, J., Organic Chemistry, 3rd ed., Brooks/Cole, Pacific Grove, CA, 1992.

Newspaper Association of America (NAA), Newsprint Recovery Continues to Climb, 2002. http://www.naa.org.

Paper Technology, The Recovered Paper Market — Global Influences, April 2001, 2001, p. 16.

Pfeffer, J.T., Solid Waste Management Engineering, Prentice-Hall, Englewood Cliffs, NJ, 1992.

Pilkington, The Chemistry of Glass, St. Helens, UK, 2003. See: http://www.pilkington.com/corporate/english/education/chemistry/default.htm

Rhyner, C.R., Schwartz, L.J., Wenger, R.B., and Kohrell, M.G., Waste Management and Resource Recovery, Lewis Publishers, Boca Raton, FL, 1995.

Roberts, J.C., The Chemistry of Paper, Royal Society of Chemistry, Cambridge, UK, 1996.

Rubber Manufacturers Association, no date, Scrap Tire Characteristics. See: http://www.rma.org/scraptires/characteristics.html#anchor135840

Rubberecycle, 2001. http://www.rubberecycle.com.

Russell, C.S. and Vaughan, W.J., Steel Production: Processes, Products, and Residuals, John Hopkins University Press. Baltimore, MD, 1976.

Schwarz, S.C. and Brunner, C.R., Energy and Resource Recovery from Waste, Noyes Data Corp, Park Ridge, NJ, 1983.

Scrap Tire Management Council, Scrap Tire Use/ Disposal Study: 1998–1999 Update, September 1999. See: http://www.rma.org/scrap_tires/scrap_tire_markets/tire_disposal_study_exec_summary.pdf

Sims Group Ltd., No date, Cadillac Faces Shredder Jaws. See: http://www.unitednotions.com.au/prjob_7.html

Smook, G.A., Handbook for Pulp and Paper Technologists, Canadian Pulp and Paper Association, Montreal, Quebec, Canada, 1982.

Society for the Plastics Industry, Processing Methods, 1999. http://www.socplas.org.

Tchobanoglous, G., Theisen, H., and Vigil, S., *Integrated Solid Waste Management: Engineering Principles and Management Issues*, McGraw-Hill, New York, NY, 1993.

Testin, R.F. and Vergano, P.J., Plastic Packaging Opportunities and Challenges, Society of the Plastics Industry, Inc., Washington, DC, 1992.

Tire Retread Information Bureau, 2002. http://www.retread.org

U.S. Environmental Protection Agency, Markets for Scrap Tires, EPA/530-SW-90-074A, U.S. Government Printing Office, Washington, DC, 1991.

U.S. Environmental Protection Agency, Air Emissions from Scrap Tire Combustion, EPA-600/R-97-115, Office of Research and Development, Washington, DC, 1997.

U.S. Environmental Protection Agency, Characterization of Municipal Solid Waste in the United States: 1998 Update, EPA 530, Office of Solid Waste and Emergency Response, Washington, DC, 1999.

U.S. Environmental Protection Agency, Municipal Solid Waste in the United States: 1999 Facts and Figures, EPA 530-R-01-014, Office of Solid Waste and Emergency Response, Washington, DC, 2001.

U.S. Environmental Protection Agency, Tires. Product Stewardship, 2002. See: http://www.epa.gov/epr/products/tires.html.

World-Aluminum, Alumina Refining, 2000. See: http://www.worldaluminium.org/production/refining/index.html

SUGGESTED READINGS

Aadland D.M. and Caplan, A.J., Household valuation of curbside recycling. *J. Environ. Plann. Manag.*, 42, 781–799, 2000.

Ackerman F. and Mirza, S., Waste in the inner city: asset or assault? *Local Environ.*, 6, 113–120, 2001.

Benhart J.E. and McCartney, D., Solid waste management in southcentral Pennsylvania, *Pennsylvania Geographer*, 22, 1–7, 2000.

Creason J. and Podolsky, M.J., Economic impacts of municipal recycling, *Rev. Regional Stud.*, 31, 149–164, 2002.

Derksen L. and Gartrell, J., The social context of recycling, *Am. Sociol. Rev.*, 58, 434–442, 2000.

Everett J.W. and Peirce, J.J., Curbside recycling in the USA: convenience and mandatory participation, *Waste Manag. Res.*, 11, 49–61, 2000a.

Everett J.W. and Peirce, J.J., Measuring the success of recycling programs, *Resour. Conserv. Recycling*, 6, 355–370, 2000b.

Fleschner, E., Crombie, G., and Moreau, T., A city shifts into full-scale recycling, *BioCycle*, 33, 38–42, 2000.

Gamba R.J. and Oskamp, S., Factors influencing community residents' participation in commingled curbside recycling programs, *Environ. Behav.*, 26, 587–612, 2000.

Hadjilambrinos, C., The USA plastics recycling industry: A survey of manufacturers and vendors of recycled plastic products, *Environ. Conserv.*, 26, 125–135, 2000.

Highfill, J., McAsey, M., and Weinstein, R., Optimality of recycling and the location of a recycling center, *J. Regional Sci.*, 34, 583–597, 2000.

Hyde, J., Multifamily communities, multiple approaches, *BioCycle*, 32, 50–53, 2000.

Lave L.B., Hendrickson, C.T., Conway-Schempf, N.M., and McMichael, F.C., Municipal solid waste recycling issues, *J. Environ. Eng.*, 125, 944–949, 2000.

Margai, F.L., 2000. Integrating waste recovery and reduction systems in low-income urban communities, Research in Contemporary and Applied Geography: A Discussion Series, State University of New York at Binghamton, Vd. 19, 2000, p. 13.

Newenhouse, S.C. and Schmit, J.T., Qualitative methods add value to waste characterization studies, *Waste Manag. Res.*, 18, 105–114, 2000.

Nyamwange, M., Public perception of strategies for increasing participation in recycling programs, *J. Environ. Educ.*, 27, 19–22, 1999.

Oskamp, S., Resource conservation and recycling: behavior and policy, *J. Social Issues*, 51, 157–177, 2000.

RecycleHawaii, Recycling Successes on the Big Island: Mauna Kea Resort & Hawai'i Volcanoes National
    Park, Oct 2003. See: http://www.recyclehawaii.org/sept99.htm
Salhofer, S. and Isaac, N.A., Importance of public relations in recycling strategies: principles and case studies,
    Environ. Manage., 30, 68–76, 2003.
Subramanian, P.M., Plastics recycling and waste management in the US. Resources, *Conserv.* Recycling, 28,
    253–263, 2000.
Tilman, C. and Sandhu, R., A model recycling program for Alabama, *Resourc. Conserv.* Recycling, 24,
    183–190, 1999.
U.S. Environmental Protection Agency, Municipal Solid Waste: Recycling, Oct. 2003. See: http://www.
    epa.gov/epaoswer/non-hw/muncpl/recycle.htm.

QUESTIONS

1. A material is not truly considered "recycled" until it has proceeded through several distinct steps, and is ultimately purchased by the consumer. True or false? Justify your answer. List and discuss the steps.
2. Why have many community recycling programs failed over the past decade? How could they have been planned, operated and financed to have been more successful?
3. What are the primary approaches to recycling MSW? Which is superior in terms of producing a clean, quality product? Which method is often preferred on account of its convenience to the consumer? How do total costs differ between the different approaches?
4. How do source reduction, reuse, and recycling differ?
5. Why is price volatility in recycling markets a "given" in the industry?
6. How does purity affect the quality of a recycled product? What can be done (by the consumer, the municipality) in order to improve overall purity?
7. List the three major methods of pulping virgin fiber. How might each method affect the quality of recycled paper?
8. What is the major limitation on the amount of waste paper that can be recycled in a given year?
9. How many lifetimes do office paper or newsprint have before it can no longer be effectively recycled? Aluminum? Steel? Glass?
10. List some of the alternative recycling markets for paper, glass, and plastics.
11. Why does aluminum container manufacture from UBCs save substantially more energy and produce less pollution than using raw materials? Be specific.
12. What are the primary contamination concerns with recycled glass? Recycled aluminum?
13. "Due to the positive net value of scrap aluminum, there is virtually no waste of aluminum containers in the United States." True or false? Discuss.
14. List the main manufacturing processes used to transform polymers into a useful form. Are any of these processes preferable for recycling polymers?
15. What are the two most commonly recycled polymers? In what types of products are they used?
16. The city of Pristine, Illonois, will develop a comprehensive waste management program in order to divert waste materials from the county landfill. The city will employ curbside collection of glass, paper, aluminum, and PETE. What actions can the municipality undertake that will ensure the success of the recycling program? Consider legislative initiatives, educational programs, marketing, and other constructive efforts.
17. Suppose the city was to avoid working with a materials broker and instead work directly with material buyers. What agreements should be specified in a proposal from a recycled materials buyer?
18. The city is allowing a new recycling company to acquire the abandoned Hi-Jinx Chemical Company warehouse building for use as a recycling center. List five practical issues which the company must consider before accepting the building.

19. "As of the late 1980s, recycling in the United States reached its maximum potential." True or false? Justify your answer.
20. Identify the materials that are currently being recycled in your community. What other materials could potentially be recycled?
21. If your community is engaged in a recycling program or waste reduction program, is the program voluntary or mandatory? How is the public involved-for example, are educational programs available? Which agency or office is responsible for managing the program? Are there areas in which the program could be improved?
22. In your home, which waste materials do you now separate? What other components could potentially be separated for eventual recycling or reuse?
23. How does resource recovery affect the overall cost of solid waste management?
24. At your university, what efforts have been undertaken to reduce the volume of solid waste? Describe any resource recovery or waste reduction programs in place.

7 Municipal Solid Waste Processing; Materials Recovery Facilities

Sooty, swarthy smiths, smattered with smoke,
Drive me to death with the din of their dents.
Such noise at night no man heard, never;
With knavish cries and clattering of knocks!
... They spit and sprawl and spill many spells;
They gnaw and gnash, they groan together
And hold their heat with their hard hammers
Heavy hammers they have that are hard-handled,
Stark strokes they strike on a steely stump

Anonymous, ca. 1400
The Blacksmiths

7.1 INTRODUCTION

The ideal resource recovery scenario in any community would include thorough segregation of individual waste components by each homeowner, commercial establishment, industry or municipal institution (i.e., at the *source*). Subsequently, the individual *products* (e.g., aluminum cans, paper, glass, and plastics) are collected on a regular basis, stored in separate bins within the collection vehicle, and transported to a facility for further processing (densifying, shredding). These slightly processed, clean materials would then be sold to and removed by buyers for eventual reprocessing. The above scenario would result in a clean and highly marketable resource, thus decreasing the capital expenses for purchase of large separation equipment, energy costs, and labor. Since the above approach for materials separation from MSW is often not feasible due to lack of information, lack of participation, and insufficient support from local and state governments, another approach is often be needed.

7.2 THE MATERIALS RECOVERY FACILITY

The materials recovery facility (MRF) is a relatively recent approach to MSW management, but its utility has become obvious and its popularity is increasing. In 1898 the first MRF was built in New York City. The facility processed the waste of over 116,000 residents and recovered up to 37% (by wt) of the wastes. More recently, the first modern MRF was established in the 1980s in Groton, Connecticut. Despite facing a volatile market for materials, the number of MRFs has grown markedly over the past decade. In 1991, a total of 40 projects were planned or operated. Two years later, this number had quadrupled to 166. In 1995, another doubling occurred to 307 projects. By 1997 the trend had slowed; about 40 new or expanded projects came on line between 1995 and 1997 (Berenyi, 2001). Since 1995, when the implementation of curbside recycling collection programs had spread across the country, MRFs began to be found in nearly equal proportion by region.

The primary reason for the increased interest in mechanized facilities for waste processing is that as MSW disposal costs rise, a greater incentive develops in favor of recycling, and convenient and rapid methods of separation and processing develop. For example, in areas where landfill tipping fees are below $30 to 40 per ton, recycling the waste stream may not appear economically attractive to municipalities and the waste industry. However, with tipping fees in some areas exceeding $100 per ton, cities and waste management companies clearly see the advantage to serious investment in recycling.

The two major configurations of MRFs are:

- Facilities which handle source-separated materials ("clean MRFs")
- Facilities which handle mixed (commingled) wastes ("dirty MRFs")

In many parts of the United States, markets exist for most materials recovered from the waste stream. In those markets the specifications for separated materials will vary. Some of the forms and conditions applicable to finished products are shown in Table 7.1.

TABLE 7.1
Some Forms and Conditions Applicable to Products to be Recycled

Paper
- Separated by grade (laser-quality white, mixed colored paper, old newspaper, corrugated, etc.)
- Baled or loose
- Dry
- Clean (or not weathered)

Ferrous Containers
- Flattened, unflattened, shredded
- Labels removed
- Clean or limited food contamination
- May not include bi-metal
- Loose, baled, or densified into biscuit form

Aluminum Containers
- Flattened, shredded, baled, or densified into biscuit form
- Free of moisture, dirt, foil, plastic, glass, oil, other foreign substances

PETE and HDPE
- Baled, granulated
- Separated by color or mixed
- Without caps

Glass
- Separated by color or mixed
- Size of cullet specified
- Nature and amount of allowable contamination

General
Available markets for secondary materials typically specify the means of packaging and shipping each product. The specifications depend upon location and end-use. The specifications often include the following:
- Skids or pallets
- Bundles, bins, boxes, cartons, or drums
- Trailer loads
- Roll-offs
- Rail cars

Source: U.S. EPA, EPA/6025/6-91/031, 1991.

7.2.1 Unit Operations

Unit operations in a centralized facility include screening, magnetic separation, shredding, and air classification. The unit operations for the separation and processing of wastes are designed to accomplish the following (Tchobanoglous et al., 1993):

- Modify the physical characteristics of the waste so that the components can be removed more easily.
- To remove specific, useful components from the waste stream.
- To remove contaminants from the waste stream.
- To process and prepare the separated materials for subsequent uses.

Common unit operations for MSW separation are shown in Table 7.2.

Even though many MRF systems are highly mechanized, human labor is still needed to carry out a number of duties. For example, the removal of hazardous wastes (batteries, paint cans, and pesticide containers) from MSW can only be accomplished by manual sorting. The same is true for 2 L soda bottles made from PETE. Additionally, nearly all MRFs that sort glass by color must rely on the human eye and hand.

TABLE 7.2

Common Unit Operations and Facilities for the Separation and Processing of Separated and Commingled MSW

Unit Operation	Function and Material Processed
Shredding	
Hammer mills	Size reduction
Flail mills	Size reduction, also used as bag breaker
Shear shredder	Size reduction, also used as bag breaker
Glass crushers	Size reduction, glass
Wood Grinders	Size reduction, yard trimmings, and wood wastes
Screening	Separation of over and under-sized material; trommel also used as bag breaker
Cyclone separator	Separation of light combustible materials from air stream
Air classification	Separation of light combustible materials from air stream
Magnetic separation	Separation of ferrous metal from commingled wastes
Densification	
Balers	Compaction into bales or paper, cardboard, plastics, textiles, aluminum
Can crushers	Compaction and flattening; aluminum and tin cans
Weighing facilities	
Platform scales	Operational records
Small scales	Operational records
Handling moving and storage facilities	
Conveyer belts	Materials transport; all types of materials
Picking belts	Manual separation of waste material; source-separated and commingled MSW
Movable equipment	Materials handling and moving; all types of waste
Storage facilities	Materials storage; all types of recovered materials

Source: Tchobanoglous, G., et al., *Integrated Solid Waste Management: Engineering Principles and Management Issues*, McGraw-Hill, New York, 1993. Data reproduced with kind permission of the McGraw-Hill Companies, Inc.

7.2.2 WEIGH STATION

Weighing is an obvious and necessary step for any MRF. Scales of various types are used to weigh the amount of materials delivered, recovered, and removed from the facility. Scale types vary from small units for weighing modest amounts brought in by individuals, to the large, platform scales which are suited to handle the largest collection trucks.

Trucks typically are required to enter a weigh station immediately upon entry to the facility property. The station often consists of a small office with series of platform scales that can handle a truck of any weight (Figure 7.1). The gross weight of the truck is measured. After tipping its load at the facility receiving area, the truck returns to the weigh station for final weighing and calculation of the net weight of the waste. These data are used to bill the waste hauling company. Some weigh stations are equipped with magnetic card readers. Vehicles may be provided with magnetic cards that are inserted into the reader. Information on tonnages is thus collected and calculated automatically.

The weigh station provides other useful data such as the rate at which waste is processed by the facility. The input tonnage is important for calculating certain facility operations, including possible requirements for additional storage space, greater equipment capacity, and a larger workforce. The weigh station also provides data for determining the total waste production for a particular collection area. Collection trucks may be identified by route, and the quantities of wastes delivered from a particular neighborhood or town can be determined. Such data are beneficial in the planning of improved collection routes or other services.

7.2.3 RECEIVING AREA

After the initial weigh-in, collection vehicles transport their loads to a receiving (tipping) area for temporary storage and initial processing (Figure 7.2). The facility must be designed to create an optimum flow of collection trucks. In other words, full trucks should not interfere with the prompt exit of just-emptied vehicles from the tipping area.

7.2.4 STORAGE AREA

Storage of MSW at an MRF is a major practical concern, from the standpoint of both efficiency and of safety. Storage encompasses sufficient space for the raw, in-coming commingled MSW as well as the sorted, cleaned, baled product that is to be collected by a buyer.

FIGURE 7.1 Weigh station at a materials recovery facility.

FIGURE 7.2 Receiving area at an MRF.

A very common storage system at an MRF is the tipping floor, also known as "slab storage." In this scenario, either collection trucks or front-end loaders deposit the wastes onto the floor (Figure 7.3). The waste may be stacked, if necessary, by a front-end loader to as high as 6 to 8.5 m (approximately 20 to 25 ft). Front-end loaders will load the wastes on to a conveyor that feeds into the processing system. The slab is partly surrounded by a "push wall", a reinforced concrete wall designed to withstand the force of a large front-end loader pushing wastes against it to load the bucket (Pfeffer, 1992).

The size of the tipping floor must consider the number of trucks that will unload in a given period. Calculations for floor size should take into account large pulses of waste deliveries. For example, collection trucks tend to arrive at the MRF in large numbers late in the day after the last loads are collected. Similarly, Monday deliveries may result in accumulation of substantial volumes of MSW, as there may have been no waste processing over the weekend.

Another storage system, although more common for incinerators, is the standard pit with an overhead crane. The pit may be 6 to 12 m (20 to 40 ft) feet deep. Waste collection trucks back up to the edge of the pit and dump their loads directly into the pit. The overhead crane is used to retrieve the solid waste and also to spread the waste in the pit area. The crane can drop the wastes into a feed chute or onto a conveyor belt. Slab storage is clearly less expensive than pit storage, especially when storage requirements are modest.

The design of storage facilities requires knowledge of materials flow; however, a means of experimentally evaluating the flow rate of MSW in a storage area is also useful. This will address problems such as how quickly (or how slowly) materials are moving out of the tipping area. A number of potentially effective techniques include stereophotogrammetry, radio pills (i.e., transmitters that move with the solids in the chamber), radiological tagging and x-ray methods. With heterogeneous materials such as MSW, the radio pill or stereophotogrammetry is applicable (Resnick, 1976; Vesilind et al., 2002).

The materials recently processed and separated at the MRF should be stored away from in-coming wastes and machinery. These items should also be protected from weather. In some facilities, storage areas for processed wastes are physically separated from storage areas for incoming wastes. Such a separation will facilitate movement of trucks. Also, the separated wastes can be placed on display for potential buyers in a clean, orderly location.

FIGURE 7.3 Tipping floor of an MRF.

Common units for storage of processed wastes include:

- Enclosed warehouse space
- Open-sided, roofed structures (i.e., pole barns)
- Roll-off containers
- Shipping containers

Shipping containers tend to serve as an economical storage system. The materials buyer may provide these containers. This method is popular for the shipping of various grades of paper and cardboard to overseas markets on container ships (Tchobanoglous et al., 1993).

EXAMPLE 7.1

Consider a 400 MT/day resource recovery facility. Assume all MSW is received in 15 m³ (20 yd³) compactor trucks, providing an average density of 900 kg/m³. During a routine work day and assuming normal equipment operation, how many trucks can be accommodated per hour?

Truck capacity = (15 m³ × 900 kg/m³)/1000 kg/MT = 13.5 MT/truck

The facility processes MSW 12 h/day, 5 days/week.
MSW processing rate = (400 MT/day)/12 h/day = 16.7 MT/h = 16.7/5 = 3.3 truckloads/h
Waste collection takes place 5 days/week, 1 shift/day, and only about 7 h/shift are actually used for on-the-tipping floor processing. Therefore, actual receiving rate = (400 × 5 days)/(5 days × 7 h/day) = 57 MT/h

(57 MT/h)/13.5 MT/truck = 4.2 truckloads/h

In addition to the above, it would be useful to consider hourly peaking factors and seasonal peaking factors in calculations. (Adapted from Schwarz and Brunner, 1983.)

7.2.5 Mobile Equipment in the MRF

Front-end loaders and forklifts are universal in the routine operations of the MRF. For example, when a collection truck tips its load on to the slab, the front-end loader will promptly arrive on the scene to lift the material onto a conveyor for subsequent processing (Figure 7.4). Front-end loaders also move wastes after processing. Forklifts are used to move baled materials from balers to storage areas, and from storage areas onto trucks. Mobile equipment which may be found in a MRF include (U.S. EPA, 1991):

- Bins
- Containers
- Forklift
- Front-end loader
- Manulift
- Skid steer loader
- Steam cleaner
- Vacuum or sweeper or magnetic pickup
- Yard tractor

7.2.6 Fixed Equipment in the MRF

The various types of fixed equipment that may be included in an MRF are presented in Table 7.3. The types of equipment in use in a single facility are highly variable, and selection is based on goals for the number of commingled articles that are to be separated, their purity, the original characteristics of the waste, space considerations, the throughput capacity of the facility, and, of course, economics. The various fixed components are discussed in the following sections.

7.2.7 Conveyors

The conveyor, a system to transfer wastes from one location to another (and also for manual and mechanical removal of individual components from the waste stream), is the most common unit of

FIGURE 7.4 Front-end loader on the tipping floor.

TABLE 7.3
Fixed Equipment Which May be Used in a MRF

Size Reduction Equipment

Baler
Shredder
    Vertical or horizontal hammermill
    Rotary shear
    Flail mill
Can shredder
Can densifier or biscuiter
Can flattener
Glass crusher
Plastics granulator

Separating Equipment

Magnetic separator
Eddy current device (aluminum separator)
Trommel screen
Vibrating screen
Air classifier

Environmental Control Equipment

Dust collection
Noise suppression
Odor control

Other Equipment

Storage bins
Floor scale
Truck scale
Belt scale
Conveyors
    Belt
    Screw
    Apron
    Bucket
    Pneumatic
    Vibrating
Source: U.S. EPA, EPA/625/6-91/031, 1991.

equipment for handling materials in a MRF. Several types are available including hinge, bucket, apron, belt drag, screw, vibrating, and pneumatic (CEMA, 1995). Comprehensive engineering data are available for many types of conveyors; as a result, their performance can be accurately predicted when they are used for handling materials possessing well-known characteristics.

Factors to consider in the selection of the appropriate conveyor include (U.S. EPA, 1991):

- Capacity of the belt
- Length of travel
- Lift
- Characteristics of the material to be transported
- Overall cost

Horizontal and inclined belt conveyors, where the material is carried along the surface of the belt, and drag conveyors, equipped with crossbars to drag the input wastes, are the most commonly used conveyors for handling MSW (Figure 7.5) (Tchobanoglous et al., 1993).

The belt conveyor can be designed with idler rolls that create a concave cross-section (Figure 7.6). Such shapes will retain lighter materials along the length of the belt, thus preventing spillage. To minimize further spillage losses, skirt bands are employed at belt transfer points. The flat belt conveyor used at a MRF is of the slider belt design, in which the belt is supported by and slides on a steel supporting surface rather than on idler rolls. When used in an inclined position, it is supplied with cleats and skirt boards over its full length in order to prevent spillage (U.S. EPA, 1991).

The capacity of a conveyor belt is directly related to its cross-sectional area and belt speed. The depth is limited by the height of the sidewalls, the shape of the belt, or the angle of repose of the MSW feed. The volume flow rate on the belt can be calculated by (Pfeffer, 1992)

$$Q = AV \qquad (7.1)$$

where Q is the flow rate in m³/min, A the cross-sectional area in m², and V the belt speed in m/min. The mass flow rate can be calculated by using the density of the MSW on the conveyor belt. The thickness of the waste on the belt can be calculated using the equation

$$TW = [(LR)(1000 \text{ kg/MT})(100 \text{ cm/m})]/[(60 \text{ min/h})(VDW)] \qquad (7.2)$$

where TW is the thickness of the waste on the conveyor belt (cm), LR the loading rate of solid waste (MT/h), V the belt velocity (m/min), D the waste density on belt (kg/m³) and W the belt width (m).

Waste thickness is an especially practical consideration for handpicking operations at an MRF.

EXAMPLE 7.2

Calculate the waste thickness for a conveyor belt measuring 1.1 m with an average belt speed of 17.5 m/min. Waste loading rate is 28 MT/h and the average density of the waste on the belt is 120 kg/m³.

$TW = [(28 \text{ MT/h})(1000 \text{ kg/MT})(100 \text{ cm/m})]/[(60 \text{ min/h})(17.5 \text{ m/min})(120 \text{ kg/m}^3)(1.1 \text{ m})]$
$\quad = 20.2 \text{ cm}$

The pneumatic conveyor is sometimes used to transport shredded, lightweight materials such as newspaper, plastic, or refuse-derived fuel. Pneumatic conveying systems consist of a fan, a feed device, piping, and a discharge device, typically a cyclone. Systems can be operated under positive pressure (i.e., blowing air) or negative pressure (suction). Air velocities for processed wastes are in

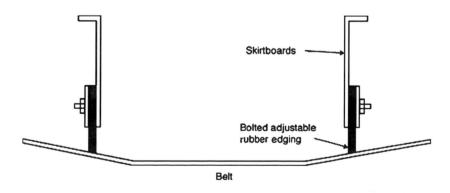

FIGURE 7.5 Cross-section of a belt conveyor (U.S. EPA, EPA/625/6-91/031, 1991).

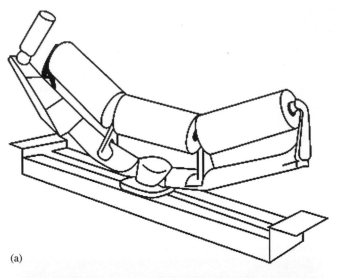

(a)

(b)

FIGURE 7.6 Base of a conveyor belt showing rollers: (a) schematic (reproduced with kind permission of CEMA, 1995); (b) photo.

the range of 1220 of 1520 m/min (4000–5000 ft/min) and the materials to air ratio is about 0.045 kg (0.1 lb) material to 0.45 kg (1 lb) air (Vesilind and Rimer, 1981; Tchobanoglous et al., 1993).

The utilization of conveyors for MSW transport has not been without its share of problems. Wastes which are too heavy or extremely sharp can be dropped on the belt, thus damaging the belt, pulleys, or other components; wastes may fall off the belts at transfer points, i.e., where one belt empties onto another (Figure 7.7); and wires and string within the wastes can become tangled around pulleys and other equipment.

7.2.8 SCALES

Scales are included inside the MRF. These are typically small models and are used to weigh objects such as bales of metal or paper, or cages of materials (Figure 7.8).

7.3 MATERIALS RECOVERY AT MRF UNIT OPERATIONS

In a hypothetical situation where one material is to be segregated from a mixture, the separation process is termed binary, as two outputs result from the operation. A binary separator receiving a

FIGURE 7.7 Transfer point along a conveyor belt. Wastes sometimes fall away from the belts during transfer to another belt.

FIGURE 7.8 Small scale for weighing bales and other small- and medium-sized objects.

mixed feed of x_0 and y_0 is shown in Figure 7.9. The goal of the unit operation is to separate the x fraction in as pure a form as possible and with the greatest total recovery possible.

One exit stream will contain the x component, the desired material, also designated the *product* or *extract*. Separation will not be perfect, however, so there will inevitably be contamination as y_1. A second stream, containing mostly the y is known as the *reject*. Note that this stream will contain some of the product x. The recovery of x can be expressed as (Hasselriis, 1984)

$$R(x_1) = (x_1/x_0) \times 100 \tag{7.3}$$

where $R(x_1)$ is the recovery of x in the first output stream (%).

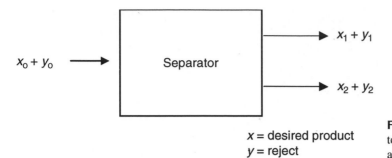

x = desired product
y = reject

FIGURE 7.9 A binary separator receiving a mixed feed of x_0 and y_0.

This equation, however, does not take into account purity of the product. If the separation device is not operational, then all the input (both the desired product as well as the reject) will pass through. In other words, $x_0 = x_1$ with the result that $R_{(x1)} = 100\%$. A second requirement, the purity of the product, is therefore necessary. The purity of the extract stream is defined as

$$P_{(x1)} = (x_1)/(x_1 + y_1) \times 100 \tag{7.4}$$

There are also difficulties with using purity alone as a descriptor of separator performance. For example, it might be possible to extract a small amount of x in a pure state, but the recovery ($R_{(x1)}$) will be very small. It is therefore necessary to describe the operation of a materials separation device by incorporating both the recovery and purity. Binary separator efficiency can be determined as (Rietema, 1981)

$$E_{(x,y)} = (x_1/x_0)(y_1/y_0) \times 100 \tag{7.5}$$

EXAMPLE 7.3

An eddy current separator (see below) is to separate aluminum product from an input stream of shredded MSW. The feed rate to the separator is 1500 kg/h. The feed is known to contain 55 kg of aluminum and 1445 kg of reject. After operating for 1 h, a total of 65 kg of material is collected in the product stream. On close inspection it is found that 46 kg of product is aluminum. Calculate the percent recovery of aluminum product, the purity of the product, and the overall efficiency of the separator.

$$R_{(x1)} = x_1/x_0 \times 100 = 46/55 \times 100 = 83.6\%$$

$$P_{(x1)} = x_1/(x_1 + y_1) \times 100 = 46/65 \times 100 = 70.8\%$$

$$E_{(x,y)} = (x_1/x_0)(y_1/y_0) \times 100 = 46/55[(1500-65)-(55-46)]/1445 \times 100 = 82.5\%$$

7.4 MATERIALS SEPARATION AND PROCESSING AT THE MRF

7.4.1 HAND-SORTING

The most basic and simple method for the separation of materials from MSW is hand-sorting. Workers take positions along a conveyor belt, either on one or both sides (Figure 7.10). Sorting takes place after the bags have been opened in a trommel screen or simple shredder. At a clean MRF the material may arrive already in loose form.

At the MRF, picking can occur at several points along the system, and the workers have two primary functions. First, they recover any items of potential value that do not need to be processed. Items such as metal and PET bottles are set aside in bins or chutes. Their second responsibility is

FIGURE 7.10 Hand-sorting along a conveyor belt. Note workers are not wearing eye or hearing protection.

to remove those items that are detrimental either to the workers downstream; to the quality of the final, separated products; or to system equipment. This could include removing toxic and potentially explosive items. According to Vesilind et al. (2002), the material along the conveyor is recognized visually ("coding") by such properties as color, reflectivity, and opacity; verified by noting its density; and removed (separated) by hand-picking.

Important factors in the design of the manual picking area are the width of the belt, the speed of the belt, and the average thickness of material placed on the belt for picking. A picking belt usually measures no more than 60 cm (24 in.) wide for one-sided picking or 120 cm (48 in.) wide for pickers on both sides of the belt. Belt speeds have varied from approx. 450 to 2700 cm/min (15 to 90 ft/min) depending on the material to be processed and the amount of any preprocessing which may have already taken place. The belt should not move faster than about 900 cm/min (30 to 40 ft/min) depending on the number of pickers (Engdahl, 1969; Vesilind et al., 2002). Belt speeds of 1800 cm/min (60 ft/min) were used in sorting facilities at the turn of the century (Hering and Greeley, 1921). The average thickness of wastes on the belt for effective picking is about 6 in. (Tchobanoglous et al., 1993).

The picking operation is best performed under natural lighting. Artificial light, for example from fluorescent bulbs, emits only a narrow band of light that makes identification (coding) of certain components difficult (Vesilind et al., 2002).

At those facilities where waste is not preprocessed, sorting is inefficient. Pickers can salvage about 450 kg (1000 lb)/person/h depending on the material density (Vesilind et al., 2002). For example, a worker removing metallic objects and wood will remove more materials by weight than a picker removing lightweight plastic containers.

Hand-picking is, obviously, dirty and dangerous work. There are significant quantities of dust and the wastes being handled are odoriferous. Wastes may be hazardous to the worker by being sharp-edged, explosive, flammable, or infected with pathogenic microorganisms. Noise from equipment can be extreme in some facilities. Heavy equipment may be routinely moving across the facility floor and the unit operations themselves are noisy. Appropriate worker safety including protection of eyes, skin and hearing, as mandated under OSHA statutes, is essential in a MRF.

7.4.2 SCREENING

Screening is a unit operation designed for the separation of waste input into *oversize* and *undersize* fractions (also labeled *reject* and *product*, respectively). In many MRFs, the oversize materials might consist primarily of old corrugated cardboard (OCC) and newspaper, depending on the composition of the incoming waste stream. Screening is carried out either wet or dry, although dry separation is most common. Screens are classified as primary, secondary, or tertiary, depending on where they are situated in the sequence of separation steps. The primary applications of screening during MSW processing include:

- Removal of oversized materials
- Removal of undersized materials
- Recovery of paper and plastics for recycling or as refuse derived fuel (RDF)
- Separation of soil, glass, and grit from combustible materials

Screens have a long history in various industries for particle separation by size. Rotary (trommel) screens have been used in the mineral industry for many years for coarse screening of ores, gravel, and rock.

As discussed in Chapter 4, different components of MSW possess characteristic size ranges. If screen size is properly selected, it is possible to create a fairly enriched stream of a particular waste component. Of course, due to the great variability of sizes of a single waste component, only partial separation is possible; additional processing is still required for further purification of the product.

There are three major types of screening used in materials recovery: trommel screening, disk screening, and vibrating shaker screening.

7.4.2.1 Trommel Screens

Of the major forms of MSW screening the trommel is by far the most popular. Trommel screening is primary screening, designated as such because it is usually placed before all other separation units in an MRF.

The trommel is a rotating perforated cylinder with a diameter ranging between 0.6 and 3 m (2 and 10 ft) with a screening surface consisting of a perforated plate or wire mesh (Figure 7.11). Some are equipped with spikes, usually within the first third of the drum, to break open plastic trash bags. The drum is inclined at a slight angle. A motor is attached at one end which rotates the drum at a rate of about 10 to 15 r/min. The waste is introduced at the elevated end via a conveyor belt. As the drum rotates, waste particles are carried up the side until they reach a certain height and then fall to the bottom. The waste which falls through the openings is collected by a conveyor or a hopper and the fraction retained within the trommel is collected on a separate belt.

A typical trommel screen is depicted in Figure 7.12. The length and diameter of the drum have a direct relationship with the efficiency of the trommel. The longer the drum, the longer the MSW will remain in contact with the screen; and, the greater the diameter, the more effective the trommel will be in breaking up large objects such as trash bags. Large trommels (2.5 to 3 m, or 8 to 10 ft in diameter, up to 15 m long) have been used to separate large OCC and newsprint from mixed paper or commingled containers (particularly from glass containers). Small trommels (0.3 to 0.6 m in diameter by 0.6 to 1.2 m long) have been used to separate labels and caps from crushed glass. These small units have been used in conjunction with an air stream to aid in separation (U.S. EPA, 1991).

Two-stage or compound trommels have been used in waste processing. In two-stage trommels the first section is set with small openings (e.g., 2 to 3 cm [approx 1 in.] diameter) which permit soil, broken glass, and other small fragments to fall through and be collected. This material is largely nonrecyclable and will probably be landfilled. The second stage is provided with larger apertures (e.g., 12 to 15 cm diameter), which allows glass, aluminum, and plastic containers to be removed from the waste stream.

Trommel screens separate waste materials based on size and do not identify the material by any other property. As a result, trommels are used as a classification step before true "separation" of materials. For example, smaller particles such as grit and broken glass can be removed early in the processing scheme to produce better quality (i.e., higher quality, greater purity) recyclables such as

FIGURE 7.11 Trommel screen.

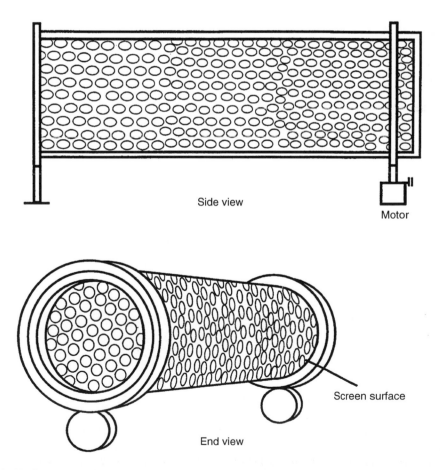

FIGURE 7.12 Schematic of a trommel screen (U.S. EPA, EPA/625/6-91/031, 1991).

paper, aluminum, and glass containers. Removal of coarse or abrasive components early in the process will reduce the load on a shredder (see below) which will therefore reduce shredder maintenance.

Waste behaves within the trommel in distinctly different ways depending on the speed of rotation. Waste rotating very slowly within the cylinder will travel only slightly up the sides and will immediately slide back, thus missing most openings. This is known as *cascading* (Figure 7.13). Waste that is rotated more rapidly within the cylinder will rise up farther, and then tumble and slide back. This *cataracting* motion causes substantial turbulence of the waste. At even higher speeds the material may adhere to the inside of the trommel and will not effectively tumble or fall by gravity through the screens, i.e., it tends to *centrifuge*. The so-called "critical speed", i.e., the frequency of rotation at which the force of the trommel on the waste holds the materials against the wall throughout a complete revolution, is given by the equation (Vesilind et al., 1988)

$$n_c = (g/4\pi^2 r)^{1/2} \tag{7.6}$$

where n_c is the critical speed (rotations/sec), g the acceleration due to gravity (cm/sec^2) and r the radius of trommel (cm).

The ideal trommel rotation speed is that immediately *prior* to the point where the waste starts to centrifuge; in other words, it climbs the side of the trommel and then falls immediately upon reaching the zenith of rotation, which is the upper limit of the cataracting type of motion. This creates the greatest opportunity for waste particles to fall through the screen openings.

EXAMPLE 7.4

Calculate the critical speed for a trommel screen having a diameter of 3.2 m.

$n_c = (g/4\pi^2 r)^{1/2}$

$\quad = (980/[4\ (3.14)^2\ (320/2)])^{1/2}$

$\quad = 0.39$ rotations/sec

Some trommels are equipped with horizontal lifter bars along the inside which help to carry waste part-way up the side of the drum. Any additional upward motion depends on the rotational speed of the drum (Pfeffer, 1992).

Separation Efficiency with Trommel Screens
The speed of rotation plays a role in the trommel's separation efficiency via agitating the waste input. The tumbling action of waste within the trommel efficiently brings about separation of individual items that may be attached to each other, or of one material contained within another. The

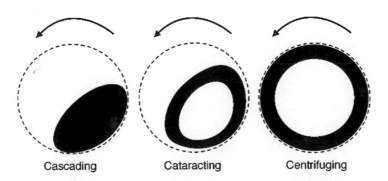

Cascading   Cataracting   Centrifuging

FIGURE 7.13 Cascading, cataracting, and centrifuging of waste input in a trommel screen (Stessel, R.I., *Recycling and Resource Recovery Engineering* 1996. Reproduced with kind permission of Springer-Verlag, Inc., Berlin).

more cycles of rising and dropping, the greater the separation efficiency. At the same time, however, the rate of throughput must be considered so that there is a limit as to how long a waste charge should remain in the trommel.

The following equation was developed for calculating trommel throughput (Sullivan et al., 1992):

$$D = [11.36 \, Qm/(d_b \, F \, K_v \, g^{0.5} \tan a)]^{0.4} \tag{7.7}$$

where D is the trommel diameter (m), Qm the trommel throughput, (kg/s), d_b the bulk specific weight of MSW (kg/m³), α the angle of trommel from base frame (deg), K_v the velocity correction factor ($K_v = 1.35$ when $\alpha = 3°$, and $K_v = 1.85$ when $\alpha = 5°$), F the fillage factor (a typical range is between 0.25 and 0.33), and $g = 9.81$ m/s².

Using data such as these shown in Figure 7.14, separation efficiency at a particular screen size can be estimated. For example, in a trommel with an aperture of 10 cm, about 90% of metals and glass will fall through and be captured as undersize material. Concurrently, about 30 to 35% of the paper and plastics will also fall through and thus contaminate the metal or glass fraction.

Many factors influence the separation efficiency of a trommel including:

- Characteristics of the incoming materials (dense, loose, fragile, and wet)
- Quantity of the incoming materials (feed rate)
- Size ranges of the cylinder screen
- Incline angle of the cylinder
- Rotational speed
- Size and number of screen openings

The primary factor influencing separation is, ultimately, retention time of the waste within the trommel. The average waste retention time in trommels ranges from 25 to 60 sec for raw waste prior to shredding, to about 10 sec for shredded, air-classified light materials. Long trommels are desirable because they achieve a more thorough screening. Optimal trommel performance was

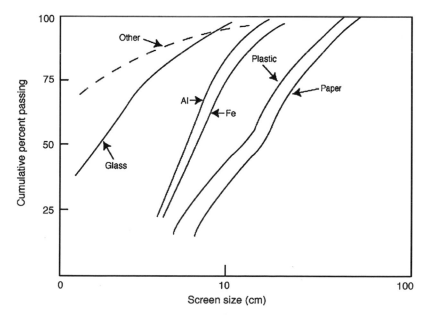

FIGURE 7.14 Hypothetical particle size distribution as relates to screening MSW. (Reproduced with kind permission of Diaz, L.W. et al., *Resource Recovery from Municipal Solid Wastes*, Vols. 1 and 2, CRC Press, Boca Raton, FL, 1982.)

found to occur with a solids retention time of 30 to 60 sec, with the material making 5 to 6 revolutions within the drum (Vesilind et al., 2002).

A practical advantage with waste separation in trommels is its avoidance of clogging. Some material may attach to the interior of the drum, but, with the continued tumbling motion and impaction by other materials, this attached material will eventually fall out and be removed.

7.4.2.2 Disk Screens

A disk screen is not a "screen" in the conventional sense, rather, it occurs as a series of rounded or lobed-shaped disks mounted on shafts (Figures 7.15 and 7.16). The disks are parallel and interlocked. The shafts rotate in one direction, carrying the waste charge along in a fashion analogous to a conveyor belt. Due to turbulence and the irregular shape of the disks, however, undersized materials fall between the spaces in the disks and are collected in one hopper while the larger particles are carried along the top to be deposited in a second hopper.

The spacing between the outer diameter of the shafts and the spacing of the disks on the shaft determine the size of separation. Particles having two dimensions less than or equal to these spacings can fall through. Most particles have a tendency to orient such that the two larger dimensions are situated horizontally. Therefore, the size separation is usually based on these two larger dimensions (Pfeffer, 1992). Varying the spacings of the disks on the drive shaft will change the desired particle size ranges.

In the event of blockage, an electronic sensor will signal for the shafts to rotate in the opposite direction to clear any materials.

7.4.2.3 Vibrating Screens

Another variation of screening is the so-called vibrating screen which consists of a mounted flat screen and undergoes a reciprocating or gyrating motion. Such flat screens are typically not used to process mixed MSW, however. They are most successful in purifying more concentrated fractions of waste that have previously been processed into a relatively fine particle size. Examples include glass, metals, and wood chips. Flat screening may be applied to remove impurities (e.g., broken glass, ceramics, and stones) from compost feedstock (Rhyner et al., 1995).

FIGURE 7.15 Disk screen.

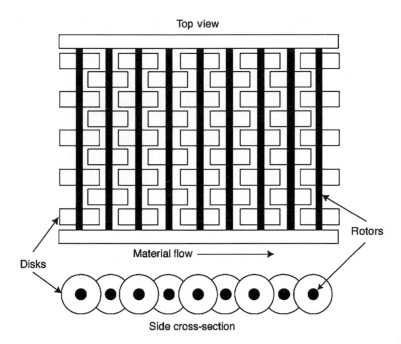

FIGURE 7.16 Schematic of a disk screen (Stessel, R.I., *Recycling and Resource Recovery Engineering.* Springer, Berlin, 1996. Reproduced with kind permission of Springer-Verlag, Inc.)

7.4.3 SIZE REDUCTION

Size reduction, a unit operation that can also be considered as volume reduction, is of great importance for various methods of treatment and disposal (e.g., composting and incineration) as well as for cost-effective transportation of materials.

There is a wide range of size reduction methods available and many types of size reduction equipment are used at a MRF, many originating from other industries. Such equipment is employed to reduce the particle size or increase the density of material in order to meet market specifications or to reduce the cost of storage and transportation. Either incoming MSW or separated and outgoing components can undergo size reduction.

7.4.3.1 Densifiers

The basic purpose of a densifier is to enhance the storability or transportability of waste components that are to be used as fuel. This includes RDF, which will be discussed in Chapter 9. Essentially, RDF is composed of the light organic fraction of MSW including paper, plastics, and some food waste.

Densification allows more MSW (by wt) to be stored in the same volume and will also result in an increase in energy value per unit volume (e.g., kJ/kg or BTU/lb). Densifiers are located at the end of the process line before the storage and retrieval system. There are six equipment types that densify MSW: pelletizers, briquetters, cubetters, extruders, compactors, and balers (Figure 7.17). Each type applies heavy force to reduce large volumes of MSW into smaller volumes. Compactors and balers are discussed below.

7.4.3.2 Compactors

Compactors became popular in the 1960s in response to increased waste hauling and disposal rates. The earliest stationary compactors compressed wastes into roll-off boxes, i.e., large metal containers usually measuring $2.5 \times 2.5 \times 6.5$ m. When the box was sufficiently filled with wastes, a transporter

FIGURE 7.17 Pelletizer (densifier) for shredded organic wastes in the manufacture of RDF.

removed the box for shipment to a sanitary landfill. This system was ideal for dry wastes. Some businesses and industries, however, such as restaurants and hospitals, disposed of liquid as well as solid wastes. Because the compactor was separate from the container, liquid wastes resulted in spills and residue. This residue left an odor and attracted animals and insects. In response to this practical problem the self-contained compactor was developed. Self-contained compactors were simply a compactor and a roll-off box housed together in the same unit. For hauling, the electrical power unit was separated from the assembly. Self-contained units typically included a liquid collection area situated directly beneath the compactor to contain any spillage (Ely, 1993).

7.4.3.3 Baler

A baler (Figure 7.18) is one of the more common components of the MRF waste processing system. Balers are used for producing bales of corrugated cardboard, newspaper, high-grade paper, mixed paper, aluminum cans, and plastic containers (Figure 7.19). Balers are available with a wide range of horsepower and levels of sophistication. Some balers have fully automated operation while others require a significant amount of work by an operator. As previously discussed, many industrial buyers have specific requirements for any secondary materials purchased; therefore, the market specifications for a particular product should be determined before a baler is selected (U.S. EPA, 1991).

Most balers are of sufficiently low force that, once released, the baled product will simply rebound to its original form. Hence, bales must be tied, either with steel wire, high-tension nylon string, or similar durable material. However, some high-power balers apply sufficient force such that a bale will maintain its shape even after the force is removed. Of course, such balers tend to be much more expensive and maintenance-intensive than the low-force models. The original waste input material will also influence the necessity of tying a bale. Bales of aluminum or other metal will hold their form better than would bales of HDPE containers or old newspaper, for example.

The performance of baling and compaction equipment is measured by calculating the percentage volume reduction and the compaction ratio. Percentage volume reduction is calculated by the equation

$$\text{Volume reduction } (\%) = (V_i - V_f)/V_i \times 100 \qquad (7.8)$$

where V_i is the initial volume of wastes before compaction (m^3), and V_f the final volume of compacted wastes (m^3)

FIGURE 7.18 Baler at an MRF.

FIGURE 7.19 Bales of OCC awaiting shipment to a recycling facility.

The compaction ratio is calculated as

$$V_i / V_f \qquad (7.9)$$

7.4.3.4 Shredders

In terms of processing raw MSW, shredding is a unit operation designed for size reduction; however, other benefits also accrue. Shredders can be applied to items including scrap metal, plastic, aluminum, wood, paper products, and possibly others. Shredders were originally developed for the crushing of stone and ores. The shredder's versatility has resulted in its application in other areas besides processing MSW for recycling. For example, shredders now serve in construction and demolition activities for breaking down concrete, steel, and other building materials.

The shredding process involves three types of action: crushing, shearing, and grinding. *Crushing* involves the reduction of particles by pounding; *shearing* involves forcing two parts of an item in different directions; and *grinding* is friction applied to the surface of an object. All shredding units employ two or more of these actions simultaneously.

Shredding can impart a number of benefits to waste. Shredded wastes are more amenable to sanitary landfilling by virtue of decreased odor and therefore less rodents and insects; it also provides for greater ease of movement of landfill equipment. Shredding is extremely useful for the manufacture of RDF as it increases the surface area of the fuel particles.

Shredding can process demolition debris and yard wastes. However, the most important application of shredders is for materials recovery. Shredded MSW provides a number of advantages to recycling and waste-to-energy systems:

- Waste volume is significantly reduced
- The waste becomes more homogenous
- Waste separation processes (e.g., Fe, paper, etc.) are facilitated
- Separation of the noncombustibles from the combustibles is made easier
- RDF is sized for convenient burning in power plant boilers

In an MRF, shredders can be installed in one or more positions along the processing scheme. Primary shredders are used to reduce incoming raw MSW, whereas secondary shredders are used to further reduce the size of the output product from the primary shredder.

Shredders are available in a variety of shapes and sizes, from portable paper shredders to huge units that shred flattened automobiles at the rate of 1/min. The three most common types of shredding units used for size reduction of MSW are the hammermill, the flail mill, and the rotary shear (Figure 7.20). The tub grinder is also used; however, it is primarily devoted to the processing of yard wastes or construction and demolition debris (Figure 7.21).

7.4.3.5 Hammermills

The most common shredder type is the hammermill. A hammermill is a large cylindrical or tapered unit equipped with a central rotor with a series of attached rapidly rotating hammers (Figures 7.20a and 7.22). Hammers are either fixed or swing on the rotating shaft to allow for rotation over bulky or very dense waste components. The rotor and hammers are enclosed within a heavy-duty housing. The housing interior may be lined with stationary breaker plates or mounted cutter bars. Shredding relies on heavy force breakage of particles by rapidly swinging hammers in the enclosed vessel. Size reduction occurs by the combined actions of tearing and impaction. Wastes are further reduced in size by being struck against breaker plates or cutting bars fixed around the inner walls of the chamber.

Feedstock can be commingled or sorted MSW; in other applications, however, entire automobiles or steel I-beams are processed for shredding.

A shredder used for MSW processing usually has a width/diameter ratio greater than 1.0, a hammer weight of 70 kg (150 lb), a hammer tip speed of 4260 m/min (14,000 ft/min), four rows of hammers, and a starting time of 30 sec (Vesilind et al., 2002). Rotational speed ranges from about 700 to over 3000 r/min with power often set at approximately 500 to 700 kW. Hammermills vary drastically in terms of horsepower, electrical needs, and type of acceptable input. For high-speed shredders, rotational speeds are usually set between 1000 and 3500 r/min. A high-speed shredder relies on brute force and is very noisy. High-horsepower motors, i.e., 50 Hp and higher, are necessary. As a result, electrical costs can be substantial for high-speed shredders.

7.4.3.6 Horizontal-Shaft Hammermills

The hammer shaft can be oriented in either the horizontal or the vertical direction. In the horizontal configuration, input is from the top and the materials flow through the machine by gravity and

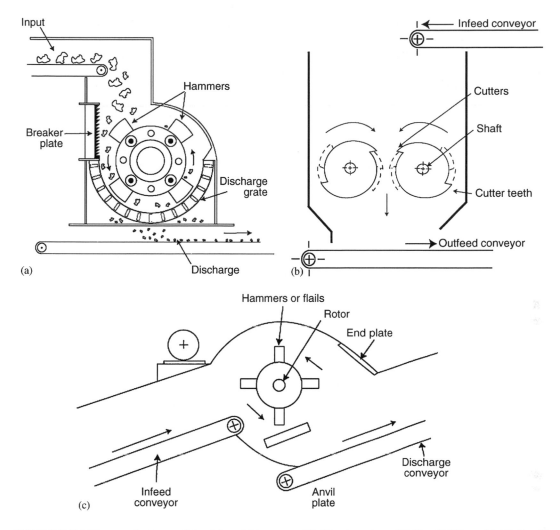

FIGURE 7.20 Three major types of shredders: (a) hammermill; (b) rotary shear (Pfeffer, 1992); and (c) flail mill (Pfeffer, J.T., *Soild Waste Management Engineering*. Prentice Hall, Englewood Cliffs, NJ, 1992). Reproduced with kind permission of Pearson Education, Inc., Upper Saddle River, NJ.

exit the bottom through a grate. Most horizontal hammermills have a grate placed across the outlet under the swinging hammers. The grate, possessing specific sized openings, may be changed depending on the desired size of the final product. This allows for more accurate control over final particle size. The hammers pound the material until it is small enough to pass through the grate openings. The size of the output material in horizontal type hammermills is therefore ultimately controlled by the size of the openings in the grate.

A disadvantage of the horizontal shaft unit is that if a durable waste such as a small engine block enters, it may remain there until it is broken to smaller sizes. This will result in considerable wear on the hammers; furthermore, excessive heat and sparks create a fire or explosion risk. Rejection portals in horizontal units are utilized in such cases.

7.4.3.7 Vertical-Shaft Hammermills

Vertical-shaft hammermills (Figure 7.23) are designed with the shaft mounted vertically. The vertical unit was originally designed in the United Kingdom for MSW processing. Large steel hammers

FIGURE 7.21 Tub grinder for yard wastes.

FIGURE 7.22 A new hammer for a horizontal hammermill.

rotate at high speeds within a large steel housing. The input enters at the top and flows downward by gravity. The housing is cone-shaped and tapers down to a narrow throat section. The rotating hammers create a vortex, or fan effect which, in addition to gravity, pulls wastes downward into the unit. Some size reduction occurs as the waste is milled in the upper portion. Farther down the rotor shaft more hammers are attached. Once the feed passes the throat section it is further reduced by the action of the lower hammers and the breaker bars on the housing walls. Adjusting the spacing between the hammers and the walls controls particle size. It is in the lower portion of the unit that most of the size reduction occurs. When the input waste reaches the bottom it has been sufficiently reduced in size such that a grate is not needed. The final shredded product is forced out through the discharge chute.

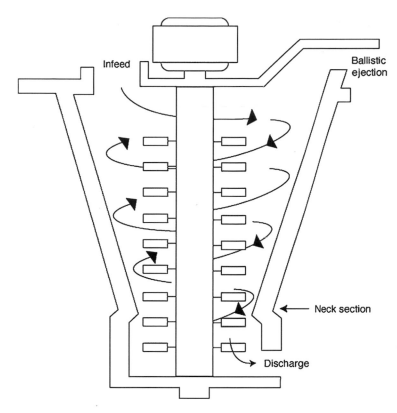

FIGURE 7.23 Vertical hammermill (Pfeffer, J.T., *Solid Waste Management Engineering*, Prentice Hall, Englewood Cliffs, NJ, 1992). Reproduced with kind permission of Pearson Education, Inc., Upper Saddle River, NJ

If the input material is difficult to break, for example a chunk of steel, the continued impact of the hammers will impart a centrifugal motion that will eventually direct the difficult item to an ejection portal. Such items will normally be ejected rather than remain in the mill and damage the hammers (Pfeffer, 1992).

The primary factors affecting particle size in the vertical-shaft hammermill include retention time in the mill and the number of impacts by the hammer. The clearance between hammers and wall of the housing in both the upper and lower portions of the machine regulates passage through the mill. By changing the number and position of the hammers, the particle size is changed.

An advantage of the vertical shaft hammermill is the ability to achieve a high degree of size reduction. Energy consumption per ton of waste processed is less for the vertical shaft hammermill than for the horizontal-shaft model. Primary disadvantages include high-energy costs and high maintenance. Operational problems encountered with the vertical shaft unit may include internal jamming and explosions.

A common waste processing problem when using a hammermill is contamination of organic materials with inorganics. For example, paper wastes are impregnated with shards of glass when the high-speed hammer shatters glass containers. The best course to pursue, therefore, is separation of potential contaminants before they enter the mill. Screening and hand-sorting are appropriate in this case.

7.4.3.8 Flail Mill

The flail mill (Figure 7.20c) is similar to the hammermill in that a rotary shaft is secured with a number of rotating appendages; however this provides only coarse shredding since the hammers are spaced farther apart. In some units, chains or knives replace the hammers. Flail mills tend to be single-pass devices whereas in the hammermill, wastes may be retained until they are small enough to pass through the gratings at the bottom of the unit. Flail mills tend to have a low power requirement (Pfeffer, 1992).

As the waste passes through the mill, the hammers strike the MSW and smash it against the anvil plate. If the particle is sufficiently small it will pass through the mill without size reduction. Typically the mill will impact larger containers, bags, boxes, and so on.

Advantages of the flail mill include low power requirements and low maintenance. The major disadvantage is the limited capability for size reduction.

7.4.3.9   Rotary Shear

The rotary shear or shear shredder contains two parallel counterrotating shafts with a series of disks mounted perpendicularly that act as cutters, working in a scissor-like fashion (Figure 7.20b). Rotary shears are low-speed devices (60 to 190 rpm) compared with hammermills. The input to be shredded is directed to the center of the rotating shafts. The size of the input is reduced by the shearing or tearing action of the cutter disks. The spacing and orientation of the shafts and the spacing between the shafts control the final particle size of the product (Pfeffer, 1993). Particle sizes can range from as low as 2.5 to 25 cm (approximately 1 to 10 in.). The shredded materials fall through or are pulled through the disks. Shutdown is not typical because even large bulky objects like railroad ties can be processed. Most are driven by hydraulic motors that can be reversed automatically in the event of an obstruction (Tchobanoglous et al., 1993).

Advantages of the rotary shear include slower speeds and the consequent lack of brute, destructive force. This will help in preventing contamination of organic wastes by broken glass — any glass containers or other small input can simply fall between the shears, given adequate spacing of the shafts. Additionally, lower speeds imply lower energy costs and also less maintenance of moving parts. As was the case for other shredders, large steel and other durable objects pose a problem for the rotary shear and should be removed prior to entering the shredder.

7.4.3.10   Hammer Wear

Due to the abrasive nature of high-speed shredding of highly heterogeneous wastes, hammers, grates, and housing walls are all subject to excessive wear. Both high- and low-speed shredders are maintenance-intensive due to their violent type of work.

Hammer damage and wear is the major maintenance issue related to MSW shredders (Figure 7.24). As the hammers wear, their effectiveness in shredding the waste decreases due to blunting the hammer tips and increased clearance between hammers and the housing or grates. Hammers are usually double-sided, and hammer wear occurs primarily on the outer edge, since this is the area of impact as the material is crushed against the grate. The wear is mostly due to abrasion although severe impact with very hard objects also contributes to wear. Maintenance involves periodically turning the hammers or knives. Once both sides are worn down, however, they must either be replaced, retipped, or resurfaced. Resurfacing involves rewelding a work surface on the hammers followed by resharpening to a cutting edge (Stessel, 1996). After a number of retippings the entire hammer must be replaced.

Hammer wear is reduced by employing special hardened facings to the hammers with abrasion-resistant alloys and by slowing the speed of shredding (Stessel, 1996; Vesilind et al., 2002). Shredding after removal of metals, concrete, glass, and ceramics greatly extends the lifetime of the hammers.

7.4.3.11   Safety Issues Related to Size Reduction

A number of hazards are possible during the operation of high-speed shredders. First, because waste materials are so heavily pulverized, dust concentrations can become sufficiently high, both in the

FIGURE 7.24 Old hammer. Compare with Figure 7.22.

shredder housing and in the shredder room or building, to cause an explosion. If a hammer strikes a metallic object and produces a spark, the dust in the atmosphere could ignite immediately. This hazard is magnified by the fact that the frictional action of the hammers and shafts produce excessive heat.

A significant danger is the inadvertent shredding of containers storing volatile or explosive compounds such as solvents, which create explosive atmospheres. A logical protection against such a hazard is a comprehensive inspection program to detect and remove explosive or flammable materials from entering the size reduction unit. Recent regulations governing the disposal of organic wastes should reduce the amount of such materials entering a MRF. A single empty solvent container is not sufficient to cause a significant explosion. A full solvent container could cause a problem if the vapors are allowed to accumulate. Dust is the most likely cause of explosions in shredders, however (Pfeffer, 1992).

Electrical switches, controls, and lighting should be installed in explosion-proof housing and conduit to avoid sparking (Tchobanoglous et al., 1993). Shredders should be installed in structurally isolated rooms separate from other processing areas. In some facilities, the MSW is conveyed to a shredder that is operated by remote control. In the event of an explosion, worker safety is enhanced by their physical separation from the mill. Keeping the MSW moist via a fine mist of water adds protection against dust hazards. Wetting will cause problems with liquid accumulation, treatment and removal, however. Water applied to wastes will also affect manufacture of RDF. A good quality ventilation system will pull dust-free, filtered air into the shredder and will greatly improve the safety of the local atmosphere (Pfeffer, 1992).

Explosion suppression devices have been manufactured which serve to reduce the force of an explosion. Sensors, designed to detect the pressure change from an explosion, are installed within the shredder housing. In response to a sudden pressure increase, an inert gas such as nitrogen or carbon dioxide is rapidly released into the chamber, thus displacing oxygen and preventing the propagation of a flame. Essential to the use of fire and explosion suppressants is adequate response time. This system works for most types of explosions but is not fast enough for supersonic releases (detonations) as occurs with TNT, gunpowder, and similar explosives. The added gases from the explosion suppression device work to fight a fire after an explosion, also by oxygen displacement. All shredders possess outlets for pressure buildup. Blast doors are situated in the roof above a shredder and a blast duct is used to direct the force of the blast upward, through the blast doors.

Blockage of materials due to material jamming between the rotor and the housing within the shredder is a frequent cause of operational problems. Should a high-speed shredder encounter an object that it cannot cut through, there is no reverse or overload setting. Either the problem material or the revolving shaft will have to give way. Preferably, the machine will simply jam up, although shafts have been known to break. In the event of a simple jam it is necessary to open the machine and remove the object. There may be substantial down time in order to remove jammed articles and to repair hammers. A hazard exists for workers entering the hammermill as there may be a buildup of pressure on the rotor and hammers caused by the obstruction.

Problems with flying objects and noise are also encountered in MSW shredding. Since raw MSW is contaminated, any dust produced contains a broad range of microorganisms and the atmosphere in the vicinity of the shredder may become a potential health hazard to facility personnel.

7.4.4 MAGNETIC SEPARATION

Magnetic separation is a relatively simple unit process designed to recover magnetic material, primarily ferrous metals, from mixed MSW. Two important reasons for removing the ferrous metals in a MRF are to increase the heat content of RDF and to recover a saleable product. On average, there is approximately 5.3% ferrous metal in the incoming MSW (U.S. EPA, 2001). Furthermore, metal removal reduces wear on subsequent processing and handling equipment and also reduces the amount of ash generated if the waste is to be incinerated. Magnetic recovery systems have also been used at landfill sites to recover product for recycling.

Magnetic separators are available in three primary configurations, i.e., the drum, the magnetic head pulley, and the magnetic belt pulley (Figures 7.25–7.27). Either permanent or electromagnets are employed. Magnets may be composed of exotic (e.g., rare earth) metals and are expensive. The principle of a single drum-type magnetic separator is shown in Figure 7.25. The drum is positioned under the lead pulley of a conveyor belt carrying mixed, shredded MSW. A stationary magnet is located inside the revolving drum. The ferrous metal in the MSW is attracted to the magnet against the force of gravity and is conveyed around the drum circumference until it exits the magnetic field

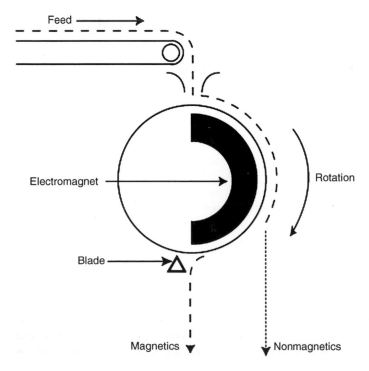

FIGURE 7.25 Drum magnet. (Vesilind, P.A., et al., *Solid Waste Engineering*, Brooks/Cole, Pacific Cole, CA, 2002. Reproduced with kind permission of Brooks/Cole, a division of Thomson Learning.)

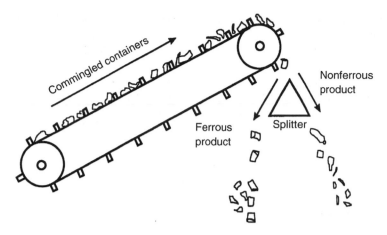

FIGURE 7.26 Magnetic head pulley (U.S. EPA, EPA/625/6-91/031, 1991).

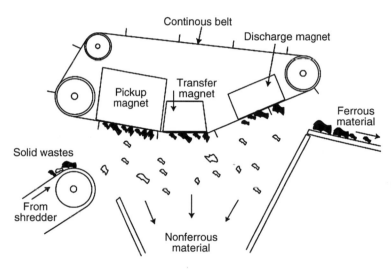

FIGURE 7.27 Magnetic belt pulley. (Tchobanoglous, G., et al., *Integrated Solid Waste Management: Engineering Principles and Management Issues.* McGraw-Hill, New York, 1993. Data reproduced with kind permission of the McGraw-Hill Companies, Inc.)

and is discharged. The drum magnet assembly can be installed for either overfeed or underfeed and directs the ferrous along a trajectory other than that taken by the nonferrous material.

The single-drum magnet tends to entrap pieces of paper and plastics. To minimize this problem a design using two-drum magnets with an intermediate belt conveyor can be used (Figure 7.28). The first drum is suspended above the end of the MSW feed conveyor and rotates in the direction of the material flow. Ferrous materials are picked up and directed forward to the intermediate belt conveyor. Most of the nonmagnetic materials fall to a conveyor located below the first drum. The second drum, which can be smaller than the first because of less material flow, is positioned over the discharge end of the intermediate conveyor, and rotates in a direction opposite to the material flow to avoid bridging or jamming. The ferrous metal is carried over the top of the drum and released on to a conveyor or bin on the far end (U.S. EPA, 1991).

The magnetic head pulley (Figure 7.26) conveyor is arranged so that material to be sorted is passed over the pulley such that the nonferrous material will fall along a different trajectory than will the ferrous material. A separator ('splitter') is positioned over the discharge end of the MSW

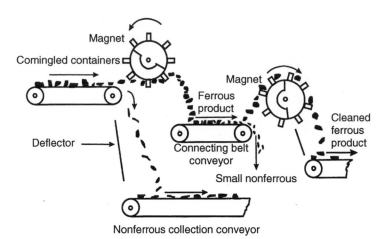

FIGURE 7.28 Two-drum magnet configuration (U.S. EPA, EPA/625/6-91/031, 1991).

feed belt. This is the simplest of the magnetic separation devices described in this chapter; unfortunately, there is a tendency of contamination by nonferrous components.

The overhead belt magnet is the most common magnet in MSW processing systems (Figure 7.27)(Stessel, 1996). The magnetic belt in its simplest form consists of a single magnet mounted between two pulleys that support a cleated conveyor belt mechanism (Figure 7.29). In placing the belt around the magnet, ferrous materials will rise upwards, and nonferrous materials will fall out of the stream by the action of gravity. The gap between the belt and the magnet permits an interval where entrained nonferrous materials can fall back onto the feed belt.

The depth of the waste stream affects the efficiency of magnetic separation. For more complete removal of ferrous, a secondary magnetic separator may be added to the processing train. In order to limit interferences, conveyor and hopper components in the vicinity of the magnetic field should be constructed of nonmagnetic materials.

Entrainment of nonferrous particles with the desired ferrous product is a common problem. One solution is to employ a dual-sequential magnet system. More commonly, an air classifier is utilized to clean the input (Stessel, 1996). In order to improve ferrous recovery a more sophisticated belt magnet has been devised. The belt is again suspended above a typical conveyor belt that is transporting processed MSW. It consists of a strong electromagnet that can recover relatively heavy pieces of ferrous metal. A belt that transports the ferrous to a recovery bin again covers the magnets. As the ferrous is transported to the main magnet, the polarity of the magnetic field is reversed, causing the metal to rotate. As the polarity changes, the metal drops a very small distance from the belt and rotates 180°. This movement allows for the entrapped nonferrous wastes to be released from the belt (Pfeffer, 1992).

Although magnetic separators have been used for numerous industrial applications, their use with MSW presents some problems. There is a tendency for nonmagnetic materials such as paper and plastic to be entrapped with the ferrous metal thereby reducing the purity of the recovered metal product. Furthermore, sharp edges on metals shorten the life of rubber belts. Although the resale value of ferrous scrap is low, it is advantageous to remove most of the ferrous materials from the waste stream early on. As noted earlier, metals will cause problems for other parts of the MRF processing train.

The effectiveness of magnetic separation depends on several variables, including:

- The height of the magnet above the conveyor belt carrying the MSW. The closer the magnet is to the MSW input, the more effective is the ferrous removal (Figure 7.30) (Vesilind et al., 2002).
- The greater the magnetic force applied, the greater the recovery of the ferrous fraction (Parker, 1983).
- Speed of the conveyor. Higher speeds will experience reduced recovery due to insufficient contact of ferrous materials with the magnet.

(a)   (b)

FIGURE 7.29 Overhead belt magnet in operation.

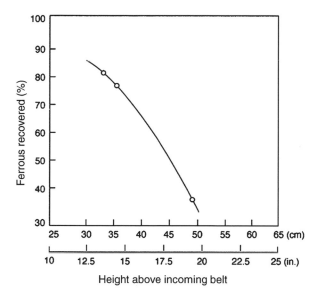

FIGURE 7.30 The height of a magnet above the waste stream affects the efficiency of ferrous recovery. Reproduced with kind permission of B.L. Parker.

- Depth of burden on conveyor. The deeper the waste on the belt, the lower recovery of ferrous.
- Material density. Dense wastes such as steel containers will sink below other wastes on a conveyor. Such settling is increased with increased time on the conveyor belt (Vesilind et al., 2002).

7.4.5 EDDY CURRENT DEVICE

The eddy current unit operation separates aluminum products from other nonmetals. An aluminum separator employs either a permanent magnetic or electromagnetic field to generate an electrical current (eddy), which causes aluminum cans (nonferrous) to be ejected and separated from other materials. Eddy current separation is based on the use of a magnetic rotor with alternating polarity, spinning rapidly inside a nonmetallic drum driven by a conveyor belt. Eddy current separation is based on Faraday's law of electromagnetic induction:

$$-dB/dt = V/A \qquad (7.10)$$

where B is the magnitude of magnetic flux density (T), V the voltage, and A the cross-sectional area normal to magnetic field (m^2).

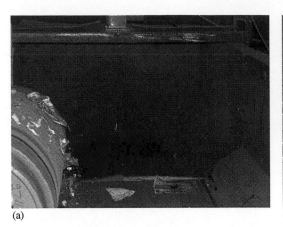

(a)   (b)

FIGURE 7.31 An eddy current separator in operation.

As nonferrous metals pass over the drum, the alternating magnetic field creates eddy currents in the particles, repelling the material away from the conveyor. While other materials drop off at the end of the conveyor, the nonferrous metals are propelled forward over a splitter for separation (Figure 7.31) (Walker Magnets, no date).

A time-varying field can be created either by rapidly reversing the voltage on an electromagnet (i.e., using alternating current) or by using strips of permanent magnets with alternating polarities.

7.4.6 AIR CLASSIFIER

Air classification is a unit operation designed to separate light waste components such as paper and plastic from heavier materials based on their differential behaviors when subjected to a stream of air. When a waste mixture is fed into an air stream of sufficient velocity, the lighter materials will be carried away with the air stream while the heavier components fall.

Air classification has been employed by industry for many years for the separation of various components in mixtures. Air classifiers are used in waste-to-energy processing lines to segregate the MSW stream into two fractions. One fraction consists of light materials (paper, plastic, wood, and dust), and the other is composed of heavy materials (metals, glass, and stones). In most MSW, the light fraction constitutes 60 to 75% of the total (see Chapter 4). Air classification concentrates the combustibles into the light fraction as a fuel product. Also, the metals and glass can be separated from the heavy fraction and sold in secondary markets. Often in the processing scheme an air classifier is situated after the magnetic separator and before the secondary shredder.

Separation is optimized through the proper design of the separation chamber, airflow rate, and material feed rate (Tchobanoglous et al., 1993). Specific variables of the input waste feed will affect material separation through air classification. Variables include:

- Particle density
- Particle size
- Particle surface area

Air classifiers may be configured in several designs of varying capacity and efficiency of separation. A schematic diagram of a typical air classifier is provided in Figure 7.32. The vertical, straight type is one of the most common and basic configurations of air classifiers. In the vertical unit, shredded MSW is dropped downward into the chute. An upward stream of air, fed by blowers, lifts lightweight materials upward for subsequent capture in a cyclone or other receptacle. There is little breakage of aggregated particles. Airflow direction is fairly uniform and the airflow rate is held constant. Variations to the simple vertical design include the installation of baffles along the length.

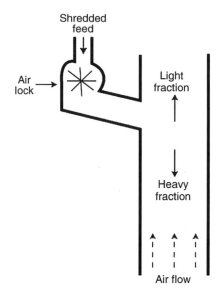

FIGURE 7.32 Schematic of a typical air classifier. (Reproduced with kind permission of Rhyner, C.R. et al., *Waste Management and Resource Recovery*, Lewis Publishers, Boca Raton, FL, 1995. Copyright Lewis Publishers, an imprint of CRC Press.)

Such appendages may be angular, thus causing turbulence within the housing and causing some of the aggregated particles to separate.

A second type of air classifier consists of a vertical column having a zig-zag internal configuration through which rapid stream of air is drawn up at a constant rate. Shredded wastes are introduced at either the top or middle of the column and air is introduced at the base. A rotary airlock mechanism is necessary to introduce the shredded wastes into the classifier housing. These classifiers use gravitational force and the impact upon the sides of the housing to break up aggregates and minimize entrapped "lights." The shape of the structure creates a vortex effect (Figure 7.33), which causes the waste to tumble and thus enhances separation of clumps. Lighter particles will follow the air stream up and heavier components fall. Although the zig-zag shape of the housing has been shown empirically to enhance separation the shape has also been shown to also enhance blockage of input wastes (Stessel, 1996).

In a third configuration, the pulsed air classifier uses a varying airflow velocity instead of a constant airflow velocity. Airflow to the column is varied with a louver valve. The pulsed airflow unit achieves a better discrimination between materials. Velocity of a falling object is a function of time until its terminal velocity has been attained. Varying the velocity of the air stream has the effect of keeping the falling particles in a velocity range such that particles with similar terminal velocities can be more completely separated (Tchobanoglous et al., 1993). A pulsed system can employ a simple straight throat as was the case for the vertical air classifier.

Horizontal air classifiers are also in use in MRFs (Figure 7.34). In the horizontal system both the light and heavy waste components are entrained with the air stream in one direction. The waste and air enter at one end of the shaft and are forced toward the other end. Separation occurs when heavier components of the waste that have hugged the bottom of the shaft fall through the opening and are collected, while the light fraction is forced beyond the opening to a separate collection area (Rhyner et al., 1995).

A similar concept to the standard air classifier is the air knife (Figure 7.35). This relatively simple separation device has been compared with throwing leaves and twigs upward into an autumn breeze. In the air knife, the airflow is forced horizontally through a vertically dropping input. Lighter particles are carried with the air stream while the heavier ones quickly drop. Another use of the air knife has been to prevent light contamination from carrying over during magnetic separation. Air is blown opposite the direction of travel of the metal under a magnet. The air flow helps in separating the lights from the metals and keeps the lights from being carried over to the metals conveyor (Vesilind et al., 2002).

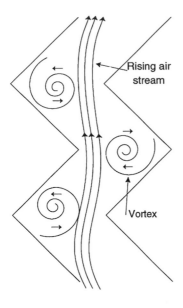

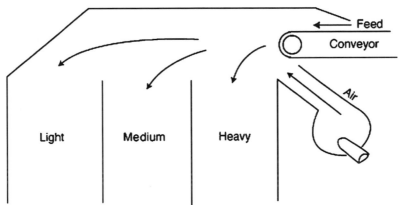

FIGURE 7.33 Vortex effect occurring within an air classifier. (Stessel, R.I., *Recycling and Resource Recovery Engineering,* Springer, Berlin, 1996. Reproduced with kind permission of Springer-Verlag, Inc.)

FIGURE 7.34 Horizontal air classifier (NASA CR-2526 as cited in Domino, 1979).

All air classifiers use one of two types of air transport to aid in the separation. A positive-pressure air transport system will push the MSW feed through the system. This is accomplished by attaching a blower to the air classifier housing and creating a higher pressure within the system relative to the ambient environment. The other method, a negative-pressure air transport system, pulls the MSW through. An exhaust fan is placed at the end of the system, creating a lower pressure within the system. This has the same effect as a vacuum.

The extracted materials must be removed from the air stream once they are separated. Following the air classifier, a cyclone separator is often used to separate the light fraction from the conveying air. Before being discharged to the outer atmosphere, the conveying air is passed through a dust collection system, typically a baghouse (Tchobanoglous et al., 1993). Alternatively, the discharge air can be recycled back to the air classifier. The light fraction is stored in bins or conveyed to another shredder for further size reduction before storage or utilization as a fuel or compost feedstock.

In the cyclone (Figure 7.36), particles and air enter the chamber at a tangent, setting up a high-velocity rotational air movement within the chamber. The solid particles, having greater mass, move outward toward the inside wall, are slowed down on contact, and eventually drop out of the bottom

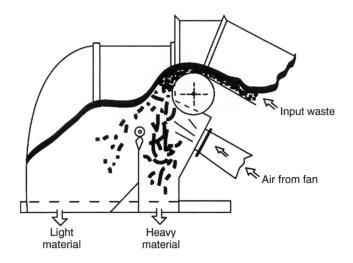

FIGURE 7.35 Air knife. (Pfeffer, J.T., *Solid Waste Management Engineering*, Prentice Hall, Englewood Cliffs, NJ, 1992. Reproduced with kind permission of Pearson Education, Inc., Upper Saddle River, NJ.)

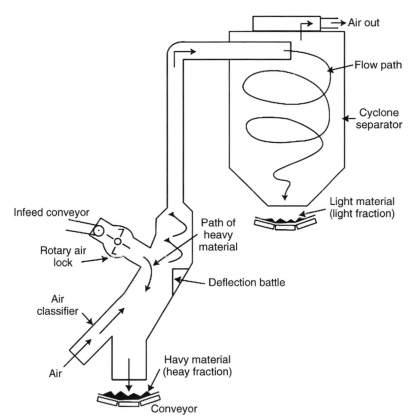

FIGURE 7.36 Complete air classification system including cyclone. (Tchobanoglous, G. et al., *Integrated Solid Waste Management: Engineering Principles and Management Issues*. McGraw-Hill, New York, 1993. Data reproduced with kind permission of the McGraw-Hill Companies, Inc.)

of the chamber under the force of gravity. The air, free of solids, exits the unit. Some suggest that a large fitter bag be installed which will capture particles while the exit air is filtered and escapes. The air can be either pushed or pulled, and a fan can be placed either before or after the cyclone (Vesilind et al., 2002).

7.4.7 EFFICIENCY OF SEPARATION IN AN AIR CLASSIFIER

As may be obvious from discussions of the various unit operations in an MRF, complete separation of one material from all others is not possible. In the case of air classification, the recovery of organics is complicated by two factors (Vesilind et al., 1988):

- Not all organics are aerodynamically light, and many inorganics (e.g., aluminum foil) are aerodynamically heavy.
- Perfect separation of heavy and light materials is difficult because of the stochastic (i.e., any phenomenon obeying the laws of probability) nature of material movement within the classifier.

In Figure 7.37, terminal settling velocity (i.e., air velocity at which the particle will just begin to rise with the air stream) is plotted against the fraction of particles of various materials. Regardless of the air velocity chosen there will never be a complete separation of the lighter organics (paper and plastic) from the heavier inorganics (steel). Figure 7.38 shows the efficiency of separation of fractions vs. the feed rate to the air classifier. With a greater loading of solids, an increasing proportion of light particles will fall into the underflow stream rather than be separated as originally planned.

The effectiveness of air classification can be estimated using published data (Figure 7.37). As the air velocity in an air classifier is increased from zero, the first material to float upwards and collect in the extract is paper. At an air velocity of about 500 cm/sec (1000 ft/min) all of the paper occurs in the extract. However, at this velocity about 50% of the plastic is also entrained. If we increase the air velocity beyond 500 cm/sec, some of the aluminum will start to become entrained with the extract. At an airflow of 1010 cm/sec (2000 ft/min), all of the paper and plastic will be collected in the extract; however, this will be contaminated by 50% of the aluminum. The steel component would not become entrained until the air velocity exceeded 1010 cm/sec (2000 ft/min). At 1525 cm/sec (3000 ft/min), virtually all of the input feed would become entrained. From the data in Figure 7.37, then, if the goal of an MRF is to produce high-quality feedstock for RDF, the air velocity is best maintained under 500 cm/sec (1000 ft/min).

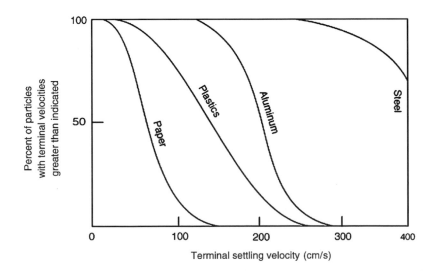

FIGURE 7.37 Terminal settling velocities of MSW components. (Vesilind, P.A. et al *Environmental Engineering*, 2nd ed., Butterworths, Boston, MA, Reproduced with kind permission of Elsevier, Oxford, U.K.)

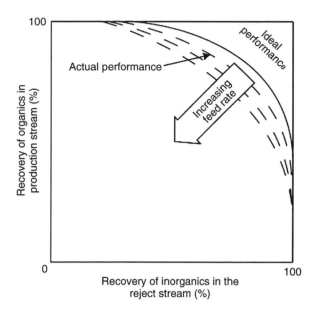

FIGURE 7.38 Actual and ideal performance of air classifiers. (Vesilind, P.A. et al *Environmental Engineering*, 2nd ed., Butterworths, Boston, MA, Reproduced with kind permission of Elsevier, Oxford, U.K.)

EXAMPLE 7.5

Shredded MSW containing equal amounts of paper, plastics, aluminum, and steel is fed into an air classifier operating with an air velocity of 175 cm/sec. Calculate the recovery of the organic product and the purity of the product. Use the terminal settling curves of the individual waste components shown in Figure 7.37.

Based on the figure, at 175 cm/sec, the fractions of the components that are captured as product are:

Paper	98%
Plastics	75%
Aluminum	20%
Steel	0%

The total organics (paper + plastic) in the product is

$$98\ (1/4) + (75)\ (1/4) = 43\% \text{ of feed}$$

The total aluminum + steel in the product is

$$(20)\ (1/4) + (0)\ (1/4) = 5\% \text{ of the feed}$$

Using equations 7.3–7.5, the recovery of organics is calculated as

$$R_{org} = 43/(25 + 25) \times 100 = 86\%$$

and the purity is

$$P_{product} = 43/(43 + 5) \times 100 = 90\%$$

7.4.8 Miscellaneous Processing

7.4.8.1 Dryers

In waste-to-energy systems, dryers reduce the moisture content of the MSW thereby increasing its heat (Btu) value. Moisture reduction also results in improved storability, possibly improved air classification, and a reduction in populations of potentially pathogenic microorganisms. The dryer is typically installed just before or just after the air classifier.

A drying system is equipped with: (1) a blower to circulate the heat and to force the MSW charge forward; (2) a combustion chamber to generate heat using RDF, (3) a rotary drum with a variable speed motor, and (4) a cyclone separator which separates the MSW from the air stream. The MSW inlet is positioned directly in front of the rotary drum. The waste and hot air stream flow through the drum in multiple stages, typically passing the length of the drum several times before exiting to the cyclone separator. As the drum rotates constantly, the waste is agitated and increased surface area is exposed to the heat (Bendersky, 1982).

7.5 MATERIALS FLOW IN THE MRF

There are a number of designs for unit operations in a MRF. As mentioned earlier, a MRF can be "clean" or "dirty." The former processes materials that have already been source-separated into various fractions. The latter type will accept bags of commingled wastes collected directly from the curbside. Considerations as to clean vs. dirty modes and what equipment to employ will vary according to factors such as initial capital costs and funding available, political pressures, and convenience to the consumer. A simplified waste separation scheme is depicted in Figure 7.39.

Placement of unit operations in an MRF varies substantially depending on the types of material desired for separation and the purity desired. For example, the placement of a trommel screen upstream of a shredder will result in the removal of stones and other small abrasive debris. This removal will lengthen the lifetime of the hammermill shredder by reducing hammer wear. Additionally, trommels will separate a large proportion of glass containers. If not removed, glass will shatter in the shredder and become embedded in paper and other potentially recyclable materials. If this paper product is combusted as fuel, there are implications for handling larger volumes of ash.

In some communities, source-separated wastes may be collected in see-through bags which are collected in the same truck along with mixed MSW. A typical process flow diagram for a MRF employing separation, manual combined with mechanical, of materials from commingled MSW and source-separated wastes is illustrated in Figure 7.39. Commingled MSW from residential and other sources are discharged in the receiving area. Hazardous items are removed immediately by hand. Recyclable and oversized materials such as lumber, white goods, and furniture are also removed in this first-stage operation. Source-separated materials in see-through plastic bags are also removed from the commingled MSW. Next, the commingled waste is loaded onto an inclined conveyor. Additional cardboard and large items are removed manually from the conveyor at the second presorting station as the waste material is transported to the bag-breaking station. The next step involves breaking open the plastic bags, which can be accomplished either manually or mechanically. In some facilities, a short enclosed trommel equipped with protruding blades is used as a bag breaker. Flail mills, shear shredders, and screw augers have also been used as bag breakers.

After the presorting steps the materials typically removed include paper, cardboard, all types of plastic, glass, and metals. In some operations, different types of plastic (e.g., PETE and HDPE) are separated simultaneously. Material remaining on the conveyor is discharged into a trommel or disc screen for size separation. The oversized material is sorted manually a second time (second-stage sorting). Mixed source-separated materials are further sorted using the second-stage sorting line. Source-separated mixed paper and cardboard are processed separately using a flow diagram such as the one given in Figure 7.40. The undersized material from the trommel screening and the material remaining after the second-stage sorting operation are hauled away for landfill disposal, processed further and combusted, or used to produce compost to be used as daily landfill cover. As shown in Figure 7.40, further processing of the residential materials usually involves shredding and magnetic separation (Tchobanoglous et al., 1993).

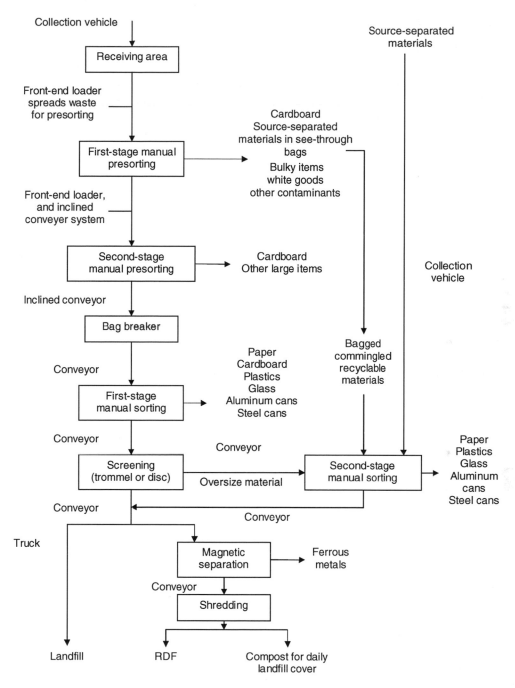

FIGURE 7.39 MRF flow scheme for a dirty MRF. (From Vesilind, P.A. et al., *Solid Waste Engineering*, 1st ed., Brooks/Cole, Pacific Grove, CA, 2002. Reproduced with kind permission of Brooks/Cole, a division of Thomson Learning: www.thomsonrights.com.)

7.6 THE CONTAMINATION ISSUE

Depending on the design of the MRF, the quality of the separated product will vary significantly. Mixed MSW received at the facility poses the greatest challenge to ensuring separation of clean, quality product. For this reason it is advantageous to require some initial separation, ideally by the

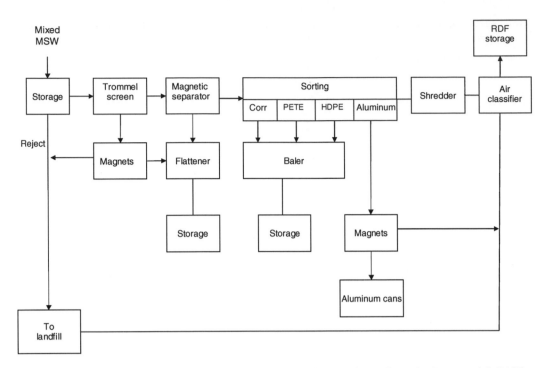

FIGURE 7.40 A second materials flow plan, dirty MRF. (Tchobanoglous, G. et al., *Integrated Solid Waste Management: Engineering Principles and Management Issues*, McGraw Hill, New York, 1993. Data reproduced with kind permission of the McGraw-Hill Companies, Inc.)

waste generator, well in advance of collection. Having recyclables picked up separately from the non-recyclable wastes best ensures adequate separation.

The success of recovery at the MRF is variable, depending on the processes used, but recovery from a dirty MRF is obviously much less than we would expect for a source-separated system.

7.7 ENVIRONMENTAL CONTROL

To protect the health and safety of facility employees as well as to meet environmental requirements of the community in which the MRF is located, it is often necessary to install equipment beyond that which normally is supplied with the material handling, separation, or size reduction equipment. Title 29 of the Code of Federal Regulations, Part 1910, presents the Occupational Safety and Health Administration (OSHA) standards which must be met to provide for adequate worker protection. Local codes often address the environmental relationship of a facility within the community. In the planning and design phase of the facility, those operations likely to cause problems for either the worker or community should be studied in order to determine methods as to how to best eliminate the problems (U.S. EPA, 1991).

7.7.1 Dust Collection

MSW brought to the tipping room floor is typically laden with soil and dust. Additionally, shredding, crushing, baling, screening, and conveying are dust-producing operations. Dust can cause several problems: it can be a vector for the transmission of pathogenic microorganisms, it can itself have a detrimental effect on health by affecting the respiratory system, and it can explode.

OSHA standards presently limit dust inhalation to 15 mg/m^3 of total dust over an 8 h day. Studies of dust production in resource recovery facilities have shown that dust levels are from

7 to 13 times higher than the OSHA standard. This finding dictates the use of facemasks while working.

Plate counts at shredding operations have indicated that the total bacterial counts during shredder operation are as much as 20 times greater than the ambient, which contains about 880 organisms/m^3 of air. In resource recovery facilities where shredders are used, coliform counts can increase from 0 to 135 per m^3 (0 to 69 per ft^3), and fecal streptococci from 0 to over 975 per m^3 (0 to over 500 per ft^3) (Diaz et al., 1976; Vesilind et al., 2002). These data show the potential danger of disease transmission by air during shredding of MSW.

The degree of dust collection is a function of the types and volumes of wastes handled, unit operations, and local climate. The solution can vary from the installation of individual dust collection units at each operation along the line, to one or two centralized dust collection systems. Dust collection systems include fans, ducts, cyclones, and baghouses. The plant worker may be required to wear a simple dust mask during operations (Drobny et al., 1971; Vesilind et al., 2002).

7.7.2 ODORS AND THEIR CONTROL

Odors can be significant in MRFs, especially the "dirty MRFs" where raw MSW is brought to the facility, stored, and subsequently processed. Odors can often be reduced or eliminated by minimizing storage time of raw materials or product followed by frequent wash-down of tipping floors.

Some of the most effective techniques involve applying negative pressure (suction) within an enclosed MRF and treating the exhaust gases. These gases can be incinerated, passed through GAC filters, or chemically (catalytically) oxidized. They can furthermore be scrubbed by passage through a gravity spray tower, i.e., an enclosed chamber containing a fine mist of water or other solution. There has been some work on biofiltering exhaust gases; i.e., forcing them through a mixture of soil and gravel. Indigenous microorganisms will act upon and oxidize a wide range of these gases, using them as substrates. In the interest of speed and cost, odors can simply be masked. A 'disinfectant spray' can be released in the working area. Its purpose is primarily to coat the nostrils of workers, thereby masking the malodorous MSW gases — it does not remove the noxious gases.

In situations with severe odors, multiple technologies may be required. Each technology may be accompanied by problems (in addition to capital and operating costs) of its own.

7.7.3 NOISE SUPPRESSION

Much of the equipment used in MRFs generate noise: conveyor belts, crushers, pumps, front-end loaders, and so on. The noise levels around a 3 ton/h hammermill range from 95 to 100 dBA with much of the noise produced being low-frequency. The unit 'dBA' is a standard method of noise measurement and stands for decibels on the A scale of the sound-level meter. This scale is an attempt to duplicate the hearing efficiency of the human ear. In addition to a high constant noise level, materials recovery facilities processing MSW produce considerable impact noise which is difficult to measure, and whose effect on human beings is poorly understood. The existing OSHA standard limits noise to 90 dBA over an 8 h working day. The corresponding limit set by the EPA is 85 dBA. Shredder operators must wear ear protection (Vesilind et al., 2002).

Engineering controls are difficult to establish in many situations. Sound muffling equipment and soundproofing at specific work locations may be installed throughout the building or specific pieces of equipment can be isolated, although this is often impractical.

7.7.4 AESTHETICS

Practical issues to be addressed in the design of MRFs include aesthetics and public health. In order to be a good neighbor to the community, it is important for a MRF to be constructed and landscaped in order to fit in well with an area zoned for commercial or industrial use. Berms and attractive vegetation can be established at the perimeter of the property. Blowing litter

FIGURE 7.41 The exterior of the MRF should be aesthetically appealing to limit NIMBY attitudes.

should quickly be removed from facility grounds. Attractive signs are to be posted at the entrance (Figure 7.41).

Odors of decomposing wastes will fill the MRF and will readily migrate beyond the facility boundaries if proper precautions are not taken. Rodents and insects may be a problem. Spontaneous combustion can occur in MSW piles. A rule of thumb is that 2 days of storage is a safe maximum, with a week being dangerous (Vesilind et al., 2002). A waste fire is difficult to extinguish and the newly formed wet wastes produced after extinguishing pose new disposal problems.

REFERENCES

Bendersky, D., *Resource Recovery Processing Equipment*. Noyes Data Corp., Park Ridge, N.J., 1982.

Berenyi, E. B., State of MRFs: 2001, *Resource Recycling*, January 2001, pp. 16–21.

Conveyor Equipment Manufacturers Association, Conveyor Terms and Definitions, CEMA No. 102-1994, Manassas, VA, 1995.

Diaz, L.F., Riley, L., Savage, G., and Trezek, G.J., Health considerations associated with resource recovery, Compost. Sci., 1976, 17, pp. 18–24.

Diaz, L.F., Savage, G., and Golueke, C.G. *Resource Recovery from Municipal Solid Wastes*, Vols. 1 and 2, CRC Press, Boca Raton, FL, 1982.

Drobny, N.L., Hull, H.E., and Testin, R.F., Recovery and Utilization of Municipal Solid Wastes, EPA-OSWMP SW-Ioc, Washington, DC, 1971.

Ely, K., Jr., Processing equipment, in *The McGraw-Hill Recycling Handbook*, Lund, H.F., Ed., McGraw-Hill, Inc., New York, NY, 1993.

Engdahl, R.B., *Solid Waste Processing*, EPA OSMP, Washington, DC, 1969.

Hering, R and Greeley, S.A., *Collection and Disposal of Municipal Refuse*, 1st ed., McGraw-Hill, Inc., New York, NY, 1921.

Parker, B.L., Magnetic Separation of Ferrous Metals from Shredded Refuse, M.S. thesis, Duke University, Durham, NC, 1983.

Pfeffer, J.T., *Solid Waste Management Engineering*, Prentice Hall, Englewood Cliffs, NJ, 1992.

Rietema, K., On the efficiency of separating mixtures of two components, *Chem. Eng. Sci.*, 7, 89, 1981.

Resnick, W., Flow Visualization Inside Storage Equipment, *Proceedings of the International Conference on Bulk Solids — Storage, Handling and Flow*, Powder Advisory Centre, London, 1976.

Rhyner, C.R., Schwartz, L.J., Wenger, R.B., and Kohrell, M.G., *Waste Management and Resource Recovery*, Lewis Publishers, Boca Raton, FL, 1995.

Schwarz, S.C. and Brunner, C.R., *Energy and Resource Recovery from Waste*, Noyes Data Corp, Park Ridge, NJ, 1983.

Stessel, R.I., *Recycling and Resource Recovery Engineering*, Springer-Verlag, Berlin, 1996.

Sullivan, J.W., Hill, R.M., and Sullivan, J.F., The place of the trommel in resource recovery, *Proceedings of the Waste Processing Conference*, American Society of Mechanical Engineers, New York, NY, 1992.

Tchobanoglous, G., Theisen, H., and Vigil, S., *Integrated Solid Waste Management: Engineering Principles and Management Issues*, McGraw-Hill, Inc. New York, NY, 1993.

U.S. Environmental Protection Agency, Municipal Solid Waste in the United States: 1999 Facts and Figures, EPA 530-R-01-014, Office of Solid Waste and Emergency Response, Washington, DC, 2001.

U.S. Environmental Protection Agency, Materials Recovery Facilities for Municipal Solid Waste, EPA/625/6-91/031, Office of Research and Development, Washington, DC, 1991.

Vesilind, P.A., Worrell, W.A., and Reinhart, D.A., *Solid Waste Engineering*, Brooks/Cole. Pacific Grove, CA, 2002.

Vesilind, P.A., Peirce, J.J., and Weiner, R.F., *Environmental Engineering*, 2nd ed., Butterworths Publishing, Boston, MA, 1988.

Vesilind, P.A. and Rimer, A.E., *Unit Operations in Resource Recovery Engineering*, Prentice Hall, Englewood Cliffs, NJ, 1981.

Walker Magnets, No date. http://www.walkermagnet.com/mag.

SUGGESTED READINGS

Boettcher, R.A., Air Classification of Solid Waste, EPA OSWMP, SW-30c, Washington, DC, 1972.

Drobney, N.L., Hull, H.E., and Testin, R.F., Recovery and Utilization of Municipal Solid Waste, EPA OSWMT SW-10c, Washington, DC, 1971.

Hasselriis, F., *Refuse-Derived Fuel Processing*, Butterworths, Boston, MA, 1984.

Makar, H.V. and DeCesare, R.S., Unit Operations for Nonferrous Metal Recovery. Resource Recovery and Utilization, ASTM STP 592, 1975, pp. 71–88.

Michaels, E. L., Woodruff, K. L., Fryberger, W. L., and Alter, H., Heavy media separation of aluminum from municipal solid waste, *Trans. Soc. Mining Eng.*, 258, 34, 1975.

Murray, D.L. and Liddell, C.L., The dynamics, operation and evaluation of an air classifier, *Waste Age*, March 1977.

Nelson, W., Kruglack, A., and Overton, M., *Trommel Screening*, Duke Environmental Center, Duke University, Durham, NC, 1975.

Rietema, K., On the efficiency in separating mixtures of two components, *Chem. Eng. Sci.*, 7, 89, 1957.

Salton, K., I., Nagano, and S., Izumin, New separation technique for waste plastics, *Resour. Recovery Conserv.*, 2, 127, 1976.

Senden, M.M.G. and Tels, M., Mathematical model of air classifiers, *Resour. Recovery Conserv.*, 2, 129, 1978.

Trezek, G.J. and Savage, G., MSW component size distribution obtained from the Cal Resource Recovery System, *Resour. Recovery Conserv.*, 2, 67, 1976.

Worrell, W.A. and Vesilind, P.A., Evaluation of air classifier performance, *Resour. Recovery Conserv.*, 4, 4, 1980.

QUESTIONS

1. The number of MRFs in the United States has increased dramatically over the past decade. Explain the causes for this increase, based on factors such as economics, NIMBY, and environmental concerns.
2. Where in the MRF is human labor essential (i.e., a machine cannot adequately do the job)?
3. What factors influence the separation efficiency of a trommel screen?
4. What is the *most critical* parameter for efficient MSW separation by a trommel screen?

5. To avoid carryover of glass fines in an air classifier, what device or unit operation may wastes first be passed through?

6. What are the advantages to using a shredder for MSW processing in a MRF? What are the disadvantages?

7. Plastic wastes can be sorted optically or mechanically using 'color screening' (true/false).

8. What is the ideal trommel rotation speed?

9. List and discuss some of the major safety hazards associated with shredding MSW.

10. All other factors being equal, are there any practical advantages to horizontal hammer-mills over vertical models? Consider efficiency of shredding, energy requirements, noise production, and jamming by rigid articles.

11. Using mechanical separation equipment for separation mixed MSW, it is not possible to achieve 100% separation efficiency. Explain and provide an example.

12. During air classification, the recovery of organics is complicated by what factors?

13. What factors influence the effectiveness of magnetic separation?

14. Calculate the waste thickness for a conveyor belt measuring 1.0 m in width with an average belt speed of 20 m/min. The waste loading rate is 25.5 MT/h and the average density of the waste on the belt is 105 kg/m^3.

15. A trommel screen measuring 9 m long and 3 m in diameter is rotating at 2.5 r/min. The feed rate is 12 MT of raw MSW per hour. Calculate the critical speed. At the above speed would the waste input be cascading, cataracting, or centrifuging?

16. If the trommel screen in Question 15 were to be adjusted to a steeper angle, how would the separation efficiency and critical speed be affected?

17. A magnetic separator is employed at a MRF for ferrous recovery from MSW. The feed rate to the separator is 1255 kg/h. The feed contains 4.2% ferrous. A total of 40 kg is collected in the product stream and 32 kg is actually ferrous. Calculate the recovery, purity, and efficiency of the unit operation.

18. List the key components of the receiving area of an MSW processing facility.

19. In terms of separation of potentially recyclable components, what is a detrimental effect of shredding raw MSW to a fine particle size?

20. Contact your local waste management authority and determine the fraction of recovered materials for recycling collected via drop-off centers vs. curbside pickup programs or MRFs.

21. Diagram a sequence of unit operations for a mechanized waste separation system. The input waste includes paper products, food waste, glass, plastic (PETE, HDPE, and others), metals (ferrous and nonferrous), and hazardous wastes. List all the equipment that would be required for adequate separation. Include at least one shredder.

22. Hazardous materials can be removed from the commingled wastes arriving at a MRF via (a) magnetic separation, (b) manually, (c) air knife, (d) froth flotation.

23. Waste paper should be removed: (a) using an air classifier, (b) after ferrous removal, (c) after shredding the waste stream, (d) after trommel screening, (e) all of the above.

24. Suppose your community has decided to develop a waste management program with a goal of substantially greater waste recycling and materials recovery. The three major alternatives are source separation, a dirty MRF, and a clean MRF. Discuss the key factors for consideration by the community in order to make an educated decision about the optimum alternative. Consider issues such as short- and long-term economics, public acceptance, environment and aesthetics, and convenience to local citizens.

8 Composting MSW

Drive my dead thoughts over the universe
Like withered leaves to quicken a new birth!

Percy Bysshe Shelley (1792-1822)
Ode to the West Wind

8.1 INTRODUCTION

Organic materials comprise the majority of municipal solid wastes (MSW) generated in the United States, accounting for about 141 million tons (67%) of the waste stream (U.S. EPA, 1997). Some organics such as newspaper, office paper, and corrugated cardboard are recovered extensively for recycling. Other potentially useful materials (e.g., yard waste, food scraps, other grades of paper), however, continue to be lost from the recycling loop and landfilled.

The organic fraction of MSW can circumvent land disposal and undergo other fates including direct conversion into fuels, fermentation as a biofuel, pyrolysis, and composting. Composting has been in practice even before Roman times, and the method has since been streamlined in terms of efficiency, time of completion, health impacts, and area requirements. Composting of MSW, agricultural wastes (plant residues, animal manures), food factory waste, and municipal wastewaster treatment solids ("biosolids") is increasingly used worldwide as a means of waste management. It is estimated that about 75 million tons of the U.S. waste stream, including 28 million tons of yard waste, 25 million tons of food waste, and 25 million tons of soiled or unrecyclable paper, is available for composting (see Table 1.3).

A range of programs has been established to divert organic materials from landfills and create beneficial uses. These include:

- Mixed waste composting at centralized processing facilities that accept MSW and separate into composting, recycling, and disposal streams.
- Residential source-separated systems using organic matcrials separated by the generator, set out for collection, and processed at a centralized facility.
- Commercial composting operations that utilize materials generated by commercial and industrial establishments.
- Yard waste composting at centralized facilities.
- Backyard composting of food and yard waste.

8.2 BENEFITS OF COMPOSTING

Composting is defined as a controlled aerobic, biological conversion of organic wastes into a complex, stable material. The final product has a number of beneficial uses, most commonly for agriculture and landscaping.

If shredded raw MSW were introduced directly into a soil to be used for agriculture, the organic component would undergo rapid transformation by soil microorganisms. A number of adverse effects would result:

- *Undesirable reactions.* Anaerobic transformations will produce ammonia (NH_3), hydrogen sulfide (H_2S), and methane (CH_4) gas. Several gaseous products are toxic to plant growth and will additionally cause odor problems.

- *Competition for plant nutrients.* The most important nutrient to most crop plants is nitrogen. With the addition of raw waste to soil, microorganisms attack the carbon (an energy source), and will also require large quantities of nitrogen for manufacturing cell biomass. Being opportunistic and fast-growing, microorganisms can incorporate and render unavailable virtually all plant-available soil N such that plants cannot obtain sufficient quantities. This situation is termed "nitrogen depression."
- *Leaching.* Potentially toxic materials (e.g., salts, metals, acids, microbial cells) are released from raw waste into soil and water.

In contrast, composting transforms the organic feedstock via:

- Mineralizing the simple, easily assimilable substances, i.e., protein, cellulose, sugars, and lipids to carbon dioxide and simple N compounds (e.g., nitrate).
- Humifying more complex compounds such as lignin to produce a more homogeneous and stable organic product.

The final humus-like product is hygienically safer, more aesthetically appealing, and substantially lower in odor than the original MSW. The finished organic product has several potential applications. A primary application of compost is for agriculture: compost serves as a soil conditioner (i.e., an organic matter source which improves water-holding capacity, increases aeration, and improves drainage), and it supplies nutrients, particularly N, P, and S, all of which occur primarily in the organic form in soils. Compost also provides a number of micronutrients including Cu, Fe, Zn, and Ni. Many such trace nutrients will occur as organic chelates and complexes which are relatively plant-available. Finally, because composts are often circumneutral in pH they moderate pH extremes of the recipient soil. Compost is also used in landfill operations as a daily cover material, for landscaping applications, and for remediation of contaminated sites and mined lands (U.S. EPA,1997, 1998).

8.3   OVERVIEW OF THE COMPOSTING PROCESS

Composting on the commercial scale occurs in three major phases. Initial processing includes size reduction to enhance microbial reactions. First, separation of inert materials (glass, plastic, metals, etc.) from the organic fraction is necessary. Size reduction and chemical or biological conditioning are extremely important at the outset if the finished product is to be used in agriculture. Next, microorganisms decompose the raw feedstock into simpler compounds, producing heat as a result of their metabolic activities. The volume of the compost pile is reduced during this stage and the heat generated destroys many pathogens. In the final stage, the compost product is "cured." Microorganisms deplete the supply of available nutrients in the pile, which, in turn, slows down their activity. As a result, heat generation diminishes and the compost mass dries. When curing is complete, the compost is considered "stabilized" or "mature." Any further microbial decomposition occurs only very slowly. Figure 8.1 provides the overall steps involved in the aerobic composting of the organic fraction of MSW.

8.4   THE ROLE OF MICROORGANISMS IN COMPOSTING

Composting is an aerobic biological process; a diverse consortium of microorganisms acting concurrently controls this process. The most active players in composting are bacteria, actinomycetes, fungi, and protozoa. These microorganisms are naturally present in most organic materials, including food waste, soil, leaves, grass clippings, and other organics.

Composting is also dependent upon a succession of microbial activities, whereby the environment created by one group of microorganisms ultimately promotes the activity of successor groups.

Different types of microorganisms are active at different phases in the composting pile. Bacteria have the most significant effect on decomposition; they are the first to become established in the pile, processing readily decomposable substrates (e.g., proteins, carbohydrates, and sugars) faster than any other group. Nitrogen-fixing bacteria are also present in the compost pile, which will fix atmospheric N for incorporation into cellular mass. Table 8.1 lists some of the major microbial types involved in composting. Commercial products are available that claim to speed the composting process via the introduction of selected strains of bacteria, but inoculating compost piles has not been found to bring about completion any more rapidly (Gray et al., 1971; Haug, 1980; Rynk, 1992; U.S. EPA, 1994).

Fungi play an important role in composting as the pile dries, since fungi can tolerate low-moisture environments better than bacteria. Some fungi also have lower nitrogen requirements than bacteria and are, therefore, able to decompose lignin and cellulose materials, which bacteria cannot. Because fungi are numerous in composting, concern has arisen over the growth of genera such as *Aspergillus*, which pose a potential health hazard.

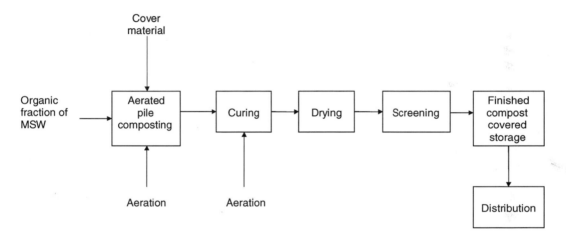

FIGURE 8.1 Flow chart showing the steps involved in the aerobic composting process.

TABLE 8.1
Microbial Populations During Aerobic Composting

| | Number per wet gram of compost | | | |
Microbe	Mesophilic Initial Temperature (40°C)	Thermophilic (40-70°C)	Mesophilic (70°C to cooler)	Species Identified
Bacteria				
Mesophilic	10^8	10^6	10^{11}	6
Thermophilic	10^4	10^6	10^7	1
Actinomycetes				
Thermophilic	10^4	10^8	10^5	14
Fungi				
Mesophilic	10^6	10^3	10^5	18
Thermophilic	10^3	10^7	10^6	16

Source: Haug, R. T., *The Practical Handbook of Compost Engineering*, Lewis Publishers, Boca Raton, FL, 1993. Reproduced with kind permission. Copyright Lewis Publishers, an imprint of CRC Press. Boca Raton, FL

The actinomycetes are often considered a middle group between the bacteria and the fungi. Most genera of actinomycetes produce slender, branched filaments that develop into a mycelium. Actinomycetes are widely distributed in soil, compost piles, river sediments, and other environments. They are second to bacteria in terms of abundance. Actinomycetes decompose aromatics, steroids, phenols, and other complex organic molecules (U.S. EPA, 1983; Eweis, *et al.*, 1998).

Macroorganisms also play a role in composting. Rotifers, nematodes, mites, springtails, sowbugs, beetles, and earthworms reduce the size of the compost feedstock by foraging, moving within the pile, or breaking up particles of the feedstock. These actions physically break down the materials, creating greater surface area and sites for microbes to attach and metabolize (U.S. EPA, 1994).

The bacteria and fungi important in decomposing MSW feedstock can be classified by optimal temperature regime as mesophilic or thermophilic. Mesophilic microorganisms experience most rapid growth at temperatures between 25 and 45°C (77 to 113°F). These are dominant within the pile early in the process when temperatures are near ambient. The mesophiles use oxygen within the interstices (pores) to oxidize carbon and thus acquire energy. End products of the reactions include carbon dioxide (CO_2) and water. Heat is also generated as chemical bonds in the substrate are broken during metabolism.

If the pile is insulated from the local environment with no aeration or turning, most of the heat generated is trapped within the pile. In the insulated center, when temperatures of the mass rise to about 45°C (112°F), the mesophiles die or are inactivated. At this time, thermophilic microorganisms, i.e., those that prefer temperatures between 45 and 70°C (112 and 158°F), are activated. These multiply and metabolize substrates and replace the mesophiles in most sections of the pile. Thermophiles generate even greater quantities of heat than mesophiles. The temperatures reached during this phase of the process are sufficiently high to kill most pathogens and also weed seeds. Many composting facilities maintain a temperature of 55°C (131°F) in the interior of the pile for 72 h to ensure pathogen destruction and to inactivate weed seeds.

The thermophiles continue decomposing the feedstock as long as nutrient and energy sources are available. As the substrates are depleted, the thermophiles die and the pile temperature falls. Mesophiles are again activated and decompose the remaining substrate until all available energy sources are utilized (U.S. EPA, 1994). Figure 8.2 provides a typical temperature pattern for composting processes, and Table 8.1 shows the density of microorganisms as a function of temperature during composting.

8.5 FACTORS AFFECTING THE COMPOSTING PROCESS

Composting is strongly influenced by many environmental factors; as a result, much research into system design and environmental controls has been conducted in attempts to optimize the process.

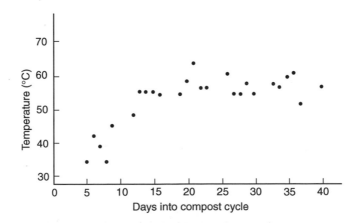

FIGURE 8.2 Temperature trends in the early stages of composting (U.S. EPA, 1984).

8.5.1 PREPROCESSING OF THE FEEDSTOCK

The preparation of the organic component of MSW for composting can be carried out in two ways:

- *Mechanical.* Composting a mixed waste material, i.e., organics commingled with inerts such as plastic and metal, is fraught with hazards and complications and should be avoided. Inerts (nonreactives) may benefit the process by acting as a bulking agent and promoting aeration of the mass; however, they end up as useless and sometimes hazardous components of the finished material. Mechanical processing involves size reduction by shredding, followed by separation of inert materials by screening, magnetic separation, and other unit operations (see Chapter 7). The resulting material possesses a higher surface area for reaction and a more available substrate for microbes.
- *Biological and mechanical.* In a combined process, waste is physically processed as described above, and then transferred to biological reactors (Figure 8.3) for 1 to 3 days. The reactors are rotating heated cylinders mounted on a slight incline. Biological activity increases significantly in the reactor and degradation of organics begins along with size reduction. After the preliminary treatment in the reactor, the feedstock is transferred to a compost pile.

8.5.2 ENVIRONMENTAL FACTORS

Microorganisms are clearly central to successful composting; therefore, the factors that affect their proliferation and activity are those which determine the rate and extent of composting. The principal environmental factors regulating the speed and degree of decomposition include nutrient levels, nutrient balance (e.g., carbon to nitrogen ratio), aeration, moisture, temperature, pH, and particle size of the feedstock material. Any shift in these factors are interdependent; a change in one parameter often results in changes in others. The closer these factors collectively approach optimum levels, the more rapid will be the rate of composting. The chemical and physical nature of the substrate, and aeration are especially important in process design.

FIGURE 8.3 Photo of a biological reactor. Reproduced with kind permission of Cornell Waste Management Institute.

8.5.3 Nutrients

The organic fraction of the MSW feedstock possesses a range of microbial substrates including proteins, lipids, sugars, starch, amino sugars, chitin, cellulose, lignin, crude fiber, and other compounds (see Chapter 4) that will vary in terms of nutrient content, energy content, and availability to microbes.

Nitrogen, P, K, Mg, S, Fe, Ca, Mn, Zn, Cu, Co, and Mo are integral to the protoplasmic structure of the microbial cell. These nutrients along with C, H, and O are essential for proper cell synthesis. Nutrients obviously must be present in sufficiently large concentrations in a substrate; however, they must also occur in a form that can be easily assimilated by the microbial cell. Availability is partly a function of the enzymatic production by the microbe. Thus, certain microbes possess enzymes that permit them to attack and utilize the organic matter within the raw feedstock, whereas others can utilize only intermediate products. The significance is that decomposition and, hence, the composting of MSW is the result of the activities of a dynamic succession of different groups of microorganisms, in which one group prepares the local environment for its successors.

Another aspect of nutrient availability in composting is that certain organic molecules are resistant to microbial attack, even to those that possess the required enzyme systems. Such *refractory* materials are therefore broken down slowly, even with all other environmental conditions set at an optimum level. Common examples of such materials are lignin (wood) and chitin (exoskeletons). Cellulose C is unavailable to most bacteria, although it is readily available to certain fungi. Nitrogen is easily available when in the amino acid form whereas N present in chitin is relatively unavailable. Many sugars and starches are readily decomposed, being a common unit of assimilable substrate. Similarly, many fats and fatty acids are relatively available to microorganisms.

The availability of nutrients is also influenced by pH of the feedstock. In the circumneutral pH range, trace metals (e.g., Cu, Ni, Zn) are typically soluble and therefore available in sufficient quantities. In contrast, excess quantities, for example under acid pH regimes, will prove toxic and inhibitory. Also, at neutral pH, phosphorus is maximally available. A pH of 5.5–8 is therefore generally considered optimal for composting (see below).

8.5.4 C:N Ratio

The ratio of carbon content to nitrogen content in the feedstock strongly affects the rate of microbial activity. With few exceptions, all other nutrients are present in organic waste in adequate amounts and ratios.

Carbon and nitrogen are used by microbes to obtain energy and for the synthesis of new cellular material. A large percentage of the carbon substrate is oxidized to CO_2 during metabolic activities. The remaining carbon is converted into cell wall or membrane, protoplasm, and storage products. The principal use of nitrogen is in the synthesis of protoplasm (e.g., proteins, amino acids, nucleic acids). Much more carbon is required than nitrogen for adequate microbial growth.

After much empirical research, the optimum C:N ratio for soil and compost microorganisms has been established at approximately 25:1. A ratio much higher than this will slow down the decomposition; if the initial ratio is over 35, the microbial consortium must pass through many life cycles, oxidizing the excess carbon to CO_2 until a more suitable ratio is attained. On the other hand, if the C:N ratio is lower than about 20:1, composting will be inhibited due to low-energy supplies and nitrogen will be lost both by leaching and volatilization as ammonia ($NH_{3(g)}$). A low C:N ratio in composts and soils is typically rare, however. Of course, these ratios may vary widely depending on the type of carbonaceous materials initially present.

If the initial C:N ratio of a waste is too high, adding a nitrogenous waste (e.g., blood meal) can bring it to acceptable levels. If the ratio is too low, a carbonaceous waste (straw, wood shavings, sawdust, shredded paper) can be added. The nitrogen contents and C: N ratios of various wastes and other materials are listed in Table 8.2.

TABLE 8.2
Carbon/Nitrogen Ratios of Various Wastes and Materials

Material	C:N
Sawdust	200-500:1
Wheat straw	125-150:1
Grass clippings	12-20:1
Corn stalks	60:1
Humus	10:1
Activated sludge	6:1
Cow manure	18:1
Horse manure	25:1
Poultry manure	15:1
Food scraps	15:1
Mixed MSW	50-60:1

Adapted from Diaz, L. F. et al., *Composting and Recycling Municipal Solid Waste*, 1994. Reproduced with kind permission of Lewis Publishers, an imprint of CRC Press. Boca Raton, FL.

8.5.5 AERATION

The atmosphere within the interstices of the composting mass will vary significantly during decomposition. When the organic feedstock is delivered to the compost site, the oxygen supply available to microbes occurs from the diffusion of ambient air, and in air originally trapped within the voids. However, the rate of diffusion of ambient air into the mass is very limited; hence, interstitial air is the major source of oxygen.

At the outset, the composition of air in voids is similar to that of ambient air (i.e., approximately 20.9% O_2 and 0.03% CO_2, v/v). Within a short time (hours to days), however, the heterotrophic pioneer communities become activated and begin decomposition of the raw organic substrates, with the concurrent increase in CO_2 content and decrease in O_2 level. With a closely monitored composting system, the oxygen content can be varied from 15 to 20% and CO_2 from 0.5 to 5% for the process to be successful.

During aerobic respiration, organic chemicals are oxidized to carbon dioxide and water or other end products using molecular oxygen as the terminal electron acceptor. Aerobic respiration occurs under highly oxygenated conditions. The reaction for the aerobic oxidation of a glucose molecule is

$$C_6H_{12}O_6 + 6O_2 \rightarrow 6CO_2 + 6H_2O + \text{energy} \tag{8.1}$$

If the O_2 concentration falls below approximately 15%, facultative anaerobic microorganisms are activated and rapidly become dominant. Fermentation and anaerobic respiration reactions take over. Undesirable products such as acetic acid, ethanol, methane, and ethane will form. These are odoriferous and may inactivate beneficial compost microorganisms:

$$C_6H_{12}O_6 \rightarrow 2C_2H_5OH + 6CO_2 + 6H_2O \tag{8.2}$$
$$\text{glucose} \qquad \text{ethanol}$$

Additional reactions under anaerobic conditions are described later.

The decomposition of organic materials is significantly faster and more complete in the presence of oxygen. The energy available in Equation 8.1 is approximately 14 times greater than that for anaerobic decomposition of glucose (Equation 8.2) (Zubay, 1983).

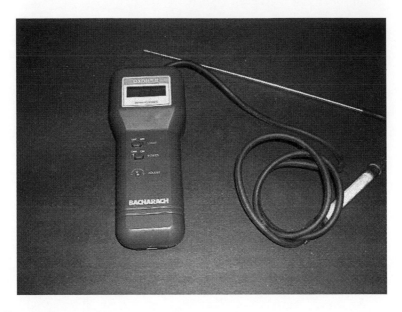

FIGURE 8.4 Portable O_2 meter.

The oxygen level of the pile can be measured using a simple portable O_2 meter (Figure 8.4). Oxygen requirements during aerobic composting can also be determined based on the composition of feed substrate and final product. This method can be applied when small pilot-scale composting studies are conducted and the final product composition is determined. Degradability of the feed substrate is thus assessed as well. Rich (1963) formulated the following stoichiometric equation:

$$C_aH_bO_cN_d + 0.5(ny + 2s + r - c)O_2 = nC_wH_xO_yN_z + sCO_2 + rH_2O + (d - nz)NH_3 \quad (8.3)$$

where $r = 0.5[b - nx - 3(d-nz)]$

$\quad s = a - nw$

The formulae $C_aH_bO_cN_d$ and $C_wH_xO_yN_z$ represent the compositions of feed substrate and final product, respectively. An elemental analysis is necessary in order to establish the subscripts.

EXAMPLE 8.1

Pilot-scale aerobic composting was conducted on 1000 kg (dry solids) of MSW feedstock determined to have an initial composition $C_{35}H_{67}O_{31}N$. By completion of composting (60 days), the initial 1000 kg of feedstock was reduced to 180 kg dry solids. The final product composition was determined to be $C_{14}H_{20}O_9N$. Calculate the stoichiometric oxygen requirement per 1000 kg of feedstock.

SOLUTION

1. The molecular weight of the substrate is

Carbon	35(12)=420
Hydrogen	67(1)=67
Oxygen	31(16)=496
Nitrogen	1(14)=14
Total	997

The kg-mol of organic feedstock at the start of the process is 1000/997=1.00.

2. Molecular weight of the compost product is

Carbon	14(12)=168
Hydrogen	20(1)=20
Oxygen	9(16)=144
Nitrogen	1(14)=14
Total	346

The kg-mol of finished compost per kg-mol at the start of the process is $n = 180 / (1.00)(346) = 0.52$
The following values are to be used in the calculations:

$a = 35$ $w = 14$
$b = 67$ $x = 20$
$c = 31$ $y = 9$
$d = 1$ $z = 1$

The r and s values are determined as

$$r = 0.5\{67-0.52\,(20)-3[1-0.52(1)]\} = 27.58$$
$$s = 35-0.52\,(14) = 27.7$$

4. From Equation 8.3, the quantity of oxygen required to complete the composting process is

$$W = 0.5\,[0.52(9) + 2(27.7) + 27.58-31]\,(1.00)(32) = 906.6 \text{ kg}$$

5. The above data can be checked with a materials balance.

Inputs	
Substrate	1000
Oxygen	907
Total in	1907 kg

Outputs		
Compost		180
CO_2	1.00(27.7)(44)	1218.8
H_2O	1.00(27.58)(18)	496.44
NH_3	[1−0.52(1)](1.00)(17)	8.16
Total out		1903 kg

Note: Since air (not O_2) is being applied to the pile, its requirement is 907/0.232=3909 kg. This value is equivalent to 3909/1000=3.909 kg air/kg substrate.

8.5.6 MOISTURE CONTENT

The preferred moisture content and oxygen availability for composting are closely interrelated. The interstices in the waste feedstock will contain either water or air, so the presence of one will directly affect the concentration of the other. The optimal moisture content for successful composting varies depending on the physical state and size of the particles and on the composting system used. Regular monitoring of the chemical and physical properties of the pile, previous experience with MSW composting, along with a review of the literature should serve as a practical guide to optimum moisture content.

Less moisture in the pile will result in dehydration, which slows biological processes. Water is required for numerous cellular processes, and properties including nutrient transport, waste removal, turgidity, and as a component in innumerable biochemical reactions. Excess water interferes with aeration by clogging pores. If the moisture content of the mass is so high as to displace most of the air from the interstices, anaerobic conditions develop within the mass. Therefore, the maximum acceptable moisture content is a level at which no nuisance conditions (e.g., anaerobiosis) will develop and at which the process will proceed satisfactorily.

Moisture content of the pile can be measured in the field using a range of analytical equipment ranging from gypsum blocks to tensiometers. Alternatively, a sample can be taken to the laboratory and measured field-moist and oven-dry (i.e., after 48 h in an oven at 105°C). These data are used to calculate the gravimetric moisture content.

Moisture is rapidly depleted from an active compost pile and must be replaced by regular additions of water or, in some cases, application of wastewater sludge (also a rich source of heterotrophic microorganisms). The optimum amount of water to be applied to a compost pile can be calculated from a mass balance equation (Vesilind et al., 2002)

$$M_p = (M_s X_s + 100 X_w)/(X_w + X_s) \qquad (8.4)$$

where M_p is the moisture content of the compost pile at the start of composting (%), M_s the moisture content of the solids, for example shredded MSW (%), X_s the mass of solids (wet metric tons) and X_w the mass of water, wastewater, or other source of water (metric tons).

EXAMPLE 8.2

At a municipal waste handling facility, a mixture of approximately 25 metric tons of food waste, yard waste, and paper waste is to be composted. The moisture content of this feedstock measures 9.5%. It has been previously determined that an ideal moisture content for the compost pile should be about 55%. Calculate the metric tons of water to be added to the solids to achieve the optimum moisture content.

SOLUTION

$$M_p = (M_s X_s + 100 X_w)/(X_w + X_s) = [(25 \times 9.5) + (100 \times X_w)]/(25 + X_w)$$

$$X_s = 25.3 \text{ MT of water to be applied to the pile.}$$

8.5.7 TEMPERATURE

There is a direct relationship between microbial activity and temperature of the pile. High temperatures result from biological activity, i.e., heat liberated from microbial respiration and the resultant breaking of chemical bonds of substrate compounds. This heat builds up within the pile; dispersal of this heat is limited due to the insulating effects of the pile.

Thermophilic vs. mesophilic temperature ranges have their own advantages and disadvantages with respect to composting. The temperatures that enhance microbial activity are in the range 28 to 55°C (48 to 131°F). The highest O_2 consumption also occurs within this range.

High temperatures are commonly considered as necessary conditions for good composting. Excessively high temperatures, however, inhibit growth of most microorganisms, thus slowing decomposition of feedstock. When the temperature rises beyond approximately 65 to 70°C (150 to 160°F), the tendency is for spore formers (e.g., *Bacillus* and *Clostridium*) to convert into spores. This transition is undesirable, because the spore-forming stage is a resting stage and therefore the rate of decomposition is reduced. Moreover, microbes incapable of forming spores are strongly inhibited or killed at those temperatures. Consequently, the maximum temperature should be kept at about 65°C (150°F).

The temperature distribution within a composting mass is affected by the surrounding climatic conditions and by the method of aeration. In static piles (see below), the highest temperatures develop at the center of the mass and the lowest temperatures occur at the edges of the pile. These temperature gradients promote a small degree of convection (i.e., natural airflow). The degree of air movement is a function of ambient conditions as well as porosity of the composting mass. The problem of temperature control is best solved, however, by either periodically turning the pile or using forced ventilation throughout the process.

8.5.8 pH

The optimum pH range in composting is so broad that difficulties due to an excessively high or low pH level are rarely encountered. Organic materials of a pH range from 3 to 11 can be composted. Optimum values, however, are between 5.5 and 8. Recall from above that bacteria are the key catalysts in organic matter transformations, and typically prefer a near-neutral pH. In contrast, fungi develop better in an acid environment.

During the early stages of composting the pH level normally decreases (perhaps to as low as 5.0) because of the production of organic acids (e.g., formic, acetic, and pyruvic). These acids serve as substrates for succeeding microbial populations. As the acids are decomposed, pH rises and often stabilizes at approximately neutral. In some cases, compost pH may reach as high as 8.5.

Because it is unlikely that the pH will drop to inhibitory levels, there is no need to buffer the feedstock by adding liming materials (e.g., limestone, calcium hydroxide). The addition of lime should be avoided because it can lead to excessive losses of ammonium nitrogen. The lime does, however, promote the formation of aggregates which in turn improves physical properties such as air and water movement.

The pH of the compost pile is commonly measured using a standard glass electrode pH meter. A known mixture, for example, a 2:1 ratio of water/solids is mixed and the resultant pH is read on the meter.

8.6 THE COMPOSTING STAGE

After MSW feedstock is preprocessed it is introduced into the composting operation. At this time, indigenous microorganisms actively decompose the feedstock; most of the physical and chemical changes to the feedstock occur during this stage.

The actual compost process can be established in a number of environments, from simple outdoor piles to sophisticated reaction vessels with controlled temperature, airflow, and humidity. Some popular composting methods are:

Open systems

- Turned piles
- Turned windrows
- Static piles using air blowing or suction

Closed systems

- Rotating drums
- Tanks

All systems are designed and operated to establish optimum conditions for composting. These conditions directly influence the growth and metabolism of the microorganisms that are responsible for the process. The factor that can be most influenced by technology, around which composting designs are developed, is the availability of oxygen.

8.6.1 TURNED PILES

Turned piles are a widely used method for composting MSW due to their simplicity of operation. As the name implies, the feedstock is mixed periodically using a front-end loader or similar equipment. Turning of the feedstock maintains oxygen, moisture, and temperature at adequate levels for microbes. The outer layers are incorporated within, where they are exposed to higher temperatures and more intensive microbial activity. The frequent turning allows for the introduction of oxygen and also releases excess heat from the center of the pile. Turning therefore promotes uniform decomposition of materials. Using the turned pile method, the composting process is completed in approximately 2 months to 1 year.

Turned piles are constructed outdoors; however, piles can also be situated under shelters. Such a cover will prevent saturation with the consequent production of anaerobic conditions as well as leachate generation. Leachate problems are further addressed by constructing piles on firm surfaces (preferably paved) surrounded by berms or trenches to collect runoff. This is discussed below.

Turning frequencies range from twice per week to once per year. The more frequently that piles are turned, the more quickly the composting process is completed. Where odor control and composting speed are a high priority, oxygen monitoring equipment can be installed to alert operators when oxygen levels fall below 10 to 15%, which is the minimum oxygen concentration required for aerobic decomposition and to minimize odor problems (Richard, 1992). Simple portable oxygen meters and long-stem thermometers can be inserted within the pile to assess these conditions.

8.6.2 TURNED WINDROWS

Turned windrows are elongated compost piles that are turned frequently to maintain aerobic conditions (Figure 8.5). Forming windrows of the appropriate size helps in maintaining the desired temperature and oxygen levels. Windrows operate most effectively at a height of 1.5 to 1.8 m (5 to 6 ft) (CRS, 1989). This height allows the feedstock to be insulated but prevents the buildup of excessive heat. Windrow heights vary, however, based on the feedstock (e.g., the tendency to compact), season, local climate, and turning equipment used. Windrow widths generally are twice the height of the piles. Land availability, operating convenience, type of turning equipment, and desired end-product quality also affect the windrow width (U.S. EPA, 1994).

If the windrow resembles a triangular shape, the volume can be determined by the equation

$$\frac{1}{2}\,(WHL) \tag{8.5}$$

where W is the average width at the bottom, H the average height of the pile, and L the length of the windrow.

For a trapezoidal shape, volume is determined by

$$\frac{1}{2}\,(W_1 + W_2)HL \tag{8.6}$$

where W_1 is the average width at the bottom, and W_2 average width at the top.

FIGURE 8.5 Composting of MSW using the windrow system.

omposition of materials. Using the turned pile method, the composting process is completed in roximately 2 months to 1 year.

Turned piles are constructed outdoors; however, piles can also be situated under shelters. Such over will prevent saturation with the consequent production of anaerobic conditions as well as hate generation. Leachate problems are further addressed by constructing piles on firm surfaces ferably paved) surrounded by berms or trenches to collect runoff. This is discussed below.

Turning frequencies range from twice per week to once per year. The more frequently that piles turned, the more quickly the composting process is completed. Where odor control and com- ing speed are a high priority, oxygen monitoring equipment can be installed to alert operators n oxygen levels fall below 10 to 15%, which is the minimum oxygen concentration required for bic decomposition and to minimize odor problems (Richard, 1992). Simple portable oxygen ers and long-stem thermometers can be inserted within the pile to assess these conditions.

.2 TURNED WINDROWS

ned windrows are elongated compost piles that are turned frequently to maintain aerobic condi- s (Figure 8.5). Forming windrows of the appropriate size helps in maintaining the desired tem- ature and oxygen levels. Windrows operate most effectively at a height of 1.5 to 1.8 m (5 to 6 ft) S, 1989). This height allows the feedstock to be insulated but prevents the buildup of excessive . Windrow heights vary, however, based on the feedstock (e.g., the tendency to compact), sea- , local climate, and turning equipment used. Windrow widths generally are twice the height of piles. Land availability, operating convenience, type of turning equipment, and desired end- duct quality also affect the windrow width (U.S. EPA, 1994).

If the windrow resembles a triangular shape, the volume can be determined by the equation

$$\frac{1}{2}(WHL) \tag{8.5}$$

re W is the average width at the bottom, H the average height of the pile, and L the length of windrow.

For a trapezoidal shape, volume is determined by

$$\frac{1}{2}(W_1+W_2)HL \tag{8.6}$$

re W_1 is the average width at the bottom, and W_2 average width at the top.

URE 8.5 Composting of MSW using the windrow system.

2. Molecular weight of the compost product is

Carbon	14(12)=168
Hydrogen	20(1)=20
Oxygen	9(16)=144
Nitrogen	1(14)=14
Total	346

The kg-mol of finished compost per kg-mol at the start of the process is $n = 180 / (1.00)(346) = 0.52$

The following values are to be used in the calculations:

$a = 35 \quad w = 14$
$b = 67 \quad x = 20$
$c = 31 \quad y = 9$
$d = 1 \quad z = 1$

The r and s values are determined as

$$r = 0.5\{67-0.52(20)-3[1-0.52(1)]\} = 27.58$$
$$s = 35-0.52(14) = 27.7$$

4. From Equation 8.3, the quantity of oxygen required to complete the composting process is

$$W = 0.5[0.52(9) + 2(27.7) + 27.58-31](1.00)(32) = 906.6 \text{ kg}$$

5. The above data can be checked with a materials balance.

Inputs	
Substrate	1000
Oxygen	907
Total in	1907 kg

Outputs		
Compost		180
CO_2	1.00(27.7)(44)	1218.8
H_2O	1.00(27.58)(18)	496.44
NH_3	[1−0.52(1)](1.00)(17)	8.16
Total out		1903 kg

Note: Since air (not O_2) is being applied to the pile, its requirement is 907/0.232=3909 kg. This value is equivalent to 3909/1000=3.909 kg air/kg substrate.

8.5.6 MOISTURE CONTENT

The preferred moisture content and oxygen availability for composting are closely interrelated. The interstices in the waste feedstock will contain either water or air, so the presence of one will directly affect the concentration of the other. The optimal moisture content for successful composting varies depending on the physical state and size of the particles and on the composting system used. Regular monitoring of the chemical and physical properties of the pile, previous experience with MSW composting, along with a review of the literature should serve as a practical guide to opti- mum moisture content.

Less moisture in the pile will result in dehydration, which slows biological processes. Water is required for numerous cellular processes, and properties including nutrient transport, waste removal, turgidity, and as a component in innumerable biochemical reactions. Excess water inter- feres with aeration by clogging pores. If the moisture content of the mass is so high as to displace most of the air from the interstices, anaerobic conditions develop within the mass. Therefore, the

Pile-turning equipment determines the size, shape, and space between the windrows. Front-end loaders are commonly used in smaller operations. Windrow turners, also known as scarab composters, straddle windrows and thoroughly mix materials as it moves over the pile. These machines are either self-propelled or mounted to front-end loaders (Figure 8.6). Self-propelled windrow turners minimize the required space between windrows.

8.6.3 AERATED STATIC PILES

In terms of operation, aerated static piles are relatively more complicated than turned piles. This approach is effective when space is limited and the composting process must be completed relatively rapidly. In this method, a series of perforated pipes is situated within or below a pile (or windrow). Air can be supplied via a negative pressure (suction) system or a positive pressure (blower) system. Fans or blowers force air through the pipes, which is then drawn through the feedstock materials (Figure 8.7). The air movement through the pipes maintains aeration within the pile, thus eliminating the need for turning.

Static piles are built to approximately 3 to 3.7 m (10 to 12 ft) in height. Topping off the pile with a layer of finished compost protects the surface from drying, insulates it from heat loss, discourages pests, and filters odors generated within the pile (Rynk, 1992). The compost is finished within 3 to 6 months.

To ensure that decomposition proceeds at high rates within aerated static piles, temperature and oxygen levels must be closely monitored and maintained. Aeration management depends on how

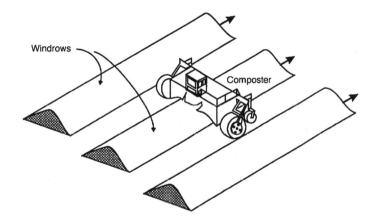

FIGURE 8.6 Schematic of windrow composting showing scarab compost turner (U.S. EPA, 1984).

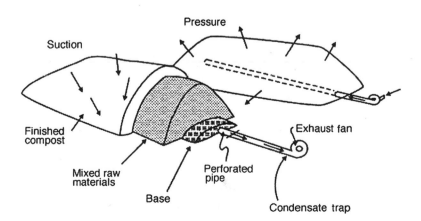

FIGURE 8.7 The aerated static pile (Rynk, R., Ed., *On-Farm Composting Handbook*, NRAES-54, 1992.)

the blower is controlled. The blower can be run continuously or intermittently. Continuous operation permits lower airflow rates because oxygen and cooling are supplied constantly. Intermittent operation of the blower is achieved with a timer or temperature feedback system.

8.6.4 IN-VESSEL SYSTEMS

In-vessel systems are relatively sophisticated units in which composting is conducted within a fully enclosed chamber. Environmental conditions are mechanically controlled and usually automated. An in-vessel system may be suitable for MSW composting if:

- The process must be finished rapidly
- Odor and leachate control are a significant concern
- Space is limited

In-vessel technologies range from simple to complex. Two categories of in-vessel technologies include rotating drums and tank systems. Rotating drums rely on a tumbling action to mix continuously feedstock materials. Figure 8.8 illustrates a rotating drum composter. The drums typically are long cylinders, approximately 3 m (10 ft) in diameter which are rotated slowly, usually at less than 10 r/min (CRS, 1989). Oxygen is forced into the drums through nozzles from air pumps. The tumbling action allows oxygen to be maintained at high and uniform levels throughout the drum.

Tank systems are available as horizontal or vertical types. These tanks are long vessels in which aeration is accomplished through the use of external pumps that force air through the perforated bottom of the tanks. Mixing is accomplished by mechanically passing a moving belt, paddle wheel, or flail-covered drum through the feedstock. The agitation breaks up clumps and maintains porosity. Solids are retained in this system for 6 to 28 days and then cured in windrows for 1 to 2 months (U.S. EPA, 1994).

8.7 THE CURING STAGE

Once the materials are adequately stable, they must be cured. Oxygen uptake and CO_2 evolution measurements will indicate the degree of maturity of compost. One method to measure pile maturity is to monitor the internal temperature of the compost pile after it is turned. If reheating of the pile occurs, then the material is not yet ready for curing.

During the curing stage, compost is stabilized as the remaining microorganisms metabolize the remaining available nutrients. For the duration of the curing stage, microbial activity diminishes as available nutrients are depleted. Curing is a relatively passive process compared with the primary composting operation, so less intensive methods and operations are used. In general, materials are formed into piles or windrows and left until the specified curing period has ended (Figure 8.9).

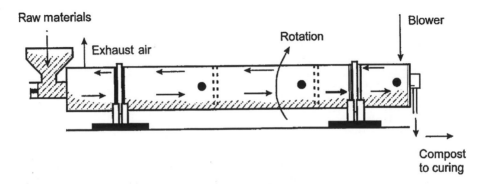

FIGURE 8.8 A drum composter (Bedminster Bioconversion Corporation).

FIGURE 8.9 A cured pile ready to go to market.

Since curing piles undergo slow decomposition, monitoring is important during this period so that piles do not become anaerobic.

The curing piles should be small enough to permit adequate natural air exchange. A maximum pile height of 2.4 m (8 ft) is suggested. If compost is intended for high-quality uses, curing piles should be limited to 1.8 m (6 ft) in height and 4.8 to 6.1 m (15 to 20 ft) in width (Rynk, 1992).

The C:N ratio of finished compost should not exceed 20:1. As mentioned earlier, C: N ratios that are too low can result in the production of phytotoxins within the pile, which are dispersed into the recipient soil when the compost is land-applied. One group of phytotoxins is produced when microorganisms are not capable of utilizing all the excess nitrogen. The free nitrogen is biologically transformed, resulting in the release of NH_3 and other chemicals that are toxic to plant roots and inhibit growth.

8.8 ENVIRONMENTAL CONCERNS DURING COMPOSTING

Composting at the municipal level involves the treatment of tons of potentially odoriferous and wet material containing a wide range of microorganisms. Homes and businesses may be located near the facility. It is therefore essential to control carefully the composting process at all times in order to limit or prevent environmental concerns such as air and water pollution, odor, noise, vectors, fires, and litter. These concerns can be minimized through proper design of the facility and conscientious daily operations.

8.8.1 AIR QUALITY

Air pollution is typically not a major concern at composting facilities, with the exception of natural odor problems. Dust can be a problem, particularly in the dry summer months. Dust is generated from dry, uncontained organic materials, especially during screening and shredding operations and from vehicle traffic. Dust carries bacteria and fungi that can affect facility workers and possibly facility neighbors. Dust may also clog certain equipment and filters.

8.8.2 ODOR

Most stages of the composting process can release odors. The feedstock itself will contain odorous compounds; odors can be produced during collection, transport, and storage of the feedstock. Improper composting procedures, for example, not providing adequate O_2 and allowing anaerobiosis, will encourage the formation of malodorous compounds (Kissel et al., 1992). Anaerobic conditions encourage generation of organic acids, mercaptans, alcohols, amines, hydrogen sulfide gas, and other sulfur compounds (Diaz, 1987; Williams and Miller, 1992). Ammonia is released under anaerobic conditions and infrequently in some aerobic conditions (for example, if the C:N ratio is less than 20:1) (Kissel et al., 1992). The compounds commonly linked to odor production at composting facilities are listed in Table 8.3.

Hellman and Small (1973) formulated the odor index (OI) to measure the potential of a chemical compound to become an odor problem. The OI is defined as (Haug, 1993)

$$OI = \text{vapor pressure/odor recognition threshold (100\%) (ppm)} \qquad (8.7)$$

The OI is a measure of the potential of a particular odorant to cause odor problems under evaporation conditions. The OI takes into account the vapor pressure of a compound, which is the measure of the potential of the compound to occur in the gas phase, and the odor recognition threshold, which is a measure of the strength of the odorant.

The boiling points of ammonia, hydrogen sulfide, ethyl mercaptan, dimethyl sulfide, and acetaldehyde are all lower than the temperatures associated with composting. Thus, these compounds will 'boil off' into the vapor phase if they are generated during composting. Other compounds have boiling points near or above the temperatures common to composting. Nevertheless, they may possess significant vapor pressures and can evaporate into the vapor phase. This evaporation can be enhanced by the large airflow rates employed in many composting systems (Haug, 1993).

8.8.3 NOISE

Noise is generated by trucks entering and leaving the facility and by equipment used in composting operations. Hammermills and other shredding machines are the noisiest of these equipment,

TABLE 8.3
Threshold Odor Concentrations and Boiling Points for Selected Odorous Compounds

Compound	TOC (ppmv)		Boiling point (°C)
	Detect	Recognition	
Ammonia	0.037	47	-33
Hydrogen sulfide	0.00047	0.0047	-62
1-Butene	0.069		-6
Methyl mercaptan	0.0011	0.0021	8
Ethylamine	0.026	0.83	17
Dimethyl amine	0.047	0.047	
Acetaldehyde	0.004	0.21	20
Ethyl mercaptan	0.002		
1-Pentene	0.0021		30
Dimethyl sulfide	0.001	0.001	36
Dimethyl disulfide	0.001	0.0056	
Diethyl sulfide	0.0008	0.005	88
Butyl mercaptan	0.0005		65
Acetic acid	0.008	0.2	63
α-Pinene, oil of pine	0.011		37

Source: Haug, R. T., *The Practical Handbook of Compost Engineering*, 1993. Reproduced with kind permission. Copyright Lewis Publishers, an imprint of CRC Press. Boca Raton, FL.

generating about 90 dB at the source. Many states have mandated noise control that limit noise at the property line. Measures that can reduce noise emanating from the facility include (U.S. EPA, 1994):

- Providing a sufficient buffer zone around the facility with plenty of trees.
- Including noise reduction features in facility design, such as noise hoods, when procuring equipment.
- Properly maintaining mufflers and other equipment components.
- Coordinating hours of operations with adjacent land uses.
- Limiting traffic to and from the facility.

8.8.4 TOXINS WITHIN THE PILE

Many lawns, golf courses, farm fields, and other vegetated areas in the United States receive copious quantities of herbicides, pesticides, and other biocides. Some have been found to persist after composting. The ten most commonly used agricultural pesticides in the United States include seven herbicides and three fumigants (Table 8.4) (U.S. EPA, 1997a). Herbicides are also commonly used for residential and commercial or industrial applications.

A study conducted in Illinois tested for the presence of 21 pesticides in yard waste and compost from 11 landscape composting facilities (Table 8.5) (Miller et al., 1992). Concentrations of all pesticides found in feedstocks and compost samples were below the Maximum Allowable Tolerance for Raw Agricultural Commodities. In a Portland, Oregon study, a total of 19 pesticides were monitored in yard waste compost (Gurkewitz, 1989). Only four pesticides were detected, and at extremely low levels. The testing program was expanded to include 27 pesticides. Low concentrations of pentachlorophenol and chlordane were consistently detected in yard waste compost. Dieldrin, DDT, DDE, toxaphene, aldrin, chlorpyrifos, and dinoseb were detected in only a limited number of samples. In leaf compost in Westchester County, New York, 200 pesticides were tested for (Richard and Chadsey, 1989). Chlordane, lindane, captan, and 2,4-D were the only ones detected. Mean concentrations of all pesticides, except chlordane, were well below the minimum USDA tolerance level for food.

TABLE 8.4
Pesticides most Commonly Applied in the United States by Agricultural, Residential, and Commercial Users, 1994-1995 (based on Active Ingredient Applied)

Rank	Agriculture Common Name	Type	Home and Garden Common Name	Type	Commercial, Industrial, Government Common Name	Type
1	Atrazine	H	2,4-D	H	2,4-D	H
2	metolachlor	H	Glysophate	H	chlopyrifos	I
3	Metam	SF	Dicamba	H	glysophate	H
4	methyl bromide	SF	MCPP	H	methyl bromide	SF
5	Dichloroprop	SF	Diazinon	I	copper sulfate	F
6	2,4-D	H	chlopyrifos	I	MSMA	H
7	glysophate	H	Carbaryl	I	diazinon	I
8	cyanazine	H	Benfluralin	H	diuron	H
9	Pendimethalin	H	DCPA	H	malathion	I
10	trifluralin	H				

F=fungicide, H=herbicide, I=insecticide, SF=soil fumigant.
Source: U.S. EPA, 1997.

TABLE 8.5
Pesticides Monitored in Yard Waste and Compost in Illinois Study

	Pesticide	Average Levels in Yard Waste (ppm)	Average Levels in Finished Compost (ppm)	MAT (ppm)[a]
Herbicides	2,3,4-T	0.788	1.15	
	2,4-D	1.04	0.268	300
	Alachlor	0.749	0.304	3
	Atrazine	4.61	3.03	15
	Dichlobenil	0.0144	0.0133	0.15
	Metolachlor	1.06	0.972	30
	Trifluralin	0.142	0.156	2
Organochlorine insecticides	Chlordane	0.526	0.4	
	DDD	0.0641	0.0505	
	DDE	0.0516	0.0807	
	Dieldrin	0.00992	0.00834	
	Heptachlor	0.00942	ND	
	Heptachlor epoxides	0.0216	0.0151	
	Lindane	0.495	0.314	7
	Methoxychlor	0.314	0.507	100
Organophosphate insecticides	Chlorpyrifos	0.00996	0.0077	15
	Diazinon	0.991	0.587	40
	Fonofos	0.0112	0.00538	
	Malathion	0.313	0.169	135
	Parathion	0.235	0.104	5
Carbamate insecticides	Carbaryl	22.5	11.0	100

[a] Maximum Allowable Tolerance for Raw Agricultural Commodities (U.S. EPA 40 CFR); ND=not detected; NA=not available.

Source: Miller, T. L. et al., Illinois Department of Energy and Natural Resources, Springfield, IL.

The Washington State University (WSU) composting facility and the Spokane Regional Compost Facility discovered traces of persistent herbicides in their composts including clopyralid and picloram (Bezdicek et al., 2001). In both cases, the compost damaged sensitive plants in gardens and nurseries. Clopyralid contamination has since been reported in other facilities in Washington, Maine (Maine Department of Environmental Protection, no date), Pennsylvania, and New Zealand. Rose and Mercer (1968) investigated the fate of pesticides during composting of fruit and vegetable processing wastes. DDT, dieldrin, parathion, and diazinon were applied to a mixture of processing residues and rice hulls, and the mixture was composted using either a batch system with minimal turning, or a system with frequent turning that achieved thermophilic temperatures. Over 120 days, the pesticides degraded faster in the thermophilic system, except for dieldrin.

Most herbicides break down rapidly after application. Buyuksonmez et al. (1999, 2000) reported that herbicides generally break down during normal composting. However, some of those in the pyridine carboxylic acid group such as clopyralid break down very slowly, especially during composting (Bezdicek et al., 2001). Monitoring incoming feedstock to remove pesticide containers and other foreign materials can help in reducing the occurrence of synthetic chemicals in compost.

8.8.5 LEACHATE

Leachate is produced in uncovered piles exposed to excessive quantities of precipitation. The leachate released from the pile can have elevated biochemical oxygen demand (BOD) and phenols, resulting from the natural decomposition of organic material. Nitrates (NO_3^-) are also generated by composting grass clippings along with leaves. Leachate composition from a compost pile is shown in Table 8.6.

Leachate can also contain potentially toxic synthetic compounds, including chlorinated organic compounds from treated wood, pesticides, polycyclic aromatic hydrocarbons (PAHs), and combustion products of gasoline, oil, and coal. Chlorinated organics and PAHs are resistant to biodegradation and tend to persist after composting (Gillett, 1992). Microorganisms can partly degrade some PAHs during composting; however, the resultant compounds can be more toxic than the original PAHs (Menzer, 1991; Chaney and Ryan, 1992).

Leachate generation can be reduced or prevented by monitoring and correcting the moisture levels in the composting pile. In some facilities, windrows or piles are installed under a roof to limit excessive moisture levels arising from precipitation. If the compost feedstock (MSW or yard trimmings) contains excess moisture, leachate will be released during the first few days of composting regardless of any rainfall event. Following this initial release, the volume of leachate generated will decrease as the compost product matures and humifies, thus improving its water-holding capacity.

The age of the pile will also affect leachate composition. In a mature pile, microorganisms have decomposed complex compounds and released or consumed substantial carbon and nitrogen. If the

TABLE 8.6
Croton Point, New York, Yard Waste Compost Leachate Composition

	Compost Leachate (16 samples)	
	Average (mg/L)	Standard Deviation (mg/L)
Cd	ND	
Cu	ND	
Ni	ND	
Cr	ND	
Zn	0.11	0.13
Al	0.33	0.38
Fe	0.57	0.78
Pb	0.01	0.02
K	2.70	0.99
NH_4-N	0.44	0.35
NO_3-N	0.96	1.00
NO_2-N	0.02	0.02
Phosphorus	0.07	0.08
Phenols (total)	0.18	0.45
COD	56.33	371.22
BOD	>41[a]	>60
PH	7.75	0.36
Color	ND	
Odor	ND	

ND=not determined; COD=chemical oxygen determined.

[a] Includes three samples above detection limit of 50 mg/L.

Source: Richard, T. and Chadsey, M., *Biocycle*, April, 31: 42-46, 1990. Reproduced with kind permission of The J.G. Press, Inc.

C:N ratio is maintained within the desired range, little excess N will leach from the pile since this is rapidly utilized by microorganisms for growth (U.S. EPA, 1994).

The installation of a concrete pad for a compost base is valuable for collection and control of any leachate produced. A simple method to handle leachate is to collect all liquids from the pad and reintroduce them into the pile. Such leachate recycling should not be conducted once the compost pile has completed the high-temperature phase, as any harmful microorganisms that were inactivated by the high heat can be reintroduced with the leachate (CC, 1991). Leachate can also be transported to a municipal wastewater treatment plant. If the contaminant levels in the leachate are too high, an on-site wastewater pretreatment system could be installed (U.S. EPA, 1994).

Measures to control leachate include:

- Diverting from the compost curing and storage areas to a leachate holding area
- Installing liner systems made of low-permeability soils such as clay or synthetic materials
- Using drain pipes to collect the leachate for treatment
- Curing and storing compost indoors to eliminate infiltration of leachate into the ground (Wirth, 1989)

8.8.6 RUNOFF

Runoff is caused by heavy precipitation, by components within the feedstock, and by practices at the facility that use water. For example, the water used to wash trucks can contribute to runoff. Polluted water can be spilled in the tipping area of composting facilities when packer trucks from restaurants, grocery stores, and food processors are emptied. Operations that compost MSW and yard waste can produce runoff containing measurable quantities of inorganic nutrients and other pollutants.

For both yard waste and MSW composting facilities, water that has come into contact with incoming raw materials, partially processed materials, or compost should not be allowed to run off the site. Figure 8.10 shows several options for diverting water from composting piles and for containing runoff. Provisions for isolating, collecting, treating, or disposing of water that has come in contact with the composting feedstock can include (U.S. EPA, 1994):

- Maintaining sealed paving materials in all areas
- Grading facility areas (1 to 2% grade) where contaminated water will be collected
- Installing containment barriers to prevent contaminated water from contacting adjacent land and waterways

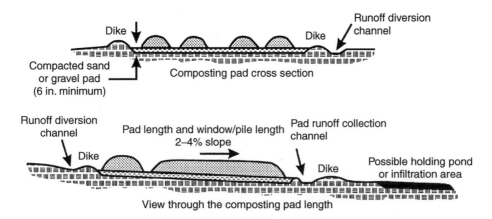

FIGURE 8.10 Methods to divert runoff water from a compost pile for eventual collection. (With kind permission of Rynk, R., Ed., *On-Farm Composting Handbook*, NRAES-54, 1992.)

- Covering compost beds and cured piles
- Percolating contaminated water through soil so as to absorb and break down organic compounds
- Building retention ponds to prevent discharge of runoff to surface water

8.8.7 VECTORS

Vectors are small animals or insects that carry disease. Mice, rats, flies, and mosquitoes may occur at composting facilities. Rodents can be attracted by the food and shelter available at composting facilities (particularly MSW composting operations) and can be difficult to eliminate. Flies, which can transmit salmonella and other food-borne diseases, are often carried in with the incoming material and are attracted to piles that have become anaerobic. All life stages of the housefly are killed by the temperatures attained in the compost pile (Golueke, 1977). Mosquitoes, which can also transmit disease, breed in standing water. Keeping the processing area neat can control insects; also, maintaining aerobic conditions and proper temperatures in the piles and grading the area properly to prevent ponding will limit mosquito breeding.

8.8.8 FIRES

If the compost material dries out and becomes too hot, spontaneous combustion may occur in the pile (Figure 8.11). Organic material can ignite spontaneously at a moisture content of between 25 and 45%. This is unlikely, however, unless the material reaches temperatures higher than 93°C (199°F). The site must be designed for access by firefighting equipment including clear aisles between piles or windrows, and must have an adequate water supply (Richard et al., 1990).

Key conditions that lead to spontaneous combustion are biological activity, relatively dry materials or dry pockets, large well-insulated piles, limited airflow, and sufficient time for the temperature to build up. Other contributing factors may include short circuiting of airflow, a nonuniform mix of materials, poor moisture distribution, and inadequate monitoring of temperature and other variables. These conditions tend to be more common within large undisturbed piles containing raw feedstocks, curing compost, and finished compost than in the active composting system. Actively

FIGURE 8.11 Compost fire presumably initiated by spontaneous combustion. Reproduced with kind permission of Winchester News-Gazette, Winchester, IN.

composting piles and vessels tend to be monitored and controlled for temperature, moisture, and aeration while storage and curing piles may be neglected (Rynk, 2000).

8.8.9 LITTER

Although not a hazard *per se*, litter from the facility is an aesthetic problem and a possible source of complaints from nearby residents. Litter originates from MSW brought into the facility, plastic, and paper blowing from piles, and rejects (such as plastic) blowing away during screening. Litter can be controlled by (Wirth, 1989: U.S. EPA, 1994):

- Requiring loads of incoming material to be covered
- Using moveable fencing or chain link fences along the site perimeter as windbreaks and to facilitate collection of litter
- Enclosing receiving, processing, and finishing operations
- Collecting litter as soon as possible before it becomes scattered off-site
- Removing plastic bags before windrowing
- Collecting leaves and woody materials in paper bags, plastic bins, or in bulk

8.9 OCCUPATIONAL HEALTH AND SAFETY CONCERNS DURING COMPOSTING

Potential health and safety problems at composting facilities include accidents with heavy equipment, exposure to excessive noise, and exposure to bioaerosols and potential toxic chemicals. Proper siting, design, and operation of the facility and adequate worker training and education can minimize these problems.

8.9.1 BIOAEROSOLS

A variety of biological aerosols (bioaerosols) can be generated during composting. Bioaerosols are suspensions of particles in the air consisting partially or wholly of microorganisms. These aggregates can remain suspended in the air for long periods, retaining their viability (infectious nature). The bioaerosols of concern during composting include actinomycetes, bacteria, viruses, molds, and fungi. *Aspergillus fumigatus* is a common fungus that is naturally present in decaying organic matter. The fungal spores can be inhaled or enter the body through cuts and abrasions in the skin. The fungus is not considered a hazard to healthy individuals (U.S. EPA, 1994). *A. fumigatus* is readily dispersed from dry compost piles during and after mechanical turning. The levels of *A. fumigatus* decrease rapidly a short distance from the pile or a short time after composting activity ceases (Epstein and Epstein, 1989).

Endotoxins are another concern at composting facilities. Endotoxins are toxins produced within a microbial cell and released upon cell destruction. These compounds are carried by airborne dust particles. The level of endotoxins in the air at one yard waste composting facility ranged from 0.001 to 0.014 mg/m^3 (Roderique and Roderique, 1990).

Because bioaerosols and endotoxins are both carried as dust, dust should be controlled at all times at the facility. Steps to minimize dust generation may include (U.S. EPA, 1994):

- Keeping compost piles and feedstock moist.
- Moistening compost during the final pile teardown and before being loaded onto vehicles, taking care not to over-wet the material (which can produce leachate or runoff).
- If the facility is enclosed, proper ventilation is required via engineering controls such as collection hoods, negative air pressure at dust generation points, and the use of baghouse filtration.

Workers should also be informed that disease-producing microorganisms are present in the work environment. Precautions should be followed for personal protection and include (U.S. EPA, 1994):

- Wear dust masks or respirators under dry and dusty conditions, especially when the compost is being turned.
- Cuts should receive prompt attention to prevent contact with incoming loads or feedstock.
- Individuals with asthma, diabetes, or suppressed immune systems should be advised not to work at a composting facility because of their greater risk of infection.

8.9.2 POTENTIALLY TOXIC CHEMICALS

Compounds such as benzene, chloroform, and trichloroethylene can present potential risks to workers at MSW composting facilities (Gillett, 1992). Certain solvents, paints, and cleaners contain volatile organic carbon compounds (VOCs). The combination of forced aeration (or periodic turning in the case of window systems) and elevated temperatures serve to release VOCs from the compost pile into the local atmosphere.

To avoid worker exposure to VOCs, adequate ventilation is needed. Control technologies developed for odor control also apply to VOC control. The best method of controlling VOC emissions, however, is to limit their presence in the feedstock in the first place. Limiting MSW composting to residential and high-quality commercial feedstocks, instituting source separation, and implementing effective household hazardous waste collection programs will minimize the amount of VOCs in MSW.

8.10 FACILITY SITING

As discussed, compost feedstock is originally derived from MSW and is therefore odoriferous. It is thus logical and practical to locate a composting site in proximity to a solid waste transfer station, landfill, wastewater treatment plant, or similar disposal facility in an area zoned for industry or commercial use. Some of the major factors in facility siting include (U.S. EPA, 1994):

- Location to minimize hauling distances
- Adequate buffer between the facility and nearby residents
- Suitable site topography and soil characteristics
- Sufficient land area for the volume of material to be processed

Current federal guidelines prohibit siting any solid waste facility, including composting facilities, within 10,000 ft of an airport. This is to prevent birds, which could be attracted to the site by potential food sources, from interfering with airplanes (see Chapter 10).

Local residents may be concerned about potential odors and other nuisance conditions. Locating a site with an extensive natural buffer zone, planted with trees and shrubs, is an effective way to reduce such concerns. Artificial buffers might also need to be constructed. Visual screens such as berms or landscaping can be installed to protect the aesthetics of the surroundings. Figure 8.12 shows a suggested field plan for a large-scale composting facility.

8.10.1 TOPOGRAPHY

The composting site should be graded to avoid standing water, runoff and erosion. The land at a composting site should be sloped at least 1% and ideally 2 to 4% (Rynk, 1992).

The type and structure of the soil present at the site should be assessed to control runon and runoff. A firm base is preferred in order to capture and control liquids and prevent groundwater contamination. If the site is unpaved, the soil should be permeable enough to ensure that excess water

Buffer Zone

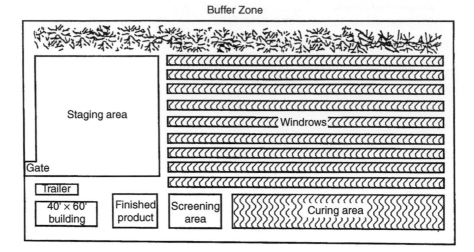

FIGURE 8.12 Suggested layout of a compost facility (From Appelhoff, M. and McNelly, J., 1988. Reproduced with kind permission of the Michigan Department of Environmental Quality.)

is absorbed during periods of heavy precipitation and that the upper layers of the soil do not become waterlogged. If the soil is impermeable or the site is paved, drainage systems are necessary in order to divert precipitation from the composting pad and storage areas.

Proximity to certain water sources should also be considered. Floodplains, wetlands, surface waters, and groundwater all need to be protected from runoff or leachate from the site. The water table should be no closer than 60 cm (24 in.) below the surface. Otherwise, leachate from the compost may percolate downward and contaminate groundwater (Richard et al., 1990).

8.10.2 LAND AREA REQUIREMENTS

To operate efficiently, a composting facility must allot sufficient space to the preprocessing, processing, and postprocessing compost stages as well as to the surrounding buffer zone.

8.10.3 OTHER FACTORS AFFECTING SITING DECISIONS

Other factors to consider when siting a composting facility include:

- *The existing infrastructure.* The presence of utility hookups, storage space, and paved access roads.
- *Zoning issues.* The construction of composting facilities is permitted in locations as directed by local zoning laws.
- *Nearby land users.* Sites near public parks, schools, or residential areas could cause objections from citizens concerned about odor and noise (U.S. EPA, 1994).

8.11 DESIGN

The following items must be incorporated in the design of a large-scale composting facility:

- Preprocessing area
- Processing area
- Postprocessing area
- Buffer zone

- Access and onsite roads
- Site facilities and security

While designing the facility, the possibility for future expansion should be considered.

8.11.1 PREPROCESSING AREA

A preprocessing area provides space to receive feedstock and to sort and separate materials. Receiving materials in a preprocessing area eliminates the need for delivery trucks to unload directly into piles in inclement weather. The size and design of the preprocessing area depends on the amount of incoming material and the way the materials are collected and sorted. The tipping area (i.e., where incoming feedstocks are unloaded) is often under a roof to avoid the effects of severe weather.

8.11.2 PROCESSING AREA

The processing area includes the composting pad and the curing area. The pad surface should be paved to prevent infiltration. Adequate drainage is also essential. Precipitation collected on the pads can be diverted through the use of drains. Poor drainage will result in ponding of water, saturated composting materials, muddy and unsightly site conditions, odor production, and excessive runoff and leachate from the site (Rynk, 1992).

Proper ventilation is required in enclosed preprocessing and processing areas because the air within the structure can be sources of bioaerosol, odors, dust, and excess moisture. Air filters can be installed to clean the exhaust air and biofilters to absorb odor-producing compounds. Vents can be situated over preprocessing equipment (e.g., conveyor belts, trommels) to reduce dust and odors.

A curing area is used to store the compost for the last phase of the composting process, to allow the material to stabilize. The material should be fairly stable and therefore runoff, groundwater contamination, and other siting issues should be of less concern. The curing area needs less space, about one quarter of the area of the compost pad (University of Connecticut, 1989; Richard et al., 1990).

8.11.3 BUFFER ZONE

The larger the buffer zone, the greater the acceptance of the facility among residents. The buffer zone installed at a composting facility depends on the type of feedstock being composted and the level of technology (i.e., monitoring and odor control) employed at the facility. State and local regulations frequently require minimal buffer zone sizes or specify the distances that the composting operations must be from property lines, residences, or adjacent businesses and from surface water or water supplies.

The buffer zone must be larger than the composting pad, particularly when the operation is adjacent to residential areas or businesses. Enclosed facilities may function adequately with a smaller buffer zone since operations are more closely controlled.

When designing the facility, prevailing wind direction should be considered. The buffer zone should be extended in this direction. This will help in minimizing the transport of odor and bioaerosols downwind of the facility.

8.11.4 SITE FACILITIES AND SECURITY

Composting operations might require several buildings to house various site functions, from maintenance and administrative work to personnel facilities. Access to the site must be controlled to prevent vandalism, especially arson and illegal dumping. At a minimum, the access roads must be secured with a fence, cable, locked gate, or similar barrier.

8.12   MSW COMPOSTING BY ANAEROBIC PROCESSES

Anaerobic digestion of low-solids (4 to 10%) wastewater has been carried out for decades at publicly owned treatment works and in industrial facilities. A number of waste management facilities in the United States and Europe, however, now employ the so-called high-solids reactors, containing up to 30% or greater solids content. This technology allows for the anaerobic digestion of high-solids MSW, specifically the organic fraction.

Anaerobic digestion is described by the following equation:

$$\text{organic MSW} + H_2O \rightarrow CO_2 + CH_4 + NH_3 + H_2S + \text{mixed solids} + \text{new cell biomass} \quad (8.8)$$

The desired end products of this process include methane and sludge water. Other products are carbon dioxide and trace quantities of ammonia and hydrogen sulfide. The sludge water is dewatered to produce a supernatant and a filter cake. The filter cake serves as a soil conditioner. The filtrate can be mixed with the organic MSW to create a slurry feedstock or it can be fed directly to the digester. The liquids can also be used as fertilizer. A simple schematic of anaerobic digestion of MSW appears in Figure 8.13.

High-solids anaerobic digestion (HSAD) of the MSW organic fraction occurs in three phases:

1. *Hydrolysis*. High-molecular-weight compounds are converted into low-molecular weight compounds by microbial action (e.g., hydrolyzing bacteria); for example, polysaccharides are hydrolyzed to monosaccharides, lipids to fatty acids, proteins to amino acids, and nucleic acids to purines and pyrimidines. These products subsequently serve as substrate for new populations of microorganisms.

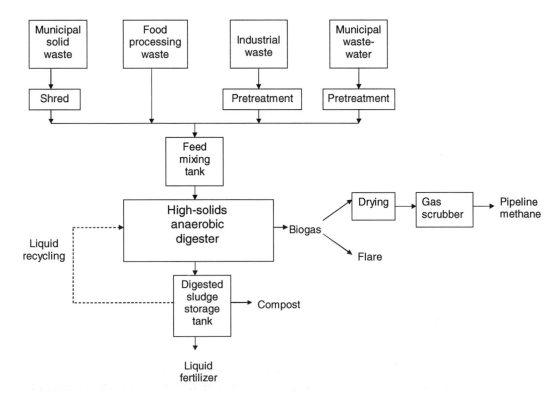

FIGURE 8.13 Schematic for a high-solids anaerobic digestion facility.

2. *Acid generation.* The low-molecular-weight amino acids, fatty acids, and monosaccharides are converted into lower molecular weight intermediate compounds by the action of nonmethanogenic, acetogenic bacteria, also known as acid formers. Acidogens are facultative and obligate anaerobes. Products include carbon dioxide and hydrogen and a number of organic acids and alcohols including acetic acid (CH_3COOH), propionic acid (CH_3CH_2COOH), butyric acid ($CH_3CH_2CH_2COOH$), and ethanol (C_2H_5OH).

An example of an acid-forming reaction is shown below:

$$C_6H_{12}O_6 \rightarrow 2C_2H_5OH + 2CO_2 \qquad (8.9)$$

3. *Methanogenesis.* The acids and alcohols produced in phase 2 are converted into methane and carbon dioxide by the action of methanogenic bacteria, which are strict anaerobes.

Methanogenesis reactions include the following:

$$CH_3COOH \rightarrow CH_4 + CO_2 \qquad (8.10)$$

$$2C_2H_5OH + CO_2 \rightarrow CH_4 + 2CH_3COOH \qquad (8.11)$$

$$CO_2 + 4H_2 \rightarrow CH_4 + 2H_2O \qquad (8.12)$$

As we shall see in Chapter 10, these identical reactions occur in a cell of a sanitary landfill, with the potential for production of enormous volumes of methane gas.

8.12.1 PROCESS DESCRIPTION

There are four basic steps involved in high-solids anaerobic digestion of MSW to produce methane:

1. Initial processing. Commingled MSW is received at the facility, sorted to obtain the organic fraction, then shredded.
2. Preparation and conditioning. In order to prepare a material suitable for biological processing, moisture and nutrients are added, the input is blended, pH is adjusted to near-neutral, and the slurry mix is heated to between 55 and 60°C.
   Moisture and nutrients are added to the wastes in the form of wastewater sludge or animal manure. Additional nutrients may also have to be added, depending on the chemical composition of the added wastewater or manure.
   Anaerobic digestion is typically conducted in a continuous-flow reactor whose contents are completely mixed. In some operations, a series of batch reactors are used. Foaming and the formation of surface crusts have caused problems in MSW digestion; therefore, adequate mixing is extremely important in system operation.
3. Recovery, storage, and separation of the gas components.
4. Dewatering and disposal (or application) of the digested sludge.

8.12.2 SUBSYSTEMS

Several subsystems have been developed to treat MSW anaerobically. The categories include (Chiang Mai University, 1998):

Batch vs. continuous

- Batch — The reactor vessel is loaded with raw feedstock and inoculated with digestate from another reactor. It is sealed and allowed to react until degradation is complete. The digester is emptied and a new input of organic mixture is added.

- Continuous — The reactor vessel is fed continuously with digestate material. Fully degraded material is continuously removed from the bottom of the reactor.

Single step vs. Multi-step

- Single step — All digestion processes occur in one reactor vessel.
- Multi-step — Several reactors operate simultaneously. In some cases the acid-forming stage is separated from the methane-forming stage. This results in increased efficiency as the two microbial populations are separated in terms of nutrient needs, growth capacity, and ability to cope with environmental stress. Some multistage systems also use a preliminary aerobic stage to raise the temperature and increase the degradation of the organic material. In other systems, the reactors are separated into a mesophilic stage and a thermophilic stage (Chiang Mai University, 1998).

Co-digestion with animal manure. The organic fraction of the MSW is mixed with animal manure and the two are co-digested. Such mixing improves the carbon/nitrogen ratio and improves gas production.

Systems in Use. There are a number of HSAD systems either commercially in use or at the pilot scale. The Dranco (Dry Anaerobic Composting) system was developed in Gent, Belgium. Feed is introduced into the top of the reactor. Digested materials are removed from the base of the unit as feed is added. A portion of the digestate is recycled as an inoculum and the remainder is de-watered, resulting in a usable compost product. No mixing occurs within the reactor. Solids content ranges from about 15 to 40%. Retention time in the reactor varies between 15 and 30 days and the operating temperature ranges between 50 and 58°C. The Dranco system can yield between 100 and 200 m^3 of gas per ton of MSW input. The gas content is 55% CH_4 (Six and DeBaere, 1992).

The Valorga system, developed in France, combines four mesophilic HSAD reactors. Mixing of feed within the reactor is carried out by circulation of a portion of the biogas under pressure. The biogas product has about 55 to 60% CH_4 content. The process operates with a solids content of 25 to 50% with residence times ranging between 18 and 25 days.

The BIOCEL process is a mesophilic dry anaerobic batch digestion system (ten Brummeler, 2000). In this process, net energy production is achieved by converting the biogas into heat and power. The first full-scale plant started up in Lelystad, The Netherlands, in 1997. This plant processes 50,000 tons of MSW per year. Anaerobic digestion with the BIOCEL process has been found to inactivate several important groups of plant and animal pathogens. The mechanism that causes the inactivation is not yet fully understood, but may be related to the relatively high volatile fatty acid concentration during the first 2 weeks of digestion.

The Kompogas system is a thermophilic digestion system developed in Switzerland. The reaction vessel is a horizontal cylinder where feed is introduced daily, and movement through the cylinder is accomplished via a horizontal plug-flow. An agitator is installed within the cylinder to mix the material intermittently. Digested material is removed from the end of the cylinder after about 20 days. The digestate is dewatered, and some of the press water is either used as an inoculum or is sent to a wastewater treatment facility to produce more biogas (University of Southampton, 2002).

The Wabio Process was developed by Ecotechnology of Finland (Chhabria, no date). There are two Wabio Process plants in operation. A plant at Vaasa, Finland, has been in operation since 1991 and a facility has been operating in Bottrop, Germany, since 1995. The Wabio process is a single-stage process operating in the mesophilic temperature range. At the treatment plant, feed preparation tanks receive the screened material and a slurry of 15% solids concentration is prepared. The slurry is then pumped to bioreactors. Digestion occurs at 30°C and the retention time of the material is 15 to 20 days. The process can also be operated in the thermophilic range at 55°C. The supernatant liquid is recirculated to make the slurry. Biogas is stored in the gas holder and a portion is used for mixing the contents in the bioreactors. From the gasholder the gas is transferred to engines to produce electricity. The digested slurry is pasteurized at ~7°C for 30 min to produce a material that is safe for spreading on agricultural land.

The feasibility of applying anaerobic digestion for stabilization of solid wastes generated during space missions was investigated (Chynoweth et al., 2003). High-solids leachbed anaerobic digestion involved solid-phase fermentation with leachate recycling between new and old reactors for inoculation, wetting, and removal of volatile organic acids during startup. Anaerobic biochemical methane potential assays were run on several waste feedstocks expected during space missions. The methane yields ranged from 0.23 to 0.30 L/g of volatile solids added.

REFERENCES

Appelhoff, M. and McNelly, J., Yard waste composting, *Guidebook for Michigan communities*, Michigan Department of Natural Resources, Lansing, MI, 1988.

Bezdicek, D., Fauci, M., Caldwell, D., Finch, R., and Lang, J., One year later: persistent herbicides in compost, *BioCycle,* July 2001, p. 25, 2001. See: http://www.jgpress.com/BCArticles/2001/070125.html

ten Brummeler, E., Full scale experience with the BIOCEL process, *Water Sci. Technol.*, 41, 299–304, 2000.

Buyuksonmez, F., Rynk, R., Hess, T. F., and Bechinski, E., Occurrence, degradation and fate of pesticides during composting, Part I, *Compost Sci. Util*, 7, 66–82, 1999.

Buyuksonmez, F., Rynk, R., Hess, T. F., and Bechinski, E., Occurrence, degradation and fate of pesticides during composting, Part II, *Compost Sci. Util*, 8, 61–81, 2000.

Chaney, R.L. and Ryan, J.A., Heavy metals and toxic organic pollutants in MSW composts: research results on phytoavailability, bioavailability, fate, etc., As cited in Hoitink, H.A.J. et al., Eds., *Proceedings of the International Composting Research Symposium*, 1992.

Chiang Mai University, *Review of Current Status of Anaerobic Digestion Technology for Treatment of Municipal Solid Waste*, Regional Information Service Center for South East Asia on Appropriate Technology, Institute of Science and Technology Research and Development, November 1998. See: http://www.ist.cmu.ac.th/riseat/documents/adreview.pdf

Chhabria, N.D., No date, Wabio Anaerobic Digestion Process to Produce Energy from Garbage, United Nations Development Programme. See: http://www.undp.org.in/Programme/GEF/september/page16-20.htm

Chynoweth, D.P., Haley, P., Owens, J., Teixeira, A., and Welt, B., Anaerobic Composting for Regenerative Stabilization of Wastes During Space Missions, Bioastronautics Investigators' Workshop, Jan. 13–15, 2003, Galveston, TX. See: http://www.dsls.usra.edu/dsls/meetings/bio2003/pdf/Environmental/2198 Chynoweth.pdf

Composting Council, Compost facility planning guide, Composting Council, Washington, DC, 1991.

Diaz, L.F., Air emissions from compost, *BioCycle*, 28, 52–53, 1987.

Diaz, L.F., Savage, G.M., Eggerth, L.L., and Golueke, C.G., *Composting and Recycling Municipal Solid Waste*, Lewis Publishers, Boca Raton, FL, 1994.

Eco Technology JVV Oy, Espoo, Finland.

Epstein, E. and Epstein, J.I., Public health issues and composting., *BioCycle*, 30, 50–53, 1989.

Eweis, J.B., Ergas, S.J., Chang, D.P.Y., and Schroeder, E.D., *Bioremediation Principles*, McGraw-Hill, Boston, MA, 1998.

Gillett, J.W., Issues in risk assessment of compost from municipal solid waste: occupational health and safety, public health, and environmental concerns, *Biomass Bioenergy*, 3, 145–162, 1992.

Golueke, C.G., *Biological Reclamation of Solid Wastes*, Rodale Press, Emmaus, PA, 1977.

Gray, K., Sherman, K., and Biddlestone, A.J., A review of composting, Part 2 — the practical process, *Process Biochem.*, 6, 22–28, 1971.

Gurkewitz, S., Yard debris compost testing, *Biocycle*, 30, 58–59, 1989.

Haug, R.T., *Compost Engineering Principles and Practice*, Ann Arbor Science Publishers, Ann Arbor, MI, 1980.

Haug, R.T., *The Practical Handbook of Compost Engineering*, Lewis Publishers, Boca Raton, FL, 1993.

Hellman, T.M. and Small, F.H., Characterization of petrochemical odors, *Chem. Eng. Prog.*, 69, 75–77, 1973.

Kissel, J. C., Henry, C.H., and Harrison, R.B., Potential emissions of volatile and odorous organic compounds from municipal solid waste composting facilities, *Biomass Bioenergy*, 3, 181–194, 1992.

Maine Department of Environmental Protection, no date, Persistent Herbicides in Leaf & Yard Compost. See: http://www.state.me.us/dep/rwm/compost_herbicides.htm

Menzer, R.E., Water and soil pollutants. As cited in Amdur, M. O., Doull, J., and Klaassen, C.D., Eds., *Casarett and Doull's Toxicology The Basic Science of Poisons*, 4th ed., Pergamon Press, New York, NY, 1991.

Miller, T.L., Swager, R.R., and Adkins, A.D., Selected Metal and Pesticide Content of Raw and Mature Compost Samples from Eleven Illinois Facilities, Illinois Department of Energy and Natural Resources, Springfield, IL, 1992.

Rich, L.G., *Unit Processes of Sanitary Engineering*, Wiley, New York, NY, 1963.

Richard, T.L., Municipal solid waste composting. Physical and biological processing, *Biomass Bioenergy*, 3, 195–211, 1992.

Richard, T. and Chadsey, M., Environmental impact of yard waste composting, *BioCycle*, 31, 42–46, 1990.

Richard, T. and Chadsey, M., Croton Point Compost Site Environmental Monitoring Program, Westchester County Solid Waste Division, White Plains, NY, 1989.

Richard, T., Dickson, N., and Rowland, S., Yard waste management, A planning guide for New York State, New York State Energy Research and Development Authority, Cornell Cooperative Extension, New York State of Environmental Conservation, Albany, NY, 1990.

Roderique, J.O. and Roderique, D.S., *The Environmental Impacts of Yard Waste Composting*, Gershman, Brickner and Bratton, Inc., Falls Church, VA, 1990.

Rose, W.W. and Mercer, W.A., Fate of Insecticides in Composting Agricultural Wastes, National Canners Association, Washington, DC, 1968.

Rynk, R., Fires at composting facilities: causes and conditions, *BioCycle*, 41, 2000. http://www.environmental-expert.com/magazine/biocycle/january2000/article4.htm.

Rynk, R. (Ed.), *On-Farm Composting Handbook*, NRAES-54; Natural Resource, Agriculture, and Engineering Service (NRAES), Cooperative Extension, Ithaca, NY, 1992, www.nraes.org.

Six, W. and DeBaere, L., Dry Anaerobic Conversion of Municipal Solid Waste by Means of the Dranco Process at Brecht, Belgium, *Proceedings of International Symposium on Anaerobic Digestion of Solid Waste*, in Venice, April 14-17, 1992, Cecchi, F., Mata-Alvarez, J., and Pohland, F.G., Eds., Int. Assoc. Wat. Poll. Res. Control, 1992, pp. 525–528.

University of Connecticut, Leaf Composting, A Guide for Municipalities, University of Connecticut Cooperative Extension Service; State of Connecticut Department of Environmental Protection, Local Assistance and Program Coordination Unit, Recycling Program, Hartford, CT, 1989.

University of Southampton, Anaerobic Digestion. Environment Division, Civil and Environmental Engineering, 2002. See: http://www.soton.ac.uk/~env/research/wastemanage/anaerobic.htm#plants 4 organic MSW

U.S. Environmental Protection Agency, Windrow and Static Pile Composting of Municipal Sewage Sludges, Project Summary, EPA-600-S2-84-122, Engineering Research Laboratory, Cincinnati, OH, 1984.

U.S. Environmental Protection Agency, Composting Yard Trimmings and Municipal Solid Waste, EPA 530-R-94-003, Office of Solid Waste and Emergency Response, Washington, DC. 1994.

U.S. Environmental Protection Agency, Innovative Uses of Compost for Erosion Control, Turf Remediation, and Landscaping, EPA530-F-97-043, Office of Solid Waste and Emergency Response, Washington, DC, (5306W), 1997b.

U.S. Environmental Protection Agency, Pesticide Industry Sales and Usage, 1994 and 1995 Market Estimate, 733-R-97-002, Authored by Aspelin, A., U.S. EPA Office of Prevention, Pesticides and Toxic Substances, Washington, DC, 1997a.

U.S. Environmental Protection Agency, An Analysis of Composting as an Environmental Remediation Technology, EPA530-B-98-001. Office of Solid Waste and Emergency Response, Washington, DC, (5306W), 1998.

Vesilind, P.A., Worrell, W., and Reinhart, D., Solid Waste Engineering. Brooks/Cole, Pacific Grove, CA, 2002.

Wellinger, A., Wyder, K., and Metzler, A.E., Kompogas - a new system for the anaerobic treatment of source separated waste, in: Checchi, F., Mata-Alvarez, J., and Pohland, F.G., Eds., *Anaerobic Digestion of Solid Waste*, Stamperia di Venezia, Venice, 1992.

Williams, T. O. and Miller, F.C., Odor control using biofilters, Part I, *BioCycle*, 33, 72–77, 1992.

Wirth, R., Introduction to Composting, Minnesota Pollution Control Agency, St. Paul, MN, 1989.

Zubay, G., *Biochemistry*, Addison-Wesley, Reading, MA, 1983.

SUGGESTED READINGS

Anderson, J., Ponte, M., Biuso, S., Brailey, D., Kantorek, J., and Schink, T., Case study of a selection process, *The Biocycle Guide to In-Vessel Composting*, JG Press, Inc., Emmaus, PA, 1986. pp. 39–47.

Barlaz, M.A. and Ham, R.K., Methane production from municipal refuse: A review of enhancement techniques and microbial dynamics, *Crit. Revi. Environ. Control*, 19, 557–584, 1990.

Broda, P., Biotechnology in the degradation and utilization of lignocellulose, *Biodegradation*, 3, 219–238, 1992.

Daniels, L., Biological methanogenesis: physiological and practical aspects, *Trends Biotechnol.*, 2, 91–98, 1984.

DeBaere, L., Van Meenen, P., Debroosere, S., and Verstraete, W., *Anaerobic fermentation of refuse, Resour. Conserv.* 14, 295–308, 1987.

Gillis, A.M., Shrinking the trash heap, *BioScience*. 42, 90–93, 1992.

Goldstein, N., States regulations on MSW composting, *Biocycle*, 30, 50–53, 1989.

Goldstein, N. and Steuteville, R., Biosolids composting strengthens its base, *Biocycle*, 35, 48–57, 1994.

Horan, N. J., *Biological Wastewater Treatment Systems*, Wiley, New York, NY, 1990.

Kashmanian, R.M. and Taylor, A.C., Costs of composting yard wastes vs. landfilling, *Biocycle*, 30, 60–63, 1989.

Segall, L., In-vessel composting for low volume generators, *Biocycle*, 35, 61–66, 1994.

Segall, L. and Alpert, J., Compost market strategy, *Biocycle*, 31, 38, 1990.

QUESTIONS

1. It is undesirable to land-apply raw solid wastes to soil because undesirable reactions may occur which could inhibit plant growth. Explain.
2. When a waste possessing a high (200:1) C:N ratio waste is land-applied:
   (a) microbial growth is relatively unchanged; (b) agricultural plants cannot compete with soil microbes for soil N; (c) N is converted into ammonia gas (NH_3) and lost to the atmosphere; (d) nitrogen is converted into N_2 gas
3. During composting of MSW, a series of complex N transformations, including immobilization, nitrification, mineralization, and others occur. Explain how the C:N ratio declines during composting. What are the fates of N and C? Provide specific reactions and compounds.
4. Microbial succession is important to bring the composting process to completion. Explain.
5. Explain how fire could be generated in an actively composting pile. How could such a scenario be prevented?
6. Compare the dynamics of microbial populations and oxygen levels over time with the turned pile method of composting vs. aerated static piles.
7. Discuss the pH requirements for optimizing composting. Why, from a biochemical or microbiological perspective, is this pH range most effective?
8. Composting of sewage sludge poses different management concerns compared with composting of MSW. Describe how the process may differ in terms of pathogen control, odor control, leachate control, and aeration.
9. Explain how anaerobic reactions may occur in a compost pile that contains 15-20% oxygen in the interstitial spaces.
10. What is the relationship between pile temperature and microbial growth and activity? What is an ideal compost temperature range? Why is this range considered optimal?
11. The city of Pristine, Illinois, in developing their comprehensive waste management program, will establish a composting facility adjacent to the transfer station. What attributes should be considered when screening potential locations for a composting site? Consider size of area, soils, drainage and slope, land-use compatibility, and controls for runon and runoff.
12. The land area of the selected compost site is smaller than optimum. Based on practical issues, the following compost method should be used: (a) turned pile; (b) static pile with

forced aeration; (c) sheet composting; (d) on-the-shelf heated bins; (e) avoid composting altogether and directly land-apply the organic component of the wastes. Explain your choice.

13. Is there a large-scale MSW composting program in operation in your community? What are the feedstock materials? Where is the facility located, for example, adjacent to the transfer station or landfill? On privately-owned land? How are odors and leachate production managed?

14. List and discuss the possible uses for finished compost. What are the benefits of MSW composting in a community integrated waste management program? Given the time, space requirements, energy and labor requirements, is composting economically justified for a community?

15. For over a decade, the market value of compost has been quite low. However, many communities continue to support yard waste composting programs. Explain why this is so.

16. Bench scale tests of aerobic composting were conducted on a compost feedstock with the starting empirical formula $C_{28}H_{46}O_{22}N$. Pilot tests indicated that 1000 kg dry solids of the feedstock decreased to 245 kg dry solids by completion. The final product empirical formula was determined to be $C_{12}H_{16}O_6N$. Determine the stoichiometric oxygen required to complete the aerobic decomposition per 1000 kg of feed.

17. At a waste handling facility, a mixture of approximately 70 metric tons of food waste and yard waste is to be composted. The moisture content of this feedstock measures 5.5%. It has been previously determined that an ideal moisture content for the compost pile should be about 58%. Calculate the metric tons of water to be added to the solids to achieve the optimum moisture content.

EXCEL EXERCISE

SOIL AND GROUNDWATER QUALITY AT A COMPOSTING FACILITY

FILE NAMES: COMPOST_SOIL.XLS

COMPOST_GW.XLS

The Situation

In 1992, a municipality in the eastern United States installed a MSW and food waste composting facility. A local farmer contracts some of his property for the facility. Surficial materials are silt loam or silty clay loam. At a depth of approximately 40 cm (16 in.) below ground surface, loam is the predominant texture, and beyond this horizon there are occasional lenses of coarse sand and gravel. Three small streams occur within 0.5 to 3 mi of the facility. There were no siting regulations at the time of the installation, and therefore no concrete pad was installed during construction. Rainfall ranges from 42 to 48 in. per year.

The data for this exercise can be located at www.crcpress.com/e_products/downloads/download.asp?cat_no=3525

MSW arrives at the compost site from two municipalities ('MSW1' and 'MSW2'); both have previously processed their wastes via shredding and magnetic separation. A third facility tips significant quantities of vegetable processing wastes including plant waste and washings. These three wastes are composted separately for a research project (below).

MSW1 contained large proportions of yard waste whereas MSW2 did not. Additionally, MSW2 contained a significantly higher proportion of inerts such as glass, stone, and some metals.

Research plots were established on the property (see map) by a university group to test the effects of the composts (if any) on soil and groundwater properties. Maize, soybeans, and pasture hay were grown on the plots. Plot setups are shown below.

The materials applied at the East Fork site were as follows:

Material	Plot Designation
MSW1	1A, 1B, 1C
MSW2	2A, 2B, 2C
Inorganic fertilizer	3A, 3B, 3C
Food processing waste	4A, 4B
No treatments	Control

For the Powder Creek site:

Material	Plot Designation
MSW1	5A, 5B, 5C
MSW2	6A, 6B, 6C
Inorganic fertilizer	7A, 7B, 7C
Food processing waste	8A, 8B
No treatments	Control

Tasks

1. Determine the direction of groundwater flow and draw directional arrows.
2. Observe the data for groundwater quality in the site wells. Are there any constituents that are in excess of MCLs (see Table 10.1, also the Code of Federal Regulations)?
3. Based on groundwater data, can you suggest possible plots and waste types that may be contributing to the highest contaminant data?
4. Do you observe any correlation between concentrations of any of the contaminants in groundwater?
5. In the groundwater, does the EC correlate with data for any elements or compounds?
6. As mentioned above, MSW2 contained relatively higher concentrations of metals compared with other feedstocks. Why were these metals detected only in very small amounts in the soil or groundwater?
7. If we were to assume that the soils were similar from both sites, which compost feedstock results in the lowest NO_3 concentrations in groundwater? The highest concentrations?
8. From the data, what is the general relationship between soil data for NO_3^- and groundwater data for NO_3^-? For NH_4^+? For metals?
9. This study was not a strictly scientific one; however, conduct an Analysis of Variance on soil data and determine whether any of the treatments is significantly different in terms of NO_3^- content, P, or K levels.
10. Conduct an ANOVA on groundwater data and determine whether any of the treatments is significantly different in terms of NO_3^- contamination of groundwater.
11. What corrective measures would you propose in order to control the excess leaching of nutrients from any of these sites?

EXAMPLE 9.2

A carbonaceous waste given by the empirical formula $C_{65.5} H_{102.3} O_{40.8} N_{1.1}$ is to be incinerated. Proximate and elemental analyses of the waste are as follows:

Proximate Analysis	%	Elemental Analysis	%
Moisture	4.8	Carbon	47.36
Noncombustibles	6.2	Hydrogen	6.25
		Oxygen	39.25
		Nitrogen	0.85
		Sulfur	0.19
		Ash	6.10

Calculate the following: (a) the gross heat value and net heat value of this waste as received; (b) the volume of air needed for the complete combustion of 1000 kg (i.e., 1 metric ton) of the input material.

SOLUTION

(a) The higher heat value (HHV) and lower heat value (LHV) of the waste can be calculated using Equations 4.6 and 4.7:

HHV = 0.339 (C) + 1.44 (H) – 0.139 (O) + 0.105 (S) MJ/kg

HHV = 0.339 (47.36) +1.44 (6.25) – 0.139 (39.25) + 0.105 (0.19) MJ/kg,

HHV = 19.61 MJ/kg

LHV = HHV (in MJ/kg) – 0.0244 (W + 9H) MJ/kg

LHV = 19.61 MJ/kg – 0.0244 (4.8 +9 (6.25) MJ/kg

LHV = 18.12 MJ/kg

(b) When computing the oxygen requirements, the chlorine and sulfur components may be neglected. Given that $a = 65.5$, $b = 102.3$, $c = 40.8$, $d = 0$, $e = 0$, $f = 0.85$, $g = 0$, the combustion equation is as follows:

$$C_{65.5}H_{102.3}O_{40.8}M_{0.85} + 71.1O_2 \rightarrow 65.5CO_2 + 51.2H_2O + 1.1NO$$

The formula mass of the waste is then calculated:

Carbon	$12 \times 65.5 = 786$	
Hydrogen	$1 \times 102.3 = 102.3$	
Oxygen	$16 \times 40.8 = 652.8$	
Nitrogen	$14 \times 0.85 = 11.9$	
Total	1553	

Therefore the molar mass of the material is 1553 or 1.55kg.

Of the 1000 kg of the material, 890 kg (i.e., 1000 kg minus 48 kg moisture and 62 kg inert material) is combustible. This quantity corresponds to 890 kg/1553 kg/mol = 573 mol.

From the equation, we see that 1 mol of the material requires 71.1 mol of O_2. Therefore 573 mol of material requires $573 \times 71.1 = 40,746$ mol of O_2.

At standard temperature and pressure (i.e., 0°C and 1atm), 1 mole of a gas occupies $22.4 \times 10^{-3} m^3$. Consequently,

volume = 40,746 mol of $O_2 \times 22.4 \times 10^{-3}$ m³/mol of O_2

volume = 913 m³ of O_2

Air contains about 21% oxygen, therefore 4348 m³ of air is required to supply this volume of oxygen. This value converts to 4.35 m³ of air per kg of dry combustible material.

There are many other formulas which have been devised to calculate the air required for waste combustion. For example, Dvirka (1986) devised the following:

$$W_a = 0.0431[\ 2.667C + 8H + S - O\ ] \text{ kg of air/kg of waste} \tag{9.2}$$

where W_a is the mass of dry stoichiometric air (at STP) required to burn 1 kg of combustible waste and C, H, S, and O are the mass % of carbon, hydrogen, sulfur, and oxygen, respectively, of the moisture- and ash-free material.

Other critical factors influencing the completeness of combustion are temperature, time, and turbulence, labeled the "three T's of combustion." Each combustible substance has a minimum ignition temperature that must be attained in the presence of oxygen for combustion to be sustained. Above the ignition temperature, heat is generated at a higher rate than it loses to the surroundings, which makes it possible to maintain the elevated temperatures necessary for sustained combustion. The residence time of the input wastes in the high-temperature region of the combustion zone should exceed the time required for combustion to take place. Such a requirement will affect the size and shape of the furnace. Turbulence (i.e., the thorough mixing of MSW as it passes through the combustion chamber) will expose particle surfaces to oxygen and high temperatures and will speed the evaporation of liquids for combustion in the vapor phase. Inadequate mixing of combustible gases and air in the furnace will lead to the generation of PICs, even from a unit containing sufficient oxygen.

9.3 THE MASS-BURN INCINERATOR

Mass burning is unquestionably the most straightforward incineration technology, involving combustion of MSW as received from the collection vehicle. The only processing involved is simple blending of wastes and removal of large bulky items such as white goods (stoves and washing machines), bulky combustible items (mattresses, furniture, etc.), and hazardous items. The crane operator often accomplishes the removal in the waste storage pit. Therefore, a major benefit of mass-burn systems, beyond its relative simplicity, is the avoidance of capital and operating costs associated with extensive waste processing. Some incinerators may utilize shredding equipment for reducing bulky items to workable sizes. As will become apparent, the convenience of mass burn is matched by a number of significant health and environmental concerns.

The major components of the mass burn incinerator include:

- Tipping area or receiving floor
- Storage pit
- Equipment for charging, or loading the waste into the incinerator hopper. This is often a crane or front-end loader
- The combustion chamber
- Energy recovery system
- Pollution control equipment
- Flue

A typical mass burn system is shown in Figure 9.1.

Mass-burning incineration can be divided into four broad areas (Hickman, 1984):

- Incineration without energy recovery
- Incineration using modular furnaces
- Incineration using refractory furnaces with heat recovery boilers
- Incineration using waterwall furnaces

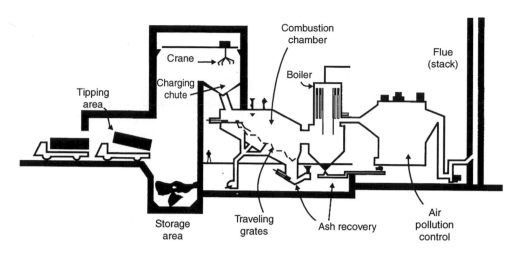

FIGURE 9.1 Cross section of a typical mass-burn incinerator. From Holmes, J.R., *Refuse Recycling and Recovery*, Wiley, New York, 1981. Copyright John Wiley and Sons Limited. Reproduced with permission.

All four methods are in use in both the United States and Europe; however, waterwall incinerators have proven to be the superior means of recovering energy from MSW (Hickman, 1984).

MSW is dumped ("tipped") by the collection vehicle directly into a storage pit. The pit must allow for storage of sufficient volumes of waste for steady uniform operation and should provide for 24 h/day, 7 day/week operation. The MSW charge is next transferred into loading hoppers by crane, which then falls into the furnace by gravity. The temperature of the combustion zone will vary with furnace type and is usually maintained between about 815 and 1095°C (1500 and 2000°F). Within this temperature range combustion is optimized and the production of odoriferous compounds is minimized. These temperatures are also adequate to protect the refractory linings of the combustion chamber. The waste is conveyed through the combustion chamber by a system of agitating grates. There are a limited number of grate types in use (Figure 9.2), all with the functions of transporting waste through the firebox, agitation, and conducting underfire air upwards. The rocking or turning action of the grate agitates the MSW for more complete combustion. There are openings in the grates that allow for the ash to fall through into a collection bin. This residue is the so-called "bottom ash." Additional unburned residue is carried to the end of the grates and is collected and combined with other bottom ash.

During mass burn of MSW the charge is spread out several inches thick on the grate surface. During agitation the waste mixes with air, which is forced over the grates ("overfire air"). The overfire air assists in completing combustion of the fuel gas and any MSW-generated gases and particulate matter rising from the grates. Air is also directed under the grates. This underfire air (about 40 to 60% of the total air entering the furnace) feeds the combustion process and cools the grates. If there is too low a flow of underfire air, grate temperatures will increase and ash will soften and clog the grates, resulting in damage to the grates and less than optimum combustion.

The combustion gases transfer heat to boilers or waterwalls. Boilers are defined as enclosed units using controlled combustion and whose primary purpose is the recovery and export of useful thermal energy in the form of hot water, saturated steam, or superheated steam. The principal components of a boiler are a burner, a firebox, a heat exchanger, and a means of creating and directing gas flow through the unit. A boiler's combustion chamber and primary energy recovery sections are usually of "integral design", i.e., the combustion chamber and the primary energy recovery sections, such as waterwalls and superheaters, are manufactured as a single unit (U.S. EPA, 2002). Figure 9.3 illustrates a cross section of a typical boiler.

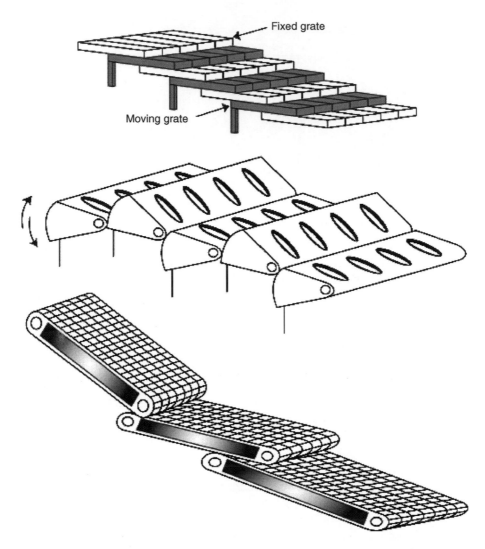

FIGURE 9.2 Common grate types for a MSW combustor. Underfire air is forced upwards through the grates, and overfire air is passed over the top of the burning MSW. (From Vesilind, P. A., *Solid Waste Engineering*, 1st ed., 2002. Reproduced with kind permission of Brooks/Cole, a division of Thomson Learning: www.thomsonrights.com.)

MSW can be combusted for the production of steam, which is useful for driving turbines and subsequent electricity generation. The remaining steam has little industrial use, however, unless it is located sufficiently close to other buildings to be applied for space and water heating. Often the residual steam is condensed to liquid water which is either cooled and used again in the power plant or released into the local environment. Usually, boiler water is treated and reused because it is too expensive to be used only once. Small amounts, less than 10%, are blowdown (i.e., fresh water added to the system) to minimize the concentration of dissolved solids (Vesilind et al., 2002).

If hot water is discharged directly into a body of surface water, it will create adverse effects in streams, rivers, and estuaries; as a result, heat discharges are regulated by federal and state codes. The typical limit on heat discharges is that the temperature of the receiving water cannot be raised by more than 1°C. Given this restriction on the temperature of water returned to local bodies of water, the heated water must be cooled prior to discharge. Various means are used for dissipating

occurring within the original waste is simply converted into either a gaseous form or ash. Under ideal conditions, carbonaceous wastes are converted into innocuous products such as CO_2 and H_2O along with the release of heat energy. The actual composition of flue gases, however, is highly complex and a function of the composition of the original MSW, furnace design, and combustion conditions. Many of the substances emitted from incinerator flue gases are known to negatively impact human health. The air pollutants of concern from MSW combustion are primarily particulates, acid gases, and trace gases.

9.4.4 PARTICULATES

Particulates, also known as "fly ash", occur as either solid particles or liquid droplets composed of organic or inorganic substances. A number of terms are used to describe atmospheric particles; the more important of these are summarized in Table 9.1. Particulate matter makes up the most visible and obvious form of air pollution.

Atmospheric aerosols are solid or liquid particles smaller than 100 μm in diameter. Particles in the 0.001 to 10 μm range are commonly suspended in the air near sources of pollution such as industrial facilities and power plants. Aerosols consist primarily of carbonaceous material, metal oxides and glass, dissolved ionic species (electrolytes), and ionic solids. The predominant constituents are carbon, water, sulfate, nitrate, ammonium nitrogen, and silicon. The composition of aerosol particles varies significantly with size. Smaller particles tend to be acidic and often originate from gases such as from the conversion of SO_2 into H_2SO_4. Larger particles tend to consist of materials generated mechanically (Manahan, 1994).

Particulates are a public health concern because large quantities that occur in the respirable fraction (approximately 15 μm in diameter) are commonly emitted from MSW incinerators. In addition, heavy metals, chlorinated dibenzodioxins, and other trace elements are attached to fly ash. The rates of particulate emissions from a mass-burn incinerator depend on:

- Ash content, i.e., the percentage of noncombustible materials in the waste. If particle size is sufficiently small, these can become entrained in the gases passing through the system.
- Furnace design. Some systems create greater degrees of agitation of the waste, thus releasing particles. Furthermore, entrainment of particulates can result from too much excess air. Optimal use of overfire and underfire air are important to limit particulate emissions.
- Temperature. A low-temperature zone occurring in the combustion chamber can result in the formation of incompletely burned residuals that are often lightweight and easily transportable.

TABLE 9.1
Terminology Associated with Atmospheric Particles

Term	Definition
Aerosol	Colloidal-sized atmospheric particle
Condensation aerosol	Formed by condensation of vapors or reactions of gases
Dispersion aerosol	Formed by grinding of solids, atomization of liquids, or dispersion of dusts
Fog	Term denoting high level of water droplets
Haze	Denotes decreased visibility due to the presence of particles
Mists	Liquid particles
Smoke	Particles formed by incomplete combustion of fuel

Source: Manahan, S.E., *Environmental Chemistry*, 6th ed., CRC Press, Boca Raton, FL. Reproduced with kind permission. Copyright Lewis Publishers, an imprint of CRC Press. Boca Raton, FL.

9.4.5 ACID GASES

As shown in Equation 9.1, the combustion of an organic material results in the generation of carbon dioxide and water and other components. Gases such as SO_x, NO_x, and HCl may be produced by incinerators at rates of several pounds per ton of waste charged. These gases are collectively considered "acid gases" because they dissolve readily in water to form the corresponding strong acids.

Sulfur occurs in tires, wallboard, and plant tissue (yard waste). During combustion, sulfur is converted into the corresponding oxides. Sulfur dioxide is a primary pollutant as it is emitted directly from MSW burning and concurrent sulfur oxidation,

$$S + O_2 \rightarrow SO_2 \tag{9.3}$$

SO_2 can cause direct respiratory irritation and damage materials such as stone and metal. Sulfur emissions are converted into a secondary pollutant when sulfur dioxide reacts with water vapor and oxygen in the atmosphere, producing sulfur trioxide:

$$2SO_2 + O_2 \rightarrow 2SO_3 \tag{9.4}$$

The SO_3 combines with water to form sulfuric acid,

$$SO_3 + H_2O \rightarrow H_2SO_4 \tag{9.5}$$

This product is corrosive to skin and mucosa and is linked with several respiratory ailments. Sulfuric acid is the primary component of acid rain, and will damage materials and is hazardous to biota. A wide range of values of SO_2 has been measured in stack emissions. As much as 0.68 to 1.4 kg (1.5 to 3 lb) SO_2 has been measured per ton of MSW charged, which can substantially alter the pH of local precipitation. Natural, uncontaminated rain has a pH of about 5.6 to 5.7, but the pH of acid rain can be as low as 2. The deposition of atmospheric acid on freshwater aquatic systems prompted the U.S. EPA to recommend a limit of 10 to 20 kg SO_4^{-2}/ ha/year.

Nitrogen (N) also occurs in food and yard waste. The product of N combustion is nitrogen oxides, NO_x:

$$N_2 + O_2 \rightarrow 2NO \tag{9.6}$$

$$2NO + O_2 \rightarrow 2NO_2 \tag{9.7}$$

Nitrogen dioxide is an important component of photochemical smog. The formation of photochemical smog begins with the production of nitrogen oxides whether from automobiles, industrial facilities, or MSW combustion. Hydrocarbons are also emitted into the atmosphere from various sources such as automotive and industrial sources (including incineration). The various constituents react with sunlight to yield ozone (O_3), a secondary pollutant, which in turn reacts with hydrocarbons to form a wide range of compounds including aldehydes and organic acids. Table 9.2 lists some of the major reactions involved in the formation of photochemical smog.

The concentrations of NO_2 generated from incineration, however, are often low due to the relatively low temperatures at which incineration occurs. Nitrogen can also be converted into HNO_3, another component of acid rain, by a series of reactions abbreviated below:

$$2NO_2 + \frac{1}{2}H_2O \rightarrow \rightarrow 2HNO_3 \tag{9.8}$$

Chlorine occurs in MSW in paints, dyes, polyvinylchloride (PVC)-based products, and bleached paper. During combustion gaseous hydrogen chloride, HCl, is produced which condenses with

TABLE 9.2
Simplified Reaction Scheme for Photochemical Smog

NO_2 + light	$\rightarrow$	NO + O
O + O_2	$\rightarrow$	O_3
O_3 + NO	$\rightarrow$	NO_2 + O_2
O + $(HC)_x$	$\rightarrow$	HCO^o
HCO^o + O_2	$\rightarrow$	$HCO_3{}^o$
$HCO_3{}^o$ + HC	$\rightarrow$	Aldehydes, and ketones .
$HCO_3{}^o$ + NO	$\rightarrow$	$HCO_2{}^o$ + NO_2
$HCO_3{}^o$ + O_2	$\rightarrow$	O_3 + $HCO_2{}^o$
$HCO_x{}^o$ + NO_2	$\rightarrow$	Peroxyacetyl nitrates

Adapted from Vesilind, P.A. et al., *Solid Waste Engineering,* 1st Ed., Brooks/Cole, Pacific Grook, CA, 2002. Reproduced with kind permission of Brooks/Cole, a division of Thomson Learning: www.thomsonrights.com.

water to form the corresponding hydrochloric acid. This corrosive liquid affects eyes, skin, and mucosa, and is linked with acid rain.

$$HCl\ (g) + H_2O \rightarrow HCl\ (aq) \tag{9.9}$$

9.4.6 TRACE GASES

In this category are gases that may occur at levels of a few parts per million (ppm), yet they may still exert a hazardous effect on living systems. Polychlorinated dibenzodioxins (PCDDs) and polychlorinated dibenzofurans (PCDFs) (Figure 9.5), some of which are highly toxic, are now known to form during the combustion of chlorine-containing wastes. There are 75 possible isomers of PCDD and 135 of PCDF (Lisk, 1988). The 2,3,7,8-tetrachlorodibenzodioxin isomer (2,3,7,8-TCDD) (Figure 9.5) is an animal teratogen and by far the most toxic, but its toxicity varies over 5000-fold among species.

A draft report released for public comment in September 1994 by the U.S. EPA described PCDDs as a serious public health threat. The EPA report confirmed that PCDDs are a cancer hazard to humans; that exposure to PCDDs, even at extremely low levels, can cause severe reproductive and developmental problems; and that PCDDs can cause immune system damage and interfere with regulatory hormones. The International Agency for Research on Cancer of the World Health Organization declared in 1997 that 2,3,7,8-TCDD is a Class 1 carcinogen, i.e., it is a known human carcinogen.

Various isomers of PCDD and PCDF have been detected at parts per billion (ppb) levels in fly ash and ng/m^3 concentrations in emissions from incinerators in many countries. The concentrations of several isomers of PCDD and PCDF in fly ash samples from MSW incinerators are shown in Table 9.3. Formation of PCDDs and PCDFs in the combustion chamber itself is unlikely due to the high temperatures present; however, these compounds form in the cooling gases as they exit the flue. Three possibilities have been proposed to account for the presence of PCDDs and PCDFs in MSW incinerator emissions (Hutzinger et al., 1985; Lisk, 1988):

- They are already present in the refuse to be burned and are not completely destroyed during incineration.
- They are produced from chlorinated precursors such as PCBs, chlorophenols, and chlorobenzenes contained in the refuse.
- They result from the cracking of complex organic substances (such as lignin to produce phenol) and are subsequently synthesized in the presence of chlorine at high temperatures, perhaps catalyzed by metal ions. Formation of chemically unrelated chlorinated organics such as PVC after pyrolysis is also possible.

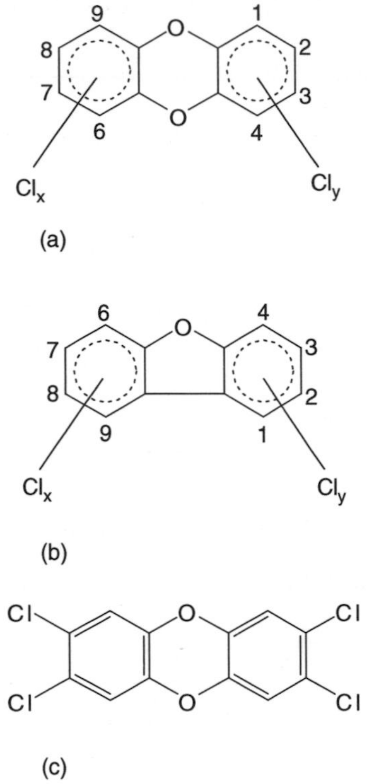

(a)

(b)

(c)

FIGURE 9.5 Structure of (a) a generic PCDD molecule, (b) a generic PDCF molecule, and (c) 2,3,7,8 - tetrachlorodibenzodioxin.

TABLE 9.3

Concentrations of Tetrachloro through Octachloro Group Isomers of PCDD and PCDF in Fly Ashes from Five North American MSW Incinerators

Incinerator	Cl_4	Cl_5	Cl_6	Cl_7	Cl_8
PCDD (ng/g)					
1	85	213	354	184	97
2[a]					
3	2.7	6.6	11.6	5.7	3.5
4	12.9	37.5	75.6	41.9	35.2
5	2.4	7.9	9.7	9.1	2.1
PCDF (ng/g)					
1	209	549	1082	499	24
2[a]					
3	7.0	17.8	32.1	10.9	0.7
4	8.2	19.8	38.7	20.6	4.0
5	4.4	21.0	21.6	16.6	

[a] < 0.5 ng/g

Source: Lisk, D.J., *Sci. Total Environ.*, 74, 39–66, 1988. Reproduced with kind permission from Elsevier, Oxford, UK.

PCDDs form in incinerators at temperatures of approximately 500°C and are destroyed at a minimum of 900°C. PCDD and PCDF formation and persistence are favored by low combustion temperature, wet MSW, insufficient or excess oxygen, and inadequate residence time (Lisk, 1988). High temperatures and well-oxygenated multistage combustion zones are incorporated in modern furnace designs to optimize conditions for more complete destruction of PCDDs and PCDFs. Some newer incineration facilities employ auxiliary burners utilizing fossil fuels to maintain the temperature in the combustion zone sufficiently high at critical times, e.g., when burning wet MSW or when starting-up and shutting-down operations. The U.S. EPA recently issued new guidelines for MSW incinerator emissions, and optimum operational parameters to meet these emission standards have been published. Degradation of PCDDs and PCDFs requires sufficient oxygen, ample turbulence in the combustion zone to avoid quench zones, and adequate residence time of the compounds in the combustion zone. About 7 to 10% oxygen or 50 to 100% excess air and a residence time of at least 1s are estimated to be required for adequate destruction.

Early theories of PCDD formation from MSW incineration centered on the content of PVC, which typically accounts for 50% of the chlorine content of the original waste; however, later studies found that if the temperature, oxygen, turbulence, and residence time parameters are optimized for the destruction of PCDDs and its precursors, the quantities of PCDDs emitted in the flue gas are independent of the PVC content of the original MSW. PCDDs are known to form during wood burning, so the chlorine content of wood is apparently sufficient to combine with precursors (e.g., phenols) released in the combustion process (Choudry et al., 1982; Olie et al., 1983). Removing PVC from MSW before incineration in order to reduce PCDD emissions, therefore, may be of questionable benefit.

It has also been hypothesized that PCDDs and PCDFs may be formed in the pollution control devices or in the stack well beyond the combustion zone. These compounds may be produced by chlorination of precursors adsorbed to the surface of fly ash particles and subsequently desorbed for release into the flue gases. Since PCDDs and PCDFs condense on to fly ash particles beyond the combustion zone and are tightly adsorbed, these compounds may be removed in the adsorbed form by conventional gas cleaning technologies. For example, electrostatic precipitators (see below) efficiently trap large fly ash particulates but do not consistently remove fine particles (< 2 μm in diameter) unless sophisticated multistage plates are incorporated (Lisk, 1988). Baghouses (fabric filters; see below) are also highly efficient for particulate removal from the flue gas stream. The efficiency of baghouses for fly ash removal can be improved by use of a dry scrubber upstream. It is suggested that scrubbers, which remove acidic constituents by introduction of alkaline (e.g., lime) slurry into the flue gas, will increase agglomeration of fly ash particles, thus further improving collection efficiency by baghouses.

Measurement of PCDDs during waste combustion is difficult and expensive. The emissions of PCDDs from a stack can be roughly estimated by measuring the emission of carbon monoxide. According to Hasselriis (1987), the generation of PCDDs is proportional to the CO concentration as

$$PCDDs = (CO / A)^2 \qquad (9.10)$$

where CO is the concentration of carbon monoxide in the flue gas as percent of total gas, and A is a constant, a function of the operating system. PCDD concentrations in the off-gases are expressed as ng/m^3.

The emission of PCDDs increases with increasing CO emissions, both of which are regulated by the amount of excess air used and the combustion temperature. From empirical evidence, several quantitative relationships have been developed that are good predictors of PCDD and PCDF formation (Vesilind et al., 2002).

For modular incinerators,

$$PCDDs + PCDFs = 2670.2 - 1.37T + 100.06CO \qquad (9.11)$$

For waterwall incinerators,

$$PCDDs + PCDFs = 4754.6 - 5.14T + 103.41CO \qquad (9.12)$$

where T (°C) is the temperature in the secondary chamber for modular combustors, and the furnace temperature in waterwall incinerators, respectively.

PCDDs and PCDFs may be produced, albeit in extremely low quantities, in all incineration processes, and not only those involving MSW. Their formation has been reported during the combustion of paper, wood, vegetable wastes, chlorophenols, polychlorinated biphenyls (PCBs), and from coal- and gasoline-powered engines (Lisk, 1988).

Berlincioni and di Domenico (1987) monitored vapor and smoke emissions from a MSW incinerator for PCDDs and PCDFs and found that the fraction of the compounds associated with fly ash accounted for less than 10% of the total emitted. They also sampled soils up to 1 km in several directions from the incinerator and found a maximum PCDD concentration of 7×10^4 ng/m^2 of soil surface. These compounds were not confined to the top 5 cm of soil and may have reached deeper layers by leaching or plowing. The more highly chlorinated isomers accumulated to the greatest degree. The authors postulated that the less chlorinated isomers were less persistent owing to their high vapor pressure and reactivity with light (photolability).

Other chlorinated and organic compounds arising from MSW combustion include PCBs and polycyclic aromatic hydrocarbons (PAHs). Compounds in the latter category include pyrene, benzo[a]pyrene, and chrysene, among others (Figure 9.6). PAHs such as benzo[a]pyrene are carcinogenic. Similar to PCDDs, PAHs are produced as a result of incomplete combustion and have been reported in gaseous emissions and fly ash from MSW incinerators. For example, the incomplete combustion of saturated hydrocarbons can form PAHs. At temperatures exceeding about 500°C (950°F), carbon–hydrogen and carbon–carbon bonds are broken, resulting in the formation of free radicals. The radicals are dehydrogenated and combine to form aromatic rings that are resistant to thermal degradation (Figure 9.7) (Manahan, 1994).

Concentrations of PAHs on fly ash from a MSW incinerator were found to vary markedly from day to day and the variations were consistent with those of the total concentrations of organic compounds on fly ash (Eiceman et al., 1981). Colmsjö et al. (1986) found that the concentration of PAHs in stack gases from a MSW incinerator increased by more than 1000-fold during cold start-up of the plant. Large PAH molecules were strongly adsorbed to fly ash particles. Pierce and Katz (1975) found the highest concentrations of PAHs on the smallest particulates, those < 5 μm in diameter, and

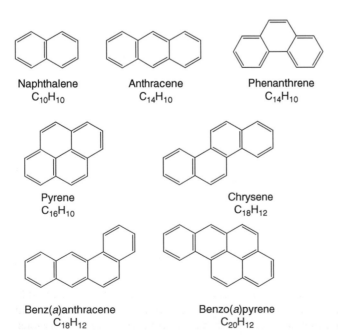

Naphthalene
$C_{10}H_{10}$

Anthracene
$C_{14}H_{10}$

Phenanthrene
$C_{14}H_{10}$

Pyrene
$C_{16}H_{10}$

Chrysene
$C_{18}H_{12}$

Benz(a)anthracene
$C_{18}H_{12}$

Benzo(a)pyrene
$C_{20}H_{12}$

FIGURE 9.6 Structures of some common polycyclic aromatic hydrocarbons.

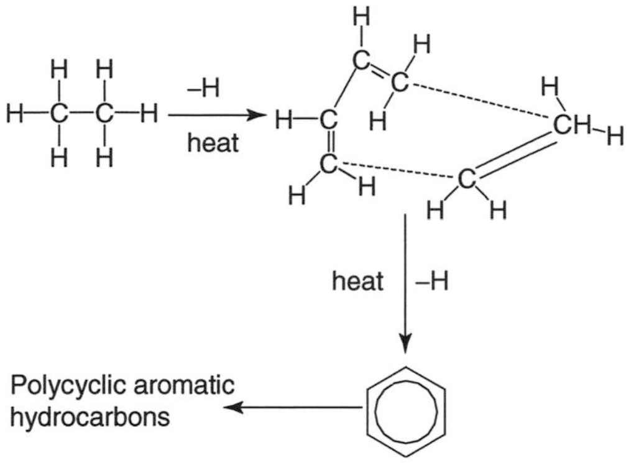

FIGURE 9.7 Formation of a PAH molecule from a simple aliphatic compound. (From Manahan, S.E., *Environmental Chemistry*, 6th ed., CRC Press, Boca Raton, FL. Reproduced with kind permission. Copyright Lewis Publishers, an imprint of CRC Press. Boca Raton, FL.)

therefore in the respirable range. Davies et al. (1976) reported that the PAHs in the stack gases from a MSW incinerator were mainly the more volatile compounds and that an electrostatic precipitator and a spray tower were capable of removing them. Incinerators operated under conditions for optimum destruction of PCDDs and PCDFs should also markedly reduce PAH emissions. Concentrations of 21 PAH compounds from the gaseous and particulate phases in incinerator flue gas are shown in Table 9.4.

Similar to PCDDs and PCDFs, PAHs are also produced in other combustion processes including wood burning, operation of gasoline engines, sewage sludge incineration, and coal burning.

9.4.7 TOXIC METALS

The escape of heavy metals with emission gases is another significant concern with combustion of MSW. Mercury, cadmium, and lead have been the most studied, represent the metals of most likely health concern, and are presently regulated under the Clean Air Act.

Mercury as an atmospheric pollutant is especially difficult to control because it readily volatilizes and escapes in the incinerator flue gases. Furthermore, different species of mercury have different physical and chemical properties and thus behave quite differently in air pollution control equipment and in the atmosphere. Emissions of mercury from waste incinerators are approximately 10 to 20% elemental mercury (Hg^o) and 75 to 85% divalent mercury (Hg^{2+}), which may be predominantly $HgCl_2$. In comparison, emissions of mercury from coal combustion sources are approximately 20 to 50% Hg^o and 50 to 80% divalent mercury (Carpi, 1997). The emission of mercury from combustion facilities depends not only on input composition but also on the species in the exhaust stream and the type of air pollution control equipment used at the source. The partitioning of mercury in flue gas between the elemental and divalent forms may be dependent on the concentration of particulate carbon, HCl, and other pollutants in the stack emissions. In a study by Nishitani et al. (1999), the proportion of $HgCl_2$ (i.e., $HgCl_2$/total Hg) increased with increasing HCl concentration.

A number of elaborate technologies are in use for the removal of Hg from stack gases; however, all are very expensive. Air pollution control equipment for mercury removal at combustion facilities includes activated carbon injection, sodium sulfide injection, and wet lime or limestone flue gas desulfurization. While Hg^{2+} is water-soluble and may be removed from the atmosphere by wet and dry deposition close to combustion sources, the combination of a high vapor pressure and low water solubility facilitate the long-range transport of Hg^o in the atmosphere. Elemental mercury is eventually removed from the atmosphere by dry deposition onto surfaces and by wet deposition after oxidation to water-soluble, divalent mercury (Carpi, 1997). The change in mercury speciation upon passing through a dust collector was investigated by Nishitani et al. (1999). A portion of Hg^o was converted into $HgCl_2$ when the mercury in the flue gas passed through a fabric filter. Clearly, however, the best solution to reduce the quantities of Hg in incinerator flue gas is to prevent its entry into the waste stream. Household battery collection programs and the virtual elimination of mercury from batteries in the early 1990s have resulted in a substantial decrease in atmospheric mercury emissions.

TABLE 9.4
Concentrations of 21 PAH Compounds from the Gaseous Phase and Particulate Phase in Incinerator Flue Gas

Compound	Gaseous phase ($\mu g/nm^3$)	Particulate phase ($\mu g/nm^3$)	Total ($\mu g/nm^3$)
Naphthalene	1086	3.61	1090
Acenaphthylene	111	0.689	112
Acenaphthene	3.96	0.228	4.19
Fluorene	4.39	0.079	4.47
Phenanthrene	25.0	0.203	25.4
Anthracene	23.7	0.66	24.4
Fluoranthene	3.77	0.53	4.27
Pyrene	1.42	1.29	2.71
Cyclopenta[c,d]pyrene	0.003	0.006	0.009
Benz[a]anthracene	0.402	4.65	5.05
Chrysene	0.075	0.544	0.618
Benzo[b]fluoranthene	0.070	0.920	0.989
Benzo[k]fluoranthene	0.170	1.47	1.64
Benzo[e]pyrene	0.684	3.03	3.71
Benzo[a]pyrene	0.754	2.53	3.28
Perylene	0.944	1.85	2.79
Ideno[1,2,3,-c,d]pyrene	0.024	0.055	0.79
Dibenzo[a,h]anthracene	0.306	1.24	1.54
Benzo[b]chrycene	0.069	0.163	0.232
Benzo[ghi]perylene	0.119	0.991	1.11
Coronene	0.461	2.35	2.81
Total PAHs	1260	27.1	1290

Source: Lee, W.J. et al., *Atmos. Environ.,* 36, 781–790, 2002. Reproduced with kind permission from Elsevier, Oxford, UK.

Cadmium is another toxic metal that may be volatilized and therefore mobilized in a mass-burn incinerator. In a study by Zhang et al. (2001), average Cd losses from a laboratory-scale system combusting assorted waste types were 69 and 74% at 850 and 1000°C, respectively. Twenty other metals were additionally lost to the atmosphere. At 500°C, Sn was emitted; at 850°C K, Mg, Na, Bi, Cr, Ge, Li, Pb, Sn, Tl, and Zn were lost, and nine more metals, Al, Be, Cs, Nb, Sb, Sr, Th, Y, and Zr were lost at 1000°C. It was speculated that the released metals were transferred to the combustion flue gas mainly in the forms of metallic chloride compounds, e.g., $CdCl_2$, $SnCl_4$, $SnCl_2$, $ZnCl_2$, and $PbCl_2$. No significant losses for Ca, Fe, Ag, Ba, Co, Cu, Ga, Hf, Mn, Mo, Ni, Rb, Sc, Ta, Ti, U, V, and W were reported. Transformations of inorganic substances during MSW combustion are depicted in Figure 9.8.

Several studies have been conducted on the effects of atmospheric emissions from MSW incinerators on local soils and plants. For example, Kukkonen and Raunemma (1984) found that the concentrations of Br, Ca, Cl, Cr, Fe, Ni, Pb, Si, Ti, V, and Zn on birch leaf samples showed a strong inverse correlation with distance from a MSW incinerator in Finland. Fewer elements showed this relationship when grass was sampled.

9.4.8 AESTHETICS

Noise is inevitable during MSW incineration. Waste collection vehicles, processing equipment, the combustion process itself, air pollution control (e.g., operation of pumps), and production of steam

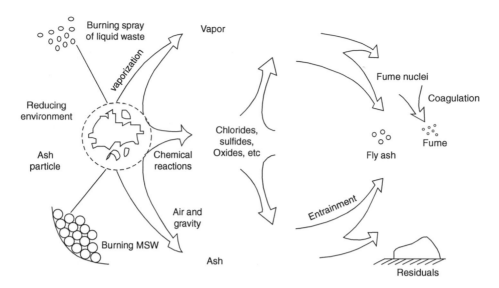

FIGURE 9.8 Transformations of inorganics during combustion of MSW (U.S. EPA, EPA/625/R-93/008, 1993).

or other energy all produce noise. Therefore, worker safety must be addressed and the facility should be sited such that local populations are considered.

MSW is odoriferous. Organics generated by decomposition of putrescibles can be detected at low concentrations in ambient air over substantial distances. The most significant sources of odor are the tipping floor, storage pits, and shredders. Sometimes the flue gases themselves will carry a strong odor. The extent of odor production and dispersal (and therefore effects on workers and local populations) is related to air temperature, barometric pressure, humidity, wind direction, and wind speed.

The ideal odor control scenario is to contain odors within the facility boundary. One possible solution is to apply negative air pressure (suction) within the tipping area and recycle the withdrawn air into the incinerator. The gases can also be passed through a charcoal filter system which will capture foul-smelling organic vapors. An additional precaution to reduce odor production is to require collection vehicles to keep compartments closed except only when tipping wastes. One of the best preventative actions for odor impact, however, involves proper siting of the facility. Residential neighborhoods and other sensitive areas must be avoided in siting. The facility is best sited in an area zoned for heavy industry.

9.5 AIR POLLUTION CONTROL

There is a wide range of incinerator air pollution control devices available, ranging from a series of simple baffles to trap particulates, to scrubbers designed to remove certain acid gases. Many of these technologies, although high in capital costs, are extremely effective in removing specific air pollutants. The proper choice of equipment depends not only on desired emission quality and quantity but also on conditions outside the incineration system. For example, a lack of local water supply will restrict the use of wet scrubbers.

9.5.1 THE ELECTROSTATIC PRECIPITATOR

Many large municipal incinerators employ the electrostatic precipitator (ESP) for flue gas cleaning, specifically for the removal of particulate matter (Figure 9.9). The ESP can remove particles down to fractions of a micron and are about 99% effective.

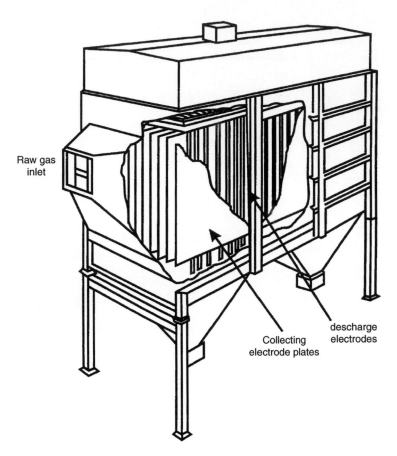

Raw gas inlet

Collecting electrode plates

descharge electrodes

FIGURE 9.9 Electrostatic precipitator (U.S. EPA, Air Pollution Engineering Manual, 2nd ed., 1973).

The stream of "dirty" gas is passed through a series of discharge electrodes (Figure 9.10). The electrodes are negatively charged, usually in the range of 1000 to 6000 V. At this voltage a *corona*, or cloud of charge, is generated. Most particles passing through this corona, regardless of initial composition, will develop a negative charge. A grounded (positive) surface, or collector electrode, is situated near the discharge electrode. The negatively charged particulates will be attracted to and collect on the grounded surface. Lastly, the particulate matter is removed from the collector surface by cutting off the voltage to each electrode and then striking them with rappers at regular intervals or by wetting down the plates. The collected particulates are then disposed. In some cases, particulates may physically or chemically resist changing charge. These will pass through the ESP without being captured.

The advantages of using the ESP for flue gas cleaning include:

- Highly efficient removal of particulates
- Relatively insensitive to high effluent gas temperatures
- No wastewater treatment requirements

Disadvantages include:

- High capital costs (a simple model will cost several million dollars)
- Large space requirements
- The ESP is often sensitive to the chemical composition of flue gas. Acid gases will corrode metallic components

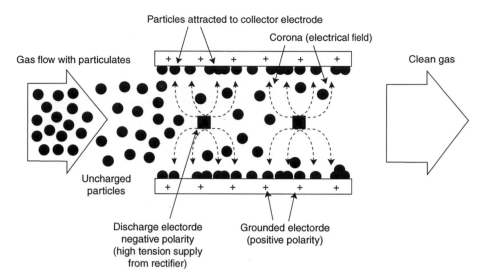

FIGURE 9.10 Electrodes within the ESP. (From Williams, 1998. Reproduced with kind permission of John Wiley & Sons, Inc.)

- Equipment is needed for collection of the captured materials
- Extensive electrical equipment is needed

9.5.2 FABRIC FILTERS (BAGHOUSES)

The baghouse is one of the oldest, simplest, and most efficient methods for removing solid particulate contaminants from gas streams, using filtration through fabric media. The baghouse is constructed as a simple series of permeable bags that capture particulate matter but allow the passage of gases (Figure 9.11). The filter fabric is composed of heat-resistant materials, ranging from cotton to nylon to glass fibers (Table 9.5). The choice of the fabric is a function of the operating temperature range, chemical composition of the flue gas, moisture, and the physical and chemical properties of the particles being collected.

The fabric filter bags are usually tubular or flat. The structure in which the bags hang is referred to as a baghouse, and the number of bags in a baghouse may vary from less than ten to several thousand. The baghouse system can be operated continuously, with airflow to some bags turned off for cleaning and maintenance. In bottom-feed units, flue gases are introduced through the baghouse hopper at the base and then to the interior of the bag. In top-feed units, dust-laden gas enters the top of the filters.

The baghouse filter fabric is typically woven with relatively large spaces, about 50 μm across. However, these filters are capable of capturing particulates measuring < 1 μm; obviously, more than simple sieving is taking place. Capture of particulates apparently occurs as a result of electrostatic attraction as well as entrapment within the fabric weaving. With woven fabrics composing the bag, a dust cake eventually forms, which, in turn, acts as an effective sieving mechanism. When felted fabrics are used, this dust cake is minimal or nonexistent and the primary filtering mechanisms are a combination of inertial forces and impingement (Vesilind et al., 2002).

As particles are collected, the pressure decreases across the fabric filtering media; therefore, the filter must be cleaned at predetermined intervals. Dust is removed from the fabric by gravity or mechanical means. When large number of bags are involved, the baghouse is compartmentalized so that one compartment may be cleaned while others are still in service.

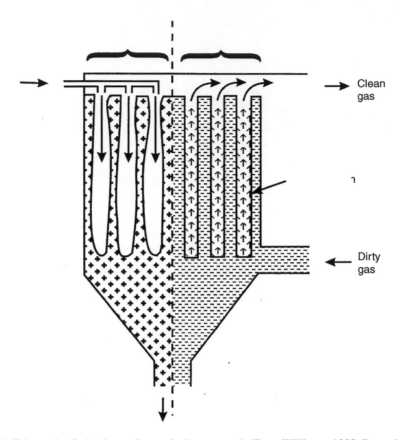

FIGURE 9.11 Schematic of a baghouse for particulate removal. (From Williams, 1998. Reproduced with kind permission of John Wiley & Sons, Inc.)

TABLE 9.5
Baghouse Fabric Ratings

Fabric	Recommended Maximum Temperature, (°C)	Chemical Resistance	
		Acid	Alkali
Cotton	81	Poor	Fair
Wool	103	Good	Poor
Nylon	103	Good	Poor
Dacron	134	Excellent	Good
Glass	285	Excellent	Excellent

Source: U.S. EPA, *Air Pollution Engineering Manual*, 2nd ed., 1973.

EXAMPLE 9.3

A baghouse holds a total of 50 bags for particulate removal from a mass-burn incinerator unit. A single bag is cylindrical in shape and measures 25 cm diameter and 6 m in length. What is the filtering area of the bag?

If the baghouse unit (50 bags) is to treat 15,000 m³/h of flue gas, calculate the effective filtration velocity in meters per minute and the mass of particles collected daily if the inlet loading is 120 g/m³ and the unit operates at 99.99+% collection efficiency. Note that 1 kg of collected residue ~ 3150 g.

SOLUTION

The total area of the bag is calculated as (Reynolds et al., 2002)

$$A = A_{\text{curved surface}} + A_{\text{flat top}} \tag{9.13}$$
$$= \pi\, Dh + \pi\, D^2/4$$
$$= \pi\, [(25\ \text{cm}/100)(6\ \text{m}) + \pi\, (25\ \text{cm})^2\,/\,100]/4$$
$$= 4.76\ \text{m}^2$$

where D is the bag diameter and h is the length (or height).

The combined area for the 50 bags is

$$A = (50)\,(4.76) = 238\ \text{m}^2$$

The filter velocity is then

$$V = qG/A \tag{9.14}$$
$$= [15,000\ \text{m}^3/60)]\,/\,238$$
$$= 1.05\ \text{m/min}$$

If we assume 100% collection efficiency, the mass collected daily is

$$= q_G C_i = (15,000)(24)(120)/3150$$
$$= 13,714\ \text{kg/day}$$

The advantages of the baghouse for cleaning flue gases are:

- High particle removal efficiencies over a wide range of particle sizes
- Variations in loading and flow rates do not affect removal efficiency
- Corrosion and rusting is minimized because the bags are manufactured from resistant materials
- Simple operation and maintenance
- Flexible designs are possible

Disadvantages include:

- The adhesion and accretion of hygroscopic material. These accretions will block the filter pores and waste energy
- High temperatures, acids, and alkalis within the flue gas tend to shorten fabric life
- There is the potential for fire and explosion if oxidizable particulates accumulate in the bags
- Gases are not removed

9.5.3 GAS WASHING

Wet scrubbers have become popular for cleaning contaminated gas streams because of their ability to remove effectively both particulate and gaseous pollutants. Wet scrubbing involves bringing a contaminated gas stream into intimate contact with a liquid that is introduced into the scrubbing device as a finely atomized mist. The most common low-energy scrubbers are gravity spray towers in which liquid droplets, often simply cold water or a dilute alkaline solution, are made to fall through rising exhaust gases and are drained at the bottom of the chamber into a wastewater collector (Figures 9.12 and 9.13). The droplets are usually formed by liquid atomized in an array of spray nozzles. The hot flue gas enters from the bottom of the unit and rises. The vertical gas velocity ranges from 75 to 150 cm/s (2 to 5 ft/s). For higher velocities, a mist eliminator must be used at the top of the tower (Figure 9.12) (Vesilind et al., 2002). Particulate matter is wetted immediately upon entering the chamber and falls out by gravity. Gases such as H_2SO_4, HNO_3, and HCl readily dissolve in the mist, forming the corresponding aqueous acids which also fall out by gravity. The spray water continuously washes the walls of the chamber.

Sulfur dioxide is one of the most common gaseous pollutants from MSW combustion and from other sources such as coal combustion. For decades, coal burning utilities and other large-scale emitters of SO_2 have condensed SO_2 to sulfuric acid as the primary means to remove sulfur dioxide from stack gas. SO_2 is fairly soluble in water; once dissolved, the acidic liquid is collected and treated for disposal. The reactions for SO_2 capture are identical to those for acid rain formation, given earlier:

$$SO_2 + \frac{1}{2}\ O_2 \rightarrow SO_3 \qquad\qquad (9.15)$$

$$SO_3 + H_2O \rightarrow H_2SO_{4(aq)} \qquad\qquad (9.16)$$

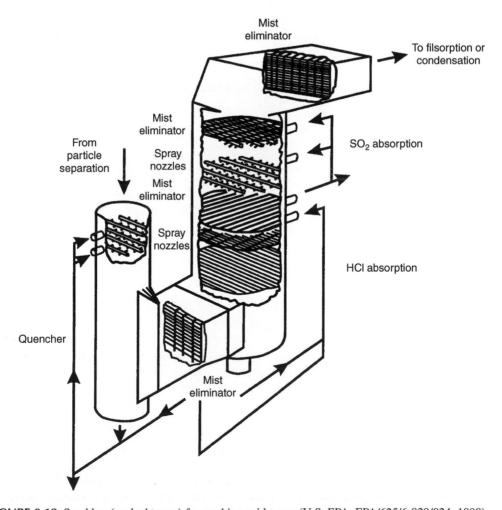

FIGURE 9.12 Scrubber (packed tower) for washing acid gases (U.S. EPA, EPA/625/6-829/024, 1990).

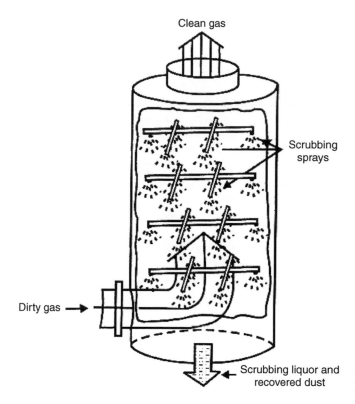

Clean gas

Scrubbing sprays

Dirty gas →

Scrubbing liquor and recovered dust

FIGURE 9.13 Gravity spray tower scrubber for washing acid gases (U.S. EPA, EPA/625/6-829/024, 1990).

A quicklime or limestone solution can also be formulated to absorb the SO_2. The reaction with quicklime is

$$SO_2 + CaO \rightarrow CaSO_3\,(s) \tag{9.17}$$

The reaction with limestone is

$$SO_2 + CaCO_3 \rightarrow CaSO_4\,(s) + CO_2 \tag{9.18}$$

Lime materials can be injected directly into the scrubber or added to the combustion chamber. If limestone is injected into the furnace, it quickly reacts to form quicklime:

$$CaCO_3 \rightarrow CaO + CO_2 \tag{9.19}$$

Both calcium sulfite and calcium sulfate are solids of low solubility that can be separated in gravity settling tanks. The calcium salts formed create a new problem, however, of disposal of enormous quantities of solid waste, actually a high water content slurry. In some facilities, the flue gas desulfurization sludge is simply stored in ponds on company property. There is much ongoing research in the utilization of this sludge. Applications under study include use as an agricultural amendment (Carlson and Adriano, 1993), for reclamation of coal mine spoils (Stehouwer et al., 1995), roadway construction, and for use in building panels (drywall) (EPRI, 1995). Based on the discussions above, flue gases may contain heavy metals and other inorganic pollutants; treatment of the scrubbing medium must consider these contaminants as well. At a MSW incinerator in Germany, Reimann (1995) used a combination of lime and trimercaptotriazine in a two-stage process to remove mercury, other heavy metals, and salts from the scrubbing medium. The treated effluent met Germany's stringent discharge requirements.

Advantages of gas washing include:

- Simultaneous removal of acid gases as well as particulates. It is recommended, however, that particulates be removed first (e.g., using a baghouse). Particle removal allows for greater efficiency of gas removal.
- The system, fed by a liquid mist, accommodates high-temperature flue gas streams.
- The scrubbing medium can be modified to increase removal efficiencies. For example, a dilute NaOH solution can be used in place of water for treatment of especially acidic flue gases.

Disadvantages:

- There is a high input of energy, primarily to pump liquids in and out of the system.
- A constant supply of water is necessary.
- Since aqueous acids and alkalis are formed, equipment corrosion is inevitable. Therefore, maintenance costs may be substantial.
- There will be large quantities of wastewater produced. This liquid must be treated prior to discharge into a receiving body of water.
- Acid gas removal is more efficient if particulates are removed first from the gas stream.

9.6 ASH QUALITY FROM MASS BURN

Incinerator residues consist of noncombustible materials such as metal, glass and stones, and also incompletely burned combustibles. MSW incinerators produce two types of ash: (1) bottom ash, the large, dense debris that falls through the grates by gravity and collects at the base of the combustion chamber; and (2) fly ash, the fine particles transported out of the combustion chamber with the air stream, which are removed by air pollution control devices. Most facilities combine the two types of ash before disposal.

Incinerator fly ash and bottom ash are often a hazard to health and the environment. If preprocessing of the MSW does not occur, there are a number of health and environmental concerns regarding the generation, storage, and ultimate disposal of MSW ash. These are outlined in the following sections.

The composition of a typical MSW ash sample is shown in Table 4.17. The predominant health and environmental concern with MSW incinerator ash is the presence of heavy metals. Table 9.6 lists a representative array of heavy metals found in combined fly ash and bottom ash from a MSW waste-to-energy unit.

Based on chemical composition and the leachability of certain components, MSW ash may technically be classified as a hazardous waste by the U.S. EPA. As discussed below and in Chapter 11, the Toxicity Characteristic Leaching Procedure (TCLP) is an extraction procedure used to determine whether a solid waste may be declared hazardous. If fly ash alone is tested, its constituents often fail the test and it may be classified as hazardous. Combined with the bottom ash, however, the mixture often meets the requirements for a nonhazardous waste (Vesilind et al., 2002).

9.6.1 METALS

Ash may contain high concentrations of a number of toxic metals such as Cd, Pb, As, Be, V, and Hg and other, comparitively less toxic metals, such as Cu, Zn, Fe, and Al (Table 9.6). Fly ash contains several thousand times more lead and cadmium than bottom ash; however, the levels in bottom ash still greatly exceed the amounts found in uncontaminated soils. These metals are concentrated in the ash via the incineration process. Incineration removes the matrix materials such as paper and plastics that contained the metals and had restricted their release to the biosphere. Once in the form of ash,

TABLE 9.6

Concentration Ranges of Elements in MSW and Bottom Ash, Fly Ash, and Suspended Particulates from MSW Incineration

Element	MSW (Combustible Fraction)	Bottom Ash	Fly Ash	Suspended Particulates	Possible Carcinogens
Ag	<3–7		52–220	84–2000	
Al(%)	0.54–1.17	2.6–14.2	9.0–14.2	0.58–4.8	
As			9.4–74	81–510	X
Ba	47–447	80–9000	1600–360	40–1700	
Be	<2				X
Bi	<15–30				
C(%)		1.0–28.7	1.7–7.4	1.8–2.2	
Ca(%)	0.59–1.65	3.6–11.2	3.3–8.6	0.66–5.3	
Cd	4—22	3.8–442	<1–477	520–2100	X
Cl		0.2–1.0	0.12–1.12	9.29	
Co	<3–5		25–54	3.8–28	
Cr	22–96		730–1900	122–1800	X
Cu	79–877	630–4281	69–2000	3000	
F	140–200	130–250	1500–3100	990–6800	
Fe(%)	0.10–0.35	2.1–32	2.4–8.7	0.17–1.8	
Hg	1–4.4	0.03–3.5	0.09–25	20–2000	
K(%)	0.09–0.21	0.42–2.41			
Mg(%)	0.09–0.21	0.04–0.86	0.5–2.1	0.31–2.8	
Mn(%)	0.005–0.02	0.08–39	0.20–0.85	0.03–0.57	
Mo					
N(%)		0–0.35	0		
Na(%)	0.18–0.74	2.3–14.2	1.12–1.94	5.1–9.8	
Ni	9–90	110–210	38.6–960	65–440	X
P(%)		0.04–0.83			
Pb(%)	0.01–0.15	0.04–0.80	0.06–0.54	2.5–15.5	X
S(%)		0.27–1.0	1.9–3.6	0.001–0.01	
Sb	20		139–760	610–12000	
Se			1.4–13	7.0–122	
Si(%)		4.7–9.4			
Sn(%)	<0.002–0.004	0.01–0.1	0.12–0.26	0.4–1.51	
Sr	11–35		110–220		
Ti(%)	0.14–.31	0.04–0.90	2.5–4.2	0.13–1.29	
Tl				150	
V			110–166	6–60	
Zn(%)	0.02–0.25	0.35–3.61	0.08–2.6	4.7–24	

Source: Lisk, D.J., *Sci. Total Environ.,* 74, 39–66, 1988. Reproduced with kind permission from Elsevier, Oxford, UK.

metals become much more bioavailable. Metals such as lead and cadmium are readily leachable from ash at levels that frequently exceed federal limits for defining a hazardous waste.

Many consumer products contribute toxic metals to the municipal waste stream. Arsenic may originate in paint, ceramics, and old insecticides. Chromium may originate from metal plating, and occurs in plastics, inks, and paints. Mercury occurs in flashlight batteries, fungicides, newspapers, paints, and plastics. Batteries, plastics, and other pigment uses are major contributors of lead and cadmium to MSW. The major sources of lead and cadmium in MSW appear in Tables 4.13 and 4.14, respectively. Recycling of batteries via specialized collection systems and prohibition on disposal are practical approaches to reducing the amount of toxic metals in the waste stream. Such approaches are more difficult to implement for plastics and pigment uses, however.

9.6.2 HEALTH EFFECTS OF METALS IN INCINERATOR ASH

Ash may be dispersed into the workplace or the local environment at all stages of ash management, including during on-site handling and storage, transport, and handling at the disposal site. At each step, there is potential for airborne and water-borne dispersal of ash. Most of the metals of concern have the ability to be adsorbed by soils and sediments, and many can accumulate in living tissue; therefore, heavy metals persist in the biosphere. Thus, long-term releases even at low levels can substantially increase metal levels in the environment. Figure 9.14 presents possible exposure routes to air emissions from MSW incinerators and incinerator ash.

Many heavy metals of concern have well-defined health effects. Many are carcinogenic; however, they also exert neurological, hepatic, renal, hematopoietic, and other adverse effects, both in humans and in other biota. As a very brief overview, arsenic, cadmium, beryllium, and lead are carcinogenic metals; arsenic, lead, vanadium, cadmium, and mercury are neurotoxins; zinc, copper, and mercury are acutely toxic to aquatic life (Denison and Rustin, 1990). More detailed effects of these and other metals are discussed in several excellent works (see Suggested Readings). Total

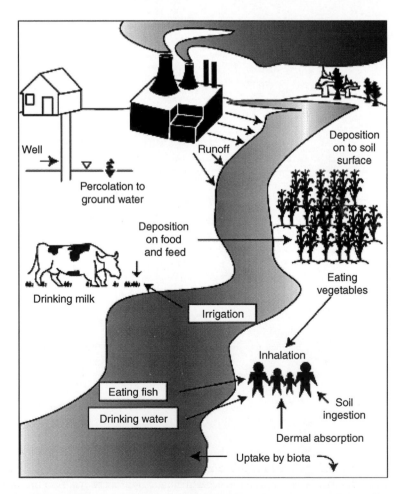

FIGURE 9.14 Exposure pathways for emissions from a MSW incinerator (U.S. EPA, 1986).

metal concentrations provide only a partial picture of the relative risk associated with heavy metals. Other methods are necessary to better understand the behavior of these metals in the biosphere, for example, in a landfill, surface impoundment, or as fine dust attached to plant tissue.

9.6.3 LEACHABILITY OF METALS IN INCINERATOR ASH

In the high-temperature zone of the incinerator, metals are vaporized. For example, cadmium and mercury boil at 765°C (1412°F) and 355°C (674°F), respectively, temperatures that commonly occur within the combustion chamber. As the combustion gases cool, these metals condense on to the surface of fly ash particles. The concentrations of these condensed metals will increase with decreasing ash particle size.

If ash containing metal-coated particulates comes into contact with ambient moisture, for example, within the leachate of a landfill, some metals will become mobilized. The small particle size of ash increases the surface area exposed to leaching, and the presence of metals at or near the surface of such particles also enhances their leachability. Another factor contributing to metal leachability in ash involves conversion into soluble salts. MSW contains significant amounts of chlorine from plastics, bleached paper, and other sources. During incineration, the chloride ion combines with metals to form metal chlorides:

$$M^{2+} + 2Cl^- \rightarrow MCl_{2\,(aq)} \tag{9.20}$$

where M^{2+} represents the cationic metal species.

These salts are typically much more soluble in water than the original metallic form.

The leachability of metals in incinerator ash is one estimate of hazard, and relates specifically to the potential for groundwater or surface-water contamination. Under federal law (40 CFR Part 261), leachability is assessed using the TCLP. To perform the test, a sample of ash is exposed to a dilute acid solution for 16 to 20 h, and the concentrations of eight metals that leach into solution are then measured. Most metals become more soluble under conditions of increasing acidity. If a minimum permissible limit of a metal in solution is exceeded, the ash (or other test material) must be managed as a hazardous waste.

The slightly acidic conditions of the TCLP are designed to simulate conditions found in a typical sanitary landfill. The majority of incinerator ash is disposed of in sanitary landfills along with unburned MSW. As will be seen in Chapter 10, landfill leachate can become quite acidic as a result of microbial decomposition processes. Metals in intimate contact with leachate will become solubilized to some extent. Additionally, large amounts of incinerator ash are managed by even less-controlled means such as being used as landfill cover, fill material in marshy areas, and deicing grit on winter roads.

The leachability of metals in incinerator ash is a function of numerous variables, for example, the species of the metal. Lead occurring as $PbCl_2$ is much more soluble in water than PbO or $Pb(OH)_2$, for example (see Appendix). As mentioned above, a smaller ash particle size results in a corresponding higher particle surface area. Metal occurring on a particle exterior will more rapidly solubilize than the same metal occurring as a whole, intact particle, all other factors being equal. The most important variable affecting metal mobility in ash, however, is the pH of the solution into which the metal is leached. In leaching tests of ash from several U.S. incinerators, lead and cadmium leached at high levels, often in excess of regulatory standards defining a hazardous waste. Such leaching occurred even when the tests were conducted using water rather than dilute acid required in the TCLP (Vesilind et al., 2002).

9.6.4 ASH MANAGEMENT

As discussed above, ash is hazardous because it contains high levels of toxic metals and can also contain chlorinated dibenzodioxins and polycyclic aromatic hydrocarbons. Ash mobility and toxicity concerns have focused on the leachability of heavy metals in ash; in other words, how quickly will a particular metal leach from ash and enter the groundwater supply or some other environmental receptor? Some have questioned the "leachability methods", however, for being incomplete in terms of assessing toxicity. The total concentrations of both metals and PCDDs must be considered in assessing ash toxicity, because humans and the environment can be exposed to ash through many routes in addition to contaminated groundwater. For example, humans can inhale ash particles into the lungs after which toxins on the particles are directly absorbed into the tissue or bloodstream. In addition, ash particles may be ingested, either directly or through contaminated food or water. Because these exposure routes can be highly significant, a full assessment of the hazards posed by ash must include data of its *total* chemical composition.

Reducing the hazard relating to metals and other toxins in ash requires several actions (Denison and Rustin, 1990):

- Keeping toxic metals out of products that may enter the waste stream
- Keeping metal-containing materials out of incinerators
- Chemically or physically treating ash prior to disposal (e.g., mixing with Portland cement and allowing to set)
- Disposing of all ash in secure facilities that do not contain other types of waste (this practice is termed "monofilling")
- Compacting the ash prior to or during landfilling

Noncompacted MSW ash may have a density of 900 kg/m^3 (1500 lb/yd^3). If the ash is compacted the density increases to as high as 1980 kg/m^3 (3300 lb/yd^3). At this density the ash is highly impermeable; permeability may be as low as 1×10^{-9} cm/sec (Vesilind et al., 2002). As more and more ash is being produced and landfill space is becoming more difficult to acquire, alternative uses for ash are being sought. Some other uses include:

- Road base material
- Structural fill
- Gravel drainage ditches
- Capping strip mines
- Mixing with cement to make building (construction) blocks

In addition, ash from MSW combustion contains metals that can be reclaimed, especially steel and aluminum.

9.7   MSW INCINERATION IN THE UNITED STATES: THE FUTURE

A number of factors have been operating over the last several years to expand the use of incineration in managing MSW. These include (Denison and Rustin, 1990):

- Diminishing landfill capacity, especially in heavily urbanized areas of the Northeast, along with rising landfill costs and difficulties with siting.
- An aggressive marketing campaign conducted by incinerator vendors.
- A public perception that waste-to-energy is cleaner and more conserving of resources than landfilling.
- A perception that the convenience that incineration provides is preferred, albeit more expensive, to implement than are municipal-level recycling programs.

However, a number of arguments continue regarding the suitability of incineration. Factors restricting the development of incineration include:

- Intense public opposition
- Unresolved risk issues relating to air emissions and ash residues

- Uncertainty over regulatory requirements
- Major long-term economic risks
- Initially high capital cost tag and frequent cost overruns
- Concern over the effect that a long-term commitment to incineration may have on recycling and other conservation efforts

9.8 REFUSE-DERIVED FUEL

During mass burning of MSW, no processing or separation of the fuel occurs other than simple mixing by the tipping vehicle. In contrast, the refuse-derived fuel (RDF) technology employs a two-stage production–incineration system. Wastes are processed to produce a higher quality and more homogeneous fuel product than raw MSW. The input waste is usually shredded to reduce particle size. Ferrous metals are recovered using magnetic separators. Glass, stones, and soil may be removed by trommel screening. In some RDF plants, additional equipment is employed to process the fuel product further, eliminating additional noncombustible materials. The final stages of processing may involve air classification to remove the lightest fractions of the charge followed by, in some cases, densification to produce a fuel in pellet or briquette form. Ideally, the separated, mostly organic fraction is composed of paper products and nonhalogenated plastics only; however, PVC, food, and yard wastes also occur. The RDF is either marketed to outside customers or burned on-site in a so-called "dedicated" furnace.

The practice of selling a solid fuel derived from the physical processing of MSW dates only to the early 1970s. Since that time a number of processes have been developed for RDF production and utilization. Several plants are in operation both in the United States and Europe. RDF facilities represent the smallest portion of the various waste-to-energy facilities studied in a DOE (2001) report. The RDF-only facility came on line in 1975 and gradually increased in number through 1991. Reaching the peak in 1990/1991 (29 plants), their numbers have since declined. Of the 38 facilities operating since 1975, 15 were still operating in 1998. In addition to combusting RDF only, however, many more plants burn RDF as a co-fuel, i.e., in a mixture with other solid fuels such as coal. Co-firing with coal in a ratio of approximately 75:25 coal/RDF is a fairly typical ratio.

The overall benefits of RDF utilization include diverting potentially useful organic materials out of the landfill, energy recovery from solid wastes, and reduction in a number of gaseous pollutants compared with the combustion of coal alone.

9.8.1 OVERVIEW OF RDF PREPARATION

Although other methods have been tested, the so-called "dry separation" of RDF from municipal wastes is by far the most common method of initial preparation. The method may be adapted to produce various qualities of RDF, depending mostly on the extent of separation of inorganic materials and the putrescible (i.e., rapidly decomposable and potentially malodorous) components from the paper and the plastics in the organic fraction. Because waste may undergo processing for other reasons, such as separation of potentially recyclable metals and other inorganics, dry separation is a logical fit for fuel production.

In order to separate the organic fraction from metals, glass, and other dense components, it is first necessary to pass the waste through a trommel screen. The incorporation of a screen early in the waste flow can be used either to avoid the need for air classification of the process stream or to remove stones, dirt, and putrescible materials before air classification, so that the RDF contains mostly paper and plastics. Screening may be followed by pulverization in a hammermill or other shredding device. Separation of ferrous metals by magnetic extraction is a next logical step (see Chapter 7). Most systems for RDF production include an air classifier to separate the heavy inorganic components from the largely organic RDF. In some cases, bundled newspaper and cardboard are manually removed from the incoming waste. Most RDF systems in the United States incorporate all organics into the fuel, while many developed in Europe separate mainly paper and plastics.

9.8.2 GRADES OF RDF

Different grades of RDF can be produced from MSW. The higher the fuel quality, the lower is the total yield of fuel. For example, if a MRF simply shreds the incoming waste and passes it under a magnetic separator to remove the ferrous component, the fuel yield may be 90 to 95%, while the average Btu value may approximate raw MSW. Conversely, producing a pelletized fuel of paper and plastic may yield 50% of fuel based on the total incoming waste. However, the heating value may be as much as 14,000 to 15,650 kJ/kg (6500 to 7000 Btu/lb), which is approximately two thirds the heat value of many Midwest bituminous coal samples. Industry-wide specifications for RDF do not exist, but RDF has been classified according to the type and degree of processing and the form of fuel produced. The properties of RDF to consider and incorporate into supply contracts include the proximate analysis (moisture content, ash content, volatiles, and fixed carbon); ultimate analysis (C, H, N, O, S, and ash percentage); HHV; and content of chlorine, fluorine, lead, cadmium, and mercury (Liu and Liptak, 2000).

The types of RDF produced are functions of equipment design, sequence of separation steps, and operation. The RDF forms fall into the following broad groups: coarse RDF, fluff RDF, powder RDF, and densified RDF (d-RDF). Details of the various RDF categories are provided in Table 9.7. A photograph of d-RDF is shown in Figure 9.15.

9.8.3 PROPERTIES OF RDF

The chemical analysis of RDF samples gives an indication of the combustion performance that might be expected. Heat content (Btu/lb) is obviously one of the top priorities for RDF production.

TABLE 9.7
Major Categories of RDF

RDF-1 (MSW)	Raw MSW with minimal processing to remove oversize bulky waste
RDF-2 (c-RDF)	MSW processed to a coarse particle size with or without ferrous metal separation such that 95% (by wt) passes through a 6 in. square mesh screen
RDF-3 (fluff RDF)	Shredded fuel derived from MSW processed for the removal of metal, glass, and other entrained inorganics; particle size of this material is such that 95% (by wt) passes through a 2 in. square mesh screen
RDF-4 (p-RDF)	Combustible waste fraction processed into powdered form such that 95% (by wt) passes through a 10 mesh screen (0.035 in. square)
RDF-5 (d-RDF)	Combustible waste fraction extruded (densified or compressed) into pellets, cubettes, briquettes, or similar forms. This form has become increasing popular owing to the advantages of ease and cost of transportation and storage as well as of adaptability to certain types of firing
RDF-6	Combustible waste fraction processed into a liquid fuel
RDF-7	Combustible waste fraction processed into a gaseous fuel

Source: ASTM. Reproduced with kind permission from the American Society for Testing and Materials.

The total sulfur, nitrogen, and chlorine contents are a guide to possible gaseous emissions problems. Similarly, the contents of heavy metals will help to predict the chemical properties of the resultant ash. The total ash content will help direct ash handling protocols. It may also help to predict particulate generation rates. Moisture content provides an indication of burnability as well as ease of handling and shipping. Variations in physical and chemical properties of RDF due to differing sources, time of year, and methods of waste sorting make it difficult to present average analysis values; however, Table 9.8 provides a comparison between RDF and coal as fuels.

RDF has a calorific value of 50 to 60% and bulk density of 65 to 75% that of coal (Table 9.8). As a result, considerably larger amounts of RDF must be burned to obtain performance similar to that obtained with coal. Optimization of fuel feeding and firing parameters will also be necessary. The higher ash content of RDF and lower ash fusion temperatures may require modifications to ash storage and removal procedures.

The sulfur content of RDF is often significantly lower than that of coal; in some cases, total S of RDF is less than 1/100 that of coal. Thus, there are obvious atmospheric benefits with co-combustion of RDF with the latter fuel. On the other hand, the chlorine content of RDF is higher than that of typical

FIGURE 9.15 Two types of densified RDF, pellets and cubettes.

TABLE 9.8
Analysis of Fuel Used in Boiler Tests

	RDF Pellets	Bituminous Coal
Calorific value (Btu/lb)	8,110	14,600
Moisture (%)	9.6	9.2
Ash (%)	8.1	6.2
Carbon (%)	45.6	69.0
Hydrogen (%)	6.3	4.8
Sulfur (%)	0.3	1.8
Chlorine (%)	1.8	0.02

coals. The higher Cl contents result from the presence of PVC in MSW as well as the presence of Cl in paper waste from the bleaching process and other Cl-containing materials, for example NaCl.

9.8.4 UTILIZATION OF RDF: PRACTICAL ISSUES

The so-called "densified" or d- RDF is available for use as a fuel immediately after processing, or mixed with coal in the field (Figure 9.16) and into a loading hopper of a boiler equipped to handle solid fuels (Figure 9.17). For RDF utilization to be successful, whether burned alone or as a co-fuel with coal, however, various potential difficulties must be addressed. For example, many coal-burning plants have experienced problems in handling, storing, and conveying materials; for starters, RDF is less dense than coal. It has been found that when coal and RDF are mixed and stored for long periods, the more dense coal tends to sink to the bottom of the mixture. If the mix continues to remain separated at the boiler hopper, uneven combustion will take place as the RDF is burned first

FIGURE 9.16 Piles of coal and RDF to be mixed in the field prior to shipment to heating plant.

FIGURE 9.17 Truck loading a coal–RDF mixture for combustion at a heating plant.

and the coal second. Such uneven burning will cause fluctuations in steam production. This segregation problem can be partly alleviated by mixing the fuels in the field, immediately before burning (Figure 9.16).

The RDF may cause problems in storage. It is fibrous, carbonaceous, and of relatively low density. Contact with rainfall will rapidly alter the physical and chemical properties of the RDF. Pelletized RDF will decompose and lose its physical strength and will no longer be easily handled. Second, a wet material will rapidly undergo anaerobic reactions. The RDF has many fine pores which will tenaciously retain moisture. Foul odors will be produced and the RDF will cause the growth of mold and other undesirable organisms. The best precautions against this scenario are to store it indoors or in a covered facility in the field. Furthermore, storage of RDF should not be prolonged; ideally, it should be burned within 24 h of its production. Poslusny et al. (1987) found that addition of a $Ca(OH)_2$ binder to the pellets during initial processing was successful in lengthening the storage lifetime of the pellets.

Dust production is inevitable with storage and handling of dry RDF; therefore, dust control equipment must be provided within both the combustion and storage areas. Forced ventilation combined with air filters are strongly recommended.

It should be clear by now that mechanical separation of MSW components is by no means 100% effective; therefore, contamination by food and yard waste, and other undesirable components will occur. As a result, odor production is inevitable in stored RDF, particularly in the warmer months. The RDF must therefore not be stored in a boiler building for extended periods; rather it should be loaded into the building daily for combustion.

In a study by Fiscus et al. (1978), total airborne bacteria concentrations were measured in a number of waste handling facilities including a RDF plant, incinerator, landfill, transfer station, waste collection vehicle, and wastewater treatment plant. The highest airborne bacterial concentrations were found in the RDF plant. Mahar (1999) studied the atmospheres in several locations within two RDF facilities. The data for particulate matter appear in Table 9.9, and that for total bioaerosols and endotoxins are in Table 9.10. The particulates detected were primarily in the nonrespirable size range. Biologically derived particulates were found to a greater extent in areas where the waste had been processed as opposed to stored.

9.8.5 GASEOUS EMISSIONS AND CORROSION ISSUES

When combusting a fuel containing 100% RDF, emissions of acid gases and hydrocarbons were consistently lower compared with bituminous coal alone (Pichtel, 1991). Oxides of sulfur and NO_x measured approximately 700 and 200 mg/m^3 compared with 1600 and 550 mg/m^3, respectively, for

---

TABLE 9.9
Particulate Comparisons[a] (mg/m^3) in Different Areas of an RDF Plant

Location	Inhalable Particles	Total Particles	Respirable Particles
Floor	2.24 (7)	1.15 (6)	0.09 (7)
Loadout	0.52 (2)	0.15 (2)	0.04 (2)
Lunchroom	0.13 (3)	0.06 (3)	0.07 (3)
Magnetic separator	3.06 (4)	1.26 (4)	0.10 (4)
Processing	0.73 (4)	0.38 (3)	0.16 (4)

[a]Geometric mean, n

Source: Mahar, S., *Waste Manage. Res.,* 17, 343–346, 1999. Reproduced with
        kind permission of the International Solid Waste Association.

TABLE 9.10

Comparisons of Bioaerosol (10^{-6}/m³) and Endotoxin (EU/m³) Concentrations Within an RDF Plant

Location	Total Bioaerosols[a]	Total Endotoxin[a]	Respirable[a] Endotoxin
Floor	0.08 (7)	38.1 (6)	0.70 (7)
Loadout	0.15 (2)	7.81 (2)	3.70 (2)
Lunchroom	0.13 (3)	1.02 (3)	0.89 (3)
Magnetic separator	3.22 (4)	72.0 (4)	12.9 (3)
Processing	0.58 (4)	2.80 (3)	3.09 (4)

[a]Geometric mean, n

Source: Mahar, S., *Waste Manage. Res.*, 17, 343–346, 1999. Reproduced with kind permission of the International Solid Waste Association.

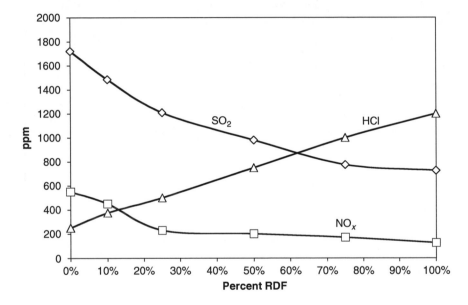

FIGURE 9.18 Concentration of gaseous SO_2, NO_x, and HCl with varied coal/RDF ratios.

coal. When increasing proportions of RDF were added to a mixture with Illinois bituminous coal, the concentrations of atmospheric SO_2 and NO_x consistently declined (Figure 9.18) (Pichtel, 1991). In contrast, total HCl concentrations increased in the flue emissions. These emissions data correlated with fuel compositions.

As is the case with mass-burn incineration of MSW, the production of undesirable gaseous organic compounds from RDF combustion is of significant concern. A number of studies (U.S. EPA, 1985a, 1985b; Poslusny et al., 1987; Pichtel, 1991) have demonstrated that the concentrations of PCDDs and PCDFs have been below detectable limits. Concentrations of PAHs, however, have been variable. In one study (U.S. EPA, 1985b), large quantities of PAHs were detected in combustion experiments where shredded and pelletized RDF were combusted. Poslusny et al. (1987) found that naphthalene was the major component of PAH emissions and tetra- and penta-chlorinated biphenyls were the major components of PCB emissions. In a study by Henstock et al. (1983), concentrations of vinyl chloride were in the ppb range.

9.8.6 PROPERTIES OF RDF ASH

As mentioned earlier in this chapter, ash quality from mass-burn incinerator ash is of concern to regulators and the general public due to the common presence of heavy metals, chlorinated dibenzodioxins, furans, and other toxins. With processing via trommel screens, shredding, magnetic separation, and air classification, the chemical properties of the waste charge are drastically changed. Composition of the ash is also correspondingly improved over that of mass-burn incinerator ash. The total elemental analysis of RDF ash is similar to that for coal ash but greatly improved compared with that of MSW ash.

When leached by the TCLP (U.S. EPA, 1986), concentrations of all TCLP metals and nonmetals as well as reactive sulfide and cyanide are typically well below RCRA limits (Table 9.11; Pichtel, 1991). The pH of ash tends to vary slightly; however, the majority are in the alkaline range, a result of the presence of Ca, Mg, Al, Na, and other basic cations in the RDF. In addition, the concentration of TCLP extractable and volatile organics, chlorinated dibenzodioxins and furans, and PCBs are well below the regulatory limits. RDF ash, therefore, can be disposed with much less concern about possible environmental and health impacts as compared with mass-burn incinerator ash.

9.8.7 PLASTICS-DERIVED FUELS

As discussed in Chapter 4, plastics are derived from petroleum and natural gas and are composed mostly of C, H, and O. Given the high-energy content of plastics, there is interest in using segregated postuse plastics and paper in industrial fuel applications. When such materials derived from residential, commercial, and industrial sources are used as an industrial fuel, they have been labeled as process engineered fuel (PEF). Conventional PEF contains 70 to 90% paper and the remaining percentage is plastic (APC, 1999).

Common reasons for encouraging a higher plastics content include the effect on densified PEF integrity and high heat value in comparison with conventional fuels. The American Plastics Council has explored formulations containing in excess of 30% plastics. In the United States, the estimated amount of plastics currently used in PEF is less than 4.54×10^7 kg/year (Fisher and Tomczyk, 1999).

TABLE 9.11
Analysis for Ignitability, Corrosivity, Reactivity, and TCLP Metals and Nonmetals in RDF Ash

Analyte	RCRA Limit[a]	Result
Flash point, °F	140°F	Negative
Corrosivity, pH	2–12	9.48
Reactive cyanide	250	<10
Reactive sulfide	500	25
Silver	5.0	<0.1
Arsenic	5.0	<0.2
Barium	100.0	0.1
Cadmium	1.0	<0.1
Chromium	5.0	<0.1
Mercury	0.20	<0.05
Lead	5.0	<0.1
Selenium	1.0	<0.2

[a] All units are in mg/L unless otherwise noted.

PEF can be produced in either shredded (fluff) or densified form. Preselected postuse plastic waste as well as wood, sawdust, or scrap paper is freed of glass and metal contaminants and ground to uniform size by a primary grinder. In some cases, a secondary grinder further processes these materials which are then densified. The most common methods for densifying PEF are cubing and pelletizing. Densification makes the final fuel product easier to transport and handle and assures consistent heating value. A typical commercial product is a densified pellet about 5 to 8 cm (2 to 3 in.) long and about $\frac{3}{4}$ in. in diameter.

Compared with conventional RDF, PEF is a more refined, low ash, low moisture, high heat value fuel. The high heating value of PEF is due to its plastics content. Plastics contribute to heating values in the range of 15,650 to 35,750 kJ/kg (7,000 to 16,000 Btu/lb). Conventional RDF has a heating value of about 14,000 to 15,650 kJ/kg (6,500 to 7,000 Btu/lb), and most coals have in the range of 20,000 to 27,000 kJ/kg (9,000 to 12,000 Btu/lb). As is the case with RDF, using PEF as a supplement for coal can also reduce some types of environmental emissions, particularly sulfur dioxide.

9.8.8 Tire-Derived Fuel (TDF)

In recent years, tire burning for fuel recovery has increased (U.S. EPA, 1991). Facilities such as cement kilns and pulp and paper mills use scrap tires (shredded or whole, depending on the industry) as a combustion fuel, burning approximately 42% of all scrap tires generated annually and utilizing the energy. Tires are shredded into small particles, the steel is removed magnetically, and the particles are often shredded a second time to produce crumb rubber. The fine rubber particles are mixed with coal, typically less than 10 to 20% (by wt), and fed directly into the combustion chamber. The so-called "tire-derived fuel" or TDF possesses the equivalent amount of energy per unit weight as oil and comparitively more energy than coal (average 32,500 kJ/kg or 14,000 Btu/lb). New technologies and pollution control equipment allow facilities to burn tires at high temperatures, reducing air emissions.

When tires are burned as fuel during controlled combustion, atmospheric emissions are similar to those emitted when coal or petroleum are burned. As with other fossil fuels, emissions include "criteria" pollutants such as particulates, carbon monoxide (CO), sulfur oxides (SO_x), oxides of nitrogen (NO_x), and "noncriteria" hazardous air pollutants such as PAHs, PCDDs, PCDFs, and trace metals. When operated properly, the burning of tires for fuel is a reasonably safe and economical practice that has been approved by the U.S. EPA. Air emissions usually are improved with the incorporation of TDF with coal due to its low sulfur and nitrogen content.

The California Integrated Waste Management Board (CIWMB) (1996) stated:

> In general, test results have shown that tire derived fuels have no additional adverse effect on emissions when compared to coal. In fact, test results indicate a net reduction of SO_x (sulfur oxides), NO_x (nitrous oxides) and particulate matter.

The EPA's Office of Research and Development published a study in 1997 entitled "Air Emissions From Scrap Tire Combustion", which stated:

> TDF has been used successfully in properly designed combustors with good combustion control and appropriate add-on controls, particularly particulate controls, such as electrostatic precipitators (ESPs) or fabric filters. Air emissions characteristic of TDF combustion are typical of most solid fuels, such as coal and wood. The resultant air emissions can usually satisfy environmental compliance limits even with TDF representing up to 10 to 20% of the fuel requirements.

Each facility that uses TDF must, after being permitted, pass an initial compliance test to ensure that it complies with emissions limits and operating conditions stated in their permit. Following the initial test, most coal-fired boilers are equipped with continuous emission monitors that are subsequently used for monitoring. This ensures that the facility is in regulatory compliance at all times.

REFERENCES

American Plastics Council, Information on Processed Engineered Fuels (PEF)/Plastics Derived Fuels, June 1999. See: http://www.plasticsresource.com/recycling/recycling_backgrounder/bk_fuels.html.

Berlinicioni, M. and di Domenico, A., Polychlorodibenzo-*p*-dioxins and polychlorodibenzofurans in the soil near the municipal incinerator of Florence, Italy, *Environ. Sci. Technol.,* 21, 1063–1069, 1987.

California Integrated Waste Management Board (CIWMB), Effects of Waste Tires, Waste Tire Facilities, and Waste Tire Projects on the Environment, Sacramento, CA, 1996.

Carlson, C.L. and Adriano, D.C., Environmental impacts of coal combustion residues, *J. Environ. Qual.,* 22,227–247, 1993.

Carpi, A., Mercury from combustion sources: a review of the chemical species emitted and their transport in the atmosphere, *Water Air Soil Pollut.,* 98, 241–254, 1997.

Chang, N.B. and Huang, S.H., Statistical modeling for the prediction and control of PCDDs and PCDFs emissions from municipal solid waste incinerators, *Waste Manage. Res.,* 13, 379–400, 1995.

Choudry, G. G., Olie, K., and Hutzinger, O., Mechanisms in the thermal formations of chlorinated compounds including polychlorinated dibenzo-*p*-dioxins and polychlorodibenzofurans in urban incinerator emissions, in *Chlorinated Dioxins and Related Compounds, Impact on the Environment,* Pergamon Press, New York, 1982.

Colmsjö, A.L., Zebühr, Y.U., and Östman, C.E., Polynuclear aromatic compounds in flue gases and ambient air in the vicinity of a municipal incineration plant, *Atmo. Enviro.,* 20, 2279–2282, 1986.

Danielson, J.A., Air Pollution Engineering Manual, 2nd ed., Air Pollution Control District, County of Los Angeles, U.S. Environmental Protection Agency, Research Triangle Park, NC, May 1973.

Davies, I.W., Harrison, R.M., Perry, R., Ratnayaka, D., and Wellings, R.A., Municipal incinerator as source of polynuclear aromatic hydrocarbons in environment, Environ. Sci. and Technol., 10, 451–453, 1976.

Denison, R.A. and Ruston, J., Recycling and Incineration, Evaluating the Choices, Environmental Defense Fund, Washington, DC, 1990.

Eduljee, G., Badsha, K., and Scudamore, N., Environmental monitoring for PCB and trace metals in the vicinity of a chemical waste disposal facility — II, *Chemosphere,* 15, 81–93, 1986.

Eiceman, G.A., Clement, R.E., and Karasek, F.W., Variations in concentrations of organic compounds including polychlorinated dibenzo-*p*-dioxins and polynuclear aromatic hydrocarbons in fly ash from a municipal incinerator, *Analyt. Chem.,* 53, 955–959, 1981.

Electric Power Research Institute, Use of FGD Gypsum and Bottom Ash in Roadway and Building Construction, EPRI TR-105236, Palo Alto, CA, 1995.

Fiscus, D.E., Gorman, P.G., Schrag, M.P., and Shannon, L.J., Assessment of Bacteria and Virus Emissions at a Refuse Derived Fuel Plant and Other Waste Handling Facilities, U.S. EPA Publication 600/2-78–152, Washington, DC, 1978.

Fisher, M.M. and Tomczyk, L., Plastics and Process Engineered Fuel (PEF): An Overview, SPE Annual Recycling Conference 1999. See: http://www.plasticsresource.com/recycling/ARC99/PEF_ARC99_PAPER_Final.htm

Gittinger, J.S. and Arvan, W.J., Waste-to-energy installations, *Steam: Its Generation and Use,* Stultz, S.C. and Kitto, J.B., Eds., 40th ed., Babcock & Wilcox, Chap. 27, Barberton, OH, 1998.

Hasselriis, F., Optimization of combustion conditions to minimize dioxin emissions, *Waste Manage. Res.,* 5, 311–325, 1987.

Hickman, H.L. Jr., *Thermal Conversion Systems for Municipal Solid Waste.,* Noyes Publications, Park Ridge, NJ, 1984.

Hutzinger, O., Blumich, M.J., Gerg, M.V.D., and Olie, K., Sources and fate of PCDDs and PCDFs: An overview, *Chemosphere,* 14, 581–600, 1985.

Kukkonen, J. and Raunemaa, T., Dispersion studies on a solid waste refuse incinerator, *Int. J. Environ. Stud.,* 23, 235–247, 1984.

Lee, W.-J., Liow, M.-C., Tsai, P.-J., and Hsieh, L.-T., Emission of polycyclic aromatic hydrocarbons from medical waste incinerators, *Atmos. Environ.,* 36, 781–790, 2002.

Lisk, D.J., Environmental implications of incineration of municipal solid waste and ash disposal, *Sci. Total Environ.,* 74, 39–66, 1988.

Liu, D.H.F. and Liptak, B.G., *Hazardous Waste and Solid Waste*, Lewis Publishing, Boca Raton, FL, 2000.

Mahar, S., Airborne particulates in refuse-derived fuel plants, *Waste Manage. Res.,* 17, 343–346, 1999.

Manahan, S.E., *Environmental Chemistry,* 6th ed., CRC Press, Boca Raton, FL, 1994.

National Academy of Sciences, Waste Incineration and Public Health, Committee on Health Effects of Waste Incineration, Board on Environmental Studies and Toxicology, National Research Council, Washington, DC, 2000.

Nishitani, T., Nomura, T., Fukunaga, I., and Itoh, H., The relationship between HCl and mercury speciation in flue gas from municipal solid waste incinerators, *Chemosphere* 39, 1–9, 1999.

Olie , K., Berg, M.V.D., and Hutzinger, O., Formation and fate of PCDD and PCDF from combustion processes, *Chemosphere,* 12, 627–636, 1983.

Pichtel, J., Assessment of Air Quality and Ash Residues from Combustion of Refuse-Derived Fuel, Council of Great Lakes Governors, Chicago, IL, 1991.

Pierce, R.C. and Katz, M., Dependency of polynuclear aromatic hydrocarbon content on size distribution of atmospheric aerosols, *Environ. Sci. Technol.,* 9, 347–353, 1975.

Poslusny, M., Moore, P., and Daugherty, K., Organic emission studies of full-scale cofiring of pelletized RDF/coal, *AIChE Symposium Series,* Vol. 84, No. 265, 1987, pp. 94–106.

Reimann, D., Treatment of waste water from refuse incineration plants, *Waste Manage. Res.,* 5,147–157, 1987.

Reimann, D.O., Future Gas Cleaning Systems in Accordance with EC Regulations, Solid Waste Management: Thermal Treatment and Waste to Energy Technologies. VIP-53, Proc. Int. Spec. Conference sponsored by the Air & Waste Management Assoc., Washington, DC, April 18–21, 1995, pp. 37–49.

Reynolds, J.P., Jeris, J.S., and Theodore, L., *Handbook of Chemical and Environmental Engineering Calculations*, Wiley, New York, NY, 2002.

Stehouwer, R.C., Sutton P., and Dick, W.A., Minespoil amendment with dry flue gas desulfurization by-products: Plant growth. *J. Environ. Quality*, 24, 861–869, 1995.

U.S. Department of Energy, Renewable Energy 2000: Issues and Trends, DOE/EIA-0628. Energy Information Administration, Office of Coal, Nuclear, Electric and Alternate Fuels, Washington, DC, 2001.

U.S. Environmental Protection Agency, Air Pollution Engineering Manual, 2nd ed., AP-40, NTIS PB-225132, Office of Air Quality Planning and Standards, Research Triangle Park, NC, 1973.

U.S. Environmental Protection Agency, Assessment of organic contaminants in emissions from refuse-derived fuel combustion, Project Summary, EPA/600/S2-85/115, Hazardous Waste Engineering Research Laboratory, Cincinnati, OH, 1985a.

U.S. Environmental Protection Agency, Emissions assessment refuse-derived fuel combustion. Project Summary, EPA/600/S2-85/116, Hazardous Waste Engineering Research Laboratory, Cincinnati, OH, 1985b.

U.S. Environmental Protection Agency, Markets for scrap tires. EPA/530-SW-90-074A, U.S. Government Printing Office, Washington, DC, 1991.

U.S. Environmental Protection Agency, Methodology for the Assessment of Health Risks Associated with Multiple Pathway Exposure to Municipal Waste Combustor Emissions, Office of Air Quality Planning and Standards, Research Triangle Park, NC, 1986.

U.S. Environmental Protection Agency, Operation and Maintenance of Hospital Medical Waste Incinerators, EPA/625/6-89/024, Center for Environmental Research Information, Cincinnati, OH, January 1990.

U.S. Environmental Protection Agency, Operational Parameters for Hazardous Waste Combustion Devices, EPA/625/R-93/008, Office of Research and Development, Cincinnati, OH, 1993.

U.S. Environmental Protection Agency, Industrial-Commercial-Institutional Boilers MACT Standards Development, 2002. See: http://www.epa.gov/ttn/atw/combust/boiler/boilback.html.

Vesilind, P.A., Worrell, W.A., and Reinhart, D.A., *Solid Waste Engineering*, Brooks/Cole, Pacific Grove, CA, 2002.

Williams, P.T., Waste Treatment and Disposal, McGraw-Hill, New York, NY, 1998.

Wilson, D.C., *Waste Management: Planning, Evaluation, Technologies,* Clarendon Press, Oxford, England, 1981.

Zhang, F.-S., Kimura, K., Yamasaki, S., and Nanzyo, M., Evaluation of cadmium and other metal losses from various municipal wastes during incineration disposal, *Environ. Pollut.,* 115, 253–260, 2001.

SUGGESTED READINGS

Bartley, D. A., Vigil, S. A., and Tchobanoglous, G., The Use of Source-Separated Waste Paper as a Biomass Fuel, *Biotechnology and Bioengineering Symposium*, Wiley, New York, NY, 1980, p. 67.

Brunner, C.L., *Incinerator Systems Selection and Design,* Van Nostrand Reinhold, New York, NY, 1984.

Bumb, R.R. et al., Trace chemistries of fire: a source of chlorinated dioxins, *Science*, 210, 385, 1980.

Cooper, C.D. and Alley, F.C., *Air Pollution Control: A Design Approach,* PWS Publishers, Boston, MA, 1986.

Cross, F.L. and Hesketh, H.E., *Controlled Air Incineration,* Technomic Press, Lancaster, PA, 1985.

Czuczwa, J.M. and Hites, R.A., Airborne dioxins and dibenzofurans: sources and fates, *Environ. Sci. Technol.,* 20, 195, 1986.

Cruczwa, J.M. and Hites, R.A., Environmental fate of combustion-generated polychlorinated dioxins and furans, Environ. Sci. Technol., 18, 444, 1984.

Drobny, N.L., Hull, H.E., and Testin, R.F., Recovery and Utilization of Municipal Solid Waste, SW 10c, U.S., Environmental Protection Agency, Washington, DC, 1971.

Finnis, P., Heat Recovery Dry Injection Scrubbers for Acid Gas Control, *Proceedings of the ASME Asian–North American Solid Waste Management Conference (ANACON),* Los Angeles, CA, 1998.

Fricilli, P.W. Impact of EPA's Air Pollution Control, *Proceedings of the 84th AWMA Meeting*, Vancouver, BC, Paper No. 91-26.2, Air and Waste Management Association, June 16–21, 1991.

Getz, N., *How does Waste-to-Energy Stack Up? Municipal Waste Combustion*, 1993, pp. 951–965.

Gleiser, R., Nielsen, K., and Felsvang, K., Control of Mercury from MSW Combustors by Spray Dryer Absorption Systems and Activated Carbon Injection, *Municipal Waste Combustion,* 1993, pp. 106–120.

Gorman, P.G., Markus, M. et al., Project Summary-Environmental Assessment of Waste-to-Energy Process: Union Carbide's Purox Process, EPA-600/S7-80-161, U.S. Environmental Protection Agency, Cincinnati, OH, December, 1980.

Harrison, B. and Vesilind., P.A., *Design and Management for Resource Recovery, Volume 2: High Technology — A Failure Analysis,* Ann Arbor Science, Ann Arbor, MI, 1980.

Hasselriis, F., *Refuse Derived Fuel*, Ann Arbor Science, Butterworth, Boston, MA, 1966.

Hasselriis, F., Optimization of Combustion Conditions to Minimize Dioxin Emission, *Waste Manage. Res.* 5, 311, 1987a.

Hasselriis, F., Optimization and Combustion Conditions to Minimize Dioxin, Furan, and Combustion Gas Data from Test Programs at Three MSW Incinerators, *J. Air Pollut. Control Assoc.,* 37, 1451, 1987b.

Hasselriis, F., Variability of Metals and Dioxins in Stack Emissions of Three Types of Municipal Waste Combustors over Four Year Period, Paper No. 95-RP147B.03, Air and Waste Management Association, San Antonio, TX, 1995.

Kenna, J.D. and Turner, J.H., *Fabric Filter-Baghouses I: Theory, Design and Selection*, ETS International, Inc, Roanohe, VA, 1989.

Kisker, J.V.L., A Comprehensive Report on the Status of Municipal Waste Combustion, *Waste Age*, November 1990.

Licata, A., Babu, M., and Nethe, L-P., Acid Gases, Mercury and Dioxin from MWCs,*Proceedings of the ASME National Waste Processing Conference*, Atlantic City, NJ, 1994.

Licata, A., Schuttenhelm, W., and Klien, M., Mercury Control for MWCs Using the Sodium Tetrasulfide Process, 8th *Annual North American Waste-to-Energy Conference,* Nashville, TN, 2000.

Orning, A.A., Principles of combustion, in *Principles and Practices of Incineration*, Corey, R. C., Ed., Wiley-Interscience, New York, NY, 1969.

Orvacic, V. et al., Emissions of chlorinated organics from two municipal incinerators in ontario, *J. Air Pollut. Control Assoc.,* 35, 849, 1985.

Rigo, H.G. and Chandler, J., Metals in MSW — Where are They and Where Do They Go in an Incinerator? *ASME National Waste Processing Conference*, Boston, MA, 1994.

Rigo, H.G., Chandler, J., and Sawell, S., Debunking Some Myths About Metals, *Municipal Waste Combustion*, 1993, pp. 609–627.

Shaub, W.M. and Tsang, W., Dioxin formation in incinerators, *Environ. Sci. Technol.,* 17, 721, 1983.

Turner, J.H., Lawless, P.A., et al., Sizing and costing of electrostatic precipitators: Part 1-sizing considerations, *J. Air Pollut. Control Assoc.,* 38, 458, 1988.

U.S. Environmental Protection Agency, Risk Assessment for the Waste Technologies Industries (WTI) Hazardous Waste Incineration Facility (East Liverpool, Ohio), EPA-905-R97-002a, U.S. Environmental Protection Agency Region 5, Chicago, IL, 1997.

U.S. Environmental Protection Agency, Emissions Test Results from the Stanislaus County, California, Resource Recovery Facility, *International Conference on Municipal Waste Combustion*, Hollywood, FL, U.S. Environmental Protection Agency Washington, DC, 1989.

Walsh, P., O'Leary, P., and Cross, F., Residue disposal from waste-to-energy facilities, *Waste Age*, April 1988.

WASTE, Waste Analysis, Testing and Evaluation: The Fate and Behavior of Metals in Mass Burn Incineration, A.J Chandler Associates Ltd., Willowdale, ON, 1993.

QUESTIONS

1. MSW combustion involves physical and chemical transformations in which solid materials are converted into gases and some solid residues. What factors affect the types of gases produced? What factors will influence the amount of solid residues, both carbonaceous and inorganic?
2. Compare the operation of a mass-burn incinerator with that of RDF, in terms of: fuel types; waste processing and equipment; convenience; resource recovery; energy utilization.
3. What are the functions of combustion chamber overfire and underfire air? How can they be adjusted to optimize incineration?
4. List the engineering and design factors that serve to enhance MSW combustion in an incinerator.
5. Define stoichiometric air and heat value.
6. Discuss the major types of gaseous emissions from a mass-burn incinerator and how each may be effectively removed from the flue.
7. SO_2 production may be controlled during mass-burn incineration by the addition of limestone to the combustion chamber. What are the advantages and disadvantages of this procedure over flue gas desulfurization?
8. An incinerator operating at a sufficiently high temperature and air inflow rate may still generate PICs. Explain how such a phenomenon may occur.
9. How do PCDDs and PCDFs form during mass-burn incineration (given that the firebox temperature is sufficiently high, for example, > 1000°C, to destroy virtually all organic compounds)? In what physical form(s) are these compounds emitted?
10. Describe "particulate matter" as relates to MSW combustion. What are its chemical and physical properties? What size range of particulates are the most potentially damaging when inhaled? How do certain toxins (e.g., metallic vapors, chlorinated hydrocarbons) react with particulate matter to increase their risk of exposure?
11. Generation of atmospheric pollutants is a function of MSW charge rate and combustion chamber conditions, among other factors. Explain.
12. Explain how the following air pollutants can be removed from stack gases: SO_2, particulates, mercury, and PCDDs.
13. Explain why most MSW incinerators in the United States are mass-burn rather than RDF-fired.
14. How do electrostatic precipitators and cyclone separators differ in terms of efficiency of removal of particulate matter, SO_2, and PCDDs.
15. Cd occurring in raw MSW can become significantly more soluble (and hence more leachable) following MSW combustion in a mass-burn incinerator. Explain.
16. Discuss the major concern(s) with RDF storage, both indoors and outdoors.
17. Which of the following is a significant concern when considering RDF production and utilization with coal: (a) dust production; (b) odor production; (c) separation of RDF and coal during handling; (d) some plants are unable to market the RDF; (e) all of the above.

18. For RDF to produce the same amount of heat as coal, more ash will probably be produced. Explain.

19. Compare and contrast RDF and raw MSW in terms of fuel properties. How do they differ in terms of heat content, moisture content, density, and ash content?

20. How do RDF and coal differ in terms of emissions of SO_2, NO_x, and HCl? How do they differ in ash composition?

21. A materials recovery facility is being installed in Pristine, Illinois (pop = 110,000). The MRF will be receiving mixed MSW. RDF will be produced. For maximum efficiency and ease of mixing with coal, what form (e.g., fluff, wet-pulped, densified, etc.) of RDF is recommended?

22. How will the RDF be stored: (a) in the customers' yard for easy utilization; (b) in a covered pole barn; (c) no need to store the RDF as it will be immediately sent to market; (d) in 50:50 mixtures with coal in customers' yard.

23. An incinerator burns 120 MT/h of MSW with the formula $C_{285}H_{455}O_{235}N_4S$. How much air is needed to combust this waste? A rate of 35% excess air is used during combustion.

24. Calculate the heating value for the waste discussed in question 23.

EXCEL EXERCISE

ANALYSIS OF RDF BEFORE AND AFTER COMBUSTION

FILE NAME: RDF.xls

Background

Data are provided from actual field-test burns of densified RDF. An MRF had recently begun operations and was producing d-RDF with the expectation of selling it to local markets. The tests outlined below were conducted in order to satisfy concerns of the state regulatory agency as to any potential hazards with gaseous emissions or ash.

The data for this exercise can be located at www.crcpress.com/e_products/downloads/download.asp?cat_no=3525

Some of the analyses and experiments were as follows:

- Ultimate analysis of fuel as received (three RDF and two local coal samples)
- Varying mixtures of coal and d-RDF were combusted in a coal-fired boiler. Steam production, ash generation, and acid gas production were documented.
- A field-scale experiment was conducted in which pellets were manufactured containing fixed ratios of paper and assorted plastics. These pellets were combusted in a coal-fired boiler and gaseous emissions were monitored.

Tasks

1. Determine the relationship between percentage of plastic in the RDF pellets and acid gas production. Graph the data for SO_2, NO_x, and HCl. What is the trend for HCl with increased ash? How could this trend occur?
2. Graph the concentrations of SO_2, NO_x, and HCl for the coal and RDF mixtures. What trends do you discern? On a separate page, graph steam production with percentage of coal/RDF. Also graph ash generation with percentage of coal/RDF. Discuss the trends, if any.
3. Analyze the TCLP data for the RDF ash. Compare the values generated with TCLP limits. Based on the TCLP limits in the table, are any of the ash samples considered "hazardous waste"?
4. Calculate the mean and standard deviations of the ultimate analysis data for the three RDF samples. How precise are the data? Within one standard deviation or greater? What does this tell us about the sampling and analysis procedures, and about the inherent chemical variability of a natural material such as southern Indiana coal?
5. The data provided are from actual pilot tests conducted at a facility considering the use of RDF alone or coal and RDF mixtures for steam production. The RDF supplier placed a bid of $29/ton of RDF. The coal was $33/ton. Based on steam production data, SO_2 data, and increased ash handling requirements, under what percentage of coal/RDF ratio would the facility accrue maximum savings?

10 The Sanitary Landfill

This is Anacreon's grave.
Here lie the shreds of his exuberant lust,
but hints of perfume linger by his gravestone still.

Antipater of Sidon (c. 130 BCE)
This is Anacreon's Grave

10.1 INTRODUCTION

Prior to enactment of Resource Conservation and Recovery Act (RCRA) (1976), what Americans had referred to as 'landfills' were typically not much more than open dumps (Figure 10.1). There was no requirement for a daily layer of soil, for example, which is important in deterring vectors and preventing other hazards or nuisance conditions. As a result, insect and rodent infestations were common at pre-RCRA facilities and there were frequent fires. These facilities were typically constructed without protective liners; therefore, contents readily leached into subsurface formations, including those which store groundwater. Many were sited at locations thought to be convenient, without regard to subsurface geology or groundwater features. There was no requirement for impermeable substrata below the landfill unit that could have prevented migration of liquids.

As a result of the RCRA regulations, modern sanitary landfills must meet stringent requirements for siting, construction, operation and maintenance, and final closure. The RCRA regulations apply to all municipal solid waste (MSW) landfills that are active (i.e., receiving waste) and do not apply to landfills that stopped accepting MSW before October 1991. Because of the complex technology required, the federal requirement for installing groundwater monitoring systems was phased in over a period of 5 years. To protect drinking water sources, landfills nearest to groundwater sources must comply before those that are sited farther away. By April 9,1994, landfill owners and operators were required to demonstrate the ability to pay the costs of closure, postclosure care, and clean-up of any known releases (U.S. EPA, 1993a).

10.2 RELEVANT DEFINITIONS UNDER THE RCRA REGULATIONS

Municipal solid waste landfill (MSWLF): A discrete area of land or an excavation that receives household waste, and is not a land application unit, surface impoundment, injection well, or waste pile. A MSWLF unit may also receive other types of wastes as defined under Subtitle D of RCRA, such as, commercial solid waste, nonhazardous sludge, small quantity generator waste, and industrial solid waste. Such a landfill maybe publicly or privately owned.

Existing unit: A MSWLF unit that is receiving solid waste as of October 9, 1993. Waste placement in existing units must be consistent with previous operating practices or modified practices to ensure good management.

Lateral expansion: A horizontal expansion of the waste boundaries of an existing unit.

New unit: Any MSWLF unit that has not received waste prior to October 9, 1993.

(a)

(b)

FIGURE 10.1 Landfill operated and closed prior to the RCRA regulations. Subsidence has occurred, creating a toxic wetland. Leachate is exiting freely from the sides of the landfill.

10.3 SITING THE LANDFILL

The Code of Federal Regulations (40 CFR) has instituted six restrictions as related to landfill siting in order to limit hazards to the local public and certain sensitive environments. Landfill owners and operators are required to demonstrate that their units meet all these criteria.

10.3.1 AIRPORT SAFETY

As many landfills are cited along coastal areas, seagulls and other scavenging bird populations often occur in high numbers. Concentrations of birds increase the likelihood of bird and aircraft collisions that damage aircraft and injure its users. As a result, RCRA requires that all new and existing MSWLF units, and lateral expansions located within 10,000 ft of an airport runway end used by turbojet aircraft (or within 5,000 ft of an airport runway end used by only piston-type aircraft), must not pose a bird hazard to aircrafts (40 CFR Part 258.10). In other words, the facility must be sited far enough from an airport to prevent excessive bird populations from the airspace. Similarly, a thick daily soil cover over the landfill cell will limit the attraction to birds.

10.3.2 FLOODPLAINS

Floodplains under the RCRA connotation are lowland and relatively flat areas adjoining inland and coastal waters that are inundated by a 100 year flood, defined as one with a magnitude that equaled or exceeded once in 100 years. Operators of landfills and lateral expansions located in 100 year floodplains must demonstrate that the unit will not restrict the flow of the 100 year flood or result in washout of any deposited MSW and therefore pose a hazard to health and the environment (40 CFR Part 258.11). "Washout" refers to the carrying away of solid waste by waters of the flood.

10.3.3 WETLANDS

The U.S. government regulatory definition of wetlands, as per Section 404 of the 1977 Clean Water Act Amendments, is (33 CFR 328.3[b]) (also 40 CFR 232.2[r]):

> those areas that are inundated or saturated by surface or ground water at a frequency and duration sufficient to support, and that under normal circumstances do support, a prevalence of vegetation typically adapted for life in saturated soil conditions. Wetlands generally include swamps, marshes bogs and similar areas.

New or expanding municipal landfills may not be built or expanded within wetlands. However, exceptions can be made for units when the owner can show:

- No siting alternative is available.
- Operation will not violate applicable regulations on water quality or toxic effluent; threaten endangered or threatened species or sensitive habitats; or violate protection of a marine sanctuary.
- The unit will not cause or contribute to significant degradation of wetlands.
- Steps have been taken to achieve no net loss of wetlands (e.g., restoring damaged wetlands or creating man-made wetlands).

The landfill operator must demonstrate the integrity of the landfill unit and its ability to protect local natural resources by addressing the following factors:

- Erosion, stability, and migration potential of native wetland soils used to support the unit
- The volume and chemical nature of the waste managed in the unit
- Impacts on aquatic and terrestrial wildlife and their habitat from release of the solid waste
- The potential effects of terrestrial catastrophic release of waste to the wetland and the resulting impacts on the environment

10.3.4 FAULT AREAS

New units or lateral expansions are prohibited within 200 ft of fault areas that have shifted since Holocene time (the most recent epoch of the Quaternary period, extending from the end of the Pleistocene Epoch to the present) (40 CFR Part 258.13).

10.3.5 SEISMIC IMPACT ZONES

RCRA requires that landfills must not be sited in a seismic impact zone. In the event of siting in a potentially unstable area, the landfill must be designed to withstand the effects of surface motion due to an earthquake. All containment structures including liners, leachate collection systems, and surface-water control systems must be designed to resist the maximum horizontal acceleration in earth material.

10.3.6 UNSTABLE AREAS

The landfill must be designed to ensure that the integrity of the unit will not be disrupted during destabilizing events such as (U.S. EPA, 1993):

- Flows of debris from heavy rains
- Fast-forming sinkholes caused by excessive withdrawal of groundwater

- Rock falls set off by explosives
- Sudden liquefaction of soil after a long period of repeated wet–dry cycles

Unstable areas are susceptible to events or forces that may impair the integrity of landfill structural components (e.g., liners, leachate collection systems, final covers, and run-on and runoff systems). Unstable areas include those susceptible to mass movements (landslides, avalanches, debris slides, and rock fall) and karst topography. Karst topography has developed from the dissolution of limestone or other soluble rock. Common physiographic features present in karst terranes include sinkholes, sinking streams, and caves. Such conditions may impair foundation conditions and result in inadequate support for the components of a landfill unit.

10.4   REQUIREMENTS OF OPERATION

The operating requirements for MSWLFs, provided in RCRA Subpart C, came into effect in 1993. These detailed requirements were formulated to ensure the safe daily operation and management at MSWLF units and include:

- Detection and exclusion of hazardous waste from the facility
- Use of appropriate cover material for daily cells and the closed landfill
- Disease vector control
- Explosive gas control
- Air monitoring
- Facility access
- Run-on and runoff control systems
- Surface-water requirements
- Restrictions on liquids entering the cells
- Record-keeping requirements

The Subpart C requirements are by no means, however, the sole determinants of landfill operation. Operators must comply with a host of other federal laws. For example, discharges from a MSWLF into surface waters must be in conformance with sections of the Clean Water Act. In addition, burning of MSW (on those infrequent occasions when it is permitted) is regulated under the Clean Air Act.

10.4.1   RECEIPT OF HAZARDOUS WASTE

A key concern of regulators, site owners, and lenders is the possible transformation of a sanitary landfill designed to accept only municipal and commercial wastes to a contaminated site. Unfortunately, prior to the enactment of Subtitle D, several Superfund sites had their origins in this manner. Long before the enactment of RCRA, some sanitary landfills accepted substantial volumes of industrial wastes, some of which were hazardous and many were in liquid form. The hazardous composition of the waste combined with their proximity to populations and other sensitive receptors contributed to a high ranking on the National Priorities List. In order to prevent such situations occurring in the future, operators of MSWLFs were required to implement a program for detecting and preventing the disposal of hazardous wastes and polychlorinated biphenyl (PCB) wastes at their facility (40 CFR Part 258.20). This program includes random inspections of incoming loads, training of facility personnel to recognize hazardous wastes and PCB wastes, and notification of regulatory authorities if a hazardous waste is discovered at the facility.

According to Subpart D of 40 CFR Part 261, a "solid waste" is deemed a "hazardous waste" if it: (1) is listed; (2) exhibits a characteristic of a hazardous waste (ignitability, corrosivity, reactivity, and toxicity) (see Chapter 11, Identification of Hazardous Waste); or (3) is a mixture of a listed hazardous waste and a nonhazardous solid waste. PCBs are regulated under the Toxic Substances Control Act (TSCA). Commercial or industrial sources of PCBs include:

- Oil and dielectric fluids
- Transformers and other electrical equipment containing dielectric fluids
- Contaminated soil, dredged material, sewage sludge, and other debris from a release of PCBs
- Hydraulic machines

10.4.2 INSPECTIONS

An inspection is a visual observation of incoming waste loads by trained personnel. Ideally, all loads should be screened; however, it is impractical to inspect all incoming loads. Random inspections, therefore, are often the only feasible technique to control the receipt of inappropriate wastes. Loads should be inspected prior to disposal at the working face of the landfill unit to provide the opportunity to refuse the wastes if necessary. Inspections can be conducted on a tipping floor of a transfer station before shipping to the disposal facility. Inspections may also occur inside the site entrance, at the disposal facility tipping floor, or as a last resort, near the working face of the landfill unit.

Inspections may be accomplished by tipping the vehicle load in an area designed to contain hazardous wastes. The waste should be spread on to the surface using a front-end loader. Facility personnel should be trained to identify questionable wastes. Suspicious wastes may be identified by a number of clues, including:

- Placards or markings indicating hazardous contents
- Presence of sludges or liquids
- Presence of powders or dusts
- Bright or unusual colors of the contents
- Drums or commercial size containers
- Significant chemical odors

The receiving facility must always be aware that containers may arrive at the facility with suspicious contents. Only trained personnel should open an unmarked 55 gal drum. OSHA regulations, as promulgated in 29 CFR 1910, provide clear guidelines as to how to handle and open drums having questionable contents. If the waste is deemed acceptable, then it may be transferred to the working face for disposal.

Testing of questionable wastes should include the Toxicity Characteristic Leaching Procedure (TCLP) and other tests for characteristics of hazardous wastes including corrosivity, ignitability, and reactivity (see Chapter 11). Wastes that are suspected of being hazardous should be handled and stored as a hazardous waste until a determination is completed. If the operator discovers hazardous waste while still in the possession of the transporter, the operator can refuse to accept the waste at the facility. Thus, the waste remains the responsibility of the transporter.

If wastes delivered to and stored at the site are determined to be hazardous, the landfill owner or operator is now responsible for the management of a hazardous waste. Management includes requirements for packaging, storage, runoff control, documentation, and other detailed practices. If the wastes are to be transported from the facility, the waste must be:

- Stored at the landfill in compliance with all requirements of a hazardous waste generator (see Chapter 12)
- Manifested
- Transported by a licensed transporter (i.e., having a U.S. EPA identification number)
- Shipped to a permitted treatment, storage or disposal (TSD) facility for final disposal

Operators of landfills should be prepared to handle hazardous wastes that are inadvertently received at the facility. This may include having 55 gal drums available and keeping a list of the nearest companies licensed to transport hazardous waste. Hazardous waste may be stored at the landfill for 90 days, provided that the following procedures are followed (40 CFR Section 262.34):

- The waste is stored in tanks or containers. Both terms are defined in the federal regulations.
- The date of receipt of the waste is noted on each container
- The container is marked with the words 'Hazardous Waste'
- An employee is designated for coordinating any emergency response measures

If the landfill facility transports the wastes off-site, it must comply with 40 CFR Part 262 or the analogous state requirements, which include:

- Obtain an EPA identification number (the landfill is now a 'generator of hazardous waste')
- Package the waste as per Department of Transportation (DOT) regulations (49 CFR Parts 173, 178, and 179)
- Manifest the waste designating a permitted facility to treat, store, or dispose of the waste

If the landfill decides to treat, store (for more than 90 days), or dispose of the hazardous waste on-site, they are legally defined as a hazardous waste treatment, storage and disposal facility, and must comply with state or federal requirements for such facilities. This typically requires a permit. The major requirements for generators, transporters, and TSD facilities are discussed in greater detail in Chapters 12 through 17.

PCB wastes detected at a landfill must be stored and disposed according to 40 CFR Part 761. The operator is required to:

- Obtain an EPA PCB identification number
- Properly store the waste
- Mark containers with the words 'Caution: Contains PCBs'
- Manifest the waste for shipment to a permitted incinerator, chemical waste landfill, or high-efficiency boiler for disposal

Clearly, it is to the facility's advantage to detect and remove any potential hazardous waste before it ever enters the tipping floor. Preventing the entry of these wastes may be accomplished through other methods. For example, facilities may receive only household wastes and processed (shredded or baled) wastes that are adequately screened for the excluded wastes.

10.4.3 TRAINING

Landfill operators must ensure that personnel are trained to identify hazardous and PCB wastes. The training program should emphasize methods to identify containers and labels typical of hazardous and PCB wastes. Training also should address hazardous waste handling procedures, safety precautions, and record keeping. Again, OSHA regulations as detailed in 29 CFR 1920.120 are extremely useful in providing proper protocols for investigations and worker safety.

10.4.4 LANDFILL DESIGN

According to RCRA, the criteria for landfill design apply only to new units and lateral expansions; existing landfills are not required to retrofit systems such as liners. The criteria provide for two basic design options. The first design option consists of a composite liner and a leachate collection system. Landfills in states without EPA-approved programs must use this design. The composite liner system consists of an upper synthetic geomembrane liner (also known as a flexible membrane liner, FML) and a lower layer of compacted soil at least 0.61 m (2 ft) thick with a hydraulic conductivity of not greater

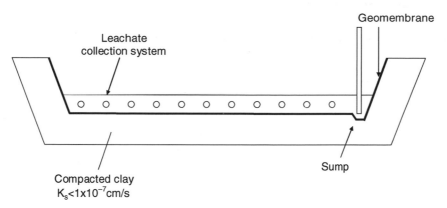

FIGURE 10.2   The layers beneath a sanitary landfill as required under RCRA (U.S. EPA, 1994). Note: Not to scale.

than 1×10^{-7} cm/sec (Figure 10.2). The geomembrane liner must measure at least 30 mil thick (1 mil = 0.001 in.) except for high-density polyethylene (HDPE) geomembranes, which must be at least 60 mil thick. The geomembrane liner minimizes the exposure of the compacted soil liner to leachate. A leachate collection and removal (LCR) system is situated above the composite liner to collect, divert, and remove liquids during landfill operation and well after closure. The LCR must be designed to limit the depth (hydraulic head) of the leachate above the liner to less than 30 cm (U.S. EPA, 1993b).

Second, in states with EPA-approved programs, landfills may be constructed to comply with a design approved by the state regulatory agency. In approving the design, the agency must ensure that maximum contaminant levels (MCLs) will not be exceeded in the uppermost aquifer at a 'relevant point of compliance.' This point is determined by the agency, but it must be located not further than 150 m from the landfill boundary. The U.S. EPA MCLs for a number of solid waste constituents are shown in Table 10.1. In planning such performance-based landfill designs, other factors must also be considered, such as the hydrogeologic characteristics of the facility and surrounding land, the local climate, and the amount and composition of the leachate (U.S. EPA, 1993b).

It must be emphasized that, in the design of Subtitle D landfills, the U.S. EPA provides minimum design standards only. Several states require double-composite liner systems in the design of Subtitle D landfills.

10.4.5 CLAY LINERS

Clay is an extremely important component of soil liners because it tends to be relatively available and amenable of mechanical and other stresses. Clay materials, being natural, incorporate readily with native soil materials and are obviously very durable. Additionally, the clay fraction of the soil ensures low hydraulic conductivity. The U.S. EPA requires that soil liners be constructed so that hydraulic conductivity is less than 1×10^{-7} cm/sec (Figure 10.3). To meet this requirement, certain characteristics of the soil materials must be met. First, the soil should contain at least 20% fines (i.e., fine silt and clay-sized particles). Second, the plasticity index (PI) must be greater than 10%. Third, coarse fragments should be screened to no more than about 10% gravel-size particles. Soils with a greater percentage of coarse fragments can contain areas that have high hydraulic conductivities. Finally, the material should not contain rocks larger than 2.5 to 5 cm (1 to 2 in.) in diameter (U.S. EPA, 1989).

Many different clay types exist with variations in surface area, external and internal charge, and interlayer cations. These differences in chemical and physical properties influence swelling behavior, potential for cracking and liquid transmission, and ultimately determine their possible utility in landfill liners.

TABLE 10.1
Maximum Contaminant Levels for MSW Constituents

Chemical	MCL[a] (mg/L)
Arsenic	0.01
Barium	2.0
Benzene	0.005
Cadmium	0.005
Carbon tetrachloride	0.005
Chromium (hexavalent)	0.05
2,4-Dichlorophenoxy acetic acid	0.07
1,4-Dichlorobenzene	0.075
1,2-Dichloroethane	0.005
1,1-Dichloroethylene	0.007
Endrin	0.002
Fluoride	4.0
Lindane	0.0002
Lead	0.015
Mercury	0.002
Methoxychlor	0.04
Nitrate	10
Selenium	0.05
Silver	0.05
Toxaphene	0.003
1,1,1-Trichloroethane	0.2
Trichloroethylene	0.005
2,4,5-Trichlorophenoxy acetic acid	0.05
Vinyl chloride	0.002

[a] Not to be exceeded in the uppermost aquifer under a MSWLF

Source: 40 CFR §§258.40.

The clays of importance are the so-called silicate clays, those possessing a crystalline structure composed of two relatively simple constituents, i.e., a silica tetrahedron (SiO_4) and an aluminum octahedron ($Al_2[OH]_6$). Different clay minerals result as these basic units are stacked upon each other. In many cases the central metal (Si or Al) is replaced by other metals of similar diameters, thus imparting a significant electrical charge to the clay units. Also, different ions may bind the clay units together. Some important clay properties are listed in Table 10.2. The smectite group is known for substantial swelling upon wetting; water molecules are easily inserted between the layers, which results in expansion. As a result, smectites (in particular bentonite clays) have been popular for landfill liners and caps and also for the installation of slurry walls, i.e., vertical barriers that restrict horizontal liquid migration.

10.4.6 HYDRAULIC CONDUCTIVITY

Vertical seepage of leachate and consequent contamination of groundwater is an important consideration in design of a Subtitle D landfill. During routine landfill operations, leachate will collect at the base of a landfill, typically from inputs of natural precipitation and the presence of moisture within the waste. It is of great practical importance, therefore, to appreciate the behavior of liquids such as water or leachate in a saturated soil (or clay) column.

FIGURE 10.3 Installation of a layer of low-permeability clay. The clay is spread in two lifts and then rolled. (Reproduced with kind permission of the Town of Bourne Department of Integrated Solid Waste Management, MA).

TABLE 10.2

Categories of Common Clays and Some Important Chemical and Physical Properties

Clay	Substitution	Interlayer Component	Swelling	Cation Exchange Capacity (cmol/kg)	Total Surface Area (m²/g)
Kaolinite	None	None	None	3–15	10–20
Illite	T	K^+	None	15–40	65–100
Vermiculite	T/Oc	H_2O	Moderate	100–200	600–700
Smectite	Oc/T	Cations, H_2O	High	80–150	700–800
Chlorite	—	$Mg(OH)_2$	None	10–50	75–100

T=tetrahedral layer; Oc=octahedral layer.

Henri Darcy, a 19th century French engineer, developed one of the earliest descriptions of groundwater flow. He observed a relationship between the volume of water flowing through sand and properties of the sand, and formulated the equation

$$Q/t = KA \, dH/dL \tag{10.1}$$

where Q is the volume of flow per unit time t through a column of a given cross-sectional area of flow A. The flow is under a pressure gradient dH/dL, and the change in water level over a given length is L. K is the saturated hydraulic conductivity, a proportionality constant. The difference in elevation of the water table, dH or (h_2-h_1) over the length (L) is the slope of the water table or the hydraulic gradient. Darcy's law calculates the volumetric flow rate through a unit cross section of the aquifer. The hydraulic conductivity of saturated clays is dependent upon grain size and particle sorting and is relatively stable over time. Hydraulic conductivity in unsaturated clay is influenced by grain size and sorting, and also by the water content in the pores.

EXAMPLE 10.1

Calculate the volumetric flow rate through a compacted clay liner at a landfill measuring 2.5 ha. Liner thickness is 1 m and the saturated hydraulic conductivity is 10^{-8} cm/cc. Assume 0.3 m water ponded on the liner.

SOLUTION

$$Q = KA\mathrm{d}h/\mathrm{d}L$$
$$Q = (K\,AH)/L$$

Convert all distances to meters, so 10^{-8} cm/sec $= 10^{-10}$ m/sec. Also, 2.5 ha $= 2.5 \times 10^4$ m^3

$$Q = (10^{-10}\ \mathrm{m/sec} \times 2.5 \times 10^4\ \mathrm{m}^3 \times 1.3\ \mathrm{m})/1\ \mathrm{m}$$
$$= 3.25 \times 10^{-5}\ \mathrm{m}^3/\mathrm{sec}$$
$$= 2.81\ \mathrm{m}^3/\mathrm{day}\ (86{,}400\ \mathrm{sec} = 1\ \mathrm{day})$$

10.4.7 GEOMEMBRANE LINERS

Given its possible contact with a landfill leachate that is chemically complex, a geomembrane liner must provide for substantial chemical resistance and reliable seams. The polymers most commonly used in geomembranes are HDPE, linear low-density polyethylene (LLDPE), polyvinyl chloride (PVC), flexible polypropylene (fPP), and chlorosulfonated polyethylene (CSPE) (Table 10.3 and Figure 10.4) (Qian et al., 2002). A number of factors must be considered for successful geomembrane liner design and installation, including (U.S. EPA, 1994):

- Selection of proper membrane polymer materials
- Proper subgrade preparation
- Membrane transportation, storage, and placement
- Proper installation conditions (appropriate weather, temperature)
- Seaming and testing
- Use of construction quality assurance

TABLE 10.3
Types of Geomembranes and their Approximate Formulations

Type	Resin	Plasticizer	Filler	Carbon Black	Additives or Pigment
HDPE	95–98	0	0	2–3	0.25–1.0
VLDPE	94–96	0	0	2–3	1–4
PVC	50–70	25–35	0–10	2–5	2–5
CSPE	40–60	0	40–50	5–40	5–15

Source: U.S. EPA, 1993c.

HDPE = high density polyethylene; PVC = polyvinyl chloride; CSPE = chlorosulfonated polyethylene.

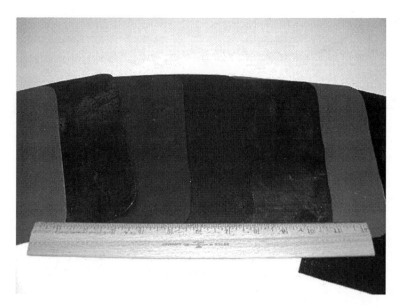

FIGURE 10.4 Geomembrane materials.

The thickness of geomembrane liners range from 30 to 120 mil. The recommended thickness for all geomembrane liners is 30 mil (0.75 mm) with the exception of HDPE, which should be set at 60 mil (1.5 mm) (Qian et al., 2002).

10.4.8 COMPATIBILITY OF LINERS WITH WASTES

The chemical compatibility of a geomembrane with waste leachate is a critical consideration regarding choice of material. Materials used in landfill construction must be expected to withstand a wide range of natural stresses for very long periods. Many materials deteriorate over time when exposed to chemicals occurring in leachate. Landfill owners and operators must anticipate the composition of leachate that a site will generate and select the appropriate liner materials. The chemical resistance of any geomembrane materials as well as LCR pipes should be thoroughly assessed prior to installation.

The EPA Method 9090A test (U.S. EPA, 1986b) is used to evaluate the chemical compatibility of synthetic materials used in liner and leachate collection and removal systems. A primary objective of chemical compatibility testing is to ensure that liner materials will remain intact during the operating lifetime of a landfill, and also through the postclosure period and beyond. EPA Method 9090A is used to predict the effects of leachate under field conditions. The test is performed by immersing a synthetic membrane in a chemical mixture for 120 days at two different temperatures, i.e., ambient and elevated. Samples are removed every 30 days and evaluated for changes in physical properties. Tests performed on geomembranes are listed in Table 10.4.

Intuitively, the results of a 120 day test under controlled conditions should be considered to be of limited predictive capability to a real-world landfill situation. Method 9090A has been verified, however, with limited field data. The U.S. EPA conducted a 5 year study of the impact of MSW on common liner materials and measured little if any deterioration within that period. In other studies, however, the results of chemical exposure on geomembranes can range from minor effects such as discoloration to more serious problems such as swelling. In extreme cases, the liner may dissolve, or tearing, cracking, or puncturing may occur. The waste may react with the liner causing degradation of the polymer or its additives or the waste may dissolve into the liner, resulting in swelling of the membrane without degrading it (Butler et al., 1995).

TABLE 10.4
Selected ASTM Tests for Geomembrane Integrity

Number	Title
D4437-99	Standard Practice for Determining the Integrity of Field Seams Used in Joining Flexible Polymeric Sheet Geomembranes
D4545-86	Standard Practice for Determining the Integrity of Factory Seams Used in Joining Manufactured Flexible Sheet Geomembranes
D4716-01	Test Method for Determining the (In-plane) Flow Rate per Unit Width and Hydraulic Transmissivity of a Geosynthetic Using a Constant Head
D4759-88	Standard Practice for Determining the Specification Conformance of Geosynthetics
D4833-00e1	Standard Test Method for Index Puncture Resistance of Geotextiles, Geomembranes, and Related Products
D4885-01	Standard Test Method for Determining Performance Strength of Geomembranes by the Wide Strip Tensile Method
D5262-97	Standard Test Method for Evaluating the Unconfined Tension Creep Behavior of Geosynthetics
D5321-92	Standard Test Method for Determining the Coefficient of Soil and Geosynthetic or Geosynthetic and Geosynthetic Friction by the Direct Shear Method
D5322-98	Standard Practice for Immersion Procedures for Evaluating the Chemical Resistance of Geosynthetics to Liquids
D5323-92	Standard Practice for Determination of 2% Secant Modulus for Polyethylene Geomembranes
D5397-99	Standard Test Method for Evaluation of Stress Crack Resistance of Polyolefin Geomembranes Using Notched Constant Tensile Load Test
D5494-93	Standard Test Method for the Determination of Pyramid Puncture Resistance of Unprotected and Protected Geomembranes
D5496-98	Standard Practice for In Field Immersion Testing of Geosynthetics
D5514-94	Standard Test Method for Large Scale Hydrostatic Puncture Testing of Geosynthetics
D5514-94	Standard Test Method for Microscopic Evaluation of the Dispersion of Carbon Black in Polyolefin Geosynthetics
D5617-99e1	Standard Test Method for Multiaxial Tension Test for Geosynthetics
D5641-94	Standard Practice for Geomembrane Seam Evaluation by Vacuum Chamber
D5721-95	Standard Practice for Air-Oven Aging of Polyolefin Geomembranes
D5747-95a	Standard Practice for Tests to Evaluate the Chemical Resistance of Geomembranes to Liquids
D5820-95	Standard Practice for Pressurized Air Channel Evaluation of Dual Seamed Geomembranes
D5884-01	Standard Test Method for Determining Tearing Strength of Internally Reinforced Geomembranes
D5886-95	Standard Guide for Selection of Test Methods to Determine Rate of Fluid Permeation Through Geomembranes for Specific Applications
D6214-98	Standard Test Method for Determining the Integrity of Field Seams Used in Joining Geomembranes by Chemical Fusion Methods
D6364-99	Standard Test Method for Determining the Short-Term Compression Behavior of Geosynthetics
D6365-99	Standard Practice for the Nondestructive Testing of Geomembrane Seams using the Spark Test
D6392-99	Standard Test Method for Determining the Integrity of Nonreinforced Geomembrane Seams Produced Using Thermo-Fusion Methods
D6434-99	Standard Guide for the Selection of Test Methods for Flexible Polypropylene (fPP) Geomembranes
D6455-99	Standard Guide for the Selection of Test Methods for Prefabricated Bituminous Geomembranes (PBGM)
D6495-02	Standard Guide for Acceptance Testing Requirements for Geosynthetic Clay Liners
D6496-99	Standard Test Method for Determining Average Bonding Peel Strength Between the Top and Bottom Layers of Needle-Punched Geosynthetic Clay Liners
D6497-02	Standard Guide for Mechanical Attachment of Geomembrane to Penetrations or Structures
D6574-00	Test Method for Determining the (In-Plane) Hydraulic Transmissivity of a Geosynthetic by Radial Flow
D6636-01	Standard Test Method for Determination of Ply Adhesion Strength of Reinforced Geomembranes
D6693-01	Standard Test Method for Determining Tensile Properties of Nonreinforced Polyethylene and Nonreinforced Flexible Polypropylene Geomembranes
D6706-01	Standard Test Method for Measuring Geosynthetic Pullout Resistance in Soil

10.4.9 SURVIVABILITY TESTS

Several tests are available to determine the survivability of unexposed polymeric liners. For example, puncture tests are useful to estimate the survivability of geomembranes in the field. During a puncture test, a 5/16 in. steel rod with rounded edges is pushed down through an anchored membrane (Figure 10.5). A very flexible membrane that is shown to have high strain capacity under tension may have great survivability in the field. High-density polyethylenes provide a high penetration force but experience high brittle failure. Thus, puncture data may not adequately predict field survivability (U.S. EPA, 1989).

10.4.10 PERMEABILITY

Even if a liner is installed correctly, i.e., without punctures and defects, liquid will inevitably diffuse through. However, such rates are extremely low. U.S. EPA data (1988) for water vapor transmission across various geomembranes is given in Table 10.5. Permeability of a geomembrane is evaluated using ASTM E96, the Water Vapor Transmission test (ASTM, 2000). A sample of the membrane is attached to the top of a small aluminum cup containing a known volume of water. The cup is then placed in a chamber of controlled humidity and temperature. The chamber is typically set to 20% relative humidity while the humidity in the cup is 100%; thus, a concentration gradient is set up across the membrane. Moisture diffuses through the membrane and with time the liquid level in the cup will fall. From these measurements, the rate at which moisture is moving though the membrane is measured and the permeability of the membrane is calculated with a simple diffusion equation (Fick's first law)

$$J = -D(dC/dx) \tag{10.2}$$

where J is the flux (mol/cm^2 sec), D the diffusion coefficient (cm^2/sec), C the concentration (mol/cm^3), and x the length in the direction of movement (cm).

FIGURE 10.5 Puncture apparatus for geomembrane testing.

TABLE 10.5
Water Vapor Transmission for Different Geomembranes

Geomembrane	Thickness		Vapor Transmission Rate	
	mm	mil	g/m²/day	gal/acre/day
PVC	0.75	30	1.9	2.03
CPE	1.0	40	0.4	0.43
CSPE	1.0	40	0.4	0.43
HDPE	0.75	30	0.02	0.021
HDPE	2.45	98	0.006	0.0064

Source: U.S. EPA, 1988.

CPE=chlorinated polyethylene

It follows that Fick's first law controls leakage through a synthetic liner. The diffusion process is similar to the rate of flow governed by Darcy's law, except that the former is driven by concentration gradients as opposed to hydraulic head. Diffusion rates in membranes are very low in comparison with hydraulic flow rates, even in clays. Of course, if a synthetic liner is installed with incomplete seams or small holes, the amount of leachate that leaks through will increase substantially.

Data revealing problems with synthetic membranes have been documented in the water industry, where contamination of drinking water due to permeation of trace organic contaminants from soil through plastic pipes has occurred. Laboratory studies have also demonstrated the transport of solvents through membranes. Haxo and Lahey (1988) demonstrated the transport of trichloroethylene and toluene through a membrane. Park and Nibras (1993) measured diffusion parameters for a range of volatile organic compounds (VOCs) in HDPE liner materials, and demonstrated that this might be a significant source of releases from lined landfills. Diffusive mass transport could, in theory, have a significant environmental impact by allowing the release of organic solvents though intact membranes at rates comparable with those of leakage through a defect (Butler et al., 1995). Also of possible concern is solvent gas transmission through the membrane. Very light gases, such as methane (CH_4), will rise from the waste cell and contact the membrane. Methane gas transmission rates for several geomembranes are shown in Table 10.6.

Factors in a landfill cell that might affect leachate diffusion rate through an intact liner include temperature, pressure, and elongation due to tensile stress. As the temperature increases, diffusion will increase due to greater thermal motion in the polymer chains, thus producing more voids through which leachate can escape (Butler et al., 1995).

Of more practical importance, however, is the occurrence of holes in the liner caused by improper placement and positioning over sharp stones. Giroud and Bonaparte (1989) state that with good quality control, 2.5 holes/ha of geomembrane (1 hole/acre) is a fairly typical occurrence. With poor quality control, we can expect 75 holes/ha (30 holes/acre). Most defects tend to be small (<0.1 cm²), but larger holes do occasionally occur (Qian et al., 2002). Table 10.7 provides data for estimated losses from geomembranes having holes.

The Bernoulii equation can be used to estimate the flow rates through holes in geomembranes, assuming that the size and shape of the holes are known:

$$Q = C_b \, a \, (2gh)^{0.5} \tag{10.3}$$

where Q is the flow rate through a geomembrane (cm³/sec), C_b the flow coefficient with a value of about 0.6 for a circular hole, a the area of circular hole (cm²), g acceleration due to gravity (981 cm/sec), and h the liquid head acting on the liner (cm).

TABLE 10.6
Methane Transmission for Different Geomembranes

Geomembrane	Thickness		Methane Transmission Rate (ml/m² day atm)
	mm	mil	
PVC	0.25	10	4.4
PVC	0.5	20	3.3
LLDPE	0.45	18	2.3
CSPE	0.8	32	0.27
CSPE	0.85	34	1.6
HDPE	0.6	24	1.3
HDPE	0.85	34	1.4

Source: U.S EPA, 1988.

LLDPE = Linear low- density polyethylene

TABLE 10.7
Calculated Flow Rates through a Geomembrane with a Liquid Head of 0.3 m (1 ft)

Size of Hole (cm²)	Number of Holes		Flow Rate	
	hole/ha	hole/acre	L/m²/day	gal/acre/day
No holes	0	0	9.4×10^{-6}	0.01
0.1	2.5	1	0.31	330
0.1	75	30	9.4	10,000
1	2.5	1	3.1	3,300
1	75	30	94	100,000
10	2.5	1	31	33,000

Source: U.S. EPA, 1991.

The above equation applies to a geomembrane that has one or more holes that are widely spaced, such that leakage through each hole acts independent of the other holes, that the leachate head *h* is constant, and that the soil that underlies the geomembrane has a relatively large hydraulic conductivity (Qian et al., 2002).

10.4.11 STRESS

Stress considerations are especially critical for the design of side slopes and base of a landfill. For side slopes, the weight of the membrane itself and waste settlement can place severe tensile strains on the geomembrane. The primary geomembrane must be able to support its own weight on the side slopes. In order to calculate self-weight, the specific gravity, friction angle, thickness, and yield stress of the geomembrane must be known. Waste settlement is an additional stress consideration. For the bottom of the facility, localized settlement must be considered in the design. As waste settles in the landfill a downward force acts upon the primary geomembrane. A low friction component between the geomembrane and underlying material prevents the force from being transferred to the underlying material, thus limiting tension on the primary geomembrane (U.S. EPA, 1989).

10.4.12 GEOMEMBRANE LINER HANDLING AND PLACEMENT

The surface of the compacted soil liner must be smooth and sufficiently strong to provide continuous support for the geomembrane liner. The soil surface must be relatively free of rocks, roots, and excess water. EPA studies (U.S. EPA, 1988) show that nonangular stones present at the surface and smaller than ¾ in. in diameter will not damage most geomembrane liners.

Geomembrane liners composed of PVC are commonly prefabricated into large panels, folded, and shipped on pallets. The liners manufactured from HDPE and PP must not be folded and are shipped to the site in rolls. Once delivered to the site, liners should be stored to avoid direct contact with the ground surface. A protective surface such as a geotextile (see below) may be placed over the ground, or the geomembrane liner rolls could be wrapped in plastic at the factory. The stored geomembrane liner should also be protected from exposure to excessive heat, dust, and water (U.S. EPA, 1994).

At the time of installation, the geomembrane liners are rolled out or spread out over the soil liner with each sheet overlapping the adjacent sheets. The geomembrane liners are then seamed together to create a single impermeable layer. A number of methods are available to create strong seams including extrusion, fusion, chemical, and adhesive seams (Figure 10.6). Thermal seaming is the most common method of attaching the sheets. It requires both proper weather conditions and a clean surface on both membrane surfaces. If the surface of a membrane is wet, water can vaporize and produce bubbles within the seam that reduces seam strength and may ultimately result in leakage. Ambient temperature must also be considered during installation. Thermal seaming should be performed when the ambient temperature is between 4.4 and 40°C (40 and 104° F). Another practical concern in geomembrane liner seaming is the presence of dust; therefore, dust control during the seaming process is critical (U.S. EPA, 1994). As geomembrane liner seaming is a critical aspect in maintaining membrane integrity, a seam testing program should be established for quality control.

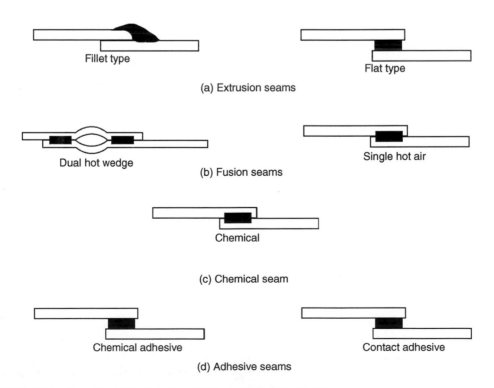

FIGURE 10.6   Seaming methods for landfill liners (U.S. EPA, 1994).

10.4.13 CONSTRUCTION QUALITY ASSURANCE

To minimize holes in a geomembrane liner (whether caused by product defects, transportation, installation, or seaming) and to meet required standards, a construction quality assurance (CQA) program should be established for liner installation (U.S. EPA, 1994, 1992a). The CQA program is a planned system of activities performed by landfill owners to ensure that the cells and associated facilities are constructed as specified in the design. The program should be developed during the design stage, and the state regulatory agency should review a facility CQA program before a permit is issued for construction.

10.4.14 DAILY OPERATIONS AND ISSUES

10.4.14.1 Filling Sequences

After weighing the truck at the weigh station, the waste is brought to the so-called 'working face' of the landfill. Waste is deposited into daily cells, i.e., small units of land that are filled with MSW and then covered at the end of the day by a layer of soil or similar material (e.g., compost and shredded tires). At the base of a new landfill cell, waste must be placed in order to prevent compactor wheels from contacting the leachate collection systems, liner, and other sensitive layers installed below. Filling continues with the placement of successive lifts, typically starting in a corner and moving outward. The filling sequence is established at the time of landfill design and permitting. The working face must be large enough to accommodate several vehicles unloading simultaneously, approximately 4 to 6 m (12 to 20 ft) per vehicle (Vesilind et al., 2002).

The maximum working area for a landfill can be calculated by (Kiely, 1997)

$$A_{max} = (0.1 \ W)/R \tag{10.4}$$

where A_{max} is the maximum working area, R the average annual rainfall (m), and W the average annual waste input (metric tons).

The above equation assumes an absorption capacity of the waste of 0.1 m^3/metric ton.

EXAMPLE 10.2

Compute the maximum working area of a landfill cell if W is 12,000 MT/year and R is 1.1 m/year (approx. 43 in./year). Comment on the result.

SOLUTION

$$A_{max} = (0.1 \times 12,000)/1.1 = 1090 \ m^2$$

If the cell measures 75 m in length, then the average width of the working area would be 14.5 m.

As waste is placed in the landfill, heavy equipment drives over and compacts the waste (Figure 10.7). The degree of compaction is related to several factors including thickness of the waste layer (see Figure 10.8), the number of passes made over the waste (see Figure 10.9), slope (flatter slopes and steeper slopes compact better by landfill compactors and track-type tractors, respectively and moisture content (wetter waste compacts more effectively than dry waste) (Vesilind et al., 2002).

EXAMPLE 10.3

Determine the area required for a new sanitary landfill with a projected lifetime of 25 years. The landfill will serve a population of 250,000 persons, generating 28 kg (62 lb) per household per week. Waste density in the landfill averages 550 kg/m^3. Landfill height is not to exceed 25 meters. Assume four persons per household.

FIGURE 10.7 Compactor vehicle at a sanitary landfill.

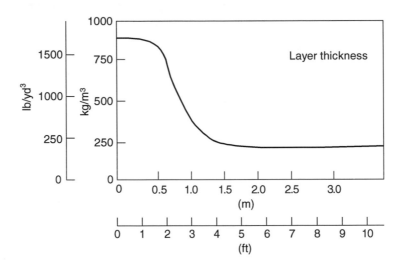

FIGURE 10.8 Density of MSW in a landfill cell as related to thickness of the waste layer. (From Vesilind, et al., *Solid Waste Engineering*, 1st ed., © 2002. Reproduced with kind permission of Brooks/Cole, a division of Thomson Learning:www.thomsonrights.com.)

SOLUTION

MSW generated $= 250,000/4 \times 28$ g/$10^3 = 1750$ MT/week
$$= 91,000 \text{ MT/year}$$

Volume of landfill space needed $= (91,000$ MT/year $\times 10^3$ kg/MT$)/550$ kg/m^3
$$= 165 \times 10^3 \text{ m}^3/\text{year}$$

For a maximum height of 25 m, the required land area
$$= (165 \times 10^3 \text{ m}^3/\text{year})/25 \text{ m} = 6618 \text{ m}^2$$
$$= 0.7 \text{ ha or } 1.63 \text{ acres}$$

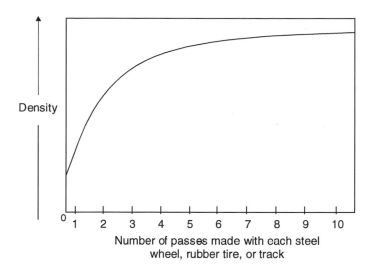

FIGURE 10.9 Density of MSW in a landfill cell as related to the number of passes by machinery over the waste. (From Vesilind et al., *Solid Waste Engineering,* 1st ed., © 2002. Reproduced with kind permission of Brooks/Cole, a division of Thomson Learning:www.thomsonrights.com.)

The above value, however, should be increased by about 50% to allow for use of daily cover, final cover, receiving areas, roads, fencing, and other structures.

Required land area for 25 years = 0.7 ha × 25 years × 1.5

= 26.25 ha or 64.8 acres

In the above example, what is the daily waste generation rate per capita? How does this figure compare with the estimated national average (Chapter 1)?

10.4.14.2 Cover Materials

At the end of each operating day, active landfill cells must be covered with at least 6 in. of soil or similar material to control disease vectors, fires, odors, blowing litter, and scavenging by animals (40 CFR Part 258.21) (Figure 10.10). More frequent application of soil may be required if a specific nuisance or hazard condition occurs at the facility, for example, strongly malodorous or gas-producing wastes.

In some instances local soil may not be readily available as a cover material; therefore, alternative materials may be required. The alternative material must be shown to control all relevant hazards and nuisances (disease vectors, fires, and odors). Alternative cover materials may be natural or commercially produced and must not pose a threat to human health and the environment. Some useful materials are those that may be considered waste; therefore, using these materials is an efficient use of landfill space. Examples of alternative materials include (U.S. EPA, 1992b):

- Fly ash and bottom ash from utilities and municipal waste incinerators
- Composted MSW or sewage sludge
- Foundry sands
- Yard waste (lawn clippings, leaves, and tree branches)
- Construction and demolition debris
- Shredded automobile tires
- Some commercially available cover materials include:
  - Foam that is sprayed onto the working face
  - Slurry products (e.g., fibers from recycled newspaper and wood chip slurry) (U.S. EPA, 1992b).

Some of the commercial alternatives may require specially designed application equipment.

FIGURE 10.10   Application of daily cover prevents pest infestations and controls odors.

10.4.14.3   Disease Vector Control

Vectors include rodents, flies, mosquitoes, or other organisms that transmit diseases to humans. Putrescible waste attracts vectors, acting as a food source. Application of cover at the end of the operating day is typically sufficient to control disease vectors; however, other practices may be necessary (40 CFR Part 258.22). These include (U.S. EPA, 1993c):

- Reducing the area of the working face
- Increasing the thickness of the daily cover
- Changing cover type, for example, to a material less permeable to air and water
- Application of repellents, insecticides, and rodenticides
- Composting of organic wastes prior to disposal
- Use of predators for control of insect, bird, and animal populations

Standing water serves as a potential breeding ground for mosquitoes. Water collects in depressions, open containers, leachate storage ponds, and siltation basins. To control mosquitoes, standing water should be removed and an insecticide possibly applied (U.S. EPA, 1993c). Table 10.8 lists insecticides used at sanitary landfills. In order to control rodent populations, various birds of prey, for example, hawks, falcons, and owls can be introduced.

10.4.14.4   Biological Control of Pests

Insecticides may serve as an effective means of combating insect pests at landfills; however, genetic resistance to applied chemicals is always a consideration. Biological control methods may be a viable alternative to chemical control at landfills. A number of workers have surveyed the parasitoids and predators active in municipal wastes and animal manures. Rueda et al. (1997) recorded five species of wasp parasitoids in the pupae of house flies and two from blow flies. Sulaiman et al. (1990) found nine species of pupal parasitoids of flies breeding in municipal wastes and on poultry farms. Hoyer (1986) determined 22 species of fly parasitoids. In Washington State, biological control was used in preference to chemical control of flies associated with manure pits (Guhlke, 1985). Flies had developed resistance to insecticides. Parasitic chalcid wasps had the

TABLE 10.8
Insecticides Approved for Use at UK MSWLFs, their Chemical Grouping and Target Pest

Pest	Active Ingredient	Approved Insecticide	Chemical Group
Ants, beetles, bugs, cockroaches, crickets, earwigs, fleas, mites, moths, silverfish	Fenitrothion	Antec Kurakil EC-Kill	Organophosphorus
Flies	Chloropyrifos-methyl	Smite	Organophosphorus
	Diflubenzuron	Dimilin Flo	Insect growth regulator
	Fenitrothion	Antec Durakil EC-Kill	Organophosphorus
	Pyrethrin	Multispray Pybuthrin[33]	Pyrenthroid
	Trichlorfon	Dipterex 80[a]	Organophosphorus
	Bioresmethrin	Blade	Pyrethroid
General insect control	Alphacypermethrin	Fendona ASC	Pyrethroid
	Bendiocarb	Various products	Carbamate
	Bioalletrin+bioresmethrin	Pybuthrin 33 BB	Pyrethroid
	Bioresmethrin	Biosol RTU Safe kill RTU	Pyrethroid
	Chlorpyrifos-methyl	Smite	Organophosphorus
	Chlorpyrifos-methyl+ Permethrin+pyrethrins	Multispray Pyrethroid	Organophosphorus+
	Cypermethrin	Various products	Pyrethroid
	Deltamethrin+ S-bioallethrin	Crackdown Rapide	Pyrethroid
	Diazinon	Knox Out 2 FM	Organophosphorus
	Fenitrothion	Various products	Organophosphorus
	Fenitrothion+resmethrin+ Permethrin	Turbair beetle killer	Organophosphorus+ Pyrethroid
	Fenitrothion+tetramethrin	Killgerm	Organophosphorus+ Pyrethroid concentrate
	Permethrin	Various products	Pyrethroid
	Permethrin+bioallethrin	Various products	Pyrethroid
	Permethrin+S-bioallethrin	Various products	Pyrethroid
	Pirimiphos-methyl	Actellic 25 EC Actellic Dust	Organophosphorus
	Propoxur	Killgerm Propoxur 20 EC	Organophosphorus
	Pyrethrins	Various products	Pyrethroid
	Tetramethrin	Killgerm Py-kill W	Pyrethroid
	Tetramethrin+d-phenothrin	Various products	Pyrethroid
	Tetramethrin+resmethrin	Swat A	Pyrethroid

[a] Off-label approval — although approved, off-label uses are not endorsed by the manufacturers and such treatments are made entirely at the risk of the user.

Reprinted with kind permission of the British Crop Protection Council, 1999.

advantage of providing continuous control of house fly populations. Costs for biological control were a fraction of what they would have been for chemical control. In addition, the reduced use of insecticides allowed populations of other naturally occurring parasites and predators to build up (Ellis and Blood-Smyth, no date).

10.4.14.5   Generation of Landfill Gases

Gases occurring in active and closed landfills include methane (CH_4), carbon dioxide (CO_2), carbon monoxide (CO), hydrogen (H_2), hydrogen sulfide (H_2S), ammonia (NH_3), nitrogen (N_2), and oxygen (O_2). The most common gases produced within landfills are CH_4 and CO_2, with lesser amounts of the other gases occurring, typically in concentrations below 1% (v/v) (Table 10.9). Data on molecular weight and density of specific gaseous components are presented in Table 10.10. Virtually all landfill gases are produced from the microbial decomposition of solid waste. Figure 10.11 shows the major phases of MSW decomposition in a landfill cell and the resultant gaseous products. Table 10.11 lists many biodegradable organic costituents of MSW that are related to gas production.

TABLE 10.9
Common Landfill Gases and their Concentrations

Component	Percent (dry volume basis)
Methane	45–60
Carbon dioxide	40–60
Nitrogen	2–5
Oxygen	0.1–1.0
Sulfides, disulfides, and mercaptans	0–1.0
Ammonia	0.1–1.0
Hydrogen	0–0.2
Carbon monoxide	0–0.2
Trace constituents	0.001–0.6

Characteristic	Value
Temperature, °C (°F)	38–50 (100–120)
Specific gravity	1.02–1.06
Moisture content	Saturated
High heating value, kJ/m^3 (Btu/ft^3)	900–1100 (400–500)

Source: Tchobanoglous, T. et al., *Integrated Solid Waste Management: Engineering Principles and Management Issues*, McGraw-Hill, New York, 1993. Data reproduced with kind permission of the McGraw-Hill Companies, Inc.

TABLE 10.10
Selected Chemical and Physical Properties of Gases Found in a Sanitary Landfill

Gas	Formula	Molecular Weight	Density (g/L)
Air		28.97	1.2928
Ammonia	NH_3	17.03	0.7708
Carbon dioxide	CO_2	44.00	1.9768
Hydrogen	H_2	2.016	0.0898
Hydrogen sulfide	H_2S	34.08	1.5392
Methane	CH_4	16.03	0.7167
Nitrogen	N_2	28.02	1.2507
Oxygen	O_2	32.00	1.4289

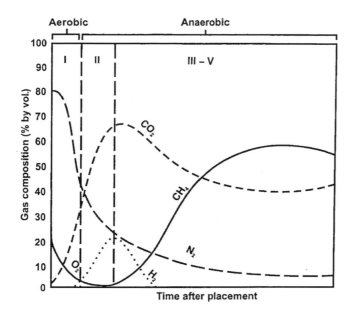

FIGURE 10.11 The stages of decomposition in a sanitary landfill (U.S. EPA, 1994).

TABLE 10.11
Rapidly and Slowly Biodegradable Organic Constituents in MSW

Organic Waste Component	Rapidly Biodegradable	Slowly Biodegradable
Food wastes	x	
Newspaper	x	
Office paper	x	
Cardboard	x	
Yard wastes	x	x[a]
Textiles		x
Rubber		x
Leather		x
Wood		x

[a] Branches, twigs, and other woody portions of yard wastes.

Phase I: The aerobic phase. During Phase I, the biodegradable components of MSW undergo microbial decomposition immediately after placement in a landfill cell. Initially, oxygen occurs in sufficient quantities in the interstices (voids) to allow for aerobic decomposition of the organic waste fraction. The sources of the heterotrophic microbial populations responsible for waste decomposition, both aerobic and anaerobic, are the waste itself and the soil material used as a daily cover. Wastewater treatment plant sludge, disposed of in many MSWLFs, and recycled leachate are other sources of organisms.

The reaction for the aerobic oxidation of a glucose molecule was shown in Equation 8.1. Common products of Phase I include CO_2, H_2O, NO_3^-, and other oxygenated compounds. Oxygen is rapidly depleted in the covered landfill cell by the action of the heterotrophic aerobic microorganisms. Diffusion of oxygen into the void spaces is negligible; once the O_2 level drops below 10 to 15% (v/v), anaerobic microorganisms are activated (Phase II).

Phase II: First anaerobic phase. By the onset of Phase II, anaerobic conditions have already begun to develop. Nitrate and sulfate ions serve as electron acceptors for anaerobic heterotrophs and are reduced to N_2 and H_2S (see Equations 10.5 and 10.6). The extent of anaerobic conditions can be monitored by measuring the oxidation and reduction potential of the waste. Reducing conditions sufficient to support the reduction of nitrate and sulfate occur at about -50 to -100 mV.

$$\frac{1}{5}NO_3^- + \frac{1}{4}\{CH_2O\} + \frac{1}{5}H^+ \rightarrow \frac{1}{10}N_2 + \frac{1}{4}CO_2 + \frac{7}{20}H_2O \tag{10.5}$$

$$SO_4^{2-} + 2\{CH_2O\} + 2H^+ \rightarrow H_2S + 2CO_2 + 2H_2O \tag{10.6}$$

In Phase II, the pH of landfill liquids decreases due to the formation of organic acids and the effect of the elevated concentrations of CO_2 within the voids, which may partly dissolve and form carbonic acid, H_2CO_3 (Tchobanoglous et al., 1993).

$$CO_2 + H_2O \rightarrow H_2CO_3 \tag{10.7}$$

Phase III: Second anaerobic phase. In Phase III, also known as the *acid phase*, anaerobic microbial activity is accelerated with the concomitant production of copious amounts of organic acids and modest amounts of H_2 gas. The process is the result of enzyme-mediated hydrolysis of high-molecular-weight compounds such as lipids, polysaccharides, and proteins into smaller compounds. Microbial populations subsequently convert these compounds into a number of organic acids such as acetic acid (CH_3COOH), butyric acid ($CH_3CH_2CH_2COOH$), and lactic acid ($CH_3CH(OH)COOH$), and small concentrations of fulvic acid and other complex organic acids. Carbon dioxide is the principal gas generated during Phase III.

The pH of landfill liquids drops to ~5 due to the presence of the organic acids and the relatively high concentrations of CO_2 within the void spaces. There is no methane production during this period, as methanogenic (methane-producing) bacteria cannot tolerate acidic conditions. The biochemical oxygen demand (BOD_5), the chemical oxygen demand (COD), and the conductivity of the leachate increase significantly during Phase III due to the dissolution of the organic acids in the leachate. Also, because of the low pH values, metals and other inorganic constituents are solubilized during this phase.

Phase IV: Methane fermentation. In Phase IV, a second consortium of anaerobic microorganisms becomes prominent, which converts the organic acids and H_2 gas formed previously into CH_4 and CO_2. The microorganisms responsible for this conversion are strict anaerobes and are labeled methanogenic. In Phase IV, both methane and some acid formation proceed simultaneously. Many of the acids have already decomposed, however, so the pH rises and stabilizes at about 6.8 to 8 (Tchobanoglous et al., 1993). Consequently, metals which were previously soluble now precipitate. The concentration of BOD_5, the COD, and the conductivity also decline.

Phase V: Maturation. Phase V occurs after the readily available biodegradable organic material has been converted into CH_4 and CO_2 (Phase IV). The rate of gas generation declines significantly because most of the available nutrients have been removed with the leachate during the previous phases and the substrates that remain in the landfill are only slowly biodegradable. The principal landfill gases evolved are CH_4 and CO_2. During the maturation phase, landfill liquids often contain humic and fulvic acids, which are complex and highly stable compounds.

The duration of each of the phases outlined above vary as a function of the distribution of the organic components in the landfill cell, the availability of nutrients, the moisture content of the waste, and the degree of initial compaction. For example, the generation of landfill gas will be limited if sufficient moisture is not available. Increasing the density of the material in the landfill may prevent adequate water movement to all parts of the waste thereby reducing the rate of biological reactions and subsequent gas production. Data on the distribution of gases occurring in a newly closed landfill cell as a function of time are shown in Table 10.12.

The volume of the gases released during anaerobic decomposition can be estimated in a number of ways. For example, if the individual organic constituents found in MSW (excepting plastics) are represented with a generalized formula of the form $C_aH_bO_cN_d$, then the total volume of gas can

---

TABLE 10.12
Landfill Gas Composition Over the First 48 months after the Closure of a Landfill Cell

Time since Closure (months)	Average (% by vol.)		
	N_2	CO_2	CH_4
0–3	5.2	88	5
3–6	3.8	76	21
6–12	0.4	65	29
12–18	1.1	52	40
18–24	0.4	53	47
24–30	0.2	52	48
30–36	1.3	46	51
36–42	0.9	50	47
42–48	0.4	51	48

Source: Tchobanoglous, T. et al., *Integrated Solid Waste Management: Engineering Principles and Management Issues*. McGraw-Hill, New York, 1993. Data reproduced with kind permission of the McGraw-Hill Companies, Inc.

---

be estimated using Equation 10.8, assuming the complete conversion of the biodegradable organic waste into CO_2 and CH_4:

$$C_aH_bO_cN_d + (4a - b - 2c + 3d)/4H_2O \rightarrow (4a + b - 2c - 3d)/8CH_4$$
$$+ (4a - b + 2c + 3d)/8CO_2 + dNH_3 \quad (10.8)$$

The organic component of MSW can be divided into two categories: (1) those materials that will decompose rapidly (3 months to 5 years) and (2) those materials that will decompose slowly (up to 50 years or more) (Vesilind et al., 2002). The rapidly and slowly decomposable components of the organic fraction of MSW are listed in Table 10.11.

10.4.14.6 Predicting Gas Production

It is essential for landfill operators to estimate the volume of gas produced from an operating or closed landfill. Similarly, the composition of the gas (e.g., methane, moisture, and sulfur content) is important to energy users. Engineers use mathematical models to predict landfill gas generation. Models are designed based on population data, per capita waste generation, waste composition, waste moisture content, and expected gas yield per unit dry weight of waste. Mathematical models are also used to model gas recovery systems including layout, equipment type, operation parameters, and failure simulation (Vesilind et al., 2002). The following parameters must be known if gas production is to be accurately estimated: gas yield per unit weight of waste, lag time prior to gas production, shape of the gas production curve, and duration of gas production.

In theory the biological decomposition of 1 ton of MSW produces 442 m^3 (15,600 ft^3) of landfill gas containing 55% CH_4 and a heat value of 19,730 kJ/m^3 (530 Btu/ft^3). Only a portion of the waste converts into CH_4 due to the presence of inaccessible waste and non-biodegradable fractions; therefore, the actual average methane yield is closer to 100 m^3/MT (3,900 ft^3/ton) of MSW. Significant variation in gas production data has been noted at landfills across the United States due to differences in climate, waste types, and landfill management. The methane generation usually is between 0.06 and 0.12 m^3/kg (1 to 2 ft^3/lb) of waste on a dry basis over 10 to 40 years. Gas yields based on waste generation have been predicted using assumptions such as:

- 50% of the organic material placed in the landfill will actually decompose
- 50% of the landfill gas generated is recoverable
- 50% of landfills are operating within a favorable pH range

Once the expected yield is determined, a model is utilized to show the pattern of gas production over time. The U.S. EPA has published a model called The Landfill Gas Emissions Model (LandGEM) based on the following equation (U.S. EPA, 1998):

$$Q_T = \sum_{j=1}^{n} 2kL_0 M_i e^{-kt_i} \tag{10.9}$$

where Q_T is the total gas emission rate from a landfill (volume and time), n the total time period of waste placement, k the landfill gas emission constant ($time^{-1}$), L_0 the methane generation potential (volume/mass of waste), t_i the age of the ith section of waste (time) and M_i the mass of wet waste, placed at time i.

In this model, the gas generation rate is based on a first-order decomposition model, which uses two parameters: L_0, the potential methane generation capacity of the waste and k, the methane generation decay rate. The methane generation rate is assumed to be maximal upon MSW placement in the landfill. This model allows the user to enter L_0 and k values using test data and landfill-specific parameters or use default L_0 and k values derived from research data (U.S. EPA, 1998).

The amount of MSW in the landfill is calculated for this model using site-specific characteristics entered by the user, such as the years the landfill has been in operation, the amount of MSW in place in the landfill, and landfill capacity. Emission rates are estimated for CH_4, CO_2, nonmethane organic compounds (NMOCs), and air pollutants expected to be emitted from landfills based on test data compiled in the U.S. EPA compilation of air pollutant emission factors, AP-42 (U.S. EPA, 1997a).

EXAMPLE 10.4

A landfill cell receives about 225,000 metric tons of MSW per year. Calculate the gas production for the first year, given a landfill gas emission constant of 0.0335 $year^{-1}$ and a methane generation potential of 175 m^3/metric ton.

SOLUTION

For the first year,

$$Q_T = 2\,(0.0335)\,(175)\,(225,000)\,(e^{-0.0335(1)}) = 2,551,067 \ m^3$$

Note. In the second year, this same cell will produce less total gas; however, the new layer for the second year will produce gas, and the yields of the two cells will be combined to calculate the total gas production for the second year, and so on.

The lag period prior to CH_4 generation may range from a few weeks to a few years, depending on landfill conditions. The duration of gas production is also influenced by environmental conditions within the landfill.

An example utilizing the full LandGEM model appears in the Appendix to this chapter.

EXAMPLE 10.5

Using data from the table below, estimate the chemical composition and the amount of gas that can be derived from the organic constituents in MSW. Determine the chemical composition and the amount of gas that can be derived from the rapidly and slowly decomposable organic constituents in MSW. Note that some of the yard wastes will decompose rapidly (e.g., grass clippings) while others will be more stable (branches).

SOLUTION

1. Determine the distribution of the major elements within the waste.

Component	Wet Weight, (kg)	Dry Weight, (kg)	Compostion, (kg)					
			C	H	O	N	S	Ash
Rapidly Decomposable Constituents								
Food wastes	4.2	1.9	0.59	0.08	0.47	0.04	0.00	0.73
Paper	19.0	17.0	6.26	0.86	6.30	0.05	0.04	3.49
Cardboard	2.5	2.2	0.89	0.08	1.02	0.01	0.00	0.20
Yard wastes	6.1	2.5	0.85	0.12	0.75	0.05	0.00	0.72
Total	31.8	23.6	8.58	1.14	8.54	0.15	0.05	5.14
Slowly Decomposable Constituents								
Textiles	1.0	0.8	0.45	0.05	0.25	0.04	0.0	0.01
Rubber	0.2	0.2	0.17	0.02	0.00	0.00	0.0	0.00
Leather	0.22	0.21	0.11	0.01	0.02	0.02	0.0	0.05
Yard wastes	3.5	1.9	0.63	0.08	0.51	0.04	0.0	0.63
Wood	0.9	0.7	0.36	0.05	0.29	0.00	0.0	0.01
Total	5.82	3.81	1.71	0.22	1.08	0.10	0.0	0.7

2. Compute the molar composition of the elements neglecting the ash.

	C	H	O	N	S
lb/mol	12.01	1.01	16.0	14.01	32.06
Total moles					
Rapidly decomposable	0.7144	1.1262	0.5337	0.0109	0.0015
Slowly decomposable	0.1423	0.2139	0.0673	0.0071	0.0001

Calculate an approximate chemical formula, excluding sulfur. Determine mole ratios of all components.

Component	Mole. Ratio (Nitrogen=1)	
	Rapidly Decomposable	Slowly Decomposable
Carbon	65.4	20.0
Hydrogen	103.1	30.1
Oxygen	48.9	9.5
Nitrogen	1.0	1.0

The chemical formulas for the waste mixtures (excluding sulfur) are:

Rapidly decomposable

$$C_{65}H_{103}O_{49}N + 16H_2O \rightarrow 57\ CH_4 + 24\ CO_2 + NH_3$$

g/mol　　　　1684.1　　　288　　　912　　　1056　　17

Slowly decomposable

$$C_{20}H_{30}O_{10}N + 9H_2O \rightarrow 11CH_4 + 9CO_2 + NH_3$$

g/mol   436   162   176   396   17

Note. The equations do not balance exactly due to rounding.

Calculate the volume of CH_4 and CO_2 that can be generated from the different waste fractions.

Rapidly decomposable

$CH_4 = [(912)(23.6 \text{ kg})]/[(1684.1)(0.718 \text{ kg/m}^3)] = 17.8 \text{ m}^3$ at STP

$CO_2 = [(1056)(23.6 \text{ kg})] / [(1684.1)(1.98 \text{ kg/m}^3)] = 7.5 \text{ m}^3$ at STP

(density of $CH_4 = 0.718 \text{ kg/m}^3$; density of $CO_2 = 1.98 \text{ kg/m}^3$)

Slowly decomposable

$CH_4 = [(176)(3.81 \text{ kg})] / [(436.4)(0.718 \text{ kg/m}^3)] = 2.14 \text{ m}^3$

$CO_2 = [(396)(3.81 \text{ kg})] / [(436.4)(1.98 \text{ kg/m}^3)] = 1.75 \text{ m}^3$

Determine the total theoretical quantity of gas generated per unit weight of organic matter.

Rapidly decomposable

volume/kg $= (17.83 \text{ m}^3 + 7.5 \text{ m}^3)/23.6 \text{ kg} = 1.07 \text{ kg/m}^3$

Slowly decomposable

volume/kg$=(2.14 \text{ m}^3 + 1.75 \text{ m}^3)/3.81 \text{ kg} = 1.02 \text{ kg/m}^3$

(Adapted from Tchobanoglous et al., 1993)

10.4.14.7 Control of Explosive Gases

Landfill gas emissions contribute to local smog and cause unpleasant odors and trigger complaints from neighbors. Methane is the primary concern in evaluating landfill gas generation because it is highly combustible. Methane accumulation in structures near a landfill may result in fire and explosions. Methane hazards can be prevented through monitoring of landfill gas and corrective action.

Although methane is lighter than air and carbon dioxide is heavier, these gases tend to remain mixed. They migrate as a function of the density of the mixture and other gradients such as temperature and partial pressure (U.S. EPA, 1994; Tchobanoglous et al., 1993). In an ideal situation, landfill gas would simply diffuse to the surface of the unit and disperse into the atmosphere. Unfortunately, there are a number of circumstances that will force landfill gas to migrate laterally rather than vertically. Landfill gas will migrate through the path of least resistance. The direction of migration is controlled in part by the permeability of the soil and fill material. This is especially relevant in pre-RCRA landfills, which may lack a complete subsurface liner. Coarse, porous soils such as sand and gravel adjacent to the landfill will promote greater lateral transport of gases than would fine-grained soils. If the unit is closed, landfill gas will migrate laterally if the final cover is dense or impermeable and if the side slopes do not contain a gas barrier. Additionally, with an increase in soil moisture content near the surface, upward landfill gas flow is inhibited. A cell with a frozen surface will similarly promote lateral migration. The effects of geology and surface conditions on gas migration are shown in Figure 10.12. Lateral gas migration is more common in older facilities that lack both liners and gas control systems.

In order to ensure safety to humans and structures, landfill gas must be regularly monitored. Methane concentrations must not exceed 25% of the lower explosive limit (LEL) in facility structures, and must not exceed the LEL at the perimeter of the facility property. The LEL is defined as the lowest percent by volume of a mixture of explosive gases in air that will propagate a flame at 25°C and atmospheric pressure (see Figure 10.13). Methane is explosive when present in the range of 5 to 15% (by vol.) in air. At methane concentrations greater than 15%, the gas mixture will not

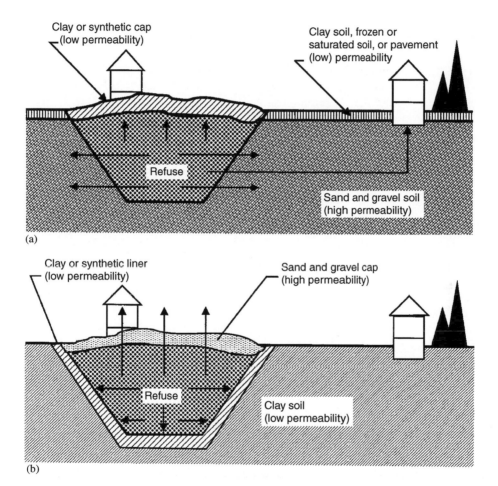

FIGURE 10.12 Effects of surrounding geology and surface features on landfill gas migration (U.S. EPA, 1994).

FIGURE 10.13 The lower explosive limit, upper explosive limit and explosive range for a hypothetical gas.

explode, as the gaseous mixture is considered 'rich.' This 15% threshold is the Upper Explosive Limit (UEL) defined as the maximum concentration of a gas, above which the substance will not explode when exposed to a source of ignition. The explosive hazard range occurs between the LEL and the UEL. It must be noted that at methane concentrations above the UEL, fire may still be possible and asphyxiation will occur. In addition, a sudden dilution of the methane in the local atmosphere can bring the mixture back within the explosive range.

Methane is generated from MSW only when the moisture content of the waste exceeds 40% under anaerobic conditions. For example, if a landfill is holding wastes having 15% moisture, the waste will be 'fossilized', i.e., it will not decay and therefore will produce very little methane (Vesilind et al., 2002).

The frequency of landfill gas monitoring is determined based on soil conditions, surface hydrology, hydrogeology, and location of facility structures. If methane gas levels exceed established limits, a remediation plan must be prepared within 60 days of detection. Air must be sampled within facility structures where gas may accumulate and in soil at the property boundary. Other monitoring methods may include sampling gases from probes within the landfill unit. A typical gas-monitoring probe installation is shown in Figure 10.14. The frequency of monitoring should be sufficient to detect landfill gas migration based on subsurface conditions and changing landfill conditions. Monitoring must be conducted at least quarterly (40 CFR 258.23). The number and location of gas probes is site-specific and dependent on subsurface conditions, land use, and location and design of facility structures. At the facility and in neighboring properties, structures with basements or crawl spaces are more susceptible to landfill gas infiltration and must also be monitored.

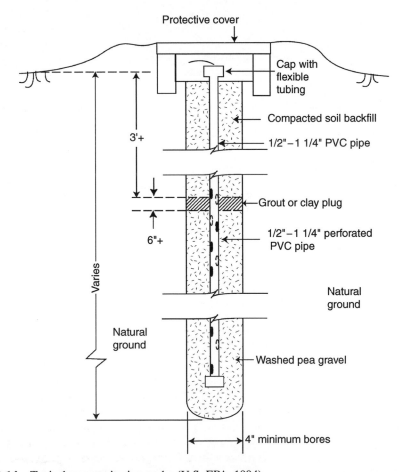

FIGURE 10.14   Typical gas monitoring probe (U.S. EPA, 1994).

FIGURE 10.15 Portable organic vapor analyzer.

Methane measurements are made in the field with a portable methane meter or organic vapor analyzer (Figure 10.15). Gas samples may also be collected and brought to the laboratory for analysis. Measurements, for example with gas chromatography-mass spectrometry, can confirm the identity and concentrations of landfill gas. In addition to measuring gas composition, other indications of gas migration may be observed. These include odor (described as either a strongly 'sweet' or a rotten egg [H_2S] odor), septic soil, and audible or visual venting of gases, especially in standing water. Stressed vegetation is a useful indicator of gas migration. Landfill gas in soil pores creates anaerobic conditions by displacing oxygen. Plant roots require sufficient oxygen to carry out normal respiration processes, and methane gas acts as a simple asphyxiant to roots (Flower et. al., 1982).

10.4.15 TRACE GASES

Table 10.13 lists many trace gaseous compounds detected at most MSWLFs. The amounts of these gases in leachate is a function of their initial concentrations and their solubility in aqueous liquids. The occurrence of significant concentrations of volatile organic compounds (VOCs) in landfill gas is associated with older landfills that had accepted industrial and commercial wastes containing VOCs. In newer landfills in which the disposal of hazardous waste has been banned, concentrations of VOCs in landfill gas have been very low.

10.4.16 LANDFILL GAS CONTROL

Landfill gas may vent naturally or be directed to the atmosphere by engineered controls. Systems used to control or prevent gas migration are categorized as either passive or active. Passive systems provide preferential flow paths by means of natural pressure, concentration, and density gradients. Active systems use mechanical equipment to direct or control landfill gas by providing pressure gradients, in essence, forcing landfill gas out by applied convective forces. The choice of system is based on the design and age of the landfill unit and on the soil and hydrogeologic conditions of the facility and surrounding environment. In other words, the degree of potential gas hazard plays a role in the choice of the particular system.

TABLE 10.13

Typical Concentrations of Trace Compounds Found in Landfill Gas at 66 MSWLFs, California

Compound	Mean Concentration	ppb, v Maximum
Acetone	6,838	240,000
Benzene	2,057	39,000
Chlorobenzene	82	1,640
Chloroform	245	12,000
1,1-Dichloroethane	2,801	36,000
Dichloromethane	25,694	620,000
1,1-Dichloroethene	130	4,000
Diethylene chloride	2,835	20,000
trans-1,2-Dichloroethane	36	850
Ethylene dichloride	59	2,100
Ethyl benzene	7,334	87,500
Methyl ethyl ketone	3,092	130,000
1,1,1-Trichloroethane	615	14,500
Trichloroethane	2,079	32,000
Toluene	34,907	280,000
1,1,2,2-Tetrachloroethane	246	16,000
Tetrachloroethylene	5,244	180,000
Vinyl chloride	3,508	32,000
Styrenes	1,517	87,000
Vinyl acetate	5,663	240,000
Xylenes	2,651	38,000

Source: Tchobanoglous, T. et al., *Integrated Solid Waste Management: Engineering Principles and Management Issues*, McGraw-Hill, New York, 1993. Data reproduced with kind permission of the McGraw-Hill Companies, Inc.

10.4.16.1 Passive Systems

Passive gas control systems rely on natural pressure and convection to vent landfill gas to the atmosphere. Passive systems involve 'high-permeability' or 'low-permeability' techniques. High-permeability systems incorporate pathways such as trenches, vent wells, or perforated vent pipes surrounded by coarse material to vent landfill gas to the surface. Low-permeability systems block lateral migration via barriers such as synthetic membranes and clayey soils. Passive systems may be incorporated into a landfill design or may be installed later for corrective purposes. They may also be installed within a landfill unit along the perimeter, or between the landfill and the facility property boundary (U.S. EPA, 1985).

At the time of landfill closure, a passive system may be incorporated into the final cover system. This may consist of perforated collection pipes and high-permeability soils located directly below the impermeable cover. Passive vent systems may be connected to header pipes located along the perimeter of the landfill unit. Figure 10.16 illustrates two passive systems.

Some practical problems have been associated with passive systems. For example, snow and soil may accumulate in vent pipes thus preventing gas migration and venting. Biological clogging of pipes and soil pores is also common in passive systems.

10.4.16.2 Active Systems

Active gas control systems employ some mechanical means to remove landfill gas and consist of either positive pressure (air injection) or negative pressure (extraction) systems. Negative pressure

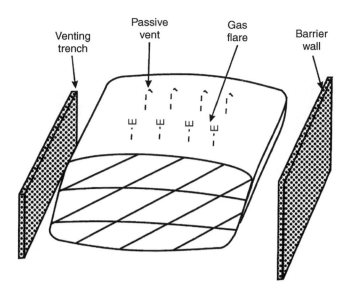

FIGURE 10.16 Passive gas control systems showing venting to the atmosphere by convective forces, and a barrier wall (U.S. EPA, 1994).

systems are more commonly used and extract gas from a landfill by using a blower to pull gas out. The gas may be recovered for energy conversion, treated, or combusted in a flare system (Figures 10.17a–c)(U.S. EPA, 1985). Gas extraction wells may be installed within the landfill cells or beyond the landfill, in nearby extraction trenches (Figure 10.18). Active systems are not as sensitive to freezing or saturation of cover soils as are passive systems.

The capital, operation, and maintenance costs of active gas systems are clearly higher than for passive systems. These costs continue throughout the postclosure period. It is possible to convert active gas controls into passive systems when gas production diminishes.

When designing the gas control system, several other practical issues must be taken into account. For example, construction materials may be indirectly affected by the elevated temperatures within a landfill unit as compared with the relatively cooler ambient air. Leachate water containing corrosive and toxic constituents may condense within the plumbing and adversely affect construction materials. Provisions for managing condensate should be incorporated to prevent accumulation. The condensate can be returned to the landfill.

10.4.17 GAS UTILIZATION

Current EPA regulations under the Clean Air Act require many larger landfills to collect and combust landfill gas. Several compliance options are available, including flaring the gas or installing a landfill gas recovery and utilization system. There are a number of environmental and economic benefits to recovering landfill gas. Gas recovery systems reduce landfill gas odor and migration, reduce the danger of explosion and fire, and may be used as a source of revenue that may help to reduce the cost of closure. Raw landfill gas, requiring removal of only water and particulates, may be used for heating small facilities. A fairly concentrated and cleaned gas can be used for both water and space heating as well as lighting, electrical generation, co-generation, and as a fuel for industrial boilers. Landfill gas is also upgraded to pipeline standards and can be sold to local utilities (SWANA, 1992).

According to the U.S. EPA (2002), 0.9 million MT (1 million tons) of MSW in a landfill generates about 8.5 m^3 min (300 ft^3/min, cfm) of landfill gas that can then generate 7,000,000 kWh (kilowatt hours) per year, energy sufficient to power 700 homes for a year. In a broader environmental sense, utilizing 300 cfm/year of landfill gas yields the same reduction in greenhouse gases as removing 6100 automobiles from the road for 1 year.

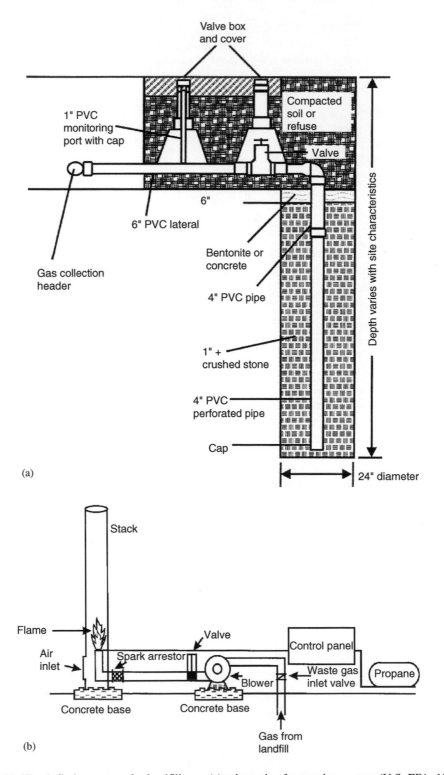

FIGURE 10.17 A flaring system for landfill gas: (a) schematic of extraction system (U.S. EPA, 1994); (b) surface features (Reproduced with kind permission of MACTEC Engineering and Consulting, Inc., formerly known as ABB Environmental Services); (c) photo.

(c)

FIGURE 10.17   (continued)

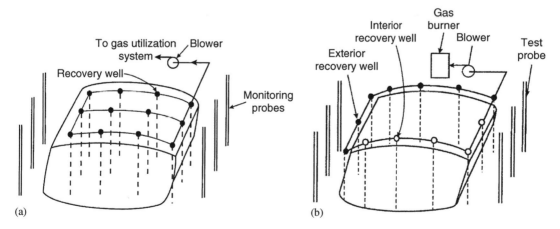

FIGURE 10.18   Gas recovery systems for active landfill gas removal: (a) interior wells; and (b) perimeter wells (U.S. EPA, 1994).

Generally, landfills closed for less than 5 years are ideal for energy recovery because of their content of relatively fresh and moist MSW. With time, the ability of a landfill to generate gas decreases. Under optimum conditions, a landfill might produce gas for 15 years or more, depending on the rate of gas generation, the water content of the waste, and the manner in which the landfill was closed. Current closure requirements for landfills are intended to restrict the entry of moisture into the landfill. These requirements will lead to greatly reduced gas generation after closure.

The heat value of unprocessed landfill gas is 18,600 kJ /standard m³ (500 Btu/standard cubic foot), about one half of natural gas, mainly because only one half of landfill gas is methane (Table 10.9). At a small landfill, gas with this heat value can be used to run a modified internal combustion engine or a generator to convert gas into electrical energy. At a larger landfill, moisture and carbon dioxide removal (through scrubbing and gas polishing with carbon or polymer adsorption) enables the gas to be used to run boilers and turbine generators for energy recovery. Purification of landfill gas to pipeline quality involves greater investment in terms of removal of impurities. In a chemical sense, landfill gas and pipeline-quality natural gas differ substantially in composition and energy content. Landfill gas has a lower Btu content, combusts at a lower temperature, is more corrosive, and contains much greater concentrations of undesirable gases (CO_2, H_2S, O_2, and N_2) than pipeline-quality natural gas. Extensive purification is therefore necessary to remove all components from landfill gas except methane. The required cleanup protocol includes nearly complete CO_2 removal. Processing increases the heat value of the gas to approximately 1000 Btu/scf. Conversion of landfill gas into natural gas is sufficiently costly such that only large landfills can attain the economies of sale necessary to support operations. Fewer than ten U.S. landfills convert gas for pipeline use (Vesilind et al., 2002)

Municipalities are using landfill gas to produce electricity, heat, or steam for industrial use. Local governments are discovering that use of landfill gas can reduce electrical demand for local utilities, delaying the need for building new power plants. The Rhode Island Solid Waste Management Corporation (RISWMC), responsible for managing the 62 ha (154 acres) Central Landfill, captures the landfill gas to supply as much as 12.3 MW of electrical power, capacity sufficient to serve roughly 17,000 households. The company sells this electricity to a local subsidiary of New England Power and pays the RISWMC $50,000 per month in royalties for the rights to the gas.

Lane County, Oregon, which includes the cities of Eugene and Springfield, constructed a landfill gas-to-electricity facility at the Short Mountain Landfill in 1992. The facility generates 1.6 MW of electrical power (sufficient for about 800 homes) and will increase its capacity to 4 MW by 2010. There is a roughly 1 cent / kWh profit for the local consumer-owned utility on the electricity sold to the Bonneville Power Administration. The total annual revenues from gas sales were about $150,000. Lane County receives a minimum royalty of $15,000 per year and avoids having to build and operate an expensive collection system (U.S. DOE, 1994).

10.4.18 OTHER AIR CRITERIA

Open burning of solid waste, except for the infrequent burning of agricultural wastes, land-clearing debris, diseased trees, or debris from emergency cleanup operations, is prohibited at all MSWLFs (40 CFR Part 258.24).

10.4.19 PUBLIC ACCESS

The general public and other unauthorized persons may be unaware of the hazards associated with landfills. Potential hazards include:

- The inability of equipment operators to see individuals during equipment operation
- Direct exposure to waste materials (e.g., sharp objects, pathogens)
- Falls
- Exposure to fires
- Earth-moving activities

Operators of MSW landfills must control public access and prevent unauthorized traffic and illegal dumping of wastes (40 CFR Part 258.25). This is accomplished by constructing natural or artificial barriers. Specific measures include installation of gates and fences, trees, hedges, berms, ditches, and embankments. Chain link, barbed wire added to chain link, and open farm-type fencing are examples of appropriate fencing. Access to facilities should be controlled through gates that can be locked when the site is unsupervised.

10.4.20 CONTROL OF RUN-ON AND RUNOFF

The landfill operator is required to prevent run-on onto the active portion of the landfill and to collect runoff as well (40 CFR Part 258.26). Run-on and runoff control systems must be designed based on the volume of water anticipated from a 24-h, 25-year storm. The purpose of a run-on control system is to collect and redirect surface waters to minimize the amount of water entering landfill cells. As discussed below, minimizing the volume of water entering a landfill will limit the volume of leachate generated. Run-on control is accomplished by constructing berms and swales up-gradient of the fill area in order to redirect water to stormwater control structures.

If stormwater enters the landfill unit and contacts waste, the stormwater, according to regulations, becomes leachate and must be managed as leachate. Such leachate generation will increase costs and can overload leachate treatment systems.

Runoff control systems must collect and handle runoff from the active portion of the landfill, including areas that contact MSW. Runoff control can be accomplished through stormwater conveyance structures that divert runoff and leachate to a storage system for eventual treatment. Other structures used for run-on and runoff controls include seepage ditches, seepage basins, and sedimentation basins (Figures 10.19 and 10.20). U.S. EPA (1985) discusses each of these structures in detail.

After a landfill unit has been sealed with a final cover, stormwater runoff is managed as stormwater and not leachate. Therefore, waters running off the final cover system of closed areas may not require treatment and can be combined with run-on waters. Run-on and runoff must be managed in accordance with the discharge requirements of the Clean Water Act including the National Pollutant Discharge Elimination System (NPDES) (U.S. EPA, 1994).

10.4.21 MANAGEMENT OF SURFACE WATER

MSWLFs are required to prevent any discharge of pollutants into surface water, including wetlands (40 CFR Part 258.27). The facility should determine if it is in conformance with requirements of the Clean Water Act and the NPDES requirements under the Clean Water Act. The EPA and approved states have jurisdiction over discharge of pollutants into U.S. waters including wetlands. Landfills discharging pollutants into the U.S. waters require a Section 402 (NPDES) permit.

Landfill units that have a point source discharge must have a NPDES permit. Point source discharges from landfills include:

- The release of leachate from a leachate collection or on-site treatment system into water
- Disposal of solid waste into water

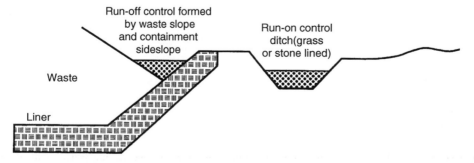

FIGURE 10.19 Schematic of run-on and runoff controls at a landfill (Reproduced with kind permission of MACTEC Engineering and Consulting, Inc., formerly known as ABB Environmental Services).

FIGURE 10.20   Landfill runoff ready for treatment.

10.4.22   Restrictions on Liquids in Landfills

As mentioned previously, many older land disposal facilities in the United States had become environmental nightmares because of the disposal of toxic wastes, often in liquid form. As a result, the RCRA regulations require that bulk or noncontainerized liquid wastes not be placed in MSWLFs unless (40 CFR Part 258.28):

- The waste is household waste.
- The waste is leachate derived from the landfill and the unit is designed with proper liners and a leachate collection system.

The restriction of free liquids is intended to limit the generation of leachate. Liquid waste refers to any waste material that is determined to contain free liquids as defined by SW-846 (U.S. EPA, 1986b) Method 9095 — Paint Filter Liquids Test. The paint filter test is performed by placing a 100 mL sample of waste in a conical, 400 μm paint filter. The waste is considered a liquid waste if any liquid from the waste passes through the filter within 5 min. The apparatus used for performing the paint filter test is illustrated in Figure 10.21. Due to concerns over cost and practicality, it is impossible to regulate household waste for its content of liquids. Containers holding liquid waste may not be placed in a landfill unless the waste is a household waste.

If the waste is considered a liquid waste, absorbent materials may be added to render it a 'solid' material (i.e., a waste or absorbent mixture that no longer fails the paint filter liquids test). Sludges are a common waste stream that may contain significant quantities of liquid. Sludge is a mixture of water and solids that has been produced during water and wastewater treatment in commercial or industrial operations. Sludge disposal is acceptable provided the sludge is nonhazardous and passes the paint filter test.

10.4.23   Leachate Formation

Landfill leachate production and migration are among the most acute concerns for operators of municipal sanitary landfills. Leachate is the liquid generated by the action of water (rainwater or infiltrating groundwater) and liquids present within the initial waste percolating through the stored waste within a landfill cell. As the absorbent components of the waste begin to fail to absorb the liquids, free leachate forms.

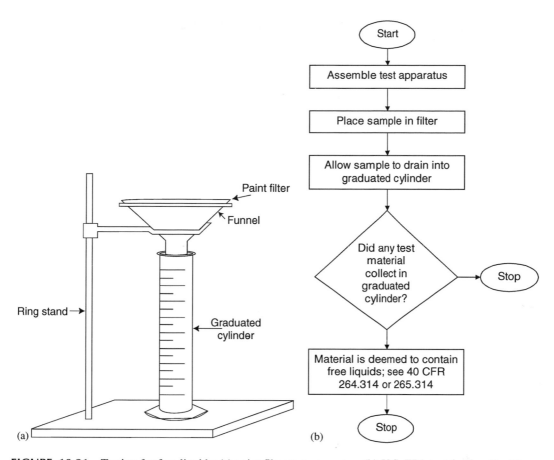

FIGURE 10.21   Testing for free liquids: (a) paint filter test apparatus; (b) U.S. EPA test for free liquids.

Many factors are involved in the chemical, biochemical, and physical variations of landfill leachate. The chemical complexity of leachate reflects the extreme heterogeneity of the input wastes. Furthermore, as the waste within a landfill cell ages, the chemical properties of the leachate will change. Climate, season, and moisture are other factors that affect composition. Chemical properties of concern include pH, BOD_5, suspended solids, N content, salt content, and the presence of trace toxic compounds. Some significant chemical components of landfill leachates are listed in Table 10.14. The major classes of microbes present within leachate include bacteria, actinomycetes, fungi, and protozoa. Microbes may be aerobes, and facultative and obligate anaerobes. Biochemical, chemical, and instrumental techniques are used to measure the types and numbers of microbial populations present. The physical characteristics of the leachate are a function of the quantity of dissolved and suspended solids, temperature, color, and the amount of inorganic solids such as iron and lead. Leachate sampling parameters are listed in Table 10.15.

10.4.23.1   Estimation of Leachate Volume Produced

The volume of leachate generated in a modern sanitary landfill cell is primarily influenced by the volume of precipitation entering the site. Assuming the facility is permitted under RCRA Subtitle D and therefore must possess a liner, it is necessary for engineers to estimate leachate production and to determine relevant factors for its collection, for example, the spacing of the leachate collection pipes (discussed below) at the base of the landfill.

TABLE 10.14

Municipal Landfill Leachate Data: Indicator Parameters, Inorganic, and Organic Compounds

Indicator Parameters	Leachate Concentration Reported (mg/L)	
	Minimum	Maximum
Alkalinity	470	57,850
Ammonia	0.39	1,200
Biological oxygen demand	7	29,200
Calcium	95.5	2,100
Chemical oxygen demand	42	50,450
Chloride	31	5,475
Fluoride	0.11	302
Iron	0.22	2,280
Phosphorus	0.29	117.18
Potassium	17.8	1,175
Sulfate	8	1,400
Sodium	12	2,574
Total dissolved solids	390	31,800
Total suspended solids	23	17,800
Total organic carbon	20	14,500
Inorganic Compounds		
Aluminum	0.01	5.8
Antimony	0.0015	47
Arsenic	0.0002	0.982
Barium	0.08	5
Beryllium	0.001	0.01
Cadmium	0.0007	0.15
Chromium (total)	0.0005	1.9
Cobalt	0.04	0.13
Copper	0.003	2.8
Cyanide	0.004	0.3
Lead	0.005	1.6
Manganese	0.03	79
Magnesium	74	927
Mercury	0.0001	0.0098
Nickel	0.02	2.227
Vanadium	0.009	0.029
Zinc	0.03	350
Organic Compounds		
Acetone	8	11,000
Acrolein	270	270
Benzene	4	1,080
Bromomethane	170	170
Butanol	10,000	10,000
1-Butanol	320	360
2-Butanone (methyl ethyl ketone)	110	27,000
Butyl benzyl phenol	21	150
Carbon tetrachloride	6	397.5
Chlorobenzene	1	685

(continued)

TABLE 10.14 (*Continued*)

Indicator Parameters	Leachate Concentration Reported (mg/L)	
	Minimum	Maximum
Chloroethane	11.1	860
Bis(2-chloroethyoxy)methane	18	25
2-Chloroethyl vinyl ether	2	1,100
Chloroform	7.27	1,300
Chloromethane	170	400
Bis(chloromethyl)ether	250	250
2-Chloronaphthalene	46	46
p-Cresol	45.2	5,100
2,4-D	7.4	220
4,4′-DDT	0.042	0.22
Dibromethane	5	5
Di-*N*-butyl phthalate	12	150
1,2-Dichlorobenzene	3	21.9
1,4-Dichlorobenzene	1	52.1
Dichlorodifluoromethane	10.3	450
1,1-Dichloroethane	4	44,000
1,2-Dichloroethane	1	11,000
cis-1,2-Dichloroethane	190	470
trans-1,2-Dichloroethylene	2	4,800
1,2-Dichloropropane	0.03	500
1,3-Dichloropropane	18	30
Diethyl phthalate	3	330
2,4-Dimethyl phenol	10	28
Dimethyl phthalate	30	55
Endrin	0.04	50
Ethanol	23,000	23,000
Ethyl acetate	42	130
Ethyl benzene	6	4,900
Bis(2-ethylhexyl) phthalate	16	750
2- Hexanone (methyl butyl ketone)	6	690
Isophorone	4	16,000
Lindane	0.017	0.023
4-Methyl-2-pentanone (methyl isobutyl ketone)	10	710
Methelene chloride (dichloromethane)	2	220,000
Naphthalene	2	202
Nitrobenzene	4	120
4-Nitrophenol	17	17
Pentachlorophenol	3	470
Phenol	7.3	28,000
1-Propanol	11,000	11,000
2-Propanol	94	26,000
1,1,2,2-Tetrachlorothane	210	210
Tetrachloroethylene	2	620
Tetrahydrofuran	18	1,300
Toluene	5.55	18,000
Toxaphene	1	1
1,1,1-Trichloroethane	1	13,000
1,1,2-Trichloroethane	30	630

(continued)

TABLE 10.14 (*Continued*)

Indicator Parameters	Leachate Concentration Reported (mg/L)	
	Minimum	Maximum
Trichloroethylene	1	1,300
Trichlorofluormethane	4	150
1,2,3-Trichloropropane	230	230
Vinyl Chloride	8	61
Xylenes	32	310

Source: U.S. EPA, 1988.

TABLE 10.15
Leachate Sampling Parameters

Physical	Organic Constituents	Inorganic Constituents	Biological
Appearance	Organic chemicals	Suspended solids total dissolved solids	Biochemical oxygen demand
pH	Phenols	Volatile suspended solids volatile dissolved solids	Coliform bacteria (total, fecal, fecal streptococci)
Oxidation–reduction potential	Chemical oxygen demand	Chloride	Standard plate count
Conductivity	Total organic carbon	Sulfate	
Color	Volatile acids	Phosphate	
Turbidity	Tannins, lignins	Alkalinity and acidity	
Temperature	Organic-N	Nitrate-N	
Odor	Ether-soluble (oil and grease)	Nitrate-N	
	Methylene blue active substances	Ammonia-N	
	Organic functional groups as required	Sodium	
	Chlorinated hydrocarbons	Potassium	
		Calcium	
		Magnesium	
		Hardness	
		Heavy metal (Pb, Cu, Ni, Cr, Zn, Fe, Mn, Hg, Ba, Ag)	
		Arsenic	
		Cyanide	
		Fluoride	
		Selenium	

Source: Tchobanoglous, T. et al., *Integrated Solid Waste Management: Engineering Principles and Management Issues*, McGraw-Hill, New York, 1993. Data reproduced with kind permission of the McGraw-Hill Companies, Inc.

One estimate of leachate quantity generated, after steady state is attained, can be developed using a simple water balance method. The contributing factors to the volume of water that enters the waste (infiltration) are the amount received via precipitation, run-on of surface water, and water entering through the sides and base of the cell. Water leaving the site from surface runoff, evaporation, and transpiration by plants are subtracted from this balance. Water loss via evaporation from the soil and from plant uptake and transpiration are typically combined into a single term, evapotranspiration. Design requirements under RCRA make the quantities of surface water run-on and water entering through the sides and bottom of the fill negligible.

The hypothetical water balance is represented by the equation

$$L = P = R_{on} + U - E - R_{off} \qquad (10.10)$$

where

L = leachate

P = precipitation

R_{on} = run-on surface water

U = underflow of groundwater into the cell

E = evapotranspiration

R_{off} = run-off surface water

If the landfill is designed and operated properly, surface water will be diverted from the waste and therefore R_{on} = 0. Additionally, a landfill constructed above the water table and possessing an impermeable liner will give U = 0 (i.e., there is no underflow).

The equation can thus be simplified to:

$$L = P - E - R_{off}$$

The integration of these concepts is shown in Figure 10.22.

The amount of runoff depends upon the soil permeability, the slope of the surface, the type of vegetation, duration and frequency of precipitation, and whether the precipitation is in the form of rain or snow. The fraction of precipitation that becomes runoff is expressed by a runoff coefficient. The fraction of precipitation that is converted into runoff is in the range 0.05 to 0.35 (Table 10.16). In utilizing the water-balance approach for predicting leachate at landfills, a number of references are available (Thornwaite and Mather, 1957; Fenn et al., 1975; Bagchi, 1994); for estimating evapotranspiration rates.

EXAMPLE 10.6

Calculate the annual volume of leachate generated per hectare for a sanitary landfill located in the northcentral United States. The climate is temperate, average annual rainfall is 1.07 m/year (42 in./year), and evapotranspiration is estimated at 55%. The wastes are covered with soil and

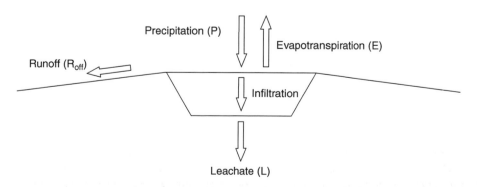

FIGURE 10.22 Mass balance of moisture in a sanitary landfill.

TABLE 10.16
Runoff Coefficients for Various Slopes and Soil Permeabilities

Surface	Slope	Runoff Coefficient
Grass, sandy soil	0–2%	0.05–0.10
	2–7%	0.10–0.15
	>7%	0.15–0.20
Grass, heavy soil	0–2%	0.13–0.17
	2–7%	0.17–0.25
	>7%	0.25–0.35

Source: Fenn, D. G., U.S. EPA, 1975.

runoff from the site is 10%. There is no run-on of surface water; similarly, there is no underflow of groundwater into the cell (i.e., $R_{on}=0$ and $U=0$).

SOLUTION

For these calculations it is necessary to convert the depth of precipitation to a volume. This is easily accomplished because the depth of water is received over a known area (units of 1 ha). The precipitation depth is thus converted into 1.07 h m. The quantity of leachate is then calculated using the equation

$L = P - E - R_{off}$

$L = 1.07$ ha m $- (0.55)(1.07$ ha m$)-(0.10)(1.07$ ha m$)$

$L = 0.37$ ha m

  $= 3750$ m^3 (1 ha$=10,000$ m^2)

  $= 3.75 \times 10^6$ L or 990,645 gal

10.4.23.2 The Leachate Collection and Removal System

A leachate collection and removal (LCR) system is designed to collect leachate and convey it out of the landfill for eventual treatment. The LCR system is constructed so that less than 30 cm of leachate (the amount of leachate the liner is designed to maintain according to Subtitle D) accumulates above the composite liner. When designing and constructing a LCR system, the following components must be considered (U.S. EPA, 1994):

- Area collector — the drain that covers the liner and collects leachate
- Collection laterals – the pipe network that drains the area collector
- Sump – the low point where the leachate exits the landfill
- Stormwater and leachate separation system

10.4.23.3 Area Collector

The area collector, also known as the blanket drain, covers the entire surface of the geomembrane liner and collects the leachate. The area collector system is typically constructed with at least a 30 cm (12 in.) layer of sand having a hydraulic conductivity greater than 10^{-2} cm/sec. An alternative type of blanket drain can be constructed using geosynthetic drainage nets (geonets), porous synthetic materials applied over the geomembrane liner. A brief discussion of geosynthetic materials for landfill construction follows.

10.4.24 GEOSYNTHETIC MATERIALS

Geosynthetic materials comprise a wide group of polymer-based mats, sheets, grids, nets, and composite materials that have found extensive use not only in modern landfills but in water management and other engineering applications. Depending on design, geosynthetics perform five major functions: separation, reinforcement, filtration, drainage, and containment (Koerner, 1998). Several common geosynthetics will be discussed below.

Geonets (also known as geospacers) are used to convey liquids (Figure 10.23a). Geonets are plastics formed into a rather tight, grid-like configuration. A geonet can be defined as (ASTM, 2002):

> a geosynthetic material consisting of integrally connected parallel sets of ribs overlying similar sets at various angles for biaxial or triaxial drainage of liquids or gases. Geonets are often laminated with geotextiles on one or both surfaces and are then referred to as drainage geocomposites.

Geonets are manufactured with layers of intersecting ribs designed so that liquid can flow within the open spaces. Geonets vary in thickness from 4.0 to 6.9 mm (U.S. EPA, 1993). Geonets require the installation of a geotextile above them.

Geotextiles (also known as filter fabrics) are defined as (ASTM, 2002):

> A permeable geosynthetic comprised solely of textiles. Geotextiles are used with foundation, soil, rock, earth, or any other geotechnical engineering-related material as in integral part of a human-made project, structure, or system.

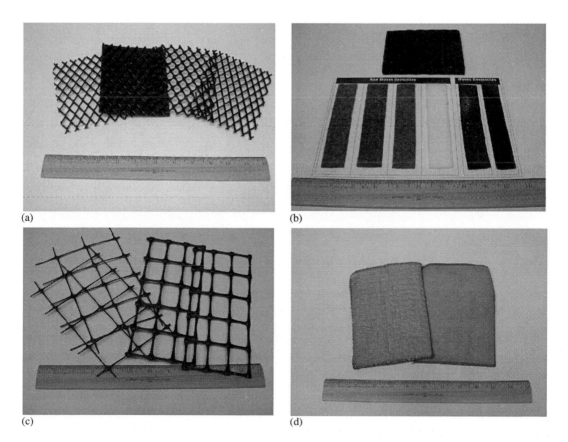

(a)

(b)

(c)

(d)

FIGURE 10.23 Common geosynthetic materials: (a) geonet; (b) geotextile; (c) geogrid; and (d) geosynthetic clay liner.

Geotextiles (Figure 10.23b) are manufactured from synthetic polymers such as polypropylene, polyester polyethylene, and nylon. Approximately 75% of all geotextiles are made of polypropylene resin, 20% are polyester, and the remainder are nylon (U.S. EPA, 1993c; Oweis and Khera, 1998). The polymers are formed into fibers and then into a woven or nonwoven fabric. When placed in the ground these fabrics are termed geotextiles. They are porous to liquid flow across their planes and also within their thickness, but to widely varying degrees. There are at least 100 specific application areas for geotextiles; however, they always perform at least one of five discrete functions (Koerner, 1998):

- Separation
- Reinforcement
- Filtration
- Drainage
- Containment (barrier, when impregnated)

Since geotextiles are composed of synthetic fibers rather than natural ones such as cotton, wool, or silk, biodegradation is not a concern.

Geogrids (Figure 10.23c) are designed to function as reinforcement materials. Koerner (1998) defines a geogrid as:

> a geosynthetic material consisting of connected parallel sets of tensile ribs with apertures of sufficient size to allow strike-through of surrounding soil, stone, or other geotechnical material.

Geogrids are plastics formed into a very open, grid-like configuration. Geogrids contain large open spaces called *apertures,* which are typically 10 to 100 mm between the ribs. The ribs themselves can be manufactured from a number of different materials. The primary function of geogrids is reinforcement. There are many application areas.

Geosynthetic clay liners (Figure 10.23d) are used as a composite component beneath a geomembrane or by themselves in environmental and containment applications as well as in transportation, geotechnical, and hydraulic applications (Koerner, 1998).

Geocomposites consist of various combinations of geotextiles, geogrids, geonets, geomembranes, and other materials. The major functions embrace the entire range of functions listed above for geosynthetics: separation, reinforcement, filtration, drainage, and containment. The general reason for the existence of geocomposites is the higher performance attained by combining the characteristics of two or more materials.

Geomembranes are discussed earlier in this chapter and shown in Figure 10.4. The materials themselves are 'impervious' polymeric materials used primarily for linings and for covers of liquid or solid storage facilities. Thus, the primary function in containment is as a liquid or vapor barrier. The range of applications, however, is broad. According to ASTM D4439, a geomembrane is defined as

> A very low permeability synthetic membrane liner or barrier used with any geotechnical engineering related material so as to control fluid migration in a himan-made project, structure, or system.

Geomembranes are manufactured from continuous polymeric sheets, but they can also be made from the impregnation of geotextiles with asphalt or elastomer sprays or as multilayered bitumen geocomposites (Koerner, 1998).

10.4.25 COLLECTION LATERALS

According to the U.S. EPA (1994), the regulatory limit of a 30 cm maximum liquid head above the liner cannot be achieved using an area collector alone; therefore, collection laterals are needed. Collection laterals are perforated pipes that direct leachate to sumps for removal. During landfill operation, leachate passes through the area collector, into collection laterals, and drains to a sump where it is removed from the landfill.

A number of materials are appropriate for the manufacture of leachate collection systems. Polymeric pipes are by far the most common. HDPE and PVC are used almost exclusively and are available as either profiled or smooth wall construction (Qian et al., 2002).

The design of perforated leachate collection pipes must address the following issues (Qian et al., 2002):

- The required flow
- Maximum drainage slope
- Maximum pipe spacing
- Pipe size
- Structural strength of the pipe

Spacing of the collection laterals depends on the permeability of the collection pipes, the slope of the liner, and the assumed entry rate of rainfall (Figure 10.24). The lower the permeability, the closer the pipes. The slope of collection laterals should be greater than 2% in order to achieve adequate flow velocity to help clean the pipes. A 2% slope will also ensure that MSW settlement will not reverse the slope of the pipes.

Predicting the value of the maximum leachate head on top of the landfill liner is important in landfill design. Factors affecting the leachate head include the permeability of the drainage materials, the drainage slope, the drainage length, and infiltration rate. Darcy's law and the law of continuity can be used to calculate the depth of leachate ponded on a liner (McBean et al., 1982; McEnroe, 1993). One equation that has been proposed is as follows (Richardson and Zhao, 2000):

$$Y_{max} = \frac{p}{2} \times \frac{q}{K} \left[\frac{K\tan^2\alpha}{q} + 1 - \frac{K\tan\alpha}{q} \left(\tan^2\alpha + \frac{q}{K} \right)^{1/2} \right] \qquad (10.11)$$

where Y_{max} is the maximum head on liner (cm), L the horizontal drainage distance (cm), tan α the inclination of liner from horizontal (deg), q the vertical inflow (infiltration), defined in this equation as from a 25-year, 24-h storm (cm/day), K the hydraulic conductivity of the drainage layer (cm/day), and p the distance between collection pipes, (cm).

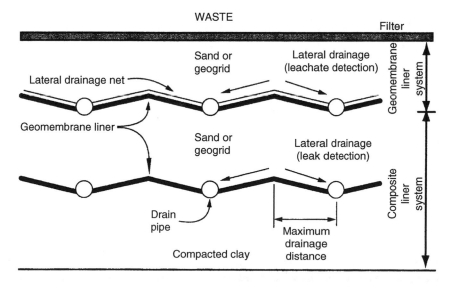

FIGURE 10.24 Schematic of the base of a landfill liner (Not to scale).

This equation can be used to calculate the maximum allowable pipe spacing based on the maximum allowable design head, anticipated leachate impingement rate, slope of the liner, and permeability of drainage materials. The equation suggests that, holding all other parameters constant, the closer the pipes are placed, the lower the head will be. A reduced head on the liner results in a lower hydraulic driving force through the liner, and the consequence of a puncture in the liner is similarly reduced.

EXAMPLE 10.7

Determine the spacing between pipes in a leachate-collection system using granular drainage material and the following properties. Assume that in the most conservative design all stormwater from a 25-year, 24-h storm enters the leachate collection system.

Design storm (25 years, 24 hours) = 8.2 in. = 0.00028 cm/s

Hydraulic conductivity = 10^{-2} cm/sec

Drainage slope = 1.5%

Maximum design depth on liner = 14.2 cm

$$P = \frac{2Y_{max}}{\dfrac{q}{K}\left[\dfrac{K\tan^2\alpha}{q} + 1 - \dfrac{K\tan\alpha}{q}\left(\tan^2\alpha + \dfrac{q}{K} \right)^{1/2} \right]}$$

$$P = \frac{2(14.2)}{\dfrac{0.0028}{0.01}\left[\dfrac{0.01(0.15)^2}{0.00028} + 1 - \dfrac{0.01(0.15)}{0.00028}\left((0.015)^2 + \dfrac{0.0028}{0.01} \right)^{1/2} \right]} = 1105 \text{ cm}$$

10.4.26 SUMPS

Sumps are situated in engineered low points in the composite liner system and are constructed to collect leachate. Figure 10.25 shows a low-volume sump. The leachate removal standpipe must be extended through the entire landfill from lowest liner to the cover, and then through the cover itself. It also must be maintained for the entire postclosure care period of 30 years or longer. Because of the difficulty in seam-testing sumps, sump areas often are designed with an additional layer of geomembrane. Figure 10.26 shows a sump being installed in a landfill.

10.4.27 LEACHATE TREATMENT

The recovered leachate is either stored in a tank until it can be safely removed, diverted directly into the sanitary sewer, or reapplied to the surface of the landfill. Leachate collection tanks should be both corrosion-resistant and able to withstand climatic extremes.

Treatment of leachate must meet the water quality standards set by regulatory authorities. There are six primary types of leachate treatment:

- Aerobic biological
- Anaerobic biological
- Land application
- Physicochemical
- Recycling leachate though the landfill
- Treatment with municipal wastewater

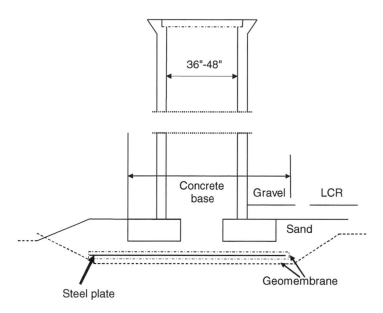

FIGURE 10.25 A low-volume sump (U.S. EPA, 1989).

FIGURE 10.26 Photograph of a sump being installed (Reproduced with kind permission of Environmental Research and Education Foundation, 2002).

10.4.28 GROUNDWATER MONITORING AND CORRECTIVE ACTION

The U.S. EPA landfill criteria establish requirements for groundwater monitoring and corrective action for all landfills. The criteria include a systematic process that requires routine groundwater monitoring ("detection monitoring"). In detection monitoring, a minimum number of indicator parameters must be tested at least annually. Figure 10.27 depicts a typical groundwater monitoring well. If statistically significant increases above background concentrations of any of the indicator parameters are detected, a more comprehensive monitoring program must be instituted. If evaluated concentrations of pollutant parameters continue or increase, the facility is then required to develop and implement a corrective action program (U.S. EPA, 1994).

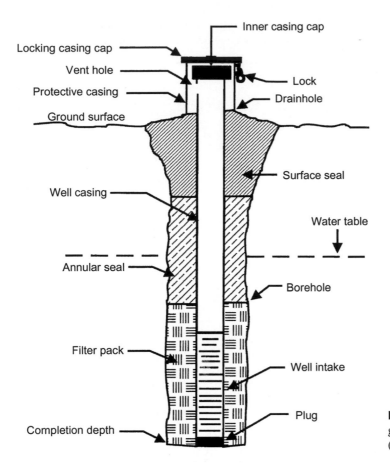

FIGURE 10.27 Typical groundwater monitoring well (U.S. EPA, 1994).

10.4.29 RECORD KEEPING

As part of routine operations, the landfill facility must retain the following information (40 CFR Part 258.29):

- Inspection records, training procedures, and notification procedures
- Gas-monitoring results from routine monitoring
- Design documentation for placement of leachate or gas condensate in a landfill
- Closure and postclosure care plans
- Any monitoring, testing, or analytical data
- Cost estimates and financial assurance documentation

10.5 CLOSURE

Subtitle D requires, at the time of landfill closure, the installation of a final cover (cap) system (40 CFR Part 258.60). The primary purposes of the cover are to minimize infiltration of rainwater (thus limiting the production of leachate) and to prevent erosion (thus protecting buried wastes from exposure and possible dispersal). The final cover system must be constructed to:

- Have a permeability less than 1×10^{-5} cm/sec
- Minimize infiltration through the landfill using a layer that contains at least 46 cm (18 in.) of soil material (the 'barrier layer')
- Minimize erosion of the final cover using an 'erosion layer' that contains a minimum of 15 cm (6 in.) of soil material capable of sustaining plant growth

Written closure plans must describe all steps necessary to close landfill units. After closure of a unit, postclosure care is required for at least 30 years. The following issues must be addressed at a minimum:

- Maintain the integrity and effectiveness of the final cover
- Maintain and operate the leachate collection system in accordance with 40 CFR 258.40
- Monitor groundwater in accordance with 40 CFR 258 and maintain the groundwater-monitoring system
- Maintain and operate the gas-monitoring system in accordance with 40 CFR 258.23

Figure 10.28 provides a schematic of a recommended top slope cap. By restricting the entry of water into the landfill by a cap system, generation of leachate is substantially minimized. However, the dry conditions that are maintained will hinder MSW biodegradation, making most landfills merely storage facilities (Vesilind et al., 2002), sometimes known as 'dry tombs.'

Components for landfill closure include (40 CFR 258.60) an infiltration (barrier) layer or an alternative barrier system, a drainage layer, an erosion control layer, and a gas venting system.

10.5.1 LOW PERMEABILITY (BARRIER) LAYER

The barrier layer for landfills consists of a compacted soil layer or a soil and geomembrane liner. Both systems are designed to reduce the rate at which surface water infiltrates into the landfill unit. An alternative barrier system may be used if approved by the state regulatory agency. The membrane material used for the final cover must be composed of a long-lasting material and must tolerate subsidence-induced strains (U.S. EPA, 1994).

10.5.2 DRAINAGE LAYER

A drainage layer is placed above the low-permeability layer and maintains the stability of cover slopes by eliminating pore water. A drainage layer in the cover system is not required under RCRA Subtitle D; however, large landfills possess such a layer. This layer prevents any water that infiltrates the erosion control layer from accumulating above the barrier layer. Accumulated water can generate pressure above the membrane and cause the erosion control layer to slide off the cover side slopes. The side slope drainage layer is drained to a large capacity toe drain (see Figure 10.19).

10.5.3 EROSION CONTROL LAYER

The erosion control layer consists of soil planted with vegetation to protect the cover from the effects of erosion. The minimum thickness of the erosion layer required under Subtitle D is 15 cm

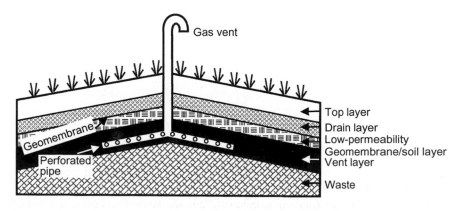

FIGURE 10.28 Cover design for a closed landfill (U.S. EPA, 1994).

(6 in.). A mixture of dense-rooted grasses and legumes is recommended for establishment on this layer. Erosion-related soil loss should not exceed 1.8 metric tons (2 tons) per acre per year to minimize long-term maintenance. To attain such a level of erosion control typically requires the construction of slopes less than 1:4 and drainage swales placed at 6 m (20 ft) vertical increments. Erosion from the effects of water is kept under control by vegetation and also by hardening the cover surface using stones or riprap (U.S. EPA, 1994).

Erosion control maintenance includes routine vegetation management (such as mowing, fertilization, liming, and replanting), repair of any areas undergoing subsidence, and run-on or runoff control. Sedimentation basins and drainage swales must be inspected after every major rainstorm and be repaired as needed.

10.5.4 Gas Collection System

A minimum of one passive gas vent per acre of cover should be installed to allow for the release of gas pressure beneath the cover. The gas venting system can use vertical gravel walls, blanket collectors (beneath the barrier layer), or gravel trench drains (also beneath the barrier layer) to collect landfill gases. The collected gases are routed through the cover using vent pipes as shown previously in Figure 10.18.

10.5.5 The Landfill Cap

Slope stability and soil erosion are important concerns in the design and installation of landfill caps. The landfill cover slope must be sufficiently stable to sustain infiltration and runoff from a 24-h, 25-year storm. Side slopes are typically 1:3 to 1:4, and the friction between adjacent layers must resist seepage forces. On slide slopes, composite liner caps (membranes placed directly above a low-permeability soil layer) are not advisable (Vesilind et al., 2002). For slopes steeper than 1:5, a drainage layer should be provided. If sliding occurs, liner systems will be damaged, soil may enter surface waterways, and the cover will need to be repaired or rebuilt.

Two different cover systems are depicted in Figure 10.29. See also the photos in Figure 10.30.

10.5.6 Subsidence Effects

Landfill subsidence can be large-scale ('global' due to uniform settlement of MSW) or localized (e.g., collapse of a large void directly below a portion of the cover). In general, global subsidence does not result in excessive tensile strains on the cover and improves the stability of the cover by reducing sliding. Localized subsidence, however, can produce depressions on the cover that can create excessive strain in cover layers and can cause ponding of water. The impact of tensile strains is minimized using a flexible geomembrane composed of PVC, low-density polyethylene, or polypropylene. Ponding of water must be avoided because it can kill or distress cover vegetation, and the weight of the water can cause expansion of a pond on the cover (U.S. EPA, 1994).

10.5.7 Weather Effects

The cover must withstand extreme weather conditions and function with minimal maintenance. The extreme weather conditions for which a final cover should be designed include heavy rains, extreme drought, and ground freezing. Cover management for heavy rains includes growing dense-rooted vegetation, maintaining a modest slope, and constructing adequate conveyances for excess runoff. Extreme drought is another consideration during the design of the erosion control layer. Certain plants are more drought-tolerant and should be included in the original seed mixture. Periodic irrigation with water and leachate may be required. Freezing of the cover is a concern because of the impact of freezing on clay permeability. Repeated cycles of freezing and thawing increase the permeability of compacted clays by causing large cracks to form. Damage to the clay layer is difficult

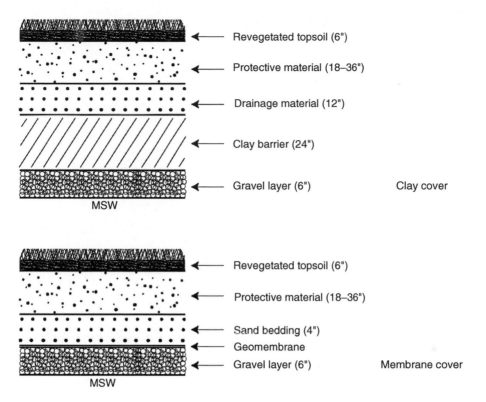

FIGURE 10.29 Two different cover types for a sanitary landfill. (From Vesilind et al., *Solid Waste Engineering* 1st Ed., 2002. Reproduced with kind permission of Brooks/Cole, a division of Thomson Learning: www. thomsonrights.com.)

if not impossible to correct in such a scenario; however, the impermeability of the clay layer can be greatly augmented by the installation of a geomembrane cap directly below or above the clay layer.

10.6 POSTCLOSURE

After a landfill cell is closed and the final cover is installed, monitoring and maintenance are necessary to ensure that the landfill remains stable. Subtitle D requires that postclosure care and monitoring be performed for at least 30 years. A postclosure care and monitoring plan is required by the state regulatory agency and must include (40 CFR 258.61):

- The start and completion dates of the postclosure period
- The monitoring plan description
- The maintenance program description
- The facility's personnel list of contacts for emergencies
- A description of the end-use plan for the site

Postclosure care activities must include (U.S. EPA, 1994):

- Maintaining the integrity and effectiveness of erosion controls
- Maintaining and operating the leachate collection system
- Maintaining and operating the gas venting system
- Monitoring groundwater for contamination

(a)

(b)

FIGURE 10.30   Landfill cap under construction showing overlapping sheets of geomembrane cap.
Photo by Theresa M. Pichtel.

After a final cover is installed, the leachate collection system will have a very small leachate load and should be relatively easy to maintain. Leachate generation should decrease to less than 9350 L/ha (1000 gal/acre) / day, which should not strain a system designed to handle storm waters. During the postclosure period, leachate production rates should be monitored to detect any decrease in production. If leachate generation falls markedly, the leachate pipes should be inspected for biological clogging. The LCR should be flushed if clogging is detected.

The vent pipes in a passive gas venting system must be inspected regularly for damage caused by mowing or other traffic. A damaged vent pipe can allow surface water to enter the gas venting system and bypass the cover. Damaged vent pipes must be repaired promptly. During the postclosure period, groundwater monitoring must continue to be conducted on a routine basis. The facility must be aware of any indications of contamination and must take necessary remedial action if contamination occurs.

10.7 THE BIOREACTOR LANDFILL

When MSW is deposited in a conventional Subtitle D landfill, certain events such as partial waste decomposition, gas production, leachate generation, and stabilization inevitably occur. Recent investigations (Rathje and Murphy, 1992) involving core sampling of sanitary landfills have revealed that wastes do not degrade significantly even after many decades, resulting in labels such as 'dry tombs' for these systems.

An innovative approach to MSW disposal, which actually encourages rapid MSW decomposition and speeds stabilization, is the bioreactor landfill. The enhanced waste degradation and stabilization carried out by the indigenous microbial populations within the waste is accomplished through the addition of liquid (typically leachate) and air. The enhanced microbiological processes within a bioreactor can transform and stabilize the decomposable organic waste within 5 to 10 years of implementation, compared with many decades for conventional Subtitle D landfills where wastes are essentially sealed off from air and moisture.

To date, there is still disagreement among scientists and engineers as to the precise definition of a bioreactor landfill. The Solid Waste Association of North America (SWANA, 2001) has defined a bioreactor landfill as:

> any permitted Subtitle D landfill or landfill cell, subject to New Source Performance Standards/Emissions Guidelines, where liquid or air, in addition to leachate and landfill gas condensate, is injected in a controlled fashion into the waste mass in order to accelerate or enhance biostabilization of the waste.

Bioreactor landfill technology has been in use for over a century. The concept originates from the systematic treatment of urban wastewater that began in the late 1800s. Bioreactor landfills can be conceptualized as an extension of anaerobic and aerobic digestion at wastewater treatment plants.

Three general types of bioreactor landfill configurations are currently in use (U.S. EPA, 2003) and are outlined below.

10.7.1 ANAEROBIC BIOREACTORS

Landfill degradation of MSW frequently is rate-limited by insufficient moisture (Campman and Yates, 2002). The average landfilled MSW has a moisture content from 15 to 40%, depending on the composition of the wastes, season of the year, and weather conditions (Emcon Associates, 1980; Tchobanoglous et al., 1993; Kiely, 1997). However, maximum methane production in landfills occurs at a moisture content of 60 to 80% wet weight (Farquhar and Rovers, 1973), suggesting that most landfills are well below the optimum moisture content for methane production.

In an anaerobic bioreactor landfill, moisture is added to the waste mass uniformly in the form of recirculated leachate, local water, or other sources to obtain optimal moisture levels. Liquid is injected into the waste via horizontal trenches, vertical wells, surface infiltration ponds, spraying, and prewetting of waste (Figure 10.31). Methods of liquid addition are addressed in detail by Reinhart and Townsend (1998). Biodegradation occurs under anaerobic conditions and produces landfill gas, primarily methane and carbon dioxide, in approximately equal proportions.

Anaerobic bioreactor landfills require careful monitoring at startup. If the waste is wetted too rapidly, a buildup of volatile organic acids might lower leachate pH, inhibiting the methane-producing bacterial population and reducing biodegradation rate. Optimal conditions for methanogenic bacteria include a pH near the neutral point. Leachate parameters such as pH, volatile organic acids, and alkalinity and gas parameters such as methane content are direct indicators of the activity of the methanogenic bacterial population. A high-volatile organic acids to alkalinity ratio (>0.25) indicates that the leachate might possess a low buffering capacity and conditions could inhibit methane generation (Campman and Yates, 2002).

When the methane content of the landfill gas exceeds approximately 40%, the methanogenic bacterial populations are considered established. A decrease in the methane gas content below 40%

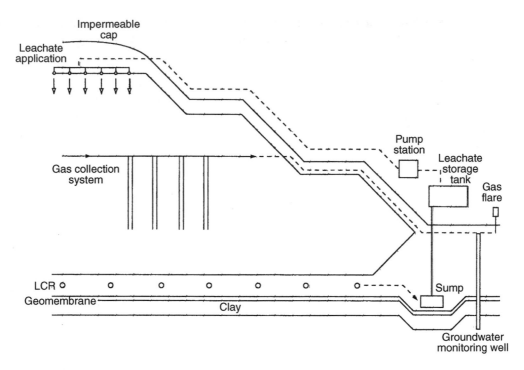

FIGURE 10.31   Schematic of a bioreactor landfill.

may indicate that the waste is too wet or dry. Once the methanogenic bacteria have become established, the rate of leachate recirculation may be increased.

Costs for piping, pumps, electricity, and equipment for increased landfill gas generation for an anaerobic bioreactor should be offset by the avoided cost of leachate treatment and landfill gas-to-energy royalties from gas usage by manufacturers or utilities (Campman and Yates, 2002).

Benefits of anaerobic bioreactor landfills include:

- Leachate storage within the waste mass
- Increased rate of landfill settlement
- More rapid waste stabilization than conventional landfills
- Increased methane generation rates (200–250% increase typical)
- Lower postclosure costs

10.7.2 AEROBIC BIOREACTOR

The aerobic bioreactor process is analogous to a composting operation in which input materials are rapidly biodegraded using air, moisture, and increased temperatures brought about by biological processes. Aerobic bioreactors operate by the controlled injection of moisture and air into the waste mass through a network of pipes.

Prior to air injection, liquid is pumped under pressure into the waste mass through injection wells in order to wet the mass to a moisture content between 50 and 70% (by wt). Once optimal moisture conditions have been reached, air is injected into the waste mass using vertical or horizontal wells. Blowers are used to force air into the waste mass through a network of perforated wells previously installed within the landfill. Leachate is removed from the base, directed to storage tanks, and recirculated into the landfill in a controlled manner. Air and liquid injection rates are similar to the air and moisture application rates used in composting systems.

Optimum temperatures for waste degradation within an aerobic bioreactor landfill are between 60 and 72°C (about 140 and 160°F) (Campman and Yates, 2002). The aerobic process continues until most of the readily degradable compounds have decomposed and the waste temperature gradually decreases during the final phase of maturation of the remaining organic matter.

Due to the substantial amounts of heat generated in an aerobic bioreactor, large quantities of leachate can evaporate. In a study of two bioreactor landfills, leachate volume was reduced by 86 and 50% (Hudgins and Green, 1999). Changing the rate of air and liquid injection will alter the temperature of the waste pile. To ensure against possible waste combustion, the waste mass is wetted adequately and air injection is uniform throughout the waste mass.

Aerobic bioreactor landfills require much greater inputs compared with their anaerobic counterparts. According to Weathers et al. (2001), the additional power required to inject air into an aerobic bioreactor was 12 times higher than the power required to extract landfill gas from an anaerobic bioreactor. However, postclosure costs should be reduced substantially due to reductions in landfill gas generation and cover settlement (Campman and Yates, 2002).

Because of the higher reaction rates, aerobic biodegradation is a more rapid process than anaerobic biodegradation. Consequently, aerobic landfills have the potential to achieve waste stabilization in 2 or 4 years as opposed to decades or longer for conventional landfills. The rapid rate of waste stabilization in aerobic landfills also offers the potential for 'mining' of the landfill waste (see below).

Aerobic bioreactors have many of the same benefits as anaerobic bioreactors, only these benefits are achieved more quickly. Since aerobic bioreactors do not produce significant quantities of methane, there is little potential to sell landfill gas for energy (U.S. EPA, 2000). The following benefits have been observed at aerobic bioreactor landfills (Hudgins and Green, 1999; Campman and Yates, 2002):

- More rapid waste and leachate stabilization
- Increased rate of landfill settlement
- Reduction of methane generation by 50 to 90%
- Capability of reducing leachate volumes by up to 100% due to evaporation
- Potential for landfill mining
- Reduction of environmental liabilities

10.7.3 ANAEROBIC VS. AEROBIC

Recall from the discussion of conventional sanitary landfills that MSW deposited in a landfill tends to undergo five phases of decomposition, starting with a brief aerobic phase, through two anaerobic phases followed by a methane generation phase and finally maturation. Aerobic bioreactor landfills attempt to sustain the Phase I activity over a longer period of time than that occurring in a conventional MSWLF. In contrast, anaerobic bioreactor landfills attempt to reduce significantly the time involved for Phase IV activities (methane generation) to possibly 5 to 10 years (a 75% reduction), with 5 to 7 years for Phase IV considered optimum (U.S. EPA, 2000).

10.7.4 HYBRID BIOREACTORS (SEQUENTIAL AEROBIC–ANAEROBIC)

The hybrid bioreactor landfill accelerates waste degradation by employing a sequential aerobic–anaerobic treatment to rapidly degrade organics in the upper lifts of the landfill and collect gas from lower sections. Operation as a hybrid results in the earlier onset of methanogenesis compared with aerobic landfills (U.S. EPA, 2003).

10.7.5 PRACTICAL OPERATIONAL ISSUES

10.7.5.1 Waste Preprocessing

Wastes may be placed directly into a bioreactor landfill after which they are compacted by heavy machinery, or wastes can be shredded prior to placement. The goal for such preprocessing is to achieve optimum exposure of waste material to the bioreaction process. There have been some concerns (U.S. EPA, 2000) regarding the absence of decomposition in bioreactor landfills when MSW is placed inside plastic bags, which may or may not be broken open during conventional compaction with heavy equipment. It may be feasible to either break open plastic bags to expose the contents (by equipment or during emplacement) or require the use of degradable bags for optimum bioreactor performance.

10.7.5.2 Daily Cover

Daily cover materials should be selected to avoid creation of low-permeability layers within the landfill cell. For example, clay can become a barrier to leachate drainage and recirculation, while foams, slurries, sludges, or reusable tarps will provide the benefits of daily cover without preventing infiltration and drainage (U.S. EPA, 2000).

10.7.5.3 Liquid Addition and Recirculation

A major landfill stability concern involves leachate (i.e., hydraulic head) buildup on the landfill liner system. Ponding of liquid on the liner can be a significant source of failure as a result of the associated hydrostatic forces. Addressing this in the initial design phase should be a straightforward issue for bioreactor landfills because liquid levels and other variables are generally known.

Considerations in addressing liquid addition to bioreactor landfills should include:

- How to differentiate between the amounts of liquid needed by different types and sizes of landfills
- Use of temperature to control liquid injection since wetting of the waste mass results in the most uniform temperature
- Timing of liquid addition, e.g., at time of waste disposal, or deposit dry waste first and add liquid later
- Determination of the desired moisture content and the amount of liquid required is necessary for the design of an effective distribution system

10.7.5.4 Alternative Liquid Sources

Alternatives to landfill leachate for liquid addition include wastewater, biosludges, biosolids from POTWs, stormwater runoff, and groundwater. Biosolids considered most suitable for bioreactor use are those in liquid form that typically undergo land application, rather than the dewatered sludge (U.S. EPA, 2002). Use of the liquid form could avoid dewatering costs, but will require more trucks to transport the larger volume of the dilute wastewater. There are concerns regarding operational health and safety issues about this material. Of particular concern is potential worker exposure to pathogens and risks to workers and nearby residential areas from aerosols resulting from biosolids application to the landfill surface.

As should be clear by now, the moisture content of wastes is critical for bioreactor operation. The addition of significant quantities of liquid may be required. Some estimates (U.S. EPA, 2000) indicate that about 50 million L (13 million gal) of liquid might be needed for 365,000 MT (400,000 tons) of waste; others estimate this requirement as 270 L of water/m³ (54 gal/yd³) of waste. Landfills in states with dry conditions may require significantly larger liquid quantities. Also, more liquid may be needed to sustain bioreactions after a low permeability cover or cap is installed because the landfill moisture may be removed via the gas collection system. Thus, the leachate generated in the landfill should not be considered sufficient to support the bioreaction moisture needs.

10.7.5.5 Fires

Active landfill gas collection systems are a source of fires. Other fire initiators include drilling operations on the landfill and lightning strikes. Potential fire hazards during drilling are easily controlled through safe work practices such as no smoking and the use of spark-free tools and equipment. Note that surface fires are much easier to control and eliminate than underground fires.

Aerobic bioreactor landfills rely on high temperatures as well as the addition of oxygen to sustain the bioreaction. For such operations, careful regulation of moisture and oxygen levels enable control of the waste mass temperature as well as fire potential.

10.7.6 OTHER CONSIDERATIONS

10.7.6.1 Leachate Strength Reduction

Bioreactor landfills decrease the strength of landfill leachate more rapidly than conventional Subtitle D landfills. Chemical oxygen demand (COD) is one common indicator of leachate strength. Reinhart and Townsend (1998) summarized measurements of COD half-lives for conventional and bioreactor landfills. The time it takes for COD to be reduced by 50% (i.e., the half-life) is about ten times faster in a bioreactor landfill than in a conventional landfill. Data up to this point, however, are limited (Campman and Yates, 2002).

10.7.6.2 Waste Mass Shear Strength

A dry waste mass may be quite strong as evidenced by modern landfills having waste heights of up to 90 m (approx. 300 ft) and slopes steeper than 3:1. However, the addition of water adds weight but not shear strength, which affects traditional landfill design factors such as the waste mass geometry. Some geometries used for dry landfills may not work with bioreactor landfills because of differences in the shear strength of the waste and elastic displacement caused by water addition (Campman and Yates, 2002).

10.7.6.3 Waste Settlement

Accelerating MSW degradation can reduce the need for new landfills by conserving volume. Conventional landfill settlement is typically around 10% of landfill height and generally occurs over a number of years as the waste decomposes (Koerner and Daniel, 1997). Settlement of the waste mass in a bioreactor landfill can be significant over time, involving 10 to 25% of the landfill height. Aerobic bioreactors might achieve this settlement within 2 to 4 years, while anaerobic bioreactors might require 5 to 10 years (Campman and Yates, 2002). Pilot scale landfill cells in Sonoma County and Mountain View, California, experienced settlement by as much as 20 and 14%, respectively, in leachate recirculation cells and approximately 8 to 10%, respectively, in the conventional dry cells (Reinhart and Townsend, 1997). Waste settlement varies greatly and is dependent on type of waste, amount of cover, and compaction. Settlement also will not be consistent across the landfill surface. Gas collection and other internal landfill systems (such as leachate collection and recirculation) must be able to shift with this settlement (U.S. EPA, 2000).

Increased rates of settlement before closure will permit additional MSW to be placed in the landfill before a cap is put in place. Such additional waste placement can therefore reduce the need for new landfills. These benefits can be realized only when waste decomposes prior to closure. Landfill operators may choose to delay closure in order to take advantage of increased space created by additional waste settlement.

10.7.6.4 Metals

The long-term fate of metals in bioreactor systems is generally unknown. Since heavy metals tend to concentrate during wastewater biosolids treatment, similar effects could be anticipated in bioreactor landfills during waste decomposition, and such changes in heavy metal concentrations could be seen in leachate quality. Issues regarding the behavior of metals in the landfill environment include:

- Microorganisms may concentrate metals
- pH and sulfides in the landfill may affect metal mobilization
- Potential for remobilization of metals if landfill conditions become anaerobic

Research conducted at Georgia Tech and elsewhere indicate that there is some potential for metal mobilization in bioreactors; however, there are multiple mechanisms for attenuation of all metals and therefore the metals generally precipitate within the waste mass. In addition, a review of data from 12 landfills indicated that heavy metals were not an issue for a fully stabilized anaerobic landfill. Over a pH range of 7 to 9, as are typically encountered in these landfills, the metals were immobilized. While metals are present in the landfill leachate, all values were below drinking water standards (U.S. EPA, 2000).

10.7.6.5   Advantages of Bioreactor Landfills

As was the case for the conventional sanitary landfill, gases emitted from a bioreactor landfill consist primarily of methane and carbon dioxide along with lesser amounts of volatile organic chemicals and hazardous air pollutants. Use of a bioreactor is expected to generate landfill gas earlier and at a higher rate compared with a conventional landfill. The bioreactor landfill gas is also generated over a shorter period of time because emissions decline as the accelerated decomposition process depletes the microbial substrates faster than in a traditional landfill. According to the U.S. EPA (2003), the bioreactor produces more landfill gas overall than the traditional landfill.

Some studies indicate that the bioreactor increases the feasibility for cost-effective landfill gas recovery, which in turn reduces fugitive emissions. This presents an opportunity for beneficial reuse of bioreactor gas in energy recovery projects. Currently, the use of landfill gas (in traditional and bioreactor landfills) for energy applications is only about 10% of its potential use. The U.S. Department of Energy estimates that if the controlled bioreactor technology were applied to 50% of the waste currently being landfilled, it could provide over 270 billion ft^3 of methane per year, which is equivalent to 1% of electrical needs in the United States. Other potential advantages of bioreactor landfills include (U.S. EPA, 2003):

- Decomposition and biological stabilization in years vs. decades in conventional landfills ('dry tombs')
- Lower waste toxicity and mobility due to both aerobic and anaerobic conditions
- Reduced leachate disposal costs
- A 15 to 30% gain in landfill space due to an increase in density of waste mass
- Reduced postclosure care

10.7.6.6   Summary of Bioreactor Landfills

Bioreactor landfills are engineered systems that incur higher initial capital costs and require additional monitoring and control during their operating life, but are expected to involve less monitoring over the duration of the postclosure period than conventional 'dry tomb' landfills.

Moisture content is the single most important factor that promotes the accelerated decomposition. The bioreactor technology relies on maintaining optimal moisture content near field capacity (approximately 35 to 65%) and adds liquids when it is necessary to maintain that percentage. The moisture content, combined with the biological action of naturally occurring microbes, decomposes the waste. Issues that need to be addressed during both design and operation of a bioreactor landfill include:

- Increased gas emissions
- Increased odors
- Physical instability of waste mass due to increased moisture and density
- Instability of liner systems
- Surface seeps
- Landfill fires

10.8 LANDFILL RECLAMATION

Landfill reclamation is a relatively new approach used to expand landfill capacity and avoid the high costs of acquiring additional land (U.S. EPA, 1997). Reclamation costs may be offset by the sale of recovered materials such as recyclables including ferrous metals, aluminum, plastic, glass, and the sale of carbonaceous wastes, which can be burned as fuel. Reclaimed soil can be used as cover material, sold as construction fill, or sold for other uses. Other benefits of landfill reclamation include the avoided liability through site remediation, reductions in closure costs, and conversion of the landfill site for other uses. Reclamation projects have been successfully implemented at MSWLFs across the United States since the 1980s (U.S. EPA, 1997).

The process of landfill reclamation is summarized as follows: The contents of the landfill cell are excavated using a bulldozer or front-end loader. A front-end loader then organizes the excavated materials into manageable stockpiles and separates out bulky material such as appliances and steel cable. A trommel screen (see Chapter 7) or vibrating screens separate soil (including old cover material) from solid waste in the excavated material. The size and type of screen used depends on the desired end use of the recovered material. For example, if the reclaimed soil typically is used as landfill cover, a 2.5 in. screen is used for separation. If, however, the reclaimed soil is sold as construction fill, a smaller mesh screen is used to remove small pieces of metal, plastic, glass, and paper.

The separated soil can be used as daily cover in a sanitary landfill. The excavated waste can be processed at a materials recovery facility to remove valuable components (e.g., steel and aluminum) or burned in a municipal waste incinerator to produce energy.

In 1986, the Collier County (Florida) Solid Waste Management Department at the Naples Landfill conducted one of the earliest landfill reclamation projects in the United States. The Naples facility, a 33 acre unlined landfill, contained MSW buried for up to 15 years. In an evaluation performed on many of the state's unlined landfills, it was discovered that the Naples Landfill, along with 27 others, posed a threat to groundwater. City officials formulated a reclamation plan with the following objectives:

- Decreasing site closure costs
- Reducing the risk of groundwater contamination
- Recovering and burning combustible waste in a proposed waste-to-energy facility
- Recovering soil for use as landfill cover material
- Recovering recyclable materials

A U.S. EPA assessment of the reclamation project found the processing techniques to be efficient for recovering soil but not for recovering recyclables of marketable quality. During a demonstration project, the county efficiently recovered a soil fraction deemed environmentally safe. The 45,000 MT (50,000 tons) of reclaimed soil were suitable for use as landfill cover material and as a medium for plant growth.

In 1990, the Lancaster County (Pennsylvania) Solid Waste Management Authority constructed a municipal solid waste incinerator to use in reducing the volume of waste entering the Frey Farm Landfill, a lined site containing MSW deposited for up to 5 years. City officials initiated a landfill reclamation project to augment the incinerator facility's supply of fresh waste with reclaimed waste. The reclaimed waste had a modest heating value (approximately 6,900 kJ/kg or 3,100 Btu/lb). To achieve a higher heating value, fresh waste, containing wood chips and discarded tires, was mixed with reclaimed waste. Approximately 220,000 m^3 (287,000 yd^3) of MSW were excavated from the landfill, and 2,400 MT (2,645 tons) of screened waste were processed per week for the incinerator. As a result, Lancaster County converted 56% of the reclaimed waste into fuel. The county also recovered 41% of the reclaimed material as soil during screening (trommeling) operations. The remaining 3% proved noncombustible and was reburied in the landfill. By the end of the project in

1996, landfill operators had reclaimed 230,000 to 305,000 m³ (300,000 to 400,000 yd³) of landfill waste. Benefits of the project at Frey Farm Landfill include:

- Reclaimed landfill space
- Supplemental energy production
- Recovered soil and ferrous metals

Drawbacks included:

- Increased generation of ash at the incinerator caused by the high soil content found in the reclaimed waste
- Increased odor and air emissions
- Increased traffic on roads between the incinerator and the landfill
- Increased wear on incinerator equipment due to the abrasive properties of the reclaimed waste

Additional drawbacks exist to landfill reclamation. During excavation, methane and other gases that result from decomposing wastes are released to the atmosphere. Excavation may also uncover hazardous materials, which are costly to manage. In addition, the excavation work may cause adjacent landfill areas to sink or collapse. To identify potential problems, engineers and landfill operators considering reclamation should conduct a site characterization study.

REFERENCES

American Society for Testing and Materials, Standard Test Methods for Water Vapor Transmission of Materials, ASTM E96-00, West Conshohocken, PA, 2000.

American Society for Testing and Materials, Standard Terminology for Geosynthetics, ASTM D4439-02, West Conshohocken, PA, 2002.

Bourne, Massachusetts, Landfill Liner System, Phase 3, Integrated Solid Waste Management Department, 2000. See: http://www.townofbourne.com/Town%20Offices/ISWM/Layer2.htm

British Crop Protection Council, Contract Report for Enventure, Ltd. Fly control on landfill: A literature review, 1999. See: http://www.enventure.co.uk/docs/Fly%20control%20on%20landfill%20sites %20-%2 0literature%20study.pdf

Buss, S.E., Butler, A.P., Johnston, P.M., Sollars, C.J., and Perry, R., Mechanisms of leakage through synthetic landfill liner materials, *J. Chartered Insti. Water Environ. Manage.*, 9, 353–359, 1995.

Butler, S.E., Butler, A.P., Johnston, F.M., Sollars, C.J., and Perry, R., Mechanisms of leakage through synthetic landfill liner materials, *J. Chartered Inst. Water and Environ. Manage.*, 9, 353–359, 1995.

Campman, C. and Yates, A., Bioreactor Landfills: An Idea whose Time has Come, MSW Management. See: http://www.coe.uncc.edu/~hhilger/CEGR%204143/Lanfill%20Presentations/Bioreactor%20Landfills.doc

Code of Federal Regulations, Vol. 40, Part 258, Criteria for municipal solid waste landfills, U.S. Government Printing Office, Washington, DC, 2004.

Code of Federal Regulations, Vol. 40, Part 761, Polychlorinated biphenyls (PCB) manufacturing, processing, distribution in commerce, and use prohibitions, U.S. Government Printing Office, Washington, DC, 2004.

Ellis and Blood-Smyth, No date, Fly control on landfill. A literature review, Contract Report for Enventure Ltd, West Yorkshire, UK.

Environmental Research and Education Foundation, Bioreactor Landfill Project Northern Oaks Landfill, 2002. See: http://www.erefdn.org/rpts_summary_ordrs/northernoaks.htm

Fenn, D.G., Hanley, K.J., and Degeare, T.V., Use of the water balance method for predicting leachate generation from solid waste disposal sites, U.S. Environmental Protection Agency, EPA-530/SW-168, U.S. Government Printing Office, Washington, DC, 1975.

Flower, F.B., Leone, I.A., Gilman, E.F., and Arthur, J.J., *Vegetation Kills in Landfill Environs,* Cook College, Rutgers University; New Brunswick, NJ, 1982.

Giroud, J.P. and Bonaparte, R., Leakage through liners constructed with geomembranes Part I, Geomembrane liners. *Geotextiles and Geomembranes*, Vol. 8, Elsevier Science Publishers, England, 1989.

Guhlke, M.R., Management program to control flies and odor in poultry manure storage and disposal, *J. Environ. Health*, 47, 314–317, 1985.

Haxo, H. E. and Lahey, T.P., Transport of dissolved organics from dilute aqueous solutions through flexible membrane liners, *Hazardous Waste Hazardous Mater.*, 5, 275–94, 1988.

Hoyer, H., Survey of Europe and North Africa for parasitoids that attack filth flies, Miscellaneous Publications of the Entomologist Society of America, No-61, 1986, pp. 35–38.

Hudgins M. and Green, L., Innovative Landfill Gas Control Using an Aerobic Landfill System, *Proceedings of the SWANA 22nd Annual Landfill Gas Symposium*, Lake Buena Vista, FL, 1999.

Intergovernmental Panel on Climate Change, *Climate Change — The IPCC Scientific Assessment*, Houghton, J.T., Jenkins, G.J., and Ephraums, J.J., (Eds.), World Meteorological Organization/United Nations Environment Programme, Geneva, 1990.

Kiely, G., *Environmental Engineering*, The McGraw-Hill Companies, New York, NY, 1997.

Koerner, R.M. and Daniel, D.E., *Final Covers for Solid Waste Landfills and Abandoned Dumps*, ACSE Press, Reston, VA, 1997.

McBean, E.A., Pohland, F.G., Rovers, F.A., and Crutcher, A.J., Leachate design for containment landfills, *J. Environ. Engin. Div.*, ASCE 108, EE1, 204–209, 1982.

McEnroe, B.M., Maximum saturated depth over landfill liner, *J. Environ. Engin. Div. ASCE*, 119, 262–270, 1993.

Oweis, S. and Khera, R.P., *Geotechnology of Waste Management*, PWS Publishing, Boston, MA, 1998.

Park, J.K. and Nibras. M., Mass flux of organic chemicals through polyethylene geomembranes, *Water Environ. Res.*, 65, 227–237, 1993.

Qian, X., Koerner, R.M., and Gray, D.H., *Geotechnical Aspects of Landfill Design and Construction*, Prentice-Hall, Upper Saddle River, NJ, 2002.

Rathje, W. and Murphy, C., *Rubbish: The Archaeology of Garbage*, Harper Collins, New York, 1992.

Reinhart, D.R. and Townsend, T.G., *Landfill Bioreactor Design and Operation*, Lewis Publishers, New York, NY, 1997.

Richardson, G. and Zhao, A., Design of Lateral Drainage Systems for Landfills, Tenax Corp., Baltimore, MD, 2000.

Rueda, L.M, Roh, P.U., and Rya, J.L., Pupal parasites (Hymenoptera: Pteromalidae) of filth flies breeding in refuse and poultry and livestock manure in South Korea, *J. Med. Entomol.*, 34, 82–85, 1997.

Schroeder, P.R., Aziz, N.M., Lloyd, C.M., Zappi, P.A., The Hydrologic Evaluation of Landfill Performance (HELP) Model. User's Guide for Version 3, EPA/600/R-94/168a, U.S. EPA Risk Reduction Engineering Laboratory, Cincinnati, OH, 1994.

Sleats, R., Harries, C., Viney, I., and Rees, J.F., Activities and distribution of key microbial groups in landfills, in *Sanitary Landfilling: Process, Technology, and Environmental Impact, Academic Press, London*, 1989.

Solid Waste Association of North America (SWANA), A Compilation of Landfill Gas Field Practices and Procedures, Landfill Gas Division of the Solid Waste Association of North America, March, 1992.

Solid Waste Association of North America as cited in U.S. EPA, 2004.

Sulaiman, S., Omar, B., Jeffrey, J., Ghauth, I., and Busparani, V., Survey of the microhymenoptera, hymenoptera, chacidoidae, parasitising filth flies, *Diptera, Muscidae, Calliphoridae* breeding in refuse and poultry farms on peninsular Malaysia, *J. Med. Entomol.*, 27, 851–855, 1990.

Tchobanoglous, G., Theisen, H., and Vigil, S., *Integrated Solid Waste Management: Engineering Principles and Management Issues*, McGraw-Hill, Inc., New York, NY, 1993.

U.S. Department of Energy, Using Landfill Gas for Energy: Projects that Pay, DOE/CH10093-322; DE94006897, Office of Energy Efficiency and Renewable Energy, Washington, DC, May 1994, Revised October 1994.

U.S. Environmental Protection Agency, Handbook - Remedial Action at Waste Disposal Sites, EPA/625/6-85/006, Office of Research and Development, Cincinnati, OH, 1985.

U.S. Environmental Protection Agency, Air Emissions from Municipal Solid Waste Landfills – Background Information for Proposed Standards and Guidelines, EPA-450/3-90-011a. PB91-197061, National Technical Information Service, Cincinnati, OH, March, 1986a.

U.S. Environmental Protection Agency, SW-846, Method 9095 – Paint Filter Liquids Test, Test Methods for Evaluating Solid Waste, Volume 1A: Laboratory Manual Physical/Chemical Methods, Office of Solid Waste and Emergency Response, Washington, DC, 1986b.

U.S. Environmental Protection Agency, Loading point puncturability analysis of geosynthetic liner materials, PB88-235544, National Technical Information Service, Cincinnati, OH, 1988a.

U.S. Environmental Protection Agency, Summary of Data on Municipal Solid Waste Landfill Characteristics – Criteria for Municipal Solid Waste Landfills, EPA/530-SW-88-038, Office of Solid Waste and Emergency Response, Washington, DC, 1988b.

U.S. Environmental Protection Agency, Requirements for hazardous waste landfill design, construction and closure, EPA/625/4-89/022, Seminar Publication, Office of Research and Development, Cincinnati, OH, 1989.

U.S. Environmental Protection Agency, Technical Guidance Document: Quality Assurance and Quality Control for Waste Containment Facilities, EPA/600/R-92/182, Cincinnati, OH, 1992a.

U.S. Environmental Protection Agency, Alternative Daily Cover Materials for Municipal Solid Waste Landfills, U.S. EPA Region IX, San Francisco, CA, 1992b.

U.S. Environmental Protection Agency, Safer Disposal for Solid Waste, The Federal Regulations for Landfills, EPA/530 SW-91 092, March, 1993, Office of Solid Waste and Emergency Response (OS-305), Washington, DC, 1993a.

U.S. Environmental Protection Agency, Criteria for Solid Waste Disposal Facilities, A Guide for Owners/ Operators, EPA/530-SW-91-089, Solid Waste and Emergency Response (OS-305), Washington, DC, March, 1993b.

U.S. Environmental Protection Agency, Technical Manual, Solid Waste Disposal Facility Criteria. 530-R-93-182, Solid Waste and Emergency Response, Washington, DC, 1993c.

U.S. Environmental Protection Agency, Design, Operation, and Closure of Municipal Solid Waste Landfills, Seminar Publication EPA 625/R-94/008, Office of Research and Development, Washington, DC, September, 1994.

U.S. Environmental Protection Agency, Compilation of Air Pollutant Emission Factors, AP-42, 5th ed., Supplement C. Office of Air Quality Planning and Standards, Research Triangle Park, NC. U.S. Environmental Protection Agency, 1997a.

U.S. Environmental Protection Agency, Landfill Reclamation. EPA530-F-97-001, Office of Solid Waste and Emergency Response (5306W), Washington, DC, July, 1997b.

U.S. Environmental Protection Agency, User's Manual, Landfill Gas Emissions Model, Version 2.0, Authors: Pelt, R., White, C., Blackard, A., Bass, R. L., Burklin, C., Heaton Amy Reisdorph, R.E., Office of Research and Development Washington, DC, 1998.

U.S. Environmental Protection Agency, The Benefits of Utilizing Landfill Gas, 2002. See http://www.epa.gov/lmop/about.htm#lfgte.

U.S. Environmental Protection Agency, Bioreactors, 2003. See: http://www.epa.gov/epaoswer/non-hw/muncpl/landfill/bioreactors.htm#3

U.S. Environmental Protection Agency, Bioreactors, 2004. See: http://www.epa.gov/epaoswer/non-hw/muncpl/landfill/bioreactors.htm

Visilind, P.A., Worell, W.A., and Reinhart, D.A., *Solid Waste Engineering*, Brooks/Cole, Pacific Grove, CA, 2002.

Whitehead, R., The UK Pesticide Guide, CAB International, British Crop Protection Council, 1998.

SUGGESTED READINGS

Akesson, M. and Nilsson, P., Material dependence of methane production rates in landfills, *Waste Manage. Res.*, 16, 108–118, 1998.

Avervalo, P., Down in the Dumps: Contra Costa County, Calif. Landfill Controversy California Lawyer, Vol. 15, 27–28, 1995.

Augello, A., Evaluation of solid waste landfill performance during the Northridge earthquake, in Earthquake design and performance of solid waste landfills: Proceedings, ASCE Annual Convention, San Diego, CA, ASCE Geotechnical Special Publication ASCE, New York, NY, 1995, pp. 17–50.

Bagchi, A., *Design, Construction, and Monitoring of Sanitary Landfills*, Wiley, New York, NY, 1990.

Baldwin, T., Stinson, J., and Ham, R., Decomposition of specific materials buried within sanitary landfills, *J. Environ. Engin.*, 124, 1193–1202, 1998.

Barlaz, M.A., Shaefer, D. M., and Ham, R. K., Bacterial population development and chemical characteristics of refuse decomposition in a simulated sanitary landfill, *Appli. Environ. Microbiol.*, 55, 55–65, 1989.

Blight, G. E. and Fourie, A. B., Leachate Generation in Landfills in Semi-Arid Climates, Proceedings of the Institution of Civil Engineers, Geotechnical Engineering, United Kingdom, 1999.

Boltze, U. and de Freitas, M., Monitoring gas emissions from landfill sites, *Waste Manag. Res.*, 15, 463–476, 1997.

Brumner, D.R. and Keller, D.J., Sanitary Landfill Design and Operation, SW-65ts, U.S. Environmental Protection Agency, Washington, DC, 1972.

California Waste Management Board, Landfill Gas Characterization, California Waste Management Board, State of California, Sacramento, CA, October 1988.

Chandler, J.A., Jewell, W.J., Gossett, J.M., Vansoset, P.J., and Robertson, J.B., Predicting Methane Fermentation Biodegradability, *Biotechnology and Bioengineering Symposium*, No. 10, pp. 93–107, 1980.

Chanton, J., Rutkowski, C.M., and Mosher, B., Quantifying Methane Oxidation from Landfills Using Stable Isotope Analysis of Downwind Plumes, in *Environmental Impact*, Academic Press, Harcourt Brace, Jovanovich, London, 1999.

Christensen, T. H., and Kjeldsen, P., Basic Biochemical Processes in Landfills, in *Sanitary Landfilling Process, Technology and Environmental Impact,* Christensen, T.H., Cossu, R., and Stegmann, P., (Eds): Academic Press, Harcourt Brace, Jovanovich, London, 1989.

Coons, L. M., Ankeny, M., and Bulik, G. M., Alternative Earthen Final Covers for Industrial and Hazardous Waste Trenches in Southwest Idaho, *Proceedings of the 3rd Annual Arid Climate Symposium*, SWANA, Albuquerque, NM, 2000.

County of Los Angeles, Department of County Engineer, Los Angeles, and Engineering-Science, Inc., Development of Construction and Use Criteria for Sanitary Landfills, An Interim Report, U.S. Department of Health, Education, and Welfare, Public Health Service, Bureau of Solid Waste Management, Cincinnati, OH, 1969.

Crawford, J.F. and Smith, P.G., *Landfill Technology*, Butterworth, London, UK, 1985.

Davis, S.N., and DeWiest, R.J. M., *Hydrogeology*, Wiley, New York, NY, 1989.

Deipser, A. and Stegmann, R., The origin and fate of volatile trace components in municipal solid waste landfills, *Waste Manage. Res.*, 12, 129–139, 1994.

Ehrig, H.J., Leachate Quality, in *Sanitary Landfilling: Process, Technology and Environmental Impact,* Christensen, T.H., Cossu, R., and Stegmann, P., (Eds.), Academic Press, Harcourt Brace, Jovanavich, London, UK, 1989.

El-Fadel, M., Shazbak, S., Saliby, E., and Leckie, J., Comparative assessment of settlement models for municipal solid waste landfill application, *Waste Manage. Res.*, 17, 347–368, 1999.

Drobny, N.L., Hull, H.E., and Testiu, R.F., Recovery and Utilization of Municipal Solid Waste, SW-10c. U.S. Environmental Protection Agency, Washington, DC, 1971.

Fisher, S.R. and Potter, K.W., Evaluation of the Use of DUMPSTAT to Detect the Impact of Landfills on Groundwater Quality, Wisconsin Department of Natural Resources, Madison, WI, 1998.

Herrera, T.A., Lang, R., and Tchobanoglous, G., A Study of the Emissions of Volatile Organic Compounds Found in Landfills, *Proceeding of the 43rd Annual Purdue Industrial Waste Conference*, Lewis Publishers, Inc., Chelsea, MI, 1989, pp. 229–238.

Hilger, H., Wollum, A.G., and Barlaz, M., Landfill methane oxidation response to vegetation, fertilization, and liming, *J. Environ. Qual.*, 29, 2000.

Kaiser, E.R., Chemical Analyses of Refuse Compounds, in Proceedings of National Incinerator Conference, ASME, New York, NY, 1966.

Kodikara, J., Analysis of tension development in geomembranes placed on landfill slopes, *Geotextiles Geomembranes*, 18, 2000.

Koerner, R.M. and Daniel, D.E., Better cover-ups, *Civil Eng.*, 62, 1992.

Lang, R.J., Stallard, W.M., Stiegler, L.C., Herrera, T.A., Chang, D.P.Y., and Tchobanoglous, G., Summary Report: Movement of Gases in Municipal Solid Waste Landfills, prepared for the California Waste Management Board, Department of Civil Engineering, University of California, Davis, Davis, CA, 1989.

Manna, L., Zanetti, M.C., and Genon, G., Modeling biogas production at landfill sites, *Resour. Conserv., Recycling*, 26, 1999.

Mantell, C.L., (Ed.), *Solid Wastes: Origin, Collection, Processing, and Disposal*, Wiley-Interscience, New York, NY, 1975.

Mathias, S.L., Discouraging Seagulls: The Los Angeles approach, *Waste Age*, 15, 1984.

Neissen, W.R., Properties of Waste Materials, in *Handbook of Solid Waste Management,* Wilson, D.G., (Ed.), Van Nostrand Reinhold, New York, NY, 1977.

Pohland, F.G., Fundamental Principles and Management Strategies for Landfill Codisposal Practices, *Proceedings Sardinia 91, 3rd International Landfill Symposium*, Grafiche Galeati, Imola Italy, Vol. II, 1991, pp.1445–1460.

Pohland, F.G., Critical Review and Summary of Leachate and Gas Production from Landfills, EPA/600/S2-86/073, U.S. EPA, Hazardous Waste Engineering Research Laboratory, Cincinnati, OH, 1987.

Pohland, F.G., Cross, W.H., Gould, J.P., and Reinhart, D.R., Behavior and Assimilation of Organic and Inorganic Priority Pollutants Codisposed with Municipal Refuse, U.S. Environmental Protection Agency, Washington, DC, 1993.

Pohland, F.G. and Kim, J.C., Microbially mediated attenuation potential of landfill bioreactor systems, *Water Sci. Technol.*, 41, 247–254, 2000.

Qasim, S.R., and Chiang, W., *Sanitary Landfill Leachate: Generation, Control, and Treatment,* Technomic Publishing, Lancaster, PA, 1994.

Reinhart, D., Full-scale experiences with leachate recirculating landfills: Case studies, *Waste Manag. Res.*, 14, 347–365, 1996.

Report on the Investigation of Leaching of a Sanitary Landfill, California State Water Pollution Control Board, Publication 10, Sacramento, CA, 1954.

Romeo, E.J., Material Recovery Design of Ocean County, NJ, *Proceedings of the 1992 Waste Processing Conference,* American Society of Mechanical Engineers, New York, NY, 1992.

Salvato, J.A., Wilkie, W.G., and Mead, B.E., Sanitary landfill-leaching prevention and control, *J. Water Pollut. Control Fed.* 43, 2084–2100, 1971.

Savage, G.M. and Diaz, L.F., Key Issues Concerning Waste Processing Design, *Proceedings of the 1986 National Waste Processing Conference*, American Society of Mechanical Engineers, New York, NY, 1986, p. 361.

Schroeder, P.R., The Hydrologic Evaluation of Landfill Performance (HELP) Model, User's Guide for Version 1, EPA/530/SW-84-009, 1. U.S. EPA Office of Solid Waste and Emergency Response, Washington, DC, 1984.

SCS Engineers, Inc., Procedural Guidance for Sanitary Landfills: Volume I. Landfill Leachate Monitoring and Control Systems, California Waste Management Board, Sacramento, CA, April, 1989a.

SCS Engineers, Inc., Procedural Guidance for Sanitary Landfills: Volume II. Landfill Leachate Monitoring and Control Systems, California Waste Management Board, Sacramento, CA, April, 1989b.

Senior, E., *Microbiology of Landfill Sites,* Lewis Publishers, Boca Raton, FL, 1995.

Stark, T. D., Stability of Waste Containment Facilities, *Proceedings of Waste Tech '99*, National Solid Wastes Management Association, New Orleans, LA, 1999, pp. 1–24.

Stark, T.D., Arellano, D., Evans, D., Wilson, V., and Gonda, J., Unreinforced geosynthetic clay liner case history, *Geosynthetics Int. J.,* Industrial Fabrics Association International (IFAI), 5, 521–544, 1998.

Stessel, R.I., Barrett, W.M., and Li, X., Comparison of the effects of testing conditions and chemical exposure of geomembranes using the comprehensive testing system, *J. Appl. Polymer Sci.*, 70, 1998.

Sullivan, J.W., Will, R.M., and Sullivan, J.F., The Place of the Trommel in Resource Recovery, *Proceedings of the 1992 Waste Processing Conference*, American Society of Mechanical Engineers, New York, NY, 1992.

Vesilind, P.A. and Rimer, A.E., *Unit Operations in Resource Recovery Engineering*, Prentice Hall, Englewood Cliffs, NJ, 1981.

Watts, K.S. and Charles, J.A., Settlement Characteristics of Landfill Wastes, *Proceedings of the Institution of Civil Engineers*, Geotechnical Engineering, UK, October 1999.

Williams, G.M., Ward, R.S., and Noy, D.J., Dynamics of landfill gas migration in unconsolidated sands, *Waste Manage. Res.*, 17, 1999.

Winkler, P.F. and Wilson, D.C., Size characteristics of municipal solid wastes, *Compost Sci.*, 14(5), 1973.

Wright, T.D., To cover or not to cover? *Waste Age*, 17, 1986.

QUESTIONS

1. List and discuss the possible passive approaches to landfill gas removal.
2. What is the minimum landfill CH_4 concentration that poses an explosion hazard? At what concentration is the methane of possible commercial (heating) value?

3. Under RCRA, new landfills cannot be located in seismic zones unless the operator can demonstrate that all containment structures (e.g. liners) are designed to resist the maximum horizontal shifting (true or false).

4. Landfill bird hazards to airports can be limited to a certain extent by shredding and baling MSW prior to disposal. Explain.

5. Landfill sizing is affected by the desired landfill lifetime, the population served, and the shape and height of the landfill, among other factors. Explain.

6. What chemical changes occur during earliest anaerobic stage of decomposition in a sanitary landfill? Discuss in terms of pH, BOD, and reactions of metals.

7. Methanogenic (methane producing) microorganisms prefer what type(s) of environmental conditions? How do they respond to pH?

8. What are the factors that influence methane gas migration below the land surface. Consider soil texture, soil temperature, soil moisture, and barometric pressure.

9. What waste types are restricted from sanitary landfills under current federal regulations? Be specific.

10. What is the significance of saturated hydraulic conductivity (K_s) in assessing soils for landfill liners, caps, and as a soil base? How is K_s influenced by soil texture and engineering practices (e.g., compaction)? What is the RCRA limit for liner K_s?

11. RCRA calls for stringent controls in MSWLF operation for air quality, explosive gases, storm water runoff, wetlands protection, cover material, and vectors. Discuss the specific requirements for each.

12. Geomembrane installation practices significantly influence possible leachate losses from a landfill. List and discuss the factors that must be considered for successful geomembrane design and installation.

13. Explain the various phases of MSW decomposition in a closed landfill cell. How do leachates and gases differ between each phase?

14. Under what conditions is passive landfill gas control acceptable? When is active gas control a requirement?

15. How does landfill gas differ from utility-strength natural gas? How is the landfill gas treated to render it suitable for possible sale to an energy utility?

16. If stormwater enters the landfill unit and contacts waste, what is the stormwater considered to be, according to the regulations (i.e., nonhazardous, hazardous, special waste)? How is it to be managed?

17. RCRA regulations require that bulk or noncontainerized liquid wastes are not to be placed in MSWLFs, with two exceptions. What are they?

18. Leachates vary in terms of physical characteristics, inorganic composition, organic composition, microbial composition, and toxicity. How are the above variables affected by waste type and by age in the landfill cell?

19. What are the major purposes of layered sanitary landfill covers (caps)? How do they function?

20. Subtitle D requires, at the time of landfill closure, the installation of a final cover (cap) system. What are the primary purposes of the cover? Consider erosion, subsidence, and limiting leachate production.

21. What are the Subtitle D requirements for proper landfill cover design and construction?

22. How does landfill reclamation occur, i.e., what mechanical steps are required for successful reclamation? How is it beneficial in terms of extending landfill lifetime and encouraging resource recovery?

23. How does sanitary landfill operation differ when ownership is private vs. public? Discuss the advantages and pitfalls of each, in terms of economics, efficiency, and environmental concerns.

24. The 'Second Law of Thermodynamics' states, in essence, that all systems proceed toward maximum disorder (chaos, entropy). How can a landfill be affected over time in the context of the Second Law? In other words, discuss how landfill liners, LCR systems, and caps can be transformed 100 years after landfill closure. Will the landfill remain impervious *ad infinitum*?

25. Soil material is being assessed as a possible liner for a sanitary landfill. A soil core was collected and brought to the laboratory. A 10-cm-tall section of soil has 2 cm of water continuously ponded on it. The area of the core surface is 78 cm². A total of 62 mL water is collected per hour. Calculate the K_s. The upper limit for K_s for liners is 1×10^{-7} cm/sec. Is this soil suitable for a landfill liner?

26. Determine the area required for a new sanitary landfill with a projected lifetime of 20 years. The landfill will serve a population of 175,000 people. It is estimated that per capita waste generation is 1.9 kg/day (4.1 lb/day). Waste density in the landfill averages 625 kg/m³. Landfill height is not to exceed 20 m.

27. Calculate the annual volume of leachate generated per hectare for a sanitary landfill located along the east coast of the United States. The climate is temperate, average annual rainfall is 122 cm/yr (48 in./yr), and evapotranspiration is estimated at 48%. The wastes are covered with soil and runoff from the site is 10%. There is no run-on of surface water; Similarly, there is no underflow of groundwater into the cell.

A.10.1 APPENDIX: LANDFILL GAS EMISSIONS MODEL

A.10.1.1 OVERVIEW

The Landfill Gas Emissions Model (LandGEM) estimates the emissions of a number of gases from landfills. Due to differences in waste types, disposal rates, climate, and operating conditions, the rate of generation of emissions is highly variable from landfill to landfill; mathematical models may account for the variability of factors such as waste types, pH, temperature, and availability of nutrients for methanogenic microorganisms (U.S. EPA, 1998). In this exercise, you will apply the LandGEM model for estimation of generation of selected landfill gases. The model, as zipped files, can be downloaded from: http://www.weblakes.com/lakeepa7.html.

A.10.1.2 HOW TO USE THE MODEL

A.10.1.2.1 Entering MSW Data for a Site

To estimate landfill emissions for a given year, the software model must be provided with the mass of MSW ('refuse') in the landfill for that year and the age of the MSW. This information is provided when you enter MSW data into the model on an annual basis, thus indicating the mass of MSW in the landfill and the age. Because MSW data are entered on an annual basis, MSW must be assigned an age in years based on an instantaneous disposal. The model assumes that MSW is accepted at the last moment of the year and that MSW accumulates into a landfill at the first instant of the year after it is accepted. Therefore, MSW accepted into the landfill one year (e.g., 2003) will appear as MSW in place the next year (e.g., 2004). Rounding the age of the accumulated MSW to the nearest whole number results in an age of zero years to MSW accepted (e.g., MSW accepted during 2003 will be zero years old in 2004). The following year (e.g., 2005), this MSW would then be assigned an age of 1 year, and so on.

The following data entry pattern to the model is recommended:

- Select the Year Opened
- Select the final year for which information is available (Current Year)

- Make any changes to the Closure Year that are needed
- Enter the Landfill Capacity
- Enter the Acceptance Rate or MSW in Place data.

The above order of entry is not required, except that landfill capacity must be entered before any waste data.

To enter the dates for calculating the years of operation:

1. In the data entry box across the top of the landfill study document, select the Year Opened text box and type the year the landfill opened.
   The default value for the year opened will be 10 years before the current year. The current year is read from the computer's clock. Replace the default values for the year opened and the current year with the appropriate years.
2. Select the Current Year text box and type the final year for which data are available or the closure year. For example, if a landfill closes in 2005, then the Current Year is 2005 or the last year data are available prior to 2005. However, if a landfill is currently open and accepting MSW, then the Current Year is the most recent year for which information is available. The landfill study table will automatically list all the years from the year opened to the current year.
3. The design capacity of the landfill is the maximum amount of MSW, in megagrams (Mg) (equivalent to metric tons), that can be accepted by the landfill. This value can be calculated using the MSW estimator utility if only the dimensions of the landfill are known.

To enter the design capacity of the landfill:

1. Select the Capacity text box and enter the design MSW capacity of the landfill (Mg). If the value available for the landfill capacity is in tons or other units, convert the value into Mg using the unit conversion utility in the Utilities menu.
2. If no value is available for capacity, but the landfill dimensions are available, use the MSW estimator utility to estimate the MSW capacity.

To enter MSW values:

1. Select the landfill study window in which you wish to enter MSW data.
2. Select the year for which the data are to be entered in the Landfill Study table. The cell for that year will be highlighted by a bold border. The waste value you type in the data entry box will be entered into this cell of the table.
3. Enter a Waste Value in the text box. To enter data in the cell of the landfill study table, select the cell in which you want to enter a value. Highlight the contents of the Waste Value box, and then type in the value for the MSW. Press [Enter]. The value entered will appear in the active cell.

MSW values can be entered into the landfill study table as either total MSW in place in the landfill (in Mg) or an annual acceptance rate (in Mg/year). If the available MSW data for the landfill are for the amount of MSW in place, select the MSW in Place (Mg) button. If the available MSW data for the landfill are annual acceptance rates, select the Acceptance Rate (Mg/year) button. Note that the MSW in place for each successive year cannot be less than the previous year's MSW in place and must always be less than or equal to the design capacity.

Details of the model appear in U.S. EPA (1998). Landfill Gas Emissions Model, Version 2.0. User's Manual. Office of Research and Development. Washington, DC 20460 (See: http://support. lakes-environmental.com/Models/Landfill/landfill.pdf), and: http://www.weblakes.com/lakeepa7. html.

A.10.1.2.2   Forecasting Landfill Emissions

To predict emissions of landfill gases, you have to select the model parameters, identify the length of operation of the landfill, enter the MSW in place or the acceptance rates of MSW, and generate a report.

Length of Operation: Specifying the length of operation of the landfill can be more complicated when forecasting emissions. If precise years of operation (e.g., 1960, 1990, and 2005) are known, they can be entered for the time variables (e.g., Year Opened, the Current Year, and the Closure Year) and the model functions normally. However, when precise dates are not known, the length of operation of the landfill can be specified with generic year numbers, such as 0001 (Year Opened), 0015 (Current Year), and 0016 (Closure Year).

Landfill Capacity: The model algorithms require that a design capacity be entered prior to entering yearly MSW data, i.e., even for a landfill not yet in operation, the total landfill capacity must be specified before any other information about the MSW in the landfill can be entered. The MSW estimator may be used to determine the landfill capacity from estimated landfill dimensions.

Landfill Waste: For each year of operation, the amount of landfill MSW must be entered as either a MSW in place or an acceptance rate. For years in which no such data are available, estimates must be provided. The Autocalc function can assist the user in entering estimates for years in between those in which MSW acceptance rate or the amount of MSW in place is known.

A.10.1.2.3   Emissions Forecasting Example

In the current situation, only partial MSW data are available for a number of years in the life of the landfill. You need to estimate emissions for a landfill which has a capacity of 6,800,000 Mg and opened in 1957. The current year is 2003, and in 2002 the landfill was at half capacity (i.e., 3,400,000 Mg). You have MSW data for each year between 1965 and the current year, but no data before 1965. Your data show that, from 1965 to 2002, a constant rate of 80,000 Mg/year of MSW was accepted by the landfill. You need to estimate emissions for the life of the landfill to determine applicability with MSWLF regulations. This landfill has no co-disposal and is scheduled to close in the year 2015.

To begin, enter the Year Opened (1957) in the box on the command bar. Then enter the Current Year (2003) in the adjacent box, and input the landfill capacity specified (6,800,000 Mg) in the Landfill Capacity box. Applicability to landfill regulations needs to be investigated; therefore, the CAA default parameters should be used. Under the Defaults menu, select the CAA item. Under the Parameters menu, indicate, with a check mark, that No Co-disposal is selected.

To input the Acceptance Rate data from 1965 to 2002, the Autocalc function from the Edit menu can be used. Move the cursor to the acceptance rate cell for the year 1965. Input 80,000 Mg/year in the Acceptance Rate box on the command bar. Move the cursor to the cell for the acceptance rate in 2002. Enter 80,000 Mg/year in this cell as well. Then highlight all the cells between and including 1965 and 2002. The Autocalc function will input the value 80,000 Mg/year into the Acceptance Rate column for each year between 1965 and 2002.

Examine the MSW in Place value for 1995. How much waste has accumulated in the landfill to this point?

As of 2002, however, we know that the landfill was at half capacity. Therefore, the waste entered between 1957 and 1964 must total 3,400,000 – [value for 2002]=_____ Mg. We will assume that the MSW accepted was evenly distributed between 1957 through 1964.

Enter the remaining data necessary to run this scenario with the Autocalc function. Begin by entering the waste accumulated between 1957 and 1964 (i.e., the difference calculated above, in Mg) in the MSW in Place cell for the year 1965 (the year in which any waste accepted in 1964 would appear as accumulated waste in the landfill). The waste acceptance rate for 1964 will now appear to be this missing value. These data will be adjusted with the next step.

To distribute evenly the waste accepted between 1957 and 1964, highlight the MSW in Place cells for these years (i.e., highlight all cells between and including 1957 and 1964). Then choose the Autocalc function from the Edit menu. The MSW data should be evenly distributed back to 1957. You should now observe as well that the MSW in Place value for the year 1995 is now 3,400,000 Mg, or half the landfill capacity.

From the Parameters menu, select the Closure Year item. Click the User Specified diamond and enter 2010 in the adjacent box. Then choose the [OK] button. Once the closure year data have been chosen, the scenario is complete, and the study can be run by generating a report. The software model will automatically calculate the waste acceptance rates necessary to reach the landfill capacity in the specified closure year.

QUESTIONS

To examine the landfill emissions calculated by the model, a report is generated. For this example, the NMOC concentrations over the lifetime of the landfill is to be determined. From the Report menu choose the Graphics option. When the Select an Emitted Substance dialog box opens, choose NMOC and click the [OK] button. A new window displaying the graphical report will open. What is the trend of NMOC concentrations over time? When does the peak in concentration occur? What does this year coincide with?

Is it realistic to expect constant values of waste acceptance in a landfill serving urban and suburban populations, including commercial establishments? What changes would you expect over the course of 45 years in a landfill which serves your home community or university? Perform another calculation on this same landfill, with the same capacity and the same annual acceptance rate; however, the landfill is now about 75% full. When will the landfill reach full capacity? What are the methane generation rates?

PROBLEM

For a second landfill, less data are available. The capacity is 4,250,000 Mg and the landfill is expected to operate for 40 years. For a case such as this, a Year Opened value of 1 and a Current Year value of 41 can be used. The model will calculate the Closure Year automatically, and it will be assumed that the waste is accepted at a uniform rate throughout the length of operation. This study will be used to estimate the actual emissions by using the AP-42 default values for k and L_o, and the landfill will be assumed to accept both hazardous and nonhazardous waste (Co-disposal).

Select the AP-42 option from the Defaults menu, and in this situation, click on Co-disposal in the Landfill Type item in the Parameters menu. The waste acceptance rates during the 50 years of operation are assumed to be constant. Therefore, the waste should be evenly distributed over the life of the landfill.

What is the annual rate of MSW acceptance?

Graph a report of NMOC generation. When do concentrations peak?

A.10.1.3 THE HYDROLOGIC EVALUATION OF LANDFILL PERFORMANCE MODEL

The Hydrologic Evaluation of Landfill Performance (HELP) model is a user-friendly computer program that estimates water balances for municipal landfills and other land disposal systems. The HELP model was developed at the U.S. Army Engineer Waterways Experiment Station in order to assist landfill designers and environmental regulators in evaluating the hydrologic performance of landfill designs. The use of this model allows engineers and regulatory personnel to compare different designs in efforts to construct an environmentally sound land disposal facility. The HELP model is recommended by the U.S. EPA and required by most states for evaluating closure designs of both municipal (RCRA Subtitle D) and hazardous (Subtitle C) waste management facilities.

A wide range of variables are incorporated into the HELP model including weather, soil type, design feature, surface storage, snowmelt, runoff, infiltration, evapotranspiration, vegetative growth, soil moisture storage, lateral subsurface drainage, leachate recirculation, unsaturated vertical drainage, and ultimately leakage through the soil and composite liners. Use of the HELP model requires general climatic data (e.g., precipitation, relative humidity, and solar radiation) for performing each landfill design analysis. HELP has a historical database built into the program which contains 5 years of daily precipitation data for 102 cities throughout the United States. The model also contains default values for soil characteristics, total porosity, field capacity, wilting point, and saturated hydrologic conductivity. Other parameters incorporated within the model include layer thickness, area, slope and maximum drainage distance, layer description, surface characteristics, and geomembrane characteristics. The inclusion of these default values allows the user to formulate hypothetical situations to determine the impact of a particular landfill design on the local environment.

The program provides for rapid estimation of the daily, monthly, annual, and average annual amounts of runoff, evapotranspiration, drainage, leachate collection, and liner leakage that may result from the operation of a wide variety of landfill designs. The model applies to open, partially closed, and fully closed sites.

The goal of this activity is to familiarize you with the HELP model and its capabilities. You will generate results and evaluate two landfill types using data for two hypothetical land disposal facilities. The procedures for performing an analysis are provided below along with a series of questions. A list of relevant terms has also been provided.

DOWNLOADS

The HELP model can be downloaded from the U.S. Army Corps of Engineers site, http://www.wes. army.mil/el/elmodels/ Engineering documentation for Version 3 of the model is also available at this site. The complete HELP User's Guide is available at: http://www.epa.gov/cgi-bin/claritgw?op-Display&document=clserv:ORD:0790;&rank=4&template=epa

A.10.1.3.1   Using the HELP Model

Some Notes about the Use of DOS Screens
For many students, experience with DOS may be minimal. Some instructions are provided below for operating HELP in the DOS mode.

Moving between cells: By pressing the Page Down key or Page Up for selecting next/previous screen; Up and Down arrows — to move through the cells of a screen; Tab and Shift-Tab keys — to move to the right and to the left between the same line.

Moving within an input cell: Each input cell is set to a given width depending upon the type of information expected to be entered in that cell.

Terminating: Press the F9 key to quit without saving changes, return to the main menu and exit the program; Esc key and Ctrl-Break — end some options and allow to continue with other operations; F10 — save the data; Ctrl-Alt-Del —termination of input or execution (by resetting or turning off the computer).

On-line help: Available from any cell location on the screen; F1 — information about the operations and purposes of the screen; F2 – specific technical assistance for the highlighted cell; F3 – various functions of keystrokes.

System of units: The HELP model allows the student to use either: the customary system of units (a mixture of U.S. and metric units traditionally used in landfill design) or metric units.

The value for a particular variable can be input for the first time or at a later time during the program session. If an input cell is left blank, a value of 0 will be assigned to the corresponding variable. The program will warn the user when a blank or zero is an inappropriate value.

OVERVIEW OF STEPS

1. Click HELP 3 on the computer screen
2. Enter weather data
   press "Enter/Edit Weather Data": Evapotranspiration, precipitation, temperature, and solar radiation data.
   Data options: 1) default data (for precipitation only); 2) synthetic; 3) user-defined data sources, e.g., NOAA Tape, Climatedata, ASCII data, HELP Version 2 data, and Canadian Climatological data.; 4) can be entered manually. Default and synthetic weather data generation is performed by selecting the city of interest from a list of cities and specifying (optional) additional data.
3. Enter soil and design data
   press "Enter/Edit Soil and Design Data": Site information, design, soil and geomembrane liner data by layers, runoff curve number, data verification, screen for saving the soil and design data file.
   Cells for entering: project title, system of units, initial soil conditions, landfill area, layer design information, i.e., layer type, thickness, soil texture, drainage characteristics, geomembrane liner information, and runoff curve number information

The user may request that the data be checked for possible violation of the design rules.

4. "Executive Simulation" Define the data files to be used in running the simulation and select the output frequency and simulation duration.
5. "View Results" Allows seeing output results.
6. "Print Results" Printing of final results
7. "Display Guidance" Information about general landfill design procedures and the HELP model
8. "Quit" Return to DOS

A.10.1.3.2 Inputs to the Model

Climate data: General evapotranspiration data and daily values of precipitation, temperature, and solar radiation are provided in the model. The HELP model has a default evapotranspiration database for 183 U.S. cities, containing data for latitude, evaporative zone depths, leaf area indices, growing season, average wind speed, and average quarterly relative humidity. A default precipitation database is included, containing 5 years of daily values for 102 cities throughout the United States. The model also has a synthetic weather generator with coefficients for 139 cities for daily precipitation data generation and for 183 cities for daily temperature and solar radiation data generation. The user interface also contains a number of utility routines to import weather data from other databases.

Soil data: Porosity, field capacity, wilting point, initial moisture content, and saturated hydraulic conductivity of up to 20 layers of materials is included. The model contains a default soil database of characteristics for 42 types of materials (soils, waste, and geosynthetics). Design data requirements include the AMC-II runoff curve number for the site, a description of the vegetation, a description of the function of each layer of material, the thickness of each layer, the slope at the base of each drainage layer, the spacing between drainage collectors in each drain system, a description of leakage potential of each geomembrane liner, and a description of leachate recirculation, if used.

Some Requirements for Data Inputs

- A vertical percolation layer may not be placed under a lateral drainage layer
- A barrier soil liner may not be placed under another barrier soil liner
- A geomembrane liner cannot situate between two barrier soil liners
- A geomembrane liner may not be placed under another geomembrane liner
- A barrier soil liner cannot situate directly between two geomembrane liners
- In the case when a barrier soil liner or a geomembrane liner is not placed directly below the lowest drainage layer, lateral drainage for the bottom section of the landfill will not be calculated
- Barrier soil liner cannot be a the top layer of a landfill
- Geomembrane liner cannot be a the top layer of a landfill
- Only five barrier soil liners and geomembrane liners in total are allowed
- Two barrier soil liners cannot be adjacent to each other

The program checks for rule violation at the time the user saves the data.

A.10.1.3.3   Relevant Terms and Their Values in HELP

Field capacity: The soil moisture storage and content after a prolonged period of gravity drainage from saturation corresponding to the soil water storage when a soil exerts a soil suction of -1/3 bar (dimensionless number between 0 and 1 and should be greater than wilting point).

Geotextile transmissivity: The product of the in-plane saturated hydraulic conductivity and thickness of the geotextile.

Installation defect density: The number of defects (diameter of hole larger than the geomembrane thickness; hole estimated as 1 cm^2 in area) per acre resulting primarily from seaming faults and punctures during installation.

Pinhole density: The number of defects (diameter of hole equal to or smaller than the geomembrane thickness; hole estimated as 1 mm in diameter) in a given area generally resulting from manufacturing flaws such as polymerization deficiencies.

Saturated hydraulic conductivity: The rate at which water drains through a saturated soil under pressure gradient (must be greater >0).

Soil moisture storage (content): The ratio of the volume of water in a soil to the total volume occupied by the soil, water and voids.

Total porosity: The soil moisture storage and content at saturation (dimensionless number between 0 and 1 and should be greater than field capacity).

Wilting point: The lowest soil moisture storage and content that can be achieved by plant transpiration or air drying, that is the moisture content where a plant will be permanently wilted corresponding to the soil water storage when a soil exerts a soil suction of -15 bars (dimensionless number between 0 and 1).

THE PROBLEM

Small Subtitle C landfills were recently constructed by the same firm for suburban locations near two cities, Indianapolis, IN, and Miami, FL. Total landfill area for both is 40 acres. Each landfill is underlain by a 48-in.-thick clay liner. There are double geomembranes, and double leachate collection and removal systems were installed. Drainage length under the landfill was set at 50 ft with a 2% slope. On the surface of closed portions of the landfill, slopes were set to 4% with slope lengths of no more than 75 ft. Leachate is being recycled (rate of 95%).

Location 1: Indianapolis, IN

Step 1: Starting the Program

- Open the program by double clicking on the HELP icon (stoplight) on your desktop
- A main menu will appear

Step 2: Entering Weather Data

A. Precipitation Data:

- Enter 1 on the main menu screen to enter/edit weather data
- Press Page Down twice on your keyboard until you see the screen titles "Weather data-Precipitation, Temperature and Solar Radiation"
- Press <Enter> to select the highlighted default option for the precipitation data
- Use the arrows on the keyboard to select Indiana and press <Enter>
- Indianapolis will appear in the smaller box below
- Press <Enter> to select this city and return to the "Weather data" screen again

B. Temperature Data:

- Use the right arrow to select Synthetic in the Temperature box
- Press <Enter> to select the highlighted customary option, then scroll to select Indiana and press <Enter>
- Three cities will appear in the small box, scroll down to Indianapolis and press <Enter>
- A screen title "Synthetic Temperature Data" will appear. Make sure to select the default normal mean monthly temperature and press <Enter>

C. Solar Radiation Data:

- Use the right arrow to select the Synthetic solar radiation data
- Press <Enter> to select the highlighted customary option and scroll over to select Indiana
- Again choose Indianapolis and press Enter twice to confirm the default data

D. Evapotranspiration Data:

- Press Page Down (PgDn) to proceed to the "Weather Data-Evapotranspiration Data" screen
- Press F5 for the default evapotranspiration data
- Select Indiana, then Indianapolis and press <Enter> to enter these default data
- Specify an evaporation zone depth of 10 by scrolling to that cell and typing in the value
- Specify a maximum leaf area index of 4.2
- SAVE the changes that you have made by pressing F10 and Y (yes) for each of the menu items – type your name into the "file" cells
- Once you have saved these changes you have finished editing the weather data and will proceed to the next step in the simulation

Step 3: Entering the Soil Data

- Press 2 on the main menu screen to begin editing the Soil Data
- Press Page Down to create new
- Press <Enter> to select customary units
- Enter the name of your simulation project (i.e., Yourname_Simulation)
- Enter 40 acres for the landfill area and change the area to be drained to 95% and leave the latter box with the default value by pressing Page Down
- Next, you must enter data regarding the layer and soil types for a total of five layers
- To specify the layer type move your cursor to the "layer type" cell and Press F2
- The HELP program provides a description of the different layer types and relative permeabilities. Read these to familiarize yourself with the different types.
- Press <escape> to return to your table and begin entering the different layers

First Layer:
- Enter 2 for the first layer type, specifying a leachate recovery layer
- Press <Enter> to specify layer thickness and enter 24 in.
- Press F6 to find the default soil data provided by the HELP program
- Scroll down to Municipal Waste (900pcy) and press <Enter>. This will add the default data to your table
- Next scroll over (press <Enter>) and enter a value of 0.5 for the initial moisture (vol/vol), this number must be larger than the specified wilting point
- Press <Enter> to begin adding the second layer

Second Layer:
- Enter 4 for the second layer type, a geomembrane liner
- Enter a layer thickness of 1.0 in.
- Enter 15 for the soil texture number, this denotes a Bentonite mat
- Again scroll over and enter 0.4 for the initial soil moisture

Third Layer:
- This layer will be the same as the first layer, so enter the same information that you used in Layer 1 above.

Fourth Layer:
- This layer will be the same as the second layer, enter the same numbers from Layer 2 above, the geomembrane liner

Fifth Layer:
- For this last layer, enter a value of 3 for the layer type to denote a layer of relatively impermeable soil
- Make the layer thickness 48 in.
- Put a soil texture number of 15, a clay soil abbreviated C in the F2 Table
- Add a initial soil moisture value of 0.4 vol/vol
- Press Page Down to continue to the next screen for further data entry
- You will see that your layer type and saturated hydrologic conductivities are already specified
- Press Enter twice until your cursor is in the "drainage length" cell and type 50 ft for each different layer (use down arrow to move through the layers)
- Enter a 2% drainage slope for all the layers
- Next enter 95% for the leachate recirculation value for all the layers
- Enter 1 into the "recirculate to layer" cells for each layer and leave the "subsurface inflow" cells blank for all layers
- Press Page Down to go to the next screen

For geomembrane liners only (type 4)
- Enter 4 for the "geomembrane pinhole density" and 0 for the "geomembrane installation defects" cells
- Enter 3 for a "geomembrane quality" of Good, a value that assumes good field installation with well prepared, smooth soil surface and geomembrane wrinkle control to ensure good contact between the geomembrane and the surrounding soil that limits drainage rate

- Leave the "geotextile transmissivity" cells blank and Press Page Down to continue
- You have now finished editing the soil and design data and will continue to the next step in the simulation

Step 4: Specifying Runoff Curve Numbers

- Scroll down and select "HELP Model Computed Curve Number"
- Enter a slope of 4% and press <Enter>
- Enter a slope length of 75 ft and press <Enter>
- Leave the soil texture of 18
- Enter a vegetation code of 4 to denote a good stand of grass
- Again the F2 option describes the different vegetation types
- Press Enter and the HELP program will compute a runoff curve number What is the runoff curve number?

Step 5: Verifying the Soil and Design Data

- Press F10 to verify your parameters
- Scroll up and press <Enter> for each item on the menu. An "OK" should appear of you have entered all of the data appropriately
- Once you have done this scroll to the Save Soil and Design Data, again enter your name into the file space
- Press Page Down to return to the main menu

Step 6: Running the Simulation

- Now that you have entered or specified all the required data, you are now ready to execute the HELP simulation
- Select Execute Simulation from the main menu and specify the file name of the data that you saved
- Next, press Page Down to continue
- Leave the default values in the output selection and press Page Down again to begin running the simulation
- HELP has now performed the simulation using that data that you entered

Step 7: Viewing the Results

- The main menu should be displayed at this point
- Scroll to View Results and press <Enter>
- Press Page Down to proceed to the specified viewing file
- You should now see the result of your simulation displayed
- Use the Arrows or Page Up/Down to navigate through these results

QUESTIONS

Answer the following questions from the Table near the bottom of the results labeled "Average Monthly Values in Inches for Years 1974–1978"

1. Which month/year experienced the highest precipitation?
2. What was the volume of runoff in March/September?
3. How much lateral drainage was recirculated into Layer 1 in April/October?
4. When was the lateral drainage collected from Layer 1 highest? Lowest?
5. When did the most leakage occur through Layer 2? What is the volume?

6. When did the most leakage occur through Layer 5? How much?
7. How do these two layers differ with regard to type and leakage values generated?
8. Based on the results of this simulation, is the design of this landfill acceptable from an environmental standpoint? Discuss.

Location Two: Miami, FL

Step 2: Entering Weather Data

A. Precipitation Data:
- Press <Enter> to select the highlighted default option for the precipitation data
- Use the arrows on the keyboard to select Florida
- Miami will appear in the smaller box below
- Press <Enter> to select this city and return to the "Weather data" screen again

B. Temperature Data:
- Use Synthetic in the Temperature box, scroll over to select Florida, then scroll down to Miami

C. Solar Radiation Data:
- Use the right arrow again to select the Synthetic solar radiation data
- Select Florida and then Miami

D. Evapotranspiration Data:
- Select Florida, then Miami and press Enter to enter these default data
- Specify an evaporation zone depth of 10 by scrolling to that cell and typing in the value
- Also specify a maximum leaf area index of 4.2
- SAVE the changes that you have made by pressing F10 and Y (yes) for each of the menu items — type your name into the "file" cells

Step 3: Entering the Soil Data
- Press 2 on the main menu screen to begin editing the Soil Data
- Press Page Down to create new
- Enter 40 acres for the landfill area and change the area able to be drained to 95% and leave the latter box with the default value by pressing Page Down
- Next you must enter data regarding the layer and soil types for a total of three layers

First Layer:
- Enter 2 for the first layer type, a leachate recovery layer
- Press <Enter> to specify layer thickness and enter 20 in.
- Press F6 to find the default soil data provided by the HELP program
- Scroll down to Municipal Waste (900pcy) and press Enter — this will add the default data to your table
- Next scroll over (press Enter) and enter a value of 0.5 for the initial moisture (vol/vol)

Second Layer:
- Enter 4 for the second layer type, a geomembrane liner
- Enter a layer thickness of 8 in.

- Enter 17 for the soil texture number
- Again scroll over and enter 0.4 for the initial soil moisture

Third Layer:
- For this last layer, enter a value of 3 for the layer type to denote a layer of relatively impermeable soil
- Make the layer thickness 36 in.
- Put a soil texture number of 15, a clay soil abbreviated C in the F2 Table
- Add a initial soil moisture value of 0.4 vol/vol
- Press Enter twice until your cursor is in the "drainage length" cell and type 50 ft for each different layer (use down arrow to move through the layers)
- Enter a 2% drainage slope for all of the layers
- Next enter 50% for the leachate recirculation value for all of the layers
- Enter 1 into the "recirculate to layer" cells for each layer and leave the "subsurface inflow" cells blank for all layers
- Press Page Down to go to the next screen

For geomembrane liners only (type 4)
- Enter 25 for the "geomembrane pinhole density" and 5 for the "geomembrane installation defects" cells
- Enter 4 for a "geomembrane quality" (poor installation)
- Leave the "geotextile transmissivity" cells blank and Press Page Down to continue

Step 4: Specifying Runoff Curve Numbers
- Scroll down and select "HELP Model Computed Curve Number"
- Enter a slope of 4% and press <Enter>
- Enter a slope length of 75 ft and press <Enter>
- Leave the soil texture of 18
- Enter a vegetation code of 1 to denote bare ground
- Again the F2 option describes the different vegetation types
- Press Enter and the HELP program will compute a runoff curve number

What is the runoff curve number?

Step 5: Verifying the Soil and Design Data
- Press F10 to verify you parameters
- Scroll up and press Enter for each item on the menu
- Once you have done this scroll to the Save Soil and Design Data, again enter your name into the file space
- Press Page Down to go back to the main menu

Step 6: Running the Simulation
- Now that you have entered or specified all the required data, you are now ready to execute the HELP simulation

Step 7: Viewing the Results
- In the main menu, scroll to View Results and press <Enter>
- Press Page Down to proceed to the specified viewing file

QUESTIONS

Answer the following questions from the Table near the bottom of the results labeled "Average Monthly Values in Inches for Years 1974–1978"

9. What month had the highest precipitation?
10. How much runoff occurred in March/September?
11. How much lateral drainage recirculated into Layer 1 in April/October?
12. When was the lateral drainage collected from Layer 1 the highest? The lowest?
13. How much leaked through Layer 3 at the highest leak event (i.e., how many cubic feet)?
14. Is the current design for the Miami facility adequate in terms of environmental concerns? How does it compare with the Indianapolis landfill in terms of leachate penetration through the upper geomembrane and through the clay liner? Given current conditions such as weather and soils, what changes could be made to the design to decrease leachate formation?

EXCEL EXERCISE

GROUNDWATER QUALITY AT A MIDWEST LANDFILL

FILE NAME: LANDFILL.XLS

The Situation

In the 1970s, a gravel pit located in the Midwest United States was converted into a landfill operation. At that time, no state regulations existed which would prohibit such a conversion. Therefore, although the land was completely unsuitable for such use, landfilling began. Landfill operation was relatively small (30 acres), and in 1977, the owners expanded operations to a total of 55 ha. In 1979 the state began to regulate landfill activities and, where the site did not meet the instituted state regulations, it was permitted to continue operations under a "grandfather clause."

The data for this exercise can be located at www.crcpress.com/e_products/downloads/download.asp?cat_no=3525

In 1982 a new operator purchased the facility. During the first year of operations the new owner had problems with leachate, runoff, and trash released from the site. The company decided to institute sound landfill management practices. They also began to buy land surrounding the original site. Since that time, the operation of the landfill improved, meeting the requirements of the state regulatory agency. Trees were planted to improve the aesthetics of the operation. The perimeter of the facility is patrolled and the company conducts business in the community. The landfill currently employs over 30 people from the community.

In 1995 the owners began petitioning the Area Planning Commission to expand the land that was permitted for landfill operation. The land for which they are seeking zoning is an additional 100 acres.

Practical Issues of the Landfill

1. Contamination from the current landfill has the potential to contaminate drinking water, not only of the immediate community, but also of other nearby communities. This potential results from two conditions:
   (a) The landfill is sited on a drainage divide. The drainage to the south of the landfill enters one river basin, and the drainage to the north enters a second river basin. Therefore, any contamination from the landfill will pollute two water systems, which pass through as many as 18 counties.
   (b) The subsurface of the land in question contains at least three aquifers. Test wells drilled show that water occurs at depths of 20 to 25 ft. In addition, a moraine, which lies beneath the landfill, conducts underground water away from the landfill in several directions. Data for a range of inorganic contaminants in test wells appears in the spreadsheet, "GW_LF", Sheet 1.
2. Limited tests of water taken from wells at the perimeter of the landfill have shown contamination from total organic carbon. The presence of certain halogenated organic compounds was also identified. Other volatile organic compounds such as trichloromethane and dichloroethane were found. These data appear in Sheet 2.
3. The effects of leachate contamination from special waste and other wastes disposed at the landfill are not presently known. The potential for adverse environmental, health, safety, and social impacts must be considered.
4. Now that the number of landfills in the state have decreased significantly, it is probable that this facility will handle even greater amounts of wastes in the coming decade.

TASKS

Using the monitoring well data from the tables, determine whether or not groundwater in the landfill environs is contaminated. Indicate where contaminants are occurring and if they are in excess of regulatory limits.

1. Determine the direction of groundwater flow and draw directional arrows.
2. From the contaminated wells identified, define the extent of any plume(s) and draw the outline on the map.
3. Discuss any observable trends in migration of any contaminants over the study period.
4. Based on the behavior of the metals, what can you conclude about the pH of the leachate?
5. Do the data for chlorinated organics indicate a possible human health hazard? What could be the possible source of these chemicals? Can any arise from natural decomposition of disposed wastes?
6. Based on the need for additional landfill space and groundwater data, is an extension of this landfill facility justifiable? Give reasons.

Part III

Hazardous Waste Management

This section covers the management of those wastes that are considered to pose a significant threat, both now and in the future, to human health and the environment when improperly managed. Wastes considered hazardous are those which are ignitable, corrosive, reactive, and toxic. It follows that such wastes cannot be handled or disposed without following special precautions. Wastes designated as "hazardous" are generated by a wide range of industries of varying sizes. As we shall see in this section, specific requirements for waste management by a generator will vary as a function of the amounts generated over a specified time frame.

Unfortunately, regulations addressing hazardous waste management were few prior to 1976, when the Resource Conservation and Recovery Act (RCRA) was enacted. Under RCRA the U.S. EPA was granted specific authority to regulate the generation, transportation, and disposal of hazardous waste. Topics in this section will adhere to requirements of RCRA (and, to a lesser extent, other important regulatory and legal frameworks such as those of the Department of Transportation and the Clean Air Act) and include identification of hazardous waste; hazardous waste generator requirements; hazardous waste transportation; treatment, storage and disposal facility requirements; incineration; hazardous waste treatment; and land disposal of hazardous waste.

Management of nuclear (radioactive) wastes is not handled in this book; radioactive wastes are not addressed by RCRA but by other laws, for example the Nuclear Waste Policy Act of 1982.

11 Identification of Hazardous Waste

I hate facts. I always say the chief end of man is to form general propositions — adding that no general proposition is worth a damn.

Oliver Wendell Holmes, Jr.
The Mind and Faith of Justice Holmes

11.1 INTRODUCTION

As discussed in Chapter 3, the Resource Conservation and Recovery Act (RCRA) was the first truly significant step in the comprehensive management of hazardous as well as municipal wastes in the U.S. The ultimate goal of RCRA is to promote the protection of public health and the environment and to conserve material and energy resources. RCRA requires the U.S. EPA to promulgate and enforce regulations regarding the management of hazardous waste. These regulations established mandatory procedures and requirements for compliance with RCRA. RCRA has remained current with waste management issues and problems by being amended several times. The most sweeping set of amendments was included in 1984 as the Hazardous and Solid Waste Amendments (HSWA).

RCRA has nine subtitles, each of which addresses some aspects of resource conservation and waste management. Subtitle C is the primary component that deals with management of hazardous waste. Its goal is to identify a hazardous waste and set standards for the accumulation, storage, transportation, treatment, and disposal of hazardous waste. The provisions of Subtitle C apply to a waste from the moment it becomes hazardous until it is no longer a hazardous waste. This embraces the so-called "cradle-to-grave" approach to the regulation of hazardous waste.

Hazardous waste management regulations are published in the *Federal Register*, which is published daily. The *Federal Register* provides a system for making regulations and legal notices issued by federal agencies available to the public.

11.2 THE RCRA SUBTITLES

U.S. EPA regulations are compiled in Title 40 of the Code of Federal Regulations (40 CFR), Protection of the Environment. The topics are as follows:

Subtitle	Topic
A	General Provisions
B	Office of Solid Waste, Authorities of the EPA administrator
C	Hazardous Waste Management
D	State or Regional Solid Waste Plans
E	Duties of the Secretary of Commerce in Resource Recovery
F	Federal Responsibilities
G	Miscellaneous Provisions
H	Research, Development, Demonstration, and Information
I	Underground Storage Tanks

Subtitle A declares that the production of hazardous waste is to be reduced; furthermore, land disposal is to become the least favored method of hazardous waste disposal. Wastes are to be handled in order to minimize the threat to human health and the environment. Subtitle A includes a set of objectives to achieve these goals, including:

- A prohibition of open dumping of waste
- State control of RCRA programs
- Promotion of research and development activities for sound waste management
- Encouragement of waste recovery, recycling, and treatment as alternatives to waste disposal

A summary of federal regulations implementing the hazardous waste management requirements of RCRA is shown in Table 11.1. Parts 124, 260 through 268, 270, 273, and 279 specifically address the management of hazardous wastes.

Part 124 contains EPA procedures for issuing, modifying, revoking, and reissuing or terminating all RCRA permits.

Part 261 identifies the wastes that are subject to regulation as hazardous waste. This part defines the terms "solid waste" and "hazardous waste," identifies those wastes that are excluded from regulations, and establishes special management requirements for hazardous waste produced by conditionally exempt small quantity generators and for hazardous waste which is recycled. Part 261 identifies characteristics and contains the various lists of hazardous wastes.

Part 262 contains the rules with which generators of hazardous waste must comply. This part requires a facility to evaluate all wastes generated on-site to determine if they meet the definition of hazardous waste. It also explains the conditions under which a hazardous waste manifest must be used, describes a generator's transportation requirements, and details the record keeping and reporting requirements.

Part 263 establishes standards that apply to people transporting hazardous wastes within the United States. In promulgating the regulations, the U.S. EPA has adopted regulations of the Department of Transportation (DOT) governing the transportation of hazardous materials. These regulations pertain to container labeling, marking, placarding, using proper containers, and reporting discharges of hazardous waste.

TABLE 11.1
Summary of Federal Regulations Implementing the Hazardous Waste Management Requirements of RCRA

40 CFR Part	Coverage of the Regulations
124	Public participation
260	General requirements, definitions, petitions
261	Identification and listing of hazardous waste
262	Generators of hazardous waste
263	Transporters of hazardous waste
264	Permitted hazardous waste facilities
265	Interim status hazardous waste facilities
266	Certain specific hazardous wastes and facilities
268	Land disposal restrictions
270	EPA administered permits
271	State hazardous waste programs requirements
273	Universal hazardous waste
279	Standards for the management of used oil

Part 264 presents the requirements that apply to facilities that treat, store, or dispose of hazardous waste. It contains general standards by which all hazardous waste treatment, storage, and disposal facilities must be operated as well as specific requirements for surface impoundments, waste piles, landfills, incinerators, land treatment facilities, and facilities with containers and tank systems used for storing or processing hazardous waste.

Part 265 establishes minimum standards that apply to facilities that treat, store, or dispose of hazardous waste and have interim status. Part 265 regulations apply to facilities that were operating before the RCRA regulations were finalized and have not yet received a final permit to operate their facility or have closed but are under EPA orders to correct some problems on-site. This part also contains the requirements for training, preparedness and prevention, and contingency planning.

Part 266 contains standards for the management of specific hazardous wastes and specific types of hazardous waste management facilities. This part includes regulations that apply to recyclable materials, hazardous waste burned for energy recovery, recyclable materials utilized for precious metal recovery, and spent lead-acid batteries being reclaimed.

Part 268 identifies hazardous wastes that are restricted from land disposal and defines those circumstances under which a restricted waste may continue to be land disposed.

Part 270 covers basic EPA permitting requirements for hazardous waste management facilities such as the information to be included in the permit application, monitoring and reporting requirements, and the conditions under which permits can be transferred or modified.

Part 271 specifies the minimum requirements with which a state must comply to receive authorization to administer and enforce its own hazardous waste management program *in lieu* of the federal programs.

Part 273 includes the management system for hazardous wastes batteries, pesticides, and thermostats. This program is referred to as the Universal Waste Program. These regulations cover the standards for universal waste handlers, transporters, and destination facilities.

Part 279 establishes minimum management standards that apply to used oil generators, collection centers, aggregation points, transporters, transfer facilities, processors, re-refiners, burners, and marketers of used oil fuel. This part also places limitations on the use of used oil as a dust suppressant and on the disposal of used oil.

11.3 SUBTITLE C: THE HAZARDOUS WASTE MANAGEMENT PROGRAM

11.3.1 DEFINITION OF A SOLID WASTE

Before a regulatory agency and a potential generator (i.e., a facility) address the issue of whether or not a specified waste is hazardous, they must first determine whether or not the waste is a solid waste. According to 40 CFR 261.2, a solid waste is any *discarded material* that is not excluded from the regulations. Going further into the definition, a discarded material is any material which is abandoned, recycled, "inherently waste-like," or military munitions identified as a solid waste in 40 CFR 266.202.

An *abandoned* material is one which is (1) disposed of; (2) burned or incinerated; or (3) accumulated, stored, or treated (but not recycled) by being disposed or burned.

Materials are solid wastes if they are *recycled*, accumulated, stored, or treated before recycling. This includes being: (1) used or placed on the land in a manner that constitutes disposal; (2) burned for energy recovery; (3) reclaimed; or (4) accumulated speculatively.

Some of the major types of materials that are recycled include (40 CFR 261.2):

- *Spent material*. These are materials that have been used and as a result of contamination can no longer serve the purpose for which it was produced without processing.
- *Sludges*. Solid, semisolid, or liquid waste generated from a municipal, commercial, or industrial wastewater treatment plant; water supply treatment; or air pollution control facility.

- *By-products*. A material that is produced as part of a production process but is not a primary product of the process. An example is process residue such as slag.
- *Scrap metal*. Metal parts (e.g., bars, rods, sheets, and wire) which when worn out or no longer needed, can be recycled.
- Discarded commercial chemical products, off-specification species, container residues, and spill residues.

Inherently waste-like materials are those which have no other possible fate except disposal. For example, hazardous waste designated by the numbers F020, F021 F022, F023, F026, and F028 (all chlorinated hazardous wastes), when they are recycled in any manner, fall into the inherently waste-like category. The F-listing and other hazardous waste listings are described below.

11.3.2 What is a Hazardous Waste

If a solid waste does not qualify for an exemption, it is declared hazardous waste if it is listed by EPA in 40 CFR Part 261, Subpart D, or if it exhibits any of the four hazardous waste characteristics identified in 40 CFR Part 261, Subpart C. The complete hazardous waste lists are available on the Internet at http://www.access.gpo.gov/nara/cfr/waisidx_03/40cfr261_03.html.

11.3.3 Exemptions and Exclusions

The EPA regulations automatically exempt certain solid wastes from the 'hazardous waste' designation under Subtitle C. There are three categories of exclusions: wastes excluded from the definition of solid waste, wastes excluded from the definitions of hazardous waste, and hazardous wastes that are partially excluded provided that they are managed in accordance with specific requirements. Table 11.2 lists the wastes contained under these exclusions.

TABLE 11.2
Exclusions from Subtitle C of RCRA

Excluded from the Solid Waste Definition	Excluded from the Hazardous Waste Definition	Excluded Materials Requiring Special Management
Domestic sewage	Household wastes	Product storage wastes
Mixture of domestic sewage and wastes going to POTW	Agricultural wastes used as fertilizers	Waste identification samples
Industrial point source-discharges under 402 CWA	Mining overburden returned to site	Treatability samples
Irrigation returns flows	Discarded wood treated with arsenic	Empty containers
Sources, special nuclear, or by-product material under AEA	Chromium wastes	Small-quantity generator wastes
	Underground storage tank cleanup wastes	
In-situ mining waste	Specific ore processing wastes	Farm wastes (pesticides)
Reclaimed pulping liquors	Specific utility wastes	
Regenerated sulfuric acid	Oil and gas exploration, development, and production wastes	
Secondary materials returned to the original process under certain conditions	Cement kiln dust	

Source: 40 CFR Part 261.4. With permission.

11.3.4 Hazardous Waste Lists

11.3.4.1 Hazardous Waste from Nonspecific Sources

Wastes in this category are placed on the so-called F-list. These wastes are determined to be hazardous; however, they are not generated by a specific industry or manufacturing process. Wastes on the F-list include certain solvent wastes, plating wastes, metal-treating wastes, wood-preserving wastes, petroleum refinery oil–water–solids separation sludge, leachate from treatment, storage or disposal facilities, wastes from the manufacture of certain chlorinated compounds, and treatment residue from the incineration or thermal treatment of soil contaminated with certain chlorinated compounds. The general categories of F-listed wastes are as follows:

- Solvent wastes (F001–F005)
- Electroplating wastes (F006–F009)
- Metal-treating wastes (F010–F019)
- Wood-preserving wastes (F032–F035)
- Petroleum refining wastes (F037–F038)

In addition, process wastes, discarded unused formulations, and incineration residues from the production of certain chlorinated aliphatic hydrocarbons, trichlorophenol, tetrachlorophenol, pentachlorophenol, and tetra-, penta-, or hexachlorobenzenes are included in the list of nonspecific source wastes.

11.3.4.2 Hazardous Wastes from Specific Sources

Wastes placed on the K-list originate from specific sources or industries that EPA has determined to be hazardous. For example, sludge from the treatment of wastewaters by the wood-preserving industry would fall into this category (K001). K048 and K052 include certain petroleum refining wastes, for example. Wastes on the K-list include those generated by the following industries:

- Wood preservation
- Inorganic pigment production
- Organic chemical production
- Inorganic chemical production
- Pesticide production
- Explosives manufacturing and production
- Petroleum refining
- Iron and steel production
- Primary copper production
- Primary lead production
- Primary zinc production
- Primary aluminum production
- Ferroalloy production
- Secondary lead smelting
- Veterinary pharmaceutical production
- Ink formulation
- Coking industries

11.3.4.3 Discarded Commercial Products, Off-Specification Materials, Container Residues, and Spill Residues

Materials on the P- and U-lists are classified as acute hazardous waste and as toxic waste, respectively. These wastes include certain commercial chemical products having the generic names listed on the P- and U-lists located in 40 CFR 261.33 when they are discarded.

Examples of commercial chemical product hazardous wastes include products from hospitals (e.g., pharmaceuticals past their expiration date, and unused reagents), research laboratories (expired or unused reagents intended for disposal), photography laboratories, and analytical laboratories. These items become hazardous waste when it is decided that they must be disposed. Some products, however, can be tested in order to determine if their expiration date can be extended. If there is another use for the material, it can be stored or used for that purpose without being classified as a hazardous waste.

11.3.5 THE CHARACTERISTIC TESTS

New chemical products are regularly available. As a result, new types of wastes are being produced as well. Since many wastes are chemical newcomers, they will obviously not be found listed in the Code of Federal Regulations. Other methods are needed to determine potential hazards of a waste. The U.S. EPA has established the four so-called 'characteristic tests' to determine whether or not a waste is hazardous.

11.3.6 IGNITABILITY

The ignitability characteristic indicates those wastes that pose a fire hazard during routine handling, for example, storage, transport, processing, or disposal. Specifically, a solid waste exhibits the characteristic of ignitability if a sample possesses any of the following properties (40 CFR Part 261.21):

- It is a liquid, other than an aqueous solution, containing less than 24% alcohol (by vol.) and has flash point less than 60°C (140°F), as determined by a Pensky–Martens Closed Cup Tester (U.S. EPA, 1986; ASTM Standard D-93-79 or D-93-80) or a Setaflash Closed Cup Tester (ASTM Standard D-3278-78).
- It is not a liquid and is capable, under standard temperature and pressure, of causing fire through friction, absorption of moisture or spontaneous chemical changes, and, when ignited, burns so vigorously and persistently that it creates a hazard.
- It is an ignitable compressed gas as defined in 49 CFR 173.300.
- It is an oxidizer as defined in 49 CFR 173.151.

Examples of characteristic (D-list) hazardous wastes include:

- Solvents used for parts cleaning or degreasing
- Paint thinners and paint removing compounds
- Carbon remover and nail polish remover solutions
- Organic-solvent-based wheel strippers

A solid waste that exhibits the characteristic of ignitability is given the EPA hazardous waste number D001.

11.3.7 CORROSIVITY

Corrosive wastes occur at extremes in pH. Wastes with very low or high pH values can corrode standard drums, oxidize skin and other living tissue, and mobilize components from certain wastes. Examples of corrosive wastes include acid wastes and alkali wastes. A solid waste exhibits the characteristic of corrosivity if a sample has either of the following properties (40 CFR Part 261.22):

- It is aqueous and has a pH $\leq$ 2 or $\geq$ 12.5 as determined by a pH meter (Method 9040, U.S. EPA, 1986).
- It is a liquid and corrodes steel at a rate greater than 6.35 mm (0.25 in.) per year and at a temperature of 55°C (130°F) as determined by the test method specified in the National

Association of Corrosion Engineers Standard TM-01-69 as standardized in U.S. EPA (1986).

Wastes generated from the following processes are examples of corrosive hazardous wastes:

- Parts-cleaning operations using highly alkaline cleaning solutions
- Alkaline strippers used to strip paint
- Acidic wastes generated from electroless metal plating lines
- Battery acid and other waste acids
- Phenol wastes

A solid waste that exhibits the characteristic of corrosivity is given the EPA hazardous waste number D002.

11.3.8 REACTIVITY

Wastes possessing the characteristic of reactivity are often unstable, and pose hazards of explosion and release of toxic gases during routine management. Examples of reactive wastes include picrate salts (derived from picric acid, 2,4,6-trinitrophenol), and certain epoxides and peroxides.

Other wastes generated from the following processes are examples of reactive hazardous wastes:

- Cyanide bearing electroplating solutions (unless they are listed in 40 CFR 261.31, F-list)
- Ordinances and explosives listed by DOT as Division 1.1, 1.2, or 1.3 explosive, or forbidden explosives

The characteristic of reactivity in a waste sample is often difficult to determine quantitatively in the laboratory. A waste exhibits the characteristic of reactivity if a representative sample possesses any of the following properties (40 CFR Part 261.23):

- It is normally unstable and readily undergoes violent changes without detonating.
- It reacts violently with water.
- It forms potentially explosive mixtures with water.
- When mixed with water, it generates toxic gases, vapors, or fumes in a quantity sufficient to present a danger to human health or the environment.
- It is a cyanide or sulfide bearing waste which, when exposed to pH conditions between 2 and 12.5, can generate toxic gases, vapors, or fumes in a quantity sufficient to pose a danger to human health or the environment.
- It is capable of detonation or explosive reaction if it is subjected to a strong initiating source or if heated under confinement.
- It is capable of detonation or explosive reaction at standard temperature and pressure.
- It is forbidden explosive as defined in 49 CFR 173.51, a Class A explosive as defined in 49 CFR 173.53, or a Class B explosive as defined in 49 CFR 173.88, DOT regulations.

A solid waste that exhibits the characteristic of reactivity is given the EPA hazardous waste number D003.

11.3.9 TOXICITY

The toxicity characteristic leaching procedure (TCLP) was formulated to simulate environmental conditions in an exposed landfill. The intent of the test is to determine whether potentially toxic components of the waste could leach to groundwater and soil if exposed to acidic precipitation.

The TCLP replaced the EP toxicity test in 1990. The new test includes 25 organic compounds as well as the eight metals and six pesticides in the EP test. In the TCLP, a representative sample is shaken in dilute acetic acid for 16 to 20 h, filtered, and the filtrate is analyzed for the required metals and organic compounds. Where the waste contains less than 0.5% filterable solids, the waste itself is considered to be the extract. Details of the method are provided in TCLP, Method 1311 (U.S. EPA, 1986).

If the extract from a representative waste sample contains any of the contaminants listed in Table 11.3 at a concentration greater than or equal to the regulatory level, the waste exhibits the

TABLE 11.3

Details on the Toxicity Characteristic Compounds (TCLP)

EPA Hazardous Waste Number	Contaminant	CAS No.	Regulatory Level (mg/L)
D004	Arsenic	7440-38-2	5.0
D005	Barium	74401-39-3	100.0
D018	Benzene	71-43-2	0.5
D006	Cadmium	7440-43-9	1.00
D019	Carbon tetrachloride	56-23-5	0.5
D020	Chlordane	57-74-9	0.03
D021	Chlorobenzene	108-90-7	100.0
D022	Chloroform	67-66-3	6.0
D007	Chromium	7440-47-3	5.0
D023	o-Cresol	95-48-7	200.0
D024	m-Cresol	108-39-4	200.0
D025	p-Cresol	106-44-5	200.0
D026	Cresol		200.0
D016	2,4-D	94-75-7	10.0
D027	1,4-Dichlorobenzene	106-46-7	7.5
D028	1,2-Dichloroethane	107-06-2	0.5
D029	1,1-Dichloroethylene	75-35-4	0.7
D030	2,4-Dinitrotoluene	121-14-2	0.13
D012	Endrin	72-20-8	0.02
D031	Heptachlor (and its epoxide)	76-44-8	0.008
D032	Hexachlorobenzene	118-74-1	0.13
D033	Hexachlorobutadiene	87-68-3	0.5
D034	Hexachloroethane	67-72-1	3.0
D008	Lead	7439-92-1	5.0
D013	Lindane	58-89-9	0.4
D009	Mercury	7439-97-6	0.2
D014	Methoxychlor	72-43-5	10.0
D035	Methyl ethyl ketone	78-93-3	200.0
D036	Nitrobenzene	98-95-3	2.0
D037	Pentrachlorophenol	87-86-5	100.0
D038	Pyridine	110-86-1	5.0
D010	Selenium	7782-49-2	1.0
D011	Silver	7440-22-4	5.0
D039	Tetrachloroethylene	127-18-4	0.7
D015	Toxaphene	8001-35-2	0.5
D040	Trichloroethylene	79-01-6	0.5
D041	2,4,5-Trichlorophenol	95-95-4	400.0
D042	2,4,6-Trichlorophenol	88-06-02	2.0
D017	2,4,5-TP (Silvex)	93-72-1	1.0
D043	Vinyl chloride	75-01-4	0.2

Source: 40 CFR Part 261.24. With permission.

toxicity characteristic. The following wastes are examples of common toxicity characteristic wastes (40 CFR Part 261.24):

- Paint waste containing metals such as lead, chromium, silver, or cadmium
- Metal strip baths used to remove paint and chrome plating
- Mercury waste from analytical instruments, dental amalgam, and batteries
- Wastewater and sludge from fabric finishing containing tetrachloroethylene
- Oily wastes and sludge from the petroleum marketing industry containing benzene

A solid waste that exhibits the characteristic of toxicity has the EPA Hazardous Waste Number specified in Table 11.3.

The U.S. EPA has assigned specific hazardous waste numbers and codes to both characteristic and listed wastes. Each listed hazardous waste will have one or more codes associated with it (Table 11.4). Many hazardous wastes meet the requirements of more than one waste type. During a waste determination, all applicable waste codes must be identified and documented.

To summarize, if a solid waste is not a listed hazardous waste and does not exhibit one of the above four characteristics, it is not in the RCRA system. Even if this is the case, however, the waste may still be subject to regulation; for example, it may fall under state regulations.

11.3.10 Mixtures of Hazardous Wastes with Other Materials

In addition to a waste being defined as hazardous waste if it is specifically listed or exhibits a hazardous characteristic, a waste is also classified as hazardous if it is: (1) a mixture of a listed hazardous waste and a solid waste (i.e., the mixture rule), (2) a listed hazardous waste contained within another material (the contained-in rule), or (3) a solid waste generated from the treatment, storage, or disposal of a listed hazardous waste.

11.3.11 The Mixture Rule

A mixture of any amount of hazardous waste and a solid (nonhazardous) waste is considered a hazardous waste (40 CFR 261.3). There is no *de minimis* concentration that qualifies for an exclusion from the mixture rule except for certain mixtures in wastewater treatment systems. For example, if an employee mixes spent ethyl ether (F003) with an absorbent clay to reduce the liquid content, the entire mixture is classified as F003.

An exception to the mixture rule is as follows: if the mixture is hazardous solely because it exhibits a characteristic and the resultant mixture no longer exhibits the same characteristic, it is not considered a hazardous waste. An example is a paint waste that is an ignitable hazardous waste. A mixture of the paint waste and a nonignitable, nonhazardous waste (e.g., machine oil) would

---

TABLE 11.4
Codes for Hazardous Wastes under RCRA

Waste Type	EPA Number	EPA Code
Ignitable	D001	I
Corrosive	D002	C
Reactive	D003	R
Toxicity characteristic	D004–D043	E
Toxic	F-, K-, and U-lists	T
Acutely hazardous	F- and P-lists	H

Source: 40 CFR Part 261. With permission.

---

become nonhazardous, provided the mixture no longer exhibits the ignitability characteristic. It must be emphasized, however, that such wastes become nonhazardous only by the inadvertent, unavoidable mixing occurring during standard processes at the facility. In other words, a facility cannot deliberately mix a nonhazardous waste with hazardous waste to render it nonhazardous. Treating a hazardous waste to render it nonhazardous may require a permit (40 CFR 262.34).

11.3.12 THE CONTAINED-IN RULE

The contained-in rule (40 CFR 261.3) relates to the incorporation of typically natural materials (e.g., soil, groundwater) with a hazardous waste. For example, if a surface impoundment leaks a listed hazardous waste into local groundwater, the resulting contaminated groundwater is to be managed as a hazardous waste.

11.3.13 THE DERIVED-FROM RULE

The derived-from rule presented in 40 CFR 261.3 states that any solid waste generated from the treatment, storage, or disposal of a hazardous waste including any sludge (pollution control residue), spill residue, ash, emission control dust, or leachate (not including precipitation runoff) is a hazardous waste. Thus, in the case of residues generated from the treatment of a listed waste, all residues remain hazardous unless specifically delisted. The generator is required to prove that the waste is no longer hazardous through the delisting process, or the treatment residues are managed as hazardous waste. A facility that treats F-listed hazardous wastes, for example via incineration, must manage the incinerator ash as hazardous waste although the toxicity of the waste may be greatly reduced. This rule also applies to treatment of hazardous wastes during a corrective action.

11.4 GENERATION OF HAZARDOUS WASTES

Table 11.5 presents recent national hazardous waste generation totals according to the percentage of characteristic wastes, listed wastes, or a mixture of both. Wastes categorized as only characteristic wastes represented 52% (20.9 million tons) of the national generation total, while listed-only wastes comprised 18% (7.3 million tons), and wastes with both characteristic and listed waste codes constituted 29% (11.8 million tons) of the national total.

TABLE 11.5
Tons of Generated Waste that were Only Characteristic Waste, Only Listed Waste, or Both Characteristic and Listed Waste, 1999

Only Characteristic Wastes		Only Listed Wastes		Both a Characteristic and a Listed Waste	
Ignitable only	681,936	F code only	2,213,492		
Corrosive only	1,075,431	K code only	3,695,803		
Reactive only	247,748	P code only	80,396		
D004-17	2,379,016	U code only	496,466		
D018-43	4,464,793				
More than one characteristic code	12,082,405	More than one listed code	845,353		
Total	20,391,330	Total	7,331,509	Both characteristic and listed	11,760,240

Note: All quantities are in tons.

Source: U.S. EPA, 2001.

11.4.1 QUANTITIES OF TOXICS RELEASE INVENTORY CHEMICALS IN WASTE BY INDUSTRY

The Toxics Release Inventory (TRI) is a publicly available database that contains information on waste management activities reported annually by certain industries and federal facilities. There are nearly 650 toxic chemicals and toxic chemical categories on the list of chemicals that must be reported to EPA and the states. These chemicals do not always correspond exactly to RCRA wastes; however, they provide a useful overview of the types and amounts of toxic chemicals generated in the United States.

According to the TRI, the chemical manufacturing industry reported the largest quantity of toxic chemicals in production-related waste managed in 2001, with 10.69 billion pounds or 40% of the total reported by all industries (Table 11.6). The primary metals industry reported the second largest quantity of toxic chemicals in production-related waste managed in 2001, with 3.10 billion pounds or 12% of the total. The metal mining industry reported the third largest quantities of toxic chemicals and the largest quantity released on- and off-site in 2001. With 2.87 billion pounds, metal mining facilities accounted for 11% of toxic chemicals in production-related waste managed by all

TABLE 11.6

Quantities of TRI Chemicals in Waste by Industry, 2001

Industry	Total Production-Related Waste Managed (lb)
Metal mining	2,869,626,395
Coal mining	16,481,994
Food	1,090,077,820
Tobacco	5,309,980
Textiles	40,714,851
Apparel	1,803,460
Lumber	69,202,953
Furniture	15,058,736
Paper	1,374,388,098
Printing	370,693,356
Chemicals	10,688,079,114
Petroleum	878,617,109
Plastics	205,625,949
Leather	8,407,158
Stone, Glass or Clay	653,183,379
Primary metals	3,100,713,329
Fabricated metals	661,029,051
Machinery	130,427,694
Electrical equipment	592,437,820
Transportation equipment	260,042,330
Measure or Photo	62,566,249
Miscellaneous	40,766,891
Multiple SIC codes 20 to 39	966,557,214
No SIC codes 20 to 39	27,510,857
Electric utilities	1,561,124,192
Chemical wholesale distributors	41,079,498
Petroleum bulk terminals or bulk storage	46,278,144
Hazardous waste or solvent recovery	957,788,017
Total	26,735,591,638

Source: U.S. EPA, 2001. With permission.

TABLE 11.7
Top 20 Chemicals with the Largest Total Production-Related Waste, 2001

CAS Number	Chemical	Total Production-Related Waste Managed (lb)
67-56-1	Methanol	2,331,011,667
108-88-3	Toluene	1,787,944,977
7647-01-0	Hydrochloric acid	1,504,105,058
—	Zinc compounds	1,355,504,817
—	Copper compounds	1,263,772,355
74-85-1	Ethylene	1,256,806,620
7440-50-8	Copper	1,088,001,030
110-54-3	n-Hexane	970,193,833
—	Lead compounds	965,794,108
98-82-8	Cumene	832,570,075
7664-41-7	Ammonia	800,432,076
115-07-1	Propylene	797,566,959
—	Nitrate compounds	701,130,070
7664-93-9	Sulfuric acid	583,305,201
107-21-1	Ethylene glycol	565,972,276
107-06-2	1,2-Dichloroethane	561,860,469
7782-50-5	Chlorine	552,091,471
1330-20-7	Xylene (mixed isomers)	479,477,559
—	Manganese compounds	477,625,043
7697-37-2	Nitric acid	411,681,261
	Subtotal (top 20 chemicals)	19,286,846,925
	Total (all chemicals)	26,735,591,638

U.S. EPA, 2001. Data are from Section 8 of Form R.

industries in 2001. Electric utilities reported the fourth largest amount of toxic chemicals in 2001 with 1.56 billion pounds, representing 6% of the total managed. This industry reported the second largest amount released on- and off-site, with 1.06 billion pounds or 17% of the quantity released by all industries in 2001.

11.4.2 QUANTITIES OF TRI CHEMICALS IN WASTE BY CHEMICAL

Table 11.7 lists the 20 TRI chemicals managed in production-related waste in 2001 in the largest quantities. Production-related waste managed from the top 20 TRI chemicals totaled 19.29 billion pounds, i.e., 72% of all toxic chemicals in production-related waste.

The TRI chemical with the largest quantity released on- and off-site was copper compounds (1.01 billion pounds), accounting for 16% of all releases on- and off-site in 2001. Copper compounds ranked fifth for total toxic chemicals managed in production-related waste. Zinc compounds were released on- and off-site in the second largest amounts (982.9 million pounds), and ranked fourth for total toxic chemicals managed in production-related waste. Hydrochloric acid was released on- and off-site in the third largest amount (587.6 million pounds) and ranked third for total toxic chemicals managed in production-related waste and second for on-site treatment.

REFERENCES

Code of Federal Regulations, Vol. 40, Part 261, Identification and Listing of Hazardous Waste, U.S. Government Printing Office, Washington, DC, 2004.

U.S. Environmental Protection Agency, Toxics Release Inventory (TRI), Public Data Release, Executive Summary, 2001. See: http://www.epa.gov/tri/tridata/tri01/pdr/index.htm

U.S. Environmental Protection Agency, Method 9040. pH Electrometric Measurement. Test Methods for Evaluating Solid Waste, Physical/Chemical Methods, EPA SW-846, Washington, DC, 1986a.

U.S. Environmental Protection Agency, Pensky–Martens Closed-Cup Method for Determining Ignitability, Method 1010, Test Methods for Evaluating Solid Waste, Physical/Chemical Methods, EPA SW-846, Washington, DC, 1986b.

SUGGESTED READINGS AND WEB SITES

Environment, Health and Safety Online, State Government Downloads and Links, 2003. See: http://www. ehso.com/stategov.php

Environmental Health and Safety Online, Identifying Hazardous Waste, 2002. See: http://www.ehso.com/ hazwaste_ID.htm

U.S. Department of Energy, Overview of the Identification of Hazardous Waste Under RCRA, 1999. See: http://tis.eh.doe.gov/oepa/guidance/rcra/define.pdf

U.S. Environmental Protection Agency, Final Paint Listing, 2002. See: http://www.epa.gov/epaoswer/ hazwaste/id/paint/index.htm

U.S. Environmental Protection Agency, Hazardous Waste Identification Studies, 2002. See: http://www.epa.gov/ epaoswer/hazwaste/id/studies.htm

U.S. Environmental Protection Agency, Hazardous Waste Management System; Identification and Listing of Hazardous Waste; Chlorinated Aliphatics Production Wastes; LDRs for Newly Identified Wastes; and CERCLA Hazardous Substance Designation and Reportable Quantities, Final Rule, Nov. 08, 2000, 2002. See: http://www.epa.gov/epaoswer/hazwaste/id/chlorali/index.htm

U.S. Environmental Protection Agency, Hazardous Waste Management System; Identification and Listing of Hazardous Waste: Dye and Pigment Industries; LDRs for Newly Identified Wastes; CERCLA Hazardous Substance Designation and Reportable Quantities; Proposed Rule - July 23, 1999, 2002. See: http://www.epa.gov/epaoswer/hazwaste/id/dyes/index.htm

U.S. Environmental Protection Agency, Hazardous Waste Characteristics Scoping Study, 1996. See: http:// www.epa.gov/epaoswer/hazwaste/id/char/scopingp.pdf

U.S. Environmental Protection Agency, 1998 RCRA §3007 Survey of the Inorganic Chemicals Industry, 1998. See: http://www.epa.gov/epaoswer/hazwaste/id/inorchem/index.htm

U.S. Environmental Protection Agency, Petroleum Refining Process Wastes Listing Final Rule, August 6, 1998, 2002. See: http://www.epa.gov/epaoswer/hazwaste/id/petroleum/index.htm

U.S. Environmental Protection Agency, RCRA Hazardous Waste Delisting: The First 20 Years, 2002. See: http://www.epa.gov/epaoswer/hazwaste/id/delist/index.htm

QUESTIONS

1. What is the top priority for hazardous waste management (*general method*) under RCRA? What is the lowest priority?
2. Which of the following under RCRA is (are) not excluded from the rules and regulations applicable to hazardous waste generators, treatment, storage, disposal, and transportation? (a) domestic sewage; (b) spent nuclear or by-product material; (c) household waste; (d) spent halogenated solvents.
3. What are acute hazardous wastes? Provide an accurate technical definition.
4. Which of the following property(ies) is (are) *not* characteristics that define a RCRA hazardous waste? (a) radioactive; (b) corrosive; (c) ignitable; (d) reactive; (e) biohazard.
5. What pH range does noncorrosive waste display?
6. What is the name of the test method used to determine if a waste is *toxic*? Describe the details of the method.
7. Can a generator legally mix a listed hazardous waste with sufficient nonhazardous solid waste to the point where it can be diluted and therefore is no longer classified as a hazardous waste? Discuss.

8. The Hi-Jinx Metalworks Corp. has produced several gallons of a spent paint stripper. Based solely on the Material Safety Data Sheet (following pages), could this waste be a RCRA hazardous waste?

9. Outside one of the Hi-Jinx warehouses (which has stored paint stripper in the past), several drums are discovered and there is a sweet solvent odor. Is the contaminated soil considered a hazardous waste? Explain.

10. The environmental safety officer at the Hi-Jinx plant identifies dozens of drums containing a reddish filter cake near the site's electroplating wastewater treatment plant. After analyzing the filter cake, the level of chromium in the sludge is determined to be 75 mg/kg and the TCLP test measures 2 mg/L in the resulting leachate. What can the facility conclude regarding the waste and its proper management?

11. Can a waste be both a listed hazardous waste and a characteristic hazardous waste? Explain.

12. An automobile body shop and painting facility is in operation. Metal parts are reworked and repainted; engine components are cleaned and reworked; fiberglass and metal body parts are repaired, replaced, and painted. List at least ten different types of waste generated at the facility. Separate into solid (nonhazardous) and hazardous (listed and characteristic) wastes.

13. List potential sources of hazardous waste generated in your university. Name any listed hazardous wastes. To what list(s) do they belong? If not listed, what are the specific characteristics that render these wastes hazardous?

14. Write in the name of each list on the table below:

Name of List	Waste Types Covered
——List	-Nonspecific sources
	-Solvents
Provide examples of wastes your university generates on this list:	-Electroplating wastes
	-Wood preserving wastes
	-Chlorinated aliphatic hydrocarbons
	-Certain pesticide wastes
——List	-Specific source wastes
	-Wood preserving
Provide examples of wastes your university generates:	-Chemical manufacturing
	-Petroleum refining
	-Explosives manufacturing
	-Metal processing
——List	Acutely hazardous commercial chemical products
Provide examples of wastes your university generates:	
——List	Toxic commercial chemical products
Provide examples of wastes your university generates:	

15. For the four hazardous waste characteristics, list their hazardous waste numbers in the table:

Characteristic	EPA Waste Number
Ignitable	
Corrosive	
Reactive	
Toxicity characteristic	

MATERIAL SAFETY DATA SHEET

Product Identification: METHYLENE CHLORIDE INDUST GRADE

ITEM DESCRIPTION

Item Name: DICHLOROMETHANE, TECHNICAL
    Type of Container: DRUM

HAZARDS IDENTIFICATION, EMERGENCY OVERVIEW

Health Hazards Acute & Chronic: ACUTE:IRRITATION OF EYES, SKIN AND RESPIRA-TORY TRACT, CNS
EFFECTS, DIZZINESS, WEAKNESS, FATIGUE, NAUSEA, HEADACHE, G. I. TRACT DISTURBANCES, NAUSEA, VOMITING, DIARRHEA. CHRONIC: INCREASE CO LEVEL IN BLOOD CAUSING CARDIOVASCULAR STRESS, CNS EFFECTS.
Signs & Symptoms of Overexposure:
EYE, SKN: IRRT, INHL: IRRT, DIZZ, WEAK, FATIGUE, NAUS, HEAD, UNCONSC. INGEST: ALSO GI IRRT, NAUS, VOMIT, DIARR.
Medical Conditions Aggravated by Exposure:
PRE-EXISTING CONDITIONS MAY BE WORSENED.
Route of Entry Indicators:
Inhalation: YES
Skin: YES
Ingestion: YES
Carcenogenicity Indicators
NTP: NO
IARC: YES
OSHA: N/P
Carcinogenicity Explanation: METHYLENE CHLORIDE IS SUSPECTED CARCINOGEN BY IARC OR ACGIH; LIVER ABNORMALITIES, LUNG DAMAGE (AMONG LAB ANIMALS).

FIRST AID MEASURES

First Aid:
INHAL: RMV TO FRESH AIR. IF NOT BRTHNG GIVE CPR; IF BRTHNG DIFF GIVE OXYGEN. EYE: IMMED
FLUSH W/PLENTY OF WATER. SKIN: WASH W/SOAP & WATER. RMV CONTAM CLTHG & SHOES.
INGEST: DO NOT INDUCE VOMIT. NOTHG BY MOUTH IF UNCONSC. GET MEDICAL ATTN.

FIRE FIGHTING MEASURES

Fire Fighting Procedures:
SELF-CONTAINED BREATHING GEAR, W/FULL FACE SHIELD
Unusual Fire or Explosion Hazard:
EMITS CARBON MONOXIDE, CARBON DIOXIDE, HYDROGEN CHLORIDE, AND PHOSGENE WHEN BURNED.
Extinguishing Media:
WATER FOG, CARBON DIOXIDE, DRY CHEMICAL

Flash Point: Flash Point Text: NONE
Autoignition Temperature:
Autoignition Temperature Text: N/R
Lower Limit(s): 13
Upper Limit(s): 23.0

ACCIDENTAL RELEASE MEASURES

Spill Release Procedures:
　　ABSORB SPILL. STOP SPILL AT SOURCE. DIKE AREA. USE PROTECTIVE EQUIP. WHEN IN AREA. FOR LARGE SPILLS PUMP LIQUID TO HOLDING TANK

EXPOSURE CONTROLS & PERSONAL PROTECTION

Repiratory Protection:
　　SELF-CONTAINED WITH FULL FACE SHIELD-OSHA/MESA APPROVED
　　Ventilation: MECHANICAL OR LOCAL AS NEEDED TO KEEP BELOW TLV
　　Protective Gloves: IMPERVIOUS
　　Eye Protection: CHEM SPLASH GOGGLES
　　Other Protective Equipment: IMPERVIOUS CLOTHING, EYE-WASH FACILITIES, BOOTS.
　　Work Hygienic Practices: AVOID CONTACT WITH EYES AND SKIN; DO NOT BREATHE VAPORS/MIST; WASH
　　THOROUGHLY AFTER USE; DO NOT USE CONTAMINATED CLOTHES.

PHYSICAL & CHEMICAL PROPERTIES

Boiling Point: Boiling Point Text: 104F, 40C
　　Melting/Freezing Point: Melting/Freezing Text: N/A
　　Decomposition Point: Decomposition Text: N/A
　　Vapor Pressure: 355 MMHG Vapor Density: 2.9
　　Percent Volatile Organic Content:
　　Specific Gravity: 1.322
　　pH: N/P
　　Evaporation Weight and Reference: 1.8 (ETHYL ETHER=1)
　　Solubility in Water: SLIGHT
　　Appearance and Odor: CLEAR, COLORLESS LIQUID. ETHER-LIKE ODOR.
　　Percent Volatiles by Volume: 100

STABILITY & REACTIVITY DATA

Stability Indicator: YES
　　Materials to Avoid: ALUMINUM, STRONG ALKALAIS
　　Hazardous Decomposition Products: CARBON MONOXIDE, CARBON DIOXIDE, HYDRO-GEN CHLORIDE, PHOSGENE
　　Hazardous Polymerization Indicator: NO

DISPOSAL CONSIDERATIONS

Waste Disposal Methods:

   PLACE ABSORBED MATERIAL IN CONTAINERS SUITABLE FOR SHIPMENT TO DIS-
POSAL AREAS. ENVIRONMENTAL LAWS TAKE PRECEDENCE. LIQUID WASTES MAY
BE DESTROYED BY LIQUID INCINERATION WITH OFF GAS SCRUBBER

DEPARTMENT OF TRANSPORTATION INFORMATION

DOT Proper Shipping Name: DICHLOROMETHANE
   Hazard Class: 6.1
   UN ID Number: UN1593
   DOT Packaging Group: III
   Label: KEEP AWAY FROM FOOD
   Non Bulk Packaging: 203
   Bulk Packaging: 241
   Maximimum Quanity in Passenger Area: 60 L
   Maximimum Quanity in Cargo Area: 220 L
   Stow in Vessel Requirements: A

12 Hazardous Waste Generator Requirements

"Always do right; this will gratify some people and astonish the rest".

Mark Twain

12.1 INTRODUCTION

RCRA regulations call for cradle-to-grave management of hazardous wastes, i.e., wastes are to be tracked from the point of their initial generation through storage and transportation, to final treatment and disposal. As a first step in this management framework (and as discussed in Chapter 11), the waste generator is required to determine if any solid wastes generated at their facility are hazardous so that the wastes will be managed and tracked properly. Secondly, a waste generator's responsibilities regarding storage, transport, and disposal options depend upon the volume of waste generated per calendar month. Hazardous waste generators are classified as large quantity generators (LQGs), small quantity generators (SQGs), and conditionally exempt small quantity generators (CESQGs), based on these monthly volumes.

12.2 DETERMINING THE GENERATOR CATEGORY

According to RCRA (40 CFR Part 262), the generator must measure (count) the quantity of hazardous waste generated per calendar month. Wastes that must be counted include those:

- Accumulated on-site before disposal or recycling
- Placed into a treatment or disposal unit on the facility site
- Collected as sludges and removed from product storage tanks

In the early days of RCRA, counting requirements had resulted in some confusion on the part of both generators and regulators. In some situations, for example, the regulations were interpreted such that the same waste was counted several times. The requirements have since been fine-tuned. Basic principles of waste counting are as follows:

- Materials generated on-site that are either listed or characteristic hazardous wastes must be counted.
- Materials are not counted until they are removed from the production process. For example, plating baths which are being used and reused, or a spent solvent still in the production process, are not counted until they are removed from the process.
- Waste is counted only once in a calendar month. In some cases, for example, a waste may be used more than once a month by recycling within the facility. Under the current requirements, only the initial quantity is counted.
- Wastes discharged to a publicly owned treatment works, in compliance with Clean Water Act standards, are not covered under the RCRA system.

12.3 GENERATOR TYPES

12.3.1 THE LARGE QUANTITY GENERATOR

Facilities that generate more than 1000 kg (2204 lb) of hazardous waste per calendar month or more than 1 kg of acutely hazardous waste per month, are designated large quantity generators (LQGs). In 1999, a total of 20,083 LQGs reported that they generated 40 million tons of RCRA hazardous waste (U.S. EPA, 2001a). A comparison of the 1997 data with the 1999 data shows that the number of LQGs decreased by 233 and the quantity of hazardous waste generated decreased by 650,000 tons or 1.5%.

The five states that contributed most to the national hazardous waste generation total in 1999 were Texas (14.9 million tons), Louisiana (4.4 million tons), Illinois (2.9 million tons), Tennessee (2.2 million tons), and Ohio (1.6 million tons). The LQGs in these states accounted for 65% of the national total quantity generated. Fourteen of the top 50 generators are located in Texas, the top-ranked state in hazardous waste generation. These 14 LQGs accounted for 92% of the state's generation total and 34% of the national generation total. The six LQGs in Louisiana, the state ranked second in hazardous waste generation, accounted for 92% of the state's generation total and 10% of the national generation total. Ten of the largest generators are located in Illinois, Tennessee, and Ohio, the states ranked third, fourth, and fifth, respectively, in hazardous waste generation. These LQGs accounted for 12% of the national total quantity generated (U.S. EPA, 2001a).

Table 12.1 illustrates the relationship between hazardous waste generation quantity ranges and the number of generators occurring within each range. Most of the LQGs (13,096 or 65% of the national total) generated between 1.1 and 113.2 tons of hazardous waste in 1999. Only 50 LQGs (< 1% of all LQGs) generated over 111,113.2 tons, but these LQGs accounted for 77% of the national total quantity generated. In 1999 95% of all LQGs generated 1,113 tons or less (U.S. EPA, 2001a).

TABLE 12.1

Quantity of RCRA Hazardous Waste Generated and Number of Hazardous Waste Generators by State, 1999

State	Hazardous Waste Quantity Generated		Large Quantity Generators	
	Tons	%	Number	%
Alabama	491,178	1.2	274	1.4
Alaska	1,335	0.0	42	0.2
Arizona	39,016	0.1	193	1.0
Arkansas	970,995	2.4	241	1.2
California	427,302	1.1	1,850	9.2
Colorado	49,190	0.1	163	0.8
Connecticut	92,201	0.2	391	1.9
Delaware	26,071	0.1	76	0.4
District of Columbia	1,167	0.0	30	0.1
Florida	272,387	0.7	366	1.8
Georgia	209,206	0.5	384	1.9
Guam	696	0.0	3	0.0
Hawaii	1,456	0.0	37	0.2
Idaho	851,764	2.1	38	0.2
Illinois	2,907327	7.3	1,006	5.0
Indiana	984,895	2.5	586	2.9
Iowa	46,828	0.1	188	0.9
Kansas	1,594,119	4.0	224	1.1
Kentucky	214,842	0.5	340	1.7
Louisiana	4,351,245	10.9	440	2.2

(continued)

TABLE 12.1 (*continued*)

State	Hazardous Waste Quantity Generated		Large Quantity Generators	
	Tons	%	Number	%
Maine	4,374	0.0	102	0.5
Maryland	80,256	0.2	289	1.4
Massachusetts	1,191,465	3.0	448	2.2
Michigan	1,385,375	3.5	823	4.1
Minnesota	56,573	0.1	262	1.3
Mississippi	1,598,642	4.0	136	0.7
Missouri	158,682	0.4	312	1.6
Montana	23,986	0.1	30	0.1
Navajo Nation	89	0.0	6	0.0
Nebraska	43,224	0.1	85	0.4
Nevada	11,473	0.0	102	0.5
New Hampshire	11,082	0.0	168	0.8
New Jersey	650,534	1.6	1,071	5.3
New Mexico	238,558	0.6	41	0.2
New York	548,928	1.4	2,647	13.2
North Carolina	74,757	0.2	508	2.5
North Dakota	2,675	0.0	16	0.1
Ohio	1,644,029	4.1	1,181	5.9
Oklahoma	417,460	1.0	147	0.7
Oregon	81,270	0.2	208	1.0
Pennsylvania	417,477	1.0	965	4.8
Puerto Rico	86,630	0.2	105	0.5
Rhode Island	37,622	0.1	145	0.5
South Carolina	14,761	0.0	347	1.7
South Dakota	1,074	0.0	21	0.1
Tennessee	2,218753	5.5	396	2.0
Texas	14,923,520	37.3	907	4.5
Trust Territories	827	0.0	4	0.0
Utah	80,427	0.2	91	0.5
Vermont	5,275	0.0	65	0.3
Virgin Islands	12,511	0.0	1	0.0
Virginia	121,787	0.3	332	1.7
Washington	91,245	0.2	545	2.7
West Virginia	92,503	0.2	139	0.7
Wisconsin	159,174	0.4	540	2.7
Wyoming	4,746	0.0	22	0.1
CBI Data	1,066	N/A	4	N/A
Total	40,026,050	100.0	20,083	100.0

Note: Columns may not sum due to rounding. Percentages do not include CBI data.

Source: U.S. EPA, 2001a. With permission.

Wastes types generated by larger industries include bottoms or residues, dusts, discarded or off-specification chemicals or by-products, lab packs, slags, sludges or slurries, spent liquors, waste packages, and wastewaters. Selected examples of manufacturing operations that generate these types of waste are shown in Table 12.2.

12.3.2 THE SMALL QUANTITY GENERATOR

Facilities that generate more than 100 kg (220.4 lb) but less than 1000 kg per calendar month, or less than 1 kg of acutely hazardous waste per month are designated small quantity generators (SQGs).

TABLE 12.2
Examples of Hazardous Wastes Generated by Large-Quantity Generators

Category	Examples of Wastes
Chemical reprocessing options	Nonhalogenated solvents, cupric chloride, pyrophosphate, acids, caustics, others
Coking operations	Ammonia, benzene, phenols, cyanide
Degreasing operations	Perchloroethylene, trichloroethylene, methylene chloride, 1,1,1-trichloroethane, carbon tetrachloride, chlorinated fluorocarbons
Distillation operations	Chlorobenzene, trichloroethylene, perchloroethylene, aniline, cumene, *o*-xylene, naphthalene, others
Electroplating processes	Cyanides, nickel, copper, acids, chrome, cadmium, gold
Ink formulation	Solvents, caustics, chromium- or lead-containing pigments and stablizers
Leather tanning	Tannic acid, chromium
Painting operations	Methylene chloride, trichloroethylene, toluene, methanol, turpentine
Petroleum processes	Arsenic, cadmium, chromium, lead, halogenated solvents, flammable oils, distillate products
Primary metal processes	Cyanides, salt baths, heavy metals such as chromium and lead
Pulp and paper operations	Chlorine, sodium sulfite, sodium hydroxide, dioxins, furans, phenols
Textile finishing	Solvents, solutions of dyes
Weapons manufacture	Trinitrotoluene, nitroglycerin, uranium alloys, plutonium
Wood-preserving processes	Creosote, pentachlorophenol, other creosote and chlorophenolic formulations, copper, arsenic, chromium

Source: Woodside, G., Hazardous Materials and Hazardous Waste Management: A Practical Guide, John Wiley and Sons, New York, NY, 1993. Reproduced with kind permission of John Wiley & Sons, Inc.

SQGs comprise a wide range of commercial and industrial activities ranging from equipment repair (degreasing and rust removal), to construction (paint preparation), to consumer service shops (e.g., auto repair). The range of SQGs along with the types of wastes produced are listed in Table 12.3.

12.3.3 EPA WASTE CODES FOR COMMON SQG WASTES

Some of the more common wastes generated by SQGs are discussed below. The U.S. EPA waste codes are also provided for these wastes.

Solvents, spent solvents, solvent mixtures, or solvent still bottoms are often hazardous. The following are some commonly used hazardous solvents (also see ignitable wastes for other hazardous solvents, and 40 CFR 261.31 for most listed hazardous waste solvents):

Benzene	F005
Carbon disulfide	F005
Carbon tetrachloride	F001
Chlorobenzene	F002
Cresols	F004
Cresylic acid	F004

TABLE 12.3

Typical Hazardous Waste Generated by Small-Quantity Generators

Typical Type of Business	How Generated	Types of Wastes	Waste Codes
Dry cleaning and laundry plants	Commercial dry cleaning processes	Still residues from solvent distillation, spent filter cartridges, cooked powder residue	D001, D039, F002
Furniture/Wood manufacturing and refinishing	Wood cleaning and wax removal, refinishing/stripping, staining, painting, finishing, brush cleaning and spray brush cleaning	Ignitable wastes, toxic wastes, solvent wastes, paint wastes	D001, F001–F005
Construction	Paint preparation and painting, carpentry and floor work, other speciality contracting activities, heavy construction, wrecking and demolition, vehicle and equipment maintenance for construction activities	Ignitable wastes, toxic wastes, solvent wastes, paint wastes, used oil, acids/bases	D001, D0002 F001–F005
Laboratories	Diagnostic and other laboratory testing	Spent solvents, unused reagents, reaction products, testing samples, contaminated materials	D001, D002 D003, F001–F005, U211
Vehicle maintenance	Degreasing, rust removal, paint preparation, spray booth, spray guns, brush cleaning, paint removal, tank cleanout, installing lead-acid batteries	Acids/bases, solvents, ignitable wastes, toxic wastes, paint wastes, batteries	D001, D002, D006, D008, F001–F005
Printing and allied industries	Plate preparation, stencil preparation for screen printing, photo processing, printing, clean-up	Acids/bases, heavy-metal wastes, solvents, toxic wastes, ink	D002, D006, D008, F001–F005
Equipment repair	Degreasing, equipment cleaning, rust removal, paint preparation, painting, paint removal, spray booth, spray guns, and brush cleaning	Acids/bases, toxic wastes, ignitable wastes, paint wastes, solvents	D001, D002, D006, D008, F001–F005
Pesticide end-users/application services	Pesticide application and clean-up	Used/unused pesticides, solvent wastes, ignitable wastes, contaminated soil (from spills), contaminated rinse water, empty containers	D001, F001–F005, U129, U136, P094, P123
Educational and vocational shops	Automobile engine and body repair, metal working, graphic arts-plate preparation, woodworking	Ignitable wastes, solvent wastes, acids/bases, paint wastes	D001, D002, F001–F005

Source: U.S. EPA, 2001b. With permission.

o-Dichlorobenzene	F002
Ethanol	D001
2-Ethoxyethanol	F005
Ethylene dichloride	D001
Isobutanol	F005
Isopropanol	D001
Kerosene	D001
Methyl ethyl ketone	F005
Methylene chloride	F001, F002
Naphtha	D001
Nitrobenzene	F004
2-Nitrobenzene	F004
Petroleum solvents (flashpoint < 140°F)	D001
Pyridine	F005
1,1,1-Trichloroethane	F001, F002
1,1,2-Trichloroethane	F002
Tetrachloroethylene (perchloroethylene)	F001, F002
Toluene	F005
Trichloroethylene	F001, F002
Trichlorofluoromethane	F002
Trichlorotrifluoroethane (Valclene)	F002
White spirits	D001

In the dry cleaning industries, filtration residues such as cooked powder residue (perchloroethylene plants), still residues, and spent cartridge filters containing perchloroethylene or valclene are hazardous and have the waste code "F002." Still residues containing petroleum solvents with a flashpoint less than 60°C (140°F) are considered hazardous and have the waste code "D001."

Acids, bases, or corrosive mixtures (40 CFR 261.22) have the waste code "D002." The following are some of the more commonly used corrosives:

Acetic acid
Ammonium hydroxide
Oleum
Chromic acid
Hydrobromic acid
Hydrochloric acid
Hydrofluoric acid
Nitric acid
Perchloric acid
Phosphoric acid
Potassium hydroxide
Sodium hydroxide
Sulfuric acid

Heavy metals and other inorganic wastes are considered hazardous if the extract from a representative sample of the waste (see discussion of TCLP, Chapter 11) has any of the specific constituent concentrations as shown in 40 CFR 262.24 (see Table 11.3). Waste sources include dusts, solutions, wastewater treatment sludges, paint wastes, and waste inks. The following are common heavy metals and inorganics:

Arsenic	D004
Barium	D005
Cadmium	D006
Chromium	D007

Lead	D008
Mercury	D009
Selenium	D010
Silver	D011

Ink sludges containing chromium and lead include solvent sludges, caustic sludges, water sludges from cleaning tubs, equipment used in the formulation of ink from pigments, driers, soaps, and stabilizers containing chromium and lead. All ink sludges have the waste code "K086."

Examples of ignitable wastes are spent solvents, solvent still bottoms, epoxy resins and adhesives, and waste inks containing flammable solvents. Unless specified, all ignitable wastes have the waste code "D001."

Acetone	F003
Benzene	F005
n-Butyl alcohol	F003
Chlorobenzene	F002
Cyclohexanone	F003
Ethyl acetate	F003
Ethyl benzene	F003
Ethyl ether	F003
Ethylene dichloride	D001
Methanol	F003
Methyl isobutyl ketone	F003
Petroleum distillates	D001
Xylene	F003

Used lead-acid batteries should be reported only if they are not recycled. Specific wastes from used batteries include:

Lead dross	D008
Spent acids	D002
Lead-acid batteries	D008

Pesticides, wastewaters, sludges, and by-products from pesticide formulators are another category of wastes produced by SQGs. The pesticides listed below are hazardous and those marked with an asterisk (*) have been designated acutely hazardous (40 CFR 261.32).

Aldicarb*	P070
Amitrole	U011
1,2-Dichloropropene	U084
Heptachlor*	P059
Lindane	U129
Methyl parathion*	P071
Parathion*	P089
Phorate*	P094

Reactive wastes (40 CFR 2612.23) all have the waste code "D003." The following materials are examples of wastes commonly considered to be reactive:

Acetyl chloride
Chromic acid
Cyanides
Hypochlorites
Organic peroxides
Perchlorates
Permanganates
Sulfides

Spent plating and cyanide wastes contain cleaning solutions and plating solutions with caustics, solvents, heavy metals, and cyanides. Cyanide wastes may also be generated from heat treatment operations and pigment production. Plating wastes typically have the waste codes F006 to F009, F007 and F009 designating wastes containing cyanide. Cyanide heat-treating wastes generally have the waste codes F010 to F012 (40 CFR 261.31).

Sludges from wastewater treatment operations for wood-preserving wastes are considered hazardous. Bottom sediment sludges from the treatment of wastewater processes that use creosote and pentachlorophenol have the waste code K001. In addition, wood-preserving compounds may include:

Chromated copper arsenate	D004
Creosote	U051
Pentachlorophenol	F027

12.3.4 THE CONDITIONALLY EXEMPT SQG

Facilities that generate 100 kg or less of hazardous waste or 1 kg or less of acutely hazardous waste are designated conditionally exempt small quantity generators (CESQGs). The total number of CESQGs in both manufacturing and nonmanufacturing sectors nationwide is approximately 455,000. The total waste volume generated by all CESQGs nationwide is about 183,000 metric tons (201,600 tons) per year. In a survey published by the U.S. EPA (1994) for 22 industry groups, approximately 80% of CESQGs are in the nonmanufacturing sector and generate approximately 88% of the CESQG waste volume. The remaining 20% of establishments are in the manufacturing sector.

The major CESQG waste types for the industry groups surveyed are spent lead-acid batteries, spent solvents and still bottoms, perchloroethylene and photographic wastes. For the industry groups surveyed, approximately 80% of CESQG waste is managed off-site. The predominant off-site management methods include (U.S. EPA, 1994):

- Recycling (73% or 69,000 tons/year managed off-site)
- Disposal at a nonhazardous solid waste landfill (10% of waste managed off-site or 9300 tons/year)
- Disposal at a permitted Subtitle C landfill (2% of the waste managed off-site or 2000 tons/year)

The predominant on-site management methods for the 22 industries surveyed include (U.S. EPA, 1994):

- Disposal in the sewer or septic system (56% of the waste managed on-site or 14,600 tons/year)
- Disposal in a nonhazardous solid waste landfill (2% of the waste managed on-site or 509 tons/year)

The vehicle maintenance industry is the largest CESQG industry (from 22 industry groups surveyed) both in terms of number of CESQGs (54%) and waste volume (71%). Of the other major CESQG waste-generating industries (U.S. EPA, 1994):

- Metals manufacturing generates the second highest amount of CESQG waste for the industries surveyed, approximately 6.1%.
- Laundries generate 4.8% of total CESQG waste volume.
- Printing and ceramics generate 4.8%.
- Pesticide end users and application services generate 2.1%.
- Construction generates 1.9%.
- Photography generates 1.8%.

TABLE 12.4
Major CESQG Industries and Waste Types

Major CESQ Generating Industries	Major CESQG Waste Types
Vehicle maintenance	Lead-acid batteries (61%)
Metals manufacturing	Spent solvents/still bottoms (18%)
Laundries	Dry cleaning filter residues (5%)
Printing/ceramics	Photographic wastes (4%)
Pesticide users/appliers	Formaldehyde (3%)
Construction	Acids and alkalis (2%)
Stone, clay, glass, and concrete	
Food and kindred products	
Primary steel and iron	
Textile manufacturing	
Pulp and paper	

Source: U.S. EPA, 1994. With permission.

The CESQG is exempt from most hazardous waste management requirements. A facility meeting the test for a conditionally exempt generator (generating < 100 kg per month and < 1 kg of acute hazardous waste) is out of the RCRA cradle-to-grave system, provided the waste is sent to a facility that is at least state-approved. Details of the major CESQG industries and waste types are listed in Table 12.4.

12.3.5 EPISODIC GENERATORS

Depending on the type of business and the amount of hazardous waste generated monthly, a facility might be regulated under different rules at different times. If, for example, a metal plating firm generates between 100 and 1000 kg (220 and 2200 lb) of hazardous waste during January, it would be considered a SQG for that month and its waste would be subject to the hazardous waste management requirements for SQGs. If, in June, it generates more than 1000 kg (2200 lb) of hazardous waste, its generator status would change, and it would be considered a LQG for June. Its waste for that month would then be subject to the management requirements for LQGs. For such generators it is to the firm's advantage to maintain all records, management protocols for storage, transportation, and so on, as a LQG.

12.4 REQUIREMENTS FOR LQGS AND SQGS

Once the waste is determined to be hazardous and is counted, the LQG must comply with the full spectrum of federal hazardous waste regulations under 40 CFR as well as 49 CFR (Department of Transportation; see Chapter 13). The SQG is subject to less stringent requirements. The LQG and SQG must notify the EPA and state regulatory agency of hazardous waste activity and obtain a U.S. EPA ID number.

12.4.1 THE EPA IDENTIFICATION NUMBER

Identification numbers are required for facilities that generate or manage hazardous waste, including LQGs and SQGs, transporters, and treatment, storage, and disposal facilities (TSDFs) (Chapter 14). Once the state regulatory authority is contacted, the generator will be sent EPA Form 8700-12, Notification of Regulated Waste Activity (Figure 12.1). An EPA identification number will be subsequently provided for each facility location.

OMB#: 2050-0175 Expires 12/31/2003

MAIL THE COMPLETED FORM TO: The Appropriate State or EPA Regional Office.	United States Environmental Protection Agency **RCRA SUBTITLE C SITE IDENTIFICATION FORM**	

1. Reason for Submittal (See instructions on page 23) MARK CORRECT BOX(ES)	**Reason for Submittal:** ☐ To provide Initial Notification of Regulated Waste Activity (to obtain an EPA ID Number for hazardous waste, universal waste, or used oil activities). ☐ To provide Subsequent Notification of Regulated Waste Activity (to update site identification information). ☐ As a component of a First RCRA Hazardous Waste Part A Permit Application. ☐ As a component of a Revised RCRA Hazardous Waste Part A Permit Application (Amendment #_____). ☐ As a component of the Hazardous Waste Report.
2. Site EPA ID Number (See instructions on page 24)	**EPA ID Number:** ⎵⎵⎵ ⎵⎵⎵⎵ ⎵⎵⎵ ⎵⎵⎵
3. Site Name (See instructions on page 24)	**Name:**

4. Site Location Information (See instructions on page 24)

Street Address:	
City, Town, or Village:	State:
County Name:	Zip Code:

5. Site Land Type (See instructions on page 24)	**Site Land Type:** ☐ Private ☐ County ☐ District ☐ Federal ☐ Indian ☐ Municipal ☐ State ☐ Other

6. North American Industry Classification System (NAICS) Code(s) for the Site (See instructions on page 24)

A.	B.
C.	D.

7. Site Mailing Address (See instructions on page 25)

Street or P. O. Box:	
City, Town, or Village:	
State:	
Country:	Zip Code:

8. Site Contact Person (See instructions on page 25)

First Name:	MI:	Last Name:
Phone Number:		Phone Number Extension:

9. Legal Owner and Operator of the Site (See instructions on pages 25 to 26)

A. Name of Site's Legal Owner:	Date Became Owner (mm/dd/yyyy):
Owner Type: ☐ Private ☐ County ☐ District ☐ Federal ☐ Indian ☐ Municipal ☐ State ☐ Other	
B. Name of Site's Operator:	Date Became Operator (mm/dd/yyyy):
Operator Type: ☐ Private ☐ County ☐ District ☐ Federal ☐ Indian ☐ Municipal ☐ State ☐ Other	

FIGURE 12.1 Notification of Regulated Waste Activity form.

12.4.2 MANAGING HAZARDOUS WASTE ON-SITE

LQGs are permitted to accumulate any quantity of waste in containers, tanks, and containment buildings for up to 90 days without a permit. Other forms of on-site storage (e.g., in a lined pond) usually require a permit. Generators must mark the date when the accumulation begins on each hazardous waste storage container so that it is visible for inspection. If the LQG facility accumulates wastes for more than 90 days it is considered a TSDF and must follow regulations described in 40 CFR Parts 264 and 270. Designation of a hazardous waste generator as a TSD is undesirable for a generator, resulting in a long list of new requirements and conditions along with increased costs for compliance.

In order to provide for more cost-effective shipments, SQGs can accumulate up to 6000 kg (13,228 lb) of hazardous waste on-site for up to 180 days without a permit. The wastes can be accumulated for up to 270 days if they must be transported more than 200 miles away for recovery, treatment, or disposal. There are limited circumstances in which the state administrator may grant extensions beyond 270 days. If the regulatory limits are exceeded, the generator is designated a TSDF and must obtain an appropriate operating permit. Special storage requirements apply to

OMB#: 2050-0175 Expires 12/31/2003

EPA ID No. [ ][ ][ ][ ][ ][ ][ ][ ][ ][ ][ ][ ]

10.   Type of Regulated Waste Activity (Mark the appropriate boxes for activities that apply to your site. See instructions on pages 26 to 30)

A. Hazardous Waste Activities

1. Generator of Hazardous Waste
(Choose only one of the following three categories.)

☐ a. LQG: Greater than 1,000 kg/mo (2,200 lbs./mo.) of non-acute hazardous waste; or

☐ b. SQG: 100 to 1,000 kg/mo (220 - 2,200 lbs./mo.) of non-acute hazardous waste; or

☐ c. CESQG: Less than 100 kg/mo (220 lbs./mo.) of non-acute hazardous waste

In addition, indicate other generator activities. (Mark all that apply)

☐ d. United States Importer of Hazardous Waste

☐ e. Mixed Waste (hazardous and radioactive) Generator

For Items 2 through 6, mark all that apply.

☐ 2. Transporter of Hazardous Waste

☐ 3. Treater, Storer, or Disposer of Hazardous Waste (at your site) Note: A hazardous waste permit is required for this activity.

☐ 4. Recycler of Hazardous Waste (at your site) Note: A hazardous waste permit may be required for this activity.

5. Exempt Boiler and/or Industrial Furnace

☐ a. Small Quantity On-site Burner Exemption

☐ b. Smelting, Melting, and Refining Furnace Exemption

☐ 6. Underground Injection Control

B. Universal Waste Activities

1. Large Quantity Handler of Universal Waste (accumulate 5,000 kg or more) [refer to your State regulations to determine what is regulated]. Indicate types of universal waste generated and/or accumulated at your site. (Mark all boxes that apply):

	Generate	Accumulate
a. Batteries	☐	☐
b. Pesticides	☐	☐
c. Thermostats	☐	☐
d. Lamps	☐	☐
e. Other (specify) _____	☐	☐
f. Other (specify) _____	☐	☐
g. Other (specify) _____	☐	☐

☐ 2. Destination Facility for Universal Waste
Note: A hazardous waste permit may be required for this activity.

C. Used Oil Activities (Mark all boxes that apply.)

1. Used Oil Transporter - Indicate Type(s) of Activity(ies)

☐ a. Transporter

☐ b. Transfer Facility

2. Used Oil Processor and/or Re-refiner - Indicate Type(s) of Activity(ies)

☐ a. Processor

☐ b. Re-refiner

☐ 3. Off-Specification Used Oil Burner

4. Used Oil Fuel Marketer - Indicate Type(s) of Activity(ies)

☐ a. Marketer Who Directs Shipment of Off-Specification Used Oil to Off-Specification Used Oil Burner

☐ b. Marketer Who First Claims the Used Oil Meets the Specifications

11.   Description of Hazardous Wastes (See Instructions on page 31)

A. Waste Codes for Federally Regulated Hazardous Wastes. Please list the waste codes of the Federal hazardous wastes handled at your site. List them in the order they are presented in the regulations (e.g., D001, D003, F007, U112). Use an additional page if more spaces are needed.

FIGURE 12.1 (Continued)

liquid hazardous wastes containing polychlorinated biphenyls (PCBs). The requirements for the management of PCBs appear in 40 CFR Part 761.

Both LQGs and SQGs must accumulate waste in tanks or containers. The U.S. EPA defines "container" as:

Any portable device in which a material is stored, transported, treated, disposed of, or otherwise handled.

The most common example of a container is a 55 gal drum. A "tank" is defined as:

A stationary device, designed to contain an accumulation of hazardous waste constructed primarily of nonearthen materials (e.g., wood, concrete, steel, and plastic), which provide structural support.

Storage tanks and containers must be managed according to EPA requirements summarized below. Each container holding hazardous waste must:

- Be labeled with the words "Hazardous Waste."
- Be marked with the date the waste was first generated.

OMB#: 2050-0175 Expires 12/31/2003

EPA ID No. [ ][ ][ ][ ][ ][ ][ ][ ][ ][ ][ ][ ]

B. Waste Codes for State-Regulated (i.e., non-Federal) Hazardous Wastes. Please list the waste codes of the State-regulated hazardous wastes handled at your site. List them in the order they are presented in the regulations. Use an additional page if more spaces are needed for waste codes.

12. Comments (See Instructions on page 31)

13. Certification. I certify under penalty of law that this document and all attachments were prepared under my direction or supervision in accordance with a system designed to assure that qualified personnel properly gather and evaluate the information submitted. Based on my inquiry of the person or persons who manage the system, or those persons directly responsible for gathering the information, the information submitted is, to the best of my knowledge and belief, true, accurate, and complete. I am aware that there are significant penalties for submitting false information, including the possibility of fine and imprisonment for knowing violations. **(See Instructions on page 31)**

Signature of owner, operator, or an authorized representative	Name and Official Title (type or print)	Date Signed (mm/dd/yyyy)

FIGURE 12.1 (Continued)

- Be constructed of, or lined with, a material that is compatible with the waste. This precaution will prevent the waste from reacting with the container, causing leakage, or creating a hazardous condition, such as the evolution of toxic or explosive vapors.
- Not store incompatible wastes together.
- Be kept closed during storage, except when adding or removing waste.
- Not be opened, handled, or stacked in a way that would cause containers to fail.
- Be located more than 50 ft from the facility property line if the waste is ignitable or reactive. This requirement does not apply to SQGs, whose ignitable or reactive wastes are to be located as far as practicable from the property line.

The generator, whether LQG or SQG, must:

- Inspect container storage areas at least weekly.
- Maintain the containers in good condition. If a container is found to be leaking, the waste must be transferred to another container immediately.

For tank systems, the generator must:

- Label each tank with the words "Hazardous Waste."
- Mark each with the beginning of the accumulation period.
- For those tanks equipped with an automatic waste feed, a feed cutoff or bypass system must be installed in the event of an overflow.
- Inspect monitoring equipment and the level of waste in uncovered tanks at least once per day. Inspect the tanks and surrounding areas for leaks and corrosion at least weekly.
- Use the National Fire Protection Association (NFPA) buffer zone requirements for covered tanks containing ignitable or reactive wastes.
- Not mix incompatible wastes.
- Provide at least 2 ft of freeboard (i.e., space at the top of each tank) in uncovered tanks, unless the tank is equipped with a containment structure.
- Report spills from a tank system to the state regulatory agency.

12.4.3 REQUIREMENTS FOR NEW TANK SYSTEMS

All new tank systems are required to be equipped with secondary containment with interstitial monitoring (Figure 12.2). This precaution should immediately alert the operators of a leak from the primary tank, thus allowing for rapid corrective action. An independent, qualified, and registered Professional Engineer must certify the design and installation of new tanks.

12.4.4 CONTAINMENT BUILDINGS

In limited circumstances, hazardous wastes may be stored in piles within containment buildings. Such storage is a permitted process that typically falls under the direction of the EPA as opposed to the state regulatory agency. Requirements for the proper operation of containment buildings are as follows. The building is to be:

- Certified by a registered Professional Engineer.
- Completely enclosed to prevent exposure to the elements.
- Of sufficient strength to support the waste and any personnel and heavy equipment that operate within the unit.
- Equipped with secondary containment and a collection system if liquid wastes are present in the building.
- Designed and operated to prevent fugitive dust emissions. Special precautions (e.g., negative air pressure to prevent releases to the outside) may be required.
- Routinely inspected.

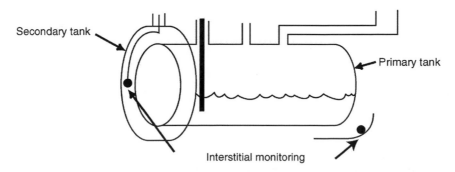

FIGURE 12.2 Schematic of a tank showing interstitial monitoring capability (U.S. GAO, 1992).

12.4.5 EMERGENCY PREPARATION

Waste-generating facilities must be operated to minimize the possibility of a fire, explosion, or any release of hazardous wastes to the environment.

All LQG facilities must usually be equipped with the following:

- An internal communications or alarm system
- A telephone or two-way radio for contacting local police and fire departments, or emergency response teams
- Fire extinguishers, fire control equipment (e.g., foam, inert gas, and dry chemicals), spill control equipment, and decontamination equipment
- A water supply for hoses, foam-producing equipment, or automatic sprinklers

Aisle space at LQG facilities must be maintained to allow the movement of personnel, fire protection equipment, spill control equipment, and decontamination equipment to any area within the facility in an emergency (Figure 12.3). The facility must familiarize police, fire departments, and other emergency response teams with the layout of the facility, the types of hazardous waste handled and associated hazards, places where personnel are working, entrance roads to the facility, and evacuation routes. It is the generator's responsibility to familiarize local hospitals with the properties of the hazardous wastes handled at the facility and the types of injuries that could result from fire, explosions, or releases. Some hospitals do not have the capability of treating persons exposed to certain chemical hazards; therefore, the correct hospital and emergency medical service team must be identified and documented.

12.4.6 EMERGENCY RESPONSE

LQGs are also responsible for preparing a thorough written contingency plan, along with training employees on hazardous waste management and emergency response. The LQG must have a written contingency plan for their facility in the event of an emergency. The plan must be designed to minimize hazards from fires, explosion, or any release of hazardous wastes.

The contingency plan must include:

- Actions in response to fire, explosion, or release of hazardous waste at the facility
- Arrangements with local police departments, hospitals, and emergency response teams that provide emergency services

FIGURE 12.3 Aisle space must be sufficient to allow the movement of persons and equipment (*Note*: this facility is *not* in compliance). Photo by Theresa M. Pichtel.

- Names, addresses, and phone numbers of all persons qualified to act as emergency coordinator
- All emergency equipment at the facility (e.g., fire extinguishing systems, spill control equipment, communication and alarm systems, and decontamination equipment)
- Evacuation plan, which must also include alternative evacuation routes

An up-to-date copy of the contingency plan must be maintained at the facility and be on file with the police and fire department, hospitals, and emergency response teams. The contingency plan must be reviewed and amended whenever the regulations are revised, the plan fails in an emergency, the facility changes its design and operation, or the facility changes the response necessary in an emergency.

At all times, the facility must designate an employee (either on facility premises or on call) with the responsibility for coordinating emergency response measures. This coordinator must be familiar with the facility contingency plan, all operations at the facility, the location and characteristics of waste regularly handled and the facility layout.

12.4.7 PERSONNEL TRAINING

As an additional safety precaution at LQGs, facility personnel must be trained in hazardous waste management protocols within 6 months of starting a new position and must be trained annually thereafter. The training program must ensure that personnel are able to respond to emergencies by becoming familiar with emergency procedures and equipment, including:

- Procedures for using, inspecting, repairing, and replacing emergency and monitoring equipment
- Automatic waste feed cutoff systems
- Communications or alarm systems
- Responses to groundwater contamination incidents
- Shutdown of operations

12.4.8 EMERGENCY RESPONSE REQUIREMENTS FOR SQGs

Emergency response requirements are less stringent for the SQG. As was the case for the LQG, there must be one employee either on the premises or on-call with the responsibility for coordinating emergency response measures. In contrast to LQGs, however, the SQG does not require a full contingency plan, but must post relevant emergency response information next to office telephones, including:

- Name and telephone number of the emergency coordinator
- Location of fire extinguishers, spill control material, and fire alarm
- Telephone number of the fire department

The emergency coordinator is responsible for responding to emergencies on-site. Certain specific responses are as follows:

- In the event of a fire, contact the fire department or attempt to extinguish it.
- In the event of a spill, contain the flow of hazardous waste, and clean up the waste and contaminated materials.
- In the event of a fire, explosion, or other release that could threaten human health outside the facility, immediately notify the National Response Center (1-800-424-8802).

The SQG must ensure that all employees are familiar with the proper waste handling and emergency procedures relevant to their responsibilities during facility operations and emergencies. No paper plan is required for the SQG (U.S. EPA, 2001b)

12.4.9 REPORTING

The LQG is responsible for submitting a biennial report to the state regulatory agency. Reports submitted for off-site shipping must include the facility's EPA identification number, information regarding the transporter and permitted TSDF, a description and quantity of the wastes generated, and actions taken to reduce the volume and toxicity of the waste. These reports can be used to encourage waste reduction (pollution prevention) within the facility. Some states require the LQG to report annually to the regulatory agency.

12.4.10 SHIPMENT OF WASTES OFF-SITE

The generator must package, label, and mark all waste containers and placard vehicles that carry the wastes according to Department of Transportation requirements (49 CFR Parts 172, 173, 178, and 179) (see Chapter 13). Additionally, a Uniform Hazardous Waste Manifest must accompany all hazardous waste shipped off-site.

12.4.11 THE UNIFORM HAZARDOUS WASTE MANIFEST

The Uniform Hazardous Waste Manifest is a paper form initiated by any generator that removes hazardous waste off-site for treatment, recycling, storage, or disposal. Both DOT (see Chapter 13) and EPA require the manifest. Each party that handles the waste, including the generator, all transporters, and TSDFs, signs the manifest and retains a copy. Once the waste reaches its final destination the TSDF returns a signed copy of the manifest to the generator. This cycle of signatures and paperwork confirms that the waste has been received and the loop is closed. Several states require the TSDF to forward a copy of the completed manifest to the state regulatory agency as well. When completed, the manifest contains information on the type and quantity of the waste that was transported, instructions for handling the waste, and signatures of all parties involved in the management process (Figure 12.4). CESQGs are not required to use a manifest when shipping wastes off-site. Detailed instructions for completing the manifest are given in Table 12.5.

If the LQG facility does not receive a signed manifest from the TSDF after 35 days (60 days for SQGs), it must attempt to locate the hazardous waste by contacting the TSDF. If there is no response after 45 days the LQG must submit an Exception Report to the state regulatory agency. The exception report (40 CFR Part 262.42) notifies the agency of a potential problem in the cradle-to-grave tracking process. The exception report contains a copy of the original manifest and a cover letter describing efforts made to locate the shipment. A flow chart for the manifest paper trail is shown in Figure 12.5. A completed albeit questionable uniform hazardous waste manifest is shown in Figure 12.6.

The manifest requirements outlined above apply only to domestic shipments of hazardous waste by road. Domestic shipments by rail or water are subject to other manifest requirements. Hazardous wastes exported from the United States are subject to additional regulatory requirements (40 CFR Part 261, 262).

12.4.12 RECORD KEEPING

The LQG must retain the following records at the facility for at least 3 years: signed manifests, biennial and exception reports, test results, and waste analyses. The 3-year-period is automatically extended in the event of an on-going enforcement action (40 CFR 262.40).

Please print or type   (Form designed for use on elite (12 - pitch) typewriter)   Form Approved OMB No. 2050 - 0039 Expires 9 - 30 - 91

UNIFORM HAZARDOUS WASTE MANIFEST	1 Generator's US EPA ID No.	Manifest Document No.	2. Page 1 of	Information in the shaded areas is not required by Federal law

3. Generator's Name and Mailing Address	A. State Manifest Document Number
	B. State Generator's ID
4. Generator's Phone ( )	
5. Transporter 1 Company Name 6. US EPA ID Number	C. State Transporter's ID
	D. Transporter's Phone
7. Transporter 2 Company Name 8. US EPA ID Number	E. State Transporter's ID
	F. Transporter's Phone
9. Designated Facility Name and Site Address 10. US EPA ID Number	G. State Facility's ID
	H. Facility's Phone

11. US DOT Description (Including Proper Shipping Name, Hazard Class, and ID Number)	12. Containers No. Type	13. Total Quantity	14. Unit Wt/Vol	I. Waste No.
G E N E R A T O R a.				
b.				
c.				
d.				

J. Additional Descriptions for Materials Listed Above	K. Handling Codes for Wastes Listed Above

15. Special Handling Instructions and Additional Information

16. **GENERATOR'S CERTIFICATION:** I hereby declare that the contents of this consignment are fully and accurately described above by proper shipping name and are classified, packed, marked, and labeled, and are in all respects in proper condition for transport by highway according to applicable international and national government regulations.

If I am a large quantity generator, I certify that I have a program in place to reduce the volume and toxicity of waste generated to the degree I have determined to be economically practicable and that I have selected the practicable method of treatment, storage, or disposal currently available to me which minimizes the present and future threat to human health and the environment; OR, if I am a small quantity generator, I have made a good faith effort to minimize my waste generation and select the best waste management method that is available to me and that I can afford.

Printed/Typed Name	Signature	Month Day Year

T R A N S P O R T E R

17. Transporter 1 Acknowledgement of Receipt of Materials		
Printed/Typed Name	Signature	Month Day Year
18. Transporter 2 Acknowledgement of Receipt of Materials		
Printed/Typed Name	Signature	Month Day Year

F A C I L I T Y

19. Discrepancy Indication Space

20. Facility Owner or Operator: Certification of receipt of hazardous materials covered by this manifest except as noted in item 19.		
Printed/Typed Name	Signature	Month Day Year

EPA Form 8700 - 22 (Rev. 9 - 88) Previous editions are obsolete.

FIGURE 12.4 The Uniform Hazardous Waste Manifest. Reproduced with kind permission of Environment, Health and Safety Online.

12.4.13 MANAGEMENT OF EMPTY CONTAINERS

Empty containers must be managed to comply with EPA regulations and to prevent contamination from residues left in the containers. Under EPA regulations and most state environmental rules, certain empty containers (and any residues within) are not subject to hazardous waste regulations.

TABLE 12.5
Details of Steps for Completion of the Uniform Hazardous Waste Manifest

Generators

Item 1. Generator's U.S. EPA ID Number — Manifest Document Number

Enter the generator's U.S. EPA 12-digit identification number and the unique 5-digit number assigned to this manifest (e.g., 00001) by the generator.

Item 2. Page 1 of —

Enter the total number of pages used to complete this manifest, i.e., the first page (EPA form 8700-22) plus the number of Continuation Sheets (EPA form 8700-22A), if any.

Item 3. Generator's Name and Mailing Address

Enter the name and mailing address of the generator. The address should be the location that will manage the returned manifest forms.

Item 4. Generator's Phone Number

Enter a telephone number where an authorized agent of the generator may be reached in the event of an emergency.

Item 5. Transporter 1 Company Name

Enter the company name of the first transporter who will transport the waste.

Item 6. U.S. EPA ID Number

Enter the U.S. EPA 12-digit identification number of the first transporter identified in item 5.

Item 7. Transporter 2 Company Name

If applicable, enter the company name of the second transporter who will transport the waste. If more than two transporters are used to transport the waste, use a Continuation Sheet(s) (EPA form 8700-22A) and list the transporters in the order they will be transporting the waste.

Item 8. U.S. EPA ID Number

If applicable, enter the U.S. EPA 12-digit identification number of the second transporter identified in item 7.

If more than two transporters are used, enter each additional transporter's company name and U.S. EPA 12-digit identification number in items 24 to 27 on the Continuation Sheet (EPA form 8700-22A). Each Continuation Sheet has space to record two additional transporters. Every transporter used between the generator and the designated facility must be listed.

Item 9. Designated Facility Name and Site Address

Enter the company name and site address of the facility designated to receive the waste listed on this manifest. The address must be the site address, which may differ from the company mailing address.

Item 10. U.S. EPA ID Number

Enter the U.S. EPA 12-digit identification number of the designated facility identified in item 9.

Item 11. U.S. DOT Description [Including Proper Shipping Name, Hazard Class, and ID Number (UN or NA)]

Enter the U.S. DOT proper shipping name, hazard class, and ID number (UN or NA) for each waste as identified in 49 CFR 171 through 177.

If additional space is needed for waste descriptions, enter these additional descriptions in item 28 on the Continuation Sheet (EPA form 8700-22A).

Item 12. Containers (No. and Type)

Enter the number of containers for each waste and the appropriate abbreviation from the table below for the type of container.

DM = metal drums, barrels, kegs
DW = wooden drums, barrels, kegs
DF = fiberboard or plastic drums, barrels, kegs
TP = tanks portable
TT = cargo tanks (tank trucks)
TC = tank cars
DT = dump truck
CY = cylinders
CM = metal boxes, cartons, cases (including roll-offs)
CW = wooden boxes, cartons, cases
CF = fiber or plastic boxes, cartons, cases
BA = burlap, cloth, paper or plastic bags

(continued)

TABLE 12.5 *(Continued)*

Item 13. Total Quantity
Enter the total quantity of waste described on each line.
Item 14. Unit (wt. or vol.)
Enter the appropriate abbreviation from the table below for the unit of measure.

G = gallons (liquids only)
P = pounds
T = tons (2000 lb)
Y = cubic yards
L = liters (liquids only)
K = kilograms
M = metric tons (1000 kg)
N = cubic meters

Item 15. Special Handling Instructions and Additional Information
Generators may use this space to indicate special transportation, treatment, storage, or disposal information or Bill of Lading information. States may not require additional, new, or different information in this space. For international shipments, generators must enter in this space the point of departure (City and State) for those shipments destined for treatment, storage, or disposal outside the jurisdiction of the United States.
Item 16. Generator's Certification
The generator must read, sign (by hand), and date the certification statement. If a mode other than highway is used, the word "highway" should be lined out and the appropriate mode (rail, water, or air) inserted in the space below. If another mode in addition to the highway mode is used, enter the appropriate additional mode (e.g., rail) in the space below.
In signing the waste minimization certification statement, those generators who have not been exempted by statute or regulation from the duty to make a waste minimization certification under section 3002(b) of RCRA are also certifying that they have complied with the waste minimization requirements.

Transporters
Item 17. Transporter 1 Acknowledgement of Receipt of Materials
Enter the name of the person accepting the waste on behalf of the first transporter. That person must acknowledge the acceptance of the waste described on the manifest by signing and entering the date of receipt.
Item 18. Transporter 2 Acknowledgement of Receipt of Materials
Enter, if applicable, the name of the person accepting the waste on behalf of the second transporter. That person must acknowledge acceptance of the waste described on the manifest by signing and entering the date of receipt.
Note: International Shipments — Transporter Responsibilities.
Exports — Transporters must sign and enter the date the waste left the United States in item 15 of form 8700-22.
Imports — Shipments of hazardous waste regulated by RCRA and transported into the United States from another country must upon entry be accompanied by the U.S. EPA Uniform Hazardous Waste Manifest. Transporters who transport hazardous waste into the United States from another country are responsible for completing the manifest (40 CFR 263.10(c)(1)).

Owners and Operators of Treatment, Storage, or Disposal Facilities
Item 19. Discrepancy Indication Space
The authorized representative of the designated facility's owner or operator must note in this space any significant discrepancy between the waste described on the Manifest and the waste actually received at the facility.
Owners and operators of facilities located in unauthorized States (i.e., the U.S. EPA administers the hazardous waste management program) who cannot resolve significant discrepancies within 15 days of receiving the waste must submit to their Regional Administrator (see list below) a letter with a copy of the manifest at issue describing the discrepancy and attempts to reconcile it (40 CFR 264.72 and 265.72).
Owners and operators of facilities located in authorized states should contact their state agency for information on State Discrepancy Report requirements. The U.S. EPA Regional Administrators are listed in Appendix A to 40 CFR Part 262.
Item 20. Facility Owner or Operator: Certification of Receipt of Hazardous Materials Covered by This Manifest Except as noted in Item 19
Print or type the name of the person accepting the waste on behalf of the owner or operator of the facility. That person must acknowledge acceptance of the waste described on the manifest by signing and entering the date of receipt.

(continued)

TABLE 12.5   (*Continued*)

The remainder of this form must be used as a continuation sheet to U.S. EPA form 8700-22 if:

- More than two transporters are to be used to transport the waste.
- More space is required for the U.S. DOT description and related information in Item 11 of U.S. EPA form 8700-22.

Generators of Hazardous Waste

Item 21. Generator's U.S. EPA ID Number — Manifest Document Number

Enter the generator's U.S. EPA 12-digit identification number and the unique 5-digit number assigned to this manifest (e.g., 00001) as it appears in item 1 on the first page of the manifest.

Item 22. Page —

Enter the page number of this Continuation Sheet.

Item 23. Generator's Name

Enter the generator's name as it appears in item 3 on the first page of the manifest.

Item 24. Transporter — Company Name

If additional transporters are used to transport the waste described on this manifest, enter the company name of each additional transporter in the order in which they will transport the waste.

Item 25. U.S. EPA ID Number

Enter the U.S. EPA 12-digit identification number of the transporter described in item 24.

Item 26. Transporter — Company Name

If additional transporters are used to transport the waste described on this manifest, enter the company name of each additional transporter in the order in which they will transport the waste.

Item 27. U.S. EPA ID Number

Enter the U.S. EPA 12-digit identification number of the transporter described in item 26.

Item 28. U.S. DOT Description Including Proper Shipping Name, Hazardous Class, and ID Number (UN/NA)

Refer to item 11.

Item 29. Containers (No. and Type)

Refer to item 12.

Item 30. Total Quantity

Refer to item 13.

Item 31. Unit (wt. or vol.)

Refer to item 14.

Item 32. Special Handling Instructions

Generators may use this space to indicate special transportation, treatment, storage, or disposal information or Bill of Lading information. States are not authorized to require additional, new, or different information in this space.

Transporters

Item 33. Transporter — Acknowledgement of Receipt of Materials

Enter the same number of the Transporter as identified in item 24. Enter also the name of the person accepting the waste on behalf of the Transporter (Company Name) identified in item 24. That person must acknowledge acceptance of the waste described on the manifest by signing and entering the date of receipt.

Item 34. Transporter — Acknowledgement of Receipt of Materials

Enter the same number as identified in item 26. Enter also the name of the person accepting the waste on behalf of the Transporter (Company Name) identified in item 26. That person must acknowledge acceptance of the waste described on the manifest by signing and entering the date of receipt.

Treatment, Storage, and Disposal Facilities

Item 35. Discrepancy Indication Space

Refer to item 19.

Source: 40 CFR Part 262 Appendix, with permission; IDEM, State of Indiana Hazardous Waste manifest Guidance Manual, Indianapolis, IN, 1994. With permission.

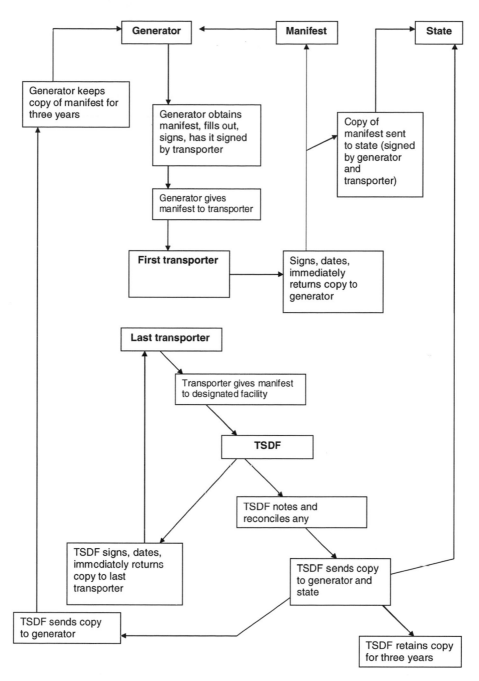

FIGURE 12.5 Flow chart for the manifest paper trail (EHSO, 2003). Reproduced with kind permission of EHSO, http://www.ehso.com.

A container or an inner liner that held hazardous waste is considered "empty" and is not regulated as a hazardous waste if (40 CFR Section 261.7):

- All waste has been removed that can be feasibly removed
- No more than 1 in. of residue remains on the bottom of the container
- No more than 3% (by wt) of the total capacity of the container remains if the container measures less than 110 gal, or no more than 0.3% remains if the container is larger than 110 gal

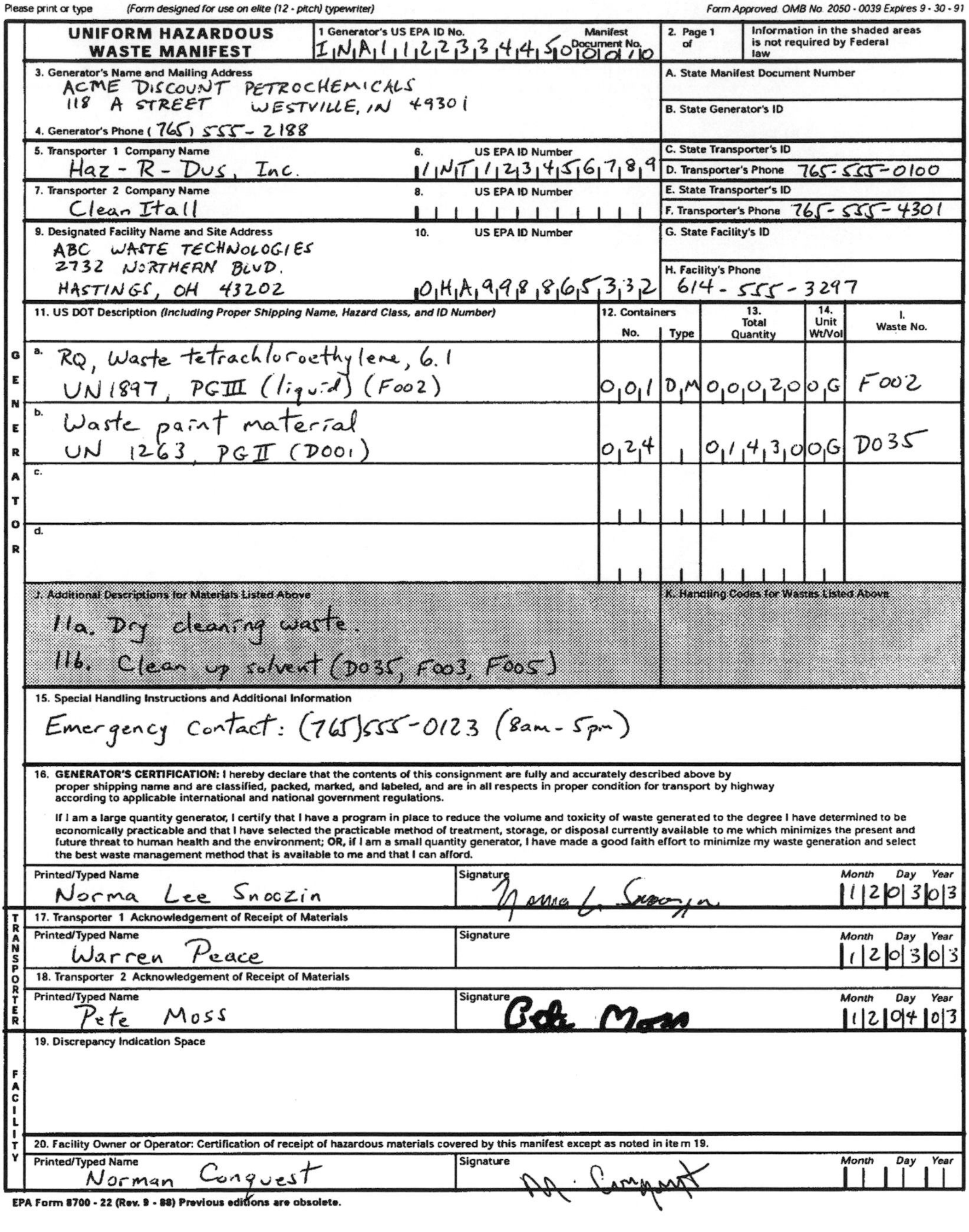

FIGURE 12.6 Questionable Uniform Hazardous Waste Manifest.

If the container held an acute hazardous waste, one of the following conditions must be met for it to be considered empty:

- The container or inner liner must be triple-rinsed with a solvent capable of removing the waste.
- The inner liner must be removed from the container.

There are several management options for used drums. They may be reconditioned and reused as shipping containers, processed for steel scrap recycling, crushed and buried at a permitted TSDF, or sent to a solid waste disposal facility. Reconditioning or recycling is preferred for handling empty containers because of the cost involved in the disposal of empty drums at a TSDF.

12.4.14 HAZARDOUS WASTE AT SATELLITE ACCUMULATION POINTS

A satellite accumulation point is a location within the generator's facility in which wastes are generated; however, it is not the primary accumulation point. Accumulation is permitted at satellite points in order to increase the efficiency of waste collection and reduce overall disposal costs.

Wastes may be collected at the satellite point indefinitely until 55 gal of hazardous waste or 1 quart of acutely hazardous waste is accumulated. There is no need to ship partially full drums of waste off-site because of accumulation time restrictions. All waste at satellite accumulation points is under the control of the operator of the process generating the waste. The U.S. EPA has established the following management standards for wastes collected at satellite accumulation points:

- Containers must be marked with the words "Hazardous Waste" or with other words that identify their contents.
- Containers must be maintained in good condition.
- Wastes must be compatible with the container.
- The container must always be kept closed during accumulation except when adding or removing waste.

12.5 REQUIREMENTS OF CESQGs

The CESQG is only required to determine which wastes are hazardous and to treat or dispose of wastes on-site or deliver wastes to an off-site TSDF that is permitted under 40 CFR 270, and is authorized to manage hazardous waste by an authorized state. Such a disposal site is not necessarily a hazardous waste facility; it might be a materials recovery facility or a solid waste management facility. The state, however, may require that CESQGs dispose their hazardous wastes at a hazardous waste facility.

REFERENCES

Code of Federal Regulations, Vol. 40, Part 261, Identification and Listing of Hazardous Waste, U.S. Government Printing Office, Washington, DC, 2004.

Code of Federal Regulations, Vol. 40, Part 262, Standards Applicable to Generators of Hazardous Waste. Appendix to Part 262 – Uniform Hazardous Waste Manifest and Instructions (EPA forms 8700-22 and 8700-22 and Their Instructions), U.S. Government Printing Office, Washington, DC, 2004.

Code of Federal Regulations, Vol. 40, Part 761, Polychlorinated biphenyls (PCB) manufacturing, processing, distribution in commerce, and use prohibitions, U.S. Government Printing Office, Washington, DC, 2004.

Environment, Health and Safety Online, Diagram of the Current Hazardous Waste Manifest System, 2003. See: http://www.ehso.com/Hazwaste/hazwaste_Manifest_process.htm

Indiana Department of Environmental Management, State of Indiana Hazardous Waste Manifest Guidance Manual, Indianapolis, IN, 1994.

U.S. Environmental Protection Agency, Generation and Management of CESQG Waste, Office of Solid Waste, Municipal and Industrial Solid Waste Division, Washington, DC, 1994.

U.S. Environmental Protection Agency, Hazardous Waste Requirements for Large Quantity Generators, EPA530-F-96-032, Solid Waste and Emergency Response (5305W), Washington, DC, 1996.

U.S. Environmental Protection Agency, The National Biennial RCRA Hazardous Waste Report (Based on 1999 Data), EPA530-S-01-001, Solid Waste and Emergency Response PB2001-106318 (5305W), June 2001a.

U.S. Environmental Protection Agency, Managing your hazardous wastes. A Guide for Small Businesses, EPA-530-K-01-005, Solid Waste and Emergency Response (5305W), Washington, DC, 2001b.

U.S. Environmental Protection Agency, Land Disposal Restrictions: Summary of Requirements, EPA-530-R-01-007, Solid Waste and Emergency Response and Enforcement and Compliance Assurance, Washington, DC, 2001c.

U.S. Environmental Protection Agency, Notification of Regulated Waste Activity. Instructions and Forms, EPA form 8700-12 (Revised 5/2002), Office of Solid Waste, Washington, DC, 2002.

Woodside, G., *Hazardous Materials and Hazardous Waste Management: A Practical Guide,* John Wiley and Sons, New York, NY, 1993.

SUGGESTED READINGS AND WEB SITES

Business and Legal Reports, Federal Hazardous Waste Generators, 2003. See: http://enviro2.blr.com/topic.cfm/topic/179/state/155

California Environmental Protection Agency, Department of Toxic Substances Control, Hazardous Waste Generator Requirements, 2002. See: http://www.coastal.ca.gov/ccbn/Apndx11.pdf

Massachusetts Department of Environmental Protection, A Summary of Requirements for Small Quantity Waste Generators of Hazardous Waste, Bureau of Waste Prevention, Boston, MA, 2000. See: www.state.ma.us/dep.

U.S. Environmental Protection Agency, Generate and Transport, 2003. See: http://www.epa.gov/epaoswer/osw/generate.htm

U.S. Environmental Protection Agency, RCRAINFO File Specification Guide, 2001 Hazardous Waste Report Submissions, 2001. See: http://www.epa.gov/epaoswer/hazwaste/data/brs01/8-01spec.pdf

U.S. Environmental Protection Agency, RCRA Environmental Indicators Progress Report: 1995 Update, Office of Solid Waste, Washington, DC, June 1996. See: http://www.epa.gov/epaoswer/hazwaste/data/ei/env-ind.pdf

Washington State Department of Ecology, No date, Hazardous Waste Generator Checklist.

Washington's Hazardous Waste and Toxics Reduction Program, See: http://www.ecy.wa.gov/pubs/9112b.pdf

QUESTIONS

1. Define "cradle-to-grave responsibility" for a hazardous waste generator. Discuss the regulatory requirements for packaging, storage, transportation, and manifesting. How long is the generator responsible for this waste?
2. Why is it advisable to use a separate container for each type of hazardous waste being accumulated? Provide at least two reasons.
3. After a satellite container becomes full, within what time period must it be removed and transferred to the central accumulation area or shipped to an off-site facility?
4. List the criteria required for a satisfactory secondary containment unit or device.
5. A "small quantity generator," as defined under RCRA, generates at least __ pounds of hazardous waste in a calendar month, whereas a large quantity generator generates at least __ pounds of hazardous waste in a month.
6. The hazardous waste accumulation period normally begins at what point?
   (a) when wastes are first added to the container or tank; (b) when the container is full; (c) when the hazardous label is affixed to the drum; (d) when the waste is determined to be listed hazardous waste.
7. How often is a facility hazardous waste report to be submitted to the U.S. EPA?
8. List instances when containers of hazardous waste can be opened at the generator's facility.
9. Under RCRA, up to what height should drums containing ignitable wastes be stacked (i.e., 2 drums, 3 drums, etc.)?
10. An LQG may ship its hazardous wastes to a landfill that is approved by the state to handle municipal or industrial waste (true or false).

11. Large quantity generators must have double liners on all new containers for hazardous wastes stored on-site (true or false).
12. Which of the following are acceptable management practices for a satellite accumulation area? (a) stores a maximum of 55 gal of hazardous waste; (b) stores a maximum of 1 quart of acute hazardous waste; (c) it is under the continuous control of the personnel who generate the wastes
13. What are the requirements for an acceptable container being used in a satellite accumulation area?
14. As a result of a new manufacturing process, the Hi-Jinx Chemical Company is generating a new waste stream. The waste has the following properties: pH = 10; flashpoint = 75°F; high concentrations of Cd and Cr (exceeds TCLP limits); approximately 2500 lb/month. Based on the *waste alone*, is the generator a LQG, SQG, or CESQG? Must the generator comply with DOT regulations for packaging the wastes and placarding the vehicles? For how long can the generator stockpile waste on-site without a permit? Is there a requirement for the submission of a written emergency response plan to the state environmental agency?
15. Based on question 14, is the waste to be classified as corrosive, ignitable, or toxic?
16. Based on the waste properties given in question 14, can the waste be stored in a single-lined tank on the premises? How far must it be stored from the property line? To what type of facility (e.g., state-approved, EPA-approved) must the waste be shipped for final disposal?
17. If the waste in question 14 could be recycled into another process at another facility, must it be considered a RCRA "hazardous waste"?
18. The waste in question 14 may possibly serve as a boiler fuel and therefore need not be managed as a waste at all (true or false).
19. At Bogus Metalworks, hazardous liquid wastes were recycled in a second process. The 2500 lb. of waste liquid produced were sent to a machine degreasing booth and reused a total of four times in September. How much waste must be counted at the end of the month?
20. At the Hi-Jinx facility, pentachlorophenol, a listed hazardous waste, was mixed with clean soil during a small spill. Is the new mixture still a hazardous waste? Justify your answer.
21. Also at the Hi-Jinx plant, the tops of several drums of the pentachlorophenol are damaged and severely rusting. Since wastes are not leaking, can they remain in these containers until they are to be transported?
22. In what specific ways are individual households subject to the requirements of RCRA?
23. According to U.S. EPA regulations, when must a hazardous waste manifest copy with an original signature of the TSD facility representative be received by the generator?
24. For how long do RCRA regulations require that manifests be retained?
25. List and discuss what is incorrect about the Uniform Hazardous Waste Manifest shown in Figure 12.6.
26. If the signed return copy (closed copy) of a manifest is not received by a large quantity generator within 35 days, an investigation must be conducted to locate the return copy. When must exception reports be submitted to the state?
27. If a small quantity generator does not receive a return copy of the manifest within 60 days, what action is required?
    For some of the following questions, it may be useful to refer to 40 CFR 262, Generator Requirements.
28. A generator of hazardous waste determines that more than 100 kg of waste will be generated in a calendar month. What documentation/identification, etc., must they obtain?
29. What is the maximum quantity of hazardous waste that can be accumulated in a satellite accumulation point?
30. Accumulation points for containers must be inspected weekly. What should be inspected?
31. How frequently should hazardous waste storage tanks be inspected?

32. While in satellite accumulation points, what must all hazardous waste containers be marked with?
33. While in primary accumulation points, what must all hazardous waste containers be marked with?
34. Tanks used for the accumulation of hazardous waste must be equipped with secondary containment, leak detection, and overflow protection. How often should this equipment be inspected?

A.12.1 APPENDIX: HAZARDOUS WASTE MANAGEMENT SCENARIOS

A.12.1.1 INTRODUCTION

This section includes several situations that are intended to apply the regulatory foundation provided in this chapter. All situations are based on actual events and inspections experienced by hazardous waste regulatory personnel. Names of companies and individuals have, of course, been changed.

After reading each scenario, discuss what, if any, violations may have occurred. How would these violations be best addressed (i.e., via changes in engineering design, a modified storage or disposal program, some use of common sense, etc.)? These are open-ended situations. Please note that there is often not one "right" solution to these situations; it must be emphasized that the regulations are often open to interpretation (just ask any state regulatory inspector or company attorney).

A.12.1.2 THE SCENARIOS

1. The Hi-Jinx Stripping Company conducts a number of commercial activities, one of its most profitable being the stripping of paint from metal parts.

The working piece is immersed in a 500 gal vat filled with methylene chloride, where most of the paint is removed. After a designated time the piece is removed and allowed to air-dry. The workpiece is then blasted with crushed walnut shells at high pressure to remove any residual paint.

After the blasting the walnut shells are impregnated with paint chips and residual methylene chloride. The contaminated shells are transported directly from the air handling system (a dust collector) to a dumpster, following which they are shipped to a local Subtitle-D landfill for disposal.

The paint chips are collected from the bottom of the vat, washed with water, and air-dried. The paint chips are stored in a trailer to allow any remaining solvent to volatilize, after which they are shipped to the local Subtitle D landfill for disposal. The inspector, Ms. Rhoda Dendron, notes that the paint chips in the trailer are releasing a strong aromatic smell. Her photoionization detector (PID) reads several hundred ppm of hydrocarbon vapors.

The wash water is used to cover the open vats of the heavier methylene chloride. Mr. Roosterson, the company's president, informs the inspector that the water layer prevents volatilization losses of the methylene chloride solvent. Any leftover washwater or methylene chloride mix is stored in 55 gal drums. The heavier methylene chloride is pumped out to be returned to the process vats. The water is reused again as washwater. The methylene chloride wastes have been stored for 4 to 5 months on-site, or longer.

Relevant portions of a MSDS for methylene chloride are shown in Figure A.12.1.

2. The Hi-Jinx Company has had other run-ins with the state. A neighbor contacts the state regulatory agency to announce that Hi-Jinx has been deliberately dumping methylene chloride directly onto the soil (this has been witnessed on several occasions). Neighbors rely on well water in their homes and businesses, and an unconfined aquifer occurs at approximately 20 ft bgs (below ground surface), and a second, confined aquifer at approximately 75 ft bgs. Local soils are predominantly silty clay and other, similarly dense materials. Mr. Roosterson is certain that, if anyone had "inadvertently"

METHYLENE CHLORIDE

Section 2 - Composition/Information on Ingredients

Ingredient Name: METHYLENE CHLORIDE (SARA III)
Ingredient CAS Number: 75-09-2 Ingredient CAS Code: M
RTECS Number: PA8050000 RTECS Code: M
OSHA PEL: 500 PPM/C,1000; Z2 OSHA PEL Code: M
OSHA STEL: OSHA STEL Code:
ACGIH TLV: 50 PPM, A2; 9192 ACGIH TLV Code: M
ACGIH STEL: N/P ACGIH STEL Code:
EPA Reporting Quantity: 1000 LBS
DOT Reporting Quantity: 1000 LBS
Ozone Depleting Chemical: N

Section 3 - Hazards Identification, Including Emergency Overview

Health Hazards Acute & Chronic: N/P

Signs & Symptoms of Overexposure:
LIGHT HEADEDNESS,MENTAL CONFUSTION,NAUSEA,HEADACHE.TINGLING OR CREEPING
SKIN FEELING ON INHALATION.

Medical Conditions Aggravated by Exposure: N/P

LD50 LC50 Mixture: N/P

Route of Entry Indicators:
    Inhalation: N/P
    Skin: N/P
    Ingestion: N/P

Carcenogenicity Indicators
    NTP: N/P
    IARC: N/P
    OSHA: N/P

Carcinogenicity Explanation: N/P

Section 4 - First Aid Measures

First Aid:
REMOVE TO FRESH AIR.IF UNCONSCIOUS,USE ARTIFICIAL RESPIRATION & CALL A
PHYSICIAN,KEEP WARM & COMFORTABLE. FOR EYE CONTACT.WASH WITH WATER FOR 15
MIN. IN CASE OF SKIN CONTACT, WASH WITH SOAP & WATER .

Section 5 - Fire Fighting Measures

Fire Fighting Procedures:
WEAR SELF-CNTD BRTHG,APP H_2O SPRAY TO COOL CONTR.
Unusual Fire or Explosion Hazard:
MAY DECOMPOSE & GIVE PHOSGENE GAS.
Extinguishing Media:
CO_2, FOAM, DRY CHEM & H_2O FOG.
Flash Point: Flash Point Text: NON-FLAMMABLE

FIGURE A. 12.1 MSDS for methylene chloride.

spilled a "small quantity" of methylene chloride, this would evaporate quickly and could not possibly migrate into the soil due to its dense structure.

Soil sampling and analysis results indicate the presence of methylene chloride, acetone, tetrachloroethane and toluene.

3. A LQG of hazardous waste uses toluene as a solvent in its manufacturing operations. An inspector walks the perimeter of company property and eventually discovers several open and rusting

drums of an aromatic-smelling substance behind an old wooden shed. Weeds grow tall around the drums, which incidentally have no labels. The inspector states that this situation is a violation of waste storage regulations. Pop, the company's health and safety person, however, claims that this is not waste but is actually *product*, on hand to be used in future operations.

4. At a degreasing operation, TCE and perchloroethylene (PCE) are routinely used as solvents. Workers have been encouraged by their supervisors to "extend the lifetime" of company clothing. At lunch breaks and at the end of the work day, employees place solvent-soaked gloves and lab coats over a sink in the locker room in order to dry. Once dried, they are reused.

5. An auto manufacturing facility uses spray booths to paint individual auto parts. Solvents, characteristic for ignitability and toxicity (due to the presence of methyl ethyl ketone), are forced under pressure into the guns periodically to clean out accumulated paint. The paint–solvent mixtures are collected in a 35 gal "purge pot" situated under the floor. These purge pots are emptied by way of plumbing, which directs the liquids to a large tank outside the building (Figure A.12.2). The company claims that the purge pots should not be regulated as hazardous waste tanks; rather, the solvents and purge pots are part of the cleaning process: without removing the solvent–paint mixture daily, the painting process would cease. Incidentally, the company claims that the plumbing should not be regulated either.

6. What is wrong with this picture (Figure A.12.3)?

7. Shy Knee Automotive Coatings, Inc., is a LQG of F019 sludge (wastewater treatment sludge from the chemical conversion coating of aluminum), and F001 and F002 waste solvents and still bottoms, spent filters, and contaminated rags.

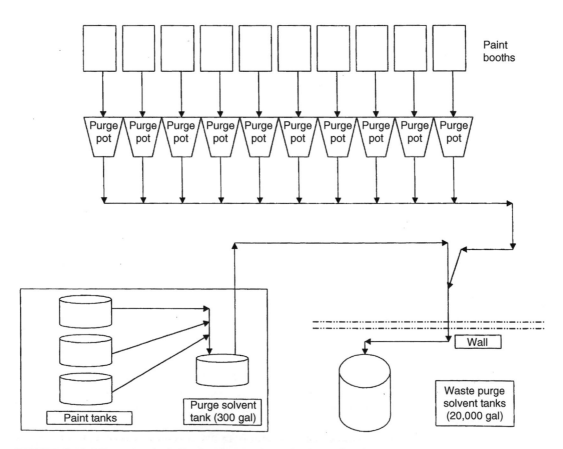

FIGURE A.12.2 Spray booths indicating waste paint and solvent storage.

FIGURE A.12.3 Drum with questionable dates on label.

FIGURE A.12.4 Roll-off container storing wastewater treatment sludge. Photo by Theresa M. Pichtel.

F019 sludge from the facility's wastewater treatment system is fed by gravity into a large (900 m³) roll-off container that is housed in a 5 × 5 m (15 × 15 ft) shed. The roll-off can be removed through a garage-style door (Figure A.12.4). An inspector notes that sludge has leaked from the roll-off on to the ground, thus constituting a violation. The plant manager, however, refutes her statement, claiming that "the building acts as secondary containment."

The inspector states that sludges are migrating outside the shed, creating a dispersal hazard (Figure A.12.5). The plant manager replies that there are drains throughout the facility, including outside the buildings, all of which empty to a wastewater treatment plant. Therefore, there is no hazard of a release to the surrounding environment.

8. The I.O. Silver Company manufactures and processes numerous metallic durable products. The company is a LQG of metallic sludges, chlorinated solvents, and other organic wastes.

There are two 90-day hazardous waste accumulation areas on the facility property. One stores F006 sludges (from electroplating operations) and the second stores assorted chlorinated solvents (F002), spent filters, lab gloves, etc. The inspector discovers that neither area has its own spill prevention equipment; rather, the equipment is kept on a large mobile cart near one processing area.

FIGURE A.12.5 Sludge migration. Photo by Theresa M. Pichtel.

The distance from the process area to the F006 sludge accumulation area is approximately 75 m, and the distance to the solvent accumulation area is almost 100 m. The inspector considers that this may be a violation, but the company environmental official claims that they can move the cart of spill equipment "quickly enough to contain any size spill."

9. At another facility, baghouse dust is generated as a result of steel production. The dust is a K061 listed waste (emission control dust from steel manufacture in electric arc furnaces) and is stored in a steel silo (tank). The inspector informs the plant manager that an integrity assessment of the silo is needed. An integrity assessment is defined and regulated as follows (40 CFR Part 264.191(a)):

> For each tank system that does not have secondary containment, the owner or operator must determine that the tank system is not leaking or is unfit for use. The owner or operator must obtain and keep on file at the facility a written assessment reviewed and certified by an independent, qualified registered professional engineer, in accordance with Section 270.11(d), that attests to the tank system's integrity. This assessment must determine that the tank system is adequately designed and has sufficient structural strength and compatibility with the wastes to be stored or treated, to ensure that it will not collapse, rupture, or fail. At a minimum, this assessment must consider the following:
>
> - Design standards according to which the tank and ancillary equipment were constructed
> - Hazardous characteristics of the wastes that will be handled
> - Existing corrosion protection measures
> - Documented age of the tank system, if available (otherwise, an estimate of the age)
> - Results of a leak test, internal inspection, or other tank integrity examination

The manager complains that since the baghouse dust is a *solid* material, an integrity assessment is not warranted; only liquid waste tanks should undergo such assessments.

10. What is wrong with this picture (Figure A.12.6.)?

11. Figure A.12.7 depicts a satellite accumulation area for a facility that manufactures truck caps. What is wrong with this picture?

12. A utility power plant combusts coal for electricity production. The facility generates large quantities of fly ash. The TCLP test fails the characteristic for arsenic. The question arises as to how to dispose appropriately this waste; if such large quantities were to be disposed in an EPA-approved secure landfill, the costs of waste disposal would be prohibitive.

13. A Midwest metal plating facility produces large quantities of strong acids (used for metal cleaning prior to plating). The spent acid waste is mixed with sodium hydroxide in a large (20,000 L) tank

(a)

(b)

FIGURE A.12.6 Hazardous waste storage area. Photo by Theresa M. Pichtel.

in order to neutralize it. This neutralized mixture of approximately pH 6.5 has been disposed on the company's property using a permitted deep-well injection system.

For financial reasons, the facility owner abandoned the company premises. State inspectors quickly detected the presence of liquids in most of the facility tanks. The owner was required to return and conduct a waste determination for all tanks. In the neutralization tank, liquids failed the TCLP test for chromium. Since the contents of these tanks have consistently been deep-well-injected, there was concern that high levels of chromium were also being injected

During an interview with an inspector, the owner stated that "We don't know how that chromium got there."

FIGURE A.12.7 Satellite accumulation area. Photo by Theresa M. Pichtel.

14. A technician at a university laboratory has discovered an aging bottle of picric acid. The bottle has never been opened, and the technician is hoping to find some inexpensive, simple means of disposal.

15. A metalworking operation uses trichloroethylene (TCE) for degreasing parts. The facility has a vapor degreaser that it uses to distill the TCE on site. The recovered TCE is used several times for degreasing. Although TCE can have an F001 or F002 listing, the facility does not consider this particular TCE to be a waste at all, as it is recycled and reused immediately; therefore, they are not a hazardous waste generator.

16. The roll-offs (containers) shown in Figure A.12.8 are storing a K-listed waste. What is the problem regarding their management?

A.12.1.3 RESPONSES TO QUESTIONS ABOUT SCENARIOS

1. (a) Although allowed to air-dry, the paint chips are by no means free of this heavy and hazardous solvent (as indicated by the aromatic smell and PID readings). The facility should have conducted a waste determination on both the paint chips and the walnut shells, i.e., determine whether the solid waste they produce is a hazardous waste.
   The paint chips are both a listed (F002) and a characteristic (toxic) hazardous waste; therefore, disposal in an ordinary MSWLF is not permitted. Similarly, the walnut shells are a listed waste according to the Mixture rule.
   (b) Given that hazardous waste is indeed generated on the premises, the generator must prepare a Uniform Hazardous Waste Manifest according to 40 CFR 262.20. Additionally, the generator must send a one-time written notification to each land disposal facility receiving the waste.
   (c) A generator may accumulate hazardous waste on-site for 90 days or less without a permit. The Hi-Jinx facility has been storing the methylene chloride for 4 to 5 months at least.
   (d) The facility was conducting "treatment" of a hazardous waste by drying the methylene chloride-contaminated paint chips in the trailer. They do not have a permit to treat hazardous wastes.
   (e) A Title V air permit is needed, as a hazardous solvent is being allowed to volatilize. The affected state has stipulated, in its own regulations, that:

   > A person may not discharge, emit, cause … or allow any contaminant or waste, including any noxious odor … into the environment.

FIGURE A.12.8 Roll-off containers. Photo by Theresa M. Pichtel.

2. (a) Regardless of soil type, groundwater contamination is possible from direct disposal onto the soil. According to the MSDS, the specific gravity for methylene chloride is 1.33; it will therefore sink once it contacts the groundwater, making any recovery and remediation operations slow and expensive.

   (b) The deliberate disposal of methylene chloride, an F002 listed RCRA hazardous waste, to the soil constitutes illegal disposal. The company owner should contact his attorney. Many states encourage voluntary notification and clean-up of spill and disposal sites rather than waiting for whistle blowers to report the problem or inspectors to discover them.

3. (a) According to 40CFR 261.2, a solid waste is any discarded material. "Discarded" means abandoned, recycled, or inherently waste-like. Based on the fact that the drums are rusted and overgrown with weeds, it is concluded that this material is abandoned and is therefore a solid waste. Based on company use of toluene, it is further concluded that these mystery liquids are probably also toluene.

   (b) Since the drums were obviously on-site for a long time, it is concluded that the facility has engaged in >90 day storage without a permit.

   (c) It is possible that wastes have been released to the surrounding soil and air. The company is responsible for assessing the degree of soil contamination; they should arrange with an outside firm to collect soil samples and analyze them for possible hydrocarbon contamination.

   (d) A waste determination should be carried out to determine the specific composition of the waste drums.

4. (a) The lab gloves and lab coats are technically hazardous waste according to The Mixture Rule.

   (b) By deliberately allowing the solvent vapors to volatilize, this facility is treating a hazardous waste. A TSDF permit is required for such an activity.

   (c) Workers are potentially being exposed to high levels of hazardous vapors, so this activity is probably a violation of OSHA regulations.

5. True, the process would likely halt if the solvent-paint wastes accumulated and overflowed into the paint booth. However, the mixture is a hazardous waste, and the purge pots and the ancillary equipment (i.e., all associated plumbing) are indeed considered a tank system. Tank systems are defined in 40 CFR 260.10 as:

   > a hazardous waste storage or treatment tank and its associated ancillary equipment and containment system.

   Regulations as to management of tank systems appear in Subpart J, Tank Systems (40 CFR 264.190-264.200).

6. For a 55 gal drum, a date is to be noted on the hazardous waste label once it has been filled; counting the days (90 for a LQG, 180 for a SQG) begins immediately thereafter. One cannot change the start date of accumulation. In the current situation (a LQG) it is probable that the drum was on-site for over the 90-day storage period. According to 40 CFR 262.34(a)(2):

> A generator may accumulate hazardous waste on-site for 90 days or less without a permit provided that:
>
> > The date upon which each period of accumulation begins is clearly marked and visible for inspection on each container;

As a side note, for a container greater than 55 gal, the date is marked on the Hazardous Waste label *at the first addition of waste*, not when the container is full.

7. (a) According to 40 CFR 265.31:

> Facilities must be maintained and operated to minimize the possibility of a fire, explosion, or any unplanned sudden or non-sudden release of hazardous waste or hazardous waste constituents to air, soil, or surface water which could threaten human health or the environment.

The building was open to the outside by way of the garage door and spaces in the walls, so there is no complete secondary containment.
(b) The inspector encouraged simple improvements in housekeeping, i.e., if the shed floor is kept free of sludge, dispersal is no longer an issue.
(c) Wastes that are tracked outside could easily be dispersed off the plant site by wind. The argument of a waste drain is therefore inadequate.

8. There is no requirement that spill prevention equipment be situated directly within the 90-day accumulation areas.

9. Although the waste is a solid hazardous waste, an integrity assessment of the tank is still required. However, the assessment does not have to be as involved as for a tank storing liquid hazardous wastes.
Incidentally, the issue over secondary containment came up between inspector and plant manager. When brought to the U.S. EPA, it was decided that secondary containment was not required.

10. This hazardous waste storage area has its share of problems. Most wastes are unidentified and unlabeled; a waste characterization is required. Incompatible wastes are stored side-by-side. Many containers are open. Containers must remain closed with few exceptions, for example (40 CFR Sec. 264.173):

> (a) A container holding hazardous waste must always be closed during storage, except when it is necessary to add or remove waste.
> (b) A container holding hazardous waste must not be opened, handled, or stored in a manner which may rupture the container or cause it to leak.

11. Although not the highest quality printing, the labeling of this drum is actually acceptable. Federal regulations (40 CFR 262.34) state that, in a satellite accumulation area, a waste drum must simply be labeled, describing the contents; there is no requirement for a specific label in a satellite area. If, however, the drum were in a 90-day accumulation area, this "label" is unacceptable. A standard yellow "Hazardous Waste" label must be affixed to the drum and dated.
This drum is situated in the middle of the aisle and must be moved.

12. The fly ash is indeed a solid waste, as it is "discarded" material. However, fly ash is exempted from designation as a hazardous waste. According to 40 CFR 261.4(b)(4), fly

ash generated from the combustion of coal or other fossil fuels is excluded from being defined as a hazardous waste.

The following solid wastes are not hazardous wastes:

> Fly ash waste, bottom ash waste, slag waste, and flue gas emission control waste, generated primarily from the combustion of coal or other fossil fuels, except as provided by Section 266.112 of this chapter for facilities that burn or process hazardous waste.

13. Unfortunately, there is no simple method to determine if Cr wastes were deep-well injected; the claim by the inspector would be difficult to prove. Corroboration by a plant employee or witnesses would have been extremely useful. This case is still under investigation.

14. Since the container has never been opened, it is considered an unused commercial chemical product. Picric acid is not, however, a U- or P-listed hazardous waste. During prolonged periods of storage, picric acid can form picrate salts, which are highly sensitive to shock and are explosive. As a result, such a waste is considered a characteristic reactive waste and must be handled with utmost care. It would be wise to contact the local emergency response team for support in handling and disposing of this container.

15. If the TCE solvent was never removed from the system before it was reused, it is not a "waste." In the current situation, however, the solvent is reclaimed (treated) before reuse and is therefore considered a solid waste. Based on the listing procedure, this solid waste is an F001 waste. The facility must manage their waste stream, including requirements for generators (counting, manifesting, and so on), according to the requirements of 40 CFR 262.

16. (a) Containers must be covered at all times. Openings in the covers allowed for rainfall to enter. The covers were not pulled tight; as a result, rain accumulated on the surface, causing the covers to collapse, thus letting in more water.

(b) The puddles under the roll-off are stained and may contain some of the released wastes. If the inspector is suspicious of a leak, he should require this facility to conduct a waste determination.

A facility must be managed to minimize the possibility of a release to the environment.

13 Hazardous Waste Transportation

From here to nowhere. And back again.
Blow foul, blow fair
All come to anchor finally in the tomb.
Passengers armed, we travel from room to room.
Whose Baggage from Land to Land is Despair

Palladas (360–430)

13.1 INTRODUCTION

Hazardous waste transportation comprises a key component of the comprehensive cradle-to-grave program for the management of hazardous wastes. Transportation most commonly takes the form of transferring from the point of generation to an EPA-registered treatment, storage, and disposal facility (TSDF). Transportation is also a significant and potential source of releases to the environment and exposure to workers and the public; hence, regulations addressing proper transportation are numerous and comprehensive.

In 1989, the National Solid Wastes Management Association reported that trucks traveling over public highways moved 98% of the hazardous waste that is treated off-site and rail transport moved the remainder. By 1993, a total of 20,800 transporters of hazardous waste were reported by the U.S. EPA (U.S. EPA, 1993). By 1997, the number of transporters declined to 18,029 (U.S. EPA, 1999). In 1999, 17,914 shippers reported shipping 8.1 million tons of RCRA hazardous waste (U.S. EPA, 2001). When comparing 1997 data with that of 1999, the number of shippers decreased by 115, and the quantity of waste shipped increased by 817 tons or 11%.

Of the 8.1 million tons of RCRA hazardous waste shipped in 1999, 5.7 million tons of waste were exported from the state of origin to other states. Between 1997 and 1999, the quantity of waste exported increased by 1.3 million tons or 30%.

When the Hazardous and Solid Waste Amendments of 1984 (HSWA) were enacted, more than 100,000 new small quantity generators (SQGs) of hazardous wastes were brought under the RCRA regulations. A majority of the SQGs had no options other than to ship hazardous wastes off-site for final treatment and disposal. As a result, HSWA initiated an increase of small shipments. In addition, the inclusion, in 1990, of 25 new chemical constituents to Table 1, 40 CFR Part 261.24, brought about 17,000 new generators under RCRA regulation (Blackman, 2001).

Hazardous waste transportation is regulated under both RCRA and U.S. Department of Transportation (DOT) hazardous materials regulations. The DOT rules are compiled in 49 CFR and were formulated to regulate the transport of *any* hazardous material, i.e., whether it exists as a new product or a waste. The DOT program had been in existence long before the promulgation of RCRA Subtitle C, and the DOT regulations were originally designed to address the transport of individual chemicals. When dealing with hazardous wastes, however, a complex waste stream, composed of many chemicals in several physical forms, may be transported. As a result, the DOT program has been modified to address classifying and managing hazardous wastes and their mixtures.

The DOT program requires the appropriate "hazard communication" (e.g., classification, labeling, marking, packaging, and placarding) of a material to be transported. If hazardous waste

427

is being transported for treatment, storage, disposal, or reclamation, the waste generator is then classified as a "shipper" by the DOT and must therefore comply with hazardous materials regulations (49 CFR Parts 171 to 179) in addition to hazardous waste (RCRA) regulations (40 CFR Part 262).

13.2   MODES OF HAZARDOUS WASTE TRANSPORTATION

Cargo tanks are the main carriers of bulk hazardous materials over roadways; however, large quantities of hazardous wastes are shipped in 55 gal drums. Cargo tanks are usually manufactured of steel or aluminum alloy, stainless steel, nickel, or titanium. Tanks range in capacity from 15,000 to 45,000 L (4,000 to 12,000 gal). Federal road weight laws usually limit motor vehicle weights to 36,000 kg (80,000 lb) gross. Table 13.1 lists DOT cargo tank specifications for bulk shipment of common hazardous materials and example cargos.

Today, rail shipments account for about 8% of the hazardous materials transported annually, with about 3000 loads transported per day. The proportion of hazardous *waste* shipments, however, is unknown (Liu and Liptak, 2000). The major classifications of rail tank cars are pressure and nonpressure (for transporting both gases and liquids). Both categories have several subclasses, which differ in terms of test pressure, presence or absence of bottom discharge valves, type of pressure relief system, and type of thermal shielding. Ninety percent of tank cars are steel; aluminum is also used. DOT tank car design specifications are covered in 49 CFR Part 179. Rail car specifications for transporting pressurized hazardous materials are in DOT 105, 112, and 114; for unpressurized shipments the numbers are DOT 103, 104, and 111. Capacities for tank cars carrying hazardous materials are limited to 131,000 L (34,500 gal) or 119,000 kg (263,000 lb) gross weight (Blackman, 2001).

Since the implementation of the RCRA regulations, most hazardous waste transporters fall into one of the following categories:

- Generators transporting their wastes to a TSDF
- Contract haulers collecting wastes from generators and transporting to TSDFs
- TSD facilities collecting wastes from generators for transport back to their facility

TABLE 13.1
Cargo Tank Table as Specified by DOT Regulations

Cargo Tank Specification Number	Types of Commodities Carried	Examples
MC-306 (MC-300,301,302,303,305)[a]	Combustible and flammable liquids of low vapor pressure	Fuel oil, gasoline
MC-307 (MC-304)	Flammable liquids, Poison B materials with moderate vapor pressure	Toluene, diisocyanate
MC-312 (MC-310,311)	Corrosives	Hydrochloric acid, caustic solution
MC-331 (MC-330)	Liquified compressed gases	Chlorine, anhydrous ammonia
MC-338	Refrigerated liquified gases	Oxygen, methane

[a]Numbers in parentheses designate earlier specifications; the older versions may continue in service but all newly constructed cargo tanks must meet current specifications.

Source: Code of Federal Regulations, Vol. 49, Parts 172.101 and 1.78.315-178.343. With permission.

Highway transportation of hazardous wastes is considered to be the most versatile. Tank trucks can gain access to most industrial generators and TSD facilities; rail shipping, in contrast, requires the installation of sidings and is suitable for only very large quantity shipments. (Blackman, 2001).

13.3 TRANSPORTATION REQUIREMENTS

13.3.1 Shipping Papers

Shipping papers are required for most shipments of hazardous materials (Figure 13.1). Such documentation provides detailed information about the materials being transported and the potential hazards that are involved. It is the transporter's responsibility to complete the shipping papers. Although the format of the papers may vary, all are required to include the following details for each hazardous material transported:

- Proper shipping name
- Hazard class or division number
- ID number
- Packing group
- Total quantity

13.3.2 The Uniform Hazardous Waste Manifest

The cradle-to-grave program requires generators to manifest all hazardous waste shipments. The Hazardous Waste Manifest System, discussed in Chapter 12, is a set of procedures and paperwork designed to track hazardous waste from the time it leaves the generator's facility until it reaches the TSDF. The manifest, a multipart paper form, allows the waste generator to verify that its wastes have been properly delivered and that all waste is accounted for.

13.3.3 Hazard Communication

13.3.3.1 Identification of the Waste being Transported

When hazardous waste is to be transported from the site of generation, the shipper must comply with regulations for its identification, classification, labeling, packaging, markings, placards, and shipping documentation, collectively known as "hazard communication." These actions are designed to protect both the health of workers and the local public who may be exposed to such wastes during transportation. Most of these requirements are contained in DOT's hazardous materials table (HMT) (49 CFR §172.101) (see below).

For purposes of safety and regulatory compliance, all materials scheduled for shipment must be fully identified. This information must also be available to the general public as well as to emergency response teams and regulators. The requirements for the proper shipping of a hazardous waste include:

- *Shipping name*: The generator (shipper) must determine the proper shipping name of the hazardous waste.
- *Hazard class*: There are 23 DOT hazard classes of shippable materials. The generator must select the appropriate hazard class.
- *Identification number*: The generator must select the identification number that corresponds to the shipping name and hazard class.
- *Label*: Generator must determine if labels are required.

U.S. GOVERNMENT BILL OF LADING
INTERNATIONAL AND DOMESTIC OVERSEAS SHIPMENTS

B/L NUMBER

TRANSPORTATION COMPANY TENDERED TO | SCAC | DATE B/L PREPARED

DESTINATION NAME AND ADDRESS | SPLC *(Dest.)* | SPLC *(Orig.)* | ORIGIN NAME AND ADDRESS

CONSIGNEE *(Name and full address of installation)* | GBLOC *(Cons.)* | SHIPPER NAME AND ADDRESS

APPROPRIATION CHARGEABLE | BILL CHARGES TO *(Dept./agency, bureau/office mailing address and ZIP code)* | AGENCY LOC CODE

VIA *(Route shipment when advantageous to the Government)*

MARKS AND ANNOTATIONS

PACKAGES		HM	DESCRIPTION OF ARTICLES *(Use carrier's classification or tariff description if possible; otherwise use a clear nontechnical description.)*	19. WEIGHTS* *(Pounds only)*	FOR USE OF BILLING CARRIER ONLY		
NO.	KIND				Services	Rate	Charges
			CLASSIFICATION ITEM NO.		TOTAL CHARGES		

TARIFF/SPECIAL RATE AUTHORITY | CARRIER WAY/FREIGHT BILL NO. AND DATE

STOP THIS SHIPMENT AT

FOR

CARRIER'S PICKUP DATE *(Year, month, and day)*

MODE

FURNISH INFORMATION ON CAR/TRUCKLOAD/CONTAINER SHIPMENTS					
SEAL NUMBERS	LENGTH/CUBE		MARKED CAPACITY		DATE FURNISHED
	ORDERED	FURNISHED	ORDERED	FURNISHED	
APPLIED BY:					
ESTIMATE	NO. OF CLS/TLS	TYPE RATE	PSC	REASON	

This U.S. Government shipment is subject to terms and conditions of 41 CFR 102-117 and CFR 102-118.

CERTIFICATE OF CARRIER BILLING -- CONSIGNEE MUST NOT PAY ANY CHARGES

DELIVERED ON *(Year, month, and day)*

FOR USE OF ISSUING OFFICE

ISSUING OFFICE *(Name and complete address)* | GBLOC | ISSUING OFFICER

CONTRACT/PURCHASE ORDER NO. OR OTHER AUTHORITY | DATED

FOB POINT NAMED IN CONTRACT

*Show also cubic measurements for shipments via air, truck or water carrier in cases where required.

AUTHORIZED FOR LOCAL REPRODUCTION

STANDARD FORM 1103 (REV. 9/2003)
Prescribed by GSA/FMR 102-118

FIGURE 13.1 Copy of a Bill of Lading.

- *Packages:* Determine the appropriate package type for the waste.
- *Markings*: Apply the required DOT and EPA markings to the package.
- *Placards*: The shipper must provide the proper placards to the transporter vehicle.
- *Shipping documentation:* The shipper must prepare a Uniform Hazardous Waste Manifest to document the identification and classification of the materials being shipped.

13.3.3.2 The Hazardous Materials Table

The HMT, found in 49 CFR §172.101, lists those materials designated by DOT as hazardous for transportation. This table provides most of the information that is needed to prepare hazardous materials for transport. The table identifies the shipping names, hazard classifications, United Nations or North American identification numbers, and references for labeling, packaging, marking, placarding, and shipping procedures. It also compiles regulations into an index that generators use to determine the appropriate procedures for waste transport.

For each hazardous material listed, the HMT identifies the hazard class or specifies that the material is forbidden in transportation, and gives the proper shipping name. In addition, the table specifies requirements pertaining to labeling, packaging, quantity limits aboard aircraft, and stowage of hazardous materials aboard vessels. A portion of the HMT appears in Figure 13.2. The specific columns are:

1. Notes (symbols) regarding requirements for shipping modes (e.g., water, air)
2. Materials descriptions and proper shipping names
3. Hazard classes and divisions
4. Identification numbers
5. Packing groups
6. Labels required
7. Special provisions
8. Packaging requirements
9. Quantity limitations
10. Requirements specifically for water shipments (e.g., vessel stowage requirements)

13.3.3.3 Hazardous Material Classes

DOT considers hazardous wastes to be a subset of the larger universe of hazardous materials. Thus, to apply the HMT for hazardous waste management, such a waste will be considered by DOT as a hazardous material subject to additional requirements. Hazardous materials are defined by DOT under 49 CFR §171.8 as products that "are capable of posing an unreasonable risk to health, safety and property when transported." A material is deemed hazardous under one of the following conditions:

- It meets one or more hazard class definitions (see below)
- It is a hazardous substance, hazardous waste, marine pollutant, or elevated-temperature material

Regardless of the above hazards, these materials can still be transported. Safe transportation is facilitated when these materials are properly classified. Classification dictates the correct handling, packaging, and emergency response actions. The DOT has established various classes of hazardous materials, established placarding and marking requirements for containers and packages, and created an international numbering system for cargo. Hazardous materials are classified for transportation into the categories shown below.

Class 1 — Explosives. An explosive is defined by DOT as "any substance or article which is designed to function by explosion (i.e., an extremely rapid release of gas and heat)." Explosives in Class 1 are divided into six divisions:

Division 1.1 — explosives that have a mass explosion hazard. A mass explosion is one that affects almost the entire load instantaneously. These explosives are among the most powerful and include bombs, mines, torpedoes, and ammunition used by the military, high explosives such as nitroglycerin and dynamite, blasting caps, detonating fuses, and rocket propellants.

Division 1.2 — explosives that have a projection hazard but not a mass explosion hazard. These substances are generally less powerful and typically function by rapid combustion rather than detonation. This class includes fireworks, flash powders, liquid or solid propellants, some smokeless powders, and certain types of ammunition.

§ 172.101 HAZARDOUS MATERIALS TABLE—Continued

Symbols	Hazardous materials descriptions and proper shipping names	Hazard class or Division	Identification Numbers	PG	Label Codes	Special provisions	Packaging (§173.***) Exceptions	Packaging Non-bulk	Packaging Bulk	Quantity limitations Passenger aircraft/rail	Quantity limitations Cargo aircraft only	Vessel stowage Location	Vessel stowage Other
(1)	(2)	(3)	(4)	(5)	(6)	(7)	(8A)	(8B)	(8C)	(9A)	(9B)	(10A)	(10B)
	Hydrocyanic acid, aqueous solutions or Hydrogen cyanide, aqueous solutions with not more than 20% hydrogen cyanide.	6.1	UN1613	I	6.1	2, B61, B65, B77, B82	None	195	244	Forbidden	Forbidden	D	40
D	Hydrocyanic acid, aqueous solutions with less than 5% hydrogen cyanide.	6.1	NA1613	II	6.1	T18, T26	None	195	243	Forbidden	5 L	D	40
	Hydrocyanic acid, liquefied, see Hydrogen cyanide, etc. *Hydrocyanic acid (prussic), unstabilized.*	Forbidden											
	Hydrofluoric acid and Sulfuric acid mixtures.	8	UN1786	I	8, 6.1	A6, A7, B15, B23, N5, N34, T18, T27	None	201	243	Forbidden	2.5 L	D	40, 95
	Hydrofluoric acid, anhydrous, see Hydrogen fluoride, anhydrous. Hydrofluoric acid, with more than 60% strength.	8	UN1790	I	8, 6.1	A6, A7, B4, B15, B23, N5, N34, T18, T27	None	201	243	0.5 L	2.5 L	D	12, 40
	Hydrofluoric acid, with not more than 60% strength.	8	UN1790	II	8, 6.1	A6, A7, B15, B110, N5, N34, T18, T27	None	202	243	1 L	30 L	D	12, 40
	Hydrofluoroboric acid, see *Fluoroboric acid.* *Hydrofluorosilicic acid,* see *Fluorosilicic acid.*												
	Hydrogen and Methane mixtures, compressed.	2.1	UN2034		2.1		306	302	302, 314, 315.	Forbidden	150 kg	E	40
	Hydrogen bromide, anhydrous	2.3	UN1048		2.3, 8	3, B14	None	304	314, 315.	Forbidden	25 kg	D	40
	Hydrogen chloride, anhydrous	2.3	UN1050		2.3, 8	3	None	304	None	Forbidden	Forbidden	D	40
	Hydrogen chloride, refrigerated liquid	2.3	UN2186		2.3, 8	3, B6	None	None	314, 315.	Forbidden	Forbidden	B	40
	Hydrogen, compressed	2.1	UN1049		2.1		306	302	302, 314.	Forbidden	150 kg	E	40, 57
	Hydrogen cyanide, solution in alcohol with not more than 45% hydrogen cyanide.	6.1	UN3294	I	6.1, 3	2, 25, B9, B14, B32, B74, T38, T43, T45	None	227	244	Forbidden	Forbidden	D	40

FIGURE 13.2 Portion of the HMT as formulated by the U.S. DOT (49 CFR 172.101).

Division 1.3 — explosives that have a fire hazard and either a minor blast hazard or a minor projection hazard, or both.

Division 1.4 — explosives that present a minor explosion hazard. The explosive effects are largely confined to the package and no projection of fragments is expected.

Division 1.5 — very insensitive explosives. This division includes substances that have a mass explosion hazard but are so insensitive that there is very little probability of initiation under routine transport. This division includes blasting agents. The material is capable of exploding under very specialized conditions.

Division 1.6 — extremely insensitive articles.

Class 2 — Gases. Gases are products that are cooled and compressed for ease in transportation. Gases pose dangers because they are stored under pressure. A compressed gas is defined as any material or mixture with an absolute pressure in a container of:

- More than 40 psa at 70°F
- More than 140 psa at 130°F

Divisions of Class 2 are as follows:

Division 2.1 — Flammable gas. A flammable compressed gas has a lower flammable limit (LFL) concentration of 13% or less by volume in air, or which has a flammable range (i.e., the difference between the LFL and UFL) of greater than 12%.

Division 2.2 — Nonflammable, nonpoisonous, compressed gas. Includes compressed gas, liquefied gas, pressurized cryogenic gas, compressed gas in solution, asphyxiant gas, and oxidizing gas.

Division 2.3 — Gas poisonous by inhalation.

Division 2.4 — Toxic gas. The term "toxic gas" may be used in place of "poison gas" for domestic shipments. The term "toxic gas" must be used for international shipments.

Class 3 — Flammable liquids. A flammable liquid refers to any liquid that has a closed-cup flash point below 37.8°C (100°F). The closed-cup test procedures are outlined in 40 CFR Part 261.21 and in EPA Method 1020B (U.S. EPA, 1986). The term "flash point" refers to a temperature at which a substance produces sufficient vapors to sustain combustion. A combustible liquid is one with flash point between 37.8 and 75.6°C (100 and 200°F).

Class 4 — Flammable solids. *Flammable solids include*: any solid material, other than one classified as an explosive, which, under conditions normally incident to transportation is liable to cause fires through friction, retain heat from manufacturing or processing, or which can be ignited readily and when ignited burn so vigorously and persistently as to create a serious transportation hazard. Spontaneously combustible and water-reactive materials are included in this class. An example of a flammable solid is aluminum hydride.

Division 4.1 — Materials that ignite easily and burn vigorously.

Division 4.2 — Spontaneously combustible material. This could include pyrophoric materials. A pyrophoric material is a liquid or solid that, without an external ignition source, can ignite after coming in contact with air.

Division 4.3 — Dangerous in the form of wet material. A material that, by contact with water, is liable to become spontaneously flammable or releases flammable or toxic gas.

Class 5 — Oxidizers and organic peroxides. *An oxidizer* "is a substance such as a chlorate permanganate, inorganic peroxide, or a nitrate, that yields oxygen readily to stimulate the combustion of organic matter." The main hazard with oxidizing agents is that contact with a combustible substance, for example organic materials (even dust) may cause the substance to ignite or explode.

Division 5.1. Releases oxygen. These will therefore promote vigorous combustion of a substance.

Division 5.2. Organic peroxides. An organic peroxide is derived from hydrogen peroxide (H_2O_2). One or more hydrocarbon groups have replaced one of the hydrogen atoms. These substances may explode under certain conditions.

Class 6 — Toxic or Poisonous materials. Poisonous materials are divided into three groups in DOT regulations according to their degree of hazard in transportation.

Division 6.1, PG I and II. Substances are "poisonous gases or liquids of such a nature that a very small amount of the gas, or vapor of the liquid, mixed with air is dangerous to life." These are highly poisonous by inhalation, ingestion, or absorption through skin. Packing groups (PG) are discussed below.

Division 6.1, PG III. These materials are moderately toxic and must be stored away from contact with food items.

Division 6.2. *An etiologic agent is* "a viable microorganism, or its toxin, which causes or may cause human disease." Such agents include infected living tissue and microbiological materials.

Class 7 — Radioactive materials. Radioactive materials are those that emit nuclear radiation. They are classified into three groups according to the controls needed to provide adequate safety during transportation.

Fissile Class I materials are the safest of these substances, do not require "nuclear criticality safety controls" during transportation, and may be shipped together in an unlimited number of packages.

Fissile Class II substances are more dangerous and can only be shipped in limited amounts when packages are shipped together.

Fissile Class III materials must be controlled to provide nuclear criticality safety in transportation.

Class 8 — Corrosive materials. The DOT defines a *corrosive material* as "a liquid or solid that causes visible destruction or irreversible alterations in human skin tissue at the site of contact, or in the case of leakage from its packaging, a liquid that has a severe corrosion rate on steel." A liquid is considered to have "a severe corrosion rate" if it dissolves >0.6 cm (0.25 in.) of a steel sample at 54°C (130°F) over 1 year.

Class 9 — Miscellaneous hazardous materials. The materials in this class have anesthetic, noxious, or other similar properties.

Other Regulated Materials — Domestic. The final DOT category is called other regulated materials (ORM). This includes a wide variety of hazardous materials shipped in limited quantities and in certain kinds of packaging. There are five classes designated ORM-A, ORM-B, ORM-C, ORM-D, and ORM-E.

The ten hazardous material classes, along with their divisions, are shown in Table 13.2. Further details regarding components of the HMT appear in Table 13.3, and the instructions for selecting the proper shipping name for a hazardous material appear in Table 13.4.

13.3.3.4 Packing Groups

After assigning a hazardous material to one of the classes listed above, the degree of risk must also be noted. Most hazardous materials are assigned to one of three packing groups (49 CFR Part 173, subpart D):

- Packing Group I (PG I) — great danger
- Packing Group II (PG II) — medium danger
- Packing Group III (PG III) — minor danger

The packing groups will determine the type of packaging that can be used for a material intended for transport. The more hazardous a material, the more stringent the packaging requirements will be. Packing groups are assigned to all hazardous materials except classes 2, 7, and ORM-D materials.

TABLE 13.2
DOT Hazard Classes and Divisions for Hazardous Materials Transportation

Label Code	Label Name
1	Explosive
1.1	Explosive 1.1
1.2	Explosive 1.2
1.3	Explosive 1.3
1.4	Explosive 1.4
1.5	Explosive 1.5
1.6	Explosive 1.6
2.1	Flammable gas
2.2	Nonflammable gas
2.3	Poison gas
3	Flammable liquid
4.1	Flammable solid
4.2	Spontaneously combustible
4.3	Dangerous when wet
5.1	Oxidizer
5.2	Organic peroxide
6.1	Poison inhalation hazard
6.2	Infectious substance
7	Radioactive
8	Corrosive
9	Class 9
ORM-D	

Source: 49 CFR Part 173. With permission.

TABLE 13.3
Details of the HMT

Column 1: Symbols
Column 1 of the table contains six symbols ("+", "A", "D", "G", "I", and "W") as follows:

A   Restricts the application to materials offered or intended for transportation by aircraft, unless the material is a hazardous substance or a hazardous waste.

D   Identifies proper shipping names which are appropriate for describing materials for domestic transportation but may be inappropriate for international transportation under international regulations (e.g., IMO and ICAO). An alternative proper shipping name may be selected when either domestic or international transportation is involved.

G   Identifies proper shipping names for which one or more technical names of the hazardous material must be entered in parentheses, in association with the basic description (see Sec. 172.203(k)).

I   Identifies proper shipping names which are appropriate for describing materials in international transportation. An alternative proper shipping name may be selected when only domestic transportation is involved.

W   Restricts the application to materials intended for transportation by vessel, unless the material is a hazardous substance or a hazardous waste.

If none of the above symbols appears in the table, the substance is regulated for all modes of transportation.

Column 2: Hazardous Materials Descriptions and Proper Shipping Names
Column 2 lists the descriptions of the hazardous materials and proper shipping names of materials designated as hazardous materials. Proper shipping names are limited to those shown in Roman type (not italics).

(Continued)

TABLE 13.3 (*Continued*)
Details of the HMT

Hazardous substances. Proper shipping names for hazardous substances are in Table 1 to Appendix A, 49 CFR §172.101.

Hazardous wastes. If the word "waste" is not included in the hazardous material description in Column 2 of the table, the proper shipping name for a hazardous waste must include the word "Waste" preceding the proper shipping name of the material. As an example, "Waste acetone" is an appropriate shipping name.

Mixtures and solutions. A mixture or solution not identified specifically by name, composed of a hazardous material identified in the table by technical name and nonhazardous material, must be described using the proper shipping name of the hazardous material and the word "mixture" or "solution."

A mixture or solution not identified in the table specifically by name, composed of two or more hazardous materials in the same hazard class, must be described using an appropriate shipping description (e.g., "Flammable liquid, n.o.s."). The name that most appropriately describes the material shall be used; for example, an alcohol not listed by its technical name in the table shall be described as "Alcohol, n.o.s.," rather than "Flammable liquid, n.o.s." Some mixtures are more appropriately described according to their application, such as "Coating solution" rather than by an n.o.s. entry. The technical names of at least two components most predominately contributing to the hazards of the mixture or solution may be required along with the proper shipping name.

Column 3: Hazard Class or Division
Column 3 contains a designation of the hazard class or division corresponding to each proper shipping name, or the word "Forbidden." A forbidden material cannot be transported. This prohibition does not apply if the material is diluted or stabilized.

Column 4: Identification Number.
Column 4 lists the identification number assigned to each proper shipping name. Those preceded by the letters "UN" are associated with proper shipping names considered appropriate for international transportation as well as domestic transportation. Those preceded by the letters "NA" are associated with proper shipping names not recognized for international transportation, except to and from Canada.

Column 5: Packing Group
Column 5 specifies one or more packing groups assigned to a material corresponding to the proper shipping name and hazard class for that material. Class 2, Class 7, Division 6.2 (other than regulated medical wastes), and ORM-D materials, do not have packing groups. Packing Groups I, II, and III indicate the degree of danger presented by the material is either great, medium, or minor, respectively.

Column 6: Labels
Column 6 specifies codes which represent the hazard warning labels required for a package filled with a material conforming to the associated hazard class and proper shipping name, unless the package is otherwise excepted from labeling by 49 CFR Part 173.
The first code indicates the primary hazard of the material. Additional label codes are indicative of subsidiary hazards. Provisions in Section 172.402 may require that additional labels other than that specified in Column 6 be affixed to the package.

Column 7: Special Provisions
Column 7 specifies codes for special provisions applicable to hazardous materials. The requirements of that special provision are as set forth in 49 CFR §172.102.

Column 8: Packaging Authorizations
Columns 8A, 8B, and 8C specify the applicable sections for exceptions, nonbulk packaging requirements and bulk packaging requirements, respectively, in Part 173. Columns 8A, 8B, and 8C are completed in a manner which indicates that "Section 173" precedes the designated numerical entry. For example, the entry "202" in Column 8B associated with the proper shipping name "Gasoline" indicates that for this material conformance to nonbulk packaging requirements prescribed in Section 173.202 is required. When packaging requirements are specified, they are in addition to the standard requirements for all packagings prescribed in Section 173.24.

Column 9: Quantity Limitations
Columns 9A and 9B specify the maximum quantities that may be offered for transportation in one package by passenger-carrying aircraft or passenger-carrying rail car (Column 9A) or by cargo aircraft only (Column 9B), subject to the following:

- "Forbidden" means the material may not be offered for transportation.
- The quantity limitation is "net" except where otherwise specified.

(Continued)

TABLE 13.3   (*Continued*)

- When articles or devices are specifically listed by name, the net quantity limitation applies to the entire article (less packaging) rather than only to its hazardous components.
- A package offered or intended for transportation by aircraft and which is filled with a material forbidden on passenger-carrying aircraft but permitted on cargo aircraft only, or which exceeds the maximum net quantity authorized on passenger-carrying aircraft, shall be labeled with the CARGO AIRCRAFT ONLY label specified in 49 CFR §172.448.

Column 10: Vessel Stowage Requirements
Column 10A (Vessel stowage) specifies the authorized stowage locations on board cargo and passenger vessels. Column 10B (Other provisions) specifies codes for stowage requirements for specific hazardous materials. The meaning of each code in Column 10B is set forth in Section 176.84 of this subchapter. Section 176.63 of this subchapter sets forth the physical requirements for each of the authorized locations listed in Column 10A.

Source: 49 CFR 172.101. With permission.

TABLE 13.4
Instructions for Selecting Proper Shipping Names for Hazardous Materials

1. Apply the most specific name listed in the Hazardous Materials Table (49 CFR §172.101). The DOT ranks the order of specificity as follows:

   - Specific chemical name (e.g., "Sulfuric acid")
   - Chemical group or family (e.g., "Acid, liquid, n.o.s.")
   - End use of material (e.g., "Pigment")
   - Generic end use (e.g., "Medicines, n.o.s.")
   - Hazard class (e.g., "Corrosive liquid, n.o.s.")

   When hazardous waste is being shipped, the word "waste" must be included in the shipping name.
2. There are to be no additions or deletions to a shipping name except as explicitly allowed by DOT. When the term "n.o.s." (not otherwise specified) is used, additional information must be included which will identify at least two major constituents present in the waste which contribute to the Hazard Class.
3. Once the most appropriate name has been selected it is assigned a specific ID number, i.e., UN or NA followed by four unique digits. The information compiled thus far, i.e., proper shipping name, Hazard Class, and the UN or NA number, comprise the U.S. DOT "Basic Description."
   The DOT description must be in the following order:

   - Proper shipping name, Hazard class, and UN or NA number
   - Additional information

4. The packing group Roman numeral is assigned.
5. Reportable Quantity (RQ) values are assigned, if applicable. The RQ value for a constituent in the material is listed in the Appendix to 49 CFR §172.101, List of Hazardous substances and Reportable Quantities. If the weight of the constituent meets or exceeds the RQ, then RQ must be added to the proper shipping name.
6. The completed form for shipping a hazardous waste is as follows:

   - RQ, proper shipping name, hazard class, UN/NA number, packing group
   - Additional information

Source: 49 CFR §172.101, Appendix A. With permission.

13.3.3.5   Identification Numbers

The DOT has assigned a four-digit identification number to each of the hazardous materials regulated in transportation. When appearing in shipping papers and other documentation, the letters "UN" or "NA" precedes these numbers. The UN numbers, such as UN1823 for solid sodium hydroxide, were assigned in cooperation with the United Nations and are used for international shipments. The NA numbers are not used for international transportation except to and from Canada (49 CFR §172.101).

Most of the numbers and the material shipping names represent specific materials. However, the DOT also permits some cargo to be identified generically. For example, the identification number UN1760 applies to Corrosive Liquid, n.o.s. The last three letters indicate "not otherwise specified." This number, therefore, does not allow for the identification of the specific material stored in the container.

13.3.3.6 Placards, Labels, and Markings

Hazard communication under DOT regulations includes the requirement of alerting regulators, emergency response personnel, and the general public of the contents of a truck, railcar, or similar vehicle. For example, DOT requires placards to appear on railroad tank cars, highway tank trucks, and other large transport vehicles (Figure 13.3). Similarly, labels must appear on packages of hazardous materials (see below). Placards and labels both have the function of warning the public and emergency response personnel of the hazards associated with a specific material. The main difference between placards and labels is that placards are larger and must be displayed on transport vehicles and other bulk packagings.

Placarding is required for the transportation of all hazardous waste with few exceptions. The generator must placard the vehicle or provide the transporter with the appropriate placards. Many transporters, however, provide their own placards. A placard is required on each side or end of the vehicle transporting hazardous materials (Figure 13.4); in the case of truck transport, the front placard may be attached to the tractor instead of the trailer. The placards must meet size, durability, color, and other requirements and must be securely affixed to the vehicle.

How a hazardous waste or material is marked, labeled, and placarded will depend upon whether it is contained in a nonbulk or bulk packaging. Nonbulk packagings are those which:

- Have a maximum capacity of 450 L (119 gal) or less (*for liquids*)
- Have a maximum net mass of 400 kg (882 lb) or less (*for solids*)
- Have a water capacity of 454 kg (1000 lb) or less (*for gases*)

Bulk packagings are those which:

- Have a capacity greater than 450 L (119 gal) (*for liquids*)
- Have a net mass greater than 400 kg (882 lb), or a capacity greater than 450 L (119 gal) (*for solids*)
- Have a water capacity greater than 454 kg (1000 lb) (*for gases*)

13.3.3.7 Hazard Warning Labels

Each package of hazardous materials must have two types of communications on the package, i.e., markings and labels. Hazard-warning labels provide an immediate warning of any hazards associated or precautions needed with a material. There are two types of labels:

- Primary labels, which indicate the most hazardous property of a material
- Subsidiary labels, which indicate other, less hazardous properties

Hazard class labels are diamond-shaped labels indicating the hazard class of the material being shipped (e.g., "flammable liquid"). Primary labels include the appropriate hazard class or division number at the bottom while subsidiary labels do not (Figure 13.5). In certain cases, both the primary hazard and a subsidiary hazard must be labeled (e.g., a flammable liquid that is also poisonous). DOT also specifies certain special precaution labels, intended to indicate an extra hazard or a special precaution to be taken during transportation. Examples include the "Dangerous When Wet" label for materials that have water-reactive properties. The HMT (49 CFR Part 172.101) specifies the type of label required for a waste. Labels should be attached to at least two sides, preferably opposite sides of the container, with at least one near the opening point of the container. Hazardous material warning labels are listed in Figure 13.6.

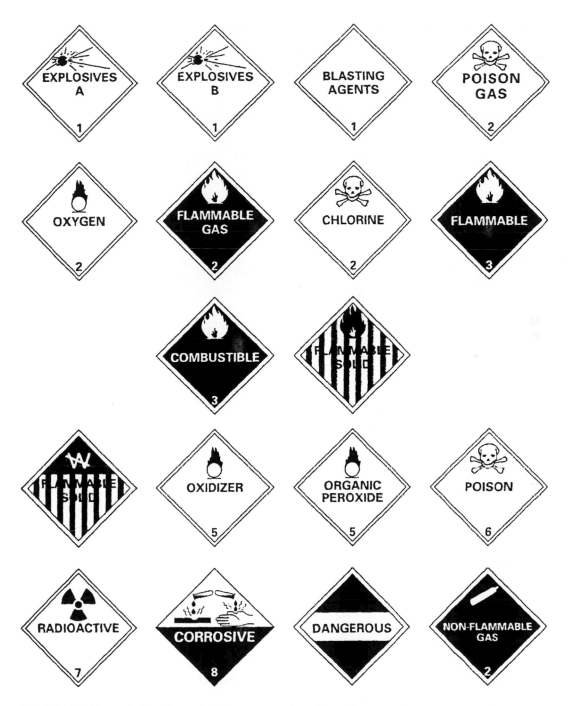

FIGURE 13.3 Types of placards required for transportation of hazardous materials.

13.3.3.8 Marking

Marking means placing one or more of the following on the outside of a shipping container: the descriptive name, proper shipping name, hazard class, identification number, instructions, cautions, and weight. Markings must be durable, in English, on a background of sharply contrasting color, and away from other markings.

FIGURE 13.4 A placard is required on each side and end of the vehicle.

FIGURE 13.5 Primary and subsidiary labels for hazardous materials.

Nonbulk Markings

The proper markings for nonbulk hazardous materials include (Figure 13.7):

- The proper shipping name
- ID number of the material
- Consignee's or consignor's name and address

Other markings may be required depending on the type of material transported. For example:

- Packagings that contain materials designated poisonous by inhalation must be marked "INHALATION HAZARD."
- Packagings that contain a consumer commodity must be marked "ORM-D" immediately following the shipping name.

Bulk Markings

All bulk packagings must be marked with the material's ID number. This is displayed across the primary hazard placard. Placement of the ID number depends on the type of packaging. For example, if the packaging has a capacity of 3785 L (1000 gal) or more and is a tube-trailer vehicle, the number

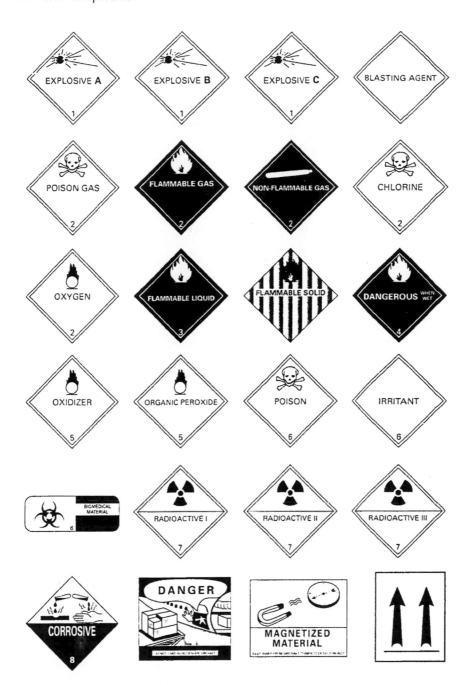

FIGURE 13.6 Hazardous materials warning labels.

must be displayed on each side and each end of the packaging. If the packaging has a capacity of less than 3785 L (1000 gal), the number must be displayed on two opposite sides. Additional markings may be required depending on the packaging and the material being transported. For example:

- Materials that are poisonous by inhalation must be marked "INHALATION HAZARD."
- Cargo tanks used to transport gases must be marked on each side and each end with the proper shipping name or common name of the gas.
- Portable tanks must be marked with the material's proper shipping name on two opposite sides, and the name of the owner.

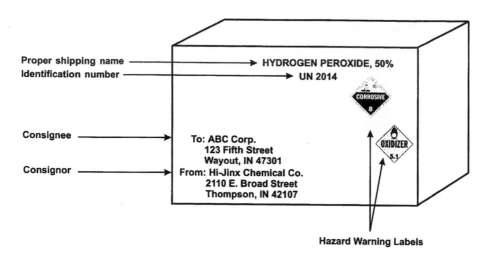

Proper shipping name ——————→ HYDROGEN PEROXIDE, 50%
Identification number ——————→ UN 2014

CORROSIVE 8

Consignee ——————→ To: ABC Corp.
　　　　　　　　　　　　　123 Fifth Street
　　　　　　　　　　　　　Wayout, IN 47301
Consignor ——————→ From: Hi-Jinx Chemical Co.
　　　　　　　　　　　　　2110 E. Broad Street
　　　　　　　　　　　　　Thompson, IN 42107

OXIDIZER 5.1

Hazard Warning Labels

FIGURE 13.7 Markings for nonbulk hazardous materials.

13.3.3.9　Packaging

There are two approaches to packaging shipments of hazardous waste. In the first approach, the generator can check with the TSD facility to determine the packaging the latter prefers for the particular waste. The TSD facility recommends using a particular DOT specification packaging. The second approach is to determine the requirements directly from the DOT regulations. Most hazardous waste will be contained in open or closed head drums or in bulk boxes. Figure 13.8 provides an example description of a drum per DOT requirements prior to the performance-oriented packaging codes (see below).

The specific section for packaging a particular material is found by reference to column 5 of the HMT. In column 8 there are three subcolumns, indicating the specific sections in Part 173 where the specific requirements are located for Exceptions (Column 8A), Nonbulk packaging (Column 8B), and Bulk packaging (Column 8C). Each of these columns is set up such that "Part 173" is the initial source, followed by the more specific subpart. For example, the entry "213" in column 8B associated with the proper shipping name "ammonium chloride" indicates that for nonbulk packaging, the requirements of 49 CFR Part 173.213 must be followed. Whenever packaging requirements are specified, these are in addition to the standard requirements for all packaging described in Part 173.

Examples of packaging
NonBulk Packagings for Liquid Hazardous Materials in Packing Group I (Section 173.201)
DOT requirements are rather specific and stringent for the selection of the appropriate packaging material for a hazardous material. For example, when a liquid hazardous material is packaged under certain DOT sections, only nonbulk packagings may be used for its transportation. Packaging must conform to the general requirements of Part 173, Part 178, and §172.101 of HMT. The following combination packagings are authorized for liquid hazardous materials in Packing Group I:

Outer packagings

Steel drum	1A1 or 1A2
Aluminum drum	1B1 or 1B2
Metal drum other than steel or aluminum	1N1 or 1N2
Plywood drum	1D
Fiber drum	1G

Plastic drum	1H1 or 1H2
Steel jerrican	3A1 or 3A2
Plastic jerrican	3H1 or 3H2
Aluminum jerrican	3B1 or 3B2
Steel box	4A

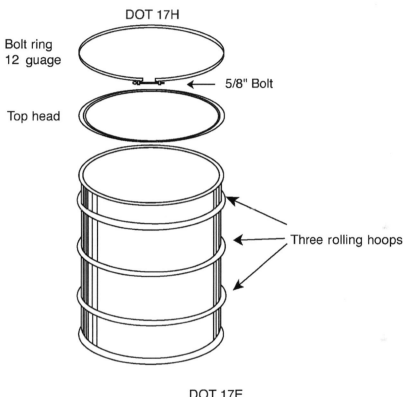

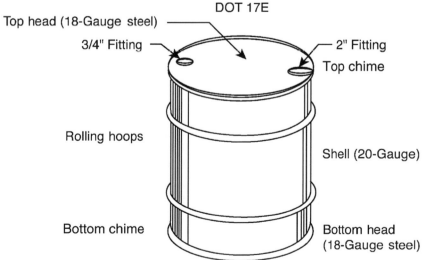

FIGURE 13.8 Drum description per earlier DOT regulations: DOT 17E (U.S. DOE, no date). Under a final rule issued September 18, 1991 (56 Federal Register 47158), DOT changed its packaging standards from specification-based (e.g., 17E) to performance-based (e.g., UN1A1).

Aluminum box	4B
Natural wood box	4C1 or 4C2
Plywood box	4D
Reconstituted wood box	4F
Fiberboard box	4G
Expanded plastic box	4H1
Solid plastic box	4H2

Inner packagings
Glass or earthenware receptacles
Plastic receptacles
Metal receptacles
Glass ampoules

Except for transportation by passenger aircraft, the following single packagings are authorized:

Steel drum	1A1 or 1A2
Aluminum drum	1B1 or 1B2
Metal drum other than steel or aluminum	1N1 or 1N2
Plastic drum	1H1 or 1H2
Steel jerrican	3A1 or 3A2
Plastic jerrican	3H1 or 3H2
Aluminum jerrican	3B1 or 3B2
Plastic receptacle in steel, aluminum, fiber, or plastic drum	6HA1, 6HB1, 6HG1, 6HH1
Plastic receptacle in steel, aluminum, wooden, plywood, or fiberboard box	6HA2, 6HB2, 6HC, 6HD2, or 6HG2
Glass, porcelain or stoneware in steel, aluminum or fiber drum	6PA1, 6PB1, or 6PG1
Glass, porcelain or stoneware in steel, aluminum, wooden or fiberboard box	6PA2, 6PB2, 6PC, or 6PG2
Glass, porcelain or stoneware in solid or expanded plastic packaging	6PH1 or 6PH2

Performance-Oriented Packaging
Performance-oriented packaging (POP) was introduced into international packaging regulations in 1989. After 1991, POP was phased into the domestic regulations (49 CFR) and has been effective since October 1, 1996. All nonbulk quantities of hazardous materials were required to be shipped in POP. POP is packaging that meets design qualification testing. Package designs are subjected to Drop, Stack, Vibration, Leakproofness, and Hydrostatic tests based upon United Nations recommendations. A United Nations Certification marking may be applied to the packagings once the design passes the performance tests (Figure 13.9).

Identification codes are set forth in the standards for packagings in 49 CFR Parts 178.504 through 178.523. A manufacturer must mark every packaging that is represented as manufactured to meet a UN standard with other DOT-specified marks. The markings must be durable, legible, and placed in a readily visible location. Packaging conforming to a UN standard must be marked as follows:

1. The United Nations symbol as illustrated in Figure 13.9 or the letters "UN." A packaging identification code designating the type of packaging and material of construction.
2. A letter identifying the performance standard under which the packaging design type has been successfully tested, as follows:

    X   for packagings meeting Packing Group I, II, and III tests
    Y   for packagings meeting Packing Group II and III tests or
    Z   for packagings only meeting Packing Group III tests

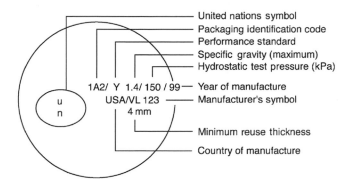

FIGURE 13.9 Performance-oriented packaging label.

3. A designation of the specific gravity or mass for which the packaging design type has been tested.
4. The last two digits of the year of manufacture. Packagings of types 1H and 3H must also be marked with the month of manufacture.
5. The state authorizing allocation of the mark. The letters "USA" indicate that the packaging is manufactured and marked in the United States.
6. The name and address or symbol of the manufacturer or the approval agency certifying compliance with DOT regulations.
7. For metal or plastic drums or jerricans intended for reuse or reconditioning, the thickness of the packaging material expressed in millimeters (rounded to the nearest 0.1 mm).
8. The rated capacity of the packaging, expressed in liters, may also be marked.

Performance-Oriented Packaging Standards for Nonbulk Packagings
Identification codes for designating non-bulk packagings consist of the following (Section 178.502):
A numeral indicating the kind of packaging:

1   drum
2   wooden barrel
3   jerricans
4   box
5   bag
6   composite packaging
7   pressure receptacle

A capital letter indicating the material of construction:

A   steel (all types and surface treatments)
B   aluminum
C   natural wood
D   plywood
F   reconstituted wood
G   fiberboard
H   plastic
L   textile
M   paper, and multiwall
N   metal (other than steel or aluminum)
P   glass, porcelain, or stoneware

A third numeral provides additional packaging details. For example, for steel drums ("1A"), "1" indicates a nonremovable head drum (i.e., "1A1") and "2" indicates a removable head drum (i.e., "1A2").

All of the above precautions and requirements have facilitated the safe transportation of hazardous materials and hazardous wastes; however, industry has expressed some frustration over the regulations. The DOT states that (49 CFR 173.24):

> Each package used for the shipment of hazardous materials under this subchapter shall be designed, constructed, maintained, filled, its contents so limited, and closed, so that under conditions normally incident to transportation — Except as otherwise provided in this subchapter, there will be no identifiable ... release of hazardous materials to the environment.

The problem with the above requirement is that two different drums may pass the same POP tests, however only one may survive a lengthy journey intact. Some argue that the problem is not with performance-based standards but with the actual tests that do not adequately mimic real-world transportation conditions in the United States. The distance factor of an average transcontinental shipment multiplies the real-world problems of abrasion between drums, vibration, shock, puncture, external corrosion, and so on. Current POP tests do not address these problems (Marshall and Andell, 1996).

13.5   EMERGENCY RESPONSE INFORMATION

The shipper is required to provide emergency response information for each hazardous material listed on the shipping paper (e.g., Bill of Lading). Information must include:

- Description of the hazardous material
- Immediate hazards to health
- Risks of fire or explosion
- Immediate precautions to be taken in the event of an accident or incident
- Immediate methods for handling small or large fires
- Initial methods for handling spills or leaks in the absence of fire
- Preliminary first aid measures

This information can be listed directly on the shipping papers. The shipper can also attach a copy of the appropriate guide from the DOT's *North American Emergency Response Guidebook* (DOT, 1996) (Figure 13.10).

13.6   SEGREGATION (49 CFR PART 177, SUBPART C)

The generator and transporters must be aware of the possible hazards if incompatible materials are loaded together. Certain materials may initiate reactions that can be dangerous to public health and property. To assist with safe loading and transport, DOT created the Segregation Table (Figure 13.11), which informs all parties as to which classes and divisions can be loaded together safely and which cannot.

The shipper is to locate the hazard classes or divisions of the material being transported, one in the vertical column, the other in the horizontal row. The codes at the intersection are defined as follows:

Code	Meaning
Blank	The materials may be loaded or stored together
X	The materials may not be loaded together
O	The materials may not be loaded together unless separated so that, in the event of leakage, there will be no mixing of the materials.
*	Class 1 (explosive) materials must be segregated in accordance with the Compatibility Table
A	An oxidizer may be loaded with Division 1.1 or 1.5 materials

If the classes or divisions are not listed in the table, there are no restrictions.

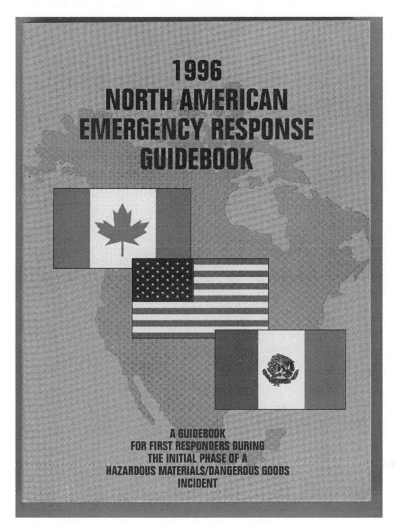

FIGURE 13.10 The North American Emergency Response Guidebook.

13.7 HAZARDOUS MATERIALS INCIDENTS (49 CFR PART 171)

From 1992 to 2001 hazardous waste incidents on the highway predominated over all other modes of transportation (Table 13.5). After 1996, the number of reported incidents began a steady downward trend. The total number of highway hazardous waste incidents totaled 4317 from 1996 to 2001. Rail incidents were second in number over this time period, with incidents ranging between 23 and 40 per year. A total of 12 air-related incidents were reported over the period 1996 to 2001 (U.S. DOT, n.d.).

Carriers of hazardous materials are required to report certain unintentional releases that occurred during transportation of hazardous materials (Figure 13.12 and Figure 13.13). The requirements appear in 49 CFR Part 171, §§171.15 and 171.16. There are also requirements in 49 CFR Parts 174 to 177 that require reporting of hazardous materials incidents (49 CFR §§174.45 [rail], 175.45 [air], 176.48 [vessel], and 177.807 [highway]). Two phases of incident reporting are required in the regulations. Section 171.15 covers immediate telephone notification following an incident and Part 171.16 outlines written reporting procedures (e.g., Incident Report Form 5800.1).

Segregation Table for Hazardous Materials

Class or Division	Notes	1.1 and 1.2	1.3	1.4	1.5	1.6	2.1	2.2	2.3 Gas Zone A	2.3 Gas Zone B	3	4.1	4.2	4.3	5.1	5.2	6.1 Liquids PG I Zone A	7	8 Liquids Only
Explosives .... 1.1 and 1.2	A	*	*	*	*	*	X	X	X	X	X	X	X	X	X	X	X	X	X
Explosives 1.3		*	*	*	*	*	X		X	O	X		X	X	X	X	X		X
Explosives 1.4		*	*	*	*	*	O		O		O			O	O	O			O
Very insensitive explosives 1.5	A	*	*	*	*	*	X		X	O	X	X	X	X	X	X	X	X	X
Extremely insensitive explosives 1.6		*	*	*	*	*			X										
Flammable gases 2.1		X	X	O	X				X	O				X	X		O		X
Nontoxic, nonflammable gases. ... 2.2		X			X														
Poisonous gas Zone A 2.3		X	X	O	X	X					X			X	X	X	X		X
Poisonous gas Zone B 2.3		X	O		X						O			O	O	O	O		O
Flammable liquids 3		X	X	O	X				X	O				X	O	O			
Flammable solids 4.1		X	X		X				X	O				X	O				O
Spontaneously combustible materials. ... 4.2		X	X	O	X				X	O				X	O	X	X		O
Dangerous when wet materials. ... 4.3		X	X		X		X		X	O	X	X	X		X	X			X
Oxidizers 5.1	A	X	X	O	X		X		X	O	X	X	X	X			X		X
Organic peroxides 5.2		X	X	O	X		X		X	O	O			X			X		X
Poisonous liquids PG I Zone A. ... 6.1		X			X				X										
Radioactive materials 7		X			X		O												
Corrosive liquids 8		X	X	O	X				X	O		O	O	X	X	X	X		

FIGURE 13.11 Segregation table for hazardous materials (49 CFR).

TABLE 13.5
Hazardous Waste Incidents in the United States Listed by Mode and Incident Year

Mode	1992	1993	1994	1995	1996	1997	1998	1999	2000	2001	Total
Air	1	1	1	0	0	2	3	2	1	1	12
Highway	377	550	519	652	424	379	381	420	325	290	4317
Railway	33	23	27	24	34	38	40	34	25	29	307
Water	0	1	0	0	0	0	0	0	0	0	1
Freight fowarder	0	0	0	0	0	0	0	0	0	0	0
Other	0	0	0	0	0	0	0	0	0	0	0
Total	411	575	547	676	458	419	424	456	351	320	4637

Source: U.S. DOT, no date. With permission

FIGURE 13.12 Hazardous materials incident on the highway.

FIGURE 13.13 Aftermath of a hazardous materials railway incident.

13.7.1 Immediate Notice of Certain Hazardous Materials Incidents (49 CFR 171.15)

Telephone notices are required immediately whenever there is a significant hazardous materials incident during transportation, during the course of transportation (including loading, unloading, and temporary storage), or storage related to transportation, in which the following occur:

1. As a direct result of hazardous materials:

    - A person is killed
    - A person receives injuries requiring hospitalization
    - Estimated carrier or other property damage exceeds $50,000
    - An evacuation of the general public occurs lasting one or more hours
    - One or more major transportation arteries or facilities are closed or shut down for 1 h or more
    - The operational flight pattern or routine of an aircraft is altered

2. Fire, breakage, spillage, or suspected radioactive contamination occurs (see also 49 CFR 174.45, 175.45, 176.48, and 177.807).
3. Fire, breakage, spillage, or suspected contamination occurs involving shipment of infectious substances (etiologic agents).
4. Release of a marine pollutant exceeding 450 L (119 gal) for liquids or 400 kg (882 lb) for solids.
5. A situation exists (e.g., a danger to life exists at the scene of the incident) such that, in the judgment of the carrier, it should be reported to the DOT.

Each notice must be given to the National Response Center by telephone (800-424-8802). Notice involving infectious substances (etiologic agents) may instead be given to the U.S. Centers for Disease Control, U.S. Public Health Service, Atlanta, Georgia (800-232-0124). Each notice must include:

- Name of reporter and of carrier
- Phone number where reporter can be contacted
- Date, time, and location of incident
- Extent of any injuries
- Classification, name, and quantity of hazardous materials involved, if such information is available
- Type of incident
- Whether a continuing danger to life exists at the scene

The NRC is staffed around the clock. The Coast Guard Duty Officer notifies concerned organizations including the Research and Special Programs Administration (RSPA), modal administrations, CHEMTREC (a U.S. and Canadian chemical tracking system), and the National Transportation Safety Board to bring about prompt resolution of serious incidents. The telephone notices received by the NRC are accumulated and transmitted daily to the Hazardous Materials Information System (HMIS) host computer which is maintained at the Volpe National Transportation System Center in Cambridge, Massachusetts. At the Volpe Center, the information is made available that day in a telephone database.

13.7.2 HAZARDOUS MATERIALS INCIDENTS REPORTS (49 CFR 171.16)

Within 30 days of the date of discovery of a hazardous materials incident, the carrier must report in writing to the DOT on Form F 5800.1 (Figure 13.14). Each incident that occurred during transportation (including loading, unloading, and temporary storage) or any unintentional release of hazardous materials from a package (including a tank) must be reported. The report identifies the mode of transportation involved, name of reporting carrier, shipment information, results of the incident, hazardous materials involved, nature of packaging, cause of failure, and narrative description of the incident. This information is available in the incident database approximately 3 months after the receipt of the report by RSPA.

If a report pertains to a hazardous waste discharge, a copy of the hazardous waste manifest for the waste must be attached to the report. Also, an estimate of the quantity of the waste removed from the scene, the name and address of the facility to which it was taken, and the mode of disposition of any removed waste must be entered in the report form (Form F 5800.1). The report is to be sent to the Research and Special Programs Administration, Department of Transportation, Washington, DC. A copy of the report must be retained for 2 years at the carrier's office.

The HMIS is a computerized information management system containing data related to the DOT program to ensure safe transportation of hazardous materials by air, highway, rail, and water. The HMIS is the primary source of national data for the federal, state, and local governmental

DEPARTMENT OF TRANSPORTATION
HAZARDOUS MATERIALS INCIDENT REPORT

Form Approved OMB No. 2137-0039

INSTRUCTIONS: Submit this report in duplicate to the Information Systems Manager, Office of Hazardous Materials Transportation, DHM-63, Research and Special Programs Administration, U.S. Department of Transportation, Washington, D.C. 20590. If space provided for any item is inadequate, complete that item under Section IX, keying to the entry number being completed. Copies of this form, in limited quantities, may be obtained from the Information Systems Manager, Office of Hazardous Materials Transportation. Additional copies in this prescribed format may be reproduced and used, if on the same size and kind of paper.

I. MODE, DATE, AND LOCATION OF INCIDENT

1. MODE OF TRANSPORTATION: ☐ AIR ☐ HIGHWAY ☐ RAIL ☐ WATER ☐ OTHER _____

2. DATE AND TIME OF INCIDENT
   (Use Military Time, e.g. 8:30 am = 0830,
   noon = 1200, 6 pm = 1800, midnight = 2400).   Date: _____/_____/_____   TIME: _____

3. LOCATION OF INCIDENT (Include airport name in ROUTES/STREET if incident occurs at an airport.)

   CITY: _____   STATE: _____

   COUNTY: _____   ROUTE/STREET: _____

II. DESCRIPTION OF CARRIER, COMPANY, OR INDIVIDUAL REPORTING

4. FULL NAME   5. ADDRESS (Principal place of business)

6. LIST YOUR OMC MOTOR CARRIER CENSUS NUMBER, REPORTING RAILROAD ALPHABETIC
   CODE, MERCHANT VESSEL NAME AND ID NUMBER OR OTHER REPORTING CODE OR NUMBER.

III. SHIPMENT INFORMATION (From Shipping Paper or Packaging)

7. SHIPPER NAME AND ADDRESS (Principal place of business)   8. CONSIGNEE NAME AND ADDRESS (Principal place of business)

9. ORIGIN ADDRESS (If different from Shipper address)   10. DESTINATION ADDRESS (If different from Consignee address)

11. SHIPPING PAPER/WAYBILL IDENTIFICATION NO.

IV. HAZARDOUS MATERIAL(S) SPILLED (NOTE: REFERENCE 49 CFR SECTION 172.101)

12. PROPER SHIPPING NAME	13. CHEMICAL/TRADE NAME	14. HAZARD CLASS	15. IDENTIFICATION NUMBER (e.g. UN 2764, NA 2020)

16. IS MATERIAL A HAZARDOUS SUBSTANCE? ☐ YES ☐ NO | 17. WAS THE RQ MET? ☐ YES ☐ NO

V. CONSEQUENCES OF INCIDENT, DUE TO THE HAZARDOUS MATERIAL.

18. ESTIMATED QUANTITY HAZARDOUS MATERIAL RELEASED (Include units of measurement)	19. FATALITIES	20. HOSPITALIZED INJURIES	21. NON-HOSPITALIZED INJURIES
22. NUMBER OF PEOPLE EVACUATED			

23. ESTIMATED DOLLAR AMOUNT OF LOSS AND/OR PROPERTY DAMAGE, INCLUDING COST OF DECONTAMINATION OR CLEANUP (Round off in dollars)

A. PRODUCT LOSS	B. CARRIER DAMAGE	C. PUBLIC/PRIVATE PROPERTY DAMAGE	D. DECONTAMINATION/ CLEANUP	E. OTHER

24. CONSEQUENCES ASSOCIATED WITH THE INCIDENT: ☐ VAPOR (GAS) DISPERSION   ☐ MATERIAL ENTERED WATERWAY/SEWER
   ☐ SPILLAGE ☐ FIRE ☐ EXPLOSION   ☐ ENVIRONMENTAL DAMAGE   ☐ NONE   ☐ OTHER: _____

VI. TRANSPORT ENVIRONMENT

25. INDICATE TYPE(S) OF VEHICLE(S) INVOLVED: ☐ CARGO TANK ☐ VAN TRUCK/TRAILER ☐ FLAT BED TRUCK/TRAILER
   ☐ TANK CAR ☐ RAIL CAR ☐ TOFC/COFC ☐ AIRCRAFT ☐ BARGE ☐ SHIP ☐ OTHER: _____

26. TRANSPORTATION PHASE DURING WHICH INCIDENT OCCURRED OR WAS DISCOVERED:
   ☐ EN ROUTE BETWEEN ORIGIN/DESTINATION ☐ LOADING ☐ UNLOADING ☐ TEMPORARY STORAGE/TERMINAL

27. LAND USE AT INCIDENT SITE: ☐ INDUSTRIAL ☐ COMMERCIAL ☐ RESIDENTIAL ☐ AGRICULTURAL ☐ UNDEVELOPED

28. COMMUNITY TYPE AT SITE: ☐ URBAN ☐ SUBURBAN ☐ RURAL

29. WAS THE SPILL THE RESULT OF A VEHICLE ACCIDENT/DERAILMENT? ☐ YES ☐ NO
   IF YES AND APPLICABLE, ANSWER PARTS A THRU C.

A. ESTIMATED SPEED:	B. HIGHWAY TYPE: ☐ DIVIDED/LIMITED ACCESS ☐ UNDIVIDED	C. TOTAL NUMBER OF LANES: ☐ ONE ☐ THREE ☐ TWO ☐ FOUR OR MORE	SPACE FOR DOT USE ONLY

FORM DOT F 5800.1 (Rev. 10/94) Supersedes DOT F 5800.1 (10/70) (9/1/76) 130-F (Rev. 10/94)   THIS FORM MAY BE REPRODUCED

FIGURE 13.14 DOT form 5800.1, Incident Report.

VII. PACKAGING INFORMATION: If the package is overpacked (consists of several packages, e.g. glass jars within a fiberboard box), begin with Column A for information on the innermost package.

ITEM	A	B	C
30. TYPE OF PACKAGING, INCLUDING INNER RECEPTACLES (e.g. Steel drum, tank car)			
31. CAPACITY OR WEIGHT PER UNIT PACKAGE (e.g. 55 gallons, 65 lbs.)			
32. NUMBER OF PACKAGES OF SAME TYPE WHICH FAILED IN IDENTICAL MANNER			
33. NUMBER OF PACKAGES OF SAME TYPE IN SHIPMENT			
34. PACKAGE SPECIFICATION IDENTIFICATION (e.g. DOT 17E, DOT 105A100, UN 1A1 or none)			
35. ANY OTHER PACKAGING MARKINGS (e.g. STC, 18/16-55-88, Y1.4/150/87)			
36. NAME AND ADDRESS, SYMBOL OR REGISTRATION NUMBER OF PACKAGING MANUFACTURER			
37. SERIAL NUMBER OF CYLINDERS, PORTABLE TANKS, CARGO TANKS, TANK CARS			
38. TYPE OF LABELING OR PLACARDING APPLIED			
39. IF RECONDITIONED OR REQUALIFIED — A. REGISTRATION NUMBER OR SYMBOL			
39. IF RECONDITIONED OR REQUALIFIED — B. DATE OF LAST TEST OR INSPECTION			
40. EXEMPTION/APPROVAL/COMPETENT AUTHORITY NUMBER, IF APPLICABLE (e.g. DOT E1012)			

VIII. DESCRIPTION OF PACKAGING FAILURE: Check all applicable boxes for the package(s) identified above.

41. ACTION CONTRIBUTING TO PACKAGING FAILURE

A B C
a. ☐ ☐ ☐ TRANSPORT VEHICLE COLLISION
b. ☐ ☐ ☐ TRANSPORT VEHICLE OVERTURN
c. ☐ ☐ ☐ OVERLOADING/OVERFILLING
d. ☐ ☐ ☐ LOOSE FITTINGS, VALVES
e. ☐ ☐ ☐ DEFECTIVE FITTINGS, VALVES
f. ☐ ☐ ☐ DROPPED
g. ☐ ☐ ☐ STRUCK/RAMMED
h. ☐ ☐ ☐ IMPROPER LOADING
i. ☐ ☐ ☐ IMPROPER BLOCKING

A B C
j. ☐ ☐ ☐ CORROSION
k. ☐ ☐ ☐ METAL FATIGUE
l. ☐ ☐ ☐ FRICTION/RUBBING
m. ☐ ☐ ☐ FIRE/HEAT
n. ☐ ☐ ☐ FREEZING
o. ☐ ☐ ☐ VENTING
p. ☐ ☐ ☐ VANDALISM
q. ☐ ☐ ☐ INCOMPATIBLE MATERIALS
r. ☐ ☐ ☐ OTHER _____

42. OBJECT CAUSING FAILURE

A B C
a. ☐ ☐ ☐ OTHER FREIGHT
b. ☐ ☐ ☐ FORKLIFT
c. ☐ ☐ ☐ NAIL/PROTRUSION
d. ☐ ☐ ☐ OTHER TRANSPORT VEHICLE
e. ☐ ☐ ☐ WATER/OTHER LIQUID
f. ☐ ☐ ☐ GROUND/FLOOR/ROADWAY
g. ☐ ☐ ☐ ROADSIDE OBSTACLE
h. ☐ ☐ ☐ NONE
i. ☐ ☐ ☐ OTHER _____

43. HOW PACKAGE(S) FAILED

A B C
a. ☐ ☐ ☐ PUNCTURED
b. ☐ ☐ ☐ CRACKED
c. ☐ ☐ ☐ BURST/INTERNAL PRESSURE
d. ☐ ☐ ☐ RIPPED
e. ☐ ☐ ☐ CRUSHED
f. ☐ ☐ ☐ RUBBED/ABRADED
g. ☐ ☐ ☐ RUPTURED
h. ☐ ☐ ☐ OTHER _____

44. PACKAGE AREA THAT FAILED

A B C
a. ☐ ☐ ☐ END, FORWARD
b. ☐ ☐ ☐ END, REAR
c. ☐ ☐ ☐ SIDE, RIGHT
d. ☐ ☐ ☐ SIDE, LEFT
e. ☐ ☐ ☐ TOP
f. ☐ ☐ ☐ BOTTOM
g. ☐ ☐ ☐ CENTER
h. ☐ ☐ ☐ OTHER _____

45. WHAT FAILED ON PACKAGE(S)

A B C
a. ☐ ☐ ☐ BASIC PACKAGE MATERIAL
b. ☐ ☐ ☐ FITTING/VALVE
c. ☐ ☐ ☐ CLOSURE
d. ☐ ☐ ☐ CHIME
e. ☐ ☐ ☐ WELD/SEAM
f. ☐ ☐ ☐ HOSE/PIPING
g. ☐ ☐ ☐ INNER LINER
h. ☐ ☐ ☐ OTHER _____

IX. DESCRIPTION OF EVENTS: Describe the sequence of events that led to incident, action taken at time discovered, and action taken to prevent future incidents. Include any recommendations to improve packaging, handling, or transportation of hazardous materials. Photographs and diagrams should be submitted when necessary for clarification. ATTACH A COPY OF THE HAZARDOUS WASTE MANIFEST FOR INCIDENTS INVOLVING HAZARDOUS WASTE. Continue on additional sheets if necessary.

46. NAME OF PERSON RESPONSIBLE FOR PREPARING REPORT	47. SIGNATURE	
48. TITLE OF PERSON RESPONSIBLE FOR PREPARING REPORT	49. TELEPHONE NUMBER (Area Code)	50. DATE REPORT SIGNED

FIGURE 13.14 (Continued)

agencies responsible for the safety of hazardous materials transportation. The industry, news media, and the general public are also entitled to use data from the system. The major components of the HMIS concern (DOT, n.d.):

- Incidents involving the interstate transportation of hazardous materials by various transportation modes
- Exemptions issued to hazardous materials regulations
- Interpretations of the regulations issued by the RSPA, as requested by concerned parties
- Approvals of specialized container manufacturers, reconditioners, and testers
- Compliance activities including inspections performed by HMS and completed enforcement proceedings
- Registrations filed by certain carriers, shippers, and other providers of hazardous materials

REFERENCES

Blackman, W.C., Jr., *Basic Hazardous Waste Management,* 3rd ed., Lewis Publishers, Boca Raton, FL, 2001.
Code of Federal Regulations, Volume 49, Department of Transportation, U.S. Government Printing Office, Washington, DC, 2004.
Indiana Department of Environmental Management, State of Indiana Hazardous Waste Manifest Guidance Manual, IDEM Office of Solid and Hazardous Waste Management, Indianapolis, IN, December 1994.
Lieberman, J.L., *A Practical Guide for Hazardous Waste Management, Administration and Compliance,* Lewis Publishers, Boca Raton, FL, 1994.
Marshall, R., and Andell, T., Hazmat packaging: Safe — but by whose standards? The move to performance-oriented packaging raises serious questions for hazmat shippers, *Transportation and Distribution,* Penton Publishing Inc., Cleveland, OH. November, 1996.
U.S. Department of Energy, Hanford Site, Test and Evaluation Document for the U.S. Department of Transportation Specification 7A Packaging, Steel drums, DOE/RL-96-57, Rev. 0-F, Vol. 1, no date, See: http://www.hanford.gov/pss/t&p/rl96-57/ch1-2/graphics/f2-9.gif.
U.S. Department of Transportation, North American Emergency Response Guidebook. A Guidebook for First Responders During the Initial Phase of a Hazardous Materials/Dangerous Goods Incident, 1996.
U.S. Department of Transportation, Incidents, no date, See http://hazmat.dot.gov/incsys.htm.
U.S. Environmental Protection Agency, The National Biennial RCRA Hazardous Waste Report (Based on 1999 Data), EPA530-S-01-001, Solid Waste and Emergency Response PB2001-106318 (5305W), June 2001.
U.S. Environmental Protection Agency, Pensky-Martens Closed-Cup Method for Determining Ignitability), Method 1010, Test Methods for Evaluating Solid Waste, Physical/Chemical Methods, EPA SW-846, Washington, DC, 1986.

SUGGESTED READING

Blackman, W.C., Jr., Environmental Impacts of Policies Toward the Rail and Motor-Freight Industries in the United States, Doctoral dissertation, Graduate School of Public Affairs, University of Colorado, Boulder, CO, 1985.
Eckmyre, A.A., Release reporting and emergency notification, in *Hazardous Materials Management Desk Reference,* McGraw-Hill, New York, NY, 2000.
Fox, M.A., *Glossary for the Worldwide Transportation of Dangerous Goods and Hazardous Materials,* CRC Press, Boca Raton, FL, 2000.
ICF, Inc., Assessing the Costs Associated with Truck Transportation of Hazardous Wastes, U.S. Environmental Protection Agency, Office of Solid Waste, Washington, DC, 1984.
Munter, F., Bell, S.W., Hillingsworth, R., Gordon, J.W., and Lovinski, C.N., Hazardous wastes, in *Accident Prevention Manual for Business and Industry — Environmental Management,* Krieger, G.R., Ed., National Safety Council, Itaska, IL, 1995.

Nicolet-Monnier, M. and Gheorghe, A.V., *Quantitative Risk Assessment of Hazardous Materials Transport Systems*, Kluwer, Norwell, MA, 1996.

Perry, D.M. and Klooster, D.J., The Maquiladora Industry: Generation, Transportation and Disposal of Hazardous Waste at the California-Baja California, U.S. – Mexico Border: Second Maquiladora Report, School of Public Health, University of California, Los Angeles, 1992.

Perry, D.M., Sanchez, R., Glaze, W.H., and Mazari, M., Binational management of hazardous waste: the maquiladora industry at the U.S.-Mexico Border, *Environ. Manage.,* 14, 441–450, 1990.

Roberts, A. I., Transport of hazardous waste, in *Transfrontier Movements of Hazardous Waste,* Organization for Economic Cooperation and Development, Paris, 1985.

U.S. Department of Transportation, Research and Special Programs Administration, 1999 *Hazardous Materials Incident Data,* Office of Hazardous Materials Safety, Washington, DC, 2000.

U.S. Office of Technology Assessment, *Transportation of Hazardous Materials,* Superintendent of Documents, U.S. Government Printing Office, Washington, DC, 1986.

QUESTIONS

1. What are the differences, in terms of content, between the Bill of Lading for hazardous wastes and the Uniform Hazardous Waste Manifest? Who is responsible for initiating each set of paperwork?
2. The uniform hazardous manifest:
   (a) Is required under RCRA/HSWA
   (b) Is to be completed where applicable by the waste generator
   (c) Is to be utilized by all LQGs and SQGs
   (d) Is essential in "cradle-to-grave" waste management
   (e) All of the above
3. The uniform hazardous waste manifest serves which of the following purposes:
   (a) Certification of treatment/disposal
   (b) Tracking document for delivery to designated TSDF
   (c) Shipping document for hazardous materials per DOT regulations
   (d) (b) and (c) only
4. List the ten classes of DOT hazardous materials. Which of the classes do not contain divisions?
5. What is the purpose of a packing group? How many packing groups exist?
6. Which of the following is *not* one of DOT's hazard classes?
   (a) Class 1 — explosive, (b) ORM-D, (c) ORM-F, (d) Class 9 — miscellaneous material.
7. According to federal regulations, what is the overall purpose of transportation placards?
8. How do hazardous placards differ from labels and markings, in terms of size, shape, and location to be affixed? Explain how the requirements for their use differ for bulk and nonbulk packaging.
9. For a truck transporting hazardous wastes, on which side(s) is (are) the placard(s) to be placed?
10. A local citizen contacts the local emergency response agency with the following information from an overturned tank truck. Which information from the truck will best help the agency identify its contents?
    (a) Company name on the truck; (b) License plate number; (c) PG II; (d) UN 1272.
11. When shipping a truckload of 25 drums of the same hazardous material, what is the number of drums that must be labeled?
12. What is the difference between primary and subsidiary hazard-warning labels on nonbulk packages of hazardous waste?
13. What details can you provide about containers with the following information stamped on its base: (a) 1A2 UN X; (b) 4G UN Z

14. The shipper is required to provide emergency response information for each hazardous material listed on the shipping paper (e.g., Bill of Lading). List the specific information that must be included on those papers.

15. Refer to the DOT waste compatibility table (Figure 13.11). State whether or not the following wastes be placed in the same shipment:
(a) Spent hydrochloric acid and a class 4.1 waste
(b) Waste classes 1.1 and 4.1
(c) Flammable gases and explosives

16. Which of the following situations does *not* require an incident report involving hazardous materials to the DOT?
    (a) A person is killed
    (b) Property damage exceeds $50,000
    (c) A person is injured but does not require hospitalization
    (d) Fire, breakage, spillage, or suspected contamination involving shipment of a biohazardous material

17. What is the name of the table that must be used to select the proper container for accumulation and transportation of hazardous waste?

18. Suppose the waste will be stored in a 55 gal metallic drum. To locate the drum in the table, under which column is the name of the waste found: the hazard class or packing group? Where is the regulation number where the drum is listed (i.e., which column)?

19. According to DOT regulations, containers used to accumulate and transport hazardous waste must be in good condition. What exactly does "good condition" mean?

20. Refer to 49 CFR 172.101 and the HMT. Find the regulatory reference for drums (i.e., nonbulk containers) used to hold waste methyl ethyl ketone (MEK).

21. Insert the packing groups authorized for containers meeting the performance standards identified below:

Performance Standard	For Packages Meeting Packing Groups
X	
Y	
Z	

22. The following labels and markings occur on a drum of hazardous waste:

RQ, Waste Corrosive Liquid, acidic, inorganic,

n.o.s., 8, UN 3264, PG I (EPA D002)

HAZARDOUS WASTE
Federal law prohibits improper
disposal. If found contact the
nearest police or public safety
authority or the US Environmental
Protection Agency

Pristine Chemical Company, Inc.
555 Quiet Blvd.
Pristine, IL 50773

EPA #1D ILD098765432

Manifest # 012345

Start Date – Oct. 3, 2003

1H1/Y 1.6/165/02
USA/AB990
6 mm

u
n

Identify the following:

Hazardous Substance Designation
DOT Proper Shipping Name
Hazard class
DOT identification number
Hazardous waste warning
Generator name

23. Based on the manufacturer's mark, answer the following:
    (a) What is the minimum drum wall thickness for reuse of the drum?
    (b) What is the limit of specific gravity of the contents?
    (c) What is the drum type?
24. Who is responsible for ensuring that containers are properly marked and labeled (i.e., hazardous waste generator, transporter, TSDF, regulatory agency)?

A.13.1 APPENDIX: HAZARDOUS WASTE MANAGEMENT SCENARIOS

A.13.1.1 INTRODUCTION

After reading each scenario, discuss what, if any, violations may have occurred. How would these violations be best addressed (i.e., via changes in engineering design, a modified storage or disposal program, some use of common sense, etc.)? These are open-ended situations.

A.13.1.2 THE SCENARIOS

1. A driver has spilled hazardous waste (baghouse dust, K061) on to the road within a secure land disposal facility (Figure A.13.1). The driver dismisses the concerns of the inspector, stating the amount released was a "minimal" quantity of hazardous waste.
2. A company truck is used to transport bulk hazardous waste to a disposal facility. What is wrong with this picture (Figure A.13.2)?
3. A machining facility uses trichloroethylene (TCE) to degrease machine parts. During lunch breaks and after work, four of the plant employees enjoy playing paintball in the back of company property, alongside Sparkling River (Figure A.13.3). They recently moved ten drums, in Moe's pickup truck, to the river's edge to use as the necessary barriers for the game. All drums are at least half-filled with TCE. One employee occasionally uses the TCE-laden drums for target practice with live ammunition (to repeat from Chapter 12, *these are true stories*).

FIGURE A.13.1 Photo by Theresa M. Pichtel.

Based on a neighbor's complaint, an inspector visits the site, and notices paint-splattered drums on the riverside both rusting and punctured by bullet holes. She instructs the company's plant manager to return the drums to the facility immediately. Later that afternoon, Moe and Larry load the ten drums on to Moe's pickup truck where they are returned to the plant floor for storage.

Back at the plant site, the inspector notices an underground storage tank vent line. The plant manager acknowledges that a UST is on the premises; however, he declares that it does not store hazardous waste, just mop water from cleaning the shop floors at the end of the working day. After some questioning, however, the manager acknowledges that spills of TCE have occurred over the past few years, and the TCE-contaminated mop water is directed into the UST opening.

A.13.1.3 RESPONSES TO SCENARIOS

1. (a) Regardless of quantity, this spill was cited as a "release to the environment", and was also a violation of the state operating permit, which required facility employees to regularly inspect their vehicles and document all inspections. The permitting issue was cited in the federal standards:

   Section 264.15 General inspection requirements.

   (a) The owner or operator must inspect his facility for malfunctions and deterioration, operator errors, and discharges which may be causing — or may lead to — (1) release of hazardous waste constituents to the environment or (2) a threat to human health. The owner or operator must conduct these inspections often enough to identify problems in time to correct them before they harm human health or the environment.

   (b) This vehicle was traveling solely on the property of the TSD facility, i.e., from the stabilization unit to the secure landfill. If this vehicle had been traveling on a public road, it would have also been subject to DOT regulations (49 CFR) for hazardous materials transport.

2. According to the facility permit, transport vehicles must be maintained in good condition. The truck shown in this photo clearly has spaces which would allow some of the hazardous cargo to disperse to air and soil. This is a permit violation.

In the preamble to Subpart A, of Standards Applicable to Transporters of Hazardous Waste (Part 263):

(a)

(b)

(c)

FIGURE A.13.2 Photos by Theresa M. Pichtel.

In these regulations, EPA has expressly adopted certain regulations of the Department of Transportation (DOT) governing the transportation of hazardous materials. These regulations concern, among other things, labeling, marking, placarding, using proper containers, and reporting discharges. EPA has expressly adopted these regulations in order to satisfy its statutory obligation to promulgate regulations which are necessary to protect human health and the environment in the transportation of hazardous waste.

3. (a) The drums are leaking alongside and possibly into the river. This is a violation of the Clean Water Act.

FIGURE A. 13.3. How *not* to manage drums of a listed hazardous waste.

    (b) Hazardous wastes are to be transported only in EPA-registered vehicles. Moe's pickup truck for transport to and from the river is clearly unacceptable and is a violation of 40 CFR 262.12(c). Since hazardous waste is being transported, a Uniform Hazardous Waste Manifest should have been completed as well. Manifesting requirements are delineated in 40 CFR 262.20.

    (c) The truck will need to be de-contaminated.

    (d) Because TCE is a F001 listed hazardous waste, the contents of the entire UST including the mop water, are also considered hazardous. The Mixture Rule, 40 CFR 261.3(b) and (d), states that hazardous waste mixed with non-hazardous waste constitutes hazardous waste.

    (e) The drums must be repacked into stable, nonrusting containers.

14 Treatment, Storage, and Disposal Facility Requirements

E'en from the tomb the voice of nature cries,
E'en in our ashes live their wonted fires.

Thomas Gray (1716–1771),
Elegy

14.1 INTRODUCTION

The generation and transportation aspects of hazardous waste management have been discussed to this point. In the final phase of the cradle-to-grave management protocol under RCRA, hazardous wastes will ultimately be treated and disposed. Treatment, storage, and disposal (TSD) technologies embrace many diverse systems. Additionally, the technologies are constantly changing and improving. With this in mind, the regulatory requirements for TSD facilities (TSDFs)are much more extensive as compared with those for generators and transporters. In this chapter, the regulatory requirements under RCRA for TSDFs will be discussed in some detail. In subsequent chapters, specific practices for hazardous waste destruction and disposal are presented.

RCRA requires a Part B permit for the treatment, storage and disposal of hazardous waste at a facility. The relevant terms are defined in 40 CFR 270.2 as follows:

Treatment: Any method, technique, or process, including neutralization, designed to change the physical, chemical, or biological character or composition of any hazardous waste so as to neutralize such wastes, or so as to recover energy or material resources from the waste, or so as to render such waste as nonhazardous, or less hazardous; safer to transport, store, or dispose of; or amenable for recovery, amenable for storage, or reduced in volume.

Storage: The holding of hazardous waste for a temporary period, at the end of which the hazardous waste is treated, disposed, or stored elsewhere.

Disposal: The discharge, deposit, injection, dumping, spilling, leaking, or placing of any hazardous waste into or on any land or water so that such hazardous waste or any constituent thereof may enter the environment or be emitted into the air or discharged into any waters, including groundwater.

According to the U.S. EPA (2001), land disposal accounted for 69% of the management methods for hazardous wastes (not including wastewaters). The quantities managed by the land disposal method include (U.S. EPA, 2001):

Deepwell or Underground injection	16.0 million tons
Landfill	1.4 million tons
Surface impoundment	705 thousand tons
Land treatment, application and farming	30 thousand tons

Thermal treatment accounted for 11% of the national nonwastewater management total. Thermal treatment methods include:

Energy recovery	1.5 million tons
Incineration	1.5 million tons

Recovery operations represented 10% of the national nonwastewater management total. The methods defined as recovery operations and the quantity managed by each method include:

Fuel blending   1.1 million tons
Metals recovery (for reuse) 720 thousand tons
Solvents recovery   368 thousand tons
Other recovery   152 thousand tons

The remaining nonwastewater management quantities (11%) were managed in other treatment and disposal units, including:

Other disposal   1.4 million tons
Stabilization   1.3 million tons
Sludge treatment   48 thousand tons

Details on hazardous waste management by method are shown in Table 14.1.

14.2 SUBPART A — GENERAL ISSUES REGARDING TREATMENT, STORAGE AND DISPOSAL

In general, all facilities involved in the treatment, storage and disposal of hazardous wastes must comply with 40 CFR 264 and 265 regulations unless they are excluded. The requirements do not apply to:

- A facility permitted to manage municipal or industrial solid waste, if the only hazardous waste the facility handles is from a conditionally exempt small quantity generator(SQG)
- A facility managing recyclable materials described in 40 CFR Part 261.6
- A generator accumulating waste on-site
- A farmer disposing of waste pesticides from his own use
- A totally enclosed treatment facility
- An elementary neutralization unit or a wastewater treatment unit
- A person engaged in treatment or containment activities during immediate response to a discharge of a hazardous waste, or an imminent and substantial threat of a discharge of hazardous waste

TABLE 14.1
Quantities of RCRA Hazardous Wastes Managed by Various Methods, 1999

Management Method	Tons Managed	Percentage of Quantity	Number of Facilities	Percentage of Facilities
Deepwell or Underground injection	16,043,912	61.0	46	8.8
Energy recovery	1,542,315	5.9	99	18.9
Incineration	1,454,403	5.5	149	28.4
Landfill	1,410,392	5.4	60	11.4
Other disposal	1,398,993	5.3	39	7.4
Stabilization	1,337,162	5.1	84	16.0
Fuel blending	1,099,687	4.2	104	19.8
Metals recovery	719,916	2.7	88	16.8
Surface impoundment	705,304	2.7	2	0.4
Solvents recovery	367,899	1.4	111	21.1
Other recovery	151,700	0.6	46	8.8
Sludge treatment	47,653	0.2	16	3.0
Land treatment, application or farming	29,873	0.1	7	1.3
Total	26,309,208	100.0	525	

Source: U.S. EPA, 2001. With permission.

- The addition of absorbent material to waste in a container
- A transporter storing manifested shipments of hazardous waste in containers at a transfer facility for 10 days or less
- Universal waste handlers and universal waste transporters (40 CFR 260.10) handling the following wastes:

  o Batteries as described in 40 CFR §273.2
  o Pesticides as described in §273.3
  o Thermostats as described in §273.4
  o Lamps as described in §273.5

14.3 SUBPART B — GENERAL FACILITY STANDARDS

Hazardous waste TSDFs are subject to a permitting system dedicated to ensuring both safe operation and adequate protection of the environment. Under the permit system, facilities must meet both general standards for proper waste management as well as requirements specific to the individual facility.

14.3.1 NOTIFICATION OF HAZARDOUS WASTE ACTIVITY

As is the case for generators of hazardous waste, every permitted TSDF is required to obtain an EPA identification number. EPA form 8700-12, Notification of Hazardous Waste Activity (see Figure 12.1), must be submitted to the U.S. EPA or state regulatory agency. This is a one-time notification. A copy of the completed form 8700-12 should be maintained at the facility.

14.3.2 WASTE ANALYSIS

Before a facility manages any hazardous wastes it must obtain a detailed chemical and physical analysis of a representative sample of the wastes. This analysis must ultimately provide all relevant information needed for the proper treatment, storage and disposal of the waste in order to comply with regulatory requirements. Permitted facilities are required to develop and adhere to a waste analysis plan (WAP), which describes the procedures conducted to ensure that sufficient information is compiled about each waste stream. The U.S. EPA or state agency must approve the WAP as part of the permit application process for facilities.

The WAP must include detailed parameters for hazardous waste analysis and the rationale for the use of these parameters, including:

- Sampling methods used to obtain a representative sample of each waste for analysis.
- The frequency with which the waste analysis is reviewed to ensure accuracy.
- Procedures used to test for parameters.
- For facilities that accept waste from off-site, the plan must document any analyses that off-site waste generators provide.

Both sampling procedures and laboratory analytical methods listed in the WAP should be the same or equivalent to those documented in Test Methods for Evaluating Solid Waste, EPA publication SW-846 (U.S. EPA, 1986).

14.3.3 SECURITY

The TSDF operator must prevent the unknowing and unauthorized entry of people onto the active portion of the facility. Security features at a facility must include:

- A 24-h surveillance system (e.g., television monitoring or surveillance by guards), which continuously monitors and controls entry to the facility

- An artificial or natural barrier (e.g., fence or berms), which completely surrounds the active portion of the facility
- A means to control entry, at all times, through the gates or other entrances (e.g., an attendant, television monitors, locked entrance, or controlled roadway access)

Signs with the warning "Danger — Unauthorized Personnel Keep Out" must be posted at each entrance (Figure 14.1). The warning must be written in English and in any other language predominant in the area surrounding the facility. Therefore, in northern Maine it may be suitable to post signs in French and in Texas to post signs in Spanish.

14.3.4 INSPECTIONS

The facility operator is required to inspect the facility for malfunctions, deterioration, and operator errors that may lead to a release of hazardous wastes or pose a threat to public health. These inspections must be scheduled often to identify and correct problems on time. The schedule must identify potential problems to be checked, the date and time of each inspection, the name of the inspector, a list of observations, and the date and types of any repairs or other actions. A written schedule for inspecting monitoring equipment, safety and emergency equipment, security devices, and operating and structural equipment must be kept on file at the facility. The inspections must be recorded in an inspection log and these records must be maintained for at least 3 years.

Inspection frequency may vary for items on the schedule; however, the frequency must be based on the rate of possible deterioration of the equipment and the probability of an environmental or public health incident if the deterioration or malfunction goes undetected between inspections. Inspection frequencies for certain equipment as specified in the regulations are listed in Table 14.2. Because of its uniqueness, each facility should develop its own inspection logs that identify the specific items to be inspected and their frequency of inspection.

14.3.5 REACTIVE AND IGNITABLE WASTES

A TSDF is required to take precautions to prevent accidental ignition or reaction of ignitable or reactive waste. This waste must be separated and protected from sources of ignition. A facility that

FIGURE 14.1 Warning signs at the entrance to a TSDF.

TABLE 14.2
Inspection Frequencies for Selected Equipment at TSD Facilities

Item	Minimum Inspection Frequency
Loading and unloading areas	Daily
Containers storage areas	Weekly
Tank systems	Daily
Surface impoundments	Weekly
Incinerators	Daily
Chemical, physical, and biological treatment units	Weekly
Containment buildings	Weekly

treats, stores, or disposes ignitable or reactive waste, or mixes incompatible wastes, must take precautions to prevent reactions that:

- Generate extreme heat, pressure, fire, or explosions
- Produce harmful quantities of toxic mists, fumes, dusts, or gases
- Produce harmful levels of flammable fumes or gases
- Damage the structural integrity of the device or facility

The operator must document compliance with the above requirements by citing published scientific literature, trial tests, waste analyses, or the results of similar treatment processes.

14.3.6 TRAINING

Training is required for facility personnel involved in the management of hazardous waste. Training records are required to document that all appropriate personnel have successfully completed their required training. In OSHA's Hazardous Waste Operations and Emergency Response (HAZWOPER) regulations (29 CFR 1910.120), all personnel who work at a permitted TSDF must complete a training program (minimum of 24 h) prior to conducting any activities that could expose them to hazardous wastes.

14.4 SUBPART C — PREPAREDNESS AND PREVENTION

Portions of new facilities where treatment, storage and disposal of hazardous waste will be conducted must not be located within 61 m (200 ft) of a fault that has experienced displacement in Holocene time. In addition, a facility located in a 100 year floodplain must be designed, constructed, and operated to prevent washout of any hazardous waste by a 100 year flood. An exemption may be permitted for surface impoundments, waste piles, land treatment units, and landfills if no adverse effects on human health or the environment will result if washout occurs (40 CFR Part 264.18). The placement of any noncontainerized or bulk liquid hazardous waste in any salt dome formation, salt bed formation, underground mine, or cave is prohibited.

A construction quality assurance (CQA) program is required for all surface impoundments, waste piles, and landfill units. The program must be developed and implemented under the direction of a CQA officer who is a registered professional engineer (40 CFR Part 264.19). The CQA program must address the following physical components:

- Foundations
- Dikes
- Low-permeability soil liners

- Geomembranes
- Leachate collection and removal systems and leak detection systems
- Final cover systems

The facility-written CQA plan must identify steps that will be used to monitor the quality of construction materials and the manner of their installation. It must also include the identification of applicable units and a description of how they will be constructed. The plan identifies key personnel in the development and implementation of the CQA plan. It also provides a description of inspection and sampling activities for all unit components to ensure that the construction materials and components meet design specifications. The description covers:

- Sampling size and locations
- Frequency of testing
- Data evaluation procedures
- Acceptance and rejection criteria for construction materials
- Plans for implementing corrective measures

The CQA program includes inspections, tests, and measurements sufficient to ensure the structural stability and integrity of all components of the unit. The program must also ensure proper construction of all components of the liners, leachate collection and removal system, leak detection system, and final cover system according to permit specifications.

14.4.1 FACILITY DESIGN AND OPERATION

Facilities must be designed, constructed, and operated to minimize the possibility of a fire, explosion, or any unplanned release of hazardous wastes to air, soil, or surface water. With limited exceptions, all TSDFs must be equipped with the following (40 CFR Part 264.32):

- An internal communications or alarm system capable of providing immediate emergency instruction (voice or signal) to facility personnel
- A telephone or hand-held two-way radio, capable of summoning emergency assistance from local police departments, fire departments, or state or local emergency response teams
- Portable fire extinguishers, fire control equipment (including special extinguishing equipment, such as that using foam, inert gas, or dry chemicals), spill control equipment, and decontamination equipment
- Water at adequate volume and pressure to supply water hose streams, foam-producing equipment, automatic sprinklers, or water spray systems

All facility communications or alarm systems, fire protection equipment, spill control equipment, and decontamination equipment must be tested and maintained as necessary to assure their proper operation in time of emergency. The facility must maintain aisle space to allow the unobstructed movement of personnel, fire protection equipment, spill control equipment, and decontamination equipment to any area of facility operation in an emergency (40 CFR Part 264.35).

14.4.2 ARRANGEMENTS WITH LOCAL AUTHORITIES

The facility must make several arrangements with local authorities in preparation for any site emergency. The facility is to familiarize police, fire departments, and emergency response teams with the layout of the facility, properties of hazardous waste handled at the facility and associated hazards, places where facility personnel normally work, entrances to and roads inside the facility, and possible evacuation routes. Where more than one police and fire department might respond to an emergency, agreements are made designating primary emergency authority to a specific police and a specific fire department, and agreements with any others to provide support to the primary emergency authority. There are also agreements with state emergency response teams, emergency response contractors, and equipment suppliers. The facility is to familiarize local hospitals with the properties of hazardous waste handled at the facility and the types of injuries or illnesses which could result from fires, explosions, or releases at the facility.

14.5 SUBPART D — CONTINGENCY PLAN AND EMERGENCY PROCEDURES

14.5.1 CONTINGENCY PLAN

Each facility must prepare and maintain a contingency plan. The contingency plan is designed to minimize hazards to human health or the environment from fires, explosions, or any unplanned releases of hazardous waste (40 CFR Part 264.51). The provisions of the plan are carried out immediately whenever there is an event that could threaten human health or the environment. The plan describes the response actions the facility personnel must take in response to a hazardous event at the facility. The plan describes arrangements agreed to by local police departments, fire departments, hospitals, contractors, and emergency response teams to coordinate emergency services. The plan lists all persons qualified to act as emergency coordinator.

The plan includes a list of all emergency equipment at the facility (such as fire extinguishing systems, spill control equipment, communications and alarm systems, and decontamination equipment). In addition, the plan includes the location and a physical description of each item and a brief outline of its capabilities. The plan must also include an evacuation plan for facility personnel. This plan describes signals to be used to begin evacuation, evacuation routes, and alternative evacuation routes, in cases where the primary routes could be blocked by releases of hazardous waste or fires.

A copy of the contingency plan and all revisions must be maintained at the facility and submitted to local police departments, fire departments, hospitals, and emergency response teams, which may be called upon to provide emergency services.

Real-world situations often do not occur according to plan; therefore, the contingency plan must be reviewed and amended whenever the following occurs (40 CFR Part 264.54):

- The facility permit is revised
- The plan fails in an emergency
- The facility changes its design, construction, or operation in a way that increases the potential for fires, explosions, or releases of hazardous waste, or changes the response in an emergency
- The list of emergency coordinators changes
- The list of emergency equipment changes

There must be at least one employee either on the facility premises or on call at all times, with the responsibility for coordinating all emergency response measures. This individual must be thoroughly familiar with the facility contingency plan, all operations and activities at the facility, the location and characteristics of waste handled, the location of all records within the facility, and the facility layout.

14.5.2 EMERGENCY PROCEDURES

Whenever there is an imminent or actual emergency situation, the emergency coordinator must immediately activate internal facility alarms or communication systems to notify all facility personnel. This individual must also notify appropriate state or local agencies if their assistance is needed. Whenever there is a release, fire, or explosion, the emergency coordinator must immediately identify the character, exact source, amount, and aerial extent of any released materials. This identification may be carried out by observation or review of facility records and by chemical analysis.

Concurrently, the emergency coordinator must assess possible hazards to human health or the environment that may result from the hazardous event. If evacuation of local areas is advisable, the coordinator must immediately notify appropriate local authorities. The coordinator must also notify either the government official designated as the on-scene coordinator for that area or the National Response Center (24-h toll free number 800/424-8802).

The report must include:

- Name and address of facility
- Time and type of incident (e.g., release, fire)
- Name and quantity of material(s) involved, to the extent known
- The extent of injuries, if any
- The possible hazards to human health or the environment, outside the facility

If the facility must cease operations in response to an emergency, the emergency coordinator must monitor for leaks, pressure buildup, gas generation, or ruptures in valves, pipes, or other equipment. Immediately after such an emergency, the coordinator must provide for the treatment, storage and disposal of recovered waste, contaminated soil or surface water, or any other material that results from such an event.

14.6 SUBPART E — MANIFEST SYSTEM, RECORD KEEPING, AND REPORTING

If a TSDF receives hazardous waste accompanied by a manifest, the facility must sign and date each copy of the manifest to certify that the hazardous waste covered by the manifest was received. It must also note any significant discrepancies in the manifest directly on the manifest. The facility then provides the transporter with at least one copy of the signed manifest. Within 30 days after the delivery, the TSDF must forward a copy of the manifest to the generator. It must also retain at the facility a copy of each manifest for at least 3 years from the date of delivery.

14.6.1 MANIFEST DISCREPANCIES

Manifest discrepancies are differences between the quantity or type of hazardous waste designated on the manifest or shipping paper, and the quantity or type of hazardous waste a facility actually receives. Significant discrepancies in quantity are (40 CFR Part 264.72):

- For bulk waste, variations greater than 10% by weight.
- For batch waste, any variation in piece count, such as a discrepancy of one drum in a truckload. Significant discrepancies in type are obvious differences that can be discovered by inspection or waste analysis.

Upon discovering a significant discrepancy, the TSDF must attempt to reconcile the discrepancy with the waste generator or transporter (e.g., via telephone contact). If the discrepancy is not resolved within 15 days after receiving the waste, the TSDF must immediately submit to the state regulatory agency a letter describing the discrepancy and attempts to reconcile it, and a copy of the manifest in question.

14.6.2 OPERATING RECORD

The facility must maintain written records of regular operations. The following information must be recorded and maintained in the operating record:

- A description and the quantity of each hazardous waste received and the method(s) and date(s) of its treatment, storage, or disposal at the facility.

- The location of all hazardous wastes within the facility and the quantity at each location. For disposal facilities, the location and quantity of each hazardous waste must be recorded on a map of each disposal area.
- Records and results of waste analyses and waste determinations.
- Summary reports and details of all incidents that require implementing the contingency plan.
- Records and results of inspections.
- Monitoring, testing or analytical data, and corrective action where required.
- Records of the quantities (and date of placement) for each shipment of hazardous waste placed in land disposal units.

All records and plans must be available for inspection by a state regulatory inspector or U.S. EPA representative.

14.6.3 BIENNIAL REPORT

The facility must submit a biennial report to the regulatory agency by March 1 of each even-numbered year. The biennial report must be submitted on EPA form 8700-13B. The report must cover facility activities during the previous calendar year and must include:

- For off-site facilities, the EPA identification number of each hazardous waste generator from which the facility received a hazardous waste during the year; for imported shipments, the name and address of the foreign generator.
- A description and the quantity of each hazardous waste the facility received during the year.
- The method of treatment, storage and disposal for each hazardous waste.
- The most recent closure cost estimate and, for disposal facilities, the most recent postclosure cost estimate.
- For generators who treat, store, or dispose of hazardous waste on-site, a description of the efforts undertaken during the year to reduce the volume and toxicity of waste generated.
- A description of the changes in volume and toxicity of waste actually achieved during the year in comparison with previous years.

14.7 SUBPART F — RELEASES FROM SOLID WASTE MANAGEMENT UNITS

Facilities must conduct a hazardous waste monitoring and response program. Whenever hazardous constituents from a regulated unit are detected at a compliance point, the facility must institute a compliance monitoring program. Also, whenever the groundwater protection standard (40 CFR Part 264.92; see below) is exceeded, the facility must institute a corrective action program. The state regulatory agency specifies in the facility permit the specific components of the monitoring and response program.

14.7.1 GROUNDWATER PROTECTION STANDARD

The facility must ensure that hazardous constituents detected in the uppermost aquifer underlying the waste management area do not exceed specified concentration limits. The state agency establishes this groundwater protection standard in the facility permit when hazardous constituents have been detected in groundwater. The regulatory agency specifies concentration limits in the groundwater for hazardous constituents. The concentration of a hazardous constituent must not exceed the background level in the groundwater, or it must not exceed the value given in Table 14.3.

TABLE 14.3
Maximum Concentration of Constituents for Groundwater Protection

Constituent	Maximum Concentration (mg/L)
Arsenic	0.05
Barium	1.0
Cadmium	0.01
Chromium	0.05
Lead	0.05
Mercury	0.002
Selenium	0.01
Silver	0.05
Endrin (1,2,3,4,10,10-hexachloro-1,7-epoxy 1,4,4a,5,6,7,8,9a-octahydro-1, 4-endo, endo-5, 8-dimethano naphthalene)	0.002
Lindane (1,2,3,4,5,6-hexachlorocyclohexane, gamma isomer)	0.004
Methoxychlor (1,1,1-Trichloro-2, 2-*bis* (*p*-methoxyphenylethane)	0.1
Toxaphene, (Technical chlorinated camphene, 67–69%chlorine)	0.005
2,4-D (2,4-Dichlorophenoxyacetic acid)	0.1
2,4,5-TP Silvex (2,4,5-Trichlorophenoxypropionic acid)	0.01

14.7.2 GROUNDWATER MONITORING REQUIREMENTS

The facility must comply with the following requirements for any groundwater monitoring program:

- The groundwater monitoring system must consist of a sufficient number of wells, installed at appropriate locations and depths to yield groundwater samples from the uppermost aquifer.
- Represent the quality of groundwater passing the point of compliance.
- Allow for the detection of contamination when hazardous waste has migrated to the uppermost aquifer.

All monitoring wells must be cased, and the casing must be screened or perforated and packed with gravel or sand, where necessary, to enable the collection of groundwater samples. The annular space (i.e., the space between the bore hole and well casing) above the sampling depth must be sealed to prevent contamination of samples and the groundwater. At a minimum, the program must include procedures and techniques for sample collection, sample preservation and shipment, analytical procedures, and chain of custody control.

In detection monitoring (or in compliance monitoring), data on each hazardous constituent specified in the permit are collected from background wells and wells at the compliance points. The facility must determine an appropriate sampling procedure for each hazardous constituent listed in the permit.

14.7.3 DETECTION MONITORING PROGRAM

A facility which establishes a detection monitoring program must monitor for indicator parameters (e.g., specific conductance, total organic carbon, or total organic halogen), waste constituents, or reaction products that indicate the presence of hazardous constituents in groundwater. The regulatory agency specifies the parameters to be monitored in the facility permit after considering factors such as the types, quantities, and concentrations of constituents in wastes managed at the regulated unit; the mobility, stability, and persistence of waste constituents or their reaction products in the unsaturated zone beneath the waste management area; and the detectability of indicator parameters and waste constituents in groundwater. The facility must also install a groundwater monitoring system at the point of compliance (40 CFR Part 264.98).

The state agency will specify the frequencies for collecting samples and conducting statistical tests to determine whether there is statistically significant evidence of contamination for any hazardous constituent. A sequence of at least four samples from each well (background and compliance wells) must be collected at least semiannually during detection monitoring.

If significant evidence of contamination is detected at any monitoring well at the compliance point, the facility must notify the state regulatory agency within 7 days. The TSD must also immediately sample all monitoring wells and determine whether constituents are present and in what concentration. Within 90 days, the TSDF must submit to the state regulatory agency an application for a permit modification to establish a compliance monitoring program.

14.7.4 CORRECTIVE ACTION PROGRAM

A TSDF required to establish a corrective action program has numerous responsibilities; for example, it must take corrective action to ensure that regulated units are in compliance with the groundwater protection standard under §264.92. The state regulatory agency will specify the groundwater protection standard in the facility permit, including a list of the hazardous constituents identified, concentration limits for each of the hazardous constituents, the compliance point, and the compliance period.

The facility must implement a corrective action program that prevents hazardous constituents from exceeding concentration limits at the compliance point by removing the constituents or treating them in place. The permit will specify the specific measures to be taken.

14.8 SUBPART G — CLOSURE AND POSTCLOSURE

14.8.1 CLOSURE PLAN (264.112)

The TSDF must prepare a written closure plan. The plan must be submitted with the permit application and approved by the state regulatory agency as part of the permit issuance procedure. The approved closure plan will become a condition of any RCRA permit.

The plan must identify steps necessary to perform partial or final closure of the facility at any point during its active life. The closure plan must include, at least:

- A description of how each hazardous waste management unit (e.g., landfill cell) at the facility will be closed.
- An estimate of the maximum inventory of hazardous wastes on-site over the active life of the facility and the methods to be used during closure, including methods for removing, transporting, treating, storing, or disposing of all hazardous wastes, and identification of the type(s) of the off-site hazardous waste management units to be used.
- A detailed description of the steps needed to remove or decontaminate all hazardous waste residues and contaminated containment system components, equipment, structures, and soils during partial and final closure, including procedures for cleaning equipment and removing contaminated soils.
- Methods for sampling and testing surrounding soils, and criteria for determining the extent of decontamination.

- Other activities necessary, including groundwater monitoring, leachate collection, and run-on and runoff control.
- A schedule for closure of each hazardous waste management unit and for final closure of the facility. The schedule must include the total time required to close each hazardous waste management unit.

The facility must notify the state regulatory agency in writing at least 60 days prior to the date on which it will begin closure of a surface impoundment, waste pile, land treatment or landfill unit, or final closure of a facility with such a unit. The facility must notify the state agency in writing at least 45 days prior to the date of beginning final closure of a facility with only treatment or storage tanks, container storage, or incinerator units, industrial furnaces or boilers.

REFERENCES

Code of Federal Regulations, Volume 40 Part 264, *Standards for Owners and Operators of Hazardous Waste Treatment, Storage, and Disposal Facilities*, U.S. Government Printing Office, Washington, DC.

U.S. Environmental Protection Agency, National Analysis, The National Biennial RCRA Hazardous Waste Report (Based on 1999 Data), EPA530-R-01-009, Solid Waste and Emergency Response PB2001-106313 (5305W), June 2001.

U.S. Environmental Protection Agency, Test Methods for Evaluating Solid Waste, EPA Publication SW-846, Washington, DC, 1986.

SUGGESTED READING AND WEB SITES

Environment, Health and Safety Online, The Hazardous Waste Permitting Process, 2003. http://www.ehso.com/ehshome/hazwaste_permitting.htm

Environment, Health and Safety Online, No date, RCRA Hazardous Waste Treatment Storage & Disposal Facilities.See: http://www.ehso.com/tsdfs.htm

Oregon Department of Environmental Quality, No date, Procedure and Criteria for Hazardous Waste Treatment, Storage or Disposal Permits. See:http://www.deq.state.or.us/wmc/hw/factsheets/HazardousWastePermits.pdf

Oregon Department of Environmental Quality, No date, RCRA Treatment, Storage, and Disposal Facility Permits. See: http://www.deq.state.or.us/pubs/permithandbook/hwrcra.htm

New York State Department of Environmental Conservation, No date, Hazardous Waste Program. See: http://www.dec.state.ny.us/website/dshm/hzwst.htm

U.S. Department of Energy, Location Standards for RCRA Treatment, Storage, and Disposal Facilities (TSDFs), EH-231-040/1093, October 1993, See: http://tis.eh.doe.gov/oepa/guidance/rcra/locate.pdf

U.S. Environmental Protection Agency, Treat, Store, and Dispose of Waste, 2002. See: http://www.epa.gov/epaoswer/osw/tsd.htm#facts

U.S. Environmental Protection Agency, Currently Operating Mixed Waste TSDFs, 2002. See: http://www.epa.gov/radiation/mixed-waste/mw_pg11a.htm

U.S. Environmental Protection Agency, RCRA, Superfund & EPCRA Call Center Training Module, EPA530-K-02-021I, Solid Waste and Emergency Response (5305W), Washington, DC, October 2001.

U.S. Environmental Protection Agency, The State of Federal Facilities, An Overview of Environmental Compliance at Federal Facilities, FY 1997-98. 300-R-00-002, Enforcement and Compliance, Assurance, Washington, DC, January 2000.

U.S. Environmental Protection Agency, RCRA Organic Air Emission Standards for TSDFs and Generators, EPA530-F-98-011, Solid Waste and Emergency Response, Washington, DC, July 1998.

U.S. Environmental Protection Agency, Protocol for Conducting Environmental Compliance Audits of Treatment, Storage and Disposal Facilities under the Resource Conservation and Recovery Act, EPA-305-B-98-006, Enforcement and Compliance Assurance (2224A), Washington, DC, December 1998.

U.S. Environmental Protection Agency, *Waste Analysis at Facilities that Generate, Treat, Store, and Dispose of Hazardous Wastes, A Guidance Manual.* OSWER 9938.4-03, Solid Waste and Emergency Response (OS-520), April 1994.

QUESTIONS

1. Which permitted TSDFs are required to obtain an EPA identification number? To whom is EPA form 8700-12, Notification of Hazardous Waste Activity to be submitted? How frequently is the form to be sent to this agency?
2. Is a WAP required for all TSDFs? Which agency must approve the WAP as part of the permit application?
3. Provide the specific components of a facility WAP.
4. At the Hi-Jinx Corporation Central Waste Incineration facility, how often should the following areas be inspected by plant personnel: (a) loading areas; (b) tank systems; (c) incinerator units; (d) container storage buildings.
5. For what types of TSDFs is a CQA program required? Under whose direction is the program developed and implemented? Which physical components must the CQA program address ?
6. The TSDF must develop and implement a written CQA plan. What is the overall purpose of the CQA plan? What must it include?
7. TSDFs must be equipped with internal communications systems and fire control equipment. What additional emergency equipment is required?
8. In preparation for any site emergency, the facility must make numerous arrangements with local authorities. List these arrangements.
9. Whenever there is a release, fire, or explosion, what specifics regarding the released materials is the emergency coordinator to identify immediately? How is the identification to be carried out? How are the findings to be reported?
10. What types of equipment must the TSDF be equipped with in order to minimize the possibility of a fire, explosion, or release of hazardous wastes which could threaten human health or the environment?
11. What is the general purpose of a contingency plan? What are the major components of the plan? To whom in the community should copies of the plan be given ?
12. TSDF contingency plans should be revised as needed. Under what conditions are contingency plans often revised?
13. The TSDF is the final stage of cradle-to-grave hazardous waste management. What are the responsibilities of the TSDF regarding the Uniform Hazardous Waste Manifest? For how long is the manifest to be kept on the premises? How are manifest discrepancies handled?
14. The groundwater monitoring program at a TSDF must include consistent sampling and analysis procedures that ensure monitoring results that provide a reliable indication of groundwater quality below the waste management area. List the minimum procedures and techniques to be included in the program.
15. In your community is there a comprehensive household hazardous waste management program currently in place? How are household hazardous wastes collected, treated and disposed? What agency or company, is responsible for administering such a program?

A.14.1 APPENDIX: HAZARDOUS WASTE MANAGEMENT SCENARIOS

A.14.1.1 INTRODUCTION

After reading each scenario, discuss what, if any, violations may have occurred. How would these violations be best addressed (i.e., via changes in engineering design, a modified storage or disposal program, some use of common sense, etc.)? These are open-ended situations.

A.14.1.2 THE SCENARIOS

1. The ABC Lead Company processes off-specification and waste automobile batteries for eventual recycling. At the facility the batteries are crushed, and the lead plates and groups are

removed and smelted to recover and purify the lead. The plastic battery cases are shredded to approximately 1 × 1 cm (0.5 × 0.5 in.) and washed with a dilute acid to remove lead. These plastic chips are shipped by truck to a battery manufacturing facility in another state to be used as feedstock for new battery housing. Upon arrival at the receiving facility, the chips are washed a second time.

An inspector claims that the plastic chips are not completely cleaned of lead and are still wet. At least one truck, filled with plastic chips, is leaking a reddish liquid, probably indicating corrosion of the truck's interior (Figure A.14.1).

The company claims that the plastic chips are cleaned of lead and are therefore not hazardous waste. They furthermore claim that the chips cannot be considered waste, as they are *feedstock* for the manufacture of new batteries.

2. At a metal foundry, baghouse dusts are enriched with cadmium. They clearly fail the TCLP test (175 mg Cd/L, whereas the RCRA limit is 1 mg/L (CFR Part 261.24)). In order to avoid the substantially higher costs associated with hazardous waste disposal, the company decides to apply a proprietary fixative agent to their process before the materials enter the furnace. The baghouse dust which exits the furnace has an average Cd value just below the TCLP limit. The inspector suggests that this procedure is treatment of a hazardous waste. The company, however, argues that the fixative compound was added to *product*, not to waste; therefore, they are not engaging in treatment of a hazardous waste.

3. Storage area for bulk hazardous wastes, Haz-R-Dus Chemical Company. What is wrong with this picture (Figure A.14.2)?

4. Bogus Pesticides, Inc. manufactures and stores a range of pesticides and fumigants. A waste hauler collects solid waste from their facility in a conventional loader truck. The driver immediately notices a strong and unpleasant odor while driving and quickly returns to the solid waste transfer station. Upon arrival he was instructed to tip the wastes immediately upon the tipping floor. By this time, the waste was smoldering and began to burn.

The fire department was called in and attempted to extinguish the small blaze. Upon contact with water, the waste pile reacted violently, spewing flames and releasing unknown gas. Eventually, the emergency response team was called in to handle the incident.

State inspectors visited Bogus Pesticides, where it was determined that containers of aluminum phosphide pesticide were placed in ordinary trash. This was the material that subsequently decomposed

FIGURE A.14.1 Truck filled with plastic chips from battery casings. Photo by Theresa M. Pichtel.

in transit and during wetting by the fire department. An inspector was informed by the plant manager that this formulation is water-reactive. The reaction with water is as follows:

$$AlP + 3H_2O \rightarrow Al(OH)_3 + PH_{3(g)}$$

The product, phosphine gas (PH_3), is highly toxic.

The inspector walked over the facility to find several aluminum phosphide containers stored in a large shed with a poor roof (water was dripping inside the building from melting snow).

The company manager made arrangements to repack the containers and transport them to a friend's farm field. There, it was planned, the fumigants would be reacted with water and allowed to decompose. The remaining residue would be landfilled. The manager contends that such a practice is acceptable as the fumigant is a *product* and not a *waste*.

(a)

(b)

FIGURE A.14.2 Storage area for bulk hazardous wastes. Photo by Theresa M. Pichtel.

A.14.1.3 Responses to Scenarios

1. (a) If the chips could be used in exactly the same form as when they were removed from the original battery casings, they are not necessarily a "waste". But, once they are processed they become a waste. In the current situation, the chips must be washed (i.e., "processed, treated") at least twice.

    (b) If the chips are indeed contaminated with lead (and, they were subsequently found to be contaminated), the chips are also hazardous waste (The Mixture Rule). The transporter must therefore possess an EPA identification number; the shipments must also be manifested (Uniform Hazardous Waste Manifest) and the chips must be sent to an EPA-approved treatment, storage, and disposal facility.

    The chips can be easily tested for the characteristic of toxicity using the TCLP. If the lead concentration in a representative extract exceeds 5 mg/L lead, the chips have failed the toxicity test.

2. "Treatment" is defined in 40 CFR 270.2 as:

    Any method, technique, or process, including neutralization, designed to change the physical, chemical, or biological character or composition of any hazardous waste so as to neutralize such wastes, or so as to recover energy or material resources from the waste, or so as to render such waste as non-hazardous, or less hazardous; safer to transport, store, or dispose of; or amenable for recovery, amenable for storage, or reduced in volume.

    Given that the new baghouse dust measures "just below" the TCLP limit, a number of samples were required by the inspector for TCLP determination. Chemists within the agency carefully assessed all data. As of this writing, this argument continues. The U.S. EPA is also being asked for guidance to settle this matter.

3. Drums are not labeled (they require the yellow "Hazardous Waste" label); several drums are open. Additionally, drums should be covered or stored indoors to limit the effects of weather on the containers.

4. Given that the fumigants were stored for long periods, the containers were in very poor condition and the inspector required immediate removal, these materials are indeed a solid waste. Furthermore, aluminum phosphide is a P006 hazardous waste (40 CFR Part 268.40).

    This waste is covered under the EPA Land Ban (see Chapter 17) so it is not to be land-disposed.

    If the fumigants could be used immediately, for example to fumigate several warehouses, the inspectors may have been willing to allow this; however, there was little practical use for pesticides and fumigants in the middle of winter.

    Given that this is a hazardous waste, there are the obvious DOT and RCRA requirements for transportation (proper transporters, labeling, packaging, etc.). Furthermore, taking this material into a field and reacting it is a form of treatment, and Bogus Pesticides would need a permit to do so. "Treatment" in the farmer's field would have been impractical; at temperatures below 49°F, it would take at least 14 days for aluminum phosphide to decompose. Given winter temperatures averaging 20 to 25°F, such decomposition will last much longer. According to 40 CFR 268.50, the appropriate treatment would be chemical oxidation, chemical reduction, or controlled incineration.

    As of this writing, Bogus Pesticides has been cited for illegal disposal and the case is being sent to the State Office of Enforcement.

15 Incineration of Hazardous Wastes

Troops harnessed in bright armor marched three times
In parade formation, and the cavalry
Swept about the sad cremation flame
Three times, while calling out their desolate cries.

Virgil (70–19 BCE)
The Pyres

15.1 INTRODUCTION

Since the enactment of the Resource Conservation and Recovery Act (RCRA) in 1976, incineration technologies for the destruction of solid, liquid, and gaseous hazardous wastes have become increasingly effective. Aerobic thermal processes detoxify a wide range of organic compounds, such as chlorinated pesticides, munitions wastes, nerve gas, polymer residues, and many other petrochemical wastes. Incineration can be employed for the destruction of contaminated soil and water; thus, the technology is not limited strictly to the treatment of organic residuals from a single production process. Furthermore, hazardous waste destruction under RCRA is not limited solely to dedicated incineration facilities; for example, thermal destruction of hazardous wastes is permitted in industrial boilers and furnaces, with the resultant recovery of heat.

As was the case for the combustion of municipal solid waste (Chapter 9), incineration of hazardous waste is defined as the *controlled* burning of a substance, where "controlled" refers to clearly defined temperature ranges, oxygen input, turbulence, atmospheric pressure, firebox design, and other aspects of the combustion environment. The regulatory definition of an incinerator is (40 CFR Part 260.10):

> any enclosed device that uses controlled flame combustion and does not meet the criteria for classification as a boiler, sludge dryer, carbon regeneration unit, or industrial furnace.

Typical incinerators include rotary kilns, liquid injectors, controlled air incinerators, and fluidized-bed incinerators. The definition of incinerator also includes the infrared incinerator or plasma arc incinerator. An infrared incinerator is a device that uses electric-powered resistance as a source of heat. A plasma arc incinerator uses a high-intensity electrical discharge as a source of heat (40 CFR Part 260.10). These two incinerator types will not be discussed here.

Other waste incineration devices include boilers and industrial furnaces. A boiler is composed of two primary components: the combustion chamber used to heat the hazardous waste and collection tubes that hold a fluid (usually water) used to produce energy (e.g., steam). Industrial furnaces are enclosed units installed within a manufacturing facility and use thermal treatment to recover materials or energy from hazardous waste. The following devices fulfill the definition of an industrial furnace (U.S. EPA, 2002):

- Cement kiln
- Aggregate kiln

- Coke oven
- Smelting, melting, and refining furnace
- Methane reforming furnace
- Pulping liquor recovery furnace
- Lime kiln
- Phosphate kiln
- Blast furnace
- Titanium dioxide chloride process oxidation reactor
- Halogen acid furnace

15.2 COMBUSTION AND ITS RESIDUES

For incineration to be an effective method of eliminating the hazardous properties of a waste, combustion must be complete. Three critical factors ensure the completeness of combustion in an incinerator: (1) temperature of the combustion chamber, (2) length of time wastes are maintained at high temperatures, and (3) turbulence, or degree of mixing, of the wastes and air. These parameters are often labeled "The Three Ts of Combustion." In each incinerator permit, the operating conditions are clearly stipulated (see below) in order to ensure that all of these factors are optimized, ultimately assuring complete combustion of the waste feed.

During a controlled burn, wastes are fed continuously or in batch mode into the incinerator combustion chamber (firebox). As the wastes are heated, they are physically converted from solids and liquids into gases. These gases, mostly organic, become sufficiently hot so that chemical bonds break. The atoms that are released combine with oxygen and hydrogen to form stable gases, primarily carbon dioxide and water, which are subsequently released from the system via the flue. In reality, however, the combustion of organic substances is a rather complex sequence of reactions that results in simple products. The combustion of ethane, a simple alkane, is as follows:

$$2C_2H_6 + 7O_2 \rightarrow 4CO_2 + 6H_2O \tag{15.1}$$

Aromatic hydrocarbons are combusted in a similar fashion, as demonstrated by the reaction for xylene:

$$C_6H_4(CH_3)_2 + 11.5O_2 \rightarrow 8CO_2 + 5H_2O \tag{15.2}$$

Incineration of halogenated hydrocarbons will result in the formation of the corresponding halogen acids, which must be treated within the flue gas prior to release. An example of combustion of a chlorinated hydrocarbon is the reaction of dichloroethane:

$$2C_2H_4Cl_2 + 5O_2 \rightarrow 4CO_2 + 2H_2O + 4HCl \tag{15.3}$$

Depending on waste composition, various quantities of sulfur oxides, nitrogen oxides, and other gases are formed. Also, if combustion is not complete, compounds such as elemental carbon (C), polychlorinated biphenyls (PCBs), benzopyrenes, and others may be emitted. This latter group is collectively referred to as products of incomplete combustion (PICs). RCRA regulations place strict limits on acceptable amounts of selected pollutants released from the flue (40 CFR Part 264).

Another significant product of waste combustion is ash, an inert solid material composed primarily of salts, metals, and some carbon. During combustion, a large proportion of the ash is collected by gravity at the base of the combustion chamber (i.e., "bottom ash"). When this ash is removed from its hopper, it may be considered hazardous waste via the "Derived-From" Rule or because it exhibits one of the four hazard characteristics (ignitable, corrosive, reactive, and toxic) (see Chapter 11). A significant fraction of the ash may be very lightweight and become entrained with gases as particulate matter. These particles, collectively referred to as "fly ash," are collected in the pollution control devices in accordance with RCRA regulations.

As a hazardous waste management practice, incineration has several unique attributes. If conducted under optimum conditions, controlled incineration permanently destroys toxic organic compounds within waste by converting them into stable molecules. Second, incineration reduces the volume of hazardous waste. Land disposal of ash, as opposed to the disposal of untreated hazardous waste, is therefore safer and should reduce long-term liability to the waste generator. Incineration, however, does not destroy inorganic compounds within the waste such as metals. In fact, the residue becomes more concentrated with the various nonburnables after the organic component has been destroyed. Ash from incinerators is subject to applicable RCRA standards and may need to be treated for metals or other inorganic constituents prior to land disposal.

15.3 OVERVIEW OF REGULATORY REQUIREMENTS

Emissions from hazardous waste combustors are regulated under two statutory authorities, RCRA and the Clean Air Act (CAA). Relevant RCRA regulations include 40 CFR Part 264, Subpart O and Part 265, Subpart O for incinerators, and 40 CFR Part 266, Subpart H for boilers and industrial furnaces (BIFs). RCRA requirements for these units are provided in 40 CFR Part 270. All of these units are subject to the general treatment, storage, and disposal facility (TSDF) standards under RCRA. Hazardous waste incinerators and hazardous waste burning cement kilns and lightweight aggregate kilns are also subject to the CAA maximum achievable control technology (MACT) emission standards. The MACT standards set emission limitations for polychlorinated debenzodioxins (PCDDs), polychlorinated dibenzofurans (PCDFs), metals, particulate matter, total chlorine, hydrocarbons, carbon monoxide, and destruction and removal efficiency (DRE) for organic emissions. The combustion standards under RCRA and the MACT standards under the CAA are discussed below.

15.4 COMBUSTION STANDARDS UNDER RCRA, SUBPART O

To minimize potential harmful effects of incinerator gaseous emissions, the U.S. EPA has developed performance standards to regulate four pollutant categories: organics, hydrogen chloride and chlorine gas, particulate matter, and metals (40 CFR Part 264.343). Boilers and most industrial furnaces are assigned performance standards. For each category or type of emission, the regulations establish compliance methods and alternatives.

The Subpart O standards apply to facilities that destroy hazardous wastes, and regulate the emissions from incinerator combustion processes. An incinerator burning hazardous waste must be designed, constructed, and operated so that the performance standards outlined below are met. Specifically, the RCRA regulations restrict gaseous emissions of organic compounds, hydrogen chloride (HCl), particulate matter, fugitive emissions, and metals. All hazardous waste incinerators must conform to the requirements of Subpart O unless the waste is considered "low risk" (40 CFR Part 264.340). These include certain listed hazardous wastes (Subpart D) or those characterized as hazardous due to ignitability or corrosivity. Incinerators in existence from May 19, 1980 were allowed to continue burning hazardous waste if the units complied with the Part 265, Subpart O interim status standards. However, on November 8, 1989, the interim status was terminated for all existing hazardous waste incinerators unless the facility submitted a Part B permit application by November 8, 1986 (§270.73(f)). Due to this deadline, there are very few incinerators presently operating under interim status (U.S. EPA, 2000).

15.4.1 ORGANIC COMPOUNDS

To obtain an operating permit, an incineration facility must demonstrate that the emission levels for selected hazardous organic constituents are within applicable limits. The main indicator of incinerator performance designated by the U.S. EPA is Destruction Removal Efficiency, DRE. An incinerator burning hazardous waste must achieve a DRE of 99.99% for each principal organic hazardous constituent (POHC) designated in the waste stream. The DRE is determined from the following equation:

$$DRE = (W_{in} - W_{out})/W_{in} \times 100\%$$ (15.4)

where W_{in} is the mass feed rate of one POHC into the incinerator and W_{out} the emission rate of that same POHC in the exhaust.

An incinerator burning the listed hazardous wastes FO20, FO21, FO22, FO23, FO26, or FO27 must achieve a DRE of 99.9999% for each. These are chlorinated hydrocarbon wastes that have the potential to contain PCDDs (40 CFR Part 264.343); therefore, combustion conditions must be more rigorous.

For many waste generators and TSDFs, it would be impractical and very costly to monitor DRE results for every organic constituent contained within the waste stream. In response to this reality, only certain POHCs are selected for monitoring and are designated in the permit. POHCs are selected due to their high concentrations in the waste feed and in the difficulty of their destruction. In other words, constituents are more likely to be designated as POHCs if they are present in large concentrations in the waste. Similarly, organic constituents that are the most difficult to destroy by incineration are most likely to be designated as POHCs. If the incinerator achieves the required DRE for the selected POHCs, regulatory agencies conclude that the incinerator should achieve the same or better DRE for other, more easily combustible, organic compounds in the waste stream.

15.4.2 HYDROGEN CHLORIDE HCl

HCl, an acidic gas, forms when chlorinated organic compounds in wastes are burned. An incinerator burning hazardous waste cannot emit more than 1.8 kg of HCl per hour or more than 1% of the total HCl in the stack gas prior to entering any pollution control equipment, whichever is larger (40 CFR Part 264.343(b)).

Boilers and most industrial furnaces must follow a tiered system for the regulation of both HCl and chlorine gas (U.S. EPA, 2002). The facility determines the allowable feed or emission rate of total chlorine by selecting one of three approaches (tiers). Each tier differs in the amount of monitoring, and in some cases, air dispersion modeling (i.e., modeling the pathways through which air pollutants may travel) that the facility is required to conduct (Figure 15.1).

Each facility can select any of the three tiers. Factors that a facility may consider in selecting a tier include the physical characteristics of the facility and the local environs, the anticipated waste composition and feed rates, and the resources available for conducting the analysis. The main distinction between the tiers is the point of compliance, i.e., the point at which the facility must ensure that chlorine concentrations will be below EPA's acceptable exposure levels. The facility must determine if the cost of monitoring and modeling is worth the benefit of combusting waste with a higher concentration of chlorine (U.S. EPA, 2002).

15.4.3 PARTICULATE MATTER

Particulate matter is composed of minute particles, solid or liquid, organic or inorganic, which are carried along with the combustion gases to the incinerator flue. Particulates are of regulatory concern because they occur in many sizes, some of which are readily inhaled and transported deep within the lungs. At the same time, some are composed of hazardous constituents or possess hazardous coatings, for example, heavy metals that volatilize and subsequently condense on a particle exterior. These effects are discussed in Chapter 9.

The Subpart O requirements control metal emissions through the performance standard for particulates, since metals are often contained within or attached to particulate matter. An incinerator burning hazardous waste must not emit particulate matter in excess of 180 mg/dscm (milligrams per

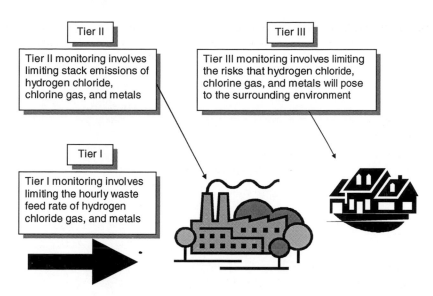

FIGURE 15.1 Performance standards for hydrogen chloride, chlorine gas, and metals (U.S. EPA, 2002).

dry standard cubic meter) (0.08 grains per dry standard cubic foot) according to the formula (40 CFR Part 264)

$$P_c = P_m \times 14 / (21 - Y) \tag{15.5}$$

where P_c is the corrected concentration of particulate matter, P_m the measured concentration of particulate matter, and Y the oxygen concentration in stack gas.

EXAMPLE 15.1

A waste mixture consisting of benzene, xylene, and chlorophenol is being incinerated. The incinerator temperature is 1075°C and the stack gas flow rate is 410.82 dscm/min. Waste feed rate is given in the table below. Determine if the unit is in compliance for each compound.

Compound	Inlet (kg/h)	Outlet (kg/h)
Benzene (C_6H_6)	245	0.015
Xylene (C_8H_{10})	442	0.061
Chlorobenzene (C_6H_5OCl)	235	0.149
HCl	—	1.1
Particulates at 8.5% O_2	—	2.775

SOLUTION

(a) Calculate the DRE for each of the POHCs:

$$DRE = (W_{in} - W_{out}) / W_{in} \times 100\%$$

Benzene $\quad DRE = (245 - 0.015) / 245 \times 100 = 99.9939\%$

Xylene $\quad DRE = (442 - 0.041) / 442 \times 100 = 99.9907\%$

Chlorobenzene $\quad DRE = (235 - 0.129) / 235 \times 100 = 99.945\%$

The DRE limit for chlorobenzene does not meet the regulatory requirement for DRE.
(b) HCl emissions. The HCl emissions shown in the table do not exceed the federal limit of 1.8 kg/h.
(c) Particulates. The outlet loading of the particulates is calculated by dividing the outlet mass rate by the stack flow rate:

$$W_{out} = [(2.775 \text{ kg/h})(10^6 \text{ mg/kg})] / [(410.82 \text{ dscm/min})(60 \text{ min/h})] = 112 \text{ mg/dscm}$$

Since the particulate concentration was measured at 8.5% oxygen, a correction factor is required:

$$P_c = P_m \times 14/(21 - Y)$$
$$= 112 \text{ mg/dscm} \times 14/(21 - 8.5)$$
$$= 125 \text{ mg/dscm}$$

This value is below the standard of 180 mg/dscm and is therefore in compliance regarding particulate release.

15.4.4 FUGITIVE EMISSIONS

Operating conditions regulated under Subpart O are also formulated to control fugitive emissions, i.e., gases that escape from the combustion chamber and do not enter pollution control devices. An example of fugitive emissions is a gas that escapes through the inlet opening of the combustion chamber. Fugitive emissions are controlled by keeping the combustion zone completely sealed, or by maintaining the combustion zone pressure lower than atmospheric pressure so that air is drawn out of the firebox and into the pollution control device.

15.4.5 METALS

For RCRA combustion units, both carcinogenic and noncarcinogenic metals are regulated under the same type of tiered system as for chlorine. The facility determines an appropriate tier for each regulated metal and must assure that the facility meets these feed rate and emission standards. A different tier may be selected for each metal pollutant (Figure 15.2) (U.S. EPA, 2002).

15.4.6 WASTE ANALYSIS

During operation, the facility must conduct sufficient waste analyses to verify that the waste feed is within the physical and chemical composition limits specified in the permit. This analysis may include a determination of a waste's heat value, viscosity, and content of hazardous constituents, including POHCs. Waste analysis is one component of the trial burn permit application (see below). The U.S. EPA stresses the importance of proper waste analysis to ensure compliance with emission limits.

15.4.7 OPERATING CONDITIONS AND THE RCRA PERMIT

Regulatory agencies must clearly delineate the operating conditions for hazardous waste incinerators to ensure compliance with the performance standards for organics, HCl, particulate matter, and fugitive emissions. The details of an incinerator permit are based upon results from trial burns of hazardous wastes (see below). The permit specifies the operating conditions that have been shown to meet these performance standards.

A RCRA permit for a hazardous waste incinerator sets operating conditions and allowable ranges for certain critical parameters and also requires continuous monitoring of these parameters.

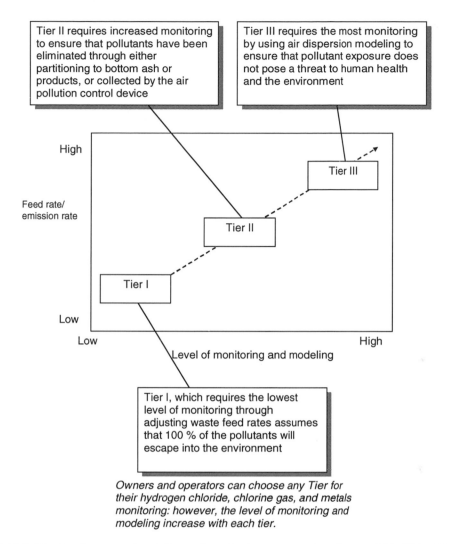

Tier II requires increased monitoring to ensure that pollutants have been eliminated through either partitioning to bottom ash or products, or collected by the air pollution control device

Tier III requires the most monitoring by using air dispersion modeling to ensure that pollutant exposure does not pose a threat to human health and the environment

Tier I, which requires the lowest level of monitoring through adjusting waste feed rates assumes that 100 % of the pollutants will escape into the environment

Owners and operators can choose any Tier for their hydrogen chloride, chlorine gas, and metals monitoring: however, the level of monitoring and modeling increase with each tier.

FIGURE 15.2 The tiered system of modeling and monitoring gaseous air pollutants (U.S. EPA, 2002).

Operation within this range ensures that combustion is performed in the most protective manner and that the performance standards are achieved. These parameters may include (U.S. EPA, 2000, 2002):

- Maximum allowable carbon monoxide levels in stack emissions
- Allowable temperature ranges
- Maximum waste feed rates
- Combustion gas velocity
- Control of the firing system
- Allowable variations of system design and operating procedures

In addition, during the startup and shutdown of an incinerator, hazardous waste must not be charged (i.e., fed into the unit) unless it is operating according to the conditions specified in the permit. The incinerator facility must stop operations when changes in waste feed, incinerator design, or operating conditions exceed any limits designated in the permit.

15.4.8 OBTAINING THE RCRA PERMIT

A facility planning to operate a new hazardous waste incinerator is required to obtain a RCRA permit before constructing the unit. The purpose of the permit is to allow the facility to establish incineration conditions that will ensure the protection of public health and the environment. The incinerator permit covers aspects of pretrial burn, trial burn, posttrial burn, and final operating conditions.

The pretrial burn phase of the permit allows the incinerator to initiate all parameters for conducting the trial burn. This may include setting charging rates, fuel and airflow rates, and installing air monitoring equipment. During the trial burn, the incinerator is prepared for operation. Operating conditions are monitored and adjusted, and gaseous emissions are measured. Test conditions are based on the operating conditions specifically indicated by the permit application. The U.S. EPA establishes conditions in the permit necessary to conduct the trial burn; in other words, the burn should represent the expected incinerator operation. Physical and chemical analysis of the waste feed is also a necessary component of the trial burn permit application. During operation, the waste must be analyzed to verify that its composition is within the limits specified in the permit. This analysis may include a determination of the content of hazardous constituents including POHCs and the heat value of the waste.

The posttrial burn period is devoted to completing the sampling, analysis and calculation of trial burn results, and the submission of the results to the U.S. EPA. During this period the EPA evaluates all data compiled during the trial burn. After reviewing the results, the EPA may modify the permit conditions again to ensure compliance with incinerator standards and the protection of health and the environment.

15.4.9 MONITORING AND INSPECTIONS

In order to ensure safe operations in compliance with all permit specifications, the operator must perform the following during the routine incineration of hazardous waste (U.S. EPA, 2000):

- Monitor waste feed rate, combustion temperature, and combustion gas velocity on a continuous basis
- Monitor carbon monoxide emissions on a continuous basis at some point downstream of the combustion zone and prior to release to the atmosphere
- Sample and analyze the waste and exhaust emissions to verify compliance with permit operating requirements
- Conduct daily visual inspections of the incinerator and associated equipment (e.g., pumps, valves, and conveyors)
- Test the emergency waste feed cutoff system and associated alarms at least once weekly (40 CFR Part 264.347)

15.4.10 MANAGEMENT OF RESIDUES

According to the Derived-From Rule, if an incinerator burns a listed hazardous waste, the ash generated is also considered a listed waste. The Derived-From Rule states that any solid waste generated from the treatment, storage or disposal of a listed hazardous waste, including any sludge, spill residue, ash, emission control dust, or leachate, remains a hazardous waste. The operator is also required to determine whether the ash exhibits any characteristics of a hazardous waste. If a facility incinerates a characteristic hazardous waste (ignitable, corrosive, reactive, and toxic), the operator must determine whether the ash exhibits any of the four characteristics, using the procedures outlined in 40 CFR Part 261. If an ash sample fails the test (i.e., exhibits a characteristic), it must be managed as a hazardous waste.

15.4.11 CLOSURE

At the time of closure, the facility operator must remove all hazardous waste and hazardous residues from the incinerator site. In addition, if the residues are hazardous waste, the operator becomes a generator of hazardous waste and must manage these residues in compliance with 40 CFR Parts 262 through 266.

15.5 MACT STANDARDS UNDER THE CAA

Hazardous waste incinerators, cement kilns, and certain aggregate kilns must also comply with other emission limitations. The MACT emission standards are established within the CAA regulations. Instead of using specific operating requirements to ensure that the unit meets performance standards, the combustion facilities subject to MACT standards are permitted to use a specific pollution control technology to achieve the stringent emission limits.

15.5.1 ORGANICS

To control the emission of organics, combustion units must comply with DRE requirements similar to the RCRA requirements for hazardous waste combustion units. MACT combustion units must select POHCs and demonstrate a DRE of 99.99% for each POHC in the hazardous waste stream. Facilities that combust F020–F023 or F026–F027 hazardous waste are required to achieve a DRE of 99.9999% for each designated POHC. Additionally, for PCDDs and PCDFs, the U.S. EPA has promulgated more stringent standards under MACT. For example, MACT incinerators and cement kilns that burn waste containing PCDDs or PCDFs must not exceed an emission limit of either 0.2 ng of toxicity equivalence per dry standard cubic meter (TEQ/dscm) or 0.4 ng TEQ/dscm at the inlet to the particulate control device. This unit of measure is based on a method for assessing risks associated with exposures to PCDDs and PCDFs (U.S. EPA, 2002).

15.5.2 HYDROGEN CHLORIDE AND CHLORINE GAS

MACT combustion units do not use a tiered system to control HCl and chlorine gas emissions; rather, facilities must ensure that the total chlorine emission does not exceed specific limits; for example, the emission limit of total chlorine for a new incinerator is 21 ppmv. The facility may achieve this emissions level by limiting or controlling the amount of chlorine-containing waste entering the incinerator (U.S. EPA, 2002).

15.5.3 PARTICULATE MATTER

The EPA developed rather stringent standards for the control of particulate matter in order to limit emissions of certain metals. For example, a new aggregate kiln cannot exceed an emission limit of 57 mg/dscm of particulate matter.

15.5.4 METALS

Hazardous waste incinerators, cement kilns, and aggregate kilns are not required to utilize the tiered approach to control the release of toxic metals into the atmosphere. The MACT rule finalized numerical emission standards for three categories of metals: mercury, low-volatile metals (arsenic, beryllium, and chromium), and semivolatile metals (lead and cadmium). Combustion units must meet emission standards for the amount of metals emitted. For example, a new cement kiln must meet an emission limit of 120 µg/dscm for mercury, 54 µg/dscm for the low-volatile metals, and 180 µg/dscm for the semivolatile metals (U.S. EPA, 2002).

15.5.5 OPERATING REQUIREMENTS

To ensure that a MACT combustion unit does not exceed MACT emission standards, the unit must operate under parameters demonstrated in a comprehensive performance test (CPT). Operating parameters such as temperature, pressure, and rate of waste feed are then established, based on the results of the CPT. Continuous monitoring systems are used to monitor the operating parameters. The facility may also use an advanced type of monitoring known as continuous emissions monitoring systems (CEMS). CEMS directly measure the pollutants that exit the stack at all times.

15.5.6 ADDITIONAL REQUIREMENTS

Because hazardous waste combustion units are a type of TSDF, they are also subject to the general TSDF standards as discussed in 40 CFR Part 264, in addition to the above combustion unit performance standards and operating requirements.

15.6 INCINERATION DEVICES

The majority of hazardous wastes occur as liquids, either as hydrocarbon or aqueous mixtures. Under RCRA, land disposal of liquid hazardous wastes has been banned; additionally, many technologies are being promoted for the chemical treatment of liquid hazardous wastes in order to render them nonhazardous (Chapter 16). Many waste streams, however, are not suitable for chemical treatment due to their inherently hazardous nature. Incineration has thus been promoted as an appropriate technology for their destruction. Some wastes are designated "hazardous" solely based on the characteristic of ignitability. Incineration can therefore serve as a means of energy generation from their destruction.

The physical form of the waste and its content of solid residues will determine the ideal type of combustion chamber. Table 15.1 provides some of the general considerations for selection of a

TABLE 15.1
Applicability of Major Incinerator Types to Physical Form of Waste

	Liquid Injection	Rotary Kiln	Fixed Hearth
Solids			
Granular, homogeneous		X	X
Irregular, bulky (pallets etc.)		X	X
Low melting point (tars etc.)	X	X	X
Organic compounds with fusible ash constituents		X	
Unprepared, large, bulky material		X	
Gases			
Organic vapor laden	X	X	X
Liquids			
High organic strength aqueous wastes	X	X	
Organic liquids	X	X	
Solids or liquids and			
Waste contains halogenated aromatic compounds (2200°F minimum)	X	X	
Aqueous organic sludge		X	

Source: Oppelt, E.T., *J. Air Pollut. Control Assoc.*, 37, 558–586, 1987. Reproduced with kind permission of the Air and Waste Management Association.

combustion chamber. The major subsystems that may occur in a hazardous waste incinerator are (Oppelt, 1987):

- Waste preparation and feeding
- Combustion chamber(s)
- Air pollution control
- Ash handling and disposal

The typical orientation of these subsystems appears in Figure 15.3. The selection of the particular system combination is a function of several variables, including the physical and chemical properties of the waste, regulatory requirements for atmospheric emissions, capital cost, and public acceptance.

15.6.1 LIQUID INJECTION

The liquid injection incinerator (Figure 15.4) is a stationary system consisting of one or more refractory-lined combustion chambers operating under high temperatures and equipped with a series of atomizing nozzles. The major units marketed are horizontally and vertically fired. The liquid injection incinerator is currently the most commonly used incinerator type for hazardous waste destruction. It is in daily use throughout the United States both at industrial facilities and at dedicated hazardous waste treatment facilities (Freeman et al., 1987).

From a combustion standpoint, liquid wastes are classified as either combustible or partly combustible. The first category includes materials having sufficient calorific value (approx. 17,900 kJ/kg [8000 Btu/lb] or higher) to support combustion in a conventional firebox. Below this value the material cannot maintain a flame. The waste often contains a high percentage of noncombustible components including water and the addition of auxiliary fuel may be necessary.

As the name of the technology implies, wastes are acceptable in a liquid injection incinerator as long as they exist as either pumpable liquids or slurries. A conventional liquid or gaseous fuel

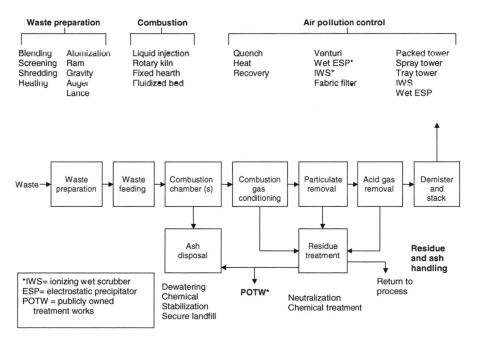

FIGURE 15.3 Schematic showing the orientation of incinerator subsystems and process component options. (From Oppelt, E.T., *J. Air Pollut. Control Assoc.*, 37, 558–586, 1987. Reproduced with kind permission of the Air and Waste Management Association.)

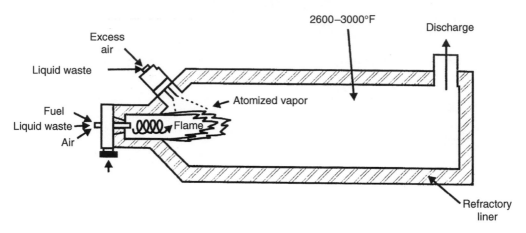

FIGURE 15.4 A liquid injection incinerator. (From Oppelt, E.T., *J. Air Pollut. Control Assoc.,* 37, 558–586, 1987. Reproduced with kind permission of the Air and Waste Management Association.)

FIGURE 15.5 Hazardous waste storage tanks. These tanks can store over 350,000 gal of hazardous waste.

(e.g., propane) preheats the system to an equilibrium temperature of approximately 815°C (1500°F) before the introduction of the waste. Liquid waste is then transferred from storage drums to a feed tank (Figure 15.5). Blending, which may be used to lower waste chlorine content or improve the pumpability or combustibility of the waste, occurs in the tank. The tank may be pressurized with nitrogen or another inert gas, and waste is fed into the incinerator using a remote valve. After waste transfer, the fuel line is purged with nitrogen to eliminate any explosion hazard.

In the combustion chamber substances will react (combust) more readily when they possess a higher surface area (e.g., finely divided in the form of a spray). Thus, atomizing nozzles are employed to inject waste liquids (Figure 15.6). Within the kiln, wastes are typically injected downstream of the fuel nozzle. If the waste possesses sufficient heat content (approx. 13,400 kJ/kg or 6000 Btu/lb), however, it can be injected directly into the fuel envelope. These wastes are said to burn *autogenously* (i.e., without the need for supplemental fuel).

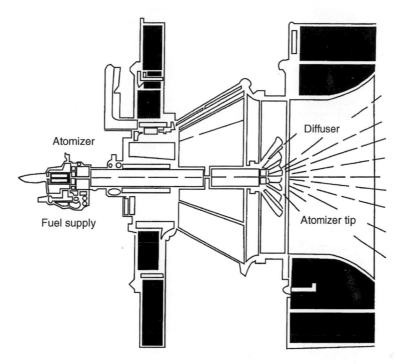

FIGURE 15.6 A mechanical atomizing burner, one of many possible nozzle designs for a liquid injection incinerator (From Brunner, C.D., *Hazardous Waste Incineration,* McGraw-Hill, New York, NY, 1993. Reproduced with kind permission of the McGraw-Hill Companies, Inc.)

Due to the utilization of nozzles there is a limit as to how viscous a waste can be for incineration. The higher the viscosity, the more difficult it is to pump, inject through a nozzle, and ultimately combust. A substance can be pumped if its viscosity is less than 10,000 Saybolt-seconds universal (SSU) (Brunner, 1993). For conventional nozzles, the viscosity should be less than 750 SSU for adequate atomization to occur. Atomization is also strongly affected by nozzle type. An ideal droplet size ranges between 40 and 100 μm, which is attained using gas–fluid nozzles and high-pressure air or steam (Wentz, 1995).

15.6.2 AIR REQUIREMENTS IN LIQUID INJECTORS

Whenever an organic material is to be incinerated, sufficient oxygen is necessary to complete combustion. Oxygen is provided through a supply of air. Air is required for several purposes, including (Brunner, 1993):

- Primary air supply, to promote combustion of the waste stream
- Secondary air supply, injected downstream of the burner
- Atomization, to promote vaporization and efficient burning

Primary combustion air is the airflow supplied at the fuel burner to combust the main fuel source. Air is supplied through a burner register, a fan-shaped unit surrounding the burner nozzle, which creates a circular motion in the airflow. The register is either fixed or adjustable. Secondary air is necessary for combustion of the waste feed and is normally introduced into the firebox downstream of the main flame. In liquid injection furnaces, the secondary air supply is often used to shape the flame and to divert the flame away from the walls. The secondary air creates turbulence within the furnace and provides a relatively cool flow on the refractory furnace surfaces, keeping them cooler than the center of the furnace. The primary and secondary airflows are also utilized to assist in fuel atomization and prevent unburned materials from contacting furnace linings (Brunner, 1993).

As the droplets vaporize and combust, any inorganics present in the waste remain in the gas stream and are carried to the air pollution control equipment. The sizes of the particles are a function of the size distribution of the original droplets. Larger droplets usually result in larger particles and vice versa. Smaller particles are less likely to be captured by air pollution control equipment and are thus more likely to escape to the atmosphere. Particle size depends on other variables, including the form of the waste burned (i.e., solid, liquid, gas) and factors specific to the combustion device (e.g., temperature, turbulence, and air flow). Only small quantities of bottom ash typically occur in the liquid injection incinerator.

As for all incinerator types, the firebox temperature, waste residence time, and overall turbulence are adjusted to optimize destruction efficiencies. The liquid injector operates within a range of temperatures, depending on the waste type. Typical combustion chamber residence times and temperature ranges are 0.5 to 2 s and 700 to 1650°C (1300 to 3000°F), respectively. Feed rates measure up to 5000 to 6000 L/h of organic wastes.

15.6.3 OVERVIEW OF LIQUID INJECTION

The type, size, and shape of a furnace are a function of waste characteristics, burner design, air distribution, and furnace wall design. The furnace can be simple in design as in a vertical, refractory-lined chamber or it can be relatively complex, involving the preheating of combustion air and the firing of multiple fuels. Liquid injection systems are capable of burning virtually any combustible waste that can be pumped. They are usually designed to burn specific waste streams and consequently are not used for multipurpose facilities. Liquid injection facilities routinely destroy a variety of wastes including phenols, PCBs, solvents, polymer wastes, herbicides, and pesticides.

The advantages in using a liquid injection incinerator include (Freeman et al., 1987):

- Fewer moving parts results in less downtime and less maintenance
- Capability to incinerate a wide range of wastes
- Low maintenance costs due to the few moving parts in the system

Disadvantages include:

- Only capable of combusting pumpable liquids and slurries
- Feed nozzles tend to clog resulting in downtime

15.6.4 ROTARY KILNS

The key component of the rotary kiln incinerator (Figure 15.7) consists of a refractory-lined rotating cylinder mounted at a slight incline from ground level. Wastes in the form of liquids, slurries, or bulk solids are fed into the entry port(s) (Figure 15.8) and agitated under elevated temperatures for a predetermined length of time depending on waste and kiln characteristics. Waste liquids may be pumped in through a nozzle, thus atomizing the charge. A screw feed mechanism may be used to inject slurries, and bulk solids may enter by a ram feed or similar mechanical system. Many rotary kilns are charged discretely; often, entire drums are fed into a kiln in a single charge (Figure 15.9). Such charging results in a cyclical temperature distribution inside the kiln.

The waste is expected to burn to ash by the time it reaches the kiln exit. Kiln rotation speed varies in the range of ¾ to 4 r/min (Brunner, 1993). Most kilns possess a smooth inner surface; however, some are equipped with internal baffles to promote turbulence of the feed. The residence time for solid wastes in the kiln is at least 30 min and is based on the rotational speed of the kiln and its angle (Wentz, 1995).

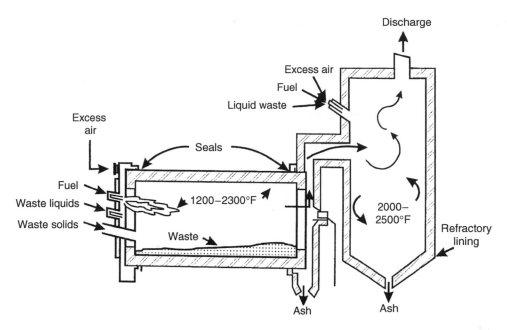

FIGURE 15.7 Rotary kiln incinerator. (From Oppelt, E.T., *J. Air Pollut. Control Assoc.*, 37, 558–586, 1987. Reproduced with kind permission of the Air and Waste Management Association.)

FIGURE 15.8 Injection ports for the introduction of liquid hazardous wastes to a rotary kiln.

A long residence time is preferred because the solids bed in the kiln is not thermally uniform. The solids retention time may be several hours, and can be estimated from (U.S. WPCF, 1988).

$$\theta = 0.19L/(NDS) \qquad (15.6)$$

where θ is the retention time (min), L the kiln length (m), N the kiln rotational velocity (r/min), D the kiln diameter (m), and S the kiln slope (m/m).

The coefficient 0.19 is based on empirical data.

(a)

(b)

FIGURE 15.9 Mechanical feeding of containerized hazardous wastes into a kiln.

EXAMPLE 15.2

Determine the waste retention time for a rotary kiln operating at 0.9 r/min with a kiln length of 5 m, a diameter of 2.2 m, and a slope of 0.1 m/m.

SOLUTION

$\theta = 0.19L/(NDS)$

$\theta = 0.19 \times 5/(0.9 \times 2.2 \times 0.1)$

$\theta = 4.8$ min

If the desired retention time is actually to be 7.5 min, what should the rotational velocity be adjusted to?

SOLUTION

Rearrange the equation to solve for r/min:

$N = 0.19L/(\theta\ DS)$

$N = 0.19 \times 5/(7.5 \times 2.2 \times 0.1)$

$\quad = 0.58$ r/min

The gas retention time for 99.99% destruction of a compound is given by (Kiely, 1996)

$$\ln t_g = (\ln 9.21/A) + (E/RT) \qquad\qquad (15.7)$$

where A is the Arrhenius preexponent frequency (s^{-1}), E the energy of activation (J/kg mol), R the universal gas constant $= 8314$ J/kg mol, and T the absolute temperature (K).

The variables A and E are typically known for a specific compound.

A source of heat is required to bring the system to operating temperatures and to maintain the desired combustion temperature. Supplemental fuel (e.g., natural gas) is injected into the kiln through a nozzle. Excess air is also provided to promote combustion. A negative pressure (i.e., suction) is applied to the kiln via an induced draft fan to remove particulate matter and noncombusted vapors.

There are two modes of kiln gas flow. In the co-current mode, the burner is installed at the entrance to the kiln, and the gas flow is in the same direction as waste flow. In the counter-current mode, the burner is placed near the kiln exit, and the gas flow is opposite to the direction of waste flow. A counter-current system has been demonstrated to be effective for the combustion of aqueous wastes. The gases exiting the kiln serve to dry the incoming aqueous waste. The entrance region is sufficiently hot for the heating and combustion of wastes entering the kiln.

Organic gases and particulates may be drawn into a second, stationary chamber, labeled an afterburner. Here temperatures are higher compared with the kiln and permit a more complete destruction of any remaining organic particles or vapors. The gases discharged from the afterburner are subsequently directed to an air pollution control system such as an electrostatic precipitator, baghouse, or scrubber.

Rotary kilns experience a high entrainment of particulate matter. Entrainment occurs because solids are continuously tumbling within the kiln and are reintroduced to the gas stream. Particle size ranges are similar for both rotary kiln incinerators and liquid injectors. In rotary kilns, entrained particles tend to be larger than 10 μm.

In some cases, it may be desirable to operate a kiln in the so-called "slagging" mode. At temperatures of approximately 1090 to 1200°C (2000–2200°F), ash will liquefy. The ash fusion temperature is a function of the composition of the waste as well as incinerator conditions (e.g., oxygen concentration). When the ash is in molten form, salt-laden wastes and metal drums are more easily incorporated into the kiln. The production of particulates is also minimized. However, the temperature range must be maintained in a higher range in a slagging kiln, resulting in higher energy costs and possible accelerated wear of components; temperatures may average 1425 to 1540°C (2600–2800°F), compared with less than 2000°F in a nonslagging kiln. The construction of a slagging kiln is more complex than for a nonslagging kiln. Finally, maintenance tends to be more frequent with a slagging kiln (Brunner, 1993).

There are several sites within the kiln in which leakage of gases may occur. Critical points include the inlet ports and the kiln seals. Efficient kiln operation requires a limited introduction of unwanted airflow into the system. During a phenomenon known as "puffing," the introduction of a volatile organic waste will result in the instantaneous production of gases with consequent expansion and a very rapid pressure increase at the inlet end of the kiln. This pressure may be sufficient to weaken kiln seals. In order to limit such a pressure increase the kiln atmosphere is maintained with a negative pressure draft.

15.6.5 OVERVIEW OF ROTARY KILN INCINERATION

The rotary kiln is one of the more popular types of incineration systems for hazardous wastes. These devices operate under a wide range of conditions, handle a wide variety of waste types, and generate a range of different emissions. There is no single temperature that is characteristic of a rotary kiln.

The advantages of the rotary kiln include (Brunner, 1993):

- Applicability for a number of waste types (liquids, slurries, sludges, and bulk solids)
- High turbulence provides for thorough mixing of the waste charge
- Minimal preprocessing of waste
- Many types of feed mechanisms available
- Readily controlled waste residence time in the kiln

Disadvantages include:

- High initial capital costs
- Significant costs for maintenance
- Separate afterburner required for destruction of volatile components
- Damage to kiln linings due to abrasion from solids such as drums
- Damage to the rotary seals
- High particulate carryover into the afterburner
- Conditions along the length of the kiln are difficult to control
- Ash production may be significant

15.6.6 BOILERS AND INDUSTRIAL FURNACES

As mentioned above, a boiler is composed of the combustion chamber used to heat the hazardous waste and tubes that hold a fluid (usually water) used to produce energy (e.g., steam). The regulatory definition of boiler requires that these two components be situated close to one another to ensure effective energy recovery. In addition, the unit must export or use the majority of the recovered energy for some beneficial purpose. Industrial furnaces are enclosed units installed within a manufacturing facility and use thermal treatment to recover materials or energy from hazardous waste. These units may use hazardous waste as a fuel to heat raw materials to make a commodity (e.g., a cement kiln manufacturing cement) or the unit may recover materials from the hazardous waste (e.g., a smelter facility which recovers silver or lead).

Not all units that meet the definition of boiler or industrial furnace are subject to the 40 CFR Part 266, Subpart H, BIF standards. Each individual unit is evaluated against a list of possible exemptions from the BIF requirements. For several reasons (e.g., to avoid duplicate regulation), the U.S. EPA exempted the following units from the BIF regulations (U.S. EPA, 2002):

- Units burning used oil for energy recovery
- Units burning gas recovered from hazardous or solid waste landfills for energy recovery
- Units burning hazardous wastes that are exempted from RCRA regulation, such as household hazardous wastes
- Units burning hazardous waste produced by a conditionally exempt small quantity generator
- Coke ovens burning only K087 decanter tank tar sludge from coking operations
- Certain units engaged in precious metals recovery
- Certain smelting, melting, and refining furnaces processing hazardous waste solely for metals recovery

15.6.7 INDUSTRIAL BOILERS

Many industrial and commercial facilities are equipped with boilers, fired by coal, number 2 heating oil, or natural gas. Boilers are high-temperature furnaces that generate heat energy from combustion. Heat is transferred by means of either a boiler adjacent to the firebox or via tubes lining the combustion chamber. The heat from the combustion gas flowing on the outside of the firebox is transferred to the water within the boiler or waterwall tubes.

Wastes designated as hazardous may be combusted in industrial boilers provided that the wastes are hazardous solely based on the characteristic of ignitability. Wastes combusted in this manner usually occur as liquids that are generated on-site. Examples include aliphatic and aromatic solvents, alcohols, and other highly volatile hydrocarbons. The U.S. EPA has required field tests of operating facilities in destroying hazardous wastes in standard boilers. The tested boilers achieved performance ratings close to 99.99% DRE.

During day-to-day operations the interior of a boiler becomes dirty due to the accumulation of particulate matter on the surface of the boiler or waterwall tubes. Such coatings result in reduced heat transfer. To address the accumulation of particles, a boiler periodically blows high-velocity air or steam into the unit to scour surfaces. This process is known as soot blowing and is an important consideration when designing a trial burn. During soot blowing, a combination of previously deposited metals, soot, and particulate matter is released. This pulse of particulates enters the air pollution control system. Because of such particulate surges, part of the trial burn must be conducted under soot blowing conditions (U.S. EPA, 1992).

The advantage of the disposal of hazardous wastes in a boiler is a reduction in cost to the waste generator compared with on- or off-site incineration. The facility obtains a fuel value from the waste, and cost savings are accrued from not having to dispose of the waste in an RCRA regulated process. Also, the waste does not have to be transported to a disposal facility. One disadvantage of hazardous waste incineration in boilers is that the process is not closely regulated and may be subject to accidents or misuse.

15.6.8 Cement Kilns

In a cement kiln, combustion conditions are more severe than those present in many waste incinerators. In the process of manufacturing cement, limestone and other additives are exposed to temperatures of 1375 to 1540°C (2500 to 2800°F) in a large rotary kiln heated with fossil fuels (Figure 15.10

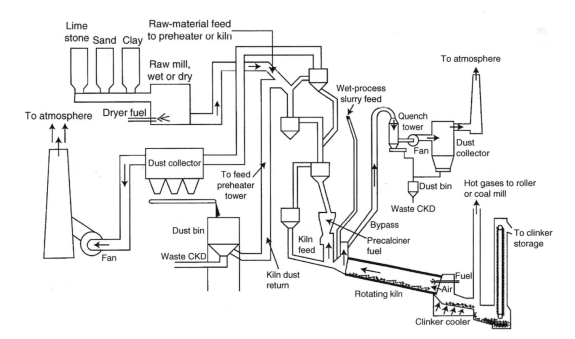

FIGURE 15.10 Cement kiln schematic. (From Chadbourne, J.F., in *Standard Handbook of Hazardous Waste Treatment and Disposal,* Freeman, H.M., Ed., McGraw-Hill, New York, NY, 1998. Reproduced with kind permission of the McGraw-Hill Companies, Inc.)

FIGURE 15.11 Dual cement kilns, each measuring 450 ft in length with a 1 in. steel shell lined with 9 to 12 in. of firebrick. This facility handles 25 million gal of hazardous waste per year.

and Figure 15.11). The gas temperature may be several hundred degrees higher. The end product of this process, and the main component of cement, is a solid material called clinker (U.S. EPA, 1993).

Cement kilns are considered to be a promising disposal option for many organic wastes. They are especially promising for the destruction of chlorinated wastes since virtually all such compounds will decompose to their component elements in this temperature range. Furthermore, the HCl produced neutralizes the clinker production process, which is normally alkaline. In test burns carried out in the United States, Canada, and Sweden, recalcitrant wastes such as PCBs have been successfully combusted in cement kilns (Mournighan, 1985).

The amount of metal in the raw materials of cement kilns can exceed that in both the waste and the fuel. Therefore, the metals in the raw materials must be considered and monitored adequately to control metals emissions. In addition, the cement matrix affects the volatility of the metals within the kiln. Cement contains a mixture of compounds that forms complexes with metals, with a resultant decrease in their volatility (U.S. EPA, 1993).

A practical and important aspect of cement kilns is their ability to recycle particulate matter. Volatile metals are also recycled as they vaporize, condense, and return to the system. As with the rotary kiln described above, cement kilns use counter-current processes in which the fuel and the air are introduced from one end, while raw materials enter the opposite end. As the hot burning fuel and air pass through the kiln, heat is transferred to the input materials. In this zone, the volatile metals vaporize, but as they are entrained with the airflow, they cool and condense onto the surface of existing particles. Any metals that escape into the gas stream are carried to a baghouse or electrostatic precipitator, which collects the majority of the particulate matter.

Advantages of cement kilns for hazardous waste incineration include:

- Destruction of organic wastes, including a number of chlorinated and recalcitrant wastes
- Reclamation of energy value of the waste
- Capacity to remove large quantities of waste

Disadvantage includes:

- A large amount of particulates are produced, requiring extensive pollution control.

15.7 AIR POLLUTION AND ITS CONTROL

A major regulatory concern associated with hazardous waste incineration (and a major justification for the NIMBY [Not in My Backyard] syndrome) is the emission of air pollutants. For a simple hydrocarbon compound (e.g., methane, propane), the primary end products from combustion, given adequate amounts of air, are carbon dioxide, water vapor, and heat. When hazardous wastes are incinerated, however, gaseous wastes posing a threat to public health or the environment often result. The types and amounts of emissions from hazardous waste incineration depend upon a number of variables including the chemical composition of the waste, waste incineration rate, incinerator type, incinerator operating parameters, and air pollution control equipment.

15.7.1 ATMOSPHERIC PRODUCTS FROM COMBUSTION

The greatest mass of air contaminants consists primarily of particulate matter and oxides of sulfur and nitrogen. Acid gases such as HCl, HBr, and HF may be produced in significant quantities, depending on feedstock. Trace levels of various other oxides, hydrocarbons (including chlorinated hydrocarbons), and heavy metals are also generated. Particulate matter consists of metal salts from the waste, metal oxides formed by combustion, fragments of incompletely burned material (primarily carbonaceous), and condensed gaseous contaminants (i.e., droplets). The metals and toxic by-products condense as or on fine particles as the exhaust gas stream cools.

Sulfur oxides, mostly as sulfur dioxide (SO_2), but also including small amounts of sulfur trioxide (SO_3), originate from sulfur or sulfur compounds present in the waste or fuel mixture. Nitrogen oxides (NO_x) originate from nitrogen in the combustion air or from organic nitrogen compounds present in the waste. HCl and chlorine (Cl_2) are derived from the incineration of chlorinated hydrocarbons such as polyvinyl chloride. Phosphorus pentoxide and phosphoric acid are formed from the incineration of organophosphorus compounds such as malathion or parathion.

15.7.2 PRODUCTS OF INCOMPLETE COMBUSTION (PICs)

Even in a well-designed incinerator, the firebox may contain areas of incomplete oxygen incorporation or other, similar, quench zones. At temperatures common to hazardous waste incinerators, hydrocarbons will not oxidize in these zones but will decompose pyrolytically, forming PICs. The primary PICs include carbon monoxide, carbon soot, hydrocarbons, organic acids, polycyclic organic matter, and any other waste constituents that escape complete thermal destruction in the incinerator. In well-designed and well-operated incinerators, these incomplete combustion products are emitted only in trace amounts.

In a study by the U.S. EPA (1986), combustion by-products were examined from 23 emissions tests at thermal destruction facilities, including eight incinerators, nine industrial boilers, and six industrial kilns. The organic emissions were compared with emissions from facilities burning coal only, and with municipal solid wastes (MSW) incinerators. A total of 28 volatile and 27 semivolatile compounds were detected in stack emissions. The compounds were emitted at rates that spanned five orders of magnitude, i.e., 0.09 to 13,000 ng/kJ of combustor heat input. Emission rates for 12 compounds emitted from the three sources are shown in Table 15.2.

Metals occurring in hazardous wastes being combusted are usually either collected as bottom ash or are emitted as particulate matter. In some cases, however, some of the more volatile elements (e.g., Hg, Cd, and Se) are emitted as vapors. Emission rates for a number of metals from hazardous waste combustion appear in Table 15.3.

TABLE 15.2
Emission Rates ng/kJ[a] of Specific Compounds for Incinerators, Boilers, and Kilns

	Incinerators		Boilers		Kilns	
	Mean	**Range**	**Mean**	**Range**	**Mean**	**Range**
Benzene	87	2–980	30	0–300	580	290–1,000
Toluene	1.6	1.5–4.1	280	0–1,200	No data	
Carbon tetrachloride	0.8	0.3–1.5	1.8	0–7.2	No data	
Chloroform	3.8	0.5–8.4	120	0–1,700	No data	
Methylene chloride	2.2	0–9.6	180	0–5,800	No data	
Trichloroethylene	1.2	2.3–9.1	1.2	0–13	1.3	0.7–2.8
Tetrachloroethane	0.3	0–1.3	63	0–780	No data	
1,1,1-Trichloroethane	0.3	0–1.3	7.5	0–66	2.4	(One value)
Chlorobenzene	1.2	1–6.0	63	0–1,000	152	33–270
Naphthalene	44	0.7–150	0.6	0.3–2.1	No data	
Phenol	7.8	0–16	0.3	0–0.8	0.02	0–0.05
Diethylphthalate	3.7	2.8–4.8	0.4	0.04–1.6	No data	

[a]Expressed as ng of emission per kJ of combustor heat input (1 ng/kJ = 2.34×10^{-6} lb/MM Btu).

Source: Oppelt, E.T., J. *Air Pollut. Control Assoc.*, 37, 558–586, 1987. Reproduced with kind permission of the Air and Waste Management Association.

TABLE 15.3
Average Stack Emissions of Metals from Five Hazardous Waste Incinerators

		Emission Rate (g/kJ)				
Metals	**Plant A**	**Plant B**		**Plant C**	**Plant D**	**Plant E**
		Uncontrolled	**Controlled**			
Sb	0.32	0.26	BDL	BDL	—	—
As	—	BDL	BDL	BDL	—	a
Be	0.052	0.19	.011	BDL	0.056	0.050
Be	—	5	BDL	BDL	—	a
Cd	0.055	0.11	0.019	BDL	0.012	0.36
Cr	0.14	0.73	0.19	2.5	a	0.094
Pb	5.4	2.3	0.64	BDL	0.24	9.0
Ni	0.024	0.50	0.087	2.7	0.052	a
Se	—	7.0	0.45	BDL	0.29	—
Ag	0.0008	BDL	BDL	0.33	0.0076	—
Ti	0.0089	BDL	BDL	BDL	—	—

[a]Some values were below detection limits; thus average not calculated.

Adapted from Oppelt, E.T., *J. Air Pollut. Control Assoc.*, 37, 558–586, 1987. (Reproduced with kind permission of the Air and Waste Management Association).

BDL = all values below detection limit.

EXAMPLE 15.3

A waste mixture of 40% xylene, 35% toluene, 23% *n*-pentane, and 2% water is to be combusted in a liquid injection incinerator at a rate of 550 kg/h. There is 18% excess air in the combustion chamber.

Properties of the waste constituents are as follows:

Compound	Chemical Formula	Molecular Weight(MW)	Heat Content (kJ/kg)
Xylene	$C_6H_4(CH_3)_2$	106.16	42,989
Toluene	$C_6H_5CH_3$	92.13	42,527
Pentane	C_5H_{12}	72.14	49,142
Water	H_2O	18.01	0

1. Calculate the total heat release in the incinerator.
2. Calculate the percent by volume of each component in the flow gas.

SOLUTION

1. Heat release in the incinerator:

   Xylene heat release $= 0.40 \times 42,989 = 17,196$ kJ/kg

   Toluene heat release $= 0.35 \times 42,527 = 14,884$ kJ/kg

   Pentane heat release $= 0.23 \times 49,142 = 11,303$ kJ/kg

   Water heat release $= 0$ kJ/kg

   Heat release/kg of mixture   43,382 kJ/kg

   Total heat release in the incinerator

   550 kg/h $\times$ 43,382 kJ/h = 23,860,100 kJ/h

2. Calculate the percent by volume of each component in the flow gas:

   (a) Xylene: $C_8H_{10} + 11.5O_2 \rightarrow 8CO_2 + 5H_2O$

   Mass of xylene $= 0.40 \times 550$ kg/h $= 220$ kg/h

   220/106 = 2.08 mol/h of xylene

Component	MW	mol/h
Xylene	106	2.08
O_2	32	23.92
CO_2	44	16.64
H_2O	18	10.4

   $0.18 \times 23.92 = 4.31$ mol/h of O_2

   $23.92 + 4.31 = 28.32$ mol/h of O_2

   Note that air is 79% N and 21% O_2, so 79/21 $\times$ 28.23 = 106.20 mol N_2

   (b) Toluene: $C_7H_8 + 9O_2 \rightarrow 7CO_2 + 4H_2O$

   Mass of toluene $= 0.35 \times 550$ kg/h $= 192.5$ kg/h

   192.5/92 = 2.09 mol/h of toluene

Component	MW	mol/h
Toluene	92	2.09
O_2	32	18.81
CO_2	44	14.63
H_2O	18	8.36

Given 18% excess air,

$0.18 \times 18.81 = 3.39$ mol/h of O_2 in addition to that calculated above.

$18.81 + 3.39 = 22.2$ mol/h of O_2

Note that air is 79% N and 21% O_2, so

$79/21 \times 22.2 = 79.72$ mol N_2

(c) Pentane: $C_5H_{12} + 8O_2 \rightarrow 5CO_2 + 6H_2O$

Mass of pentane $= 0.23 \times 550$ kg/h $= 126.5$ kg/h

$126.5/72.14 = 1.75$ mol/h of pentane

Component	MW	mol/h
Pentane	72	1.75
O_2	32	14.00
CO_2	44	77.00
H_2O	18	31.50

Given 18% excess air,

$0.18 \times 14.00 = 2.52$ mol/h of O_2 in addition to that calculated above.

$14.00 + 2.52 = 16.52$ mol/h O_2

Note that air is 79% N_2 and 21% O_2, so

$79/21 \times 16.52 = 62.15$ mol N_2

Beyond the above data for hydrocarbon combustion, there is also 2% water in the waste.

This amounts to 0.02×550 kg/h $= 11$ kg/h.

$11/18 = 0.61$ mol H_2O.

Add the moles of each component generated in the flue gas to determine the total moles.

	CO_2	H_2O	O_2	N_2
Xylene	16.64	10.4	23.92	106.20
Toluene	14.63	8.36	18.81	79.72
Pentane	77.00	31.50	14.00	62.15
Water		0.61		
Total	108.27	50.87	56.73	248.07

Total moles $= 463.94$.

Given that the mol% is equivalent to the vol.%, the flue gas contains the following:

23.34% CO_2

10.97% H_2O

12.23% O_2

53.47% N_2

15.7.3 AIR POLLUTION CONTROL

Four classes of air pollution equipment are commonly employed for particulate control in hazardous waste incinerators: electrostatic precipitators, venturi scrubbers, ionizing wet scrubbers, and baghouses. The removal of acid gases is accomplished using technologies identical to those for MSW incinerators (see Chapter 9). For example, wet scrubbers and packed-tower absorbers are highly effective for the condensation and removal of HCl and SO_x.

Most hazardous waste incinerator facilities employ one of three possible schemes for overall air pollution control:

- Venturi scrubber (for particulates) followed by a packed tower absorber (for gases)
- Ionizing wet scrubber (for particulates) combined with a packed tower absorber (for gases)
- Dry scrubber (for particulates) followed by a baghouse or an electrostatic precipitator (for particulates).

EXAMPLE 15.4

A hazardous waste incinerator is operating for the destruction of a mixed nonchlorinated solvent waste. The flue gas is passed through a lime (CaO) slurry in a dry scrubber where acid gases are partially neutralized and the gases cooled. The gases then pass through a baghouse for particulate removal and are released via the flue.

The flue gas contains 410 kg/h of SO_2 and 325 kg/h of HCl. The dry scrubber lime feed rate is 1.2 × stoichiometric rate, and it is 75% efficient in removing SO_2 and 88% efficient in HCl removal.

1. Calculate the lime feed rate in kg/h.
2. Determine how many kg/h of SO_2 and HCl will remain in the flue gas following the dry scrubbing process.

   Assume that CO_2 in the flue gas does not react with the lime.

   $CaO + SO_2 \rightarrow CaSO_3$

   $CaO + 2HCl \rightarrow CaCl_2 + H_2O$

   (MWs: CaO = 56; SO_2 = 64; HCl = 36.5)

SOLUTION

410 kg/h per 64 kg/mol	= 6.4 mol/h SO_2 in flue gas
325 lb/h per 36.5 kg/mol	= 6.71 mol/h HCl in flue gas
SO_2 requires 6.4 mol/h CaO × 1.2	= 7.68 mol/h
HCl requires 6.71 mol/h CaO × 1.2	= 8.05 mol/h
Total CaO required	= 15.73 mol/h

1. Total lime usage = 15.7 mol/h × 56 kg/mol = 879.2 kg/h
2. 0.25 × 410 kg/h = 103 kg/h SO_2 in flue gas

   0.12 × 245 kg/h = 29 kg/h HCl in flue gas

Note. Further treatment of this flue gas to remove additional acid gases is warranted.

REFERENCES

Brunner, C.D., *Hazardous Waste Incineration,* McGraw-Hill, Inc., New York, NY, 1993.

Brunner, C.D., *Incineration Systems: Selection and Design,* Incinerator Consultants, Inc., Reston, VA, 1988.

Chadbourne, J.F., Cement kilns, in *Standard Handbook of Hazardous Waste Treatment and Disposal,* Freeman, H.M., Ed., McGraw-Hill, New York, NY, 1998.

Code of Federal Regulations, Volume 40 Part 261, Identification and Listing of Hazardous Waste, U.S. Government Printing Office, Washington, DC, 1996.

Code of Federal Regulations, Volume 40 Part 264, Standards for Owners and Operators of Hazardous Waste Treatment, Storage, and Disposal Facilities, U.S. Government Printing Office, Washington, DC, 1999.

Freeman, H.M., Olexsey, R.A., Oberacker, D.A., and Mournighan, R.E., Thermal destruction of hazardous waste — a state-of-the-art review, *J. Hazard. Mater.,* 14, 103–117, 1987.

Haas, C.N., and Vamos, R.J., *Hazardous and Industrial Waste Treatment,* Prentice-Hall, Englewood Cliffs, NJ, 1995.

Manahan, S.E., *Environmental Chemistry,* 6th Ed., Lewis Publishers, Boca Raton, FL, 1994.

Mournighan, R.E., Hazardous Waste in Industrial Processes. Cement and Lime Kilns, *First International Congress on New Frontiers of Hazardous Waste Management,* EPA 600-9-85-025, Pittsburgh, PA, 1985.

Oppelt, E.T., Incineration of hazardous waste: a critical review. *J. Air Pollut. Control Assoc.,* 37, 558–586, 1987.

U.S. Environmental Protection Agency, RCRA Orientation Manual, EPA530-R-02-016, Office of Solid Waste/Communications, Information, and Resources Management Division, Washington, DC, 2002.

U.S. Environmental Protection Agency, Introduction to Hazardous Waste Incinerators (40 CFR Parts 264/265, Subpart O), RCRA, Superfund and EPCRA Hotline Training Module. EPA530-R-99-052, Office of Solid Waste and Emergency Response, Washington, DC, 2000.

U.S. Environmental Protection Agency, Operational Parameters for Hazardous Waste Combustion Devices, EPA/625/R-93/008, Office of Research and Development, Cincinnati, OH, 1993.

U.S. Environmental Protection Agency, Permitting Hazardous Waste Incinerators, EPA/625/4-87/017, Center for Environmental Research Information, Cincinnati, OH, 1987.

U.S. Environmental Protection Agency, Products of Incomplete Combustion from Hazardous Waste Combustion, Midwest Research Institute, U.S. EPA, Draft Final Report, June, 1986.

U.S. Water Pollution Control Federation, *Hazardous Waste Site Remediation,* Alexandria, VA, 1988.

SUGGESTED READINGS

Bayer, J.E., Incinerator operations, *J. Hazard. Mater.,* 22, 243–24, 1989.

Brunner, C., Industrial sludge waste incineration, *Environ. Prog.,* 8, 163–166, 1989.

Carnes, R.A., RCRA trial burns: adventures at Rollins, *J. Hazard. Mater.,* 22, 151–160, 1989.

Cooper, E. D., Incineration permitting: the long and winding road, *Hazard. Mater. Waste Manage.,* 10–13, 1986.

Cundy, V.A., Lester, T.W., Leger, C., Miller, G., Montestruc, A.N., Acharya, S., Sterling, A.M., Pershing, D.W., Lighty, J.S., Silcox, G.D., and Owens, W.D., Rotary kiln incineration — combustion chamber dynamics, *J. Hazard. Mater.,* 22, 195–219, 1989.

Cundy, V.A., Lester, T.W., Sterling, A.M., Morse, J.S., Montestruc, A.N., Leger, C.B., and Acharya, S., Rotary kiln injection, An in depth study — liquid injection, *J. Air Pollut. Control Assoc.,* 39, 63–75, 1989.

Freeman, H., Recent advances in the thermal treatment of hazardous wastes, *Manage. Hazard. Toxic Wastes Process Ind.,* 19, 34–36, 1987.

Munoz, H., Cross, F., and Tessitore, J., Comparisons Between Fluidized Bed and Rotary Kiln Incinerators for Decontamination of PCB Soils/Sediments at CERCLA Sites, *Proceedings of the National Conference on Hazardous Waste and Hazardous Materials,* Mar. 4-6, 1986, pp. 424–245.

National Research Council, *Waste Incineration and Public Health,* National Academy Press, Washington, DC, 2002.

NetLibrary, Inc., *Carbon Filtration for Reducing Emissions from Chemical Agent Incineration,* National Academy Press, Washington, DC, 1999.

Roberts, S.M., Teaf, C.M., and Bean, J.A. *Hazardous Waste Incineration: Evaluating the Human Health and Environmental Risks,* Lewis Publishers, Boca Raton, FL, 1999.

Santoleri, J.J., Rotary–Kiln Incineration Systems: Operating Techniques for Improved Performance, Hazardous and Industrial Wastes, *Proceedings of the Mid-Atlantic Industrial Waste Conference,* Technomic Publishing Company, Inc., Lancaster, PA, 1990, pp. 743–758.

Stoddart, T.L. and Short, J.J., Dioxin surrogate trial burn: full scale rotary–kiln incinerator for decontamination soils containing 2,3,7,8-tetrachlorodibenzo-p-dioxin, *Chemosphere,* 18, 355–361, 1989.

Takeshita, R. and Akimoto, Y., Control of PCDD and PCDF formation in fluidized bed incinerators, *Chemosphere,* 19, 345–352, 1989.

Theodore, L. and Reynolds, J., *Introduction to Hazardous Waste Incineration,* Wiley New York, NY, 1987.

U.S. Congress, Office of Technology Assessment, Dioxin Treatment Technologies, Washington, DC, 1991.

U.S. Environmental Protection Agency, Permitting Hazardous Waste Incinerators, EPA/530-SW-88-024, Office of Solid Waste, Washington, DC., April, 1988.

Wiley, S.K., Incinerate your hazardous waste, *Hydrocarbon Processing,* 66, 51–54, June, 1987.

Wood, R.W. and Bastian, R.E., Rotary kiln incinerators: the right regime, *Mechanical Engineering,* III, 78–81, September, 1989.

QUESTIONS

1. Which two major sets of regulations are included in the federal standards for hazardous waste thermal technologies? Under which specific acts are they administered? How do they differ in terms of regulatory coverage?

2. Combustion involves chemical transformations in which solid materials are converted to gases and solid residues. What factors, with regard to both incinerator design and operation and waste properties, affect the composition and quantities of gases produced? What factors will influence the amount of solid residues produced, both carbonaceous and inorganic?

3. Whenever an organic material is to be incinerated, air (oxygen) is necessary to complete combustion. For what other purposes is air required? Be specific.

4. List the engineering and design factors that serve to enhance combustion in an incinerator.

5. Discuss how each of the major types of gaseous emissions from an incinerator may be effectively removed from the flue.

6. Compare and contrast the rotary kiln injection system and the liquid injection incinerator in terms of overall design, efficiency and problems during use. Which is most suited to the destruction of organic sludges, organic solids, liquid solvents, and metal-enriched acidic solutions?

7. A waste mixture consisting of benzene and chlorophenol is being incinerated. Is the unit in compliance for each compound?

Compound	Inlet (kg/h)	Outlet (kg/h)
Benzene (C_6H_6)	953	0.081
Chlorobenzene (C_6H_5OCl)	950	0.149
Xylene (C_8H_{10})	442	0.061
HCl	—	0.95
Particulates (8.8% O_2)	—	48.1

Note: Flow rate = 16,250 dscfm.

8. For the data in the previous question, determine if the emissions meet federal requirements for particulates.

9. Calculate the solids retention (θ) time in a rotary kiln incinerator with the following data:
   kiln length = 6 m
   kiln rotational velocity = 0.8 r/min
   kiln diameter = 1.8 m
   kiln slope = 0.085 m/m

10. For question number 9, if the desired retention time is actually 12.0 min, what should the rotational velocity be adjusted to?

11. In a regulatory sense, how might hazardous waste incinerator ash be considered hazardous? Consider both listed and characteristic hazardous wastes.

12. Boilers and most industrial furnaces must follow a tiered system for the regulation of both hydrogen chloride and chlorine gas. How do the tiers differ in terms of monitoring requirements, dispersion modeling, and point of compliance? Which factors may be considered when a facility selects a tier?

13. Provide two examples of incinerator fugitive emissions. How can their releases to the environment be controlled?

14. A RCRA permit for a hazardous waste incinerator sets operating conditions and allowable ranges for certain critical parameters and also requires continuous monitoring of these parameters. List the important parameters.

15. Suppose there are several facilities in your county that generate liquid organic hazardous wastes. Based on economic and practical factors for the generator, incineration is considered the safest and most effective treatment. Where is the nearest hazardous waste incinerator in your state or community? Are there any special routes that the transportation vehicle must follow in order to make a delivery?

16. Flue gas from a liquid injection incinerator contains 550 kg/h of SO_2 and 475 kg/h of HCl. The dry scrubber lime feed rate is $1.5 \times$ stoichiometric rate, and is known to be about 82% efficient in removing SO_2 and 90% efficient in HCl removal. Calculate the lime feed rate in kg/h. How many kg/h of SO_2 and HCl will still be in the flue gas following dry scrubbing?

17. A liquid waste mixture of 64% xylene, 32% acetone, and 4% water is to be combusted in a liquid injection incinerator at 1275 kg/h. There is 35% excess air in the combustion chamber.

Properties of the waste constituents are as follows:

Compound	Chemical Formula	MW	Heat Content (kJ/kg)
Xylene	$C_6H_4(CH_3)_2$	106	42,989
Acetone	CH_3COOCH_3	74	13,120
Water	H_2O	18	0

(a) Calculate the total heat release in the incinerator.
(b) Calculate the percent by volume of each component in the flow gas.

16 Hazardous Waste Treatment

Because the newer methods of treatment are good, it does not follow that the old ones were bad: for if our honorable and worshipful ancestors had not recovered from their ailments, you and I would not be here today.

Confucius (551–478 BCE)

16.1 INTRODUCTION

In 1984, Congress updated RCRA by prohibiting the land disposal of certain hazardous wastes, with the consequent enactment of the Land Disposal Restrictions (LDR) program by the U.S. EPA. The LDR program required that toxic constituents within hazardous waste be adequately treated prior to disposal of the waste on land. Since the enactment of the LDR program, mandatory technology-based treatment standards have been formulated, which must be met before hazardous waste is disposed on land. These standards have been extremely important in minimizing threats to human health and the environment.

16.2 LAND DISPOSAL RESTRICTIONS

By May 8, 1990, the Hazardous and Solid Waste Amendments to RCRA (HSWA) prohibited all untreated hazardous waste from landfill disposal. Many hazardous wastes were restricted from being disposed in or on the land due to the probability of severe groundwater or soil contamination. HSWA also required the U.S. EPA to formulate treatment standards for all hazardous wastes by five specific deadlines. The treatment standards established maximum contaminant levels that a hazardous waste cannot exceed in order for it to be disposed in a hazardous waste landfill. The specific goals of the treatment standards are to (U.S. EPA, 2001c):

- Identify wastes with similar physical and chemical characteristics
- Establish treatability groups based on these characteristics
- Identify the Best Demonstrated Available Technology (BDAT) to treat a hazardous waste.

There are three types of treatment standards:

- *Concentration-based.* The waste must be treated to a level at which only allowed amounts of toxins remain in the waste
- *Technology-based.* The waste must be treated by a specific technology to below the level at which it is prohibited from landfill disposal
- No land disposal
  - o the waste can be recycled without generating a prohibited residue.
  - o the waste is not currently being disposed.
  - o the waste is no longer being generated.

Wastes that meet these treatment standards may be disposed in U.S. EPA-approved hazardous waste landfills.

The LDR program has three major components that address hazardous waste disposal, dilution, and storage. The Disposal Prohibition states that before a hazardous waste can be land-disposed, treatment standards specific to that waste material must be met. A facility may meet such standards by either (U.S. EPA, 1999):

- Treating hazardous chemical constituents in the waste to meet required treatment levels. Any method of treatment can be used to bring concentrations to the appropriate level (except dilution); or
- Treating hazardous waste using a treatment technology specified by the U.S. EPA. Once the waste is treated with the required technology, it can be land disposed.

The Dilution Prohibition states that waste must be properly treated and not simply diluted in concentration by adding large volumes of water, soil, or nonhazardous waste. Dilution does not reduce the toxicity of the hazardous constituents but only increases total volume. The Storage Prohibition states that waste must be treated and cannot be stored indefinitely. This prohibition prevents generators and treatment, storage, and disposal facilities (TSDFs) from "warehousing" hazardous waste for long periods to avoid treatment. Waste may be stored, subject to the LDR, in tanks, containers, or containment buildings, but only to accumulate quantities necessary to facilitate proper recovery, treatment, or disposal.

16.3 WASTE TREATMENT PRIOR TO LAND DISPOSAL

Chemical treatment of a hazardous waste is carried out via the application of one or a series of chemical reactions. Chemical processes may be applied for the treatment of soluble contaminants (e.g., wastewaters), or mixtures of solids and liquids (sludges) containing hazardous constituents. Table 16.1 lists the common chemical and physical processes for the treatment of hazardous wastes.

16.3.1 Neutralization

Neutralization is used for the treatment of acidic or alkaline wastes, many of which are designated as RCRA corrosive wastes. A waste that exhibits the characteristic of corrosivity as defined in 40 CFR Part 261.22 is aqueous with a pH of less than or equal to 2, or greater than or equal to 12.5, or is a liquid that corrodes steel at a rate greater than a specified rate (see Chapter 11). Some listed hazardous wastes (e.g., spent pickle liquor generated by steel finishing operations, K062) are also corrosive wastes and must be neutralized.

TABLE 16.1
Common Chemical Treatment Processes for Hazardous Wastes

Process	Specific Aapplications
Neutralization	Neutralization of acidic or basic properties of a liquid waste to reduce its corrosive properties.
Precipitation	Removal from solution of dissolved hazardous inorganic contaminants by chemical reaction.
Oxidation/reduction	Changing the valence of an element via addition or removal of electron(s). The reaction renders that element less toxic and amenable to other treatment processes.
Sorption	Physical adhesion of soluble hazardous contaminant molecules to the surface of a solid sorbent.
Stabilization	Stabilization and solidification of metal-containing waste sludges by precipitation with Portland cement, fly ash, or similar fixative agent.

It is important to conduct a waste characterization early in the design of a waste neutralization process. The overall chemical composition of the waste, including variations in strength, must be known to ensure the correct design of the treatment system. Similarly, the waste flow rate will affect the size of the treatment system. Waste characterization is a requirement for hazardous waste generators (see Chapter 12) and can be accomplished using established laboratory procedures or by considering the nature of the facility's processes. Waste strength (i.e., concentration of acidity or alkalinity) is determined by collecting representative samples of the waste and performing a simple titration.

Depending on waste properties, pretreatment may be necessary prior to neutralization. Pretreatment can include filtration, sedimentation, and equalization. Other common pretreatment steps include cyanide destruction, chromium reduction, and removal of oil and grease.

Neutralization of acidic wastes is carried out by reaction with a base, which raises the pH to an acceptable range (Figure 16.1). Neutralization is conducted on a batch basis or as a continuous-flow process. Methods of neutralizing acidic wastes include (Blackman, 2001):

- Adding appropriate volumes of strong or weak base to the waste
- Mixing acidic waste with lime slurries
- Passing acidic waste through limestone beds
- Mixing acidic waste with a compatible alkaline waste

Reagents used to neutralize acidic wastes include sodium hydroxide (caustic soda), sodium carbonate (soda ash), ammonia, limestone, and lime (Table 16.2). The choice of neutralizing agent is a function of several factors, including neutralizing ability, possible reaction products that form, and cost. For wastes having mineral acid acidity greater than 5000 mg/L, high calcium lime or caustic soda are often used, while for more dilute acid wastes, limestone treatment may be economically feasible (Camp, Dresser and McKee, 1984).

Sodium hydroxide (NaOH) is relatively expensive compared with many common neutralizing agents; however, its popularity is based on its ease of storage and delivery (i.e., low equipment

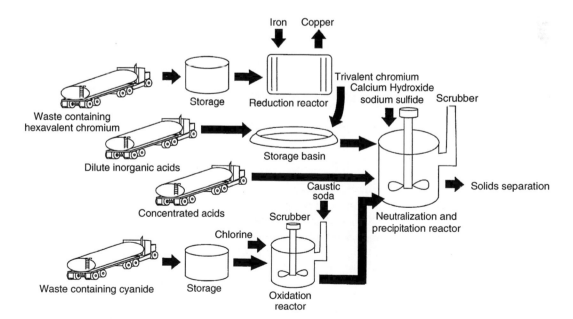

FIGURE 16.1 Chemical treatment of hazardous wastes: neutralization, precipitation, and oxidation–reduction. (From Blackman, W.J., Jr., *Basic Hazardous Waste Management,* 3rd ed., Lewis Publishers, Boca Raton, FL, 2001. Reproduced with kind permission of Lewis Publishers, an imprint of CRC Press. Boca Raton, FL).

TABLE 16.2
Common Acid Neutralizing Agents

Common Name	Chemical Formula
Calcitic limestone	$CaCO_3$
Dolomitic limestone	$Ca \cdot Mg(CO_3)$
Quickline	CaO
Hydrated lime	$Ca(OH)_2$
Soda ash	Na_2CO_3
Caustic soda	$NaOH$
Potassium hydroxide	KOH
Magnesium hydroxide	$MgOH_2$
Ammonia	NH_3
Slags, industrial blast furnace slag	Calcium silicates
Powerplant fly ash	Varied, alkaline

costs), rapid reaction rate, uniformity of composition, and relatively low volumes of sludge produced (Haas and Vamos, 1995). Sodium hydroxide is quite reactive and poses a serious hazard to workers. The reagent itself is highly corrosive to skin and materials. Rapid reactions will produce excessive amounts of heat.

Sodium carbonate (Na_2CO_3) is not a strong base and is safer to use than $NaOH$; however, it is less reactive and more expensive to use. This base is applied in slurry form because of its low solubility in water. Sodium carbonate has the advantage of buffering the pH of the waste mixture. However, the evolution of CO_2 gas can cause foaming problems (Haas and Vamos, 1995).

Ammonia (NH_3) is a strong alkali and is very reactive. Neutralization of acidic wastes with ammonia has the advantage of ease of handling and provides some buffering capacity. Ammonia is quite toxic and special precautions are required for its use. For example, neutralization reactions should be carried out in well-ventilated units with continuous atmospheric monitoring and controls. The ammonia-neutralized waste may contain high levels of dissolved nitrogen compounds.

Acidic wastes can be neutralized by mixing with limestone ($CaCO_3$), either by adding pulverized or granular material to a reaction basin or by passing the acidic liquid wastes over a granular limestone bed. Limestone-based neutralization is popular because of its availability and low cost. Unfortunately, limestone has a modest neutralizing potential compared with other alkali reagents. Also, low reactivity will result in long treatment times (45 min. or more). Another practical concern involves the production of large volumes of sludge when neutralizing with lime. The reaction of sulfuric acid with limestone is given by

$$H_2SO_4 + CaCO_3 \text{ (s)} \rightarrow CaSO_4 \cdot H_2O \text{ (s)} + CO_2 \tag{16.1}$$

The presence of sulfate in acidic wastes will result in large quantities of gypsum sludge when limestone is used as the neutralizing agent. In addition, the sludge produced in this process is difficult to settle. When treating concentrated acidic wastes, limestone particles can become coated with precipitates, thus rendering them inactive, and the coated particles will end up as sludge. In order to circumvent this problem, particle diameters of less than 0.074 mm are recommended for limestone bed neutralization. An additional difficulty in limestone bed treatment is that the carbon dioxide gas produced during neutralization reactions can gas-bind the beds (Haas and Vamos, 1995).

Acidic wastes can also be neutralized by the addition of lime slurries to the waste. A slaked lime ($Ca(OH)_2$) slurry, produced by reacting lime (CaO) with water, usually has a solids concentration of about 10 to 35%. Acid neutralization requires 15 to 30 min. retention times (Wilk et al., 1988; Haas and Vamos, 1995). Lime has an advantage over limestone in that it is a more soluble and more

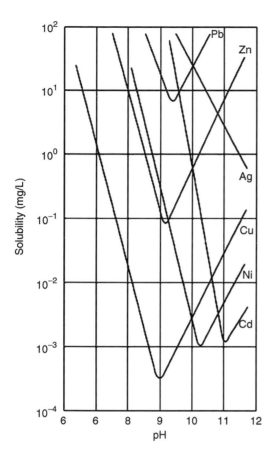

FIGURE 16.2 Precipitation of metals as a function of solution pH. (From U.S. EPA, 1989 With permission.)

concentrated neutralizing reagent. As with limestone, a disadvantage of using lime slurries is the formation of insoluble salts, especially when the waste contains sulfate. The insoluble salts can coat pH electrodes, valves, pipes, and pumps. Also, handling lime releases large amounts of lime dust.

Alkaline industry wastes can be used to neutralize an acidic waste stream. Supplemental neutralizing agents should be available to account for any incomplete reaction. Mixing alkaline wastes with metal-containing acidic wastes will produce heavy metal hydroxide sludges. Figure 16.2 shows the behavior of soluble metals as a function of solution pH. Wastes containing cyanide are not suitable for mutual neutralization processes because of the potential for the evolution of hydrogen cyanide gas (Haas and Vamos, 1995).

EXAMPLE 16.1

At a wire processing facility a precipitation system is being installed to remove copper from the processing solution. A pH meter will be used to control the feed of the hydroxide solution to the mix tank. To what pH should the instrument be set to achieve a Cu effluent concentration of 0.5 mg/L?

SOLUTION

K_{sp} of $Cu(OH)_2 = 2.0 \times 10^{-19}$ and the copper hydroxide reaction is

$$Cu^{2+} + 2OH^- \rightarrow Cu(OH)_2$$
$$K_{sp} = [Cu^{2+}][OH^-]^2$$

Converting mg/L of Cu to mol/L of Cu,

$$[Cu^{2+}] = (0.50 \text{ mg/L})/[(63.54 \text{ g/mol})(1000 \text{ mg/g})] = 7.87 \times 10^{-6} \text{ mol/L}$$

The solubility product equation above is rearranged to solve for the hydroxide concentration.

$$[OH^-]^2 = (2.0 \times 10^{-19})/(7.87 \times 10^{-6}) = 2.54 \times 10^{-14}$$
$$[OH^-] = (2.54 \times 10^{-14})^{1/2}$$
$$= 1.59 \times 10^{-7}$$
$$pOH = -\log(1.59 \times 10^{-7})$$
$$= 6.80$$
$$pH = 14 - pOH$$
$$= 14 - 6.8$$
$$= 7.20$$

Thus, the pH should be set to 7.20 to remove the soluble copper to 0.5 mg/L.

The neutralization of alkaline waste is achieved by reaction with an adequate quantity of an acid to bring the solution pH to within the desired range. Methods of neutralizing alkaline wastes include (Haas and Vamos, 1995):

- Adding appropriate amounts of strong or weak acid to the waste
- Adding compressed carbon dioxide gas to the waste
- Blowing acidic flue gas (e.g., from coal combustion or municipal solid waste [MSW] incineration) through the waste
- Mixing the alkaline waste with an acidic waste

Alkaline wastes are most commonly neutralized by reaction with mineral acids, typically sulfuric and hydrochloric acid. Sulfuric acid is popular by virtue of its relatively low cost. Neutralization residence times of 15 to 30 min. are recommended with sulfuric acid (Wilk et al., 1988; Haas and Vamos, 1995). A disadvantage of using sulfuric acid is that it will form sludges when reacted with calcium-containing alkaline wastes.

Hydrochloric acid (HCl) is more expensive than sulfuric acid; however, HCl neutralization will not produce sludges when neutralizing calcium-containing alkaline wastes. Neutralization reaction residence times of 5 to 20 min. are recommended with HCl (Wilk et al., 1988; Haas and Vamos, 1995). A disadvantage of using HCl is that it can form a corrosive and irritating acid mist upon reaction. Nitric acid (HNO_3) may also be used for neutralization; however, it is dangerous because it is a powerful oxidizing agent.

16.3.2 CHEMICAL PRECIPITATION

During the chemical precipitation of hazardous waste streams, a soluble hazardous species is removed from the solution by the addition of a precipitating reagent; an insoluble compound subsequently forms that contains the hazardous constituents (Equation 16.2). The precipitate is removed from the solution using a physical separation technique such as sedimentation or filtration. Coagulants or flocculants may be added to the mixture to enhance the separation of the precipitate from the soluble phase. Examples of common inorganic coagulants are aluminum sulfate (alum), ($Al_2(SO_4)_3 \cdot 18H_2O$) and ferric sulfate ($Fe_2(SO_4)_3$)

A schematic of a typical precipitation process is provided in Figure 16.3.

Precipitation processes are typically geared toward the removal of dissolved inorganic ions, particularly metals:

$$Cd^{2+} (aq) + HS^- (aq) \rightarrow CdS(s) + H^+ (aq) \tag{16.2}$$

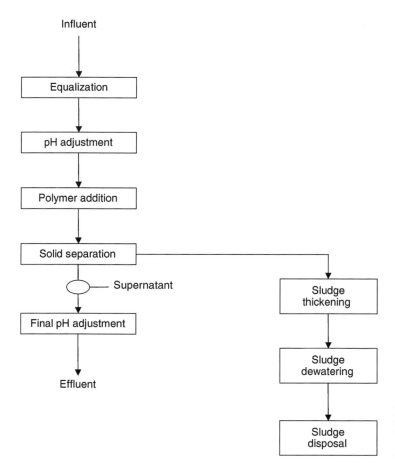

FIGURE 16.3 Schematic of a precipitation process. (Reproduced with kind permission of Water Environment Federation, Alexandria, VA.)

A number of counter anions are suitable for reaction with the metal. These anions vary widely in terms of rate of reaction, inherent toxicity, and cost. A common means of precipitating soluble metal ions is by hydroxide formation as in the example of zinc(II) hydroxide:

$$Zn^{2+} + 2OH^- \rightarrow Zn(OH)_2 \text{ (s)} \tag{16.3}$$

The hydroxide ion source can be a common alkali such as NaOH, Na_2CO_3, or $(Ca(OH)_2)$. When reacted with such alkalis, most metal ions will produce basic salt precipitates. Lime addition is the most common reagent for the precipitation of metals as hydroxides and basic salts. Sodium carbonate is used to form hydroxide precipitates ($Cr(OH)_3$), carbonates ($CdCO_3$), or basic carbonate salts ($2PbCO_3 \cdot Pb(OH)_2$) (Manahan, 1994). The carbonate anion produces hydroxide as a result of hydrolysis with water:

$$CO_3^{2-} + H_2O \rightarrow HCO_3^- + OH^- \tag{16.4}$$

The carbonate anion subsequently reacts with the metal.

Several heavy metal sulfides have extremely low solubilities; therefore, precipitation by H_2S or other sulfides serves as a very effective treatment. Unfortunately, hydrogen sulfide (H_2S) is a toxic gas and a hazardous waste, listed as a U135 waste. Another safer source of sulfide that produces metal-sulfide precipitates is iron(II) sulfide (FeS). A disadvantage of producing sulfide precipitates is that H_2S can be evolved if metal sulfide wastes come into contact with acid:

$$MS + 2H^+ \rightarrow M^{2+} + H_2S \tag{16.5}$$

Metals can also be precipitated from solution by the action of a reducing agent such as sodium borohydride (Manahan, 1990):

$$4Cu^{2+} + NaBH_4 + 2H_2O \rightarrow 4Cu + NaBO_2 + 8H^+ \tag{16.6}$$

Precipitation reactions result in the formation of the treated effluent and a sludge containing most of the contaminants originally present in the solution. After precipitation, the volume of the sludge should be substantially less than the volume of the original solution. Therefore, the precipitation process is considered a volume-reduction process and not a destruction process. The sludge may contain mostly water and must be dewatered before disposal. Dewatered sludges are often characteristic hazardous wastes based upon Toxicity Characteristic Leaching Procedure testing or are listed hazardous wastes based on the industrial process from which the wastes were generated. Such sludges may require further treatment prior to landfilling as required by the LDR.

The formation and settling of inorganic precipitates (commonly termed flocculation and sedimentation) can also entrap both dissolved and colloidal organic contaminants via physical and chemical mechanisms. The removal of organic contaminants by precipitation is viewed as a beneficial side-reaction of the process (Haas and Vamos, 1995).

Industries producing wastewaters amenable to precipitation include metal plating and finishing, steel and nonferrous, inorganic pigments, mining, and electronics. Landfill leachate and contaminated groundwater can also contain hazardous species that are suitable for chemical precipitation.

16.3.3 OXIDATION AND REDUCTION

Oxidation and reduction reactions are applied for the treatment, removal, and destruction of a variety of inorganic and organic wastes. By definition, an oxidation reaction increases the valence of an ion due to the removal of electrons; conversely, a reduction reaction decreases an ion's valence because electrons are added to its shell.

For oxidation to proceed, an oxidizing agent is reacted with the waste in question. A number of useful and common oxidizing agents are listed in Table 16.3. Some of the more common ones include ozone (O_3), hydrogen peroxide (H_2O_2), and chlorine gas (Cl_2). Ozone, employed as an oxidant gas at levels of 1 to 2 wt% in oxygen and 2 to 5 wt% in air, has been used to treat a variety

TABLE 16.3
Commonly Used Oxidizing and Reducing Agents

Oxidizing Agents
F_2
O_2
O_3
Cl_2
Permanganate (MnO_4^-)
Chromate (CrO_4^{2-})
Dichromate ($Cr_2O_7^{2-}$)
Nitric acid (HNO_3)
Perchloric acid ($HClO_4$)
Sulfuric acid (H_2SO_4)

Reducing Agents
Sodium
Magnesium
Aluminum
Zinc
Metal hydrides: NaH, CaH_2, $LiAlH_4$

of effluents and wastes including wastewater and sludges containing oxidizable constituents (Manahan, 1994). Ozone is a strong oxidant, and decomposes in a short time; it is therefore generated on-site by an electrical discharge through dry air or oxygen. Ozone must be used with caution as it is a nonselective and a rapid oxidant. Similarly, H_2O_2 and Cl_2 are nonselective and highly reactive depending on their initial concentration.

Cyanide-bearing wastewater generated by the metal-finishing industry is often oxidized with alkaline chlorine or hypochlorite solutions. In these reactions the chlorine is correspondingly reduced. In the process, the cyanide contaminant is initially oxidized to a less toxic cyanate and then to carbon dioxide and nitrogen in the following reactions (U.S. EPA, 2000a):

$$2NaOH + Cl_2 \rightarrow NaOCl + NaCl + H_2O \tag{16.7}$$

$$NaCN + Cl_2 \rightarrow CNCl + NaCl \tag{16.8}$$

$$CNCl + 2NaOH \rightarrow NaCNO + NaCl + H_2O \tag{16.9}$$

$$2NaCNO + 3Cl_2 + 4NaOH \rightarrow 2CO_2 + N_2 + 6NaCl + 2H_2O \tag{16.10}$$

Oxidation of cyanide may also be accomplished with hydrogen peroxide, ozone, and electrolysis (Dawson and Mercer 1986; Blackman, 2001).

EXAMPLE 16.2

A metal processing industry produces 95,000 L per day of a waste stream containing 325 mg/L cyanide as NaCN. Calculate the stoichiometric quantity of Cl_2 required daily to destroy the cyanide.

SOLUTION

Combining reactions 16.7 to 16.10, we obtain

$$2NaCN + 5Cl_2 + 12 NaOH \rightarrow N_2 + Na_2CO_3 + 10NaCl + 6H_2O \tag{16.11}$$

From this reaction, we can see that a total of 2.5 mol of Cl_2 are required to react completely with 1 mol of NaCN.

The total mass of NaCN generated daily is

(95,000 L/day) (1 kg / L) (325 parts /10^6 parts) = 30.88 kg NaCN/day

The kg-mol of NaCN generated per day is

(30.88 kg/day) / (49 kg/kg-mol) = 0.630 kg-mol NaCN/day

The amount of Cl_2 required daily is

2.5 (0.63) = 1.576 kg-mol Cl_2

= (1.576) (70.9 kg/kg-mol)

= 111.7 kg/day Cl_2 required

Note. Due to reaction with other contaminants, more than the stoichiometric amount of Cl_2 will be required to complete this process.

EXAMPLE 16.3

For the waste stream in Example 16.2, determine the stoichiometric amount of NaOH required to oxidize the cyanide to N_2. Refer to Equations 16.8 to 16.10 for the conversion of CN^- into N_2.

The molecular weight of $CN^- = 26$, $Cl_2 = 70.9$, and NaOH = 40.

SOLUTION

From Equation 16.9, the molar ratio of $CNO^-/CN^- = 1$. From Equation 16.10, the molar ratio of $NaOH/CNO^- = 4/2 = 2$. Therefore, the molar ratio $NaOH/CN^- = 2$. The mass ratio $NaOH/CN^- = 2 \times 40/26 = 3.08$ kg/kg. The mass of NaOH required daily is 111.7×3.08 kg/kg $= 344.0$ kg/day.

Hexavalent chromium-containing wastewater is produced in chromium electroplating and in metal-finishing operations carried out on chromium as the base material. Chromium wastes are typically treated in a two-stage batch process. In the initial stage, the highly toxic hexavalent chromium (Cr^{6+}) is reduced to the less toxic trivalent form (Cr^{3+}). There are several ways to reduce the hexavalent chromium to trivalent chromium including the use of sulfur dioxide, bisulfate, or ferrous sulfate. The Cr^{3+} can then be precipitated as chromic hydroxide and removed. Most processes use caustic soda (NaOH) to precipitate chromium hydroxide. Hydrated lime ($Ca(OH)_2$) may also be used. The key reactions are as follows:

$$SO_2 + H_2O \rightarrow H_2SO_3 \tag{16.12}$$

$$3H_2SO_3 + 2H_2CrO_4 \rightarrow Cr_2(SO_4)_3 + 5H_2O \tag{16.13}$$

Addition of NaOH will result in Cr precipitation:

$$6NaOH + Cr_2(SO_4)_3 \rightarrow 2Cr(OH)_{3(s)} + 3Na_2SO_4 \tag{16.14}$$

Some generic oxidation and reduction reactions are depicted in Table 16.4.

16.3.4 SORPTION

Sorption involves the use of a sorbent to remove a soluble hazardous contaminant (the sorbate) from an aqueous waste solution. Sorption is not a chemical process; rather, it involves the physical adhesion of molecules or particles to the surface of a solid sorbent. Sorption is solely a surface phenomenon.

One of the most popular sorbents for the removal of both organic and some inorganic substances from aqueous waste is activated carbon. Carbon possesses a high surface area and hydrophobic surface characteristics, thus making it an excellent sorbent for removing contaminants from water.

TABLE 16.4
Examples of Oxidation and Reduction Reactions Used to Treat Wastes

Waste Type	Reaction with Oxidant or Reductant
Oxidation of Organics	
Organic matter, (CH_2O)	$\{CH_2O\} + 2\{O\} \rightarrow CO_2 + H_2O$
Aldehyde	$CH_3CHO + \{O\} \rightarrow CH_3COOH$
Oxidation of Inorganics	
Cyanide	$2CN^- + 5OCl^- + H_2O \rightarrow N_2 + 2HCO_3^- + 5\,Cl^-$
Iron(II)	$4Fe^{2+} + O_2 + 10H_2O \rightarrow 4Fe(OH)_3 + 8H^+$
Sulfur dioxide	$2SO_2 + O_2 + 2H_2O \rightarrow 2H_2SO_4$
Reduction of Inorganics	
Chromate	$2CrO_4^{2-} + 3SO_2 + 4H^+ \rightarrow Cr_2(SO_4)_3 + 2H_2O$
Permanganate	$MnO_4^- + +3Fe^{2+} + 7H_2O \rightarrow MnO_2(s) + 3Fe(OH)_3(s) + 5H^+$

Reproduced with kind permission of Manahan, S.E., *Environmental Chemistry*, 6th ed., Lewis Publishers, Boca Raton, FL, 1994. Copyright Lewis Publishers, an imprint of CRC Press. Boca Raton, FL

Activated carbon is used in either granular or powdered form. Different raw materials and processing techniques result in a range of carbon types with different sorption characteristics. Activated carbon is originally prepared from coconut shells, wood, bituminous coal, or lignite. The carbon is first dehydrated by heating at 170°C; the temperature is increased further, resulting in carbonization, transforming the material into a charcoal-like substance. The final step is activation or the addition of superheated steam, which enlarges pores, removes ash, and increases the surface area. The resulting activated carbon has an extremely high surface area of approx. 1000 to 1400 m^2g (Watts, 1998). At the microscopic level, an activated carbon particle possesses a porous structure with a large internal surface.

Movement of an organic molecule to a surface site requires four separate transport phenomena: bulk fluid transport, film transport, intraparticle (or pore) diffusion, and the actual physical attachment. The driving forces that control adsorption of the organic solute include electrical attraction, a chemical affinity of the particular organic molecule for the adsorbent, van der Waal's forces (weak attractive forces acting between molecules), and the hydrophobic nature of the organic solute (Tchobanoglous et al., 1993). Most adsorption processes are physical processes resulting from van der Waal's molecular forces.

The adsorption equilibrium may be represented by the Freundlich isotherm

$$x/m = kC^{1/n} \qquad (16.15)$$

where x is the mass of solute adsorbed (g), m the mass of carbon adsorbent (g), k the empirical constant, n the empirical constant, and C the equilibrium concentration of solute (g/L)

An adsorption isotherm is essentially a plot of the amount of contaminant adsorbed per unit of mass of carbon (x/m) against the concentration of contaminant in the fluid (C). There are several different mathematical forms of isotherms. Another common isotherm to describe this type of adsorption is the Langmuir isotherm,

$$x/m = (\alpha kC)/(1 + kC) \qquad (16.16)$$

where α is the mass of the adsorbed solute required to saturate a unit mass of adsorbent, ak the a Langmuir constant, and k the b Langmuir constant

16.3.5 SORPTION SYSTEMS

Activated carbon has a number of applications; for example, in industrial and wastewater treatment, activated carbon is used to remove undesirable organic substances. Carbon sorption is also used as a polishing step for the treatment of drinking water. Some of the organic compounds amenable to sorption by activated carbon are shown in Table 16.5.

Many process configurations have been designed for the treatment of aqueous waste streams such as upflow and fluidized-bed systems; however, the most common process configuration is gravity flow through a packed-bed column (Watts, 1998). The carbon contactor consists of a lined steel column, or a steel or concrete rectangular tank in which the carbon is packed to form a filter bed. Process water is initially applied to the top of the column and contaminants are sorbed at the top of the bed. Process water exits the bottom of the column via an underdrain. Eventually, these sorption sites become saturated and the contaminants are then sorbed further down the column. The saturated zone migrates all the way down until the entire column is saturated. At this point, "breakthrough" occurs and the column loses its effectiveness.

After the carbon has been spent (i.e., exhausted), it must be regenerated. Regeneration takes place by heating it in an oxygen-rich environment. If the sorbed organic material is volatile, the carbon bed may be regenerated with the use of the steam. More typically, however, the carbon is removed from the chamber and regenerated in a controlled furnace. In a large activated carbon unit, a regeneration furnace is commonly installed as one component of the carbon treatment system.

TABLE 16.5
Organic Compounds Suitable for Sorption Treatment by Activated Carbon

Class	Example
Aromatic solvents	Benzene, toluene, xylene
Polynuclear aromatics	Naphthalene, biphenyl
Chlorinated aromatics	Chlorobenzene, PCBs, Endrin, toxaphene, DDT
Phenolics	Phenol, cresol, resorcinol, nitrophenols, chlorophenols, alkyl phenols
Aromatic amines and high-molecular weight aliphatic amines	Aniline, toluene diamine
Surfactants	Alkyl benzene sulfonates
Soluble organic dyes	Methylene blue, textiles, dyes
Fuels	Gasoline, kerosene, oil
Chlorinated solvents	Carbon tetrachloride, percholoroethyene
Aliphatic and aromatic acids	Tar acids, benzoic acids
Pesticides/herbicides	2,4-D, atrazine, simazine, aldicarb, alachlor, carbofuran

Source: U.S. EPA, 1984. With permission.

In smaller installations the carbon is removed and returned to the supplier for reprocessing (Wentz, 1995; Watts, 1998). Such regeneration processes will cause the carbon to lose some of its sorptive qualities. Furthermore, about 10% of the carbon is lost with each regeneration.

All carbon contactors must be equipped with carbon removal and loading mechanisms to allow spent carbon to be removed and virgin or regenerated carbon to be added. Spent, regenerated, and virgin carbon are typically transported hydraulically by pumping as a slurry. Carbon slurries may be transported with water or compressed air, centrifugal or diaphragm pumps, or eductors (U.S. EPA, 2000b).

Carbon adsorption is generally cost-effective only when the contaminants are present in very dilute quantities. Carbon adsorption is typically used to treat dilute aqueous streams with organics in the low parts per million range. For wastewater streams that contain a significant quantity of industrial flow, activated carbon adsorption is a proven, reliable technology to remove dissolved organics. Space requirements are low. Granular activated carbon (GAC) adsorption can be easily incorporated into an existing wastewater treatment facility. Disadvantages are also possible with the use of activated carbon for sorption. Under certain conditions, granular carbon beds may generate hydrogen sulfide from bacterial growth, creating odors and corrosion problems. Spent carbon, if not regenerated, may present a land disposal problem. Wet GAC is highly corrosive and abrasive. Lastly, carbon treatment requires pretreated wastewater with low suspended solids concentration. Variations in pH, temperature, and flow rate may also adversely affect carbon adsorption (U.S. EPA, 2000).

16.3.6 STABILIZATION

Stabilization processes are accomplished by mixing hazardous waste with a binding agent to form a crystalline or polymeric matrix that incorporates the entire waste. Inorganic binders include cement, cement kiln dust, fly ash, and blast furnace slag. Organic wastes can be immobilized by the addition of organic binders such as bitumen (asphalt) or polyethylene. Stabilization converts contaminants into a less- or a nonreactive form typically by chemical processes. Contaminants are furthermore physically immobilized within a solid matrix in the form of a monolithic block. Thus, stabilization serves to physically sorb, encapsulate, or alter the physical or chemical form of the contaminants, producing a less leachable material.

Stabilization techniques also improve the handling characteristics of the waste for transport on-site or to an off-site TSDF. Stabilization techniques designed to limit the solubility or mobility of hazardous constituents are required for RCRA hazardous wastes containing heavy metals.

The ideal stabilization agent is inert, readily available, and nondegradable. In selecting a stabilization agent, considerations should include the quantity required to eliminate free liquid, compatibility with the waste, and binding properties.

Pozzolanic materials such as fly ash form a solid monolithic mass when mixed with hydrated lime. Stabilization of waste using lime and pozzolanic materials requires that the waste be mixed with water to create an optimal consistency. Numerous treatment processes incorporate Portland cement as a binding agent along with pozzolanic materials to improve the strength and chemical resistance of the solidified waste. Soluble silicates may be added to a pozzolan–cement mixture to further enhance performance and to reduce interference from metals. Emulsifiers may be added to better incorporate organic liquids. Solidification and fixation processes are generally adjusted for the waste on a case-by-case basis.

To reduce the final volume of stabilized waste for disposal, wastes should be dewatered before stabilization reactions. Pretreatment, such as a chemical treatment to scavenge toxic materials, may contribute to more cost-effective treatment of the waste.

EXAMPLE 16.4

An aqueous sludge containing high amounts of free liquids is stabilized with a cement–slag mixture at a secure landfill. The wastes are mixed with the cement–slag using a front-end loader. The blended material is then transported to the secure landfill and spread in uniform lifts with a bulldozer. Waste properties are as follows: 45 mg/kg Cd, 1055 mg/kg Pb, and 575 mg/kg Zn. In order to optimize stabilization, the required ratio of fly ash to waste is estimated at 1.5:1. Calculate the reduction in contaminant concentrations due to dilution.

SOLUTION

$$C_o \times W_w = C_f (W_w + W_{FA}) \qquad \text{(adapted from LaGrega et al., 1994)} \qquad (16.17)$$

Where C_o is the original containment concentration, W_w the weight of waste, C_f the final concentration, and W_{FA} the weight of fly ash.

$$C_o/C_f = (W_w + W_{FA})/W_w = (1.0 + 1.5)/1 = 2.5/1 = 2.5$$
$$C_f/C_o = W_w/(W_w + W_{FA}) = 1.0/(1.0 + 1.5) = 1/2.5 = 0.40$$

There is a reduction of 60% of contaminant concentrations.

A major drawback to stabilization processes is the significant increase in the volume of material to be disposed. Waste mixtures may require several pretreatment steps that might make the process cost-prohibitive (Wentz, 1995).

REFERENCES

Blackman, W.J. Jr., *Basic Hazardous Waste Management*, 3rd ed., Lewis Publishers, Boca Raton, FL, 2001.
Camp, Dresser and McKee, Inc., Technical Assessment of Treatment Alternatives for Wastes Containing Corrosives, Prepared for U.S. EPA under contract No. 68–01–6403, Boston, MA, September 1984.
Dawson, G.W., and Mercer, B.W., *Hazardous Waste Management*, Wiley, New York, NY, 1986.

Haas, C.N., and Vamos, R.J., *Hazardous and Industrial Waste Treatment,* Prentice-Hall, Englewood Cliffs, NJ, 1995.

LaGrega, M.D. Hazardous Waste Management, McGraw-Hill, New York, 1994.

Manahan, S.E., *Environmental Chemistry,* 6th Ed., Lewis Publishers, Boca Raton, FL, 1994.

Manahan, S.E., *Hazardous Waste Chemistry, Toxicology and Treatment,* Lewis Publishers, Chelsea, MI, 1990.

Tchobanoglous, G., Theisen, H., and Vigil, S., *Integrated Solid Waste Management: Engineering Principles and Management Issues*, McGraw-Hill, New York, NY, 1993.

U.S. Environmental Protection Agency, Managing Cyanide in Metal Finishing, Capsule Report, EPA/625/R-99/009, Office of Research and Development, Washington, DC, 2000a.

U.S. Environmental Protection Agency, Granular Activated Carbon Absorption and Regeneration, Wastewater Technology Fact Sheet, EPA 832-F-00-017, Office of Water, Washington, DC, September 2000b.

U.S. Environmental Protection Agency, Corrective Action: Technologies and Applications, Seminar Publication, EPA/625/4-89/020, Center for Environmental Research Information, Cincinnati, OH, 1989.

U.S. Environmental Protection Agency, Granular Activated Carbon Systems Problems and Remedies, EPA 800/490/9198, Washington, DC, 1984.

Watts, R.J., *Hazardous Wastes: Sources, Pathways, Receptors*, Wiley, New York, NY, 1998.

Wentz, C.A., *Hazardous Waste Management*, 2nd ed., McGraw-Hill, New York, NY, 1995.

Wilk, L., Palmer, S., and Breton, M., Technical Resource Document: Treatment Technologies for Corrosive-Containing Wastes, Vol. II, EPA/600/S-87-099, U.S. Environmental Protection Agency, Cincinnati, OH, 1988.

SUGGESTED READINGS

Aieta, E.M., Reagan, K.M., Lang, J.S., McReynolds, L., Kang, J., and Glaze, W.H., Advanced oxidation processes for treating groundwater contaminated with TCE and PCE: Pilot-scale evaluations, *J. Am. Water Works Assoc.*, 80, 64–72, 1988.

Cawley, W.A., Ed., Treatability Manual, Vol III. Technologies for Control/Removal of Pollutants, 600/2-82-011c, U.S. Environmental Protection Agency, September. 1981.

Combs, G. D., *Emerging Treatment Technologies for Hazardous Waste*, Section XV, Environmental Systems Company, Little Rock, AR, 1989.

Conner, J. R., *Chemical Fixation and Solidification of Hazardous Wastes*, Van Nostrand Reinhold, New York, NY, 1990.

Eckenfelder, W.W., *Industrial Pollution Control*, McGraw-Hill, New York, NY, 1966.

George, C.A., Treatment technologies in hazardous waste management, in *Hazardous Materials Management*, Cox, D.B., and Borgias, A.P., Eds., McGraw-Hill, New York, NY, 2000.

Glaze, W.H., Kang, J. W., and Chapin, D. H., The Chemistry of water treatment processes involving ozone, hydrogen peroxide and ultraviolet radiation, *Ozone Sci*. Eng., 9, 335, 1987.

Govind, R., Kumar, U., Puligadda, R., Anita, J., and Tabak, H., *Emerging Technologies in Hazardous Waste Management 7, D*, Tedder, W. and Pohland, F. G., Eds., Plenum Press, New York, NY, 1997.

Gupta, V.K., Mohan, D., Sharma, S., and Park, K.T., Removal of chromium (VI) from electroplating industry wastewater using bagasse fly ash — a sugar industry waste material, in *The Environmentalist* Kluwer, Boston, MA, Vol. 19, 1999, pp. 129–136.

Gurol, M.D., Bremen, W.M., and Holden, T.E., Oxidation of cyanides in industrial wastewaters by ozone, *Environ. Prog.*, 4, 46–51, 1985.

Hassan, S.Q., Vitello, M.P., Kupferle, M.J., and Grosse, D.W., Treatment technology evaluation for aqueous metal and cyanide bearing hazardous wastes (F007), *J. Air Waste Manage. Assoc.*, 41, 710–715, 1991.

Murphy, J.S., and Orr, J.R., *Ozone Chemistry and Technology*, The Franklin Institute Press, Philadelphia, PA, 1975.

Palmer, S.A.K., Brenton, M.A., Nunno, T.J., Sullivan, D. M., and Suprenant, N.F., Technical Resource Document: Treatment Technologies for Metal/Cyanide-Containing Wastes, Vol. III, EPA/600/S2-87/106, U.S. Environmental Protection Agency, Cincinnati, OH, February 1988.

Patterson, J.W., Allen, H.E., and Scala, J.J., Carbonate precipitation for heavy metals pollutants. *J. Water Pollut. Control Fed.*, 49, 2397–2410, 1977.

Rice, R.G., Ozone for the Treatment of Hazardous Materials, Water-1980, *Symposium Series: American Institute of Chemical Engineers*, 209, 79–107, 1981.

Rice, R. and Browning, M., Ozone Treatment of Industrial Wastewater, Pollution Control Technology Review No. 84, Noyes Data Corp., Park Ridge, NJ, 1981.

Rice, R.G., and Netzer, A., Eds., *Handbook of Ozone Technology and Applications*, Vol. 1, Ann Arbor Science, Ann Arbor, MI, 1982.

Rosen, H.M., Use of ozone and oxygen in advanced wastewater treatment, *J. Water Pollut. Cont. Fed.*, 45, 2521, 1973.

Roy, A., Eaton, H.C., Cartledge, F.K., and Tittlebaum, M.E., Solidification/stabilization of hazardous waste: evidence of physical encapsulation, *Environ. Sci. Technol.*, 7, 1349–1353, 1992.

Sedlak, D.L., and Andren, A.W., Oxidation of chlorobenzene with Fenton's reagent, *Environ. Sci. Technol.*, 25, 777–782, 1991.

Sedlak, D.L., and Andren, A.W., Aqueous phase oxidation of polychlorinated biphenyls by hydroxyl radicals, *Environ. Sci. and Technol.*, 25, 1419–1427, 1991.

Surprenant, N., Nunno, T., Kravett, M., and Benton, M., Technical Resource Document: Treatment Technologies for Halogenated Organic Containing Wastes, Vol. 1, U.S. EPA/600/S2-87/098, Environmental Protection Agency, Cincinnati, OH, 1988.

Tomiyasu, H., Fukutomi, H., and Gordon, G., Kinetics and mechanism of ozone decomposition in basic aqueous solution, *Inorganic Chemistry*, 24, 2962, 1985.

Topudurti, K.V., Lewis, N.M., and Hirsch, S.R., The applicability of UV/Oxidation technologies to treat contaminated groundwater, *Environ. Prog.*, 12, 54–60, 1993.

Unterberg, W., Williams, R.S., Balinsky, A.M., Reible, D.D., Wetzel, D.M., and Harrison, D.P., Analysis of Modified Wet-Air Oxidation for Soil Detoxification, EPA/600/S2-87/079U.S, Environmental Protection Agency, Cincinnati, OH, 1988.

U.S. Environmental Protection Agency, Treatment of Metal Finishing Wastes by Sulfide Precipitation, EPA-600/2-77-049, Industrial Environmental Research Laboratory, Cincinnati, OH, 1985.

U.S. Environmental Protection Agency, Environmental Regulations and Technology: The Electroplating Industry, EPA/625/10-85/001, Cincinnati, OH, 1985.

U.S. Environmental Protection Agency, Technical Resource Document: Treatment Technologies for Metal/Cyanide-Containing Wastes, Vol. III, EPA/600/S2-87/106, 1988.

U.S. Environmental Protection Agency, Stabilization/Solidification of CERCLA and RCRA Wastes, Physical Tests, Chemical Testing Procedures, Technology Screening, and Field Activities, EPA/625/6-89/022, Office of Research and Development, Washington, DC, 1989.

U.S. Environmental Protection Agency. Ultrox International Ultraviolet Radiation/Oxidation Technology, San Jose, California, EPA/504/S5-89-012, Risk Reduction Engineering Laboratory, Cincinnati, OH. 1990.

Voice, T.C., Activated Carbon Adsorption, in *Standard Handbook of Hazardous Waste Treatment and Disposal*, Freeman, H. M., Ed., McGraw-Hill, New York, NY, 1989.

Wethington, B.C., Destruction of Cyanide in Wastewaters: Review and Evaluation, EPA/600/S2-88/031, Water Engineering Research Laboratory, Cincinnati, OH, 1988.

Wilson, R.D. and Thompson, C.H., Activated carbon treatment of groundwater: Results of a pilot plant program, Hazardous Materials Control, July-August 1988.

Zeff, J.D., Leitis, E., and Harris, J.A., Chemistry and Application of Ozone and Ultraviolet Light for Water Reuse—Pilot Plant Demonstration, *Proceedings of the 38th Purdue Industrial Waste Conference*, Lewis Publishers, Chelsea, MI, pp. 105–116, 1983.

QUESTIONS

1. List some of the major reagents used to neutralize acidic wastes. What factors are involved in making a choice of neutralizing agent ? What hazards may be involved in the use of certain neutralizing agents?
2. Which of the neutralizing agents discussed in this chapter have the benefit of buffering the reaction medium?
3. Precipitation processes are typically geared toward the removal of what types of compounds or elements? List the common counterions involved in precipitation.

4. What are the drawbacks to using an oxidizing agent such as O_3 to treat a waste stream consisting of numerous organic and inorganic components, only one of which is deemed hazardous?

5. What is the principle of stabilization of hazardous waste? Is stabilization solely a physical process or are chemical reactions also possible?

6. In order to remove hexavalent chromium from a waste stream, what pretreatment step is necessary?

7. A metal plating solution contains 32.2 mg/L of zinc. Calculate the concentration (mol/L) of $[OH^-]$ needed to precipitate all but 0.80 mg/L of the copper. The K_{sp} of zinc hydroxide is 7.68×10^{-17}. What is the final pH?

8. Can composting be used to treat hazardous waste? Explain.

9. At a metal plating facility, a precipitation system is being installed to remove zinc from the processing solution. A pH meter will be used to control the feed of the hydroxide solution to the mix tank. To what pH should the instrument be set to achieve a Zn effluent concentration of 1.0 mg/L?

10. Aqueous sodium cyanide wastes at a metal processing industry are to be treated with chlorine and NaOH. Calculate the stoichiometric quantity of Cl_2 required daily to destroy the cyanide from 5 kg of NaCN.

17 Land Disposal of Hazardous Waste

Farther off, farther off, the burying ground.
I wish you peace.
I wish the labyrinth of byways may one day be lost
beneath the new green of Spring.

Gu Cheng (1956–1993)
Yes, I go also

17.1 INTRODUCTION

About 23 million tons of hazardous wastes are land disposed each year (U.S. EPA, 1999). Land disposal can take place on or slightly below the ground surface, i.e., in a landfill or other land-based unit, or can occur under the Earth's surface, for example, by deep-well injection.

As will be apparent in this chapter, modern land-based disposal systems for hazardous wastes are designed and equipped with numerous safeguards and are closely regulated; however, if a hazardous waste is not properly treated before disposal, it can still contaminate local soil, groundwater and surface water. Additionally, rainwater, snowmelt, and groundwater can penetrate a landfill and disposed hazardous waste and can potentially mobilize hazardous substances.

17.2 THE LAND DISPOSAL RESTRICTIONS REVISITED

The land disposal restrictions were originally divided into five deadlines, or phases, mandated by federal law. Each phase restricted different types of hazardous waste from land disposal by a specific deadline. Phase One, for dioxin-containing wastes and spent solvents, was put into effect by 1986 (40 CFR Parts 268.30 and 268.31). Spent solvents and dioxin-containing wastes were the first wastes restricted from land disposal because they are generated in the greatest quantities and can be extremely hazardous. These wastes include F001–F005 solvents and F020–F023 plus F025–F028 dioxin wastes listed in 40 CFR Part 261.31. F-listed wastes are those from nonspecific sources and are commonly produced by manufacturing and industrial processes.

Phase Two, the so-called California List, was put in effect by 1987. This was dubbed 'The California List' because California was the first state to restrict land disposal of these hazardous wastes. Phase Two restricts land disposal of untreated, liquid hazardous waste containing the following constituents:

- Polychlorinated biphenyls (PCBs)
- Corrosives with pH < 2 or free cyanides
- Arsenic, cadmium, chromium (VI), lead, mercury, nickel, selenium, and thallium

The California List also restricts hazardous waste containing halogenated organic compounds in liquid or nonliquid form (40 CFR Part 268.32). After the Phase One wastes, California List wastes are the next most abundant and hazardous.

521

Phases Three, Four, and Five address land disposal restrictions (LDRs) for all remaining hazardous wastes listed by the EPA in the hazardous waste regulations (40 CFR Part 261). Wastes in Phases Three, Four, and Five include F, K, P, and U coded wastes. To recapitulate from Chapter 11:

- F wastes are nonspecific source wastes
- K wastes are specific source wastes
- P and U wastes are commercial chemical products waste

17.2.1 GENERAL REQUIREMENTS

LDRs apply to: (1) hazardous waste generators, including clean-up waste; (2) hazardous waste transporters; (3) hazardous waste treatment, storage, and disposal facilities (TSDFs); and (4) hazardous waste recycling facilities. The restrictions do not apply to generators of less than 100 kg of hazardous waste per month (conditionally exempt small quantity generators [CESQGs]), and to corrective action waste if there is no new placement or disposal.

Under the land disposal restrictions, hazardous waste generators, TSDFs, and recycling facilities must meet specific requirements and responsibilities for waste analysis, record keeping, notification, and certification. The Uniform Hazardous Waste Manifest for LDR wastes must be accompanied by a LDR notification. The LDR is a one-time notification form that is signed by the generator and indicates that they understand that this waste cannot be land-disposed. It is the responsibility of the generator to provide the manifest and the LDR. Most hazardous waste disposal contractors supply one or both of these forms.

Treated wastes can only be disposed in a landfill meeting the U.S. EPA minimum technology standards for hazardous waste landfills. These standards are found in the federal regulations in 40 CFR Part 264.301.

17.3 THE SECURE LANDFILL

17.3.1 REGULATORY REQUIREMENTS

According to federal requirements, a secure landfill must possess a liner system that is constructed and installed to prevent any migration of wastes out of the landfill during the active life and the closure period of the landfill; whereas a Subtitle D landfill for muncipal solid wastes (MSW) disposal is required to posses a single liner system, the requirements for a Subtitle C (i.e., RCRA hazardous waste) landfill are much more stringent; for example, there is to be a double-liner system equipped with two leachate collection and removal systems (LCRs) (Figure 17.1 and Figure 17.2). The liner system must include:

- A top liner (primary liner) designed and constructed of materials (e.g., a geosynthetic liner or flexible membrane liner) to prevent the migration of hazardous constituents during the active life and postclosure care period.
- A composite bottom liner (secondary liner) consisting of at least two components. The upper component is typically a geosynthetic liner. The lower component must be constructed of materials to minimize the migration of hazardous constituents if a breach in the upper component were to occur. The lower component must be constructed of at least 91 cm (3 ft) of compacted soil material with a hydraulic conductivity of 1×10^{-7} cm/sec or less.

A LCR system is installed immediately above the top (primary) liner to collect and remove leachate during the active life and postclosure care period of the landfill. A secondary LCR system is placed above the composite liner. This is also a leak detection system and must be capable of

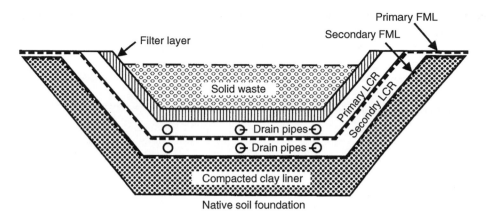

FIGURE 17.1 Schematic of a landfill base showing liners and leachate collection and removal systems (U.S. EPA, 1989) with permission. Not to scale.

FIGURE 17.2 Geosynthetic landfill liner being installed. (Reproduced with kind permission of Tenax, Baltimore, MD.)

detecting, collecting, and removing leaks of hazardous constituents through the top liner as early as possible. The state regulatory agency specifies the design and operating conditions in the permit to ensure that the leachate depth over the liner does not exceed 30 cm (1 ft). The requirements for a leak detection system include that it be (40 CFR 264):

- Constructed with a bottom slope of 1% or more.
- Constructed of granular drainage materials with a hydraulic conductivity of 1×10^{-2} cm/sec or more and a thickness of 30.5 cm (12 in.) or more; or constructed of synthetic or geonet drainage materials with a transmissivity of 3×10^{-5} m²/sec or more.
- Constructed of materials that are chemically resistant to the waste present in the landfill and the leachate expected to be generated, and of sufficient strength and thickness to prevent collapse under the pressures exerted by overlying wastes, waste cover materials, and heavy equipment used.

- Designed and operated to minimize clogging during the active life and postclosure care period.
- Constructed with sumps and pumps of sufficient size to remove liquids from the sump and prevent liquids from backing up into the drainage layer. Each unit must have its own sump. All pumpable liquids must be removed in the leak detection system sumps to minimize the head on the bottom liner.

Alternative designs and operating practices are allowed by the permitting agency if they adequately prevent the migration of hazardous constituents into groundwater or surface water. Factors to consider with alternative designs include the characteristics and quantity of the wastes and the hydrogeologic setting of the facility, including the ability of liners and soils to attenuate any contaminants which may leach from the facility. The double-liner requirement may be waived for a hazardous waste facility under certain conditions, for example if the landfill:

- Contains only hazardous wastes from foundry furnace emission controls or metal casting molding sand
- Has at least one intact liner
- Is located more than one-quarter mile from an underground source of drinking water
- Is in compliance with applicable groundwater monitoring requirements

Precautions are also required for the proper management of surface liquids including run-on and runoff. The facility must design, construct, and operate a run-on control system capable of preventing flow onto the active portion of the landfill during peak discharge from at least a 25-year storm. Also, a runoff management system must be installed to collect and control the water volume resulting from a 24-h, 25-year storm. These design aspects will be discussed below.

17.3.2 SECURE LANDFILL DESIGN, CONSTRUCTION, AND OPERATION

Many of the requirements for secure landfill design and construction are similar to those for Subtitle D (sanitary) landfills; therefore, several sections will be discussed briefly below. Details appear in Chapter 10.

17.3.3 LINER MATERIALS

17.3.3.1 Clay Liners

As discussed in Chapter 10, clay is an important component of soil liners due to its ready availability and its tendency to be accomodating to mechanical and other stresses. Clay materials, being natural, incorporate readily with native soil materials and are obviously very durable. Additionally, the clay fraction of the soil ensures low hydraulic conductivity. For the secure landfill, the U.S. EPA requires that clay liners be constructed so that hydraulic conductivity is less than 1×10^{-7} cm/sec.

17.3.3.2 Synthetic Liner Materials

Synthetic liner materials (geomembrane liners) are composed of polymers, which are either natural or synthetic compounds of high molecular weight. Different polymeric materials may be used in the construction of geomembranes (U.S. EPA, 1989), including:

- Thermoplastics — polyvinyl chloride (PVC)
- Crystalline thermoplastics — high-density polyethylene (HDPE) and linear low-density polyethylene (LLDPE)

- Thermoplastic elastomers — chlorinated polyethylene (CPE) and chlorylsulfonated polyethylene (CSPE)
- Elastomers — neoprene and ethylene propylene diene monomer (EPDM)

Additional liner materials are discussed in Chapter 10.

Membranes contain various oils and fillers that aid in geomembrane manufacture, and these additives may affect overall performance. Chemical compatibility, stress-strain characteristics, survivability, and permeability are critical issues that must be addressed while selecting a geomembrane (U.S. EPA, 1989).

At present, HDPE, very flexible polyethylene (VFPE), PVC and other geomembranes, for example CSPE and flexible polypropylene, account for much of the market for geomembranes (PGI, 1999). The commonly used formulations for HDPE and PVC geomembranes are shown in Table 17.1. U.S. EPA regulations require a thickness of 30 mil (1 mil = 1/1000 in.) for PVC and 60 mil for HDPE geomembranes. A thicker HDPE geomembrane is required for a number of reasons including the ability to weld without damage to the liner, increased strain to tensile yield, greater stress crack resistance, and less susceptibility to folding, which can lead to stress cracking.

A primary advantage of HDPE membranes is their higher chemical resistance to hydrocarbons and solvents (Vandervoort, 1992). Differences in chemical resistance between HDPE and PVC may be significant for aliphatic and aromatic hydrocarbons and chlorinated, oxygenated, and crude petroleum solvents. The semicrystalline nature of HDPE may make it more susceptible to stress cracking when tested under stress in the presence of leachate (PGI, 1999). Manufacturers continue to develop resins that are more resistant to stress cracking, chemical attack, and oxidation, and are more cost-effective.

17.3.3.3 Composite Liners

A composite liner system is one fitted with a highly impermeable liner, for example a geomembrane, situated directly above another impermeable liner such as compacted clay. A composite liner system

---

TABLE 17.1
Example Formulations for Geomembranes

Polymer

Butyl rubber
Chlorosulfonated polyethylene
Elesticized polyolefin
Epichlorohydrin rubber
Ethylene propylene rubber
Neoprene (chloroprene rubber)
Nitrile rubber
Polyethylene
   Chlorinated
   High density
   Low density
   Linear low density
   Very low density
Polypropylene
   Flexible
   Reinforced
Polyvinyl chloride
   Elasticized
Polyvinylidene fluoride

should, therefore, outperform either a geomembrane or clay liner alone. In accordance with Darcy's law (see Chapter 10), leachate that ponds directly on top of a clay liner alone will percolate down through the liner at a rate controlled by the hydraulic conductivity of the liner, the head of the leachate on top of the liner, and the liner's total area. With the addition of a geomembrane placed directly on top of the clay and sealed against its surface, leachate moving down through a hole or defect in the geomembrane does not spread out significantly between the geomembrane and the clay liner (see Figure 17.3). The composite liner system allows much less leakage than a clay liner acting alone because the area of flow through the clay liner is much smaller.

The geomembrane must be placed on top of the clay liner such that leachate liquids do not migrate along the interface between the two liners and move downward through the entire area of the clay liner. This is analogous to placing a geomembrane on a bed of highly permeable materials such as sand. Liquids moving through a small defect in the geomembrane will spread over the whole area of the clay liner and percolate down as if the geomembrane were not even present (Figure 17.4).

As discussed above, clay liners, synthetic liners, or combinations of both are required in secure landfills. Figure 17.1 presents the synthetic composite double-liner system that appears in the U.S. EPA minimum technology guidance.

17.3.4 COMPATIBILITY OF LINERS WITH WASTES

The chemical compatibility of a geomembrane with waste leachate is a critical consideration regarding choice of material. A secure landfill is considered a permanent repository for hazardous wastes; therefore, materials used in its construction must be expected to withstand a wide range of natural stresses for very long periods. Many materials deteriorate over time when exposed to chemicals occurring in both hazardous and nonhazardous leachate. Landfill owners and operators must anticipate the composition of leachate that a site will generate and select the appropri-

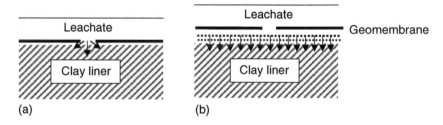

FIGURE 17.3 The FML should be placed directly upon the clay liner (a); this limits leachate infiltration into the clay (b) (U.S. EPA, 1989).

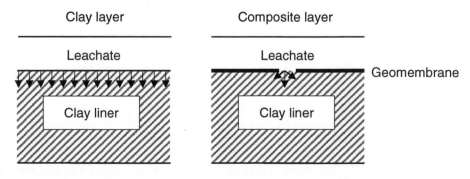

FIGURE 17.4 Leachate infiltration in a clay and a composite liner system (U.S. EPA, 1989).

ate liner materials. The chemical resistance of any geomembrane materials as well as LCR pipes should be thoroughly assessed prior to installation. Several tests for geomembranes are outlined in Chapter 10.

17.3.5 LEACHATE MANAGEMENT

A liner that is completely sealed at the base requires an efficient means of removing accumulated leachate, otherwise the head pressure will damage the geomembrane. As mentioned in Chapter 10, even the best quality material will eventually allow liquids to penetrate. Even if a liner is perfectly sealed and without holes or other disturbances, molecular diffusion will cause some liquids ponded on top of the liner system to leach through the geomembrane.

The first 'line of defense' in the collection of leachate is the primary LCR situated directly beneath the waste and above the primary geomembrane (Figure 17.1 and Figure 17.5). The primary LCR is designed and constructed on a site-specific basis in order to optimize performance. The second component of the LCR is installed between the primary and secondary liners. This secondary LCR is also known as the leak detection network. The leak detection, collection, and removal system can consist of either granular soils (i.e., gravels) or geonets. The primary purpose of this system is to determine the degree of any leakage of liquids through the primary liner. Under optimum conditions, the secondary LCR will collect negligible quantities of leachate; nevertheless, it must be designed on the basis of a worst-case scenario (U.S. EPA, 1989).

17.3.6 DRAINAGE MATERIALS

The drainage materials for the leachate management system must allow for the unimpeded flow of liquids to ensure the life of the facility. In a leachate collection system, the drains may consist of pipes, coarse soil such as gravel, or geonets (Figure 17.6). Perforated drainage pipes are in common use in a number of industrial and other (e.g., agricultural) activities. Such pipes transmit fluids rapidly. They do, however, require considerable vertical space and are susceptible to particulate clogging, biological clogging, and creep (deflection). Creep is a potential problem with both PVC and HDPE pipe materials (U.S. EPA, 1989). The crushing strength of pipes should also be tested.

FIGURE 17.5 Photograph of a LCR system. Photo by Brent Showalter. Reproduced with kind permission of Daniel Gallagher, Virginia Polytechnic University.

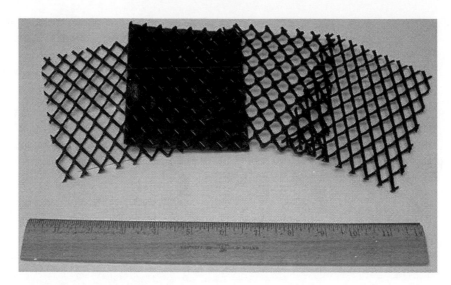

FIGURE 17.6 Sample of a geonet.

Events have been documented where landfill pipes have collapsed and failed. The ASTM D2412 test is used to measure the strength of pipe materials (ASTM, 2002).

Geonets are synthetic materials that require less space than perforated pipe or gravel, promote rapid transmission of liquids, and, because of their relatively large openings, are less likely to clog (see Chapter 10). They do, however, require geotextile filters above them and can experience problems with creep and intrusion.

Natural drainage materials should be tested to ensure that they will not dissolve in the leachate or form a precipitate that might clog the system. ASTM D2434 evaluates the ability of the materials to retain permeability characteristics (ASTM, 1968), while ASTM D1883 tests for bearing ratio, or the ability of the material to support the waste unit (ASTM, 1999).

17.3.7 LEACHATE REMOVAL SYSTEMS

A sump is needed in the lowest portion(s) of the landfill to remove accumulated leachates. The leachate removal standpipe must be extended through the entire landfill from the lowest liner to the cover, and then through the cover itself. It must also be maintained for the entire postclosure care period of 30 years or longer.

17.3.8 CLOSURE OF THE SECURE LANDFILL

Once a secure landfill or a cell is completed, it must be sealed off from its surroundings. Under the Hazardous and Solid Waste Amendments (HSWA), there are numerous requirements for preventing water entry into the landfill and for diverting and removing surface liquids.

A secure landfill cover consists of several layers. A dense clay cover is placed directly over the waste. This is compacted by heavy machinery in order to decrease pore sizes and thus slow hydraulic conductivity. After grading, a geomembrane cap is placed over the clay layer. A surface water collection and removal (SWCR) system is installed above this composite liner system. Beyond this is a soil layer that holds rooted vegetation.

17.3.9 GEOMEMBRANE CAPS

Geomembrane caps are synthetic membranes with chemical and physical properties similar to those of geomembrane liners. The geomembrane cap, however, is one component of the cover. The geomembrane cap is placed over the low permeability clay cap and below the SWCR system (Figure 17.7). Geomembrane caps function primarily by preventing surface water including precipitation from entering the landfill. In selecting materials for the geomembrane cap, operators should keep in mind some practical differences between liners and caps. Unlike a liner, a geomembrane cap is usually not exposed to leachate, so chemical compatibility is not a significant issue. Membrane caps also have lower stresses acting on them in comparison with liners. An advantage geomembrane caps have over liners is that they are much easier to repair due to their proximity to the surface of the landfill. Geomembrane caps will, however, be subject to other strains due to settlement of the waste (U.S. EPA, 1989).

17.3.10 SURFACE WATER COLLECTION AND REMOVAL SYSTEM

The surface water collection and removal (SWCR) system is installed on top of the completed unit and directly above the geomembrane cap (Figure 17.7). The purpose of the SWCR system is to prevent the infiltration of surface water into the landfill by diverting and removing any liquid that comes into contact with it. Surface water is redirected to the perimeter of the cover system. The rainwater that percolates through the topsoil and vegetative cover is carried off to an upper drainage system.

Surface water drainage systems can be composed of granular soils, geonets, or geocomposites, but the majority of drainage systems use granular soil. This is significant in frost-susceptible regions, where a 1 to 2 m (3 to 6 ft) soil layer is needed above the geomembrane liner to protect against frost penetration. In such cases, a 0.3 m (1 ft) layer of granular soil serves as the surface water collector. If good drainage materials are not available or if the site is too large, a synthetic geonet or geocomposite can be used. The advantage of drainage geocomposites is their higher flow rate capabilities compared with geonets or granular soils. All geocomposite systems are designed with polymer cores protected by a geotextile filter. Many of the polymers cannot withstand highly reactive leachates; however, in a surface drainage collector, the only contact is with water and leachate will not be encountered.

Figure 17.8 shows a typical landfill profile designed to meet the EPA minimum technology guidance requirements. The upper profile includes the soil cover, a 1 ft lateral drainage layer, and a low permeability cap of barrier soil (clay), which must be more than 2 ft thick. This three-layer system also includes a geomembrane cap and an optional gas control layer.

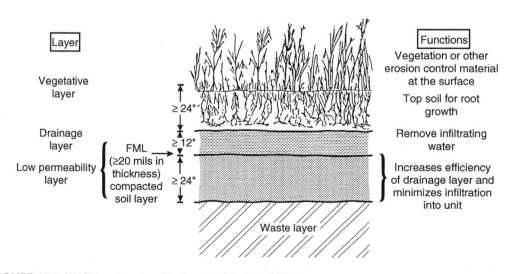

FIGURE 17.7 SWCR system, humid climate (U.S. EPA, 1989).

In designing a SWCR system, three highly practical issues must be considered: (1) cover stability, (2) puncture resistance, and (3) the ability of the system to withstand considerable stresses due to the impact of settlement. The stability of the geomembrane cap supporting the SWCR system can be affected by the materials used to construct the drainage layer and by the slope of the unit layers. Geomembrane caps must resist penetration by construction equipment, rocks, roots, and other natural phenomena. Traffic by operational equipment can cause tearing. In a Subtitle C landfill, a geomembrane is always underlain by a clay liner, which will protect against puncture. Additionally, a geotextile placed on top or beneath a membrane increases its puncture resistance by three or four times. The impact of settlement is a major concern in the design of the SWCR system. A number of facilities have settled 6 ft in a single year and 40 ft or more over a period of years (U.S. EPA, 1989).

17.3.11 Gas Control Layer

Gas collection systems are typically not used in Subtitle C landfills. Landfills dedicated to receiving only hazardous wastes are relatively new and gas is rarely detected in these systems. This is attributed to the fact that most wastes are received in stabilized or solid form, and there are no putrescible materials present as would be found in a conventional MSW landfill. Because the long-term effects of gas generation are not known and costs are minimal, the U.S. EPA recommends the use of gas collector systems. Gas collection systems are installed directly beneath the low permeability clay cap in a Subtitle C landfill.

17.3.12 Biotic Barrier

A biotic barrier is a gravel and rock layer designed to prevent the intrusion of burrowing animals into the landfill. Animals cannot generally penetrate a geomembrane cap but they can widen an existing hole or tear the material where it has wrinkled (U.S. EPA, 1989).

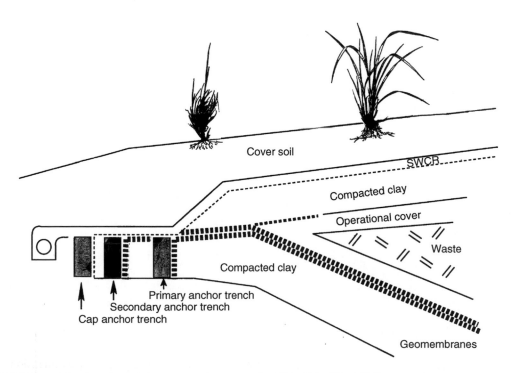

FIGURE 17.8 Profile of a Subtitle C hazardous waste landfill (U.S. EPA, 1989).

17.3.13 VEGETATIVE LAYER

The uppermost layer in the closed landfill profile is the vegetative layer. This layer is often planted with a bed of dense-rooted grasses and legumes. The vegetative layer prevents wind and water erosion, minimizes the percolation of surface water into the waste layer, and maximizes evapotranspiration, i.e., the loss of water from soil by evaporation combined with plant-induced transpiration. The vegetative layer also enhances aesthetics and promotes a self-sustaining ecosystem on top of the landfill.

There are problems in maintaining a vegetative layer on top of a landfill site, especially in arid or semiarid regions. A vegetative layer built upon a surface water collection and removal system composed of well-drained stone and synthetic material may be unable to support crops of any kind because insufficient soil moisture will be available. In arid regions a continuous sprinkler system may be needed to maintain plant growth, even if the soil is sufficiently deep and fertile.

17.4 DEEP WELL INJECTION OF HAZARDOUS WASTES

Deep well injection, also known as underground injection, involves the injection of liquid hazardous wastes into confined geologic formations deep below the surface. A geologic formation considered suitable for injection is one having sufficient permeability and porosity surrounded by thick impermeable strata.

Injection well disposal places treated or untreated liquid waste into geologic formations that are considered to have no potential for migration of contaminants into potential potable water aquifers. Injection wells have been used for the disposal of industrial and hazardous wastes since the 1950s; therefore, the equipment and methods are readily available and well known. Use of injection wells, however, continues under strict regulatory control. The overriding environmental concern regarding underground disposal of hazardous wastes is the potential for contamination of drinking water. For this reason, legislation was enacted under the Safe Drinking Water Act (SDWA) of 1974 to protect underground drinking water sources from contamination that may be caused by the disposal of hazardous liquids in injection wells (U.S. ACE, no date).

17.4.1 THE UNDERGROUND INJECTION CONTROL PROGRAM

One component of the SDWA required the EPA to report to Congress on waste disposal practices and to develop minimum federal requirements for injection practices that protect public health by preventing wells from contaminating an underground source of drinking water (USDW). A USDW is defined as an aquifer that:

- Supplies any public water system or contains sufficient water to supply a public water system
- Currently supplies drinking water for public consumption
- Contains fewer than 10,000 mg/L total dissolved solids and is not an exempted aquifer

The EPA established the underground injection control program (UIC) program to set minimum federal requirements for all injection wells that discharge hazardous and nonhazardous liquids above, below, or into USDW. The requirements affect the siting, construction, operation, maintenance, monitoring, testing and closure of injection wells. The program is also designed to provide a safe and cost-effective means for industries, municipalities, and small businesses to dispose of their wastewater, extract mineral resources, and store water for the future. All operational wells require authorization under general rules or specific permits.

Under the UIC program, a well is defined as:

- A bored, drilled, or driven shaft whose depth is greater than the largest surface dimension
- A dug hole whose depth is greater than the largest surface dimension

- An improved sinkhole
- A subsurface fluid distribution system

17.4.2 WELL SYSTEM DESIGN

A typical injection well (Figure 17.9) consists of a series of concentric pipes that may extend thousands of feet down from the ground surface into highly saline, permeable strata that are confined above and below by impermeable strata. The outermost pipe or surface casing extends below the base of any underground sources of drinking water and is cemented entirely to the surface to prevent the contamination of drinking water. Immediately inside the surface casing is a long casing that extends to the injection zone. This casing is filled with cement up to the surface in order to seal off the injected waste from the formations above the injection zone back to the surface. The waste is injected through the injection tubing inside the long casing either through perforations or in an opening below the bottom of the tube. The space between the inner casing and the injection tube, the annulus, is filled with a fluid, for example, diesel fuel under pressure, and is sealed at the bottom by a removable packer, which prevents injected wastewater from backing up into the annulus.

At the disposal facility the wastes are stored in large-capacity tanks for blending, diluting, or other processing (Figure 17.10). In some injection facilities handling very hazardous (e.g., caustic) wastes, there is no pretreatment of the waste, except simple filtering to remove particulates that may plug the well. Wastes are pumped via aboveground or underground plumbing to the pump house, which transfers the waste to the well house, a simple structure that houses the injection well head (Figure 17.11).

Common wastes for deep-well injection include caustics, volatile organic compounds (VOCs), semi volatile organic compounds (SVOCs), fuels, explosives, pharmaceutical wastes, and pesticides. However, existing permitted deep-well injection facilities are limited to a narrow range of specific wastes.

17.4.3 THE CLASSES OF INJECTION WELLS

Injection wells regulated under the UIC program are divided into five classes (Table 17.2).

17.4.3.1 Deep Wells (Class I)

Class I injection wells are designed for the disposal of industrial hazardous, industrial nonhazardous, and municipal (nonhazardous) waste. A total of 272 Class I injection facilities are operating nationwide. Of these, 51 dispose hazardous waste and 221 nonhazardous waste (U.S. EPA, 2002a). These facilities operate over 500 Class I injection wells in 19 states. The greatest concentration of Class I wells are located in the Gulf Coast, the Great Lakes, and Florida.

Class I wells dispose wastes from the following industries:

- Petroleum refining
- Metal production
- Chemical manufacture
- Pharmaceutical manufacture
- Commercial disposal
- Municipal disposal
- Food production

Examples of wastes injected include:

- Hydrochloric acid
- Sulfuric acid
- Chromic acid

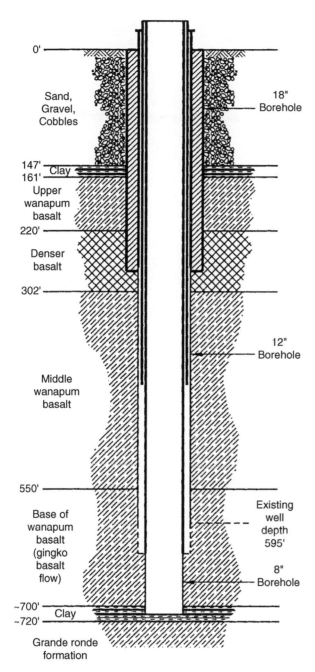

Sand, Gravel, Cobbles

18" Borehole

147'
161' Clay

Upper wanapum basalt

220'

Denser basalt

302'

12" Borehole

Middle wanapum basalt

550'

Base of wanapum basalt (gingko basalt flow)

Existing well depth 595'

8" Borehole

~700'
~720' Clay

Grande ronde formation

FIGURE 17.9 Schematic of a deep well for injecting hazardous liquid wastes (U.S. EPA, 2002c). Not to scale.

- Nitric acid
- Hydrofluoric acid
- Phosphoric acid
- Mixed acids
- Caustics
- Ammonium wastes
- Metal plating and galvanizing solutions
- CERCLA (Comprehensive Environmental Response, Compensation, and Liability Act of 1980) wastes

FIGURE 17.10 Tanks for storage and blending of liquid hazardous wastes.

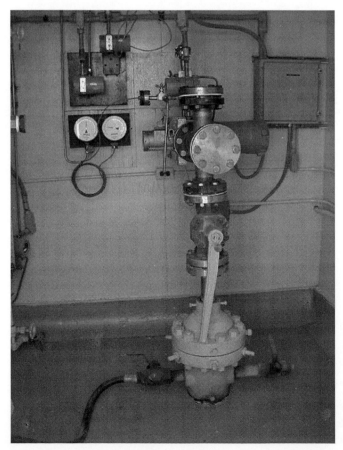

FIGURE 17.11 Well head at deep-well injection facility.

TABLE 17.2
U.S. EPA Injection Well Classification System

Well Class	Injection Well Description	Approximate Inventory
Class I	Inject hazardous wastes beneath the lowermost USDW	500
	Inject industrial nonhazardous liquid beneath the lowermost USDW	
	Inject municipal wastewater beneath the lowermost USDW	
Class II	Dispose of fluids associated with the production of oil and natural gas	147,000
	Inject fluids for enhanced oil recovery	
	Inject liquid hydrocarbons for storage	
Class III	Inject fluids for the extraction of minerals	17,000
Class IV	Inject hazardous or radioactive waste into or above a USDW	40
	This activity is banned	
	These wells can only inject as part of an authorized clean-up	
Class V	Wells not included in the other classes	500,000 to >
	Inject nonhazardous liquid into or above a USDW	685,000

Source: U.S. EPA, 2002b. With permission.

- Aqueous solutions of pesticides and pharmaceuticals
- Boiler and tank cleanouts

Although a number of toxic and hazardous wastes are acceptable for Class I injection wells, many wastes or characteristics are restricted or not accepted at all, for example:

- Reactive cyanides
- Reactive sulfides
- Flashpoint (e.g., must be greater than 100°C or 212°F)
- PCBs
- Oil content
- Radioactive wastes
- Shock-sensitive wastes
- Infectious wastes

Class I injection wells are regulated under the SDWA (UIC program) and RCRA Subtitle B (40 CFR 146.11–146.14). Class I injection wells are sited and designed such that they inject below the lowermost drinking water source with a confining zone situated above the injection zone. Injection zone reservoirs range in depth from 1700 to over 10,000 ft below the surface (U.S. EPA, 2002a). There are specific requirements regarding siting, construction, operating, monitoring and testing, reporting and record keeping, permitting and closure for all Class I wells. The two main categories of Class I wells are hazardous and nonhazardous waste wells.

Hazardous Waste Injection Wells
Injection of hazardous waste into deep wells began in the United States in the 1960s. The technology originated with the oil and gas industry. Approximately 89% of the hazardous waste that is land

disposed is through Class I wells (U.S. EPA, 2002a). Wells in this grouping are mainly used by industries such as petroleum refining and metal, chemical, and pharmaceutical manufacture.

A total of 163 Class I hazardous waste injection wells are located at 51 facilities. Most are found in Texas (78) and Louisiana (18). A small fraction of these facilities are commercial hazardous waste injection facilities. These are the only facilities that can accept hazardous waste generated off-site for injection. Ten are located in the Gulf Coast region while one is located in the Great Lakes region.

The HSWA made UIC regulations (1988) more stringent for Class I hazardous wells. This resulted in strict no-migration standards and a petition approval process for the continued operation of the wells. Of the 51 Class I hazardous waste facilities, 47 have approved no-migration petitions that cover 123 wells. To receive a no-migration petition the facility must be capable of demonstrating that injected waste will not impact groundwater or surface water for 10,000 years. Such a demonstration is often conducted using highly conservative mathematical and computer modeling in addition to scientific and engineering data.

NonHazardous Waste Injection Wells

Non-hazardous deep injection wells inject industrial, low radiation, and municipal wastewater. These wells must meet all the technical requirements of hazardous waste wells. Some states include mining wells in this group and require that these wells meet all the requirements of other deep wells.

There are 366 Class I nonhazardous injection wells nationwide. These wells occur across 19 states, and most are found in Florida (112) and Texas (110). Florida is the only state with Class I municipal waste disposal wells (104) (U.S. EPA, 2002a).

The EPA has submitted to Congress a study of Class I wells that describes the current Class I UIC program, documents past compliance incidents involving Class I wells, and summarizes studies of human health risks associated with Class I injection conducted for past regulatory efforts and policy documentation. The program is described in *Class I Underground Injection Control Program: Study of the Risks Associated with Class I Underground Injection Wells* (EPA 816-R-01-007 / March 2001).

17.4.3.2 Oil and Gas Injection Wells (Class II)

The oil and gas production industry accounts for a large proportion of the liquids injected via deep wells. When oil and gas are extracted, large amounts of salt water (brine) are often brought to the surface. About 38 L (10 gal) of brine are produced for each gallon of oil recovered. This brine is often saltier than seawater and can contain toxic metals and radioactive substances. As a result, this salt water cannot be discharged into surface water; rather, states have implemented rules that prevent the disposal of brine to surface water bodies and soils, thereby making deep well injection of this waste the prevalent form of disposal. The brine must be injected into formations similar to those from which it was extracted. Over 2 billion gallons of brine are injected daily into injection wells in the United States (U.S. EPA, 2002a).

The largest proportions of oil and gas field brines are injected into formations that contain trace quantities of extractable oil and gas. Therefore, injection of brine can actually enhance the extraction of oil and gas from the formations. Class II wells exist where there is production of oil and gas. There are approximately 167,000 oil and gas injection wells in the United States, most of which are used for this so-called secondary recovery of oil. In this process water is pumped into the formation that contains residual hydrocarbons. Some of the hydrocarbons are recovered along with the injected water by extraction wells. In a common configuration, one injection well is surrounded by four or more extraction wells. The recovered fluid is treated to remove most of the hydrocarbons in an oil–water separator.

The second type of oil and gas injection well is a disposal well. In this system, excess fluids from production and other activities directly related to the production process are injected solely for the purpose of disposal.

Class II well facilities are required to adhere to strict construction standards except when historical practices in the state and geology allow for different standards. A Class II well that follows

EPA standards is constructed with specifications similar to those for a Class I well. In 1980, Congress added Section 1425 to the SDWA that controls underground injection, relieving Class II well programs from having to meet the technical requirements in the UIC regulations. Instead, they can make a demonstration that the state has an

> ...effective program (including adequate record-keeping and reporting) to prevent underground injection which endangers drinking water sources.

Most of the oil and gas injection wells are located in the Southwest, with Texas having the largest number (53,000) and California, Oklahoma, and Kansas having about 25,000, 22,000 and 15,000 wells, respectively (U.S. EPA, 2002a). More than 700 million gallons of liquids are injected annually into Class II wells (U.S. EPA, 2002b).

17.4.3.3 Mining Wells (Class III)

A number of minerals are mined by using injection wells. The technology involves the injection of a fluid, called lixiviant, which contacts an ore containing minerals that dissolve in the fluid. The lixiviant fluid, nearly saturated with components of the ore, is pumped to the surface where the mineral is removed from the fluid. *In situ* leaching of uranium is the practice of injecting a fluid to leach out the uranium salts and pumping it back to the surface where the uranium is extracted. About 80% of the uranium extracted in the United States is produced using this technique (U.S. EPA, 2002b).

The EPA protects drinking water from contamination from mining wells by implementing regulations that set minimum standards. These regulations require mining well operators to:

- Case and cement their wells to prevent the migration of fluids into an underground drinking water source
- Never inject fluid between the outermost casing and the well bore
- Test the well casing for leaks at least once in every 5 years

17.4.3.4 Shallow Hazardous and Radioactive Injection Wells (Class IV)

Class IV wells are defined in 40 CFR 146.5(d) as:

> Wells used by generators of hazardous waste or of radioactive waste, by owners or operators of hazardous waste management facilities, or by owners or operators of radioactive waste disposal sites to dispose of hazardous waste or radioactive waste into a formation which within one quarter mile of the well contains an underground source of drinking water.

A second category includes wells that:

> ...dispose of hazardous waste or radioactive waste above a formation which within one quarter mile of the well contains an underground source of drinking water.

Class IV wells are prohibited unless the injection wells are used to inject contaminated groundwater that has been treated and is being injected into the same formation from which it was drawn. These wells are authorized by rule for the life of the well if the EPA or a state, pursuant to provisions in CERCLA or RCRA, approves emplacement of fluids.

17.4.3.5 Shallow Injection Wells (Class V)

Class V wells are a diverse grouping and are considered a 'catch-all' category; wells not included in the previous four classes that inject nonhazardous fluids into or above formations that contain USDW fall into Class V. This category came about after all the easily definable wells were placed into Classes I through IV. The EPA did not have definite information on these wells when it published the UIC regulations in the late 1970s.

The Class V designation includes shallow wells such as septic systems and dry wells, used to place nonhazardous fluids directly below the land surface. However, Class V wells can be deep, highly sophisticated wells. The EPA estimates that there are more than 650,000 Class V wells in the United States although some estimates place the number as high as one million. Class V wells are located in every state, especially in unsewered areas where the population may depend on groundwater as its source of drinking water. Aquifers serve as drinking water sources for about 90% of public water systems in the United States (U.S. EPA, 2001).

Class V wells provide a convenient and inexpensive means of disposing of a variety of nonhazardous fluids. Some examples of Class V wells are:

- Agricultural drainage wells
- Storm water drainage wells
- Large-capacity septic systems
- Sewage treatment effluent wells
- Spent brine return flow wells
- Mine backfill wells
- Aquaculture waste disposal wells
- Solution mining wells
- *In situ* fossil fuel recovery wells
- Experimental wells
- Aquifer remediation wells
- Geothermal electric power wells
- Geothermal direct heat wells
- Heat pump or air conditioning return flow wells
- Salt water intrusion barrier wells
- Aquifer recharge and aquifer storage and recovery wells
- Subsidence control wells
- Industrial wells
- Radioactive waste disposal wells other than Class IV

The majority of Class V well owners are small businesses and municipalities and the two most numerous Class V wells are stormwater drainage and large-capacity septic systems (U.S. EPA, 2002b). For facilities that generate nonhazardous wastes, Class V wells provide for disposal when there is no access to a sewer system. Class V wells are also an alternative to discharges to surface water.

The effective management of Class V wells is critical because of their large number, the wide variety of fluids discharged, and because most accessible fresh water is stored in aquifers. Regulation of Class V wells continues to pose a problem; many owners of Class V wells are not aware of the UIC regulations and furthermore may not consider their facility (e.g., a heat pump system well) to be an injection well.

17.4.4 PRACTICAL CONSIDERATIONS OF DEEP-WELL INJECTION

The primary purpose of the UIC program is to protect current and potential drinking water supplies by keeping injected liquids within the well and the intended injection zone. Injected liquids can contaminate USDWs by several major pathways (Table 17.3).

There are a wide range of factors that may limit the applicability and effectiveness of deep-well injection; some of the primary factors include the following (FRTR, 2003):

- It is not feasible in areas of seismic activity.
- Injected wastes must be compatible with the components of the injection well system and the natural formation water.

TABLE 17.3
The Major Pathways of Contamination of Underground Sources of Drinking Water

Faulty well construction	Leaks in well casing or fluid escaping between well's outer casing and well bore
Nearby wells	Fluids from pressurized area in injection zone may escape through wells in injection area.
Faults or fractures in confining strata	Fluids may leak out of pressurized area through faults or fractures in confining beds.
Direct injection	Inject fluids into or above USDWs
Diplacement	Fluid may be displaced from injection zone into hydraulically connected USDWs

Source: U.S. EPA, 2002b. With permission.

- The waste generator may be required to perform physical, chemical, biological, or thermal treatment for the removal of various constituents from the waste to modify the physical and chemical character of the waste and ensure its compatibility.
- High concentrations of suspended solids (typically >2 ppm) can cause plugging of the injection equipment and injection zone.
- Corrosive media may react with the injection well components, with the injection zone formation or with confining strata.
- Equipment will last longer if wastes are neutralized prior to injection.
- High Fe concentrations may cause fouling when conditions alter the valence state and convert soluble into insoluble species.
- The presence of significant amounts of organic carbon may result in rapid microbial population growth and subsequent fouling.
- Waste streams containing organic contaminants above their solubility limits may require pretreatment before injection into a well.
- Site assessment and aquifer characterization are required to determine the suitability of the site for wastewater injection.

REFERENCES

American Society for Testing and Materials, Standard Test Method for Determination of External Loading Characteristics of Plastic Pipe by Parallel-Plate Loading ASTM D2412-02, West Conshohocken, PA, 2002.

American Society for Testing and Materials, Standard Test Method for CBR (California Bearing Ratio) of Laboratory-Compacted Soils, ASTM D1883-99. West Conshohocken, PA, 1999.

American Society for Testing and Materials, Standard Test Method for Permeability of Granular Soils (Constant Head)ASTM D2434-68, West Conshohocken, PA, 1968.

Federal Remediation Technologies Roundtable, Deep well injection 2003. See: http://www.frtr.gov/matrix2/section4/4_57.html

Gallagher, G., Virginia Polytechnic University. See: http://www.cee.vt.edu/program_areas/environmental/teach/gwprimer/hazmat/hazmat.html, no date.

Haxo, H.E. Jr., Haxo, R.S., Nelson, N.A., Haxo, P.D., White, R.M., Dakessian, S., and Fong, M.A.,. Liner Materials for Hazardous and Toxic Wastes and Municipal Solid Waste Leachate, Noyes Publications, Park Ridge, NJ, 1985.

PGI, Comparison of 30 mil PVC and 60 mil HDPE Geomembranes, PGI Technical Bulletin, Urbana, IL, March 1999.

Tenax, No date, *Leachate Collection and Removal System* (LCRS), Baltimore, MD. See: http://www.tenaxus.com/landfills/applications/leachate.html

U.S. Army Corps. of Engineers, No date, Commercial Deep Well Injection Facilities. http://www.environmental.
    usace.army.mil/library/pubs/tsdf/sec4-3/sec4-3.html
U.S. Environmental Protection Agency, Underground Injection Control (UIC) Program,2002a. See: http:
    //www.epa.gov/safewater/uic/classii.html
U.S. Environmental Protection Agency, Protecting Drinking Water through Underground Injection Control, Pocket
    Guide #2, EPA 816-K-02-001, Office of Ground Water and Drinking Water, Washington, DC 2002b.
U.S. Environmental Protection Agency, EPA Presents Revised Proposal for Skyline Water System, Moses Lake
    Wellfield Contamination Skyline Water System, Superfund Fact Sheet, April 2002. See:http://
    yosemite.epa.gov/R10/CLEANUP.NSF/9f3c21896330b4898825687b007a0f33/2f3ffa55a450a07288
    25653100688c3a/$FILE/20020417mosesfs.pdf
U.S. Environmental Protection Agency, Safe Drinking Water Act, Underground Injection Control (UIC)
    Program, EPA 816-H-01-003, Office of Water, Washington, DC. August, 2001.
U.S. Environmental Protection Agency, Land Disposal Restrictions for Hazardous Wastes, EPA530-F-99-043,
    Office of Solid Waste and Emergency Response, Washington, DC, 1999.
U.S. Environmental Protection Agency, Requirements for Hazardous Waste Landfill Design, Construction, and
    Closure, Seminar Publication, EPA/625/4-89/022, Center for Environmental Research Information,
    Cincinnati, OH, 1989.
Vandervoort, J., *The Use of Extruded Polymers in the Containment of Hazardous Wastes*, Schlegel Lining
    Technology, Inc., The Woodlands, TX, 1992.

SUGGESTED READINGS

Ahlert, R.C. and Kosson. D.S., Treatment of Hazardous Landfill Leachates and Contaminated Groundwater,
    Project Summary, U.S. Environmental Protection Agency, Risk Reduction Engineering Laboratory,
    Cincinnati, OH, 1990.
Banta, E.R., Example Evaluation of a Permit Application for a Proposed Hazardous- Waste Landfill in Eastern
    Adams County, Colorado, U.S. Dept. of the Interior, Geological Survey: Open-File Services Section,
    Western Distribution Branch, Denver, CO., 1986.
Beval, D.L., Benthic Invertebrates in Lake Marion and Selected Tributaries in the Vicinity of a Hazardous-
    Waste Landfill near Pinewood, South Carolina, 1988, U.S. Dept. of the Interior, U.S. Geological
    Survey, Books and Open-File Reports, Columbia, SC, and Denver, CO, 1991.
Burt, R.A. Streamflow, Lake-Flow Patterns, Rainfall, and Quality of Water and Sediment in the Vicinity of a
    Hazardous-Waste Landfill near Pinewood, South Carolina, March 1987 through early January 1989, U.S.
    Dept. of the Interior, U.S. Geological Survey, Books and Open-File Reports, Columbia, SC, and Denver,
    CO, 1991.
De Rouche, J.T. and Breen., K.J., Hydrogeology and Water Quality Near a Solid-and Hazardous-Waste
    Landfill, Northwood, Ohio. Dept. of the Interior, U.S. Geological Survey, U.S. Geological Survey,
    Books and Open-File Reports, Columbus, OH, and Denver, CO, 1989.
Gilbert, P.A. and Murphy W.L., Prediction/Mitigation of Subsidence Damage to Hazardous Waste Landfill
    Covers, Project Summary, U.S. Environmental Protection Agency, Hazardous Waste Engineering
    Research Laboratory, Cincinnati, OH, 1987.
Hix, K., Leak Detection for Landfill Liners, Overview of Tools for Vadose Zone Monitoring, U.S.
    Environmental Protection Agency, Technology Innovation Office, Washington DC, 1998.
Prowell, D.C, Geology Near a Hazardous Waste Landfill at the Headwaters of Lake Marion, Sumter County,
    South Carolina, Department of the Interior, U.S. Geological Survey, Denver, CO, 1990.
Richardson, G.N., and Koerner, R.M., Geosynthetic Design Guidance for Hazardous Waste Landfill Cells and
    Surface Impoundments, U.S. Environmental Protection Agency, Hazardous Waste Engineering Research
    Laboratory, Cincinnati, OH, 1988.
Schroeder, P.R., Gibson, A.C., and Smolen, M.D., The Hydrologic Evaluation of Landfill Performance (HELP)
    Model, Municipal Environmental Research Laboratory, Office of Research and Development, U.S.
    Environmental Protection Agency, Office of Solid, Waste and Emergency Response, U.S.
    Environmental Protection Agency, Cincinnati, OH, and Washington DC, 1984.
Simonds, F.W. Simulation of Ground-water and Potential Contaminant Transport at Area 6 Landfill, Naval Air
    Station Whidbey Island, Island County, Washington, U.S. Dept. of the Interior, U.S. Geological
    Survey, Information Services, Tacoma, Washington, and Denver, Colorado, 2002.

Streissguth, T., *Nuclear and Toxic Waste*, Greenhaven Press, San Diego, CA, 2001.

Tedder, D.W., and Pohland., F.G., *Emerging Technologies in Hazardous Waste Management*, American Chemical Society, Washington DC, 1990.

Vroblesky, D.A., Hydrogeologic Framework and Simulation of Shallow Ground-water Flow in the Vicinity of a Hazardous-waste Landfill near Pinewood, South Carolina, U.S. Geological Survey, USGS Science Information Center, Open-File Reports Section, Columbia, SC, and Denver, CO, 1994.

Vroblesky, D.A., Hydrogeology and Ground-water Quality near Hazardous-waste Landfill near Pinewood, South Carolina, U.S. Dept. of the Interior, U.S. Geological Survey, Books and Open-File Reports, Columbia, SC, and Denver, CO, 1992.

QUESTIONS

1. The LDR program includes treatment standards for all hazardous wastes scheduled for land disposal. It has three major components that address hazardous waste disposal, dilution, and storage. Discuss these components.

2. What waste types are restricted from secure landfills under current federal regulations? Be specific.

3. Under what conditions may the double-liner requirement be waived for a hazardous waste facility?

4. List the advantages of clay liner materials as compared with synthetics. What are the relative disadvantages of using clay?

5. What is an advantage of HDPE liners over PVC and other polymers? Are there any significant disadvantages with the use of HDPE?

6. Discuss three tests commonly employed for assessing the integrity of a synthetic landfill liner material.

7. If a synthetic liner is perfectly sealed and without holes or other disturbances, molecular diffusion will allow some of the organics from the liquids ponded on top of the liner to leach through. True or false.

8. Is chemical compatibility with wastes of equal concern regarding installation of FMCs and FMLs? Explain.

9. What are the major purposes of layered landfill caps? How do they function?

10. What is the significance of saturated hydraulic conductivity (K_S) in assessing soils for landfill liners, caps, and as a soil base? How is K_S influenced by soil texture? By engineering practices (e.g., compaction)? What is the RCRA limit for liner K_S?

11. Geomembrane installation practices significantly influence possible leachate losses from a landfill. List and discuss those factors that must be considered for successful geomembrane design and installation.

12. The "Second Law of Thermodynamics" states, in essence, that all systems proceed toward maximum disorder (chaos, entropy, etc.). In this context, how may a landfill be affected over time with respect to the Second Law? In other words, discuss how landfill liners, LCR, caps, etc. may be transformed one hundred years after landfill closure. Will the landfill remain impervious *ad infinitum*?

13. Discuss the five major classes of injection wells. What are the primary types of wastes acceptable in each?

14. Discuss the UIC program. What is its overall scope and purpose?

15. Under what conditions (and in what class[es] of wells) are dioxin-containing wastes permitted for injection?

Part IV

Special Categories of Waste

The final section of this book serves as a "catch-all" section covering the management of those wastes that may not fit well in a regulatory or management context with either municipal or hazardous wastes. Universal Wastes, used motor oil, medical waste, construction and demolition debris, and electronics waste are included. Some of these residuals, for example, medical or infectious wastes and certain electronics wastes, may be inherently hazardous; others have the reputation of being simply nuisance materials (e.g., construction and demolition debris), which are generated in large volumes. Many of these wastes had not been adequately managed in the past. Regulations regarding the management of these special wastes have evolved rapidly over the past decade and continue to be updated with the advent of new waste types in the waste stream and new technologies for their treatment and disposal.

18 Universal Wastes

How much of the Source escapes with thee
How chief thy sessions be
For thou hast borne a universe
entirely away.

1517, Emily Dickinson (1881)

18.1 INTRODUCTION

Routine management of hazardous wastes in accordance with RCRA requirements has earned a reputation among some industries as being a costly, cumbersome and time-consuming ball-and-chain. In response to the requests of generators to ease regulatory burdens on businesses, the U.S. EPA formulated the Universal Waste Rule (40 CFR Sec. 273), first published in the May 1995 Federal Register. The primary purpose of the rule is to promote appropriate recycling and disposal of several potentially hazardous wastes produced in large quantities by businesses regulated under RCRA.

The wastes addressed in the new rule include batteries, pesticides, thermostats, and certain types of lamps. Until recently, such wastes were managed solely as RCRA hazardous wastes. As a consequence, disposal costs for many businesses were substantial. There were also the inevitable episodes of improper disposal (Figure 18.1). The streamlined recycling and disposal of the above wastes substantially reduces the amount of hazardous wastes in the municipal solid waste stream. Removing them from MSW landfills and incinerators furthermore decreases the threat to public health and the environment. In 1989, 643,000 kg (1.4 million lb) of mercury was discarded in MSW in the United States and 84% of this was landfilled (Building Green, 2004). Household batteries were by far the largest contributors of mercury in MSW in 1989 with other sources including thermostats and thermometers (3.9%), and mercury-containing lamps (3.8%). The EPA Office of Solid Waste estimates that roughly 600 million fluorescent lamps are discarded each year.

By reducing administrative requirements the Universal Waste Rule saves companies in compliance costs and reduces the amount of time spent completing paperwork. Specifically, the rule streamlines requirements related to notification, labeling of packaging, accumulation time limits, employee training, response to releases, off-site shipments, tracking, exports, and transportation. For example, under the Universal Waste Rule a waste generator may transport these wastes with a standard carrier instead of a hazardous waste transporter. In addition, the Uniform Hazardous Waste Manifest is not required.

Conditionally exempt small quantity generators (CESQGs) are not affected by the Universal Waste Rule. The U.S. EPA encourages CESQGs to participate voluntarily in collection and recycling programs by bringing universal wastes to collection centers for proper treatment and disposal. Universal wastes are also generated by households, which are not regulated under RCRA and are permitted to dispose of these items in ordinary trash. Although MSW landfills designed under RCRA can handle small amounts of hazardous wastes, these wastes are better managed in a recycling program or transferred to collection centers for proper disposal. In states that adopt the Universal Waste Rule, communities can work with both businesses and residents to facilitate proper recycling or disposal of these wastes. Municipalities can establish collection programs or participate with collection and management programs established by local businesses.

FIGURE 18.1 Clearly *not* the proper method to manage universal waste lamps.

18.2 UNIVERSAL WASTES DEFINED

Universal wastes are limited to a small group of wastes produced in relatively large quantities by businesses, institutions, and private homes. The specific wastes are defined by the 40 CFR 272 regulations and examples are provided.

"*Battery*" is defined as (40 CFR 273.0):

> a device consisting of one or more electrically connected electrochemical cells which is designed to receive, store, and deliver electric energy. An electrochemical cell is a system consisting of an anode, cathode, and an electrolyte, plus such connections (electrical and mechanical) as may be needed to allow the cell to deliver or receive electrical energy. The term battery also includes an intact, unbroken battery from which the electrolyte has been removed.

Batteries such as nickel-cadmium (Ni-Cd) and small lead-acid batteries found in household and commercial items including electronic equipment, portable computers, mobile telephones, and emergency backup lighting, are included in this definition.

"*Pesticide*" means:

> any substance or mixture of substances intended for preventing, destroying, repelling, or mitigating any pest, or intended for use as a plant regulator, defoliant, or desiccant, other than any article that:

- Is a new animal drug under FFDCA section 201(w)
- Is an animal drug that has been determined by regulation of the Secretary of Health and Human Services not to be a new animal drug
- Is an animal feed under FFDCA section 201(x) that bears or contains certain listed substances

Agricultural pesticides that have been banned from use, are obsolete, have become damaged, or are no longer needed due to changes in agricultural practices are in this category. Many such pesticides have been stored for long periods.

"*Thermostat*" means:

a temperature control device that contains metallic mercury in an ampule attached to a bimetal sensing element, and mercury-containing ampules that have been removed from these temperature control devices in compliance with the requirements of 40 CFR 273.13(c)(2) or 273.33(c)(2).

Many thermostats contain as much as 3 g of liquid mercury and are located in commercial, industrial, agricultural, community, and household buildings.

"*Lamp*", also referred to as "universal waste lamp" is defined as:

the bulb or tube portion of an electric lighting device. A lamp is specifically designed to produce radiant energy, most often in the ultraviolet, visible, and infrared regions of the electromagnetic spectrum. Examples of common universal waste electric lamps include fluorescent, high intensity discharge, neon, mercury vapor, high pressure sodium, and metal halide lamps.

Such universal waste lamps contain mercury and sometimes lead, and are found in homes and businesses.

18.3 CATEGORIES OF UNIVERSAL WASTE HANDLERS

Management of universal wastes is similar for large- and small-quantity handlers. Differences occur primarily with respect to notification, documentation, and employee training.

18.3.1 LARGE-QUANTITY HANDLERS OF UNIVERSAL WASTE

A "large-quantity handler" of universal wastes is defined as a facility that generates greater than 5000 kg (approx. 11,000 lb) of combined batteries, pesticides, thermostats, and lamps during a calendar year. A large-quantity handler must provide written notification of universal waste management to the state or federal environmental regulatory agency and must receive an EPA Identification Number. Notification must include:

* The handler's name and mailing address
* The name and business telephone number of a contact person at the facility
* The location of the universal waste management activities
* A list of all of the types and amounts of universal waste managed by the handler

18.3.2 SMALL-QUANTITY HANDLERS

A small-quantity handler of universal waste is permitted to generate not more than 5,000 kg of universal wastes in a calendar year. In contrast to large-quantity handlers, the small quantity handler does not need to provide written notification of universal waste management to the state regulatory agency. Similarly, there is no requirement to receive an EPA Identification Number. Additional details about generator types and requirements are provided below.

18.4 MANAGEMENT OF UNIVERSAL WASTES

18.4.1 UNIVERSAL WASTE BATTERIES

In 1997, 753,000 metric tons (830,000 tons) of non-ferrous meterials other than aluminum were recovered (U.S. EPA, 1999), most of this being composed of lead (Pb) recovered from batteries. Automotive batteries are considered hazardous because they contain both Pb, in the form of metallic Pb and PbO pastes, and sulfuric acid. Since 1987 the lead-acid battery has been one of the most highly recycled products in the United States. More than 93% of Pb from spent batteries, approximately 900 million kg (2 billion lb) per year, is recycled in the production of new batteries (Exide, 2002). It is estimated that 94.3% of battery lead was recovered in 1997.

In 1998 over 3 billion industrial and household batteries were sold in the United States alone. Demand for batteries in the United States is projected to increase by 5.8% every year due to technological advances in the battery industry and the convenience and portability of battery-operated equipment (Sova and Mueller, n.d.). As a consequence, demand for these metals, whether as virgin ore or recycled, will continue to increase.

Under 40 CFR 273, a large-quantity handler of universal wastes is required to manage batteries properly in order to prevent releases to the environment. Specific requirements are comparatively simple and include transferring and containing batteries that are leaking or damaged to a suitable container. The container must be compatible with battery contents and must be kept closed. In addition, a large-quantity handler is permitted to carry out the following activities with universal waste batteries:

- Sort batteries by type
- Mix battery types in one container
- Discharge batteries to expel the electric charge
- Regenerate used batteries
- Disassemble batteries into individual cells
- Remove batteries from consumer products
- Remove electrolyte from batteries

A large-quantity handler who removes electrolyte from batteries must determine whether the electrolyte fails any characteristic test for a hazardous waste (40 CFR Part 261). If a characteristic of hazardous waste is expressed, the electrolyte must be managed in compliance with 40 CFR Parts 260 through 272 (standards applicable to management of hazardous wastes). The handler is thus designated a generator of hazardous waste.

A large-quantity handler of universal waste must label waste containers to identify contents. Universal waste batteries or their containers must be labeled with one of the following: "Universal Waste — Batteries," "Waste Batteries," or "Used Batteries."

18.4.2 LEAD ACID BATTERY RECYCLING

Spent (dead) batteries are broken apart manually or mechanically in a hammermill and the lead, plastic, and acid are separated. The metallic and plastic battery pieces fall into a vat where the lead and heavy materials fall to the bottom and the plastic floats. The floating polypropylene pieces are skimmed away and the liquids are drawn off, leaving the heavier components behind.

The Pb grids, PbO, and other lead components are melted together in smelting furnaces. Molten lead is poured into ingot molds to form either large ingots weighing 900 kg (2000 lb) called "hogs" or smaller ingots weighing 30 kg (65 lb) called "pigs." Impurities (dross) float to the top of the molten lead mixture in the ingot molds. The dross is removed and the ingots are allowed to cool. The cooled ingots are removed from the molds and shipped to battery manufacturers where they are remelted and used in the production of new lead plates and other parts for new batteries (Figure 18.2).

Spent battery acid is managed in two ways. The acid is neutralized with an industrial alkali which forms a simple inert salt solution. The solution is tested for chemical composition and can be discharged to a publicly owned treatment works. Acid is also treated by conversion to sodium sulfate, an odorless and harmless salt, which is used in certain industrial and manufacturing processes. The plastic casing is shredded and the polypropylene pieces are washed, dried in a stream of air, and shipped to a plastics recycler where the pieces are melted together into a near-liquid state. The molten plastic is forced through an extruder that produces small pellets of uniform size. The pellets are finally sold to a battery case manufacturer for eventual reuse.

18.4.3 NICKEL–CADMIUM BATTERY RECYCLING

Nickel–cadmium battery recycling occurs through a high-temperature metal recovery process. First, shredding using a hammermill disassembles the Ni–Cd batteries. Consumer type Ni–Cd battery

FIGURE 18.2 Recycling of spent lead-acid batteries: (a) manual breaking of industrial batteries; (b) reverbatory furnace for lead smelting; and (c) lead ingots ('pigs') (OSHA, no date).

packs are placed in a natural gas-fired thermal oxidizer where the plastic is vaporized, leaving clean nickel–cadmium battery cells. Gases from the thermal oxidizer are sent to a rotary hearth furnace where they are combusted for heat value. The clean Ni–Cd battery cells are placed in a cadmium recovery furnace. Cadmium is reduced using carbon, vaporized and condensed, thereby producing elemental cadmium, Cd^o. This cadmium is eventually used in the production of new Ni–Cd batteries. The remaining nickel and iron become feedstocks for a metal recovery plant. The material is used to produce nickel, chromium, and iron remelt alloy that is used in stainless steel production (Sova and Mueller, n.d.).

18.4.4 Universal Waste Pesticides

A large-quantity handler must contain universal waste pesticides in either:

- Its original container
- Its container overpacked in another container
- In a tank that meets the requirements of 40 CFR Part 265
- In a vehicle or vessel

All of the above must be closed, structurally sound, and compatible with the pesticide. A container, tank, transport vehicle, or vessel that holds universal waste pesticides must include the product label, and the words "Universal Waste — Pesticides" or "Waste — Pesticides".

Many states have enacted programs to promote the drop-off and recycling of agricultural and other pesticides (Figure 18.3); and many have been highly successful in preventing improper disposal.

18.4.5 Universal Waste Thermostats

Mercury is a public health and environmental concern because it is a potent neurotoxin with the potential to accumulate in the food chain.

The Thermostat Recycling Corporation (TRC) recycles used mercury-switch thermostats in the 48 continental U.S. states. TRC is a private corporation established by thermostat manufacturers Honeywell, General Electric, and White-Rodgers. Between 1998 and 2000, TRC collected 75,000 thermostats containing 300 kg (656 lb) of mercury. Under this voluntary, industry-sponsored effort, heating and cooling contractors drop off old mercury-switch thermostats at participating wholesalers who collect them in protective bins supplied by TRC. Full bins are sent to TRC's recycling center where the switches are removed and forwarded to a mercury recycler.

TRC devotes much of its attention to heating and air-conditioning contractors and wholesalers as these entities sell and install the majority of thermostats and because the industry already has the infrastructure to support an effective recycling program. Many states encourage cooperation with the TRC program, urging contractors to sign a pledge of cooperation (Figure 18.4). Some local governments have established separate programs to manage recycling or disposal of used thermostats directly from homeowners.

A large-quantity handler managing universal waste thermostats is required to contain any thermostat that is leaking or damaged. The container must be closed and compatible with the contents of the thermostat. A large-quantity handler may remove mercury-containing ampules from thermostats provided they are handled to prevent breakage and opened over a containment device. Also, the handler must ensure that a mercury cleanup system is available and the work area is well ventilated to ensure compliance with OSHA exposure levels for mercury. The facility must train employees removing ampules with proper waste mercury handling and emergency procedures. A large-quantity handler who removes mercury-containing ampules from thermostats must determine whether any residues are characteristic hazardous wastes described in 40 CFR Part 261. If the mercury or residues are not hazardous, the waste may be managed according to solid waste regulations. Universal waste thermostats must be labeled with "Universal Waste — Mercury Thermostats," "Waste Mercury Thermostats," or "Used Mercury Thermostats."

OHIO DEPARTMENT OF AGRICULTURE
Pesticide Collection
Turn in unwanted pesticides.

Pesticides Only: A pesticide is any of the following examples:

Insecticides -	Insect Control
Herbicides -	Weed Control
Fungicides -	Control of Fungi & Disease
Rodenticide -	Control of Rodents

All Pesticides must be preregistered, just fill out the registration form below and mail it to:
Ohio Department of Agriculture, Pesticide Regulation, 8995 East Main Street, Reynoldsburg, Ohio 43068

ALL INFORMATION WILL BE KEPT CONFIDENTIAL
You will then be notified by mail where to turn in your pesticides for disposal.   There is no cost to participate.
All costs are paid by the Ohio Department of Agriculture.

**PLEASE PRINT ALL INFORMATION NEATLY. IF THE PESTICIDE LABEL IS UNREADABLE, LIST IT AS
UNKNOWN FOR THE PESTICIDE NAME.**

How Generated		Pesticide Name	Liquid (√)	Dry (√)	Aerosol (√)	Actual Pounds/Gallons	Size of Container (lbs., oz., etc.)
Home (√)	Business (√)						

If you need additional space, make copies of this form. If you have any questions, contact Larry Berger (614) 728-6392.
Only preregistered pesticides will be accepted.

CONTACT NAME:_____ COMPANY NAME:_____

ADDRESS:_____

CITY/ZIP:_____

PHONE:_____COUNTY:_____TYPE OF BUSINESS:_____
            (See other side for sample)   (For Survey Only)

FIGURE 18.3 Form printed at the state level to encourage pesticide recycling (Ohio State Pesticide Education Program). Reproduced with kind permission of the Ohio Department of Agriculture.

18.4.6 UNIVERSAL WASTE LAMPS

Fluorescent lamps work by passing an arc of electricity through mercury vapor in the lamp. The charged mercury atoms emit ultraviolet (UV) light, which is absorbed by a phosphor powder coating on the inside of the cylindrical glass lamp. The energized phosphors subsequently emit a white light. To generate the mercury vapor a small amount of elemental (liquid) mercury is added to each lamp during manufacture. The mercury instantly vaporizes when the lamp is turned on and it recondenses when the lamp is turned off. Most types of high-intensity discharge lamps, i.e., mercury vapor, metal halide, and high-pressure sodium, also contain mercury (EBN, 1997).

The Wisconsin Department of Natural Resources (WDNR) is promoting the Mercury Thermostat Reduction and Recycling Pledge Program for HVAC contractors, and wholesalers. We want you to participate.

The goals of our program are to reduce the use of mercury-containing thermostats by beginning to replace any unusable or unwanted mercury-containing thermostats with an electronic thermostat and to recycle discarded mercury-containing thermostats. All participating HVAC contractors and wholesalers will receive free promotional items to help them publicize their leadership and commitment to the environment. These items include fact sheets, a certificate for the office, and participation patches for uniforms and hats to be worn by field staff .

To achieve these goals, WDNR encourages all HVAC contractors and wholesalers to participate in this program. The pledge program consists of four parts:

▼ Increase awareness of customers on the hazards of mercury and encourage their participation in this program,

▼ Encourage the Wisconsin public to replace unusable or unwanted mercury-containing thermostats with digital or mercury-free thermostats,

▼ Ensure that mercury-containing thermostats are returned to a participating supplier for recycling with Thermostat Recycling Corporation (TRC), and

▼ Keep a supply of digital or mercury-free thermostats on hand.

(a)

The recycling portion of this program is offered by the Thermostat Recycling Corporation (TRC). The recycling of these mercury-containing thermostats is offered free-of-charge to all contractors. TRC offers mercury-thermostat collection bins to all Wisconsin HVAC wholesalers. The wholesalers who participate will pay a one-time deposit fee and receive the thermostat collection bin and directions for pick-up of a full bin. TRC will handle the fees for shipping, handling, and recycling of the mercury-containing thermostats.

WDNR is asking all contractors to collect any old, discarded mercury-containing thermostats at your shop. Collection of these thermostats as a whole unit is not inconsistent with the regulations. The key is to collect the thermostats as whole units - with the casing. DO NOT remove the mercury bulb or drain the liquid mercury from a thermostat. Collect the entire thermostat (non-damaged) and place them in some type of a sturdy, non-leaking container. Each contractor can bring their thermostats to a participating local wholesaler. TRC will handle the rest of the process to include recycling of the mercury in each thermostat. For more information, TRC has a brochure, *Answers To Dealer Questions About Recycling Mercury Containing Thermostats.*

If your company is interested in participating in the Mercury Thermostat Reduction and Recycling Pledge Program, please read and sign the Contractor Pledge Form. Once received by WDNR, the form will be signed, dated and a copy will be returned to you. In addition to the signed form, you will also receive copies of a fact sheet, a certificate, and promotional patches, if requested. Please indicate the number of each free promotional item along with your address on the reverse side of the pledge sheet. Supplies are limited. Please return the pledge form either by fax or mail. If you have further questions, contact John Shenot at 608/267-0802. Thanks for your interest in working together to protect Wisconsin's environment. Through reduction initiatives like this one, Wisconsin can become a cleaner, healthier place to live.

Mercury Thermostat Reduction and Recycling Pledge Program

Contractor Pledge

We pledge to work with Wisconsin Department of Natural Resources and our heating, ventilation, and air conditioning (HVAC) supplier to keep mercury out of the environment in order to protect the quality of Wisconsin's air, land and water.

Our company is committed to protecting the environment. Therefore, we, the undersigned Wisconsin HVAC contractor, have established these goals to minimize the impact of mercury thermostats on the environment and encourage our customers to conduct sound environmental practices.

Pledge

To achieve these goals, we pledge to participate in the Wisconsin Department of Natural Resource's Mercury Thermostat Reduction and Recycling Program. As a participant in this program we will:

▼ Increase awareness of our customers on the hazards of mercury and encourage their participation in this program,

▼ Encourage the Wisconsin public to replace unusable or unwanted mercury-containing thermostats with digital or mercury-free thermostats,

▼ Ensure that mercury-containing thermostats are returned to a participating wholesaler for recycling with the Thermostat Recycling Corporation, and

▼ Keep a supply of digital or mercury-free thermostats on hand.

Pledge Form

Company Representative

Company Name

Street Address

City

State   Zip/Postal Code

Phone

Fax

E-Mail

WDNR Representative

Today's Date (to be filled in at WDNR)

Requested Promotional Items:

Patches: _____ (Limited to Field Staff Only)

Certificate: _____ (Limited to One)

Fact Sheets: _____

Please mail or fax the signed form to:

John Shenot CO/8
P.O. Box 7921
Madison, WI 53707-7921
608.267.0802 (telephone)
608.267.0496 (FAX)
shenoj@dnr.state.wi.us

(b)   PUB-CO-102 99

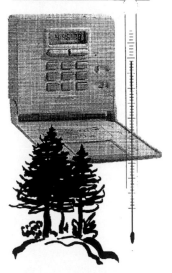

Mercury Thermostat
Reduction & Recycling Pledge Program
for Wisconsin HVAC Contractors

FIGURE 18.4 Pledge program form for mercury thermostat reduction (Wisconsin Department of Natural Resources). Reproduced with kind permission of Wisconsin Department of Natural Resources.

The amount of mercury in fluorescent lamps is highly variable depending on the type of lamp and year of manufacture. T-12 lamps (measuring 38 mm or 12/8 in. in diameter) contain significantly more mercury than T-8 lamps (25 mm or 1 in. in diameter). Additionally, the mercury content per lamp has declined substantially over the past 15 years. Most 1.2 m (48 in.) T-12 fluorescent lamps in use today contain 20 to 60 mg of mercury, although some have as much as 80 mg and new ones average 22.8 mg. Compact fluorescent lamps are highly variable in their mercury content, with an average of 4 mg (EBN, 1997).

For years, the U.S. EPA considered mercury-containing lamps to be ordinary MSW that generators could dispose of along with everyday MSW. Landfilling had been the traditional means of disposing spent lamps (NEMA, 2001). Due to their relatively low mercury content, lamps have historically accounted for only 3.8% of all the mercury deposited in municipal landfills (Truesdale et al., 1993). Debate has ensued over the years as to the extent of hazard such disposal may create. While some studies claim that there is little to no mercury in landfill leachate, others claim that there are substantial mercury releases in leachate (Massachusetts DEP, 1996; U.S. EPA, 1988).

Until recently most municipal waste incinerators were not equipped with the necessary controls to reduce mercury emissions. The incineration of mercury-containing lamps released up to 90% of mercury into the air. By the end of 2000 most incinerators added stringent new EPA-mandated mercury controls, significantly reducing the amount of mercury that incinerators released from any mercury-containing product (Truesdale et al., 1993).

In 1990 the U.S. EPA revised the test required to identify a hazardous waste, i.e., the "EP-Tox" test was replaced by the Toxicity Characteristic Leaching Procedure (TCLP) (see Chapter 11). Under the new test many spent mercury-containing lamps failed and were classified as hazardous waste. In response, the industry developed numerous types of lamps reduced in mercury content that could pass the TCLP and which were, therefore, no longer classified as hazardous waste. Federal law exempts households and CESQGs from handling spent lamps that fail the TCLP. A few states ban the disposal of all lamps in solid waste, however.

Both large- and small-quantity handlers of universal waste must contain any lamp in containers that are structurally sound, adequate to prevent breakage, and compatible with the contents of the lamps. Such containers and packages must remain closed and must show no evidence of leakage. A handler of universal waste must immediately clean up any broken lamp and place the pieces into a suitable container.

Safe recycling of fluorescent lamps involves the separation of the three primary components of the lamp: glass, aluminum end-caps, and phosphor powder (which is where most of the mercury resides in a used lamp). These components are crushed and separated, thus reclaiming nearly all of the mercury. The recovered mercury is triple-distilled and sold on the commodity market. Other recovered materials can also be marketed. The Philips Company, for example, uses substantial amounts of recycled phosphor in its new lamps (Wilson, 1997).

Recycling has the advantage of reusing potentially hazardous materials from fluorescent lamps, including mercury. Studies indicate that mercury releases into the air from well-managed lamp recycling facilities are relatively low. However, the reclaimed glass often contains some residual mercury that can be released as the glass is distributed through commerce and processed at high temperatures to make new products. In some areas of the United States the number of lamp recycling operations is still limited. In addition, the market value of the reclaimed materials from lamps is negligible, such that lamp recycling is not overly profitable and waste generators or government programs must pay for recycling (NEMA, 2001).

18.5 RESPONSE TO RELEASES

A handler of universal waste must contain releases of universal wastes and its residues. The handler must determine whether any material resulting from the release is hazardous waste, and if so, must manage the waste in compliance with 40 CFR Parts 260 through 272.

18.6   TRACKING UNIVERSAL WASTE SHIPMENTS

A large-quantity handler must maintain records of each shipment of universal waste received at the facility. The record for each shipment received must include:

- The name and address of the originating waste handler from whom the waste was sent
- The quantity of each type of universal waste received
- The date of receipt of waste shipment

A large-quantity handler must keep records of each shipment of waste sent from the handler to other facilities. The record must include:

- The name and address of the universal waste handler and destination facility to whom the universal waste was sent
- The quantity of each type of universal waste sent
- The date the shipment of universal waste left the facility

A large-quantity handler must retain the records for 3 years from the receipt of waste shipment. Similarly, the handler must retain the records of all wastes shipped off-site for at least 3 years from the shipping date.

18.7   SMALL-QUANTITY HANDLERS OF UNIVERSAL WASTE

As discussed for large-quantity handlers, the small-quantity handler of universal waste must manage universal waste batteries, pesticides, and thermostats in order to prevent releases to the environment. Operating requirements are essentially identical to those of the large-quantity handlers. A small-quantity handler may accumulate universal waste for no longer than 1 year from the date the waste is generated or received. The handler must be able to demonstrate the length of time that the waste has been accumulated.

A small-quantity handler is prohibited from sending or taking universal waste to a location other than another universal waste handler, a destination facility, or a foreign destination. If a small-quantity handler self-transports universal waste off-site, it becomes a universal waste transporter and must comply with universal waste transporter requirements. A small-quantity handler of universal waste is not required to maintain records of universal waste shipments.

A small-quantity handler must provide adequate information regarding universal waste hazards to all employees who are responsible for managing universal waste. The information must describe proper handling and emergency procedures appropriate to the types of universal waste handled at the facility.

A summary of the differences in regulatory requirements between large- and small-quantity handlers of universal waste is given in Table 18.1.

18.8   UNIVERSAL WASTE TRANSPORTERS

A universal waste transporter is responsible for the safe transport of universal waste. Restrictions on transporter operations include:

- Prohibitions from disposing universal waste
- Prohibitions from diluting or treating universal waste except when responding to releases

A universal waste transporter must comply with U.S. Department of Transportation (DOT) regulations in 49 CFR Parts 171 through 180 for the transport of any universal waste that meets the

TABLE 18.1

Differences in Regulatory Requirements for Handlers of Universal Wastes and RCRA Hazardous Wastes

	Small-Quantity Handler of Universal Waste	Large-Quantity Handler of Universal Waste	CESQG	SQG	LQG
Quantity handled by category	Accumulate <5,000 kg (11,000 lb) on site at any one time	Accumulate <5,000 kg (11,000) or more on site at any one time	Generate <100 kg (220 lb) per month; <1 kg acute per month	Generate <1,000 kg (2,200 lb) per month	Generate >1,000 kg per month >1 kg acute per month
	§273.9	§273.9	§261.5(a) and (e)	§260.10	Part 262 and §261.5 (e)
EPA identification number	Not required	Required	Not required	Required	Required
	§273.12	§273.32	§261.5	§262.12	§262.12
On-site accumulation limit	<5,000 kg	No quantity limit	<1,000 kg <1 kg acute <100 kg spill residue from acute	<6,000 kg	No quantity limit
	§273.9		§§261.5(f)(2) and (g)(2)	§262.34(d)(1)	
Storage time limit (without a storage permit)	1 year, unless for proper recovery, treatment, or disposal	1 year, unless for proper recovery, treatment, or disposal	None	<180 days or <270 days	<90 days
	§273.15	§273.25	§261.5	§§262.34(d) and (e)	§262.34(a)
Manifest	Not required	Not required but must keep basic shipping records	Not required	Required	Required
	§273.19	§273.39	§261.5	§262.20	§262.20
Personal training	Basic training	Basic training geared toward employee responsibilities	Not required	Basic training	Full training
	§273.16	§273.36	§261.5	§262.34(d)	§262.34(a)

Source: U.S. EPA, 2002. With permission.

definition of "hazardous material" in 49 CFR §. 171.8. Some universal waste materials are regulated by DOT as hazardous materials because they meet the criteria for one or more hazard classes listed in 49 CFR §. 173.2. As universal waste shipments do not require a manifest, they are not described by the DOT shipping name "hazardous waste."

A universal waste transporter is permitted to store the universal waste at a transfer facility for a maximum of 10 days. If wastes are stored for more than 10 days, the transporter becomes a universal waste handler and must comply with the requirements for handlers. A universal waste transporter must immediately contain all releases of universal wastes. It must be determined whether any of the release is hazardous waste and if so, it is subject to 40 CFR Parts 260 through 272.

18.9 DESTINATION FACILITIES

A universal waste destination facility is defined as a facility that treats, disposes of, or recycles a particular category of universal waste. The operator of a destination facility is subject to the requirements of Parts 264, 265, 266, 268, 270, and 124, and the notification requirement under section 3010 of RCRA.

The formal requirements for a Universal Waste Destination Facility are very limited in scope. This contrasts with the extensive body of requirements for RCRA-regulated TSDFs, for hazardous wastes (40 CFR Parts 265 and 265). The destination facility is prohibited from sending or taking universal waste to a site other than a universal waste handler, another destination facility, or foreign destination. If the destination facility receives a shipment containing hazardous waste that is not a universal waste the operator must notify the state regulatory office of the shipment.

The destination facility must keep a record of each shipment of universal waste received at the facility. The record for each shipment received must include:

- The name and address of the universal waste handler, destination facility, or foreign shipper from whom the universal waste was sent
- The quantity of each type of universal waste received (e.g., batteries and pesticides)
- The date of receipt of the shipment of universal waste

Records must be retained by the destination facility for at least 3 years from the receipt of the shipment.

As should be obvious at this point, the Universal Waste Rules were designed to coincide with the RCRA regulations for the management of hazardous wastes; however, the former were designed to be more business-friendly in order to promote recycling and streamline waste management. The differences between the Universal Waste and hazardous waste rules are highlighted in both Table 18.1 and Table 18.2.

18.10 RESPONSIBILITY OF THE STATES

When the U.S. EPA issues a new rule, states authorized to implement the RCRA program are expected to adopt the rule as well. A state rule-making procedure is required in order to implement

TABLE 18.2
Significant Differences Between Universal Waste and Hazardous Waste Rules

	Universal Waste Transporters (40 CFR part 273 subpart D)	Hazardous Waste Transporters (40 CFR part 263)
Compliance with DOT	Yes	Yes
	(§273.52(a) cites DOT requirements 49 CFR parts 171–180)	(§263.10 DOT requirements 49 CRF parts 171–179)
EPA ID number	None	Yes
		(§263.11)
Allow to store up to 10 days at transfer facility	Yes §273.53	Yes (§263.12)
Manifest requirements	None	Yes
		(§263.20–22)
Response to releases	Yes	Yes, with more complex requirements
	§273.54	(§263.30–31)

Source: U.S. EPA, 2002. With permission.

TABLE 18.3

Universal Wastes Specific to States. These Wastes are also Referred to as "State-only Universal Wastes"

Waste	State(s)
Aeroseol cans	CO
Antifreeze	LA, NH
Ballasts	ME, VT
Cathode ray tubes	ME, MI, NH, RI
Electronic devices and electronic components	CO
Mercury-containing devices	CO, MA, MI, ND, NH, PA, RI
Barometers	NH, RI
Gas flow regulators	NH
Intact mercury-containing ampules	NH
Manometers	NH, RI
Mercury switches and relays	MI, NH, RI
Mercury thermometers	MI
Sphygmomanometers	NH, RI
Thermocouples	NH, RI
Thermometers	NH. RI
Water meters	NH

Note: California has proposed to add cathode ray tube materials, consumer electronic devices, aerosol cans, and mercury-containing motor vehicle light switches (public hearing September 30, 2002).

Source: U.S. EPA, 2002. With permission.

FIGURE 18.5 Universal waste programs in the United States (U.S. EPA, 2002a).

the new rule. The EPA encourages state adoption of the Universal Waste Rule to encourage improved management of universal wastes. State adoption facilitates several interstate issues such as import and export between states. Similar Universal Waste programs among states facilitate the implementation of regional collection and disposal programs.

If states adopt the rule, they are also allowed to add wastes to their universal waste programs without their addition to the federal rule. States can also establish requirements that are more stringent compared with the federal regulations. Universal wastes specific to states are listed in Table 18.3, and the overall distribution of state programs is depicted in Figure 18.5.

REFERENCES

Building Green.com. Disposal of fluorescent lamps and ballasts. See http://www.buildinggreen.com/features /ds/disposal.cfm

Code of Federal Regulations, Volume 40 Part 273, Standards for Universal Waste Management, U.S. Government Printing Office, Washington, DC, 2000.

Code of Federal Regulations, Volume 40 Part 279, Standards for the Management of Used Oil, U.S. Government Printing Office, Washington, DC, 2000.

Exide Technologies, Battery Recycling and the Environment, 2002.See: http://www.exide.com/about/environment.html.

Massachusetts Department of Environmental Protection, Mercury in Massachusetts: An Evaluation of Sources, Emissions, Impacts and Controls, June 1996. See: www.state.ma.us/dep/files/mercury/appb.htm.

National Electrical Manufacturers Association, Fluorescent Lamps and the Environment: Mercury Use, Environmental Benefits, Disposal Requirements, NEMA01BR, Washington, DC, January 2001.

Occupational Health and Safety Administration, No date, Battery Breaking and Separating. See: http://www.osha.gov/SLTC/etools/leadsmelter/credits.html

Sova, C., and Mueller, H., No date, Battery Recycling Makes Sense in the Second Millennium, Battery Solutions, Inc. South Brighton, MI. http://www.batteryrecycling.com/.

The Thermostat Recycling Corporation, No date, Answers to Dealer Questions about Recycling Mercury Containing Thermostats, Rosslyn, VA.

Truesdale, R.S., Beaulieu, S.M., and Pearson, A., *Management of used Fluorescent Lamps: Preliminary Risk Assessment,* Research Triangle Institute, Raleigh, NC, May 1993.

U.S. Environmental Protection Agency, State-Specific Universal Waste Regulations, 2002a. http://www.epa.gov/epaoswer/hazwaste/id/univwast/uwsum.htm.

U.S. Environmental Protection Agency, Proposed CRT Rule, 2002b. See: http://www.epa.gov/epaoswer/osw/meeting/pdf02/goode.pdf

Wilson, A., Disposal of Fluorescent Lamps and Ballasts, Environmental Building News, Vol. 6, No. 9, October 1997. See: http://www.buildinggreen.com/features/ds/disposal.cfm

FURTHER READINGS AND WEB SITES

California Department of Toxic Substance and Control, Final Regulations: Universal Waste Rule, 2000.See: http://www.dtsc.ca.gov/LawsRegulationsPolicies/UWR/UWR_final_regs.html

Colorado Department of Public Health and Environment, Compliance Bulletin, Hazardous Waste Universal Waste Rule, June 2003. See: http://www.cdphe.state.co.us/hm/UWRgeneral.pdf

Illinois Environmental Protection Agency, Used Fluorescent and High-Intensity-Discharge Lamps as Universal Wastes, 2003. See: http://www.epa.state.il.us/land/fluorescent-lamps/

Michigan Department of Environmental Quality, Universal Wastes (Including Electric Lamps) and Lead Acid Batteries, 2003. See: http://www.michigan.gov/deq/1%2C1607%2C7-135-3585_4130_4192-12659—%2C00.html

North Carolina Division of Pollution Prevention and Environmental Assistance, No date, Fluorescent Lights and Mercury. See: http://www.p2pays.org/mercury/lights.htm

Wuf Technologies, No date, Capabilities of Recycling Universal Wastes, Concord, NH, See http://www.wuftech.com/CapabilitiesUniversals.htm

QUESTIONS

1. What is the overall rationale for the Universal Waste Rule? What are the advantages for waste generators to comply with this rule rather than Subtitle C?
2. What types of facilities must comply with the Universal Waste Rule?
3. Are universal waste handlers and universal waste transporters required to comply with 40 CFR 264 and 265 regulations? What types of operations are exempted from the regulations?
4. List and describe the different classes of universal wastes. In your opinion, and based on current data for volumes generated, are other waste types appropriate for classification as universal waste? Would personal computers fit well in this category? Give reasons.
5. How do large-quantity and small-quantity handlers of universal waste differ from large- and small-quantity generators of hazardous waste in terms of weight limitations, notification requirements, and storage requirements? Which of these categories of handler or generator requires an EPA identification number?
6. How long can a facility accumulate a universal waste? How does this differ from small- and large-quantity generators of hazardous waste?
7. When are batteries considered a universal waste and what are their handling requirements?
8. If a large-quantity handler removes electrolyte from automotive batteries, are there any requirements to determine hazardous waste characteristics?
9. How do tracking requirements differ for large-quantity handlers of universal wastes compared with large-quantity generators of hazardous waste? Is a manifest system required for the former? Explain.
10. How are mixtures of universal waste and hazardous waste regulated?
11. If a company has several locations at which universal waste is generated, how should each location be handled, i.e., is each considered a separate handler?
12. If a small-quantity handler of universal waste in 1 month out of the year generates 6250 kg of universal waste, does the handler's classification change?
13. According to the Universal Waste Rule, if a transporter stores universal waste for more than 10 days what will the transporter be classified as?
14. If an authorized state adds a waste to its universal waste category, would this designation be applicable in other states?
15. Because a manifest is not required for compliance with the Universal Waste rule, how will handlers know how and when their universal waste finally arrives at an appropriate destination facility?
16. A handler who transports universal waste is considered a universal waste transporter. Does this rule apply for *any* quantity of universal waste?

19 Management of Used Oil

The development of the superculture is the result of the knowledge explosion, which led not only to new theories and processes, but to new discoveries, especially of fossil fuels and rich ores. In 1859 the human race discovered a huge treasure chest in its basement. This was oil and gas, a fantastically cheap and easily available source of energy. We did, or at least some of us did, what anybody does who discovers a treasure in the basement — live it up and we have been spending this treasure with great enjoyment.

Kenneth E. Boulding (1910–1993)
Ecodynamics: A New Theory of Societal Evolution, 1978

19.1 INTRODUCTION

An estimated 1.4 billion gal of used oil are generated each year in the United States alone. Automotive maintenance facilities, do-it-yourself (DIY) oil-changing practices, manufacturing companies, electric-generating stations, and mining and smelter operations are among the primary sources of used oil. Prior to RCRA regulations, a significant portion of used oil was utilized and disposed of improperly; for example, used oil had routinely been applied to control road dust and kill roadside weeds. New regulations, however, have banned certain uses and encouraged others. For example, under RCRA, about 1.9 billion L (500 million gal) of used oil are burned in 30,000 boilers and industrial furnaces (Mouche, 1995). Unfortunately, large quantities of oil continue to be improperly disposed; about 200 million gal are dumped onto the ground, discarded in ordinary household trash, and poured down storm sewers and drains (U.S. EPA, 1994). The persistent components of these oils are transported via sewers and large water bodies where they sink into sediments. Pollution due to used motor oil occurs worldwide and reaches several million tons yearly.

The Used Oil Management Standards (40 CFR Part 279) impose requirements affecting the storage, transportation, burning, processing, and re-refining of used oil by commercial facilities. For facilities that generate used oil, the regulations establish storage standards. For a used oil marketer (i.e., one that sells used oil directly to a burner facility) there are additional tracking and paperwork requirements. Transporters of used oil are responsible for the chemical analysis of used oil shipments and for compliance with both RCRA and DOT requirements. Finally, recyclers and burners of used oil must comply with requirements for analysis of the oil, furnace type, and air pollution control measures.

19.2 PRODUCTION AND PROPERTIES OF LUBRICATING OILS

Petroleum refining is designed to extract a wide range of fuels and lubricating oils as a function of their respective boiling points. Within the refining tower many lubricating oils are recovered including engine oil, industrial transmission oils, hydraulic oils, heat-transfer oils, cutting oils, electrical oils, and others. Lubricating oils are distinguished from other fractions of crude oil by their high (>400°C [750°F]) boiling point and high viscosity. Compounds separated for the production of lubricating oils are hydrocarbons containing as many as 40 carbons per molecule. In these oils there is a predominance of normal (i.e., straight chain) and branched paraffins. There are also polycycloparaffins, with rings commonly condensed. Mono-, di-, and trinuclear aromatics, for example, naphthalene and phenan-threne, are the main components of the aromatic portion. Lubricating oils possess a high additive content. These compounds are included in an oil blend to improve both physical and chemical prop-erties. Additive content can be as high as 20%, the most important being detergents and dispersants (Table 19.1) (Vazquez-Duhalt, 1989). The technology used in the fabrication of different lubricat-ing oils, including the type and quantity of additives, is unique for each manufacturer.

19.3 CHEMICAL COMPOSITION OF USED MOTOR OIL

Used motor oil may originate from monograde automotive engine oil, multigrade automotive engine oil, mineral-based crankcase oil, and railway diesel oil. There is no consistent chemical composition of used motor oil since it is a function of the crude oil source, refining processes, pres-ence of additives, and length of time in use (ATSDR, 1997a). In general, used motor oil contains small amounts of gasoline, additives (detergents, dispersants, oxidation inhibitors, rust inhibitors, and viscosity improvers), N and S compounds, metals such as Pb, Zn, Ca, Ba and Mg, and a broad range of aromatic and aliphatic hydrocarbons with chain lengths ranging from C_{15} to C_{50} (Dominguez, 2002). Cotton et al. (1977) were unable to determine a single predominant organic compound in 30 samples of used motor oil examined. However, several broad classes of compounds were commonly present in the oil including saturated aliphatic, mono-, di-, tri- and polyaromatic ring products. Aliphatic compounds comprise about 73 to 80% of the total weight of used motor oil. This fraction is composed of alkanes and cycloalkanes of 1 to 6 rings. Monoaromatics and diaro-matics make up another 11 to 15% and 2 to 5% of the weight, respectively (Vazquez-Duhalt, 1989). The percentages of hydrocarbons in crankcase oil are shown in Figure 19.1.

TABLE 19.1
Typical Formulation of Engine Oils

Ingredient	% by Volume
Base oil (solvent 150 neutral)	86
Detergent inhibitor (zinc dialkyl dithiophosphate)	1
Detergent (barium and calcium sulfonates)	4
Multifunctional additive (polymethyl-methacrylates)	4
Viscosity improver (polyisobutylene)	5

Source: U.S. EPA, 1974. With permission.

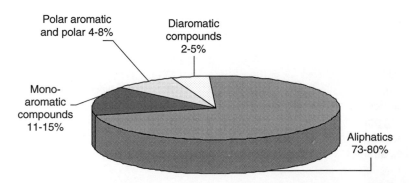

FIGURE 19.1 Chemical composition of hydrocarbons in used motor oil (Vazquez-Duhalt, R., Sci. Total Environ., 79, 1-23, 1989. Reproduced with kind permission of Elsevier, Oxford, UK).

Examples of common monoaromatic structures in used motor oil include benzene and its derivatives, toluene, xylene, and ethylbenzene (Figure 19.2). Used motor oil contains higher percentages of polycyclic aromatic hydrocarbons (PAHs) and additives compared with fresh oil (Vazquez-Duhalt, 1989; Hewstone, 1994; Dominguez, 2002). The concentration of PAHs in used oil may range from 34 to 190 times higher than that in fresh motor oil (Grimmer et al., 1982). These compounds represent a direct hazard to the environment and human health. Hydrocarbons have low solubility in water and high solubility in fat. Material Safety Data Sheets of used motor oil indicate a specific gravity of 0.9 and a low vapor pressure. Used motor oil is defined as a stable material; however, it may be incompatible with oxidizing agents.

The concentration of heavy metals in used oil is much higher than in fresh oils. Metals could originate in the crude petroleum, in motor oil additives, from engine wear, or after contact with fuel. High concentrations of lead could arise from the contact of lead-based gasoline vapors with motor oil during engine operation (Shi-Hsien Chen et al., 1994), although this is no longer an issue in the United States due to the phaseout of leaded fuels in the late 1970s. Raymond et al. (1975) found 13 metals in used motor oil with a concentration ten times higher than that in crude oil.

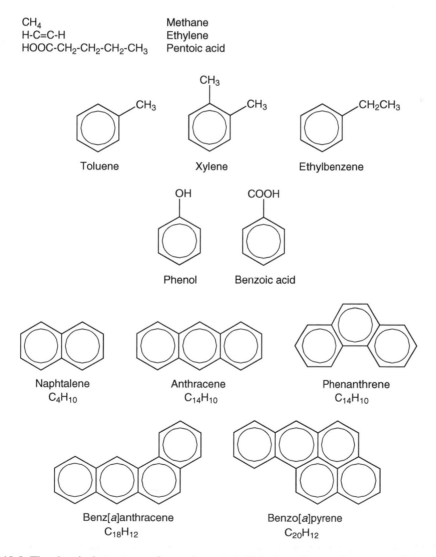

FIGURE. 19.2 The chemical structures of some important aliphatic and aromatic compounds in used oil.

19.4 CHEMICAL AND PHYSICAL CHANGES OF OIL DURING ENGINE OPERATION

In general, fresh oil is transformed by oxidation, nitration, cracking of polymers, decomposition of organometallics, and other processes. These chemical changes are brought about during motor operation from the high temperatures and mechanical strains occurring within the engine. The main chemical changes that lubricating oils undergo are caused by heating and oxidation. The minimum decomposition temperature range for most hydrocarbons in motor oils is about 300 to 315°C (575 to 600°F). Oxidation products such as acids and esters, PAHs, and resins and asphaltenes, are formed at lower temperatures. The oxidation products are more easily cracked by heat and the cracked materials are readily oxidized (Gruse, 1967). New motor oils are often light in color and become darker during use because of oxidation reactions, contamination from the combustion chamber, and wear from the piston. Generally, oxidation of oils in engines produces soluble acidic compounds and semisolid to solid insoluble materials. Laboratory studies have shown that the oxidation of paraffin is as follows (Gruse, 1967):

Paraffin hydrocarbon → hydroperoxide → water + ketone → carboxylic acid + aldehydes

Aldehydes of low molecular weight include formaldehyde, acetaldehyde, or propionaldehyde. These compounds can be oxidized, producing acids that are soluble in water. The carboxylic acid produced as a result of the reaction will have a high molecular weight and will be water-soluble; therefore, the oil will become more corrosive.

Water constitutes a contaminant of oil, entering the crankcase and fuel tanks by natural "breathing" and may condense there. About 1 gal of water is produced when a gallon of gasoline is burned in an engine (Gruse, 1967). Some water, as a vapor, will migrate to the crankcase and condensation could produce rust.

19.5 POTENTIAL HAZARDS WITH USED OIL

The presence of these cleaning and dispersing substances, along with chemical transformations, result in the production of a range of contaminants in used oil. In a study of 1000 samples of used oil, significant concentrations of toxic components such as 1,1,1-trichloroethane, trichloroethylene, tetrachloroethylene, toluene, and naphthalene were detected (Bergeson, 1985). PAHs are of particular public health and environmental relevance because of their carcinogenic properties, and a total of 140 different PAHs have been detected in used motor oil. PAHs also occur in fresh oil but in much lower quantities. They originate primarily from the fuel (Mumford et al., 1991) and the combustion process.

The mutagenic effects of used engine oil have been determined by the Ames toxicity test (Ames et al., 1973). The Ames test was developed to determine whether a specific chemical is a mutagen. The test is based on the assumption that any substance that is mutagenic to a specific strain of the bacterium *Salmonella typhimurium* may also be carcinogenic to humans. Many chemicals are not carcinogenic in themselves, but become converted into carcinogens as they are metabolized by the body. It is for this reason that the Ames test includes a mixture of liver enzymes. A large number of chemicals used in industry provide a positive Ames test.

In one study, up to 70% of the carcinogenic effects of used oil were caused by PAHs with more than three rings. This fraction represented less than 1% of the total volume of the oil. Of this fraction, 18% of the effects were caused by benzo[*a*]pyrene. There were few mutagenic effects caused by the PAH-free portion of the used oil. Schulte et al. (1993) found a significant increase in lung tumors and a dose-dependent increase in malignant lung tumors for mice exposed to PAH-enriched exhausts containing 0.05 or 0.09 mg/m^3 benzo[*a*]pyrene. Thyssen et al. (1981) showed evidence of a dose-response relationship between inhaled benzo[*a*]pyrene particles (most were between 0.2 and 0.54 μm in diameter) and respiratory-tract tumorigenesis. Respiratory-tract tumors were induced in the nasal cavity, pharynx, larynx, and trachea in a dose-related manner in hamsters exposed to 9.5

or 46.5 mg/m^3 for 109 weeks (ATSDR, 1997b). Studies have shown an increase in the carcinogenic effect of engine oil with mileage (after 25,000 km) in gasoline-powered engines (Hewstone, 1994). However, The Department of Health and Human Services (DHHS), the International Agency for Research on Cancer (IARC), and the U.S. EPA have not classified used crankcase oil with regard to its mutagenicity or carcinogenicity in humans (Hewstone, 1994; ATSDR, 1997b). Furthermore, used motor oil is not considered hazardous waste under RCRA.

As a result of friction within a metallic cylinder, metals are scoured from pistons and the cylinder walls. Therefore, used oil commonly contains high levels of heavy metals such as Pb, Zn, Cu, Cr, Ni, and Cd; all are toxic to biota. Apart from heavy metals and contaminants formed within the oil itself, used crankcase oil contains other compounds that are dangerous to the environment such as detergents and lubrication additives. Contaminants of potential concern in used oils are listed in Table 19.2.

19.6 HISTORY OF USED OIL MANAGEMENT

In 1980, the U.S. Congress directed the U.S. EPA through the Used Oil Recycling Act to determine a suitable classification for used oil in order to ensure its appropriate management. The used oil regulations are codified in 40 CFR Part 279. Used oil is defined in 40 CFR Part 279 as:

> any oil that has been refined from crude oil, or any synthetic oil, that has been used and as a result of such use is contaminated by physical or chemical impurities.

This definition does not include crude oil or fuel oils spilled onto land or water, wastes from petroleum refining operations, mineral oils, or oils derived from animal or vegetable fats. Examples of used oil according to the EPA definition are shown in Table 19.3 (EPA, 2003). Based on composition data such as that described above, there are significant concerns regarding the effects of used oil on public health and the environment. Congress therefore directed the EPA to determine whether used oil should

TABLE 19.2
Contaminants of Potential Public Health and Environmental Concern Occurring in Used Oil

Contaminant	Probable Source	Approximate Concentration Range
Aromatic hydrocarbons(μg/L)	Petroleum base stock	
PAHs		
Benzo[a]pyrene		360–62,000
Benz[a]anthracene		870–30,000
Pyrene		1,670–33,000
Monoaromatics alkylbenzenes	Petroleum base stock	900,000
Diaromatic naphthalenes		440,000
Chlorinated hydrocarbons (μg/L)		
Trichloroethanes	May be formed during use of	18–1,800
Trichloroethylenes	contaminated oil	18–2,600
Perchloroethylene		3–1,300
Metals(mg/kg)		
Ba	Additives	60–690
Zn		630–2500
Al	Engine wear	4–40
Cr		5–24
Pb	Contamination from leaded gasoline	3,700–14,000

Source: U.S. EPA, 1981. With permission.

TABLE 19.3

Examples of Used Oil according to 40 CFR Part 260.10

Used Oil	Not Used Oil
Engine oil (gasoline, diesel engine, crankcase oils, piston-engine oils for cars, trucks, boats, airplanes, locomotives, and heavy equipment)	Waste oil that is bottom cleanout waste from virgin fuel storage tanks, fuel oil spill cleanups, or other oil wastes that have not been used.
Synthetic oil (from coal, shale, or polymer-based starting material)	
Transmission fluid	Products such as antifreeze and kerosene
Refrigeration oil	Vegetable and animal oil, even when
Compressor oil	used as a lubricant.
Metalworking fluids and oils	Petroleum distillates used as solvents.
Industrial and hydraulic fluid	
Copper and aluminum wire drawing solution	
Electrical insulating oil	
Industrial process oils	
Oils used as buoyant	

Source: U.S. EPA, 2003. With permission.

be designated a RCRA hazardous waste, and to formulate regulations for its appropriate management and disposal. In 1985 the EPA proposed a rule to list all used oil as hazardous waste (U.S. EPA, 1994). Standards were proposed for recycling used oil and restrictions were established regarding the burning of used oil. The new rule raised concerns, however, that the new restrictions and associated increased costs for management may discourage recycling. By the following year, the EPA rescinded the rule (RCRA, 1999). In November 29, 1986, EPA designated used oil as a nonhazardous waste.

Based on reviews of toxicological and other data, a court decision in 1988 required the EPA to reconsider their most recent designation of used oil. In response to these conflicting factions, three options for used oil management were published in the 1991 *Federal Register*:

- To designate all used oils as listed hazardous wastes under RCRA
- To designate only certain used oils (primarily nonindustrial oils) as hazardous
- To formulate management standards for used oil and classify used oils as RCRA hazardous material when disposed (RCRA, 1999)

On May 20, 1992, a final ruling stated that used oil destined for disposal would not be listed as a hazardous waste. The EPA reasoned that used oil has the potential to be recycled or re-refined. Used oil could then be utilized as a fuel or recycled as a lubricant instead of disposed of in landfills. Under the final ruling, in addition, drained used oil filters could also be disposed of as nonhazardous waste (40 CFR Part 261.4). On September 10, 1992, the Used Oil Management Standards were published in the *Federal Register* (U.S. EPA, 1999). The standards are codified in 40 CFR Part 279. The standards regulate used oil generators, collection centers, transporters and transfer facilities, re-refiners, used oil burners that burn off-specification used oil, facilities using the oil as a dust suppressant, and facilities that dispose of used oil.

19.7 REQUIREMENTS OF THE USED OIL MANAGEMENT STANDARDS

19.7.1 USED OIL GENERATORS

Any business that produces used oil as a result of commercial or industrial operations, or collects used oil from these operations or private households is classified as a used oil generator. Examples include industrial facilities that use lubricated machinery, taxi, bus, and delivery companies; government and military motor pools; and shipyards. Household used oil generators, also known as do-it-yourself (DIY) operations, are not regulated under management standards; however, DIY collection centers are considered generators.

Used oil generators are required to engage in "good housekeeping practices in handling used oil." Some important requirements include:

- Storage containers and tanks at generator facilities must be labeled "Used Oil."
- Fill pipes used to transfer used oil into underground storage tanks must be labeled "Used Oil."
- Storage is to occur only in containers or tanks in good condition (no severe rusting, structural defects, and not leaking).
- Used oil spills and leaks must be immediately stopped, contained, and cleaned.

Used oil generators are subject to the spill prevention, control and countermeasures program (SPCC) as listed in 40 CFR Part 112. The SPCC consists of plans for emergency response in the event of a spill or release. Plans are specified by the Oil Pollution Act of 1990, which was promulgated to address spills into surface water. Industries managing used oil must comply with SPCC specifications, except facilities located in zones where a release is not expected to reach a navigable waterway. Used oil generators are also subject to the Underground Storage Tank standards (40 CFR Part 280) for storage in underground tanks. There is no requirement for generators to have secondary containment for containers or tanks of used oil.

19.7.2 Used Oil Collection Centers

A used oil collection center accepts and stores used oil collected from generators that bring in shipments of no more than 55 gal. Collection centers may also accept used oil from household DIYs. Collection centers must:

- Comply with the used oil generator standards, Subpart C (40 CFR Parts 279.20-279.24)
- Be registered, licensed or permitted by a state, county or municipal government to manage used oil

19.7.3 Marketers of Used Oil

Used oil marketers are defined as generators, processors, re-refiners, transporters, or burners of used oil. They must comply with additional management standards such as conducting analyses to ensure that the used oil has the suitable burning specifications (see below), and to maintain these analysis reports for 3 years.

19.7.4 Transporters and Transfer Facilities

Used oil transporters are subject to more stringent requirements than are generators. Transporters are required to obtain an EPA identification number when shipping used oil off-site in amounts greater than 55 gal, and are permitted to deliver used oil only to:

- Another used oil transporter, a used oil processing and re-refining facility or an off-specification used oil burner facility that has obtained an EPA identification number
- An on-specification used oil burner facility (no EPA identification number required)

A key responsibility of a used oil transporter is to determine whether or not used oil is hazardous by measuring the total halogen content. Transporters can apply the analytical Test Method for Evaluating Solid Waste (SW-846) (U.S. EPA, 1986) for total halogen content. If concentrations of individual halogenated constituents are found to be insignificant, the used oil is not considered hazardous waste. If, however, more than 1000 ppm of halogenated compounds are measured, the used oil is considered hazardous waste and must be handled under RCRA Subtitle C (standards for management of hazardous waste). Analytical reports are to be maintained by the transporter for 3 years.

The transporter is required to maintain records for shipments of used oil for 3 years. Transfer facilities are required to keep the same storage, labeling, and release response specifications as generators. In addition, tanks and storage containers must be equipped with impermeable secondary containment. This includes an impermeable floor and may include dikes, soil berms, or retaining walls.

19.7.5 DOT REQUIREMENTS

The Hazardous Materials Transportation Act (HMTA) regulates the transportation of used oil. The DOT has established its own definition of a "hazardous material;" DOT considers used oil a hazardous material if it is a combustible liquid (flash point of 38 to 75°C [100 to 200°F]) or a flammable liquid (flash point of 38°C [100°F] or less). Persons transporting used oil that meets the definition of a hazardous material in 49 CFR Part 171.8 must comply with DOT requirements in 49 CFR Parts 171 through 180.

In the event of a discharge of used oil during transportation, the transporter must take immediate action to protect human health and the environment (e.g., notify local authorities, and dike the discharge area). A state or federal official may authorize the removal of the used oil by transporters who do not have EPA identification numbers.

19.7.6 USED OIL RECYCLING

Approximately 560 million L (150 million gal) of used oil are recycled on-site in the United States by generators each year. The basis of the decision to focus the Used Oil Management Standards on recycling was to protect public health and the environment, for conserving energy resources, and for economic benefits. For example, re-refining used oil consumes one- third of the energy of refining crude oil to lubricant quality. In addition, about 3.8 L (1 gal) of used oil processed for fuel contains about 147,700 kJ (140,000 Btu) of energy. A discussion of used oil recycling technologies appears below.

Generation and handling are subject to the managements standards until the used oil is shipped away for disposal. Used oil recyclers must comply with the following requirements:

- Tanks and storage containers must be equipped with impermeable secondary containment. This includes an impermeable floor and may include dikes, soil berms, or retaining walls.
- Used oil mixed with listed hazardous waste is to be treated as hazardous waste.
- Used oil mixed with characteristic hazardous waste is under hazardous waste regulations if the mixture fails one of the four characteristic tests.
- Used oil with more than 1000 mg/kg total halogens is designated a hazardous waste, because EPA concludes that such oil has been mixed with listed hazardous waste.
- Metalworking oils or fluids containing chlorinated paraffins processed to reclaim metalworking oils or fluids are exempt.
- Used oils containing chlorofluorocarbons (CFCs) removed from refrigerators where the CFCs are reclaimed, are exempt.
- People who change their own oil and farmers generating less than 95 L (25 gal) per month are not regulated by the same requirements (40 CFR Part 279.53).

Used oil processors and re-refiners must have an EPA identification number and must develop practices that reduce the risk of fire, explosion, or any kind of release. Communications, emergency, and spill control equipment must be maintained at the facility. A contingency plan for the prevention of health and environmental damage must be formulated. Operating records must be kept on file until the re-refiner is closed. Shipment records must be maintained for 3 years.

Used oil processing and re-refining facilities must develop a written analysis plan describing the procedures to comply with the analysis requirements listed above (e.g., total halogen content). The plan must specify whether sample analyses or knowledge of the halogen content of the used oil is used to make the determination. If laboratory analyses are used to make the determination, the following information is required:

- The sampling method used to obtain representative samples
- The frequency of sampling to be performed
- Whether the analysis is performed on-site or off-site
- The methods used to analyze used oil for the necessary parameters

The analysis plan must be kept on file at the facility.

19.8 USED OIL RECYCLING METHODS

Recycling is loosely defined as the reuse of a substance in a beneficial way. The most common used oil recycling methods that are approved by the Used Oil Management Standards include:

- Re-refining for use as a base for lubricating oil
- Slipstreaming during refining of crude petroleum
- Processing
- Direct burning

The first two methods convert used oil into a lubricating oil or similar product; the last two methods are designed to recover heat energy from combustion of the used oil.

19.8.1 RE-REFINING

During re-refining used oil is subjected to a series of physical and chemical treatments to remove impurities. The resulting product is blended with virgin oil and additives to produce new lubricating oil. The recycled product is typically of a quality equivalent to that of a product derived from virgin oil. Re-refining employs vacuum distillation and hydrotreating (Figure 19.3). The used oil is

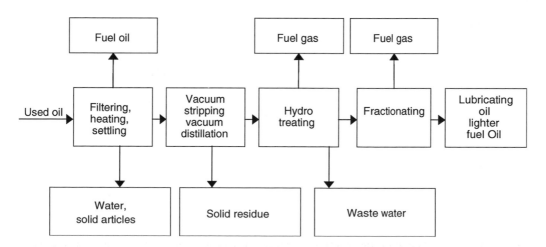

FIGURE 19.3 A vacuum distillation or hydrotreatment re-refining system for used oil (U.S. EPA, 1994).

first subjected to filtering, heating, and settling to remove water and large solids (Bryant, 1989; Arner, 1992). A vacuum is then created within a column of used oil, which strips out organic contaminants. The oil is subsequently treated with hydrogen, which bonds with certain contaminants that subsequently settle. Finally, the heavier lubricating oil is separated from the lighter fuel oil. With the re-refining process the production of hazardous waste is minimized and residuals can be burned as a fuel or used to produce asphalt (EERC, 1988).

Re-refining possesses several advantages. First, unlike other recycling options, re-refining allows used oil to be reused over and over again. In addition, re-refining used oil into lubricating oil is simpler and less expensive than refining crude oil into lubricating oil. It saves energy; approximately 50 to 85% less energy is needed to re-refine used oil than to refine virgin oil into lubricating oil (Byrne et al., 1989). A disadvantage of re-refining is that it is more complicated than other recycling options. Another disadvantage is that only a few re-refiners are currently in operation in the United States. As a result, costs of transporting used oil to a re-refiner may be substantial.

19.8.2 SLIPSTREAMING

With slipstreaming, small amounts of used oil (approximately 1% of the feed) are mixed with crude oil and introduced into the standard refining process (Arner, 1992). The used oil does not require pretreatment before it is mixed with the crude oil because the refining process removes contaminants that might impair the quality of the final product.

The major advantage of slipstreaming is that the heat or lubrication value of the used oil is realized without complicated processing methods. Slipstreamed used oil could serve as a base for several petroleum products including fuel oil, gasoline, and lubricating oil. In addition, slipstreaming poses no greater environmental risk than refining crude oil. Used oil channeled into the refining process after distillation or catalytic cracking is exempt from the Used Oil Management Standards provided that it is on-specification used oil (see below)(U.S. EPA, 1994).

19.8.3 PROCESSING

Processing involves treating used oil to improve its fuel characteristics. Relevant contaminants that must be removed include water, sediment, and ash (e.g., metals). Once removed, the quality of the used oil is similar to that of virgin fuel oil (Mueller, 1989; Arner, 1992). Physical treatment methods such as settling, filtering, and centrifuging are used to remove water or solid contaminants (e.g., metallic fragments from engine wear) (Figure 19.4). During settling, used oil is retained in large tanks for relatively long periods. Heavy contaminants eventually sink and light contaminants rise to the top. Filtering screens out solids and captures lighter particles that do not settle. During centrifugation at high speeds, the oil is separated from substances with different densities such as water and solids.

Processing also includes chemical treatment of used oil to remove contaminants and improve burning characteristics. For example, simple reagents can be added to neutralize acids. During processing, used oil can be blended with crude oil to adjust its viscosity or improve its fuel quality.

The major advantage of processing is that it improves the burning quality of used oil. Processing can also allow off-specification used oil to be upgraded to on-specification used oil so that it can be burned by a greater number of facilities (see the next section). Processing used oil is a widespread industry in the United States. Most facilities are small operations that service local markets (Nolan et al., 1990). Processing is therefore a more readily available recycling option for used oil generators than is re-refining.

19.8.4 DIRECT BURNING

Direct burning is as the name implies, i.e., burning used oil without prior processing to remove water, solids, or other contaminants.

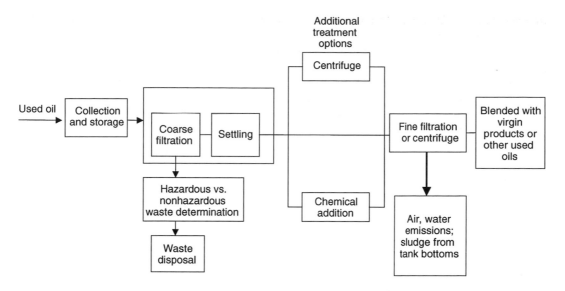

FIGURE 19.4 A used oil processing schematic (U.S. EPA, 1994).

19.8.5 OFF-VS. ON-SPECIFICATION USED OIL

Under the Used Oil Management Standards, if used oil exceeds any of the allowable levels of the constituents and properties levels listed in Table 19.4, it is designated "off-specification used oil," and there are restrictions as to its use including burning for energy recovery. In the table, limits are provided regarding concentrations of certain metals, flash point, and total halogens. Used oil that does not exceed any of these values may be burned for energy recovery and any fuel produced from used oil by processing, blending, or other treatment, is not subject to regulation. If it complies with all specifications, it is deemed "on-specification used oil."

Off-specification used oil may be burned for energy recovery in the following:

- Industrial furnaces
- Industrial boilers (such as asphalt plants and cement kilns), located on the site of a facility engaged in manufacturing processes
- Utility boilers used to produce electric power, steam, heated, or cooled air
- Hazardous waste incinerators subject to 40 CFR Parts 264 or 265
- Used oil-fired space heaters provided that (a) the used oil is generated on the facility site or is received from household DIY used oil generators; (b) the heater is designed to have a maximum capacity of 0.5 million Btu/h or less; and (c) the combustion gases are vented outside (U.S. EPA, 1992)

The used oil must be generated on-site or collected from DIYs.

Off-specification used oil is effectively burned in the above systems because their operating temperatures are much higher than space heaters and other small heaters. The heat generated causes the used oil and its organic contaminants to be combusted almost completely, thus reducing the quantity of atmospheric contaminants released. In addition, large burners are usually equipped with pollution control equipment that further reduces emissions (Nolan et al., 1990).

Burners of off-specification used oil are required to obtain an EPA identification number. Facilities must demonstrate that the used oil will be burned in a suitable device; for example, their combustors must possess equipment that adequately reduces atmospheric emissions. Storage specifications are the same as those for transporters, and burners are required to follow the same procedures to treat releases to the environment (Booz et al., 1999).

TABLE 19.4
Used Oil Specifications Limit

Constituent/property	Allowable Level
Arsenic	5 mg/kg
Cadmium	2 mg/kg
Chromium	10 mg/kg
Lead	100 mg/kg
Flashpoint	100°F (minimum)
Total halogens	4000 mg/kg

Source: 40 CFR Section 279.11. With permission.

The major advantage of direct burning is that it allows the heat value of used oil to be utilized without the need for additional processing. The advantage is greater when used oil is directly burned on-site, thus obviating any transportation costs. Another advantage is that numerous facilities in the United States are capable of directly burning used oil. Unlike re-refining, therefore, direct burning is a readily available recycling option for generators which are also off-specifications burners.

For service stations, quick-lube shops, fleet operations, DIY collection centers, and retailers, the major advantage of directly burning used oil on-site comes from utilizing the generated waste oil as a heating fuel. During seasons when heating is not required, used oil must be removed from the site unless sufficient storage capacity is available for colder months. Generators that choose to store used oil will incur increased storage costs.

One disadvantage of direct burning off-specification used oil relates to the chemical analysis (or lack thereof) of the used oil. Under the Used Oil Management Standards, generators who burn used oil in space heaters do not have to test the oil to determine if it meets burning specifications as long as the used oil is generated on-site or collected from DIYs. Consequently, these generators could potentially burn off-specification used oil. Emissions from burning off-specification used oil could, therefore, pose a hazard to public health.

On-specification used oil can be burned in space heaters, boilers, and industrial furnaces without being subject to the Used Oil Management Standards. Burning on-specification used oil is not expected to pose any greater risk to human health and the environment than burning virgin fuel oil; as a result, there are no special burning requirements (U.S. EPA, 1994). Burning used oil in small heaters is a common method of recycling. Approximately 70,000 small heaters are used every year in the United States to burn more than 450 million L (120 million gal) of used oil (Arner, 1992). In many areas of the country, facilities (service stations, quick-lube shops, fleet operations, DIY collection centers, and retailers) that wish to directly burn on-specification used oil must obtain local or state air pollution control permits.

Periodic maintenance of space heaters is necessary and the ash from space heaters must be removed and disposed of properly. Prior to disposal, a generator must determine whether the ash is regulated as a hazardous waste. The ash should be tested using the EPA characteristic tests for hazardous waste to determine ignitability, corrosivity, reactivity and toxicity of the sample in question. If the ash fails any one of these tests, it must be disposed of in a permitted facility.

19.9 DISPOSAL

Under some circumstances, costs of recycling used oil are sufficiently high such that recycling is not a practical option. For example, the used oil may be generated in such small quantities and so far away from a recycling facility that recycling is impractical. Disposal may be the only cost-effective option for the generator. The first step the generator must take when disposing of used oil is to determine if it is hazardous. The generator must test the used oil to determine if it has been mixed with a hazardous waste or if it exhibits the characteristics of a hazardous waste. The generator must determine if the used oil contains more than 1000 ppm of total halogens. The used oil must be disposed of as a hazardous waste under Subtitle C if it contains 1000 ppm or more of total halogens or if it exhibits a characteristic of a hazardous waste.

The used oil can be disposed of as a solid waste under Subtitle D if it contains less than 1000 ppm total halogens and does not exhibit the characteristics of a hazardous waste. If the oil is to be disposed of as a solid waste, the generator must identify either a Subtitle D industrial landfill or incinerator that will accept the oil. Different states have promulgated different regulations concerning whether or not solid waste facilities can accept used oil. If no Subtitle D facilities can be identified that will accept the used oil, then the generator must identify a hazardous waste facility to accept the used oil.

Irrespective of whether or not the used oil has a halogen content of 1000 ppm or more or exhibits other hazardous characteristics, it is considered a hazardous substance under DOT's HMTA regulations when transported. Consequently, the used oil generator must meet all the HMTA requirements for a hazardous waste generator. This includes preparing a Uniform Hazardous Waste Manifest, properly labeling the used oil, using DOT-approved packaging, and hiring a transporter that meets DOT requirements. Details for hazardous materials transportation are provided in Chapter 13.

19.9.1 PROHIBITIONS

Used oil is not permitted for storage or disposal in surface impoundments or waste piles. The use of used oil as a dust suppressant is prohibited except when such activity takes place in one of the states permitted by the U.S. EPA. A state may petition the EPA to allow the application of used oil as a dust suppressant. It must be demonstrated that the used oil is not mixed with hazardous waste and does not exhibit a hazardous characteristic other than ignitability. The state is also required to minimize the local environmental impacts of its use as a dust suppressant.

19.9.2 EXEMPTION FROM COMPREHENSIVE EMERGENCY RESPONSE, COMPENSATION AND LIABILITY ACT LIABILITY

Service stations, government-run DIY collection centers, and quick-lube shops can be considered "service station dealers" (defined by the Comprehensive Emergency Response, Compensation and Liability Act [CERCLA or "Superfund"]) (see Chapter 3) if DIY used oil is accepted for recycling. The service station dealer is exempted from CERCLA liability. However, the Used Oil Management Standards must be complied with. If the used oil is mixed with hazardous waste, it is not exempt from CERCLA liability (Booz et al., 1999).

REFERENCES

Agency for Toxic Substances and Disease Registry (ATSDR), Total Petroleum Hydrocarbons, 1997a. www.atsdr.cdc.gov/toxprofiles/tp123-c-3.pdf.
Agency for Toxic Substances and Disease Registry (ATSDR), Toxicological profile for used mineral-based crankcase oil, U.S. Department of Health and Human Services, Public Health Service, Atlanta, GA, 1997b. http://www.atsdr.cdc.gov/tfacts102.html
Ames, B., Lee, F., and Durston, W., An improved bacterial test system for the detection and classification of mutagens and carcinogens. *Proc. Natl. Acad. Sci.*, 70, 782–786, 1973.
Arner, R., Used oil recycling markets and best management practices in the United States, National Recycling Congress, Boston, MA, September 1992.
Bergeson, L., Legal Lookout, EPA to list used oil as a hazardous waste, *Pollut. Eng.*, 17, 58–59, 1985.
Booz, A., and Hamilton Inc, Used Oil (40 CFR Part 266, Subparts E, and Part 279), RCRA, Superfund and EPCRA, Hotline Training Module, 1999.

Bryant, C., Slick new oil re-refining process, *Res. Recycling*, 8, 36–70, 1989.

Byrne, J.P., Cody, C.A., Doyle, P.J., MacKinnon, J.S., Mayor, A.H., Reid, A.M., Rosner, S.K., and Talbot, C.J., Used Motor Oil in Massachusetts: A Prioritization of End Uses based on Human Health and Environmental Risk, Prepared for the Commonwealth of Massachusetts Department of Environmental Protection, Tufts University Department of Civil Engineering, 1989.

Code of Federal Regulations, Volume 40, Part 279, The Used Oil Management Standards, U.S. Government Printing Office, Washington, DC, 2004

Cotton F.O., Whisman, M.L., Gowtzinger, S.W., and Reynolds, J.W., Analysis of 30 used motor oils, *Hydrocarbon Process*, 56, 131–140, 1977.

Dominguez, E., Phytoremediation of Soils Contaminated by Used Motor Oil, M.S. thesis, Ball State University, Muncie, IN, 2002.

Energy and Environmental Research Corporation, and Robert H. Salvesen Associates, Evergreen Oil, Inc., Guide to Oil Waste Management Alternatives for Used Oil, Oily Wastewater, Oily Sludge, and Other Wastes Resulting from the Use of Oil Products, Final Report, Prepared for Alternative Technology Section, Toxic Substances Control Division, California Department of Health Services, in cooperation with the U.S. Environmental Protection Agency, 1988.

Grimmer, G., Naujack, K. W., Dettbarn, G., Brune, H., Deuschwenzel, R., and Mifeld, J., Studies on the carcinogenic action of use engine lubricating motor oil, *Erdol Kohle*, 35, 466–472, 1982.

Gruse, W. A., *Motor Oils: Performance and Evaluation,* Reinhold Publishing Corporation, New York, NY, 1967.

Hewstone, R. K., Health, safety and environmental aspects of used crankcase lubricating oils. *Sci. Total Environ.,* 156, 255–268, 1994.

Mouche, C., Managing used motor oil, Pollut. Eng., 27, 40 –3, 1995.

Mueller Associates, Inc., *Waste Oil: Reclaiming Technology, Utilization and Disposal,* Noyes Data Corp., Park Ridge, NJ, 1989.

Mumford, J.L., Williams, R.W., and Walsh, D.B., Indoor air pollutants from unvented kerosene heater emissions in mobile homes: Studies on particles, semivolatile organics, carbon monoxide, and mutagenicity, *Environ. Sci. Technl.,* 25, 1732–1738, 1991.

Nolan, J.J., Harris, C., and Cavanaugh, P.O., Used Oil: *Disposal Options, Management Practices and Potential Liability,* 3rd ed., Government Institutes, Rockville, MD, 1990.

Raymond R. L., Hudson, J. O., and Jamison, V. W., Oil degradation in soil, Appl. Environ. Microbiol., 31, 522–535, 1975.

RCRA, Superfund and EPCRA Hotline Training Module, Used Oil, 40 CFR Part 266, Subpart E, and Part 279, 1999.

Schulte, A., Ernst, H., Peters, L., Induction of squamous cell carcinomas in the mouse lung after long-term inhalation of polycyclic aromatic hydrocarbon-rich exhausts, Exp. Toxicolo. Pathol., 45, 415–421, 1993.

Shih-Hsien Chen, Delfino, J., and Suresh, P., Partitioning of organic and inorganic components from motor oil into water, *Sci. Total Environ.,* 1385–1400, 1994.

Thyssen, J., Althoff, J.K.G., and Mohr, U., Inhalation studies with benzo[a]pyrene in Syrian golden hamsters, *J. Natl. Cancer Inst.,* 66, 575–577, 1981.

U.S. Environmental Protection Agency, Waste Oil Recycling and Disposal, EPA-670/2-74-052, Author: K. Weinstein, Washington, DC, 1974.

U.S. Environmental Protection Agency, Listing of Oil as a Hazardous Waste Pursuant to Section (8)(2), Public Law 96– 463, Washington, DC, 1981.

U.S. Environmental Protection Agency, Method 9040, Test Methods for Evaluating Solid Waste, Physical/Chemical Methods, EPA SW-846, Washington, DC, 1986.

U.S. Environmental Protection Agency, Environmental Fact Sheet: Management Standards Issued to Control Potential Risks from Recycled Used Oil — No Hazardous Waste Listing, EPA/530-F-92-018, Office of Solid Waste and Emergency Response, Washington, DC, 1992.

U.S. Environmental Protection Agency, Environmental Regulations and Technology: Managing Used Motor Oil, EPA/625/R-94/010, Office of Research and Development, Cincinnati, OH, 1994.

U.S. Environmental Protection Agency, Used Oil Management, U.S. EPA Publications, 1999. See: http://www.ehso.com/usedoil.htm.

U.S. Environmental Protection Agency, Managing Used Oil: Advice for Small Businesses, EPA530-F96-004, 2003. See: http://www.epa.gov/epaoswer/hazwaste/usedoil/usedoil.htm#what.

Vazquez-Duhalt, R., Environmental impact of used motor oil, *Sci. Total Environ.,* 79, 1–23, 1989.

SUGGESTED READING

Alaska Department of Environmental Conservation, and Stephl Engineers, Sound Waste Management Plan Environmental Operations and Used Oil Management System, Prince William Sound Economic Development Council, Juneau, AK, 1998.

Anon, Closing the loop: Waste oil, Consulting-Specifying Engineer, 23, 82–86, 1998.

Arner, R., Safe recycling of used oil, *Biocycle*, 36, 36–38, 1995.

Blaszczak, R.J. and Sandborn, C., Used oil analysis and waste oil furnace emission study, Control Technology Center, U.S. Environmental Protection Agency, Research Triangle Park, NC, 1996.

Brinkman, D.W. and Dickson, J.R., Contamination in used lubricating oils and their fate during distillation/hydrotreatment refining, *Environ. Sci. Technol.*, 29, 81–92, 1995.

California Environmental Protection Agency, California's Used Oil Recycling Program, Integrated Waste Management Board, Sacramento, CA, 1997.

Cowan, D., Christiansen, E., It's All About Used Oil Recycling, County of Santa Barbara Public Works Dept., Solid Waste & Utilities Division, Santa Barbara, CA, 1990.

Donnelly, J.E., Numbers never lie, but what do they say? A comparative look at municipal solid waste recycling in the United States and Germany, *Georgetown Intl. Environ. Law Rev.*, 15, 29–53, 2002.

Driedger, R.J., From cradle to grave; Extended producer responsibility for household hazardous wastes in British Columbia, *J. Ind. Ecol.*, 5, 89–102, 2001.

Elbashir, N.O., Al-Zahrani, S.M., Abdul Mutalib, M.I., and Abasaeed, A.E., A Method of predicting effective solvent extraction parameters for recycling of used lubricating oil, *Chemi. Eng. Process.*, 41, 765–770, 2002.

Energy and Environmental Research Corp, Guide to Oil Waste Management Alternatives Used for Oil, Oily Wastewater, Oily sludge, and other Wastes Resulting from the Use of Oil Products, Irvine, CA, 2003.

Ferguson, J., Cement companies go toxic, *Nation*, 256, 306–309, 1993.

Grassy, J., The waste oil monster, *Garbage,* 3, 34–38, 1991.

Gourgouillon, D. and Shrive, L., An environmentally friendly process for the regeneration of used oil, *Environ. Sci. Technol.*, 34, 3469–3474, 2000.

Hamilton, D., Environmental liabilities associated with oil industry operations, *Environ. Geosci.*, 7, 209–211, 2000.

Hammett, W.S., Recycling Used Oil a Guide for Do-it-Yourselfers, N.C. Cooperative Extension Service, Raleigh, NC, 1994.

Hammett, W.S., Recycling Used Oil a Guide for Farmers, N.C. Cooperative Extension Service, Raleigh, NC, 1994.

Hegberg, B.A., Hallenbeck, W.H., and Brenniman, G.R., Used Oil Management in Illinois, University of Illinois Center for Solid Waste Management and Research, Office of Technology Transfer, School of Public Health, Chicago, IL, 1991.

Hippensteel, D.L., The RCRA exemption for oil and natural gas exploration and production wastes- What you may not know, *Environ. Geosci.*, 6, 106–110, 1999.

Hotter, D.S., Bringing used oil back to life, *Machine Design*, 68, 124–126, 1996.

Inyang, H.I., Framework for recycling of wastes in construction, *J. Environ. Eng.*, 129, 887–899, 2003.

Lohof, A., Used Oil Management in Selected Industrialized Countries, American Petroleum Institute, Washington, DC, 1991.

Nixon, W., Cleaning up dirty oil, E Magazine: *Environ. Mag.*, 5, 44–50, 1994.

Nobbe, G., Greasing the wheels of progress — With recycled oil, *Omni*, 15, 27–30, 1993.

Nolan, J.J., Harris, C., and Cavanaugh, P.O., Used Oil: Disposal Options, Management Practices, and Potential Liability, Government Institutes, Rockville, MD, 1990.

Oudenhoven, J.A. and Van, C.M., Oil Refinery Disposal Methods, Quantities and Costs; 1993 survey, Concawe, Brussels, 1995.

Rincon, J., Canizares, P., Garcia, M.T., and Gracia, I., Regeneration of used lubricant oil by propane extraction, *Ind. Eng. Chemi. Res.*, 42, 4867–4874, 2003.

Schroder, J.L., Basta, N.T., Payton, M., Wilson, J.A., Carlson, R.I., Janz, D.M., and Lochmiller, R. L., Ecotoxicological risks associated with land treatment of petrochemical wastes; Residual soil contamination and bioaccumulation by cotton rats, *J. Toxicol. Environ. Health*, 66, 305–326, 2003.

Schwarzbauer, J., Heim, S., Brinker, S., and Littke, R., Occurrence and alteration of organic contaminants in seepage and leakage water from a waste deposit landfill, *Water Res.*, 36, 2275–2288, 2002.

Sengupta, P., Sarkia, N., and Borthakur, P.C., Bricks from petroleum effluent treatment, plant sludge, properties and environmental characteristics, *J. Environ. Eng.*, 128, 1090–1095, 2002.

Sims, B.C., Regulation of hazardous waste in the oil field, The railroad commission of texas' approach, *Environ. Geosci.*, 6, 155–156, 1999.

Smith, K.P., Arnish, J.J., Williams, G.P., and Blunt, D.L., Assessment of the disposal of radioactive petroleum industry waste in nonhazardous landfills using risk-based modeling, *Environ. Sci. Technol.*, 37, 2060–2067, 2003.

QUESTIONS

1. To what aspects of used oil management (e.g., generator requirements, transportation, etc.) do the Used Oil Management Standards (40 CFR Part 279) apply? Be specific.

2. Discuss the general chemical composition of used oil in terms of the presence of aliphatics, aromatics, PAHs, metals, etc. How has the oil chemically changed compared with when it was fresh?

3. What contaminants are routinely and inadvertently *added* to oil during engine operation?

4. Although used oil cannot be utilized as a weed killer on fencerows, can it be utilized to burn brush and other unwanted vegetation? Check 40 CFR Part 279.

5. Based on the used oil regulations, is it acceptable for a DIY oil changer to place used oil in the trash?

6. Is it permissible for farmers to use oil to dip their pigs or cattle in to remove lice and mites? Check 40 CFR Part 279.

7. Used oil generators are required to engage in "good housekeeping practices in handling used oil." List and discuss some of the important requirements.

8. There are at least three bodies of federal regulations governing some aspect of used oil management. List them. Provide the appropriate citations from the Code of Federal Regulations.

9. Is it required that used oil be transported to permitted recycling facilities with EPA identification numbers?

10. What are the inspection requirements for used oil storage tanks?

11. What are the storage and secondary containment requirements for used oil storage tanks?

12. In what ways are the requirements for used oil processors and rc-refiners similar to those for TSDFs as managed under RCRA Subtitle C?

13. Discuss the most common used oil recycling methods that are approved by the Used Oil Management Standards. What are the advantages and disadvantages of each method?

14. How does "on-specification" used oil differ from "off-specification" in terms of composition? Where can off-specification used oil be burned for energy recovery? What is the rationale for the use of these systems?

15. The ash from used oil-burning space heaters must be removed and disposed of properly. Prior to disposal, is the generator responsible for any special management of the ash?

16. Under what condition(s) must used oil be disposed of as a Subtitle C hazardous waste?

17. What is an approved procedure for rendering oil filters acceptable for disposal? Can drained and crushed oil filters be disposed of in a Subtitle D landfill?

18. Where in your community can you bring used oil for recycling? How is the used oil managed (i.e., how is it stored, where is it shipped for processing, etc.)? Can used oil filters also be brought to this facility?

19. Trace the chronology of the Used Oil Management Standards. How did the waste classification for used oil evolve over time?

A.19.1 APPENDIX: SCENARIOS

This chapter includes several situations that are intended to apply the regulatory foundation provided in previous chapters. All situations are based on actual events and inspections experienced by hazardous waste regulatory personnel. Names of companies and individuals have been changed.

After reading each scenario, discuss what, if any, violations may have occurred. How may these violations be best addressed (i.e., via changes in engineering design, a modified storage or disposal program, some use of common sense, etc.)? These are open-ended situations.

1. What is wrong with this picture (Figure A. 19.1)?
2. At a wire-coating facility, an employee alleges that F002 solvent was mixed with used oil and the mixture was stored in the facility basement. An inspector found the drums in question to be leaking (Figure A. 19.2), and cracks were apparent in the floor and along the walls.

FIGURE A. 19.1 Drum storage area.

FIGURE A. 19.2 Drum storage and wire coating facility.

A.19.1.2 Responses to Scenarios

1.
    (a) Containers must be labeled "Used Oil", not "Waste Oil" (40 CFR 279.22(c)(1)).

    (b) Clearly, this is not the proper method to store drums, regardless of whether they contain hazardous wastes or used oil. Some of the drums have experienced corrosion and rust damage, and some appear to have leaked. Drums should be protected from the elements and should be kept away from water in order to prevent rusting and leaks. According to the federal regulations:

> Containers and aboveground tanks used to store used oil at generator facilities must be:
> (1) In good condition (no severe rusting, apparent structural defects or deterioration)
> (2) Not leaking (no visible leaks) (40 CFR 279.22(b))

    (c) In some facilities, hazardous wastes such as chlorinated solvents have been stored mixed with used oil. If the inspector is suspicious, a waste determination of the drum contents can be ordered.

    (d) There are no federal regulations that limit the quantity of used oil in storage at a facility; however, local fire departments or state agencies may establish their own limits.

2.
    (a) A waste determination by the generator should be made to ascertain the presence of hazardous waste solvent with the used oil.

Before the results of the waste determination were provided, the inspector stated in his report that "poor housekeeping was apparent at the facility, which creates the potential for a release to the environment." According to 264.31:

> Facilities must be designed, constructed, maintained, and operated to minimize the possibility of a fire, explosion, or any unplanned sudden or non-sudden release of hazardous waste or hazardous waste constituents to air, soil, or surface water which could threaten human health or the environment.

20 Medical and Infectious Wastes

Nearly all men die of their medicines, not of their diseases.

Moliere (1622–1672)

20.1 INTRODUCTION

Each year approximately 500,000 tons of regulated medical waste is produced in the United States by about 375,000 generators. For comparison, about 229,000,000 tons of municipal solid waste (MSW) is generated annually (see Table 1.3). The majority of medical waste is generated during the administration of healthcare or research by medical institutions and home healthcare activities. The institutions generating most of the medical waste include hospitals, laboratories, physicians, dentists, veterinarians, long-term healthcare facilities, clinics, blood banks, and funeral homes. The majority (approximately 77%) of regulated medical waste, however, is generated by hospitals. Most of the remaining classes of generators produce relatively small quantities (< 23 kg or 50 lb per month) of regulated medical waste (U.S. EPA et al., 1991). The estimated number of generators in each of these categories and volumes of waste generated are presented in Table 20.1.

Before 1988, the category of waste labeled "medical waste" received little attention by regulators or the general public. In that year, however, syringes, blood vials, laboratory rats, and other medical-related debris began washing up on the beaches of the Atlantic coast and Great Lakes. In 1990, medical waste began appearing on West Coast beaches as well. Beaches near San Francisco and San Diego closed because of potentially dangerous public health conditions. While there was little chance of this medical detritus causing illness, public fears of possible contact with hepatitis B and HIV viruses led to a corresponding collapse in local tourist industries. It was subsequently determined that much of the beach washups consisted of garbage and other debris attributed to malfunctioning solid waste management systems rather than to illegal dumping. A small portion of the waste consisted of syringes, medical vials, or other wastes of medical origin. This debris was not linked to hospitals, but more likely the result of home injections (insulin and medications), drug users, recreational boaters, cruise ships, and U.S. Navy ships (Lipman, 1992).

Federal agencies with the authority to establish medical and infectious waste regulations such as the EPA and the Centers for Disease Control often developed guidelines or best management practices rather than promulgate regulations for managing medical wastes. Before federal regulations for medical wastes were established, state and local authorities regulated most medical waste issues. Regulations varied from one locality to another and up to 1988, several states had no formal definition of medical waste. A number of states did not regulate medical and infectious wastes, thereby allowing hospitals and other medical facilities to dispose all wastes collectively as municipal waste (Lipman, 1992).

20.2 THE MEDICAL WASTE TRACKING ACT

Congress passed House Bill 3515, the Medical Waste Tracking Act (MWTA) in November 1988, which directed the U.S. EPA to develop protocols for dealing with infectious waste disposal. The EPA was required to publish an interim final rule for a 2-year demonstration medical waste

TABLE 20.1
Estimated U.S. Medical Facilities and Annual Waste Generated

Generator Category Generated (tons)	Number of Facilities	Annual Infectious Waste Generated (tons)	Annual Total Waste (tons)
Hospitals	7,000	360,000	2,400,000
Laboratories			
Medical	4,900	17,600	117,500
Research	2,300	8,300	55,500
Total	7,200	25,900	173,000
Clinics (outpatient)	41,300	26,300	175,000
Physicians' offices	180,000	35,200	235,000
Dentists, offices	98,000	8,700	58,000
Veterinarians	38,000	4,600	31,000
Long-term care facilities			
Nursing homes	18,800	29,700	198,000
Residential care	23,900	1,400	9,000
Total	42,700	31,100	207,000
Blood banks	900	4,900	33,000
Funeral homes	21,000	900	6,000
Health units in industry	221,000	1,400	9,000
Fire and rescue	7,200	1,600	11,000
Corrections	4,300	3,300	22,000
Police	13,100	<100	<1,000
Total	682,400	504,000	3,361,100

Source: U.S. EPA, 1994. With permission.

management and tracking program. The Resource Conservation and Recovery Act (RCRA) was amended by adding a Subtitle J.

In a fashion analogous to RCRA, the Act established a cradle-to-grave medical waste tracking protocol; however, this program was implemented in only a limited number of states. New York, New Jersey, and Connecticut were required to participate in the program. Additionally, the program was open to any state wishing to petition the U.S. EPA for inclusion. The Commonwealth of Puerto Rico petitioned for participation in the program. The states contiguous to the Great Lakes were designated by the MWTA to participate in the program; however, the Act provided the option for them to voluntarily withdraw. These states did indeed choose to opt out.

The medical waste tracking program had limited participation and the program expired in June 1991 without being reauthorized by Congress; however, the course of United States medical waste management changed significantly as a result of this legislation. With the American public's increasing fear of the AIDS epidemic along with washups of medical waste on beaches, United States medical waste policy shifted in response to the potential risks associated with this waste stream.

20.3   DEFINITIONS AND SOURCES OF MEDICAL WASTE

As is the case for municipal and hazardous wastes, one of the first steps in managing medical waste is to identify its sources. A logical next step is to determine the types and quantities of waste generated. Third is the determination as to whether the waste should be classified as solid, potentially infectious, hazardous, radioactive, etc. Definitions of medical waste to be managed and disposed as potentially infectious will vary depending on which regulation or guideline is chosen.

Definitions for *infectious waste* also vary widely in different federal regulations and from state to state; there is no national standard defining which wastes comprise infectious wastes. Some of the federal designations of infectious waste include those by the CDC (1985), the EPA *Guide for Infectious Waste Management* (U.S. EPA, 1986), the *Medical Waste Tracking Act of 1988* (U.S. EPA, 1988), and others. The CDC definition of infectious waste is any waste from microbiology laboratories, pathological waste, sharps, and blood or blood-product waste (CDC, 1985). The definitions provided in the EPA *Guide* were more specific (U.S. EPA, 1986):

> Isolation waste, cultures and stocks of infectious agents, human blood and blood products, pathological waste, contaminated sharps (e.g., hypodermic needles, syringes, Pasteur pipettes, scalpel blades, blood vials) and contaminated animal carcasses, body parts, and bedding all are considered infectious.

'Optional infectious waste' is also listed in the *Guide* and includes surgery and autopsy wastes, miscellaneous laboratory wastes, dialysis unit wastes, and contaminated equipment. These are not considered to pose a risk, and the decision as to whether optional waste should be handled as infectious is left to an authorized representative at the facility. The MWTA now defines regulated medical waste to include cultures and stocks of infectious agents, human pathological wastes, human blood and blood products, sharps (used and unused), contaminated animal wastes, and isolation wastes (40 CFR Part 259). These are described further below.

Based on the MWTA definition of medical waste, sources of medical and infectious wastes were identified as: (1) hospitals, (2) physicians' offices, (3) dentists' offices, (4) biomedical research facilities, (5) clinical laboratories, (6) manufacturing facilities, (7) veterinary offices and clinics, (8) funeral homes, (9) in-home medical care, (10) other healthcare and residential care facilities, (11) illicit intravenous drug use, and (12) other sources (e.g., cruise ships and naval vessels).

Because medical and infectious wastes are defined in many ways, because the general public tends to consider all medical wastes as potentially infectious, and because off-site disposal contractors may define any medical waste as potentially infectious, some institutions categorize all patient-contact wastes as potentially infectious. Each healthcare facility must formulate its own definition of medical and infectious wastes based on definitions established at the federal and state levels.

20.3.1 POTENTIALLY INFECTIOUS WASTE

A portion of the medical waste stream from healthcare and similar institutions is categorized as being potentially infectious. Other terms for infectious waste are biohazardous waste, biological waste, biomedical waste, contaminated waste, pathogenic waste, pathological waste, red-bag waste, and regulated medical waste (RMW). Regardless of regulatory definition, however, a waste is infectious when all of the following conditions are met simultaneously (U.S. EPA, 1991, 1989a, 1989b, 1989c):

- The presence of a virulent pathogen
- Sufficient concentration of the pathogen
- Presence of a host
- Portal of entry
- Host susceptibility

20.3.2 REGULATED MEDICAL WASTE (40 CFR PART 259.30)

The MWTA of 1988 defined regulated medical waste as follows (40 CFR 259.1):

> Medical waste means any solid waste which is generated in the diagnosis, treatment (e.g., provision of medical services), or immunization of human beings or animals, in research pertaining thereto, or in the production or testing of biologicals. The term does not include any hazardous waste identified or listed under Part 261 or this chapter or any household waste as defined in 261.4(b)(I) of this chapter.

Mixtures of hazardous waste and medical waste are subject to this part except as provided in 40 CFR Part 259.31. Specific classes of medical wastes are described in 40 CFR Part 259 and include:

Class 1 — cultures and stocks: Cultures and stocks of infectious agents and associated biological materials, including cultures from medical and pathological laboratories, cultures and stocks of infectious agents from research and industrial laboratories, wastes from the production of biological materials, discarded live and attenuated vaccines, and culture dishes and devices used to transfer, inoculate, and mix cultures.

Class 2 — pathological wastes: Human pathological wastes, including tissues, organs, body parts, and body fluids that are removed during surgery and autopsy or other medical procedures, and specimens of body fluids and their containers.

Class 3 — human blood and blood products: Waste human blood and products of blood, items saturated and/or dripping with human blood; or items that were saturated or dripping with human blood that are now caked with dried human blood; including serum, plasma, and other blood components, and their containers, which were used or intended for use in either patient care, testing and laboratory analysis, or the development of pharmaceuticals. Intravenous bags are included in this category.

Class 4 — used sharps: Sharps that have been used in animal or human patient care or in medical, research, or industrial laboratories, including hypodermic needles, syringes (with or without needle), Pasteur pipettes, scalpel blades, blood vials, test tubes, and culture dishes (regardless of presence of infectious agents). Other types of broken or unbroken glassware that were in contact with infectious agents, such as used slides and cover slips are also included.

Class 5 — animal waste: Contaminated animal carcasses, body parts, and the bedding of animals that were known to have been exposed to infectious agents during research (including research in veterinary hospitals), production of biological materials, or testing of pharmaceuticals.

Class 6 — isolation wastes: Biological waste and discarded materials contaminated with blood, excretion, exudates, or secretions from humans who are isolated to protect others from highly communicable diseases or isolated animals known to be infected with highly communicable diseases.

Class 7 — unused sharps: Unused, discarded sharps including hypodermic needles, suture needles, syringes, and scalpel blades.

20.3.3 EXEMPTIONS TO THE DEFINITION OF REGULATED MEDICAL WASTE

According to 40 CFR Part 259, several wastes are not to be regulated as medical waste. This ruling is in effect because some wastes are already managed under other regulations; in other cases, for example, household waste, regulation is simply impractical. The following wastes are not to be regulated as medical waste under 40 CFR Part 259:

- Hazardous waste identified or listed in 40 CFR Part 261
- Household wastes
- Ash from incineration of regulated medical waste
- Residues from treatment and disposal of medical waste
- Human remains intended for interment or cremation

20.3.4 RCRA HAZARDOUS WASTE

Hazardous wastes are defined in RCRA Subtitle C, and are either listed or meet the characteristics of ignitability, corrosivity, reactivity, or toxicity characteristic leaching procedure (TCLP) toxicity (see Chapter 11). Quantities of hazardous waste generated by hospitals compared with industry are small; however, some may be acutely toxic. Many chemotherapy wastes may be defined by RCRA as hazardous and are therefore regulated by 40 CFR Parts 260-265. Other hazardous wastes generated include antineoplastic drugs (used to treat certain forms of cancer and malignant hematological diseases), formaldehyde, solvents, mercury, and waste anesthetic gases (U.S. EPA, 1990a; 1990b).

Sources of potentially hazardous chemical wastes include clinical and research laboratories, patient-care activities, pharmacies (spills and expired items), physicians' offices (outdated items), physical plant departments, or building and grounds departments (e.g., pesticides and solvents) (U.S. EPA, 1991).

If a container holds less than 3% (by wt) of the original amount or capacity of hazardous material, it is considered 'empty' and does not require disposal as a hazardous waste (40 CFR 261.7). This exemption does not apply to seven chemotherapy drugs listed by EPA as acutely toxic (40 CFR 261.33f).

20.3.5 RADIOACTIVE WASTE

Radioactive waste, specifically low level, is produced through a number of healthcare activities including those associated with research laboratories, clinical laboratory procedures, and nuclear medicine procedures such as diagnostic and therapeutic applications. Low-level waste includes items that have become contaminated with radioactive material or have become radioactive through exposure to radiation. This waste consists of contaminated clothing, wiping rags, mop heads, filters, medical tubes, swabs, injection needles, syringes, and laboratory animal carcasses and tissues. Liquid radioactive wastes include scintillation fluids and research chemicals. Radioactivity can range from just above background levels found in nature to highly radioactive. In a study of one university hospital, radioisotopes included ^{125}I (25.5%), ^{32}P (19.1%), ^{3}H (14.5%), ^{14}C (8.7%), ^{35}S (6.2%), ^{131}I (1.1%), ^{51}Cr (0.8%), and several others (Emery et al., 1992).

Low-level waste is stored on-site by the generator either until it has decayed sufficiently and can be disposed of as ordinary trash, or until amounts are large enough for shipment to a low-level waste disposal site in containers approved by the U.S. Department of Transportation (DOT).

20.3.6 MIXTURES

Mixtures of solid waste and regulated medical waste are also to be regulated as medical waste (40 CFR Part 259.31). Similarly, mixtures of listed or characteristic hazardous waste and regulated medical waste are considered medical waste.

20.4 MANAGEMENT OF MEDICAL WASTES PRIOR TO TRANSPORT

20.4.1 HANDLING AND PACKAGING

The handling of medical wastes depends largely upon federal and state regulations, the disposal method, and location of the disposal facility. Medical wastes that are intended for transport off-site must be segregated into the following categories whenever possible: sharps (used and unused), fluids (greater than 20 mL), and other regulated medical waste (40 CFR 259.40). If other waste is placed in the container with medical wastes, the generator must label and mark the container and its entire contents. Generators must ensure that all regulated medical waste is placed in containers that are:

- Rigid
- Leak-resistant
- Impervious to moisture
- Of sufficient strength to prevent tearing or bursting under normal handling conditions
- Sealed to prevent leakage during transport

In addition, generators must package used and unused sharps in puncture-resistant packaging (Figure 20.1). Fluids are to be placed into break-resistant and tightly lidded packaging.

If medical wastes (excluding sharps and fluids) are to be incinerated off-site, they are generally packaged in plastic bags at the point of generation. Bags are usually red or labeled with a biohazard

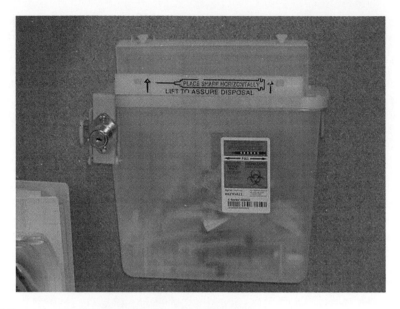

FIGURE 20.1 Puncture-resistant packaging for used sharps.

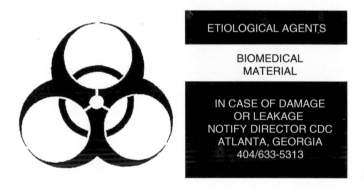

FIGURE 20.2 Biohazard symbol for the labeling of infectious wastes.

symbol for easy identification (Figure 20.2). The waste may be single- or double-bagged (U.S. EPA, 1990).

All nonrigid packaging and inner liners must be managed as regulated medical waste and cannot be reused. Any storage containers that are designated for reuse must be decontaminated after they are emptied. If contamination cannot be removed from the container, the container is to be managed as medical waste.

Medical waste must be stored in a location that maintains the integrity of the packaging and provides protection from water and the elements. The location should also be protected from entry by animals. The waste is to be maintained in a nonputrescible state (i.e., will not rot or putrify), using refrigeration if necessary. Areas that store dumpsters, sheds, and tractor-trailers are to be locked in order to prevent unauthorized entry. On-site storage areas are limited to access by facility employees only.

20.4.2 LABELING AND MARKING MEDICAL WASTE

Packages containing untreated medical wastes must have a water-resistant label attached to the outside of the containers. The label must include the words 'Medical waste' or 'Infectious waste,' or display the universal biohazard symbol (40 CFR Part 259). Red plastic bags used as inner packaging need not be labeled. The outermost surface of each waste package must be marked with a water-resistant identification tag with the following information:

- Generator name
- Generator's state permit number or address
- Transporter's name
- Transporter state permit number or address
- Date of shipment
- Identification of contents as medical waste

20.4.3 GENERATOR REQUIREMENTS

A facility that generates medical waste must determine whether that waste is a regulated medical waste (40 CFR 259.50). However, a generator that treats and destroys or disposes of regulated medical waste on the facility site (e.g., incineration) is not subject to any waste tracking requirements. A generator must determine the quantity of regulated medical waste generated in a calendar month and the quantity that is transported off facility property.

A generator that transports medical waste for off-site treatment or disposal must prepare a tracking form (see Figure 20.3). This form is analogous to the Universal Hazardous Waste Manifest that is required for shipments of RCRA hazardous waste (Chapter 12). The generator is responsible for preparing copies of the form for the generator, each transporter, any intermediate handlers, and the destination facility. The generator is to keep a copy of the form, signed by all parties involved in storage, transport, and disposal. This form is to be kept on file by the generator for at least 3 years from the date that the waste was accepted by the transporter.

A generator must contact the destination facility to determine the status of any tracked waste if a copy of the complete tracking form is not returned within 35 days. If the form is not returned after a total of 45 days, the generator must submit an exception report to the state regulatory agency. The exception form must include a copy of the tracking form and a letter signed by the generator explaining the efforts taken in locating the waste shipment. A copy of the report must be kept on file by the generator for 3 years.

Generators of < 23 kg (50 lb) of regulated medical waste per month are exempt from a number of requirements including of using a transporter who has notified EPA and the use of the tracking form. Such small quantity generators are exempt from other transportation requirements provided that certain provisions are met (Sec. 259.51(a)).

20.4.4 TRANSPORTER REQUIREMENTS

Transporters must notify the U.S. EPA and the state regulatory agency in writing as to their intent to transport regulated medical wastes (40 CFR Part 259.70). Notification must include the transporter's name and address and EPA hazardous waste identification number (if applicable); name and address of each transfer facility that the transporter will operate from; state permit to handle medical or infectious waste; and a statement of willingness to comply with 40 CFR Part 259. The EPA will subsequently provide each transporter with a unique identification number.

Waste that is transported by commercial haulers is often prepared for off-site transport in reusable bins or drums or in cardboard boxes. Transport vehicles may be compactor vehicles depending on state or local regulations. Secondary containers or additional packaging may also be required. Vehicles that transport regulated medical waste must possess a fully enclosed, leak-resistant cargo-carrying body, which must also be maintained in good sanitary condition. The transporter must ensure that the waste is not subject to mechanical stress or compaction during transit, loading, or unloading.

Transport vehicles must have the following identification on both sides and the back of the cargo body: the name of the transporter, the transporter's state identification number, and a sign with the words 'Medical waste' or 'Regulated medical waste.' Also, as per the 'Mixture Rule' for

Federal Register / Vol. 54, No. 56 / Friday, March 24, 1989 / Rules and Regulations 12383

Appendix I to 40 CFR Part 259—Medical Waste Tracking Form and Instructions

State Information Block (Name, Address, Contacts, Phone Numbers, etc.)

State Tracking Form Number (if applicable)

Medical Waste Tracking Form

GENERATOR

1. **Generator's Name and Mailing Address**
2. Tracking Form Number
3. **Telephone Number** ( )
4. **State Permit or ID No.**
5. **Transporter's Name and Mailing Address**

EPA Med. Waste ID No.

6. Telephone Number ( )
7. State Transporter Permit or ID No.
8. **Destination Facility Name and Address**
9. Telephone Number ( )
10. State Permit or ID No.

11. US EPA Waste Description	12. Total No. Containers	13. Total Weight or Volume
a. Regulated Medical Waste (Untreated)		
b. Regulated Medical Waste (Treated)		
c. State Regulated Medical Waste		

14. **Special Handling Instructions and Additional Information**

15. **Generator's Certification:**

Under penalty of criminal and civil prosecution for the making or submission of false statements, representations, or omissions, I declare on behalf of the generator that the contents of this consignment are fully and accurately described above and are classified, packaged, marked and labeled in accordance with all applicable State and Federal laws and regulations, and that I have been authorized in writing to make such declaration by the person in charge of the generator's operation.

Printed/Typed Name Signature Date

BILLING CODE 6560-50-C

INSTRUCTIONS FOR COMPLETING MEDICAL WASTE TRACKING FORM

INSTRUCTIONS

Copy 1 — GENERATOR COPY: Mailed by Destination Facility to Generator
Copy 2 — DESTINATION FACILITY COPY: Retained by Destination Facility
Copy 3 — TRANSPORTER COPY: Retained by Transporter
Copy 4 — GENERATOR COPY: Retained by Generator

As required under 40 CFR Part 259

1. This multi-copy (4-page) shipping document must accompany each shipment of regulated medical waste generated in a Covered State
2. Items numbered 1-14, must be completed before the generator can sign the certification. Items 4, 7, 10, 11c, & 19 are optional unless required by the State. Item 22 must be completed by the destination facility

For assistance in completing this form, contact your nearest State office, Regional EPA office or call (800) 424-9346.

TRANSPORTER

16. **Transporter 1** (Certification of Receipt of Medical Waste as described in items 11, 12 & 13)

Printed/Typed Name Signature Date

17. **Transporter 2 or Intermediate Handler** (name and address)
18. **Telephone Number**
19. **State Transporter Permit or ID No.**

EPA Med. Waste ID No.

20. **Transporter 2 or Intermediate Handler** (Certification of Receipt of Medical Waste as described in items 11, 12 & 13)

Printed/Typed Name Signature Date

21. **New Tracking Form Number** (for consolidated or remanifested waste)

DESTINATION

22. **Destination Facility** (Certification of Receipt of Medical Waste as described in items 11, 12 & 13)

☐ Received in accordance with items 11, 12 & 13

Printed Typed Name Signature Date

(If other than destination facility, indicate address, phone and permit or ID no. in box 14)

23. **Discrepancy Box** (Any discrepancies should be noted by item number and initials)

FIGURE 20.3 Tracking form to accompany medical waste shipments off-site.

medical wastes, a transporter cannot ship regulated medical waste mixed with solid wastes unless both are designated as regulated medical waste.

A tracking form must be provided and completed for shipments of over 50 lb of medical waste. For shipments from generators shipping < 50 lb, the transporter still needs to compile a log containing the following information for each shipment: the generator name, address, and state identification number, the quantity of waste accepted (number of packages and total weight by waste category [i.e., 'treated' and 'untreated']), and the date on which the waste was accepted.

20.4.5 TREATMENT AND DISPOSAL OF MEDICAL WASTE

There is a wide range of methods available for the treatment and disposal of medical waste based on the original chemical and physical properties of the waste, regulatory issues, public concerns, and economics, among other factors. Incineration and steam sterilization are currently the most popular technologies for the destruction of microbiological cultures. For years incineration has been the method of choice because it could be carried out on-site, and disposal involved only the removal of the residual bottom ash. Other methods, less commonly used, include chemical disinfection, pulverization with chemical disinfection, microwave (thermal) inactivation, gas-vapor sterilization, and ionizing radiation. There is also the option of conventional landfilling, which is likely to be combined with one of the methods listed above. When the waste occurs in liquid form it may be possible to dispose it directly to a publicly owned treatment works. Such disposal is, of course, contingent upon local and state regulations (Hoeltge, 1995).

Table 20.2 presents the medical waste types appropriate for treatment by each of the major medical waste treatment technologies.

20.4.6 REQUIREMENTS FOR THE DESTINATION FACILITY

The destination facility for regulated medical wastes is analogous to a treatment, storage, and disposal facility (TSDF) for RCRA hazardous wastes. When medical wastes arrive at the facility the operator must review the tracking form for any discrepancies. Discrepancies may include a variation in the count of drums, boxes, or containers, discrepancies in the number of containers for each category of medical waste, damaged packaging, regulated waste that arrives without the tracking form, or an incomplete (e.g., unsigned) tracking form (40 CFR Part 259.80). Upon discovery of the

---

TABLE 20.2
Medical Waste Types Appropriate for Treatment by Technology

Technology	Class 1	Class 2	Class 3	Class 4	Class 5	Class 6	Class 7	Radio-active	Hazardous/ Cytotoxic
Incineration	X	X	X	X	X	X	X	X[a]	X[a]
Steam autoclave	X	X[b]	X	X	X^2	X	X		
Chemical treatment	X	X[b]	X	X	X^2	X	X		
Microwave	X	X[b]	X	X	X^2	X	X		
Radiofrequency	X	X[b]	X	X	X^2	X	X		
Gamma irradiation	X	X[b]	X	X	X^2	X	X		

[a]The treatment of radioactive antineoplastic and hazardous waste which are mixed with medical wastes can be treated with incineration, however, special permits are usually required for this type of treatment. Additionally, incineration does not inactivate radioactive waste. Thus the ash from these processes may be radioactive and contain hazardous constituents.

[b]Technology not recommended for treatment of body parts because the density of the waste may prevent adequate treatment. Grinding the waste may increase treatment efficacy, however the grinding process may present aesthetically unacceptable.

Source: RTI, Final Report, 1993. Reproduced with kind permission of Research Triangle Institute.

discrepancy, it is the responsibility of the destination facility to resolve the issue with the waste generator or transporter. If the situation cannot be resolved, the destination facility must contact the state regulatory agency within 15 days of receiving the waste. The completed form must be signed and dated with a copy provided to the transporter and generator.

20.4.7 RECORD KEEPING

The destination facility must maintain several records for at least 3 years from the date the waste was accepted. Records must include copies of all tracking forms, the name and state permit number of all generators who delivered waste to this destination, and copies of all discrepancy reports. For facilities receiving smaller amounts of waste (less than 50 lb; see the above section, Generator Requirements), the following information must be maintained: the date the waste was accepted, the name and state permit number of the waste generator, the total weight of the regulated medical waste accepted, and all appropriate signatures.

20.4.8 DISPOSAL OPTIONS FOR MEDICAL WASTE

Presently, there are a limited number of treatment and disposal technologies available for managing medical wastes in the United States. As mentioned above, the primary technologies are incineration, autoclaving (steam sterilization), and shredding with chemical disinfection. Several innovative technologies are either in the test phase or are experiencing limited use in some facilities.

20.4.9 INCINERATION OF MEDICAL WASTE

Of the disposal methods available for medical wastes, incineration is the most widely used. Hospitals incinerate 2 million tons of waste per year in the United States (Key and Marble, 1997). There are several types of incinerators in use within the healthcare industry. According to the U.S. EPA there are 2400 incinerators in the United States, with more in the planning and construction stages (U.S. EPA, 2000).

20.4.10 REGULATORY REQUIREMENTS FOR MEDICAL WASTE INCINERATORS

Based on discussions in other chapters, there is a great deal of concern on the part of the general public and regulators regarding the proper operation of incineration units. Concern tends to revolve around daily facility operations, atmospheric emissions, and ash disposal. As a consequence, federal regulations were promulgated which placed strict emission limits on medical waste incinerators. Regulations have varied as a function of age of the unit. For example, the EPA formulated 40 CFR SubPart Ec — Standards of performance for hospital, medical and infectious waste incinerators for which construction is commenced after June 20, 1996. Presented below is an overview of some of the more significant regulations for the operation of a new medical waste incinerator.

20.4.11 SITING REQUIREMENTS

A medical waste incineration facility for which construction began after September 15, 1997 is required to prepare an analysis of the impacts of the affected facility (40 CFR Part 60.54c). The analysis considers air pollution control alternatives that minimize potential risks to public health or the environment. In assessing such alternatives, the analysis considers costs, energy impacts, and additional environmental impacts.

20.4.12 OPERATOR TRAINING AND QUALIFICATION REQUIREMENTS

Fully trained and qualified incinerator operators must either be accessible at all times at the incineration facility, or available within 1 h. Training is obtained by completing an operator training course that includes the following provisions.

A. Training on the following subjects:

- Environmental concerns, including pathogen destruction and types of emissions
- Basic combustion principles including products of combustion
- Operation of the type of incinerator to be used by the operator, including proper startup, waste charging, and shutdown procedures
- Combustion controls and monitoring
- Operation of air pollution control equipment
- Methods to monitor pollutants and equipment calibration procedures
- Inspection and maintenance of the incinerator, air pollution control devices, and emission monitoring systems
- Actions to correct malfunctions or conditions that may lead to malfunction
- Bottom and fly ash characteristics and handling procedures
- Applicable federal, state, and local regulations
- Safety procedures
- Pre-startup inspections
- Record keeping requirements

B. Completion of an examination.

Qualification for incinerator operation is obtained by completion of a training course combined with experience as an operator or direct supervisor of an operator. To maintain qualification, the qualified operator must pass an annual refresher course covering the following at a minimum:

- An update of relevant regulations
- Incinerator operation, including startup and shutdown procedures
- Inspection and maintenance
- Responses to malfunctions or conditions that may lead to malfunction
- Discussion of operating problems encountered

20.5 FACILITY OPERATIONS

20.5.1 Waste Management Plan

There are many documented cases in which toxic materials such as nickel–cadmium batteries have been collected and commingled with combustibles and disposed in a hospital incinerator. In order to prevent such contamination episodes, the incineration facility must prepare a waste management plan. The purpose of the plan is to identify the means of separating certain components of solid waste from the healthcare waste stream in order to reduce the amount of toxic emissions from incinerated waste. A waste management plan may include aspects such as paper, plastics, glass, battery, and metal recycling; or the purchase of recycled or recyclable products. A waste management plan may include different goals or methods for different departments at the facility. The plan should identify additional waste management measures, taking into account the costs of the additional measures, the emission reductions expected to be achieved, and any other environmental or energy impacts they might cause. The American Hospital Association publication, *An Ounce of Prevention: Waste Reduction Strategies for Health Care Facilities* (AHA, 1993) should be considered in the development of the waste management plan.

20.5.2 Compliance and Performance Testing

The facility must conduct an initial performance test (40 CFR Part 60.8) to determine compliance with emission limits (see below). All performance tests consist of a minimum of three test runs conducted under representative operating conditions. The minimum sample time is 1 h per test run.

Following the initial performance test the facility must:

- Determine compliance with opacity limits by conducting an annual performance test
- Determine compliance with particulate matter, CO and HCl emission limits by conducting an annual performance test

Table 20.3 contains the U.S. EPA Reference Methods for measurement of the above parameters.

20.5.3 EMISSION LIMITS

After an initial performance test of the medical waste incinerator is completed, the facility is not permitted to discharge any gases that contain stack emissions in excess of the limits presented in Table 20.4. Additionally, there are to be no discharges that exhibit greater than 10% opacity. There are also requirements for large incineration facilities regarding the discharge of visible emissions of combustion ash from an ash conveying system (EPA Reference Method 22, Table 20.3).

Extensive requirements apply to those facilities equipped with a dry scrubber, fabric filter, wet scrubber, or similar air pollution control device. Additionally, there are numerous requirements for the use and maintenance of sorbents for HCl, Hg, chlorinated dibenzodioxins, and furans (40 CFR Sec 60.56c).

20.5.4 MONITORING REQUIREMENTS

The medical waste incineration facility must install, calibrate (to manufacturers' specifications), maintain, and operate devices for monitoring all applicable operating parameters (Table 20.5). The facility must obtain monitoring data at all times during incinerator operation.

20.5.5 DOCUMENTATION

The incineration facility must maintain documentation at the facility that address the following:

- Summary of the applicable standards relevant to incinerator operation and emissions
- Description of basic combustion theory applicable to a medical waste incinerator

---

Table 20.3
U.S. EPA Reference Methods for Incinerator Compliance and Performance Testing

EPA Reference Method	Purpose
1	Select the sampling location and number of traverse points
3, 3A, or 3B	Gas composition analysis, including measurement of oxygen concentration
3, 5, or 29	To measure particulate matter emissions
9	To measure stack opacity
10 or 10B	To measure CO emissions
23	To measure total dioxin/furan emissions. The minimum sample time is 4 h per test run
26 or 26A	To measure HCl emissions
29	To measure Pb, Cd, and Hg emissions
22	To determine compliance with the fugitive ash emission limit under §60.52c(c). The minimum observation time is a series of three 1-h observations

Source: 40 CFR Sec. 60, Appendix A. With permission.

TABLE 20.4
Emission Limits for Small, Medium, and Large Medical Waste Incinerators

Pollutant (Units)[a]	Incinerator Size		
	Small	Medium	Large
Particulate matter (mg/scm or grains/ dscf)	69 (0.03)	34 (0.015)	34 (0.015)
Carbon monoxide (ppm [v])	40	40	40
Dioxins or furans ng/dscm)	125 (55) or 2.3	25 (11) or 0.6	25 (11) or 0.6
Hydrogen chloride (ppm or % reduction)	15 or 99%	15 or 99%	15 or 99%
Sulfur dioxide (ppm [v])	55	55	55
Nitrogen oxides (ppm [v])	250	250	250
Lead, (mg/ dscm [grains/1000 dscf] or % reduction)	1.2 (0.52) or 70%	0.07 (0.03) or 98%	0.04 (0.02) 90%
Cadmium (mg/ dscm or [grains/1000 dscf] or % reduction)	0.16 (0.07) or 65%	0.04 (0.02) or 90%	0.04 (0.02) 90%
Mercury (mg or dscm or [grains/1000 dscf] or % reduction)	0.55 (0.24) or 85%	0.55 (0.24) or 85%	0.55 (0.24) or 85%

[a]Based on 7% O_2, dry basis.

Source: 40 CFR Table 1 to SubPart Ec. With permission.

- Procedures for receiving, handling, and charging waste
- Incinerator startup, shutdown, and malfunction procedures
- Procedures for maintaining proper combustion air supply levels
- Procedures for operating the medical waste incinerator and associated air pollution control systems within the standards established
- Procedures for responding to periodic malfunction or conditions that may lead to malfunction
- Procedures for monitoring medical waste incinerator emissions
- Reporting and record keeping procedures
- Procedures for handling ash

The above information must be readily accessible to all operators. This information, along with records of training, must be available for inspection by the EPA or the state regulatory agency.

20.5.6 REPORTING AND RECORD KEEPING

The facility must maintain a range of routine operations information. For example, prior to initial startup of the incinerator, the following information should be available:

- The type(s) of waste to be combusted
- The maximum design waste burning capacity
- The anticipated maximum charge rate

As regards routine operations:

- Concentrations of any pollutant listed in § 60.52c or measurements of opacity
- Results of fugitive emissions tests
- Medical waste incinerator charge dates, times, weights, and hourly charge rates

Table 20.5
Operating Parameters to be Monitored and Minimum Measurement and Recording Frequencies

Operating parameters to be monitored	Minimum Frequency		Control System		
	Data measurement	Data recording	Dry scrubber followed by fabric filter	Wet scrubber	Dry scrubber followed by fabric filter and wet scrubber
Maximum operating parameters					
Maximum charge rate	Continuous	1 x h	x	x	x
Maximum fabric filter inlet temperature	Continuous	1 x min	x		x
Maximum flue gas temperature	Continuous	1 x min	x	x	
Minimum operating parameters					
Minimum secondary chamber temperature	Continuous	1 x min	x	x	x
Minimum dioxin or furan sorbent flow rate	Hourly	1 x h	x	x	x
Minimum HCl sorbent flow rate	Hourly	1 x h	x		x
Minimum mercury (Hg) sorbent flow rate	Hourly	1 x h	x		x
Minimum pressure drop across the wet scrubber or minimum horsepower or amperage to wet scrubber	Continuous	1 x min	x	x	
Minimum scrubber liquid flow rate	Continuous	1 x min		x	x
Minimum scrubber liquid pH	Continuous	1 x min		x	x

Source: 40 CFR Section 60 (Subpart Ec). With permission.

- Fabric filter inlet temperatures
- Amount and type of dioxin/furan sorbent used
- Amount and type of Hg sorbent used
- Amount and type of HCl sorbent used
- Secondary chamber temperatures recorded
- Horsepower or amperage to the wet scrubber
- Temperature at the outlet from the wet scrubber
- pH at the inlet to the wet scrubber
- Records indicating use of the bypass stack

Additional required data include:

- Days and durations of malfunctions, a description of the malfunction, and the corrective action taken
- Days for which emission rates or operating parameters exceeded the applicable limits, a description of the excess readings, reasons for such excess readings, and a description of corrective actions taken
- The results of performance tests conducted to determine compliance with the emission limits and to establish operating parameters
- All documentation produced as a result of siting requirements

The above records must be on hand at the facility for at least 5 years (40 CFR Part 60.58c). An annual report must be submitted by medical waste incineration facilities for the following data:

- Values for site-specific operating parameters
- The results of any performance tests
- Any use of the bypass stack, the duration, reason for malfunction, and corrective action taken

20.6 TYPES OF MEDICAL WASTE INCINERATORS

As discussed in Chapter 9, the design function of incineration is to destroy the organic component of the waste through high-temperature combustion. A secondary function is detoxification of the waste. The residual ash from incineration may be hazardous as per 40 CFR Part 261 (Identification and Listing of Hazardous Waste). It must therefore be tested via the TCLP in order to assess possible toxicity.

Three general types of incinerators are in common use to destroy medical wastes: the rotary kiln, the multiple hearth, and the controlled-air incinerator. Some innovative incinerators have been used with varying degrees of success. The major components of an incineration system are shown in Figure 20.4.

20.6.1 ROTARY KILN

The rotary kiln incinerator was described in Chapter 15. The design includes a cylindrical, refractory-lined combustion chamber that is rotated at a slight incline. Waste is loaded at the upper end and is combusted as it is agitated along the length of the cylinder. The ash is discharged at the lower end. To comply with air emissions standards, rotary kiln incinerators must be equipped with secondary combustion chambers and air pollution control equipment such as a baghouse (U.S. EPA, 1991).

Rotary kiln incinerators have long been popular for the destruction of RCRA hazardous wastes. The rotational action provides for good turbulence of the waste and allows for continuous-feed operations, so ash is regularly removed during routine operations. There are only a limited number of rotary kiln incinerators in use for medical waste incineration, however, due to their high capital, operating, and maintenance costs. Repair and maintenance costs are especially high due to damage to the refractory lining from abrasive materials. A second disadvantage is that some wastes require processing, for example, shredding, prior to incineration. As the waste is shredded there is the potential for the release of potentially infectious material from the shredder or conveyance mechanisms (U.S. EPA, 1991).

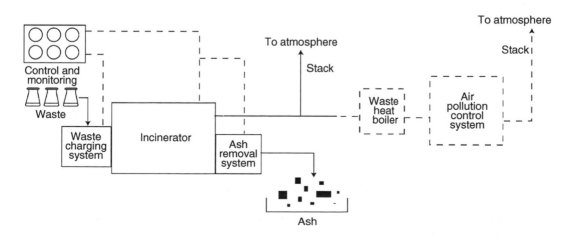

FIGURE 20.4 Major components of an incineration system for medical wastes (U.S. EPA, 1990b).

20.6.2 THE MULTIPLE-CHAMBER INCINERATOR

Hospitals and similar health-related institutions had used the multiple-chamber incinerator for destroying infectious wastes for decades. There are two basic configurations, the in-line design and the retort design (Figure 20.5). Combustion gases flow straight through in-line incinerators, turning only vertically. In the retort design, gases turn horizontally and vertically. Retort multiple-chamber incinerators are more compact and are more efficient than in-line systems at small capacities (U.S. EPA, 1991). In order to control combustion and to limit emissions, the multiple chamber systems incorporate settling chambers and are designed to operate at very high levels of excess air. The generation of gaseous and particulate emissions can be substantial with these systems.

Few multiple-chamber incinerators are now being built; however, many older systems are still in use. Some were designed with grates within the primary combustion chamber. These grates allow for noncombusted waste to fall into the ash receptacle, with the potential for exposing operators to unburned infectious waste.

20.6.3 CONTROLLED-AIR INCINERATORS

Controlled-air incinerators use two or more separate combustion chambers to combust waste (Figure 20.6). The first chamber operates under starved-air conditions to volatilize the moisture, vaporize the volatile fraction, and combust the fixed carbon in the waste. The combustion gases are then passed to the secondary chamber where excess air is provided to complete the combustion of

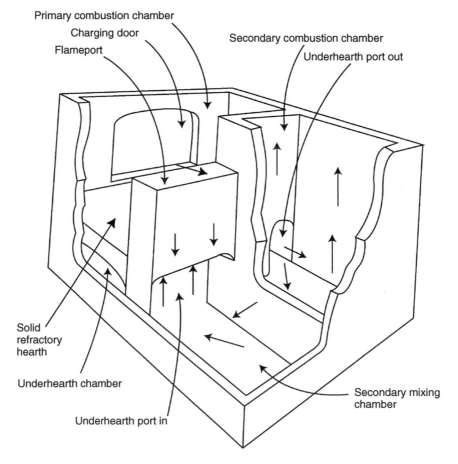

FIGURE 20.5 In-line and retort design of medical waste incinerators (U.S. EPA, 1989a).

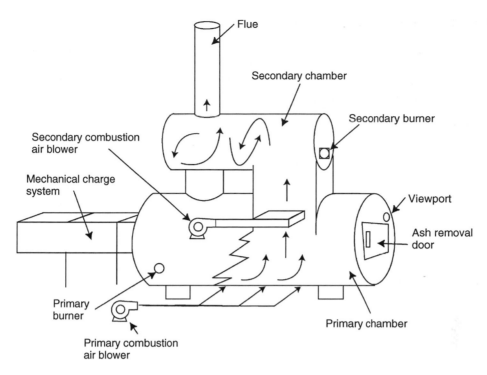

FIGURE 20.6 Controlled-air incinerator (U.S. EPA, 1989b).

the volatiles and other hydrocarbons emitted from the primary chamber. Turbulence is provided to promote mixing of the air and combustion gases. The gas/air mixture is combusted at relatively high temperatures.

Controlled-air incinerators possess several advantages over the existing multiple-chamber incineration technology. The starved-air environment of the primary chamber allows for slow, non-turbulent combustion, which minimizes the entrainment of particulates in combustion gases and thus reduces particulate emissions to the atmosphere. The lower temperatures achieved in this chamber avoid the melting and fusion temperatures of most metals, glass, and other noncombustibles, thus minimizing slagging and the formation of clinker. The high temperatures and excess air environment of the secondary chamber help to ensure more complete combustion of volatile gases, thus reducing hydrocarbon emissions. Controlled-air incinerators are comparatively low in cost and carry out clean combustion, making them popular in the hospital industry (U.S. EPA, 1991).

20.7 MODES OF INCINERATOR OPERATION

Medical waste incinerators can be operated in one of three modes: batch, intermittent duty, and continuous duty. As the name implies, batch incineration involves burning a single batch of waste, often only once per day. Waste is often loaded manually, combusted, and cooled, after which the ash is manually removed. Intermittent duty incinerators are loaded continuously and frequently with small waste batches and operate less than 24 h per day. A typical operating cycle for an intermittent duty system includes a 15- to 30-min period of cleanout of ash from the previous day, a 15- to 60-min preheat, a 12- to 14-h waste combustion period, a 2- to 4-h burndown period, and a 5- to 8-h cooldown period. Continuous duty incinerators operate 24 h per day and use automatic charging units, such as a ram-feed system (Figure 20.7), to input waste into the firebox in small, frequent batches. A mechanism will automatically remove the ash from the incinerator (U.S. EPA, 1991).

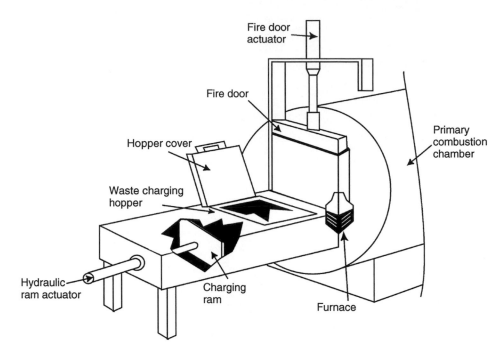

FIGURE 20.7 Hopper and mechanical feed system for a medical waste incinerator (U.S. EPA, 1990b).

EXAMPLE 20.1

A rotary kiln incinerator burns medical waste that contains 9.2% ash. The incinerator receives approximately 325 kg (about 715 lb) waste per hour over a 24 h period. Determine the amount of bottom ash to be disposed annually, if 20% of the ash is emitted as fly ash during combustion.

SOLUTION

Total output (kg/hr) as bottom ash from the incinerator is

$$(0.092) \ (325 \ \text{kg/h}) = 30 \ \text{kg/h}$$

The rate of bottom ash produced per day is calculated as

$$(30) \ (24) \ (0.80) = 576 \ \text{kg/day}$$

The rate of ash output per year is

$$(365) \ (576) = 210{,}240 \ \text{kg/yr}$$

EXAMPLE 20.2

An incinerator burns mixed hospital waste contaminated with mercury and having an ash content of 5.8%. The input waste feed rate is 1000 kg/h and the gas flow rate is 675 dscmm (approx. 23,825 dscfm). Upon analysis of the flue gas, it was found that the average Hg content in the particulates was 4.29 µg/g and the Hg concentration of the vapor was 0.20 mg/dscm. For this incinerator, the emissions meet the particulate standard of 0.1832 g/dscm (0.08 g/dscf) and there is 99.8% efficiency of particulate collection by an electrostatic precipitator. Calculate the amount of Hg bound to the fly ash which is captured in the precipitator. Also, calculate the amount of Hg leaving the stack as vapor and with the fly ash. Note that 1 kg of collected residue ~3150 g.

SOLUTION

1. The amount of ash exiting the flue is calculated as

   0.1832 g/dscm $\times$ 1 kg/3150 g $\times$ 675 dscmm/min $\times$ 60 min/h $\times$ 24 h/day = 56.5 kg

   The amount of ash collected in the ESP

   (56.5 kg/d)/(1−0.998) = 28,250 kg/day

2. Hg exiting the stack with the fly ash.

   (56.5 kg ash/day)(2.42 $\times$ 10^{-6} g Hg/0.001 kg ash) = 0.14 g Hg/day

   Hg leaving the stack as vapors

   0.2 $\times$ 10^{-3} g Hg/dscm $\times$ 675 dscm/min $\times$ 60 min/sh $\times$ 24 h/day = 194.4 g/day

   The total mercury exiting the stack = 194.4 + 0.14
   = 194.5 g/day

(Adapted from Reynolds et al., 2002)

20.8   ADVANTAGES AND DISADVANTAGES OF INCINERATION

The greatest advantage to using incineration for the disposal of medical and infectious waste is that the organic component of the waste can be reduced by up to 95% of its original volume. Therefore, significantly less waste must be handled, stored, or transferred to the disposal site (Hoeltge, 1995). Incineration will also detoxify the biological component of the waste. These factors are important for both keeping operational costs down and reducing future liability.

Medical waste incineration has its own unique problems, however. First, following the Law of the Conservation of Mass, the waste is physically transformed into a variety of solid residues, gaseous chemicals, and particulate matter. As indicated in Chapter 9, there is possible harm to public health from these residues. A pressing concern involves the release of metals such as mercury. Hospitals were among the nation's leaders in mercury emissions to the atmosphere, contributing almost 10% of all mercury from incineration in the United States (U.S. EPA, 1997). Another harmful result of incineration involves the generation of polychlorinated dibenzodioxins (PCDDs). PCDDs such as tetrachlorodibenzodioxin (TCDD) have been linked to cancer, birth defects, and a host of other health problems (Key and Marble, 1997). PCDDs can enter the food chain and bioaccumulate, making them more dangerous as they can be deposited in tissue.

All types of wastes may be treated by incineration; however, a special permit is required to incinerate low-level radioactive medical wastes. The ash from the incineration of radioactive medical waste will remain radioactive. No other treatment technology may be used for radioactive, hazardous, or cytotoxic wastes.

20.9   MICROBIAL INACTIVATION

Microbial inactivation refers to those physical or chemical processes that render microorganisms incapable of multiplication. Such processes may either kill the organisms or injure them to the extent that repair and subsequent growth of cells is not possible. The effectiveness of medical waste treatment technologies tested during a U.S. EPA investigation is presented in Table 20.6.

Level I microbial inactivation destroys most disease-causing microorganisms. There is a kill of at least 10^5 vegetative bacteria and fungi, fungal spores, and viruses (in other words, an inactivation of at least 5 Log_{10} or greater); however, Level I may be unable to inactivate mycobacteria and bacterial spores. Level I inactivation may be accomplished by several physical or chemical processes.

Level II microbial inactivation is defined as significant inactivation of all microorganisms with the exception of bacterial spores. This indicates the inactivation of at least 10^5 mycobacteria

Table 20.6

Evaluation of Level of Microbial Inactivation Achieved by Medical Waste Treatment Technologies

Waste Treatment Technology	Microbial Inactivation			
	Level I	Level II	Level III	Level IV
Steam autoclave				
Lab test results[a]	Yes	Yes	Yes	No
Field test results[b]	Yes	Yes	Yes	Yes
Microwave field test results[c]	NT	NT	Yes	No
Radio frequency field test results[d]	NT	NT	Yes	No
Chemical				
Lab test results[e]	Yes	Yes	Yes[g]	Yes[g]
Field test results[f]	Yes	Yes	No[h]	No[h]

NT = not tested.

[a]Benchtop and gravity displacement autoclaves, 121°C, 15 psi.

[b]Prevacuum system, 138°C, 30 psi; double-door gravity system, 163°C, 80 psi.

[c]Microwave treatment system (6 units at 2450 MHz each).

[d]Short-wave RF system, 11 to 13 MHz.

[e]Chemical only, sodium hypochlorite 1000 and 3000 ppm FAC prolonged exposure (≥3 hrs).

[f]Chemical/mechanical systems, sodium hypochlorite 1000, 2000, 3000 ppm FAC.

[g]Dependent on Prolonged exposure (>3 h).

[h]Not achieved under normal operating conditions (<3 h exposure).

Source: RTI, Final Report, 1993. Reproduced with kind permission of Research Triangle Institute.

in addition to Level I activation. Level II inactivation implies some measure of tuberculocidal activity on the wastes (RTI, 1993).

Level III inactivation indicates the kill of microbial life forms as evidenced by the inactivation of at least 10^4 of selected indicator spores, which possess death curves similar to human pathogenic spores. Thus, *Bacillus subtilis* spores may be used to indicate Level III microbial inactivation for moist heat treatment since they also exhibit thermal death data similar to species of the pathogenic spore-forming *Clostridium*.

Level IV indicates the kill of microbial life forms as evidenced by the inactivation of 10^6 bacterial indicator spores recognized as most resistant to the treatment process. For example, the inactivation of at least 10^6 spores of the bacterium *B. stearothermophilus*, recognized as most resistant to moist heat, is an indication of Level IV inactivation by stream autoclaving (RTI, 1993).

Nonincineration alternative treatment technologies are being relied upon increasingly as public and regulatory pressures direct the medical industry away from incineration for treating medical and infectious waste. There is a need to understand the alternative medical waste treatment system options that are available.

20.9.1 AUTOCLAVING (STEAM STERILIZATION)

Autoclave treatment utilizes moisture, heat, and pressure to inactivate microorganisms. Steam sterilization was originally developed for equipment or material sterilization, for example, glassware and microbial media, prior to use. Steam autoclaving is now an appropriate on-site or off-site treatment technology for medical and infectious wastes. Most medical wastes can be treated with this technology, with the exception of radioactive wastes, body parts, or animal carcasses. Steam sterilizers are used to treat medical waste by both small generators such as healthcare clinics and physicians' offices and commercial medical waste treatment firms treating waste for a large region.

Sterilization of medical waste involves placing contaminated waste into a sealed chamber and exposing the waste to pressurized steam of sufficient temperature for a specified length of time to render them noninfectious. For steam to penetrate the load, the air must be completely removed from the treatment chamber. Sterilization occurs primarily from the penetration of steam into the matrix. Heat conduction provides a secondary source of heat transfer.

There are three basic types of autoclaves: gravity systems, prevacuum systems, and retort systems. In the gravity system, steam replaces the air within the chamber, generally by forcing the air out through a valve located at the base of the unit as the steam is introduced into the chamber (Figure 20.8). Prevacuum systems use pumps to evacuate air from the chamber before the steam is introduced (U.S. EPA, 1986; Joslyn, 1991; Turnberg, 1996). Retort systems are designed to operate at high steam pressures (U.S. EPA, 1991).

When the steam enters the chamber the temperature increases to the desired setting. This is known as the *heat-up time* (steam penetration time). The *holding time* begins after the load has reached the minimum temperature required for achieving sterilization. The *exposure time* represents the entire period necessary to achieve sterilization and includes the sum of heat-up time and holding time plus a margin of error (Perkins, 1976; Turnberg, 1996).

20.9.2 OPERATIONAL ISSUES

Autoclaves require pressurized vessels to ensure that the waste is being exposed to the correct temperatures for the proper amount of time. Temperature and time are essential for successful steam sterilization. Air must be removed completely from the chamber (thus also from the wastes) so that

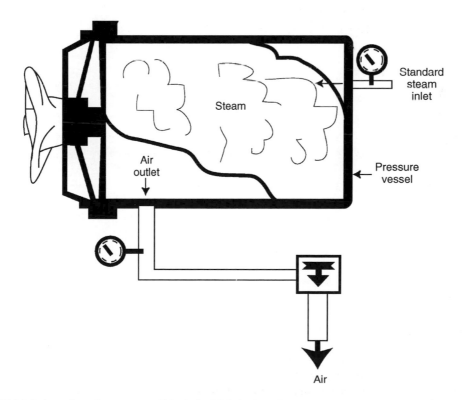

FIGURE 20.8 Autoclave from system (Block S., Disinfection, Sterlization and Preservation, 4th ed., Lea and Fibiger, Philadelphia, 1977. Reproduced with kind permission of Lippincott, Williams & Wilkins, Malvern, PA.)

steam can penetrate all areas and kill the organisms of concern. Specific factors affecting the operation of steam autoclave treatment of medical waste include:

- Temperature and pressure achieved by the autoclave
- Steam penetration of the waste
- Size of waste load
- Composition of waste
- Packaging of the waste
- Orientation of wastes within autoclave

Steam autoclaves operate most effectively when the temperature measured at the center of the waste load approaches 121°C (250°F) and there is adequate steam penetration of the load. At a given temperature, the duration of treatment is the variable that determines the heat conduction and steam penetration to the center of a load (RTI, 1993). Optimum operational temperatures are approximately 132°C (270°F), and should not go below 250°F for maximum effectiveness.

As was shown in Table 20.2, the following wastes can be steam sterilized by autoclaving: cultures and stocks, pathological wastes, human blood and blood products, sharps (used and unused), animal waste, and isolation wastes. Body parts and contaminated animal carcasses are not suitable for steam sterilization because the density of these wastes may prevent adequate steam penetration. Radioactive, hazardous, and cytotoxic wastes are also not appropriate for treatment by steam autoclaving.

Waste liquid from the steam condensate is typically permitted for discharge directly to the sanitary sewer. Wastes may be shredded after sterilization, which will greatly reduce total waste volume.

20.9.3 TESTING THE EFFICIENCY OF STERILIZATION BY AUTOCLAVING

The thermally resistant species *B. subtilis* (globigii) ATCC 9372 (10^4) and *B. stearothermophilus* ATCC 12980 are commonly employed for the verification of Level III and Level IV microbial inactivation, respectively. These organisms are available in commercial suspensions or as prepared spore strips. The test procedure is as follows: dried test spores are placed in a thermally resistant and steam-permeable container near the center of the waste load. The autoclave is operated under normal conditions. At the conclusion of the cycle the test organisms are removed from the load. To recover the test organisms, test strips are inoculated into soybean–casein digest broth medium and incubated for at least 48 h (30°C for *B. subtilis* or 55°C for *B. stearothermophilus*). At the end of the incubation, the media is examined for turbidity as an indicator of bacterial growth. Any detected growth should be subcultured on to appropriate media to confirm the identity of the organism. To establish Level III inactivation, a minimum of 10^4 *B. subtilis* spores must be completely killed. To establish Level IV, a minimum total of 10^6 *B. stearothermophilus* must be completely killed.

Effective sterilization requires the correct application of several variables. Trained autoclave operators are essential for correct and safe operation. Training should include proper autoclave operation as well as information about potential associated hazards. Processing effectiveness should be monitored to ensure that treatment has been accomplished using time temperature charts, chemical indicators that produce color changes to correspond with the necessary time-temperature relationship necessary to achieve sterilization, and biological indicators (e.g., spore strips of *B. subtilis* or *B. stearothermophilus*) to ensure inactivation of the most resistant microorganisms (Turnberg, 1996).

The ideal container for the sterilization of medical wastes is one that is corrosion-resistant, leakproof, capable of allowing complete steam penetration to its contents, and, of course, reasonably priced. In recent years, plastic bags have become the preferred container for many medical wastes. While assessing laboratory autoclave operation, however, loads contained in plastic bags have frequently been found to perform poorly. This effect was later found to be caused by inadequate steam penetration into the bags. When plastic bags are used to store medical wastes, steps should be taken to determine the necessary sterilizing time/temperature relationships and steam penetration (Everall and Morris, 1976; Turnberg, 1996).

20.9.4 HEALTH CONSIDERATIONS WITH AUTOCLAVING

Health impacts attributed to the autoclaving of medical wastes have not been documented. Operators may be exposed to hazardous constituents in the venting emissions if they come into contact with the steam, such as may occur when the autoclave door is opened at the end of the cycle. It is therefore important to exclude any waste containing potentially toxic constituents such as hazardous chemicals (e.g., RCRA waste) or radiological wastes (Turnberg, 1996).

20.9.5 ADVANTAGES AND DISADVANTAGES OF AUTOCLAVING

When properly operated, autoclaves are highly effective for the sterilization of infectious medical waste. They are most suitable for decontaminating laboratory wastes such as stocks and cultures of infectious agents, contaminated glassware, and biological tissue, but are capable of decontaminating other classes of infectious waste as well. For practical reasons, autoclaves are not appropriate for treating body parts.

An advantage of steam sterilization is that the technology has a long history of use by hospitals, laboratories, clinics, and other medical institutions for the sterilization of both medical supplies and medical and infectious waste. Steam sterilization does not raise public concerns or require the complex regulations such as those observed with medical waste incinerators. Another advantage is their greater output capacity and minimal space requirements compared with those required for on-site incinerators (Turnberg, 1996, 1989). In limited cases, the autoclaved waste will occupy less space because air has been forced out (Liberman and Gordon, 1989). Steam autoclave systems have low capital and operating costs and are comparatively easy to operate.

A major disadvantage of autoclaving is that the waste itself is not destroyed but is simply rendered nonpathogenic. The volume of the waste is not reduced after sterilization. There are also concerns about odors emanating from an autoclave in use. Drainage liquids must be stored, managed, and disposed of properly. Some bags may block air, thus limiting steam penetration and complete sterilization (Lipman, 1992). The presence of residual air within the autoclave can prevent complete sterilization by (Perkins, 1969; U.S. EPA, 1986):

- Reducing the temperature of the steam, regardless of pressure
- Causing variations in temperature throughout the chamber
- Prolonging the time needed to attain maximum temperature
- Inhibiting stream penetration into porous materials

Factors causing incomplete displacement of air include use of heat-resistant plastic bags (which trap air), deep containers (which may prevent displacement of air from the bottom), and improper loading (which may prevent free circulation of steam within chamber) (U.S. EPA, 1986).

A significant concern with an autoclave is that waste handlers are not able to determine, from looking at a product, whether it has been sterilized; a bag of autoclaved waste at a landfill may well resemble one that had not been sterilized. In some situations, landfills have refused to accept autoclaved waste due to concerns as to whether the waste was treated adequately. Some states have enacted regulations requiring not only that medical and infectious waste be decontaminated but also that it be rendered unrecognizable as medical waste. As a result, many steam sterilizers are now provided with waste shredding systems (Turnberg, 1996).

20.9.6 CHEMICAL DISINFECTION

A chemical disinfectant is an agent that destroys disease-causing agents such as pathogenic organisms. Disinfectant chemicals are registered under FIFRA (The Federal Insecticide, Fungicide, and Rodenticide Act) according to their application against specific types of pathogens.

Chemical agents have been used for disinfecting infectious waste for many years. The treatment usually incorporates some type of mechanical destruction process (i.e., shredding) that reduces the waste to a small particle size, thereby increasing particle surface area which increases contact with the chemical agent and ultimately sterilizes the waste. Disinfectants (antimicrobial agents) may be used alone or in combination with encapsulating agents.

Systems are available for small operations such as laboratories and for large operations such as hospitals. In one disinfection system (Figure 20.9; U.S. EPA, 1991), waste is loaded onto a conveyor belt that transfers it to a low-speed shredder to break bags and cardboard boxes. Beyond the low-speed shredder is a high-speed hammermill where the waste is pulverized to a fine particle size. During both the initial shredding and pulverization steps, the waste is sprayed with a sodium hypochlorite (NaOCl) solution. Solids are separated from slurry using a perforated conveyor belt. The liquids are diverted to a sanitary sewer and the solids are collected for off-site disposal. Air is drawn from the system and passed through a series of prefilters and a chlorine-resistant HEPA filter after which it is discharged to the air to control airborne contamination. The system can handle up to 675 kg (1,500 lb) of medical waste per hour (U.S. EPA, 1991).

20.9.7 TYPES OF DISINFECTION AGENTS

Classes of common antimicrobial chemicals and their advantages and disadvantages are listed in Table 20.7. Several different chemical agents are currently being marketed for use in various medical waste treatment systems. Some have been used in other applications, for example, in the treatment of drinking water. Example formulations include chlorine dioxide, sodium hypochlorite, a dry calcium oxide mixture, and peracetic acid.

20.9.8 TESTING THE EFFICIENCY OF CHEMICAL DISINFECTION

Not all microorganisms are affected to the same degree by the same chemical. In addition, genetic mutation and natural selection will result in a pattern of resistance to numerous chemicals. The general scale of resistance to chemical treatment, from least to most resistant, is:

- Vegetative bacteria
- Vegetative fungi and fungal spores

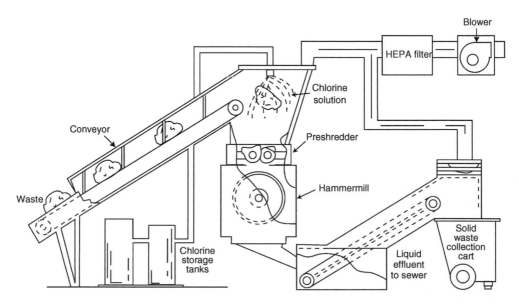

FIGURE 20.9 Combined shredding and disinfection system. Reproduced with kind permission of Medical Safe-Tec, Indianapolis, IN.

TABLE 20.7

Advantages and Disadvantages of Antimicrobial Agents

Class	Advantages	Disadvantages
Alcohols	Bactericidal	Nonsporicidal
	Tuberculocidal	Organic matter interference
	Virucidal	Incompatible with some
	Fungicidal	rubber and plastics
	Nonstaining	Highly flammable
	Nonirritating	
	Rapid action	
Quaternary	Bactericidal	Nontuberculocidal
ammonium	Virucidal (lipophilic)	Nonsporicidal
compounds	Fungicidal	Organic matter interference
	Pleasant odor	Nonvirucidal (hydrophilic)
Phenolic compounds	Bactericidal	Questionable virucide
	Fungicida	(hydrophilic)
	Tuberculocidal	Nonsporicidal
	Virucidal (lipophilic)	Skin irritant
		Unpleasant odor
		Corrosive
Iodophor compounds	Bactericidal	Prolonged exposure for
	Virucidal	tuberculocidal and
	Fungicidal	sporicidal activity
	Detergent action	Corrosive
	Storage stability	Inactivation by organic matter
Glutaraldehyde	Bactericidal	Irritant
	Virucidal	Limited shelf life
	Fungicidal	
	Tuberculocidal	
	Sporicidal	
	Lack of organic	
	matter interference	
	Noncorrosive	
Hypochlorite solution	Bactericidal	Prolonged exposure for
	Virucidal	sporicidal activity
	Fungicidal	Corrosive
	Tuberculocidal	Bleaching agent
Hydrogen peroxide	Bactericidal	Corrosive
	Virucidal	
	Fungicidal	
	Tuberculocidal	
	Sporicidal	

Adapted from RTI, Final Report, 1993. Reproduced with kind permission of Research Triangle Institute.

- Viruses
- Mycobacteria
- Bacterial spores

Several levels of antimicrobial activity are defined to indicate the types of organisms the chemical is expected to kill (Table 20.8). Antimicrobial chemicals typically include products with

TABLE 20.8
Selected Antimicrobial Efficacy Claims for Microbial Inactivation

Specific Claim	Definition
Sporicide/Sterilant	An agent intended to inactivate all living microorganisms, especially bacterial spores
Tuberculocide	An agent intended to inactivate mycobacteria, especially *Mycobacterium tuberculosis* Tuberculocidal efficacy assumes inactivation of all viruses, fungi, and vegetative bacteria
Virucide	An agent intended to destroy viruses. Virucidal efficacy may vary in regard to lipophilic and hydrophilic viruses
Fungicide	An agent that inactivates fungi including fungal spores
Bactericide	An agent that inactivates vegetative bacteria but not bacterial spores
Germicide	An agent that inactivates one or more pathogenic microorganisms (May include sporicide, tuberculocide, virucide, fungicide, or bactericide)

Adapted from EPA (1984); RTI, Final Report, 1993. Reproduced with kind permission of Research Triangle Institute.

claims such as bactericidal, fungicidal, virucidal, sporicidal, etc. The chemically resistant species *B. stearothermophilus* (ATCC 12980 or ATCC 10149) are commonly employed to demonstrate a 10^4 reduction of viable spores. These organisms may be available in commercial suspensions or as prepared spore strips.

A sufficient number of *B. stearothermophilus* spores must be added to the treatment system to demonstrate satisfactory reduction. Disinfection systems are tested by comparing samples from the procedure with and without the test chemical. The recovered samples are neutralized, filtered, and inoculated onto soybean–casein digest agar, streaked to quantify the samples, and incubated at 55°C for at least 48 h. After incubation the organisms must be quantified to confirm the appropriate level of spore reduction. The chemically treated spores should demonstrate a 10^4 reduction in comparison with spores treated with tap water.

20.9.9 OPERATIONAL ISSUES

The effectiveness of disinfection treatment is a function of (U.S. EPA, 1986; Turnberg, 1996):

- Characteristics of the disinfectant
- The concentration of the active ingredient
- Type of microorganisms in the waste
- Degree of contamination
- Characteristics of the waste
- Contact time of the disinfectant with waste sample

Other relevant factors (e.g., pH, presence of electrolytes, complex formation, and adsorption such as binding to small molecules or ions, macromolecules, or soil) will additionally influence the effectiveness of a specific disinfectant (U.S. EPA, 1986; Turnberg, 1996).

Microbial inactivation by chemical agents is a function of the active ingredient concentration and the pH of the prepared solution, among other factors. It is essential to ensure that the formulation will not be diluted during the treatment process to the point where it is no longer effective. Some antimicrobial chemicals such as quaternary ammonium compounds and halogens may be readily inactivated when in contact with organic matter (e.g., whole blood) or hard water (Ca and Mg). Depletion of the chemical disinfectant can also occur through consumption of the chemical agent by interaction with microorganisms, loss through volatilization, chemical decomposition, or metabolism by the microorganism itself (Kostenbauder, 1983; Turnberg, 1996). Such potential interference must be considered when selecting a formulation and its concentration for the treatment of medical waste (RTI, 1993).

20.9.10 FIFRA REGISTRATION

Under The Federal Insecticide, Fungicide, and Rodenticide Act (FIFRA), any chemical agent used in a treatment process may require registration with the EPA Pesticide Registration Office. If a manufacturer advertises that a chemical formulation can achieve a level of microbial inactivation (e.g., sterilant) for a specified use, that formulation would have to be registered with the EPA Office of Pesticide Registration (Turnberg, 1996).

20.9.11 ADVANTAGES AND DISADVANTAGES OF CHEMICAL DISINFECTION

The combined shredding or chemical disinfection system is relatively simple. Furthermore, this treatment train can substantially reduce waste volumes. After shredding, waste components are essentially unrecognizable. Disadvantages of chemical disinfection include relatively high capital and operating costs. There may be problems with contaminants in the slurry, concentrations of disinfectant in the work space (a potential irritant to nose, eyes, and lungs), noise levels, and bioaerosol emissions (U.S. EPA, 1991). Discharge permits may be required for the slurry. Microbes may become resistant to certain disinfectants. Finally, disinfection systems do not provide sterilization.

20.9.12 MICROWAVE IRRADIATION

During microwave treatment, waste is fed automatically to a grinding device where it is shredded and sprayed with steam to increase the moisture content to about 10%. The moist ground waste is then heated by exposure to microwave irradiation units over 2 h. Microwaves are electromagnetic waves having a frequency between infrared and radio waves. A typical industry standard radio frequency energy of 2450 MHz is absorbed by the waste to create friction in water molecules. Heat generated by this friction denatures proteins within the microbial cell thereby killing and inactivating the cell (Turnberg, 1996). Temperatures in the unit exceed 90°C. Factors affecting the microwave treatment of medical waste include the frequency and wavelength of irradiation, the duration of exposure, moisture content of the waste material, process temperature, and the mixing of the waste during treatment.

Microwave irradiation can treat most medical wastes with the exception of cytotoxic, hazardous, or radioactive wastes. Contaminated animal carcasses, body parts, and human organs are excluded from treatment by microwave irradiation for aesthetic reasons.

Thermally resistant species such as *B. subtilis* (globigii) ATCC 9372 (10^4) may be used for verification of microbial inactivation. Dried spores are placed in a steam-permeable container and added to the waste stream after the waste is ground and sprayed with steam before exposure to microwave irradiation. The microwave unit is operated under routine conditions. At the conclusion of the cycle the test strips containing the organisms should be inoculated into soybean–casein digest broth medium and incubated for at least 48 h. *B. subtilis* should be cultured at 30°C. At the end of the incubation period the media should be examined for turbidity as a sign of bacterial growth. Any growth should be cultured onto appropriate media to confirm the identity of the organism (RTI, 1993).

REFERENCES

American Hospital Association, An Ounce of Prevention: Waste Reduction Strategies for Health Care Facilities, American Society for Health Care Environmental Services of the American Hospital Association, Chicago, IL, 1993.

Block, S. (Ed.), *Disinfection, Sterilization, and Preservation*, 4th ed., Lea and Fibiger, Philadelphia, 1977.

Centers for Disease Control, Guideline for Handwashing and Hospital Environmental Control, NTIS PB85-923404, Washington, DC, 1985.

Code of Federal Regulations Vol 40 Part 259, Standards for the Tracking and Management of Medical Waste, U.S. Government Printing Office, Washington, DC, 2002.

Cole, E.C. Application of Disinfection and Sterilization to Infectious Waste Management, *Proceedings of Strategies for Improved Chemical and Biological Waste Management for Hospitals and Clinical Laboratories*, North Carolina Board of Science and Technology, Raleigh, NC, 1987.

Doucet, L.G., Infectious Waste Treatment and Disposal Alternatives, *Presented at the Symposium on Infection Control*: *Dilemmas and Practical Solutions*, Philadelphia, PA, November 1989.

Emery, R., Marcus, J., and Sprau, D., Characterization of low-level radioactive waste generated by a large university/hospital complex, *Health Phys.*, 62, 183–185,

Everall, P.H. and Morris, C.A. Failure to sterilize in plastic bags, *J. Clinical Pathol.* 29, 1132, 1976.

Hoeltge, G., Managing the Infectious Waste of a Surgery Unit and Laboratory, *Semin. Dermatol*, 14, 247–51, 1995.

Joslyn, L.J., Sterilization by heat., in *Disinfection, Sterilization, and Preservation*, Block S., Ed., 4th ed., Lea and Febiger, Philadelphia, PA, 1991.

Key, S. and M. Marble, Hospitals emit cancer-causing dioxin when they burn waste, *Cancer Weekly Plus*, 10, 19–21, 1997.

Kostenbauder, H.B., Physical factors influencing the activity of antimicrobial agents, in *Disinfection, Sterilization, and Preservation*, Block S., (Ed.), 3rd ed., Lea and Febiger, Philadelphia, PA., 1983.

Lipman, L. New rules for medical waste, *Saf. Health*, 145, 40–43, 1992.

Liberman, D. and Gordon J., (eds), *Biohazard Management Handbook*, Marcel Dekker, New York, NY, 1989.

Perkins, J.J., *Principles and Methods of Sterilization in Health Services*, 2nd ed., Charles C. Thomas, Springfield, IL, 1976.

Research Triangle Institute, Guidance for Evaluating Medical Waste Treatment Technologies, Final Report, 94U-5400-005/01-F. Research Triangle Park, NC, 1993.

Reynolds, J.P., Jeris, J.S., and Theodore, L., Handbook of Chemical and Environmental Engineering Calculations, Wiley-Interscience, New York, 2002.

Turnberg, W.L., *Biohazardous Waste. Risk Assessment, Policy, and Management*, Wiley, New York, NY, 1996.

Turnberg, W.L., Human Infection Risks Associated with Infectious Disease Agents in the Waste Stream: A Literature Review, in Washington State Infectious Waste Project — Report to the Legislature, Washington Department of Ecology, Olympia, WA, December, 1989.

U.S. Environmental Protection Agency, Efficacy Data Requirements, Office of Pesticides and Toxic Substances, Washington, DC, 1984.

U.S. Environmental Protection Agency, EPA Guide for Infectious Waste Management, EPA/530-SW-86-014, Office of Solid Waste, Washington, DC. May, 1986.

U.S. Environmental Protection Agency, Medical Waste Tracking Act of 1988. Public Law 100–582. EPA/530-SW-89-008. Washington, DC, Nov. 1, 1988.

U.S. Environmental Protection Agency, Seminar — Medical and Institutional Waste Incineration: Regulations, Management, Technology, Emissions, and Operations. Center for Environmental Research Information, CERI 89-247, Cincinnati, OH, November 1989a.

U.S. Environmental Protection Agency, Hospital Incineration Operator Training Course Manual, EPA 450/3-89-004, NTIS PB 89-189880, Research Triangle Park, NC, March 1989b.

U.S. Environmental Protection Agency, Operation and Maintenance of Hospital Medical Waste Incinerators, EPA 450/3-89-002, Control Technology Center, Research Triangle Park, NC, 1989c.

U.S. Environmental Protection Agency, Medical Waste Management in the United States — First Interim Report to Congress, Office of Solid Waste, Washington, DC., 1990a.

U.S. Environmental Protection Agency, Operation and Maintenance of Hospital Medical Waste Incinerators, EPA/625/6-89/024, Center for Environmental Research Information. Cincinnati, OH, January 1990b.

U.S. Environmental Protection Agency, Landrum, V.J., Barton, R.G., Neulicht, R., Turner, M., Wallace, D., and Smith, S., *Medical Waste Management and Disposal*, Noyes Data Corporation, Park Ridge, NJ, 1991.

U.S. Environmental Protection Agency, Medical and Institutional Waste Incineration: Regulations, Management, Emissions and Operations, EPA/625/4-91/030, Office of Research and Development, Washington, DC, 1991.

U.S. Environmental Protection Agency, Medical Waste Incinerators — Background Information for Proposed Guidelines: Industry Profile Report for New and Existing Facilities, EPA-453/R-94-042a, July 1994.

U.S. Environmental Protection Agency, Mercury Study Report to Congress: Volume I: Executive Summary, EPA-452/R-97-003, Office of Air Quality Planning & Standards and Office of Research and Development, December 1997.

U.S. Environmental Protection Agency, 2000.See: http://yosemite.epa.gov/opa/admpress.nsf/ 2000.

U.S. Environmental Protection Agency, Municipal Solid Waste in the United States: 1999 Facts and Figures, EPA 530-R-01-014, Office of Solid Waste and Emergency Response, Washington, DC, 2001.

Wastebusters. Materials management in health care 7:9, 1998.

SUGGESTED READING

Aegis Bio-Systems, The JYD-1500, Mobile Medical Waste Treatment System Using Steam Sterilization, October 2003. See: http://www.aegisco.com/techspec.htm.

Bay Area Dioxins Project, Vendor List: Medical Waste Treatment, Environmental Best Practices for Health Care Facilities, June 2003. See: http://dioxin.abag.ca.gov/pilot_projs/MW_VendorList.pdf.

Bay Area Dioxins Project, Why Are Hospitals Rethinking Regulated Medical Waste Management? Environmental Best Practices for Health Care Facilities, June 2003. See: http://dioxin.abag.ca.gov/pilot_projs/MW_Background.pdf.

BGX Technologies, LLC, Important New Technology for Medical Waste Treatment. October 2003. See: http://www.bgxonline.com/Medical%20Waste.htm.

Burke, T. and Byrns, G., Medical waste management implications for small medical facilities, *J. Environ. Health,* 55,12-14, 1992.

Boatright, D.T. and Edwards, A.J., A comprehensive biomedical waste survey, *J. Environ. Health,* 8,15-19, 1995.

California Department of Waste Services, Medical Waste Management Program, August 2003. See: http://www.dhs.cahwnet.gov/ps/ddwem/environmental/Med_Waste/medwasteindex.htm

California State University, Chico, Medical Waste Management, Office of Environmental Management Health and Safety, March 1995. See: http://www.csuchico.edu/ehs/Environmental/mwm.htm

Center for Environmental Research Information (U.S.), Medical and Institutional Waste Incineration: Regulations, Management, Technology, Emissions, and Operations, Cincinnati, OH, Center for Environmental Research Information, Office of Research and Development, United States Environmental Protection Agency, 1992.

Clean-Pro Industries, Bio Set Cement Based Solidification Process for Hazardous Medical Waste Material, 2001. See: http://www.users.qwest.net/~cleanpro/index18.html

Covell, C., Wavelengths, An Employee's Digest of Events and Issues, Division Supports the Shipboard Medical Waste Management Program, December 2000. See: http://www.dt.navy.mil/pao/excerpts%20pages/2000/medwaste12_00.html

Environmental Protection Act, SO., Biomedical Waste Management Requirements in Ontario, 1990. See: www.ene.gov.on.ca/envision/env_reg/er/documents/2001/RA01E0023_c.pdf

Fritsky, K., Kumm, J., and Wilken, M., Combined PCDD/F Destruction and Particulate Control in a Baghouse: Experience with a Catalytic Filter System at a Medical Waste, Incineration Plant, *International Thermal Treatment Technologies Conference*, Portland, OR, 2000.

Fry, B. et al., Technical Support Document to Proposed Dioxins and Cadmium Control Measure for Medical Waste Incinerators, California Air Resources Board, Sacramento, CA, 1990.

Garvin, M.L., *Infectious Waste Management: A Practical Guide*, Lewis Publishers, Boca Raton, FL, 1995.

Hasselriis, F., Corbus, D., and Kasinathan, R., Environmental and Health Risk Analysis of Medical Waste Incinerators Employing State-of-the-Art Emission Controls, P. No. 91 30.3, 84th Annual Meeting of the Air and Waste Management Association, Kansas City, MO, 1991.

Health Care Without Harm, Medical Waste, Nonincineration Medical Waste Treatment Technologies, Executive Summary, August 2001. See:http://www.noharm.org/library/docs/Non-Incineration_Medical_Waste_Treatment_Te_2.pdf.

Joint Commission on Accreditation of Healthcare Organizations, Medical Waste Management: Recycling and New Technologies, Oakbrook Terrace, IL, 1991.

Kiser, J., Hospital recycling moves ahead, *Biocycle*, 11, 30–34, 1995.

Landrum, V.J., *Medical Waste Management and Disposal*, William Andrew Inc., New York, NY, 1991.

Marcial, G., Cleaning up at Stericycle, *Bus. Week*, 3784, 107-109, 2002.

Maziuk, J., *Results of Emission Testing at a Medical Waste Incinerator Using Dry Injection/Fabric Filter APC Technology*, Church & Dwight, Princeton, NJ, 1998.

Miller, R.K., Survey on Medical Waste Management, Future Technology Surveys, New York, NY, 1991.

National Academies Press, Waste Incineration and Public Health, Committee on Health of Effects of Waste Incineration, Board on Environmental Studies and Toxicology, Commission on Life Sciences, National Research Council, 2000. See: http://books.nap.edu/books/030906371X/html/R1.html#pagetop

New Jersey Department of Environmental Protection, Alternative Medical Waste Disposal Technologies Authorized in New Jersey, June 2001. See: http://www.state.nj.us/dep/dshw/rrtp/aamwdt.htm

Office of Environment, Health & Safety, UC Berkeley, Managing and Disposing of Medical Waste, December 1997. See: http://www.ehs.berkeley.edu/pubs/helpsheets/01medwastegls.html

Perez, L., Traditional Medical Waste Treatments, Environmental Chemistry.com, 1999. See: http://environ mentalchemistry.com/yogi/environmental/medicalwaste.html

Reinhardt, P.A.and Gordon. J.G., *Infectious and Medical Waste Management*, Lewis Publishers, Boca Raton, FL, 1991.

Research Triangle Institute, Bioemissions from Medical Waste Treatment Technologies: Evaluation of Indicator Microorganism Recovery, Final Draft Report, Prepared for: Office of Solid Waste, United States Environmental Protection Agency, April 1993. See: http://www.epa.gov/epaoswer/other/medical/mwpdfs/rti2/1.pdf

Rutala, W.A. and Weber, D.J., Infectious waste, *New Eng. J. Med.*, 325, 578-583, 1991.

Sanders, D.L. et al., The New East Carolina University Medical Waste Incinerator: Combining a Wet Scrubber with Granular Carbon Bed, *International Thermal Treatment Technologies Conference*, Portland, OR, 2000.

Studnicki, J., The management of hospital medical waste, *Hosp. Top.*, 70, 11-21, 1992.

U.S. Environmental Protection Agency, Medical Waste, June 2003. See: www.epa.gov/epaoswer/other/medical/index.htm

U.S. Environmental Protection Agency, Hospital/Medical/Infectious waste Incinerators: Background Information for Federal Plan: Summary of Public Comments and Responses, EPA 456/R-00-003, Office of Air and Radiation, Office of Air Quality Planning and Standards, Research Triangle Park, NC, 2000.

U.S. Environmental Protection Agency, Medical and Institutional Waste Incineration: Regulations, Management, Technology, Emissions, and Operations, EPA/625/4-91/030, Center for Environmental Research Information, Office of Research and Development, Cincinnati, OH, 1992.

University of California, Riverside, No date, Medical Waste Management Program, University of California, Riverside Environmental, Health & Safety. See: http://www.ehs.ucr.edu/programs/bio/pdf/BIO_WAS8.PDF

Walker, B.L. and Cooper, C.D., Air pollutant emission factors for medical waste incinerators, *J. Air Waste Manage. Assoc.*, 42, 784-791, 1992.

QUESTIONS

1. List the institutions generating most of the medical waste in the United States. In which specific institution is the majority produced?
2. What are the agency(ies) that regulate wastes generated at healthcare facilities?
3. What is the status of the Medical Waste Tracking Act?
4. How is the incineration of medical waste regulated?
5. List and discuss the alternatives to incineration for treating and disposing of medical waste.
6. Which federal agencies other than the EPA regulate potentially infectious medical waste?
7. Are there regulations relating to the land disposal of medical waste?
8. What is the difference between 'infectious waste' and 'medical waste?'
9. Teeth are considered infectious waste. Under the federal regulations, is it permissible to bring an extracted tooth home from the dentist?
10. The Robust Health Care Company generates very little infectious waste (less than 20 kg, or about 50 lb/month). How are they required to manage these wastes?
11. Under federal regulations, what is the standard definition of an infectious waste? Under which body of regulations is this covered?
12. What are the exemptions to the definition of regulated medical waste?
13. What are the container requirements for regulated medical waste? How are sharps to be containerized?
14. What are the requirements regarding tracking of medical waste shipments? How are non-returned forms to be addressed?
15. What specific microorganisms are employed to test the efficiency of sterilization by autoclaving? Describe the procedure.

21 Construction and Demolition Debris

We have to discard the past and, as one builds
floor by floor, window by window,
and the building rises, so do we go throwing down
first, broken tiles, then pompous doors,
until out of the past dust rises as if to crash against the floor,
smoke rises as if to catch fire
and each new day it gleams like an empty plate

Pablo Neruda (1904–1973)
Past

21.1 INTRODUCTION

Construction and demolition (C&D) debris is generated when new structures are built and existing structures are renovated or demolished (Figure 21.1). Structures include residential and nonresidential buildings and public works projects such as streets, highways, bridges, and piers. Components of C&D debris include concrete, asphalt, wood, metals, gypsum wallboard, and roofing. State definitions of C&D debris may also be expanded to include trees, stumps, soil, and rock from clearing construction sites.

21.2 CHARACTERISTICS OF C&D DEBRIS

Sources of C&D debris include homebuilders, homeowners, commercial developers, building contractors, highway and street contractors, bridge constructors, bituminous pavement contractors, home remodelers, site grading contractors, demolition contractors, roofing contractors, drywallers, and excavating companies (Franklin Associates, 1998). The major activities that generate C&D debris from buildings include the construction, demolition, and renovation of residential and nonresidential buildings (Figure 21.2). Residential buildings include single-family houses, duplexes and high-rise multifamily housing. Nonresidential buildings include commercial, institutional, and industrial buildings. Approximately 123 million metric tons (136 million tons) of building-related C&D debris were generated in the United States in 1996 (Table 21.1). Forty-three percent of the waste (58 million tons per year) was generated from residential sources and 57% (78 million tons per year) from nonresidential sources. Building demolitions account for 48% of the total C&D waste stream, renovations account for 44%, and construction activities generated the remaining 8% (Franklin Associates, 1998).

Construction activities generally produce cleaner materials than does demolition. Demolitions generate wastes having several materials bound together or contaminated with hazardous materials such as asbestos or lead paint. Renovation projects can produce both construction and demolition type wastes. The sources of C&D debris are listed in Table 21.2.

The composition of C&D debris is highly variable and is a function of the specific activities taking place at the site. For example, concrete is the largest component of building demolition debris and wood is the largest waste component generated at construction and renovation sites. Typical components of C&D debris are listed in Table 21.3.

FIGURE 21.1 C&D debris results from construction, demolition, and renovation of residential and nonresidential structures.

Construction debris from building sites commonly consists of scraps of construction materials such as wood, sheetrock, masonry, and roofing materials. There is comparatively less concrete in construction debris than demolition debris. Scrap from residential construction sites represents between 6 and 8% of the total weight of the building materials delivered to the site, excluding the foundation, concrete floors, driveways, and patios (Franklin Associates, 1998). A comparison of the composition of construction debris vs. demolition debris is provided in Figure 21.3.

When buildings are demolished, large quantities of waste may be produced in a short period of time. Demolition techniques include imploding a structure with explosives, using a crane and wrecking ball technique, or deconstructing. A majority of demolition projects use a combination of

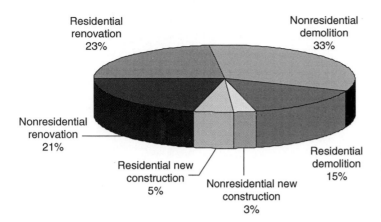

FIGURE 21.2 Generation of C&D debris from buildings (Franklin Associates, (EPA530-R-98-010, 1998). Reproduced with kind permission of Franklin Associates, Prairie Village, KS.)

TABLE 21.1
Summary of Estimated Building-Related C&D Debris Generation in1996

Source	Residential		Nonresidential		Totals	
	Thousand tons	%	Thousand tons	%	Thousand tons	%
Construction	6,560	11	4,270	6	10,830	8
Renovation	31,900	55	28,000	36	59,900	44
Demolition	19,700	34	45,100	58	64,800	48
Totals	58,160	100	77,370	100	135,530	100
Percent	43		57		100	

Source: Franklin Associates, EPA530-R-98-010, 1998. Reproduced with kind permission of Franklin Associates, Prairie Village, KS.

TABLE 21.2
Representative Sources of C&D Wastes

Site clearance materials (brush, tree, and stumpage materials)
Excavated materials (Earth, fill, and other excavated rock and granular materials)
Roadwork materials
    Concrete slabs and chunks from concrete road construction
    Asphalt chunks and millings from asphalt pavement
    Bridge/overpass construction/renovation materials
New construction materials (residential, commercial, and industrial project sources)
Renovation, remodeling, or repair materials (residential, commercial, and industrial project sources)
Demolition materials including wrecking, implosion, dismantling and deconstruction (residential, commercial, and industrial project sources)
Disaster debris

Source: Reproduced with kind permission from Gershman, Brickner and Bratton, Inc., Fairfax, VA.

the last two techniques depending on the materials used in the original project, the physical size of the structure, the surrounding buildings that cannot be impacted, and project time. The entire weight of a building, including the concrete foundations, plumbing, walls, electrical workings and so forth, may be generated as C&D debris when a building is demolished.

TABLE 21.3

Typical Components of Construction and Demolition Debris

Material	Examples
Wood	Forming and framing lumber, stumps, plywood, laminates, scraps
Drywall	Sheetrock, gypsum, and plaster
Metals	Pipes, rebar, flashing, steel, aluminum, copper, brass, and stainless steel
Plastics	Vinyl siding, doors, windows, floor tile, and pipes
Roofing	Asphalt and wood shingles, slate, tile, and roofing felt
Rubble	Asphalt, concrete, cinder blocks, rock, and soil
Brick	Bricks and decorative blocks
Glass	Windows, mirrors, and lights
Miscellaneous	Carpeting, fixtures, insulation, and ceramic tile

Source: Franklin Associates, EPA530-R-98-010, 1998. Reproduced with kind permission of
Franklin Associates, Prairie Village, KS.

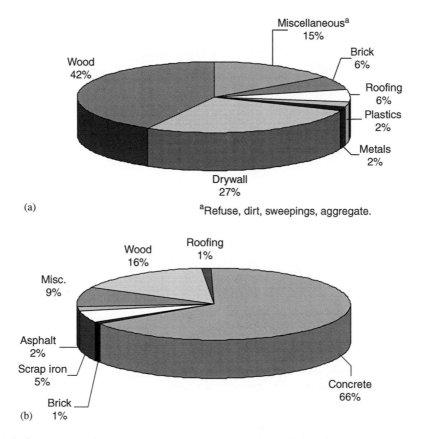

FIGURE 21.3 Comparison of wastes generated by (a) construction (Reproduced with kind permission of Franklin Associates, Prairie Village, KS); and (b) demolition (Reproduced with kind permission of R.W. Rhine, Inc., Tacoma, WA).

The quantities of C&D debris generated around the United States vary markedly from one region or community to another. This variation is partly a result of differences in state regulations, and also by the current rate of growth and the development activity of the community. The amount of C&D

debris generated in a region also depends on the overall economic status of the region, the occurrence of natural disasters, and special projects underway. In rapidly growing areas, the C&D waste stream consists primarily of construction debris, with only small volumes of demolition debris. In contrast, in older urban areas demolition debris makes up the majority of the C&D waste stream.

21.3 MANAGEMENT OF C&D DEBRIS

C&D debris is characterized as a nonhazardous waste subject to regulation under RCRA Subtitle D (Figure 21.4). As discussed in Chapter 1, other Subtitle D wastes include municipal solid waste (MSW), solids from water and wastewater treatment plants, nonhazardous wastes from industrial processes, agricultural wastes, oil and gas wastes, mining wastes, spent automobiles, and trees and brush.

C&D debris is most commonly managed via landfilling. Final disposal sites include dedicated C&D landfills, MSW landfills, and unpermitted sites. Between 35 and 45% of C&D debris was disposed in C&D landfills in 1996, and another 30 to 40% was managed on-site, at MSW landfills, or at unpermitted landfills. A survey conducted for the U.S. EPA identified about 1900 active C&D landfills in the United States (Franklin Associates, 1998). Approximately 20 to 30% of building-related C&D debris is recovered for recycling. There is a trend toward increasing recovery of C&D debris in the United States. Approximately 3500 facilities process C&D materials in the United States (Franklin Associates, 1998).

The C&D materials most frequently recovered for recycling are concrete, asphalt, metals, and wood. Metals have the highest recycling rates among the materials recovered. The recycling rate for C&D steel is about 85% (18.2 out of 21.4 million tons generated) (Franklin Associates, 1998). These numbers include scrap steel from buildings, rebar from roads, and structural steel from bridges. There are about 500 wood processing facilities in the United States that obtain wood from C&D debris. The leading states for wood processing plants are North Carolina, Oregon, and California.

21.4 MANAGEMENT OF SELECTED C&D WASTE COMPONENTS

21.4.1 AGGREGATE

Portland cement concrete and asphalt concrete are the two predominant materials used in road construction. Both consist primarily of aggregate and cement. Aggregate is composed of inert mineral

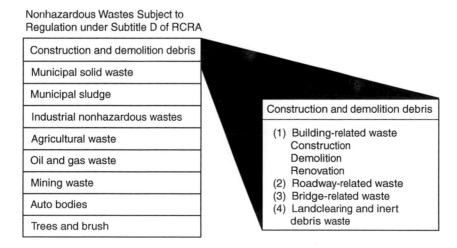

FIGURE 21.4 C & D debris as a component of Subtitle D wastes (Franklin Associates, EPA530-R-98-010, 1998. Reproduced with kind permission of Franklin Associates, Prairie Village, KS.)

materials including crushed stone, sand, gravel, slag, or rock dust. Cement and asphalt serve as binders. Aggregate is used as a replacement for road-base gravel or it is incorporated directly into asphalt or concrete.

A typical roadway is constructed in several layers: pavement, base, and sometimes subbase. The pavement is the surface layer and is composed of Portland cement concrete or asphalt concrete. The base layer supports the pavement, and is composed of aggregate base. The subbase layer supports the base and is made of aggregate subbase (Figure 21.5). The subbase layer has less strength, but is useful and cost-effective for bringing the road up to grade. The subbase layer includes a larger proportion of sand, silt and clay than the aggregate base layer (CIWMB, 2002c).

Recycled aggregate originates primarily from Portland cement concrete and asphalt concrete from road maintenance and demolition. Aggregate is recycled by crushing and sorting operations. Portland cement concrete and asphalt concrete are delivered to the processor in large chunks. Heavy crushing equipment is required to break the chunks into aggregate of manageable size. A crushing plant may include a hopper to receive the material, a jaw to break it into smaller pieces, an impact crusher to reduce further its size, a vibrating screen to sort to the required specifications, and a conveyor belt with a rotating magnet to remove metal contamination such as rebar (CIWMB, 2002). Aggregate is recycled used for many purposes, including:

- In paved roads as aggregate base, aggregate subbase, and shoulders
- In gravel roads as surfacing
- As base for building foundations
- As fill for utility trenches

The primary market for aggregate, however, is as a road base and subbase.

21.4.2 DRYWALL (SHEETROCK, GYPSUM)

Drywall, also referred to as gypsum board, wallboard, plasterboard and rockwall, is the primary material used in the United States for interior wall construction. It is composed of a sheet of gypsum covered on both sides with a paper facing and a paperboard backing. Gypsum is a naturally occurring mineral composed of calcium sulfate dihydrate ($CaSO_4 \cdot 2H_2O$). Drywall sheets are manufactured in sizes from 4×8 to 4×16 ft and range in thickness from 1/4 to 1 in.

Most drywall waste is generated from new construction (64%), followed by demolition (14%), manufacturing (12%), and renovation (10%). Approximately 15 million tons of new drywall are manufactured per year in the United States. About 12% of new construction drywall is discarded during installation.

Drywall is recycled by first separating the paper backing from the gypsum material. The paper can be recycled into new paper backing, and the gypsum is remixed and used in the manufacture

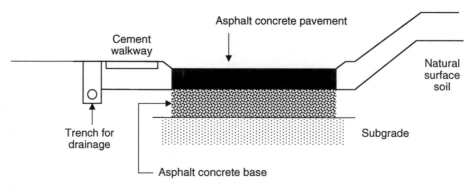

FIGURE 21.5 Cross section of a road showing the various base layers.

of new drywall. The paper content of drywall waste limits the amount of recycled gypsum allowed in new drywall because the paper content affects the fire rating. Other potential markets for drywall waste include:

- Cement plants. The gypsum is added to the clinker to control the setting time.
- Stucco additive.
- A bulking material for sludge drying and composting.
- Water treatment. Recycled gypsum can settle suspended clay particles in turbid water.
- Manure treatment. Recycled gypsum can be mixed with animal wastes to react with ammonia and reduce odor.
- Combine with wood shavings for animal bedding. Gypsum can substitute for sawdust or sand to absorb moisture.
- Grease absorption. Recycled gypsum can be sprinkled on shop floors to absorb oil and grease.
- Athletic field marker. Gypsum is used to mark lines on athletic fields.
- Agriculture.

New construction drywall can be recycled into a soil amendment. Specific applications include (CIWMB, 2002e):

- General agriculture
- Mushroom culture
- Forestry and mine land reclamation
- Nurseries
- City parks and recreation areas
- Residential lawns (sod)
- Golf courses
- Compost additives

Gypsum provides a number of practical benefits to soils; for example, it neutralizes some soil acidity and adds plant nutrients such as calcium and sulfur. Gypsum also improves water penetration and workability of an impermeable sodic (i.e., high Na^+) soil. Recycled gypsum could be used to leach out sodium salts from soil along roadways where salt is applied during winter. Gypsum can also reduce the plasticity of soils with high clay content. Regardless of reuse option, however, the following contaminants should be removed:

- Nails
- Tape
- Joint compound. This compound is made primarily of limestone or gypsum; however, if the structure was built before the mid-1970s, asbestos may occur.
- Paint usually covers demolition drywall. Structures built before 1978 may contain lead-based paint.

Drywall waste from new construction sites is freer of contaminants than demolition drywall, and it is the most commonly recycled. Grinding equipment can range from a large plant to a small mobile chipper. A hammermill is often used. The machinery grinds the drywall, producing about 93% gypsum powder and 7% (by wt) shredded paper. The gypsum can be sold as a powder, with or without paper, or molded into pellets. Drywall recycling produces dust, which is often collected via a baghouse or similar filter system. The amount of drywall waste can be reduced in construction by constructing standard-sized walls and flat ceilings, and by ordering custom-sized sheets for non-standard walls. Wastes are also reduced by finding substitutes that are reusable, such as modular "demountable partitions" for commercial buildings (CIWMB, 2002e).

21.4.3 WOOD WASTE RECYCLING

Wood waste generated at construction sites is often of better quality and purity and has a better potential for reuse than wood from demolition sites, as the former material is more easily separated from other wastes. Demolition wood is often low in demand because of contamination and due to the difficulty in separating wood from other building materials.

In wood processing facilities, wood is chipped to a fine size. The processed wood is used as mulch, a compost bulking agent, animal bedding, and fuel. Wood waste from construction or demolition is appealing as a fuel because of its low moisture content.

Treated wood in C&D debris merits special mention. Wood treatment involves the addition of chemicals either as a surface application or internally within pore spaces to increase resistance to decay by fungi, insects, and marine borers. Wood treatment chemicals include creosote, pentachlorophenol and chromated copper arsenate (CCA), among others. CCA currently comprises about 80% of the wood preservation chemicals currently in use. Copper serves as a fungicide, arsenic serves as an insecticide, and chromium is used to 'fix' the copper and arsenic into the wood (Townsend, 1991). Typical CCA-treated wood applications include telephone poles, fence posts, decks, and other components of home construction.

At C&D debris landfills, CCA-treated wood is disposed along with other components of C&D debris (e.g., untreated wood, concrete, and drywall). As long as the wood is disposed after being used for its intended purpose, CCA-treated wood is exempt from designation as a hazardous waste under RCRA. At C&D debris recycling facilities, some CCA-treated wood may be separated and recovered for reuse or shipped to a landfill for disposal. Management of treated wood from C&D sites is cumbersome because CCA-treated wood cannot be easily separated from the remainder of the wood waste stream. It is difficult to distinguish visually CCA-treated wood from untreated wood. When painted or weathered, identification of CCA-treated wood becomes even more difficult.

Processed wood from C&D debris recycling operations has been utilized as fuel for industrial boilers. Since arsenic can volatilize, combustion units that burn C&D debris containing CCA-treated wood must be equipped with the required air pollution controls.

There are approximately 315 wood processing facilities in the United States that process C&D debris. The leading states for these wood processing plants are North Carolina (44), Oregon (35), and California (34) (Franklin Associates, 1998).

21.4.4 METALS RECYCLING

Metals have the highest recycling rates among the materials recovered from C&D sites. Good markets have been in existence for ferrous metals as well as copper and brass. The recycling rate for C&D steel is about 85% (18.2 out of 21.4 million tons generated) (Franklin Associates, 1998). These numbers include scrap steel from buildings but also from streets, bridges, and highways (Heenan, as cited in Franklin Associates, 1998).

21.4.5 ASPHALT SHINGLES

Asphalt roofing shingles are composed of an asphalt-saturated felt mat with small rock grains added. The felt backing (mat) is made of organic felt, paper, or fiberglass. Organic shingles contain 30 to 36% asphalt and fiberglass shingles contain 19 to 22% asphalt. A mineral filler/stabilizer (limestone, silica, and dolomite) is added as are mineral granules (ceramic-coated natural rock, sand-sized).

The two major applications of asphalt roofing are residential (often slanted) and commercial (often flat). Asphalt shingles are used primarily on slanted residential roofs. Built-up roofing, which consists of roofing felt between layers of tar and gravel, is commonly used on flat commercial roofs. These two materials comprise the majority of the waste arising from roof replacement or repair. About two thirds of all residential roofing activities utilize asphalt shingles. Other roofing materials include wood, tile, and concrete (NAHB, 1996; Franklin Associates, 1998).

Approximately 11 million tons of waste asphalt roofing shingles are generated in the United States per year (Franklin Associates, 1998). Reroofing accounts for 10 million tons, with 1 million tons from manufacturing scrap. Old roofs are often covered over with a set of new shingles, so some tear-offs contain two layers.

To recycle shingles for use in new products, the shingles must be ground to a specified size and the contaminants removed. Primary grinding may produce fragments measuring 2 or 3 in. across. Secondary grinding may be needed to produce smaller pieces; for example, aggregate base may require ¾ in. or smaller, and asphalt pavement may require 1/2 or 1/4 in. or smaller. Depending on the desired use, the shingles may have to be sieved after grinding (CIWMB, 2002). For virtually all uses, contaminants must be removed. Possible contaminants may include metals, which can be removed by a rotating magnet, and wood.

Common uses for recycled asphalt shingles include asphalt for paving and new roofing materials. Asphalt shingle scrap can be used in a variety of products, including:

- Asphalt pavement
- Aggregate base and subbase
- Cold patch for potholes, sidewalks, utility cuts, driveways, ramps, bridges, and parking lots
- Pothole patch
- Road and ground cover
- New roofing
- Fuel oil

Meeting the specifications for paving and roofing materials limits the above applications. Preconsumer manufacturing scrap is used in hot mix asphalt; however, postconsumer scrap (8 to 10 million tons per year), which is less uniform in composition, is not as widely used in hot mix asphalt (Button, 1997; Franklin Associates, 1998).

21.4.6 ASBESTOS SHINGLES

Asbestos is a naturally occurring mineral of both the serpentine and amphibole groups. Its historical popularity has been due to its ability to withstand excess heat and its resistance to conducting electricity and chemical decomposition. Furthermore, asbestos can be woven into useful fabrics, pastes, and muds for use in construction.

The occurrence of asbestos-containing shingles in roof replacement jobs is rather low. The total asbestos content of asphalt shingles manufactured in 1963 was 0.02%; in 1977, the percentage decreased to 0.00016% (CIWMB, 2002). Asbestos is no longer used in the manufacture of asphalt roofing shingles; however, due to the practice of covering an old roof with new shingles, asbestos may still be found in the shingle waste stream.

The agencies regulating asbestos include the U.S. EPA, the U.S. Occupational Safety and Health Administration (OSHA), and city and county health departments. The U.S. EPA regulates friable asbestos over 1%. OSHA regulates friable and nonfriable asbestos over 0.1%. Processors need to contact these agencies to determine any relevant permitting and monitoring requirements (Figure 21.6).

21.5 REDUCTION OF C & D DEBRIS

As with all other waste management strategies, waste minimization must be given high priority in order to limit any adverse environmental impacts, hold down disposal costs, and minimize future liability. Regarding C&D wastes, reduction strategies should begin with the earliest choices made in the building process, including architectural design and material selection (U.S. EPA, 2002). Design concepts should include waste prevention, product durability, and recyclability of materials.

FIGURE 21.6 Buildings undergoing demolition must be assessed for possible asbestos content.

Waste prevention techniques minimize the amount of materials used during construction and renovation. For example, a technique from the homebuilding field, known as 'optimum value engineering' or 'efficient framing,' reduces the amount of wood used in the framing process without sacrificing structural integrity (U.S. EPA, 2002).

Durable products are often more expensive and contain more material than their traditional counterparts; however, they offer long-term benefits through lower maintenance and replacement costs. Additionally, products that are more durable can enhance safety and prevent other building features from deteriorating. The durability of a building is not simply a function of the actual construction materials, but also from the quality of construction. For example, constructing a roof and window frames to properly prevent water infiltration help ensure that interior walls and ceilings last longer.

21.6 DECONSTRUCTION

The amount of waste created by building demolition and removal can be minimized through waste reduction techniques including a process known as "deconstruction." This process maximizes the salvage of building materials for reuse or recycling by selectively disassembling buildings. Deconstruction concentrates on recovering items such as flooring, siding, windows, doors, bricks, plumbing fixtures, ceiling tiles, and structural components. Some sites have demonstrated that enormous amounts of demolition debris can be reused in new structures being built at the demolition site. When the Seattle Kingdome was demolished in 2000, a total of 97% of the structural waste was recycled into the new stadium complex, resulting in savings of more than $3 million. Additionally, truck traffic to and from the site was reduced by about 4500 trips. Before constructing the Four Times Square Office Tower in Manhattan, the contractor first removed six existing buildings. By salvaging all usable materials and recycling as much of the remaining C&D debris as possible, the contractor saved more than $800,000 by selling the salvaged material and reducing disposal fees.

Out of two buildings deconstructed at a demolition project at the closed Presidio Army Compound in San Francisco, workers salvaged about 85% of the first building and approximately 40% of the second building. Material sorting was performed on site. Salvaged items included timbers, dimensional lumber, flooring, windows, ceiling vent covers, iron work, plumbing fixtures, and siding. Material sales took place at the deconstruction site, avoiding the significant cost of transportation, while the remaining materials were taken to a warehouse and sold within 8 months. The

Presidio Army Compound was turned over to the National Park Service following clean-up (Regula-Thompson, 1997).

Several buildings at a former naval supply center in the port of Oakland are being deconstructed. The warehouse being deconstructing has over 400,000 board feet of lumber including old-growth Douglas fir and redwood. An open house was held at the port and invitations were sent to local architects and contractors advertising the materials available (Regula-Thompson, 1997).

Deconstruction is labor-intensive and often relies on the use of hand tools and manpower to take apart buildings and structures. When deconstructing buildings, workers should take precautions to protect themselves from exposure to substances such as lead-based paint and asbestos (U.S. EPA, 2002).

Benefits of deconstruction include (CEC, 1997):

- Lower building removal costs. Deconstruction can cost less than demolition because costs are offset by reselling materials salvaged during the process.
- Avoidance of disposal costs. The majority of materials from deconstructed structures can be salvaged, thereby decreasing debris going to landfills and avoiding shipping costs and tipping fees.
- Increased employment opportunities. Deconstruction is an opportunity to create interim jobs. Deconstruction projects have been linked to job training programs, where less skilled workers can gain experience in carpentry, construction, and materials recovery.
- Regulatory compliance. Many states now require a substantial reduction in the amount of solid wastes disposed in landfills by a specified date.
- Local business development. Salvage industries have strong growth potential.
- Cheaper and better quality building materials. Reused materials are less expensive and sometimes of higher quality than new materials. For instance, some wood structures on military bases contain old growth wood, which is excellent quality and rare lumber.
- Reduced environmental impacts. Deconstruction has fewer environmental impacts on adjacent property because it creates less dust and noise than demolition.
- Diverts valuable materials from landfills and provides consumers and builders with an alternative to buying building products made from virgin resources.

20.7 HAZARDOUS MATERIALS

A number of hazardous materials may be removed from structures during demolition and may include floor lamps, mercury thermostats, and PCB-containing lighting ballasts (Table 21.4). Mercury found

TABLE 21.4
Common Building Components Containing Hazardous Materials

Component	Hazardous Material
Flourescent light bulbs	Mercury
High-intensity discharge (HID) lamps	Mercury
Thermostats	Mercury
Lighting ballasts for fluorescent light bulbs and HID lamps	PCBs, bis(2-ethylhexyl) ester di(2-ethylhexyl) phthalate
Silent wall switches	Mercury
Batteries	Lead, cadmium
Roof vent flashings	Lead
Lead pipes and lead painted surfaces	Lead
Discarded paint, oil, pesticides, cleaning agents	Assorted
CCA-treated wood	Chromium, arsenic

in residential properties usually occurs in household appliances and lights. Items such as fluorescent and HID lamps, thermostats, water heaters, furnaces, and ranges often store mercury-containing parts (Purdue, 1996). Commercial properties usually have larger water heaters, furnaces, and appliances (depending on the type of property). Boiler rooms often contain mercury devices. Mercury switches are frequently associated with heating and cooling equipment. Some homes have mercury gas flow regulators attached to the gas meter; some larger buildings have mercury manometers to regulate gas flow. Some of these wastes are managed as Universal Wastes (see Chapter 18), whereas others such as lead components may require more stringent management and disposal.

REFERENCES

Button, J.W., Williams, D., and Scherocman, J.A., From roofing to roads: the use of recycled shingles in hot-mix asphalt. *C&D Debris Recycling,* July 1997.

Center for Economic Conversion, When is Deconstruction Cost Effective? Technical Brief #1, Deconstruction, Green Base Conversion Strategies, Techniques for Creating Environmentally Sustainable Development on Closing Military Bases, Mountain View, CA, 1997.

California Integrated Waste Management Board, Construction and Demolition Recycling Fact Sheet, Lumber Waste, 2002a. See: http://www.ciwmb.ca.gov/ConDemo/Pubs.htm

California Integrated Waste Management Board, Construction and Demolition Recycling Fact Sheet, Asphalt pavement recycling, 2002b. See: http://www.ciwmb.ca.gov/ConDemo/Pubs.htm

California Integrated Waste Management Board, Construction and Demolition Recycling Fact Sheet, Recycled aggregate, 2002c. See: http://www.ciwmb.ca.gov/ConDemo/Pubs.htm

California Integrated Waste Management Board, Construction and Demolition Recycling Fact Sheet, Asphalt Roofing Shingles Recycling, 2002d. See: http://www.ciwmb.ca.gov/ConDemo/Pubs.htm

California Integrated Waste Management Board, Construction and Demolition Recycling Fact Sheet, Drywall Recycling, 2002e. See: http://www.ciwmb.ca.gov/ConDemo/Pubs.htm

Franklin Associates, Characterization of Building-Related Construction and Demolition Debris in the United States, Prepared for the U.S. Environmental Protection Agency, EPA530-R-98-010, Prairie Village, KS, 1998.

Franklin Associates, Characterization of Building-Related Construction and Demolition Debris in the United States, Prepared for the U.S. Environmental Protection Agency. Municipal and Industrial Solid Waste Division, Office of Solid Waste, EPA530-R-98-010, June 1998.

National Association of Home Builders (NAHB) Research Center, Waste Management Update 2: Asphalt Roofing Shingles, Washington, DC, October 1996.

Purdue Research Foundation, Mercury in Buildings, West Lafayette, IN, 1996. See: http://danpatch.ecn.purdue.edu/~mercury/src/frame.

Regula-Thompson, B., Creative Reuse: Military Base Buildings: Carefully Pounding Swords Into Plowshares, Mar/Apr 1997, CalMAX Creative Reuse, 1997. See:http://www.ciwmb.ca.gov/CalMAX/Creative/default.htm

Townsend, T., Arsenic and old wood, *Recycling Today,* 2001. See: http://www.recyclingtoday.com/articles/

University of Florida, No date, Department of Environmental Engineering Sciences, Recommended Management Practices for the Removal of Hazardous Materials from Buildings Prior to Demolition. Gainesville, FL, See: http://www.ees.ufl.edu/homepp/townsend/Research/DemoHW/Guide/DHW99_12_30.PDF

U.S. Environmental Protection Agency, WasteWise Update: Building for the Future. EPA 530-N-02-003, Solid Waste and Emergency Response (5306W), Washington, DC, 2002.

SUGGESTED READING

Brickner, R H., Overview of C&D debris recycling plants, *C&D Debris Recycling*, January/February, 1997.

California Integrated Waste Management Board, Military Base Closure Handbook: A Guide to Construction and Demolition Materials Recovery, 433-96-074, Sacramento, CA, 2002.

California Integrated Waste Management Board, Wood Waste: How to Keep Wood Waste Out of Landfills, 500-94-017, Sacramento, CA, 2002.

California Integrated Waste Management Board, Designing With Vision ... A Technical Manual for Material Choices in Sustainable Construction, 431-99-009, Sacramento, CA, 2000.

California Integrated Waste Management Board, Green Building Basics, 400-99-014, Sacramento, CA, 2000.

Corson, J., *The Resourceful Renovator, A Gallery of Ideas for Reusing Building Materials*. Chelsea Green Publishers, White River Junction, VT, 1967.

IFC Inc., Construction and Demolition Waste Landfills, May 18, 1995 Draft Report, Prepared for the U.S. Environmental Protection Agency, Washington, DC, 1995.

Langewiesche, W., *American Ground: Unbuilding the World Trade Center*, North Point Press, New York, NY, 2002. P. 205.

SWANA., Construction Waste and Demolition Debris Recycling—A Primer, Publication GR-REC 300, Solid Waste Association of North America, Silver Springs, MD, 1993.

U.S. Environmental Protection Agency Region 9, Construction and Demolition (C&D) Debris, U.S. Environmental Protection Agency Region 9, 2002. Web Site www.epa.gov/region09/waste solid/debris.htm#1.

U.S. Environmental Protection Agency, Characterization of Construction and Demolition Debris in the United States, EPA/530-R-96-010, U.S. Environmental Protection Agency, Washington, DC, 1998.

U.S. Environmental Protection Agency, Draft Document, Damage Cases: Construction and Demolition, 1995. See: http://www.epa.gov/epaoswer/hazwaste/sdq/damages/dampdf/damage.pdf\

Other U.S. Environmental Protection Agency web sites, undated: Building Savings: Strategies waste reduction of construction and demolition debris from buildings.See: http://yosemite.epa.gov/ncepihom/nsCatalog.nsf/7cd1b64da384615b85256bb 8006c846e/4842c2551525e2ca85256d9f00709171?OpenDocument

Construction and Demolition Debris: The Next Frontier of Waste, EPA's Building- Related C&D Characterization Report Methodology, See: http://www.epa.gov/osw/meeting/pdf02/sandler.pdf

Environmental Aspects of the Construction and Demolition Industry: Annotated Bibliography, See: http://www.epa.gov/region01/assistance/rcra/biblio.html

Environmental Aspects of the Construction and Demolition Industry: Annotated Bibliography, See: http://www.epa.gov/NE/assistance/rcra/biblio.html

Information on Recycling of Building-Related Construction and Demolition Debris, See: http://www.epa.gov/epaoswer/non-hw/recycle/jtr/comm/construc.htm

National Compliance Clearinghouse-preview of Oregon DEQ Construction and Demolition Debris Salvage Program with Depots, Remove frame return to Clearinghouse score site. See: http://cfpub.epa.gov/clearinghouse/toolbar.cfm?RESOURCE ID=253699

U.S. EPA Region 9:Solid Waste Construction and Demolition Debris Solid and Hazardous Waste Programs, See: http://www.epa.gov/Region9/waste/solid/c&d.html

U.S. EPA Region 5: Recyclables - Construction Demolition Debris, Most Waste from Construction and Demolition Sites is Placed into Containers. See: http://www.epa.gov/RCRIS-Region-5/wptdiv/solidwaste/construction.htm

Waste Prevention, Recycling, and Composting Options — Chapter 7—Targeting Construction and Demolition Debris for Recovery, See: http://www.epa.gov/epaoswer/non-hw/reduce/recy-com/chap07.pdf

QUESTIONS

1. What are the federal regulations that exist for the management of C&D debris?
2. What C & D materials are most frequently recovered from construction and demolition sites? Of these, which has the highest rate of recycling?
3. Are land clearing debris such as stumps, rocks, and soil included in state definitions of C&D debris? Check the U.S. EPA web site, http://www.epa.gov/region09/waste/solid/c&d.html, for California's and Hawaii's definition of C&D debris.
4. Will it cost a typical contractor more to recycle or to reuse construction and demolition debris?
5. How do U.S. states regulate C&D debris? Are state programs relatively consistent? Check the U.S. EPA web site, http://www.epa.gov/region09/waste/solid/c&d.html for state regulations from Arizona, California, Hawaii and Nevada.

6. In your community, find out where C&D debris is managed. Is it landfilled or recycled? Who is in charge of managing C&D debris: private haulers or the municipality?

7. How does gypsum, $CaSO_4$, affect soil physical and chemical properties? Is it a noxious compound or does it benefit the soil?

8. How is CCA-treated wood managed under RCRA? Is it to be handled as a hazardous waste?

9. Are there any hazards associated with burning CCA-treated wood?

10. How are old asphalt shingles recycled? What are common uses for recycled shingles?

11. What federal agencies regulate asbestos-containing shingles? What are the minimum asbestos contents to be eligible for regulation?

12. Describe deconstruction. If an academic building on your campus were to be deconstructed, what would the primary recovered materials be? Consider plumbing, wiring and interior woodwork (if any), in addition to components such as stone, steel girders, and metallic doors.

13. Is deconstruction worth the effort? List the benefits of deconstruction.

22 Electronics Waste

Farewell, German radio with your green eye and your bulky box
Together almost composing a body and soul.
.... your eye would grow nervous, the green pupil widen and shrink
...your old age was announced by a cracked voice,
then rattles, then coughing, and finally blindness (your eye faded), and total silence.
Sleep peacefully, German radio

Adam Zagajewski (1945-)
Electric Elegy

22.1 INTRODUCTION

Electronics waste (e-waste) is an inevitable and unavoidable by-product of the current worldwide technological revolution. E-waste is a collective term, embracing consumer and business appliances, products, components, and accessories nearing the end of their useful life due to obsolescence, malfunction, or exhaustion (e.g., batteries). Common examples of e-waste include personal and mainframe computers, printers, televisions, VCRs, stereos, copiers, and fax machines. There is no standardized definition for e-waste; however, electronics *equipment* can be defined as those devices whose primary functions are provided by electronic circuitry and components, i.e., semiconductor devices (integrated circuits, transistors, and diodes), passive components (resistors, capacitors, and inductors), electrooptical components (CRTs, LEDs, CCDs, lasers, etc.), sensors (transducers and MEM devices), and electronics packaging (printed circuit boards, connectors) (IAER, 2002). Some major categories of electronics equipment are listed in Table 22.1.

E-wastes comprise from 2 to 5% of the United States municipal solid waste stream and is one of the fastest growing fractions of the waste stream (SVTC, 2001). Technological advances for electronic equipment, particularly personal computers, quickly render them obsolete. The average lifespan of a Pentium-class computer is currently 2 to 3 years and will gradually decline (Table 22.2). More than 20 million personal computers became obsolete in 1998 (U.S. EPA, 2002a) and more than 63 million PCs are estimated to be retired in 2003 (NSC, 1999). European studies estimate that the volume of e-waste is rising by 3 to 5% per year, almost three times faster than the MSW stream (SVTC, 2001). The U.S. EPA is currently conducting research to estimate the annual rates of generation and recovery of consumer electronics in the United States. In a recent EPA report, it is estimated that the amount of e-waste in the U.S. landfills will grow fourfold in the next few years (SVTC, 2001). New products rapidly heading to the market create additional waste problems; for example, the disposable cellular telephone is now available for marketing (Reuters, 2002). Additionally, disposable DVDs are making their appearance on the market (Truini, 2003).

The issue of how to manage e- waste is a pressing one. Many obsolete and malfunctioning electronic products can be reused, rebuilt, or recycled. However, electronics that malfunction often are not repaired due to the relatively low price of replacing them. Only 11% of discarded computers were recycled compared with 28% of MSW. Most were disposed or remain in storage (NSC, 1999). It is estimated that nearly 75% of unwanted electronics are in storage, partly because of the uncertainty as to how to manage such items.

TABLE 22.1
Categories of Electronics Equipment

User	Application
Consumer	Video: televisions, VCRs, camcorders, digital cameras, and control boxes
	Audio: stereo systems and components, CD players, and radios
	Communications: cell phones, pagers, PDAs
	Personal: computers, printers, and calculators
	Game systems
Automotive	Control systems
	Audio systems
	Instrumentation
	Computers and peripheral equipment: CPUs, monitors, printers, scanners, storage devices, servers, and networking systems
	Financial Systems: retail or check-out, banking or teller, and ATMs
	Security Systems: monitoring and detection equipment
	Entertainment: radio, television and movie production equipment, transmission systems, and sound and video projection equipment, amusements
	Office equipment: copiers, fax machines, imaging systems, and printing systems
Industrial	Telecommunications equipment: telephones, switching systems, PBXs, transmitters, receivers, and microwave systems
	Test and measurement equipment: oscilloscopes, power supplies, and signal processors
	Medical equipment: EKGs, MRIs, CAT scan, and X-ray, monitors
	Manufacturing equipment: control systems, data entry devices, workstations, instruments, and process tools
Aerospace	On-board control systems
	Communications systems
	Navigation systems
	Radar and traffic control systems
	In-flight entertainment systems
Military/defense	Weapons control systems
	Communications systems
	Navigational systems
	Security systems
	Encryption systems

Source: IAER, 2002. Reproduced with kind permission of the International Associations of Electronics Recyclers, Albany, NY.

TABLE 22.2
Link between Computing Power and the Rate of PC Disposal

Year	1971	1974	1976	1982	1986	1989	1993	1995
Chip	4004	8080	8086	80286	386DX	486	Pentium	PentiumPro
Transistors	2,300	6,000	29,000	134,000	275,000	1.2 m	3.1 m	5.5 m

Source: Reproduced with kind permission of Microprocessor Report, Scottsdale, AZ; and The Silicon Valley Toxics Coalition, San Jose, CA.

More than 3.2 million tons of electronics discards were disposed in landfills in the United States in 1997. It is believed that most households and small businesses disposing e-wastes opt to send them to landfills or incinerators rather than to recyclers. Landfilling and incineration of e-wastes pose significant health and environmental hazards (see below).

Obsolete electronic equipment is also a concern because electronics are manufactured using valuable resources such as precious metals, engineered plastics, and glass. The elemental breakdown of a personal desktop computer is listed in Table 22.3. When electronic equipment is disposed

TABLE 22.3
Elemental Breakdown of a Personal Desktop Computer

Element or Compound	Content (% of Total Weight)	Weight of Material (lb.)	% Recycling Efficiency (Current Recyclability)	Use Location
Plastics	22.9907	13.8	20	Includes organics and oxides (other than silica)
Lead	6.2988	3.8	5	Metal joining, radiation shield/ CRT, PWB
Aluminum	14.1723	8.5	80	Structural conductivity/ housing, CRT, PWB, connectors
Germanium	0.0016	<0.1	0	Semiconductor/ PWB
Gallium	0.0013	<0.1	0	Semiconductor/ PWB
Iron	20.4712	12.3	80	Structural, magnetivity/ (steel) housing, CRT, PWB
Tin	1.0078	0.6	70	Metal joining/ PWB, CRT
Copper	6.9287	4.2	90	Conductivity/ CRT, PWB, connectors
Barium	0.0315	<0.1	0	Vacuum tube/ CRT
Nickel	0.8503	0.51	80	Structural, magnetivity/ (steel) housing, CRT, PWB
Zinc	2.2046	1.32	60	Battery, phosphor emitter/ PWB, CRT
Tantalum	0.0157	<0.1	0	Capacitors/ PWB, power supply
Indium	0.0016	<0.1	60	Transistor, rectifiers/ PWB
Vanadium	0.0002	<0.1	0	Red phosphor emitter/ CRT
Terbium	<0	<0	0	Green phosphor activator, dopant/CRT, PWB
Beryllium	0.0157	<0.1	0	Thermal conductivity/ PWB, connectors
Gold	0.0016	<0.1	99	Connectivity, conductivity/ PWB, connectors
Europium	0.0002	<0.1	0	Phosphor activator/ PWB
Titanium	0.0157	<0.1	0	Pigment, alloying agent/ (aluminum) housing
Ruthenium	0.0016	<0.1	80	Resistive circuit/ PWB
Cobalt	0.0157	<0.1	85	Structural, magnetivity/ (steel) housing, CRT, PWB
Palladium	0.0003	<0.1	95	Connectivity, conductivity/ PWB, connectors
Manganese	0.0315	<0.1	0	Structural, magnetivity/ (steel) housing, CRT, PWB
Silver	0.0189	<0.1	98	Conductivity/ PWB, connectors
Antimony	0.0094	<0.1	0	Diodes/ housing, PWB, CRT
Bismuth	0.0063	<0.1	0	Wetting agent in thick film/ PWB
Chromium	0.0063	<0.1	0	Decorative, hardener/ (steel) housing
Cadmium	0.0094	<0.1	0	Battery, blue-green phosphor emitter/ housing, PWB, CRT
Selenium	0.0016	0.00096	70	Rectifiers/ PWB
Niobium	0.0002	<0.1	0	Welding alloy/ housing
Yttrium	0.0002	<0.1	0	Red phosphor emitter/ CRT
Rhodium	<0	<0	50	Thick film conductor/ PWB
Platinum	<0	<0	95	Thick film conductor/ PWB
Mercury	0.0022	<0.1	0	Batteries, switches/ housing, PWB
Arsenic	0.0013	<0.1	0	Doping agents in transistors/ PWB
Silica	24.8803	15	0	Glass, solid-state devices/ CRT, PWB

Source: Reproduced with kind permission of Microelectronics and Computer Technology Corporation, Austin, TX.

and new equipment is manufactured in its place, virgin resources are extracted, additional energy is required, valuable resources are wasted, and air and water pollution are generated.

22.2 MAJOR TYPES OF ELECTRONIC EQUIPMENT

Several major types of electronic equipment are found in the MSW streams in all U.S. states. These items and their significance are described below.

22.2.1 PERSONAL COMPUTERS

About half of the U.S. households own a PC. Most businesses have at least one compared with larger businesses having many computers and peripherals (e.g., modems, scanners, and printers). The composition of each component will vary significantly from others. Similarly, the recycling processes and economics of each component differ markedly. When a PC is resold, recycled, or scrapped, most recyclers and 'demanufacturers' (see below) will break it down into: (1) monitors; (2) central processing units CPUs (including components such as circuit boards, hard drives, and chips); and (3) peripherals (Florida DEP, 2002).

22.2.2 COMPUTER MONITORS

The monitor is the housing for the cathode-ray tube (CRT). The CRT is the 'picture tube' that produces the image on the monitor screen. The primary concern associated with monitors is the presence of lead (Pb) in the CRT.

22.2.3 COMPUTER CENTRAL PROCESSING UNITS

The CPU is the component that actually performs the computing. CPU components that possess value for recycling and demanufacturing include circuit boards, hard drives, memory chips, microprocessor chips, and video cards. These components contain a range of precious metals.

22.2.4 COMPUTER PERIPHERALS

Peripherals are devices needed to operate a computer such as the keyboard or mouse; that perform additional functions such as printers, scanners, and modems; or cables that connect the different parts of the computer.

22.2.5 HANDHELD AND DESKTOP TELECOMMUNICATIONS EQUIPMENT (TELEPHONES, FAX MACHINES)

This category of equipment includes desk telephones, fax machines, portable telephones, and communication radios. These products are found in most homes and businesses.

PCs, televisions, and desktop or handheld telecommunications equipment like cellular phones and fax machines possess components or materials having a negative or zero net value when recycled. The net value is defined as the value of recovered components or materials, minus the cost to recover the materials.

22.2.6 MAINFRAME COMPUTERS AND HIGH-END TELECOMMUNICATIONS EQUIPMENT

Mainframe computers are composed of a large and powerful central computer with a number of workstations, each consisting of a monitor and keyboard which access the mainframe via automated servers. High-end telecommunications equipment consists of switching equipment for a building or company-wide telephone equipment, and equipment used in commercial audio, video, or telephone applications.

Both mainframe computers and high-end telecommunications equipment possess high-value components or materials that can be recycled with a positive net value. As a result, mainframe computers and high-end telecommunications equipment are typically purchased by recyclers and demanufacturers, generating substantial revenue (Florida DEP, 2002).

22.2.7 TELEVISIONS

Approximately 25 million television sets are sold annually in the United States (Stanford Resources, 1998) and yearly sales have equaled or exceeded 20 million units for the last decade. Most U.S. households possess at least one television compared with about 50% of households having computers. The rate of sales growth (and obsolescence) is slower in televisions than in computers.

Over the past several decades televisions have been manufactured in a variety of sizes and styles, ranging from black-and-white to color all the way to plasma and high-definition digital quality. Television "size" is usually measured as a diagonal line across the screen. Common groupings among recyclers and demanufacturers are 19″ or less and larger than 19″. Generally, larger TVs, especially console models, are more labor intensive (and therefore more expensive) to demanufacture and recycle (Florida DEP, 2002).

Less data are available on the average lifespan of televisions or the annual volume of televisions discarded. With less than 20,000 units annually being recycled, the volume of TVs making their way into the waste stream is substantial. Further, it is anticipated that there will be a significant increase in the numbers of televisions purchased and subsequently discarded following new federal rules for high-definition televisions which became effective in 2004.

22.3 HAZARDS OF E-WASTE

Electronic equipment contains metals and other materials that are hazardous to human health and the environment when improperly managed. For example, e-waste constitutes the source of about 70% of the heavy metals (including Hg and Cd) found in landfills (SVTC, 2001).

The central component of computer monitors and televisions is the CRT, which contains leaded glass. CRTs are second only to automotive batteries in terms of Pb contribution to the U.S. MSW stream. Other primary applications of Pb in computers are for soldering of printed circuit boards and other electronic components. Consumer electronics constitute 40% of Pb found in landfills.

Cadmium (Cd) occurs in electronic components such as SMD chip resistors, infrared detectors and semiconductors. Older types of CRTs contain Cd. Cadmium is also used as a plastic stabilizer. The largest source of Cd in MSW is rechargeable nickel–cadmium (Ni–Cd) batteries, commonly found in laptop computers. Between 1997 and 2004 over 315 million computers will become obsolete, which represent about 0.9 million kg (2 million lb) of Cd (NSC, 1999).

Some manufacturers continue to utilize Cr(VI) as corrosion protection for untreated and galvanized steel plates and as a decorative coating and hardener for steel housing. Of the more than 315 million computers destined to become obsolete between 1997 and 2004, about 540,000 kg (1.2 million lb) of hexavalent Cr will accumulate (U.S. EPA, 2002a).

E-waste is also a leading source of mercury (Hg) in MSW. Approximately 22% of the yearly world consumption of Hg is used in electrical and electronic equipment (U.S. EPA, 2002a). Mercury is used in thermostats, position sensors, relays, and switches (e.g., on printed circuit boards and in measuring equipment) and discharge lamps. Furthermore, it is used in medical equipment, data transmission, telecommunications, and mobile telephones. Mercury is also used in batteries, switches/housing, and printed wiring boards. Although the amount is small for any single component, 315 million obsolete computers by the year 2004 represent more than 182,000 kg (400,000 lb) of Hg.

Based on the calculation that more than 315 million computers will become obsolete between 1997 and 2004 and that plastics make up 6.3 kg (13.8 lb) per computer on average, there will be more than 1.8 billion kg (4 billion lb) of plastic present in computer waste (SVTC, 2001). An analysis commissioned by the Microelectronics and Computer Technology Corporation (MCC) estimated that total electronics plastic scrap amounted to more than 525,000 metric tons (580,000 tons) per year. This same study estimated that the largest volume of plastics used in electronics manufacturing (26%) was polyvinyl chloride (PVC). While many computer companies have recently reduced or phased out the use of PVC, there is still a huge volume of PVC contained in the computer scrap that continues to grow — potentially up to 250 million lb per year (SVTC, 2001). PVC is a difficult plastic to recycle and it contaminates other plastics in the recycling process.

The use of PVC in computers has been mainly for cabling and computer housings. Most computer moldings are now, however, being manufactured of acrylonitrile-butadiene styrene (ABS) plastic. PVC cabling is used for its fire-retardant properties but there are concerns that if combusted, fumes can contribute to fatalities. The production and burning of PVC products generate chlorinated dibenzodioxins and furans (SVTC, 2001). Hence, there are pressures to switch to alternatives for safety reasons. Such alternatives are low-density polyethylene and thermoplastic olefins.

Brominated flame retardants are commonly used in electronic products. In PCs they are used mainly in four applications: in printed circuit boards, in components such as connectors, in plastic covers and in cables. They are also used in plastic covers of television sets and in domestic kitchen appliances. Recent research indicates that polybrominated diphenylethers (PBDE) might act as endocrine disrupters (SEPA, n.d.). PBDEs also form polybrominated dibenzofurans (PBDFs) and polybrominated dibenzodioxins (PBDDs) during the extruding process (Hoque et al., 1998; SVTC, 2001). The German chemical industry halted the production of these chemicals in 1986 (Brenner, 1986; SVTC, 2001).

22.4 CATHODE RAY TUBES

CRTs are used in computer monitors, televisions and other video display devices. Liquid crystal displays (LCDs) and plasma displays do not employ CRT technology. In picture production, x-rays are inevitably generated within the CRT; therefore, Pb is incorporated within the CRT glass to protect the viewer or user from exposure to x-rays. The average television and PC monitor CRT for the period between 1995 and 2000 measured 18.6 in. and had a Pb content ranging from 1 to 1.2 kg (2.14 to 2.63 lb) (EIA, n.d.). Table 22.4 summarizes the weight of glass and the Pb content of various-sized CRTs manufactured in the United States.

Figure 22.1 shows the locations of Pb within a typical color CRT used in a television or PC monitor. The Pb in the funnel and face plate glass is incorporated (i.e., it is *not* a simple coating) into the glass matrix as lead oxide (PbO). This Pb formulation is stable and immobile; in theory, therefore, it is not expected to leach during the Toxicity Characteristic Leaching Procedure (TCLP; see Chapter 11) or in a landfill environment. The Pb in the frit that joins or welds the face plate glass to the funnel glass is also in the form of a PbO paste. This Pb readily leaches when subjected to the TCLP. Table 22.5 lists the primary components of the PC or television monitor and their composition.

The CRT cone is an electromagnet that encircles the electron gun at the rear (narrow) end of the CRT (Figure 22.1). The cone directs the electron beams produced by the electron gun to the faceplate of the CRT to produce the image. The circuit board processes the incoming signal from the computer to control the operation of the electron gun in order to provide the correct image on the screen.

22.5 RCRA AND E-WASTES

Recall from Chapter 11 that under The Resource Conservation and Recovery Act (RCRA) a solid waste is a discarded material, which is:

- Abandoned (materials that are disposed, incinerated, or stored in lieu of disposal)
- Recycled
- Inherently waste-like

TABLE 22.4
Weight of Glass and Pb Contents for CRTs

Screen Size	Total Glass Weight (lb)	Total Pb in Lead Panel CRT (lb)	Total Pb in No-Lead Panel CRT (lb)
Portable and Table Televisions			
8" and below	12.9	1.30	1.08
9" and 10"	15.0	1.47	1.21
13"and 14"	17.2	1.75	1.46
20"	28.2	2.70	2.21
25"	43.5	4.05	3.28
29" through 31"	81.2	7.99	6.62
35"	117.4	11.28	9.27
Console Televisions			
26" and below	48.8	4.54	3.68
27"–32"	68.5	6.61	5.44
35" and above	117.4	11.28	9.27
Monitors			
14"	19.7	na	1.68
15"	19.7	na	1.68
17"	25.7	na	2.08
19" through 21"	28.0	na	2.28

Source: Reproduced with kind permission of the Electronic Industries Alliance, Arlington, VA.

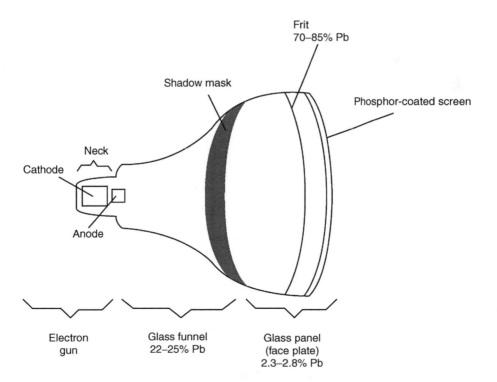

FIGURE 22.1 Cross section of a CRT showing Pb contents. Reproduced with kind permission of the Silicon Valley Toxics Coalition, San Jose, CA.

TABLE 22.5

Primary Components and Composition of a Monitor

Glass Components

Panel glass. The panel or screen makes up the front of the CRT and accounts for two thirds of the CRT mass. In late model CRTs, the panel glass contains barium oxide instead of PbO

Funnel glass. The rear portion is referred to as the bell or funnel. Funnel glass is leaded glass; most of the Pb in a CRT is in the funnel glass

Neck. The neck is the straight glass tube that surrounds the electron gun(s). The neck is made of leaded glass

Solder glass. The solder glass or frit is used to seal the CRT. Frit is 85% Pb

NonGlass Components

Banding. Metal banding reinforces the CRT. It must be removed prior to recycling

Electron gun. The stainless-steel electron gun creates the electron beam

Shadow mask. The shadow mask is a metal screen that focuses the electrons on the back of the panel

Yoke. The coils of copper wire around the neck of the CRT are known as the yoke; the yoke is an electromagnet that deflects the electrons to the proper position on the panel

Phorphor coating. Special compounds adhered to the inside of the panel; they produce light when struck by the electron beam

Source: Reproduced with kind permission of the Materials for the Future Foundation, San Francisco, CA.

When wastes are recycled rather than disposed, generators must comply with the regulations in 40 CFR Part 261 to determine if the wastes are to be regulated under RCRA. These regulations are based on the assertion that recycled secondary materials (a material that is potentially a solid waste and a hazardous waste) fit into one of five categories. The categories relevant for CRTs are:

- Spent materials. This includes obsolete computers, even if they still work.
- Commercial chemical products. This can include unused circuit boards.
- Scrap metal. If recycled, even scrap metal that is hazardous is exempt from hazardous waste regulations.

The most relevant types of recycling activities regarding CRTs are:

- Reclamation.
- Speculative accumulation. If materials are kept on-site for longer than a calendar year, they can be regulated as speculative accumulation.

By considering the category of material and type of recycling activity for a particular product or material, Part 261 is used to determine whether that material is a solid waste when recycled in that fashion. If it is determined to be a solid waste, then it could possibly be a hazardous waste.

22.5.1 HAZARDOUS WASTE

As discussed in Chapter 11, hazardous wastes are either listed or are characteristic hazardous wastes (ignitable, corrosive, reactive, and toxic). The RCRA regulations require the solid waste generator to conduct the waste determination. Research conducted at the University of Florida shows that CRTs usually fail the TCLP for toxicity. The California Department of Toxic Substances Control confirmed that the CRTs in computer monitors, television sets, ATMs, and other devices contain TCLP-soluble Pb concentrations that classify them as hazardous waste when they are discarded (Harris, 2001; SVTC, 2001). Monochrome monitors do not usually exceed the TCLP limits and could probably be segregated from color monitors for disposal. California law considers nonfunctioning CRTs from televisions and monitors to be hazardous waste.

The following generators are excluded from RCRA hazardous waste regulations; therefore, wastes from these generators are permitted in RCRA Subtitle D landfills. Some states are working to limit or end these exclusions:

- Household hazardous waste generators. Therefore, CRTs from households are not regulated as hazardous waste.
- Conditionally exempt small quantity generators (CESQGs). This exemption allows CESQGs to dispose of approximately seven to eight computers (100 kg or 220 lb) per month or store up to 1000 kg (2200 lb).
- Other generators as excluded in Section 261.4 (a) and 261.4 (b).

The Universal Waste Management Standards (40 CFR Part 273), created in 1995, streamline the hazardous waste regulations by removing barriers to collection programs, reducing complexities, and reducing the cost of compliance. As discussed in Chapter 18, there are currently four types of federal Universal Wastes: batteries, certain pesticides, mercury-containing thermostats, and lamps (e.g., fluorescent bulbs). States can add their own universal wastes and some have done so for CRTs (WEPSI, 2001).

22.5.2 THE U.S. EPA PROPOSED CRT RULE

The U.S. EPA created a rule dedicated for CRTs rather than designate them as a Universal Waste. The Electronics Subcommittee of the Common Sense Initiative (CSI) Council recommended that the EPA streamline its RCRA management requirements for CRTs. The purpose of the rule is to encourage greater reuse, recycling and better management of this rapidly growing waste stream. The agency opted to formulate a rule designed specifically for CRTs for the following reasons (U.S. EPA, 2002c):

- Computers are heavy devices, and the weight cut-offs for Universal Wastes were impractical. To repeat from above, seven or eight monitors were sufficient to exceed the limits for CESQGs.
- CRTs may resemble commodities more than wastes if handled properly, i.e., the glass can be sold to make new CRT glass, Pb can be sent to a lead smelter, and so on.

The summary of the proposed CRT rule is shown in Table 22.6.

22.5.3 CIRCUIT BOARDS

Circuit boards are expected to fail the TCLP for Pb and therefore should not be disposed in a sanitary landfill. Circuit boards tend to be recycled or handled as scrap metal, in which case they are not covered under RCRA. U.S. EPA Regions IV and V are evaluating TCLPs of other electronic components. At present, however, if material reuse is feasible, such devices are still considered to be products, not wastes. Current regulatory requirements for circuit boards are summarized as follows (U.S. EPA, 2002a):

- Whole unused circuit boards are considered unused commercial chemical products, which are unregulated.
- Whole used circuit boards meet the definition of spent materials but also meet the definition of scrap metal. Therefore, whole used circuit boards that are recycled are exempt from the hazardous waste regulations.
- Shredded circuit boards are excluded from the definition of solid waste if they are in containerized storage prior to recovery. These shredded circuit boards cannot contain mercury switches, mercury relays, Ni–Cd batteries, or lithium batteries.

TABLE 22.6
Summary of the Proposed CRT Rule

Not Regulated

Households: may send used monitors to any collector for recycling or disposal

CESQGs: Nonresidential generators of less than 100 kg per month (about seven or eight computers) are not subject to most Subtitle C standards

User who sends CRTs to a collector or reseller for potential reuse or repair: If the CRTs may be reused, they are *products*, not wastes

Intact, off-specification CRTs sent for recycling: These are products, not wastes

Non-CRT electronic materials: Whole used circuit boards and shredded circuit boards are not regulated

Regulated With Streamlined Requirements

Used Broken CRTs Sent for Recycling:

- Conditionally excluded if they comply with universal waste-type packaging and labeling requirements for storage and shipment
- No speculative accumulation

Glass Processors:

- Must store broken CRTs indoors or in accordance with universal waste-type packaging and labeling, no speculative accumulation
- May not use temperatures high enough to volatilize lead

Processed Glass:

- If sent to a glass-to-glass manufacturer or lead smelter, not regulated — treated as a product, unless speculatively accumulated
- If sent off-site for other kinds of recycling activities, must comply with universal waste-type packaging and labeling
- No speculative accumulation

Fully Regulated Under Subtitle C:

CRT disposal in a landfill or incinerator, and the disposer is not a household or a CESQG

Source: U.S. EPA, 2002c. With permission.

22.6 ELECTRONICS RECYCLING

Electronics recycling is a new but growing industry and embraces computers, telecommunications, and consumer electronics equipment (e.g., televisions and audio equipment). Over 9.7 million units (125 million kg or 275 million lb) of electronic equipment were recycled in 1998 (NSC, 1999). Computer peripherals, desktop PCs, and CRT computer monitors are the most common equipment recycled; however, the actual percentage of electronics recycled is quite low. Approximately 6% of personal computer CPUs were recycled in 1998, with televisions recycled at an even lower rate (NSC, 1999). By comparison, over 64% of white goods are recycled annually in the United States. (U.S. EPA, 1999).

Electronic products are manufactured with valuable raw materials that can be reused or recycled. In 1998 over 112 million lb of materials were recovered from electronics including steel (43.9 million lb), glass (29.2 million lb), plastic (14.4 million lb), aluminum (9.9 million lb), copper (7.9 million lb), and precious metals (2.2 million lb) (NSC, 1999; U.S. EPA, 2002a).

The number of personal computers being recycled is anticipated to reach approximately 60 million units per year within the next five years (NSC, 1999). The major sources of electronics products for recycling have been manufacturers and large users. Currently 75% of the equipment being recycled originates from electronics manufacturers and large organizations (more than 500 employees). Only a very small amount of electronics is being recovered from households (NSC, 1999; U.S. EPA, 2002a).

Recovered electronics equipment is recycled by a small group of companies due to the large capital investment and substantial infrastructure required. In the United States, the five largest recycling firms handle more waste equipment than all the other companies combined. In addition, electronics recyclers are geographically concentrated; half of all electronics recycling firms are in the mid-Atlantic and midwest regions of the United States. (NSC, 1999).

The electronics recycling industry can be considered as a chain of handlers and processors (NSC, 1999). The major steps in the chain are:

- Collection/separation
- Direct resale of computers or components
- Parts and materials recovery
- Smelting/refining of metals
- Other parties — resellers and not-for-profit organizations

Electronics recycling activity is expected to grow 18% annually between 1998 and 2007 (U.S. EPA, 2002a). This growth will be from the entry of new businesses and increased handling volume from large facilities. Estimates indicate that over 40 million units of electronic equipment will be recycled in 2007, with notebook PCs and desktop CPUs experiencing significant growth in recovery (NSC, 1999). A Carnegie Mellon report estimates that nearly 150 million computers will be recycled in 2005 (Matthews, 1997).

22.6.1 RESIDENTIAL COLLECTION PROGRAMS

In order to establish a consistent and high-volume stream of electronics discards, adequate and consistent collection is essential. The costs of residential collection and the inherently lower value of obsolete consumer electronics create economic and logistical obstacles for the public and private sectors. In recent years a number of programs have been tested to develop approaches to residential collection to overcome these obstacles. The U.S. EPA and its CSI has developed such programs, and several states have followed with programs of their own.

In the EPA pilot program, a variety of different collection models were evaluated including curbside collection, drop-off events, establishment of a permanent collection depot (Figure 22.2), and point of purchase (retail) collection. The weighted average of all electronics items collected by the various programs was as follows (U.S. EPA, 2002a):

- Televisions (36%)
- Audio and stereo equipment (16%)
- Monitors (11%)
- Computers (8%)
- VCRs (6%)

Most of the collected equipment was outdated and in poor condition; therefore, it had low scrap value and was expensive to handle. Items that contained CRTs predominated, which drove up program costs. There were some notable differences in efficiencies and costs between the program models used; for example, curbside collection yielded more pounds of material collected per resident, but high transportation costs resulted. Also, the cost per item collected was lower for 1-day collection events due to lower transportation and operating costs. Net costs for the collection program ranged from $0.17 to $0.50 per pound. The dominant costs were for transportation, demanufacturing, and disposal. Adequate planning and promotion were considered essential to the overall effectiveness of the electronics waste collection program (U.S. EPA, 2002a).

Data regarding optimizing collection programs are still lacking in a number of areas, including:

- Up-front costs
- Operational costs

FIGURE 22.2 Electronics waste drop-off area at a materials recovery facility.

- Transportation costs (to the demanufacturer/recycler)
- Economies of scale effects, which should lower program costs in the long run

Other general conclusions were as follows (U.S. EPA, 2002a):

- Transportation, demanufacturing, and disposal are the main cost considerations for collection pilot programs. New programs, however, incur a number of one-time costs. Operational costs may be more relevant in the long term.
- Recovered computers have the most material value, but CRTs, which are costly to manage, are frequently collected. The number of CRTs managed has a major impact on net cost values.
- Geographical location affects transportation costs to demanufacturers and recyclers. For example, CRT recyclers are more prevalent on the east coast, leaving smelters as more viable options for the Midwest and West regions.
- Compared with residentially collected electronics, commercial used electronics has a higher recovered value, is more uniform, and is easier to disassemble.
- The ability to sell at least a few products for positive income makes a big difference in the overall cost of a collection program.

A number of original equipment manufacturers (OEMs), including Compaq, Dell, Gateway, HP, IBM, and Micron offer electronics leasing and take-back services. While environmental considerations are a factor, demand from large corporate customers are driving the development of these services. OEMs such as HP are taking computers back from their higher volume customers but do not widely advertise these services. Many corporate customers have become aware of the implications of improper disposal such as regulatory liability and loss of proprietary information.

22.6.2 REUSE AND RESALE

Reuse and resale are potential outlets for recovered computers. Schools, nonprofit organizations, and low-income households as well as foreign markets often welcome the opportunity to acquire older computers. Certain laws such as the 21st century Classrooms Act for Private Technology Investment (a provision to the Taxpayer Relief Act of 1997) provide tax incentives for corporations to donate late model computers. However, donated equipment varies greatly in terms of quality and reusability. Also, new computers continue to be faster and cheaper, thereby reducing incentives for computer reuse. According to the Florida DEP, only one usable PC can be constructed from every three computers donated. The availability and cost of computer software for older computers has also been a barrier to computer reuse. Because of copyright requirements, valuable computer software (e.g., word processing) is often erased from PCs prior to donation (U.S. EPA, 2002a).

22.6.3 DECONSTRUCTION

As seen in Chapter 21, "demanufacturing" a product is the opposite of manufacturing it. Demanufacturers take the electronic device apart to recover usable components such as memory, disk drives and microprocessor chips, and to recover scrap materials with value such as copper cables and circuit boards. Using this approach, more value is recovered from obsolete electronic products than would be recovered if the whole product were scrapped without disassembly. Unfortunately, computers have been designed in a manner that makes disassembly difficult.

When a shipment arrives at the processing facility, pallets and containers are unloaded at a central point. Each shipment may be accompanied by an inventory of items shipped. Upon arrival contents are verified and examined for integrity and breakage. The equipment is unwrapped, sorted by type, and each type is weighed and counted. The components may then be segregated on the basis of overall condition, age, and other factors. An intact computer may first undergo simple testing of components (Figure 22.3). In many operations, usable components such as hard drives and CRTs are immediately removed for rebuilding or resale (Figure 22.4 and Figure 22.5). Computer monitors and televisions are opened and completely disassembled into glass, plastics, copper-bearing materials, and other components (Figure 22.6). Many will separate the hard drives, power supply, printed circuit boards, keypads, mice, and the steel PC housing (Figure 22.7).

22.6.4 PROCESSING AND RECYCLING E-WASTE COMPONENTS

CRT recycling produces three types of commodities: glass, metals and plastics. CRTs alone produce steel and glass; and complete monitors and televisions produce glass, various metals, and plastics. Some of the practical aspects of deconstruction and recycling are shown in Table 22.7.

FIGURE 22.3 To resell, deconstruct, or scrap? Testing PC components.

FIGURE 22.4 PC monitors approved for resale.

FIGURE 22.5 Old CRTs awaiting shipment to a recycler.

22.6.4.1 CRT Glass

Glass-to-glass recycling refers to the closed-loop process of reclaiming leaded CRT glass from discarded CRTs and using it to make new CRT glass. Glass-to-glass recycling involves collecting televisions and monitors, removing and crushing the CRTs (Figure 22.8), separating the glass from the nonglass materials, processing to meet specifications, and using the glass as an ingredient to manufacture new CRT glass. The steps in glass-to-glass recycling are as follows (MFF, 2001):

- Remove CRT from the monitor or television
- Remove all nonglass material from the exterior of the CRT

FIGURE 22.6 Disassembly of a PC.

FIGURE 22.7 PC hard drives separated for eventual recycling.

- Release the vacuum from the CRT
- Crush CRT and separate any remaining metals from glass
- Remove phosphorescent coatings and prepare uniform cullet

Most manufacturers no longer make CRT glass for monochrome CRTs; therefore, monochrome monitors are best recycled in the United States by other methods (e.g., Pb smelting).

Glass-to-glass recycling has several benefits. In addition to removing Pb from the municipal waste stream, glass-to-glass recycling avoids the environmental impacts associated with mining

TABLE 22.7
Practical Aspects of Electronics Deconstruction and Recycling

Metals   Negligible amounts of metals such as copper, nickel and steel can be recovered from CRTs. Newer models yield less precious metals than older ones. The copper wire yoke and the shadow masks are presumed to be hazardous waste. At a minimum, recyclers of this material will be classified as generators by the California DTSC

Plastics   High-value engineering plastic resins such as high-impact polystyrene, acrylonitrile-butadiene styrene and polycarbonates are sold to some domestic markets. Some companies have developed processes to recycle these plastics. However, most plastics are exported to international markets

Glass   CRT glass recyclers separate cullet into the following streams:
  • Broken Color Dirty Mix, without metals — Dirty Broken Panels
  • Broken Color Dirty Mix, with metals — Dirty Whole Panels
  • Dirty Sorted Funnels — Whole CRTs
  • Monochrome CRTs

"Mix" refers to panel and funnel glass combined; "dirty" refers to glass with coatings

Broken Color Dirty Mix, with metals is broken CRTs with the banding and masks. Broken Color Dirty Mix, without metals is color CRTs that have metal banding, gun and mask removed. Rubber stoppers under the yolk also need to be removed

Dirty sorted funnels and panels must be free of all metals except pins molded into the glass. Panel received with frit is considered Broken Dirty Color Mix, without metals

Source: Reproduced with kind permission of the Materials for the Future Foundation, San Francisco, CA. 2001

FIGURE 22.8 Unit operations for CRT recycling. A compactor unit is linked with a HEPA baghouse to capture all dust and vapors.

and processing raw Pb from ore by supplying the metal, in the form of CRT glass, for CRT manufacturing. In California, CRTs destined for glass-to-glass recycling may be managed as Universal Waste rather than hazardous waste (MFF, 2001).

22.6.5 METALS

Once electronic devices and components have been collected and segregated, recoverable metals are extracted through one of three processes (MOEA, 1995):

- Mechanical reclamation involves shredding and grinding electronic appliances and components into a fine powder. Metals are then separated from the powder. The powder is managed by means of recycling or energy reclamation.
- Chemical reclamation (hydrometallurgy) uses a bath to dissolve components and their packaging. Metals are then extracted chemically.
- Thermal reclamation (pyrometallurgy).

The technology selected depends on several factors. The choice is typically driven by which technology will provide the best return on investment. Total value extracted from electronics scrap can range from hundreds to thousands of dollars per ton depending on the kind of scrap collected and the technology used to reclaim metals (MOEA, 1995).

Precious metals such as gold, platinum, silver and palladium, and copper and steel-bearing materials are often sent to a smelter for recycling. Smelting is defined as a high-temperature metallurgical process in which mineral concentrates and fluxes such as silica and limestone are processed in one or more steps to separate molten metal from impurities. In terms of electronics wastes, Pb (from CRT tubes), copper (CRT yokes), and gold (circuit board fingers) are all recoverable by smelting. Some demanufacturing facilities are equipped with the appropriate furnaces for gold and silver recovery.

Lead is recovered in one of two types of smelters. Primary smelters process sulfide concentrates, sulfates, oxides, and metallic scrap. Primary smelting requires a sulfur (S) removal process such as a sinter or acid plant process to capture S and convert it into H_2SO_4. Secondary smelters process only Pb oxides and metallic scrap. The secondary production of Pb begins with recovery of scrap from obsolete or damaged CRTs and new scrap composed of product wastes and smelter-refinery residues (MFF, 2002).

During smelting, an oven heats the Pb to a temperature of 1260°C (2300°F). Silica is used as a fluxing agent to separate molten Pb from impurities, which form a slag. CRT glass can be used in the blast furnace as a fluxing agent. Scrap materials and fluxing agents are placed in a blast furnace where coke is burned at high temperatures. Carbon contained in the coke reacts with the hot air to form gases that chemically reduce the feeds to molten Pb and slag. The fluxing agents, limestone, and iron form a slag that floats to the top of the molten bath. Lead bullion and the slag flow out the bottom of the furnace in a continuous molten stream where the slag is separated from the Pb bullion. The molten Pb collects in pots where it is allowed to cool. A residue ("dross") forms on the surface of molten Pb as it cools. This dross is removed from the surface and the Pb bullion is further processed in a drossing furnace. The molten slag is tapped from the furnace and granulated. Once the impurities have been removed, the refined Pb is ready to cast (MFF, 2002).

The copper yoke on the CRT possesses positive scrap value and is also recovered in the smelting process. A CRT contains 0.9 to 2.3 kg (2 to 5 lb) of copper on the yoke. During copper smelting the feedstock is dried and fed into one of several different types of furnaces. There the sulfide minerals are partially oxidized and melted to yield a layer of "matte" which is a mixed copper-iron sulfide, and the slag, which is an upper layer of waste. The matte is further processed by a method known as converting. The slag is tapped from the furnace and stored or discarded. Another product of the smelting process is SO_2 which is collected, purified, and converted to H_2SO_4 for sale. Iron and S are removed during the converter process and the remaining molten copper is refined to produce high purity copper (MFF, 2002).

22.6.6 CIRCUIT BOARDS

The PC circuit board contains copper and very small amounts of gold and other precious and semi-precious metals that give the circuit board much of its scrap value. Printed circuit boards can be processed to recover the value of its raw materials. Several companies offer recycling services as an alternative to off-site disposal of obsolete printed circuit boards. These services typically process the boards by removing raw materials of value such as gold, silver, copper, and lead from the board (NFESC, 2002).

Scrap gold may include other metals including copper and zinc. Furthermore, gold may be bound with various formulations of plastic and so must be recovered and purified for eventual resale. Depending upon their composition, gold ores are treated by a variety of methods. Processes commonly include gravity separation, amalgamation, cyanidation, and flotation. Some gold ores (or scrap) may also be smelted and others may be roasted to process them for further treatment. In other refining processes gold is treated with acids or chlorine gas that react with impurities but not with the gold itself. Finally, the refined gold is melted and cast into bars or bricks.

The rest of the monitor is plastic that can sometimes be recycled; however, due to the heterogeneous plastics composition, the plastic portion usually has a negative or zero net value. A flow chart showing e-waste recycling options appears in Figure 22.9.

22.7 INTERNATIONAL ISSUES IN ELECTRONICS MANAGEMENT

There is a substantial overseas market for used PCs and CRTs. The volumes of electronics being shipped overseas are not accurately known; similarly, the uses of these waste electronics are unknown. However, the prices that are being paid for used CRTs ($8 per each) indicate that reuse, demanufacturing, and recycling are occurring (U.S. EPA, 2002b).

There is concern that a portion of U.S. e-wastes is being exported to less-developed countries for processing and disposal; however, these countries possess neither the technology nor the infrastructure to properly and safely recycle such wastes. As a result, some wastes are presumably being treated and disposed under unsafe conditions.

22.8 BARRIERS TO RECYCLING E-WASTE

The infrastructure in the United States for electronics collection, reuse, and recycling is currently insufficient to manage the huge volumes of existing, not to mention projected waste volumes. Electronic products have traditionally been designed for disposal and are difficult to disassemble, thus making them expensive to recycle. E-wastes are bulky, posing transportation and storage problems. Finally (and at the core of any recycling discussion), markets for reclaimed materials are poorly developed.

Requiring consumers and small businesses to pay the cost of recycling and disposal on the back-end has been ineffective. Reliance on back-end disposal fees reduces incentives for proper recycling, promotes "sham" recycling, and results in improper disposal which ultimately requires clean-up at a cost to taxpayers (SVTC, 2001). IBM sold more than 3 million computers in the United States last year and was the first manufacturer to establish a pay-as-you-go system for recycling obsolete computers. The results have not been impressive. According to IBM, only a small fraction of annual sales (less than 1%) have been recycled under the pay-as-you-go system (SVTC, 2001).

CRT manufacturers have enormous volumes of recycled CRT glass available to them. However, transportation costs to a glass-to-glass recycler are often prohibitively high. E-waste contains many types of highly engineered plastics which, although possessing a high resale value, are difficult to identify. Personal computers could be designed with a lower number of plastics; additionally, the manufacturing industry could standardize a selected number of plastics that can be easily identified and recycled. The addition of flame retardants to plastics also poses practical and health problems for recycling.

22.9 PRODUCT STEWARDSHIP AND ELECTRONICS

Given the complexity of most electronics products, the inherent difficulty in their recycling, and the sheer volumes produced, it is evident that improvements in their manufacture, using a more efficient

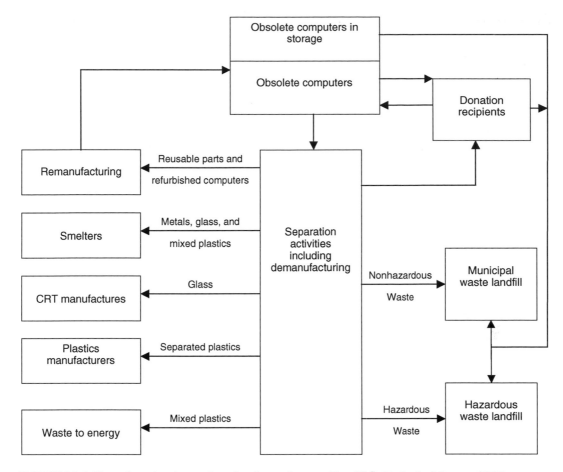

FIGURE 22.9 Flow chart showing options for electronics recycling (U.S. Geological Survey. 2001).

and environmentally appropriate context, would be welcome. The recent concept of "product stewardship" is clearly relevant at this point. Product stewardship can be defined as (NWPSC, 2001):

> whoever designs, makes, sells or uses a product takes responsibility for minimizing its environmental impact. This responsibility spans the product's life cycle — from selection of raw materials to design and production processes to its use and disposal.

Product stewardship can apply to any type of waste and addresses the environmental impact of a product at all stages of its life cycle, from design and manufacture to packaging and distribution to end-of-life management. Product stewardship transfers the responsibility for end-of-life management from the public sector (i.e., government and taxpayers) alone to a shared responsibility that includes the private sector (manufacturers and purchasers). The ultimate goal of this arrangement is to encourage environmentally beneficial design and recycling and to reduce the flow of wastes to landfills or incinerators (NWPSC, 2001). For example, it is considered appropriate to internalize the cost of waste management into the price of electronic devices at the time of purchase. Such up-front fees can be set aside and eventually allocated to cover the eventual costs for deconstruction, processing, and recycling.

Several European and Asian nations have established product stewardship models that involve numerous types of products including electronics. European nations have been the vanguard in addressing the e-waste problem by proposing an ambitious system of "extended producer responsibility." In 2001, the European Union Parliament adopted a directive that requires producers of electronics to take responsibility, financial and otherwise, for the recovery and recycling of e-waste. A second directive requires manufacturers to phase out the use of hazardous materials in electronics products.

22.10 CONCLUSIONS

Modern consumer-driven and technological society relies on an ever growing and constantly changing supply of electronic products. These products are concurrently one of the fastest growing segments of the waste stream. The National Safety Council report (NSC, 1999) estimates that a half billion computers will enter the U.S. waste stream from 1998 to 2007. This number does not include televisions or any of the other unwanted electronic products. The total volume of e-wastes is projected to increase markedly unless reuse and recycling options become more available and convenient for the user.

There are two pressing issues to consider in managing e-waste. The first is how to manage the existing waste stream. Little of this equipment was designed with any consideration for reuse or recycling. The second concern is how to manage future electronic equipment. Regulatory and public pressure will encourage innovative and recycling-oriented design strategies. By applying the principles of product stewardship, electronic equipment can be manufactured with fewer toxic constituents and designed with upgradability, durability, and recyclability in mind.

REFERENCES

Brenner, K., Formation of polybrominated dibenzofurans (PBDFs) and dioxins (PBDDs) during extrusion production of a polybutylenetenerephtalate (PBTP) glassfibre resin blended with decabromodiphenylether (PBDPE)/Sb203, Product and workplace analysis, Brenner, Kniew, BASSF, 1986.
Florida Department of Environmental Protection, Major Types of Electronic Equipment. 2002. See: http://www8.myflorida.com/waste/categories/electronics/pages/major_types.htm
Harris, P., Chief, State Regulatory Programs Division, Hazardous Waste Management Program, Department of Toxic Substances Control, in a March 20, 2001 letter to Sheila Davis of Materials for the Future Foundation, 2001.
Hoque, A., Sigurdson, A.J., Burau, K.D., Humphrey, H.E.B., Hess, K.R., and Sweeney, A.M., Cancer among a Michigan cohort exposed to polybrominated biphenyls in 1973, Epidemiology, 9, 373-378, 1998.
International Association of Electronics Recyclers, Electronics Equipment, 2002. See: http://www.iaer.org/electronicsequipment.htm
Materials for the Future Foundation, CRT Glass to CRT Glass Recycling, The Monitor of Electronics Recycling Issues, September 2001. See: http://www.materials4future.org/PUBS/g2g%20final.pdf
Materials for the Future Foundation, How Lead Smelting Works., The Monitor of Electronics Recycling Issues, January 2002. See: http://www.materials4future.org/PUBS/smelting%20final.pdf
Matthews, H., McMichael, F.C., Hendrickson, C.T., and Hart, D.J., Disposition and End-Of-Life Options for Personal Computers, Green Design Initiative Technical Report #97-10, Carnegie Mellon University, Pittsburgh, PA, 1997.
Minnesota Office of Environmental Assistance (MOEA), *Management of Waste Electronic Appliances*, Minnesota Office of Environmental Assistance, Saint Paul, MN, August 1995.
Monchamp, A., Evans, H., Nardone, J., Wood, S., Proch, E., and Wagner, T., Cathode Ray Tube Manufacturing and Recycling: Analysis of Industry Survey, Electronic Industries Alliance, 2001. See: http://www.eiae.org/chemicals/files/EIA_CRT_5-01.pdf
National Safety Council, Electronic Product Recovery and Recycling (EPR2) Baseline Report, National Safety Council — Environmental Health Center sponsored by the US Environmental Protection Agency, May, 1999. See: http://www.nsc.org/ehc/epr2/baseline.htm
Naval Facilities Engineering Service Center, Printed Circuit Board Recycling, 2002. See:p2library.nfesc.navy.mil/P2_Opportunity_Handbook/ 2_II_8.html July 2002.

Northwest Product Stewardship Council, Defining Product Stewardship, 2001. See: http://www.productstewardship.net/definingStewardship.html

Reuters, Disposable Cell Phones are Ringing, July 29, 2002.

Silicon Valley Toxics Coalition, Californians against Waste and Materials for the Future, San Jose, CA, 2001. See: http://www.svtc.org/cleancc/pubs/ppc-ttv1.pdf

Stanford Resources, Inc., Television Systems, 1998.

Swedish Environmental Protection Agency, No date, Persistent Organic Pollutants., See: http://smn.environ.se/miljonat/english/sokning/sokning.htm

Truini, J., 3,2,1: Self-Destructing DVDs may Stir Waste Issues, *Waste News*, July 21, 2003, p. 1.

U.S. Environmental Protection Agency, Characterization of Municipal Solid Waste in the United States: 1998 Update, EPA 530, Office of Solid Waste and Emergency Response, Washington, DC, 1999.

U.S. Environmental Protection Agency, Electronics Waste, 2002a. See: http://www.epa.gov/reg3wcmd/eCyclingwaste.htm

U.S. Environmental Protection Agency, Electronics, 2002b. See: http://www.epa.gov/epr/products/electronics.html

U.S. Environmental Protection Agency, Proposed CRT Rule, 2002c. See: http://www.epa.gov/epaoswer/osw/meeting/pdf02/goode.pdf

U.S. Geological Survey, Obsolete Computers, "Gold Mine," or High-Tech Trash? Resource Recovery from Recycling. Fact Sheet 060-01., By D. Bleiwas and T. Kelly, 2001.See: http://pubs.usgs.gov/fs/fs060-01/

Western Electronic Product Stewardship Initiative, RCRA and How it Applies to Electronics, 2001. See: http://www.recyclingadvocates.org/wepsi/about/subgpdocs/rcra.pdf

SUGGESTED READING AND WEB SITES

BBC News, Japanese Plant Takes on E-waste, 2003. See: http://news.bbc.co.uk/2/hi/technology/3075717. stm. July 23, 2003.

BBC News, Taxing Times for E-waste, 2003. See:http://news.bbc.co.uk/1/hi/technology/3141988.stm

California Integrated Waste Management Board, Electronic Product Management Directory, 2003. See: http://www.ciwmb.ca.gov/Electronics/Collection/

California Integrated Waste Management Board, Electronic Equipment: Reducing Waste at the CIWMB, 2003. See: http://www.ciwmb.ca.gov/Electronics/CaseStudies/CIWMB.htm

Computer Takeback Campaign, No date, Legislation and Policy, See: http://www.computertakeback.com/legislation_and_policy/e_waste_legislation_in_the_us/index.cfm

Dean, K., Recyclers Pledge to Cut E-waste, Wired.com, 2003. See: http://www.wired.com/news/technology/0,1282,57802,00.html

Guardian Unlimited, The E Waste Land, 2002. See:http://www.guardian.co.uk/weekend/story/0,3605,849530,00.html. November 30, 2002

Jacobsohn, A.P., Deleting e-waste, *Waste Age*, 2003. See: http://www.wasteage.com/ar/waste_deleting_ewaste/

Mayfield, K., E-Waste: Dark Side of Digital Age, Wired.Com, 2003. See: http://www.wired.com/news/technology/0,1282,57151,00.html

News Limited, HK Urged to Stop Exporting E-waste, 2003. See: http://www.news.com.au/common/story_page/0,4057,7010453%255E15306,00.html. August 20, 2003.

OCLC Public Affairs Information Service, Electronics Waste (June 2002). See: http://www.pais.org/hottopics/2002/June/index.stm

Platt, B., and Hyde, J., *Plug into Electronics Reuse*, Institute for Local Self Reliance, Washington, DC. 1997.

Silicon Valley Toxics Coalition, Just Say No to E-waste: Background Document on Hazards and Waste from Computers. No date. See: http://www.svtc.org/cleancc/pubs/sayno.htm

State of California Department of Toxic Substances Control, Final Regulations: Electronic Hazardous Wastes (formerly Cathode Ray Tubes), 2000. See: http://www.dtsc.ca.gov/LawsRegulationsPolicies/CRTs/CRT_final_regs.html

U.S. Environmental Protection Agency, Electronics Challenge., WasteWise, 2003. See: http://www.epa.gov/wastewise/wrr/celect.htm

QUESTIONS

1. What are the trends in generation of e-waste in the United States over the past decade? How are the majority of e-wastes managed (i.e., landfill, stockpile, reuse, etc.)? Does the current recycling rate for these products correspond with the generation rate?

2. What is the contribution of consumer electronic products to the lead input of sanitary landfills? What is the contribution to cadmium input? Mercury input?

3. How does a mainframe computer differ from a typical desktop PC in terms of recyclability? How does a laptop differ from a desktop in this respect?

4. The Pb in the funnel and face plate glass occurs as PbO and is considered stable and immobile. Under what conditions, therefore, is the lead in computer CRTs hazardous to health?

5. Where does mercury, chromium, and cadmium occur in a personal computer? List specific components where each occur.

6. What are the health effects that may be associated with PVC and brominated flame retardants in computer monitors?

7. How does desktop computer plastic differ in terms of recyclability with plastic from a nonfood container (e.g., HDPE or PVC)? You may wish to refer to Chapter 6 (Recycling).

8. What is the benefit in promulgating a separate rule for CRT recycling rather than simply relying on the Universal Waste Rule?

9. How are computer circuit boards managed under RCRA?

10. How would you expect the generation of e-waste to change over the next 10 years? What components may disappear from the waste stream? What new products may begin to appear in large quantities?

11. In your opinion, are current federal and state regulations which address e-waste management and recycling adequate to ensure environmental quality over a long term? Should there be mandatory federal and state recycling of personal computers?

12. List all the major electronic appliances and devices in your home. How long have you owned each? What is the expected lifetime of each? Which do you feel can be safely disposed in the trash when their usable lifetime is complete? Which should be recycled or shipped for special disposal?

13. In your community, how are e-wastes handled? Is there a community drop-off center or a special waste disposal event? Is there an adequate public education program so that local citizens are informed about e-waste? If there is a local collection program, check with the local waste authority as to where these wastes are shipped for recycling.

14. Contact your university inventory agent and inquire as to how old or obsolete desktop PCs are managed. How many are replaced each year? Are the obsolete PCs made available for public purchase or are they donated or recycled? If recycled, explain where they are sent and how they are managed.

15. What are the major cost considerations for a community considering establishment of an e-waste recycling program? Which steps are likely to be the most costly?

16. In some electronics demanufacturing facilities, unskilled workers may be employed. List possible hazards to workers involved in demanufacture of PCs, television sets, and similar equipment.

17. List some of the general, practical difficulties in recycling electronics waste.

18. A fire has occurred in an office which houses several dozen desktop and laptop PCs, telephones, a photocopy machine, a VCR and television monitor. List possible hazards to those persons entering the office during and immediately following the fire. You may

want to check Materials Safety Data Sheets (available on the Internet) of the electronic components to determine possible combustion products.

19. Check the Internet for sources which describe the manufacture of personal computers. What wastes may be generated during the manufacture of PC circuit boards and plastic monitor housing?

List of Acronyms and Abbreviations

API	American Petroleum Institute
ASTM	The American Society for Testing Materials
BDL	below detection limit
BIF	boiler and industrial furnace
BOD	biochemical oxygen demand
BTEX	benzene, toluene, ethylbenzene, and xylene
Btu	British thermal unit
C	carbon
°C	degrees Celsius
C-&-D	construction and demolition
CAA	Clean Air Act
CBL	Commercial Bill of Lading
CCA	chromated copper arsenate
CEMS	continuous emission monitoring systems
CERCLA	Comprehensive Environmental Response, Compensation, and Liability act of 1980
CESQG	conditionally exempt small quantity generator
CFR	Code of Federal Regulations
COD	chemical oxygen demand
CPE	chlorinated polyethylene
CWA	Clean Water Act
CO	carbon monoxide
CO_2	carbon dioxide
CRT	cathode-ray tube
CSPE	chlorylsulfonated polyethylene
d-RDF	densified refuse-derived fuel
DBA	decibels on the A scale sound level meter
DHHS	Department of Health and Human Services
DoD	Department of Defense
DOE	U.S. Department of Energy
DOT	U.S. Department of Transportation
DRE	destruction and removal efficiency
DYI	do-it-yourself
EPA	Environmental Protection Agency
ESP	electrostatic precipitator
°F	degrees Fahrenheit
FMC	flexible membrane cap
FML	flexible membrane liner
GAC	granular activated carbon
g/dscm	grams per dry standard cubic meter
GPAD	gallons per acre per day
HAZMAT	hazardous material
HCl	hydrochloric acid (or hydrogen chloride)
HDPE	high-density polyethylene
HHV	higher heat value

HMTA	Hazardous Materials Transportation Act
HMTUSA	Hazardous Material Transportation Uniform Safety Act of 1990
HSWA	Hazardous and Solid Waste Amendments
HTIS	Hazardous Technical Information Service
IARC	International Agency for Research on Cancer
ID	inside diameter
ICP	inductively coupled plasma
ISO	International Organization for Standardization
ISWM	Integrated Solid Waste Management
kg	kilogram
kPa	kilopascal
L	liter
LCS	leachate collection system
LCR	leachate collection and removal system
LDPE	low-density polyethylene
LDR	land disposal restriction
LDS	leachate detection system
LEL	lower explosive limit
LHV	lower heating value
LLDPE	linear low-density polyethylene
LQG	large quantity generator
MACT	maximum achievable control technology
MCL	maximum contaminant level
MRF	materials recovery facility
MSW	municipal solid waste
MW	megawatt
MW	molecular weight
MWTA	Medical Waste Tracking Act
NA	North American
NEPA	National Environmental Policy Act
NIMBY	Not in My Backyard
NIMEY	Not In My Election Year
ng	nanogram $(1 \text{ g} \times 10^{-9})$
NO_x	nitrogen oxides
nos	not otherwise specified
NPL	National Priorities List
NRDES	National Pollution Discharge Elimination System
OD	outer diameter
ORM	other regulated material
OSHA	Occupational Safety and Health Administration
PAH	polycyclic aromatic hydrocarbons
PCB	polychlorinated biphenyls
PCDD	polychlorinated debenzodioxins
PCDF	polychlorinated dibenzofurans
PCP	pentachlorophenol
PETE	polyethylene terephthalate
PIC	product of incomplete combustion
PM	particulate matter
POHC	principal organic hazardous constituent
POTW	publicly owned treatment works
PP	polypropylene

ppb	parts per billion
PPH	pounds per hour
ppm	parts per million
PRP	potentially responsible parties
PS	polystryene
psi	pounds per square inch
PVC	polyvinyl chloride
RCRA	Resource Conservation Recovery Act
RDF	refuse-derived fuel
SDWA	Safe Drinking Water Act
SI	System International, the metric system of units
SG	specific gravity
SO_x	sulfur oxides
SPCC	spill prevention, control, and countermeasures
SQG	small quantity generator
SSU	Saybolt-seconds universal
SVE	soil vapor extraction
SWCR	surface water collection and removal system
TCLP	toxicity characteristic leaching procedure
TDS	Total dissolved solids
TOC	total organic carbon
TPH	total petroleum hydrocarbons
TRC	Thermostat Recycling Corporation
TSCA	Toxic Substances Control Act
TSDF	treatment, storage, and disposal facility
UEL	upper explosive limit
UIC	underground injection control
µg	microgram
µm	micrometer
USDW	underground source of drinking water
UN	United Nations
U.S. EPA	United States Environmental Protection Agency
UST	underground storage tank
VFPE	very flexible polyethylene
VOA	volatile organic acid
VOC	volatile organic compound
%w/w	percent by weight

Index